Corrosion Chemistry

Chemistry Centre, London, UK

13-15 April 2015

FARADAY DISCUSSIONS

Volume 180, 2015

ROYAL SOCIETY OF CHEMISTRY

The Faraday Division of the Royal Society of Chemistry, previously the Faraday Society, founded in 1903 to promote the study of sciences lying between Chemistry, Physics and Biology.

Faraday Discussions (Print ISSN 1359-6640, Electronic ISSN 1364-5498) is published 9 times a year by the Royal Society of Chemistry, Thomas Graham House, Science Park, Milton Road, Cambridge, UK CB4 0WF.

Volume 180 ISBN-13: 978-1-78262-180-5

2015 annual subscription price: print+electronic £877, US $1635; electronic only £833, US $1553. Customers in Canada will be subject to a surcharge to cover GST. Customers in the EU subscribing to the electronic version only will be charged VAT. All orders, with cheques made payable to the Royal Society of Chemistry, should be sent to RSC Order Department, Royal Society of Chemistry, Thomas Graham House, Science Park, Milton Road, Cambridge, CB4 0WF, UK.
Tel +44 (0) 1223 432398; E-mail www.orders@rsc.org

If you take an institutional subscription to any RSC journal you are entitled to free, site-wide web access to that journal. You can arrange access via Internet Protocol (IP) address at www.rsc.org/ip. Customers should make payments by cheque in sterling payable on a UK clearing bank or in US dollars payable on a US clearing bank.

Printed in the UK

MIX
Paper from responsible sources
FSC® C013604
www.fsc.org

Faraday Discussions are unique international discussion meetings that focus on rapidly developing areas of chemistry and its interfaces with other scientific disciplines.

Corrosion Chemistry

Faraday Discussions

www.rsc.org/faraday_d

A General Discussion on Corrosion Chemistry was held in London, UK on the 13th, 14th and 15th of April 2015.

RSC Publishing is a not-for-profit publisher and a division of the Royal Society of Chemistry. Any surplus made is used to support charitable activities aimed at advancing the chemical sciences. Full details are available from www.rsc.org

CONTENTS

ISSN 1359-6640; ISBN 978-1-78262-180-5

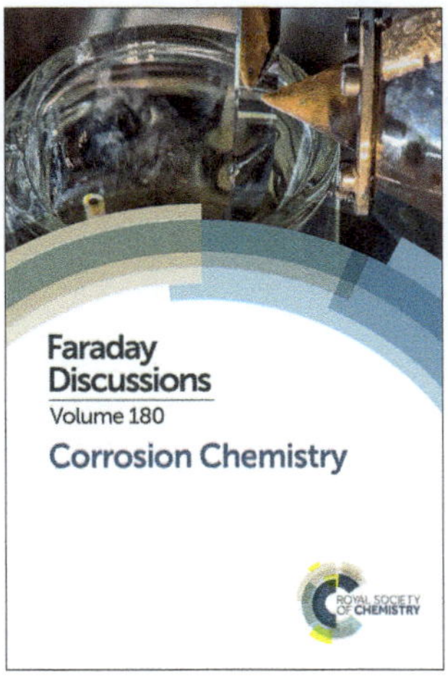

Cover
See Z. Liu, C. S. Fadley, H. Bluhm *et al.*, pp. 35–53.

Nickel thin film sample deposited on a multilayer X-ray mirror substrate in front of the conical entrance aperture of an ambient pressure XPS system. The sample is partly immersed into a 0.1 M KOH aqueous solution. Counter and reference electrodes inside the solution enable control of the electrochemical conditions in the thin KOH film that extends from the bulk of the solution to the measurement position.

Image reproduced by permission of H. Bluhm, Lawrence Berkeley National Laboratory, USA, from *Faraday Discuss.*, 2015, **180**, 35.

INTRODUCTORY LECTURE

PAPERS AND DISCUSSIONS

SPONSORS

Cogent skills
for science industries

SKT 2015
Selb, Germany
April 14th – 16th, 2015

DENS solutions

ICAM
International Centre for Advanced Materials

SUPPORTING ORGANISATION

IOM3
The Institute of Materials, Minerals and Mining

bluescientific

ec-lab

SciMed
Scientific & Medical Products Ltd

LOT
QuantumDesign

PalmSens
Compact Electrochemical Interfaces

PINE
RESEARCH INSTRUMENTATION

Metrohm

CONCLUDING REMARKS

ADDITIONAL INFORMATION

Faraday Discussions

PAPER

Introductory lecture on corrosion chemistry: a focus on anodic hydrogen evolution on Al and Mg

G. S. Frankel,* S. Fajardo and B. M. Lynch

Received 8th May 2015, Accepted 28th May 2015
DOI: 10.1039/c5fd00066a

The increase in the rate of hydrogen evolution (HE) on dissolving Mg surfaces with increasing anodic current density or potential, which is sometimes called the negative difference effect, has been the topic of much discussion in recent years. A review of the very recent contributions to this subject is given in this paper. Increased catalytic activity of the corrosion product layer, either from the accumulated impurities or from the Mg oxy-hydroxide itself, is shown to have a minor influence on the anodic HE observed on dissolving Mg at high anodic current densities and potentials. Al exhibits similar characteristics during anodic polarization in concentrated HCl, although the anodic HE rate on Al is less than on Mg. Possible mechanisms for the anodic hydrogen are provided and implications in the area of intergranular corrosion and environmental cracking are discussed.

Introduction

Corrosion, the environmental degradation of materials, involves many aspects of chemistry, as can be seen by the breadth of topics in this meeting and published volume. Details of the environment include: the concentrations of species that might be aggressive, oxidizing, passivating, inhibiting, complexing, or buffering; the temperature, pressure, and flow conditions; the electrochemical potential and distribution across the surface caused by local reactions and ohmic potential drops; atmospheric corrosion parameters such as relative humidity, pollutants, and particulates; and stress conditions such as load and load fluctuations. Details of the material include crystallinity *vs.* amorphicity, alloy composition, microstructure such as grain size and shape, second phases, inclusions, impurities, precipitates, solute depleted zones, dislocation density and surface films that are protective such as passive films or less protective deposited films. Physical descriptions of the component or test sample of relevance might include size from a massive beam to a nanoparticle, surface preparation, surface roughness, stress-concentrating defects, and physical connections that might cause galvanic

Fontana Corrosion Center, The Ohio State University, Columbus, OH 43210, USA. E-mail: frankel.10@osu.edu

coupling or crevices. Corrosion protective schemes are relevant, such as a wide variety of possible coatings and potential alteration by cathodic or anodic protection. Considerable focus has been placed on the use and development of advanced techniques for the study of corrosion, including electrochemical approaches, surface analytical methods, scanning techniques that provide spatial resolution, and powerful methods for *in situ* analysis. Finally, modeling of corrosion processes is an active field that can provide insights and predict behavior with less need for lengthy experiments. A long-term goal of this effort would be to connect the behavior at multiple length scales from atomistic to macroscopic levels in a way that would allow the prediction of component lifetime based on fundamental principles. Many of these topics can be found in the papers and discussions of this meeting and volume.

The purpose of the introductory lecture and paper in a Faraday Discussion meeting and volume, respectively, is to "set the stage." Owing to the vastness of the field, a comprehensive review is inappropriate and borders on impossible given the page constraints. It was therefore decided to limit the scope to a few issues in the area of localized corrosion and to raise some points that would initiate discussion.

Localized corrosion is a part of the corrosion field that is rich in chemical complexity.[1] Essentially all of the topics listed above can play a role or even be dominant. Pits, crevices and cracks often contain concentrated solution and steep concentration gradients, as well as potential gradients and strong sensitivity to the local potential. Microstructural features often control the severity and nature of the attack.

The introductory lecture covered four aspects of localized corrosion chemistry; this introductory paper will focus on only one of them, the accelerated production of hydrogen gas on dissolving metal surfaces, or anodic hydrogen. The topics that were presented but are not covered in this paper include the mechanism for trenching attack around cathodic particles in Al alloys (pH gradient, potential gradient or interface defects), the electrochemical properties of solute elements in a metal matrix as they combine in the initial stages of particle formation, and the effects of limited water in localized corrosion sites caused by the complexation of ions in the concentrated localized environment.

Hydrogen evolution in Al pits

Anyone who has looked closely at pits in Al has noticed that H_2 bubbles are emitted from actively growing pits and that the hydrogen evolution stops when the pits stop growing. A poster at this meeting displayed this exact phenomenon using high resolution X-ray tomography.[2] The rate of hydrogen evolution (HE) in pits can be quantified under special conditions.

HE from pits in 99.999% Al and two less pure Al alloys polarized in deaerated NaCl solution was studied by Drazic and Popic.[3] They used a rotating ring disk electrode (RRDE) in which the H_2 evolved from pits in the Al disk was oxidized on the Pt ring. Fig. 1 shows some of the results of their work. The open symbols represent the measured polarization curve on the disk and the filled symbols show the ring oxidation current. The H_2 oxidation rates on the ring matched the measured currents in the cathodic region of the curve below the corrosion potential. Al is spontaneously passive in this deaerated solution, and the HE rate

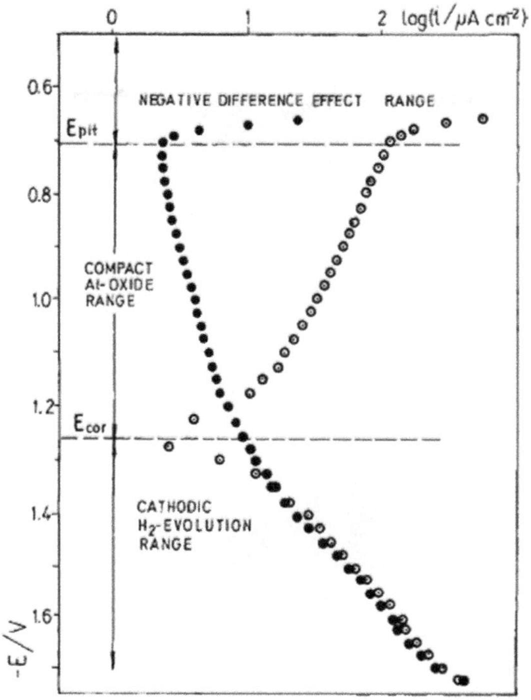

Fig. 1 HE rate during the polarization of a rotating Al disk.[3] Open symbols represent the measured current density and the closed symbols are the ring current associated with oxidation of hydrogen at the ring. With kind permission from Springer Science and Business Media.

continued to decrease as the potential increased in the passive region. However, as soon as the current increased as the result of pit initiation, a similarly sharp increase in the rate of HE was detected. The measured HE rate was about 5% of the measured current density. The authors showed similar behavior for the other Al alloys; pitting was always accompanied by a high rate of HE.

Pitting of thin film samples offers another approach for the quantification of HE during pitting. The kinetics of pits in thin film Al samples were first reported 25 years ago.[4] Sputtered Al films, about 100 nm thick deposited onto glass substrates, were tested in chloride solution. Pits in such samples very quickly penetrate the thickness of the metallic coating and reach the inert substrate. This barrier prevents further deepening, and the pits continue to grow by expanding outwards across the substrate with essentially vertical pit walls. The undermined passive film is lifted into the solution during growth by the hydrogen bubbles (more on that later) and Fig. 2a is an SEM image of a pit edge showing the undermined passive film collapsed back onto the surface. The pit growth is two-dimensional (2-D) and the pit takes the form of circles at high applied potentials. At lower potentials the pit perimeter is more convoluted owing to the partial repassivation of the growing pit front. The difference in reflectivity of the Al film and glass substrate allows for the capture of a high contrast image using incident light and a video recording device, Fig. 2b.

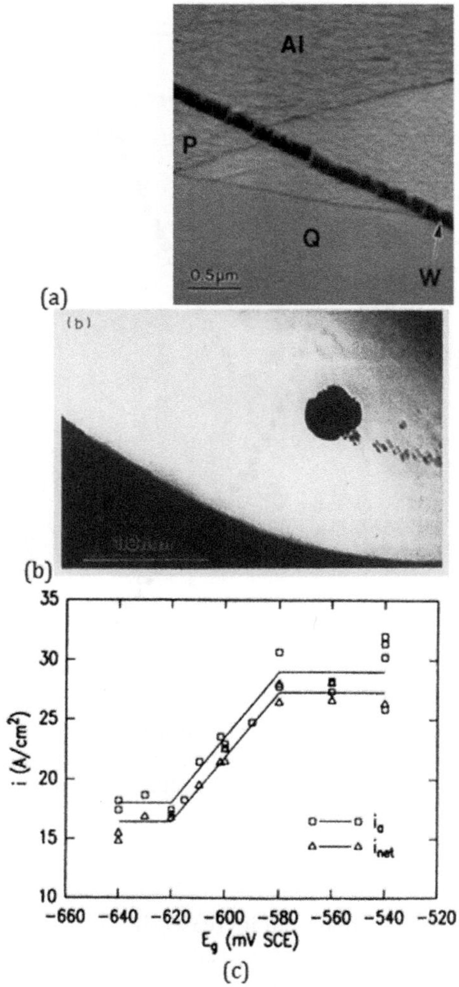

Fig. 2 Results of 2-D pit growth experiments.[4] (a) SEM image of the sample after 2-D pit growth, where Q is the quartz substrate, P is the passive film and W is the 2-D pit wall. (b) Optical image during a pit growth experiment showing hydrogen bubbles streaming upwards from the pit surface. (c) Anodic and net current density as a function of applied potential. The lines are added to aid assessment, and the difference between the two is the current density associated with HE. Reprinted with permission from Elsevier.

The original paper described how the anodic pit current density, i_{anod}, could be assessed from the increases in pit perimeter and area with time.[4] A simpler approach is to use the local velocity of the pit radius, $v = \mathrm{d}r/\mathrm{d}t$, which is constant for the thin film pits owing to their fixed pit depth:

$$i_{\text{anod}} = v\frac{\rho n F}{M} \tag{1}$$

where ρ is the metal density, $n = 3$ is the charge on the dissolved Al ion, F is the Faraday constant, and M is the molecular weight. Because v is constant, i_{anod} tends to be constant with time. Values of anodic pit current density as a function of the fixed applied potential are shown as the square symbols in Fig. 2c. The anodic current densities are high, in the order of tens of A cm^{-2} for pits in films of this thickness. This high current density reflects the very high current density of pits in bulk samples when they are small, in the order of 100 nm in depth.

Evident in Fig. 2b is a stream of H_2 bubbles emanating from the growing pit. The current density associated with the HE at thin film pits can be assessed in cases where a single pit forms. In that case, the current measured by the potentiostat is associated only with the one pit. The active pit area can be determined from the film thickness and the perimeter change with time. The measured current, $I(t)$, and the area, $A(t)$, tend to have the same dependence on time so the net current density, $i_{net} = I(t)/A(t)$, is constant with time, like i_{anod}. The HE current density for these pits can then be assessed from:

$$|i_{HE}| = i_{anod} - i_{net} \tag{2}$$

The triangular symbols in Fig. 2c represent i_{net} measured for experiments with single pits. As indicated by eqn (2), the HE current density can be assessed by the difference between the i_{anod} and i_{net}. The analysis shows that i_{HE} is huge, in the order of A cm^{-2}. Fig. 1 shows that the rate of HE on Al even at very low potentials in the cathodic region is still 4 orders of magnitude smaller than the rate in these actively growing thin film pits at a much higher potential. Clearly HE can occur in pits at extremely high rates even though Al is not known as a good catalyst for HE. It is of interest to understand how this is possible, and what implications this high rate of HE might have for corrosion and cracking of Al alloys. Before addressing this, the kinetics of HE on Mg and Al will be discussed.

Anodic HE on Mg

In recent years there has been a flurry of activity on a topic related to the high rate of HE in Al pits, *i.e.* enhanced HE on Mg during anodic dissolution, which has been called the negative difference effect (NDE). NDE is an odd term for this phenomenon that is in common use for historical reasons. It will be referred to as "anodic HE" in this paper.

In Fig. 3, the rate of HE, assessed volumetrically by hydrogen collection in a burette, is plotted as a function of applied current density during galvanostatic experiments on high purity (HP, 99.98% Mg) and ultra high purity (UHP, 99.9999% Mg) magnesium in 0.1 M NaCl.[5] As the applied negative current decreases, the measured rate of HE decreases, as expected. Although not shown, the measured potential increases as the applied negative current density decreases and then further increases as the applied anodic current density increases above the Mg open circuit potential (OCP). However, as the applied positive or anodic current density increases, the measured rate of HE increases exponentially, in contrast to the continued decrease that is expected with increasing potential for a reduction reaction such as HE. This increase in HE with increasing potential is the phenomenon of anodic HE on Mg. It should be noted that the curves in Fig. 3 look like measured polarization curves showing net cathodic and net anodic regions, but all of the data points in the figure represent HE.

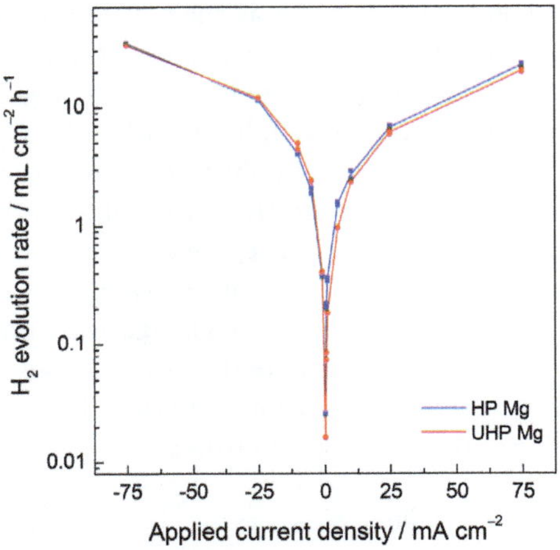

Fig. 3 HE rate determined volumetrically for high purity (HP) and ultra high purity (UHP) Mg samples as a function of the applied current density.[5] Reprinted with permission from Elsevier.

Table 1 summarizes many of the explanations for anodic HE in Mg and Mg alloys that have been suggested. A recent review paper presented a current opinion regarding the possible mechanisms of this anodic HE.[6] An overview of some of these mechanisms will be given below along with a summary of new

Table 1 Possible mechanisms for anodic HE

Mechanism	Ref.	Comment
Univalent Mg	7 and 8	Has been criticized in recent years
Crevice corrosion under H_2 bubbles	11	No evidence
Surface area increases		Cannot explain order of magnitude increases
Tafel slope changes		No strong evidence
Chunking (non-Faradaic loss of Mg metal)		No strong evidence
Deposition of impurities from solution		No strong evidence
pH change	23	Plays a role
Surface enrichment of impurities	12 and 20	Plays a role
Corrosion product coverage and catalytic activity	16, 18, 19 and 22	Plays a role
Catalytic nature of dissolving surface	5	Potentially plays a dominant role

papers published since that recent review paper was written. Indeed, this area is at the moment one of the most rapidly changing topics in the field of corrosion, with new insights being published frequently.

For many years, the most widely discussed mechanism to explain the phenomenon of anodic HE on Mg was univalent Mg, in which some fraction of Mg dissolves into solution as Mg^+, which is then further oxidized chemically with the reduction of water to H_2.[7,8] According to this theory, higher rates of Mg dissolution would result in more Mg^+ and thus more HE. This theory is consistent with much of the data and can explain apparent observations of $n < 2$ at the OCP, where n is the nominal oxidation state of the dissolved ion. A complete critique of this theory will not be attempted here; only a few relevant comments will be offered. There is still no direct evidence for the existence of Mg^+ in solution, as what had been considered the best evidence[9] has been shown to have been the result of other phenomena.[10] Furthermore, the univalent Mg theory is not consistent with apparent observations of $n < 1$, for which a theory involving isolated crevice corrosion under H_2 bubbles has been evoked.[11] If this latter mechanism is operative, it would obviate the need for the univalent Mg mechanism, as it alone could explain all observations for $n < 2$. In fact, however, the formation of surface films makes the determination of n difficult, so all of the reports of $n < 2$ should be viewed with some skepticism. Careful assessments of n by multiple methods have found that indeed $n = 2$ for Mg dissolution,[12–14] so the univalent Mg and crevice corrosion mechanisms for anodic HE, must also be considered unlikely until clear and direct evidence for their existence is developed.

An alternate explanation for the anodic HE was provided by Frankel *et al.*[15] Activation-controlled kinetics for the HE reaction (HER) can be described by the Tafel equation as the potentials at which anodic HE on Mg are observed are far below the reversible potential for the HER so the back reaction can be ignored:

$$i_{\mathrm{HER}} = i_{0,\mathrm{HER\ on\ Mg}} 10^{\left(\frac{-(E - E_{\mathrm{rev,HER}})}{b}\right)} \tag{3}$$

where i_{HER} is the HER current density, $i_{0,\mathrm{HER\ on\ Mg}}$ is the exchange current density for the HER on Mg, E is the potential, $E_{\mathrm{rev,HER}}$ is the pH-dependent reversible potential for the HER, and b is the Tafel slope. This equation predicts a decrease in the rate of HE as E increases. As shown in Fig. 3, this is observed in the cathodic region of Mg polarization, but not in the anodic region. The concept promoted by Frankel *et al.* was that the catalytic activity of the dissolving Mg surface, as embodied by i_0, increases as the rate of dissolution increases such that the increasing pre-exponential term in eqn (3) dominates the decreasing exponential term.[15] Possible reasons for increased catalytic activity were discussed but no definitive proof was given. Note that this phenomenon might be related to the high rate of HE in pits in Al, which was described above.

Recent papers have focused attention on the role of corrosion films and impurity enrichment in providing the increased catalytic activity of the dissolving surface. When Mg corrodes at the OCP and during anodic polarization, a dark corrosion film forms and this film has some special properties. A related phenomenon is the enrichment of impurity or alloying elements in or under the corrosion film. These topics will be discussed presently with a focus on recent reports.

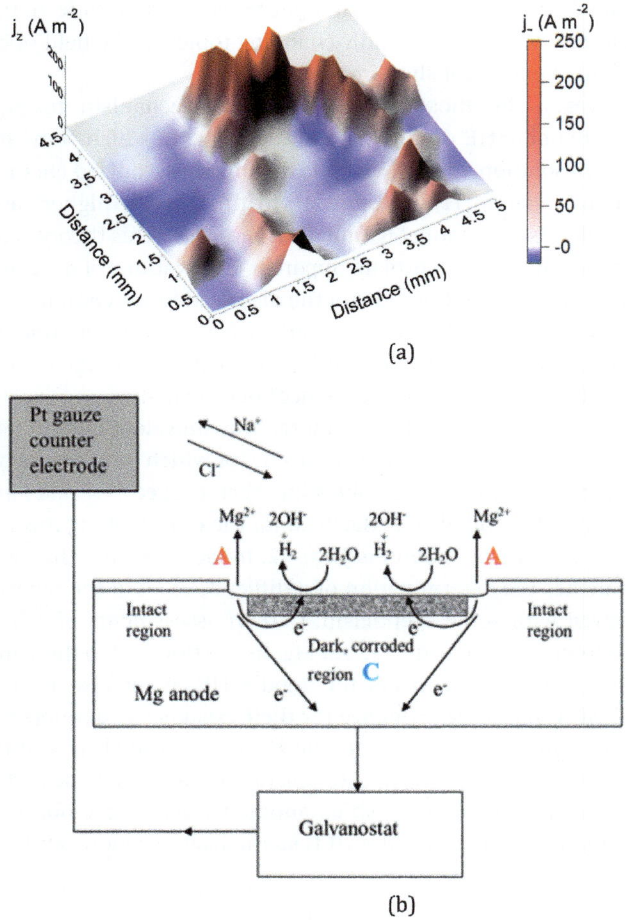

Fig. 4 (a) Current density map determined by SVET during polarization of pure Mg at a fixed anodic current density of 1 mA cm^{-2}.[16] (b) Schematic showing the proposed model for anodic HE, whereby the dark, previously corroded area acts as a net cathode even during anodic polarization. Reprinted with permission from Elsevier.

A study by Williams *et al.* used the scanning vibrating electrode technique (SVET) to map the net current flowing from a Mg surface during anodic polarization in a 2 M NaCl solution.[16] The attack progressed as dark filiform tracks or expanding rings across the surface, depending on the purity, and the active anode sites could be easily distinguished by SVET, Fig. 4a. Interestingly, the SVET could also sense regions where the net current was negative, even during anodic galvanostatic polarization at +1 mA cm^{-2}. The cathodes were the dark oxide left in the wake of the moving anodes. The integrated anodic and cathodic currents both increased with time, but summed more or less to the applied current, which was constant. The integrated cathodic current during anodic polarization increased with increasing applied current density and was found to be slightly less than the volume of hydrogen gas collected using a burette. The observation of net cathodes on an anodically polarized surface is stunning and provides evidence for the importance of the remnant corrosion layer in the anodic HE reaction. The schematic provided in Fig. 4b shows how a fraction of the anodic current can be consumed by HE occurring at the neighboring sites covered with corrosion product.[16] The area covered by the dark corrosion product increased with time in the initial stages similarly to the increase in cathodic current measured by the SVET. However, at steady state the rate of HE measured volumetrically is normally constant with time, rather than increasing with time, and it varies with the anodic current density. This constant and current-dependent rate of HE is not consistent with the dark areas being the location of the anodic HE. Furthermore, at steady state, the integrated cathodic current from the SVET is much less than the volumetric HE rate,[17] which indicates that the dark corrosion product regions, which are net cathodes, cannot be the source of most of the anodic HE under those conditions.

The robustness of the cathodic activity of the corroded surface was shown by Birbilis *et al.* using experiments in 0.1 M NaCl on pure Mg that was alternately polarized galvanostatically in the anodic direction and then potentiostatically at −1.9 V SCE, which is cathodic polarization.[18] The current measured during the cathodic polarization step increased as the prior applied anodic current density increased, Fig. 5. These findings support those of Williams *et al.*[16] regarding the catalytic nature of the corrosion product film. However, the measurements of enhanced cathodic reaction were made in the cathodic region, and the relevance to anodic polarization conditions must be confirmed.

The enhanced catalytic activity described above might be associated with the magnesium oxy-hydroxide corrosion film or with impurities enriched either within or under the film. Even the purest Mg available will have impurities that might enrich on the surface during anodic polarization owing to incongruent dissolution of Mg at potentials that are far below the reversible potentials for the impurity metal dissolution. These surface-enriched impurity elements, such as

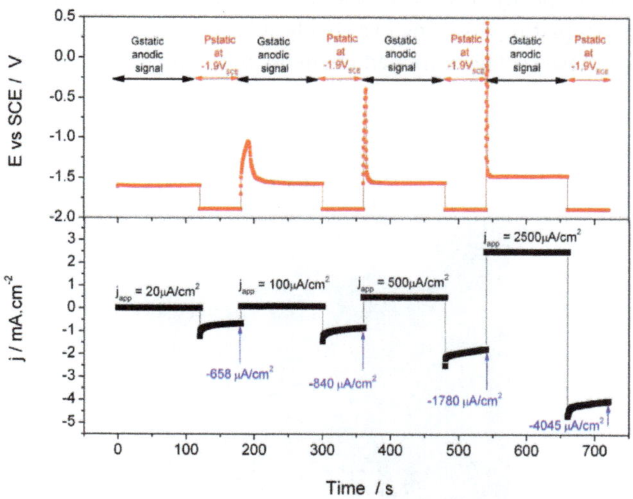

Fig. 5 Results of cycled anodic galvanostatic polarization at different current densities and cathodic potentiostatic polarization at −1.9 V SCE showing the increase in cathodic current following anodic polarization at higher currents.[18] Reprinted with permission from Elsevier.

Fe, would be polarized during Mg dissolution at potentials at which they would be expected to exhibit high rates of HE. As a result, a small enrichment of these elements on the surface, in the oxide, or under the oxide, might be the cause of the anodic HE during Mg dissolution. However, the thick corrosion product complicates the measurement of this enrichment.

Analytical transmission electron microscopy measurements on a cross section of a polarized pure Mg surface were performed by Taheri *et al.*[19] The corrosion product was found to be bilayered with a crystalline oxide inner layer and a crystalline columnar oxy-hydroxide outer layer. Very small Fe-rich particles, in the order of hundreds of nm in size, were found in the outer layer, as shown in Fig. 6. Such sites might be the source of the enhanced catalytic activity of the corrosion film described by Williams *et al.*[16] However, to do so, some of the electrons generated at the dissolving anode sites would have to conduct through the thick oxide layer to the Fe-rich particles embedded in the outer portion of the oxide, and MgO is not expected to be a good electronic conductor.

Another possibility is that the primary location of impurity element enrichment is beneath the oxide corrosion product layer. Cain *et al.* used Rutherford Backscattering Spectrometry (RBS) on pure Mg samples following anodic polarization and chromic acid cleaning to remove the oxide product layer.[12] RBS is ideally suited for the sensing of heavy metals concentrated on the surface of lighter metals. Fig. 7 shows the results of an analysis of the RBS data, which indicates an increase in the Fe/Mg ratio near the surface (under the removed oxide product layer). The authors suggest that the enrichment efficiency is low owing to possible incorporation of the Fe into a non-conducting oxide layer or release (presumably non-Faradaic) of Fe-rich particles into the solution. However, if such release occurred, the particles would reach their own open circuit potential independent of that of Mg, dissolve into solution, and replate back onto conductive surface sites as elemental Fe. The constant rate of HE with time on an anodically polarized Mg surface is not consistent with the expected continual enrichment of impurity or alloying elements over time. A low enrichment efficiency with no replating resulting in a steady-state surface concentration would be necessary for these surface enriched elements to be the site of anodic HE.

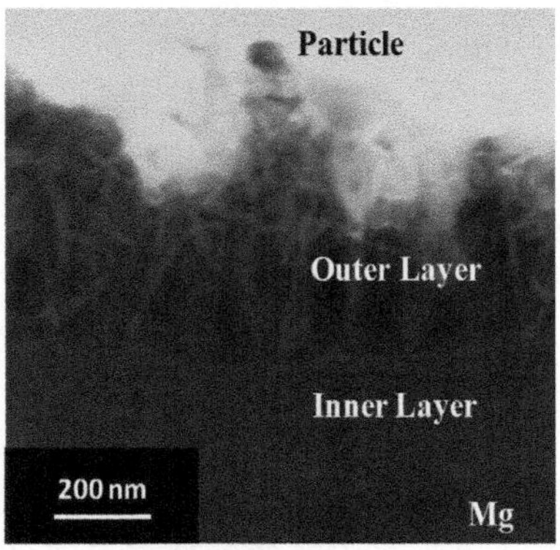

Fig. 6 TEM image of the bilayered corrosion film formed on a pure Mg sample during anodic polarization showing an example of an Fe rich particle in the outer layer.[19] Reprinted with permission from Elsevier.

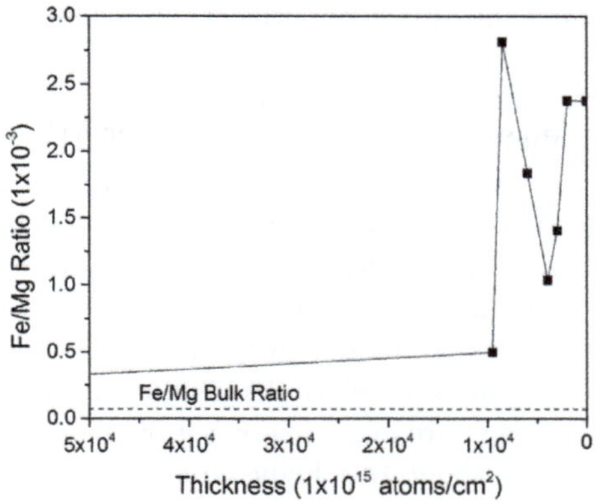

Fig. 7 Fe/Mg ratio as a function of depth from the surface as determined from RBS data for a sample polarized at −1.625 V SCE for 24 h showing Fe enrichment.[12] Reprinted with permission from the authors of ref. 12.

The cathodic behavior of prepolarized Mg samples with 40 or 200 ppm Fe was compared to that of Mg–Fe alloys with a range of Fe content to assess the extent and effect of surface enrichment.[20] The cathodic current at −1.9 V SCE in 0.1 M NaCl after prior anodic polarization was compared to the value expected if all of the Fe were retained on the surface as a result of the anodic prepolarization by comparison with the currents measured on the different alloys without prior anodic polarization. A factor termed the Fe enrichment efficiency was found to be quite low, less than 1%. The mechanisms mentioned above, incorporation into the oxide or non-Faradaic release, were offered as explanations. The authors suggest that "the enhanced HER on anodically polarised Mg is not completely dictated by the Fe concentration in the Mg matrix" and that surface enriched Fe might not be the cause of anodic HE.[20] As intuitive as it seems, the enrichment of Fe and other impurity elements apparently play a minor role in anodic HE. The similarity in anodic HE rate for Mg binary alloys with high concentrations of Fe, Li

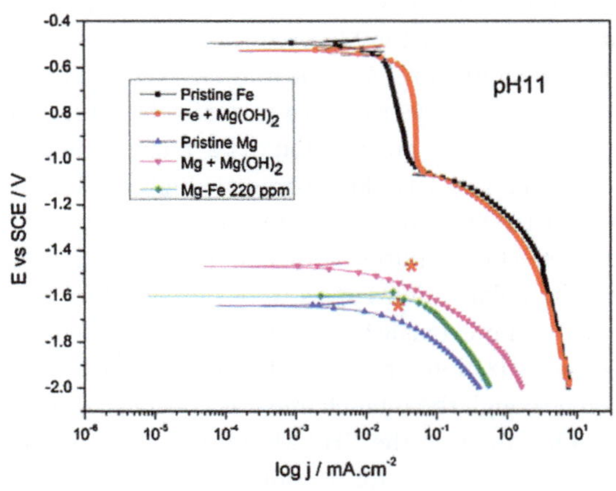

Fig. 8 Cathodic polarization curves for pure Fe and Mg as well as these materials coated with $Mg(OH)_2$ in 0.1 M NaCl solution.[22] Reprinted with permission from Elsevier.

or Ca to that of pure Mg[21] and ultra high purity Mg (99.9999% pure)[5] supports this finding.

The other aspect of the corrosion product that might be important in the anodic HE process is the Mg oxy-hydroxide layer itself. The notion that such a layer is catalytic toward the HER is, in contrast, non-intuitive. However, another report has found evidence for this type of behavior. Salleh *et al.* investigated the cathodic behavior of pure Fe and pure Mg electrodes that were hydroxide coated by immersion in a super-saturated $Mg(OH)_2$ solution.[22] The HER kinetics on uncoated and $Mg(OH)_2$-coated Fe were high and essentially the same, suggesting that the hydroxide coating does not block the HER on Fe, Fig. 8. The kinetics on Mg and $Mg(OH)_2$-coated Mg were much lower, but the kinetics on the hydroxide coated Mg electrode were about 10× higher than on the uncoated Mg. This remarkable finding takes the focus away from the impurity enrichment and places it on the Mg corrosion product. However, like the influence of impurity enrichment, this notion is not consistent with the observation of a constant HE rate with time for samples in which the corrosion sites move across the surface, resulting in an increasing surface coverage of the corrosion product.

The pH dependence of anodic HE on Mg was investigated using two different methods.[23] Volumetric collection of hydrogen was performed in chloride-free buffer solutions of pH 3, 7, and 10. The hydrogen volume increased with decreasing pH owing to the increase in the corrosion rate. Anodic HE was observed at all pH values, but the relative amount was larger at the higher pH. In pH 3 and 7 buffers, the HE rate decreased with increasing applied current just above the OCP, and anodic HE was only observed at higher anodic current densities. Also, yet higher current densities were required to observe anodic HE when the buffer strength was larger. This suggests that pH changes are important for anodic HE such that it only occurs when the buffering capacity is locally overwhelmed. Measurements were also performed in a flow cell connected to an inductively coupled plasma-mass spectrometry (ICP-MS) unit. Anodic HE was also not observed in the flow cell at low current densities, as the flowing condition prevented anodic HE at current densities for which it was observed in volumetric measurements. The pH changes seem to be related to the stability of surface films, indicating the importance of such films in the anodic HE process.

A study by the authors of this work provides evidence that neither surface-enriched impurity elements nor the Mg oxy-hydroxide corrosion product dominate the anodic HE phenomenon.[5] As shown in Fig. 3, high purity (HP, 99.98%) and ultra high purity (UHP, 99.9999% Mg) were tested by volumetric collection of hydrogen during galvanostatic polarization. The rate of anodic HE was only slightly higher on HP than on UHP Mg; extremely high rates of anodic HE were still observed on the UHP Mg, which has only 1 ppm total impurity, in line with the apparent minor effect of impurity elements on anodic HE. Following anodic polarization, cathodic polarization curves were measured in fresh 0.1 M NaCl solution, Fig. 9. As found previously,[18] the cathodic current densities were higher after the anodic prepolarization, and generally increased with increasing anodic current density, even though the total charge was constant. Also, the currents on the HP Mg were higher than on the UHP Mg for the same pretreatment. These observations provide more support for the notion of some cathodic enhancement by the corrosion film and/or enriched impurities, even though the UHP Mg had very low impurity content.

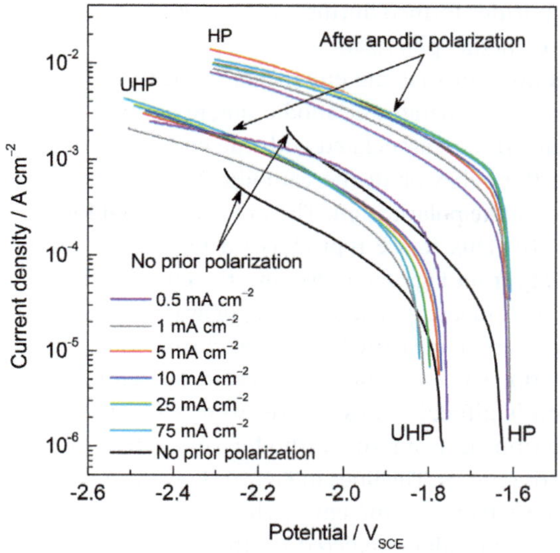

Fig. 9 Cathodic polarization curves (non-IR-corrected) measured on HP and UHP Mg samples after galvanostatic anodic polarization.[5] Reprinted with permission from Elsevier.

The previous paper then analyzed the data in Fig. 9 to show that this enhancement in the rate of the cathodic reaction plays a small role in the anodic HE phenomenon. The primary assumptions were that the corrosion film and

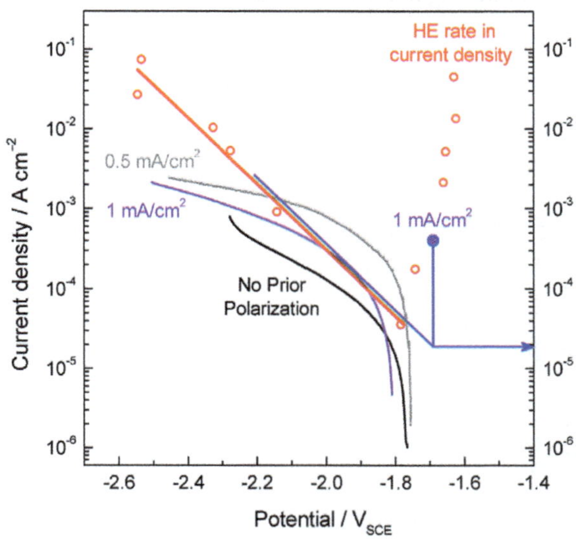

Fig. 10 Cathodic polarization curves (non-IR-corrected) from selected UHP Mg samples in Fig. 9 along with the current density calculated from the HE rate during galvanostatic polarization tests (open circles) plotted against measured and IR-corrected potential.[5] Also shown are lines describing the method for determination of the effect of the corrosion film and impurities on anodic HE. The arrow pointing to the right shows the value of HE rate from the film and impurities predicted by extrapolation for the case of 1 mA cm^{-2}.

accumulated impurities formed during anodic polarization are robust, so that the surface during cathodic polarization reflects the condition during the prior anodic polarization, and that the enhancement in the cathodic reaction is the same mechanism that operates at anodic potentials. It is therefore possible to assess the anodic HE rate associated with the corrosion film and impurities by extrapolation of the behavior in the cathodic region to the potentials reached during the prior anodic polarization. The procedure used for the extrapolation is clarified in Fig. 10. This figure reproduces a few of the cathodic polarization curves for UHP Mg from Fig. 9 and overlays the current density calculated from the volumetric HE measurements during galvanostatic polarization plotted against the IR-corrected measured potential during those measurements. The cathodic polarization curves are not IR-corrected, so there is no clear Tafel region and extrapolation is difficult. These curves were each extrapolated by using the Tafel slope determined from the cathodic volumetric measurements at the tangent point to the curves. The lines in Fig. 10 show the extrapolation for the case of polarization of UHP Mg at 1 mA cm^{-2}. The arrow pointing to the right at about 2×10^{-5} A cm^{-2} is the value predicted for the HE rate from the film and impurities at the IR-corrected potential measured during the galvanostatic polarization, about −1.69 V SCE. This value is much lower than the rate of HE measured volumetrically, which was equivalent to about 4×10^{-4} A cm^{-2}. Other approaches to extrapolation are possible, but it is clear from this figure that the values would in any case be much lower than the measured rates of HE in the anodic region.

Fig. 11 shows the predicted values of anodic HE rate from the corrosion film and accumulated impurity elements determined by the extrapolation of the cathodic polarization curves. The rates are almost independent of the applied anodic current density and are about 10× larger for HP Mg than for UHP Mg. So the impurity level of the Mg plays a role in the anodic HE process, but this role is

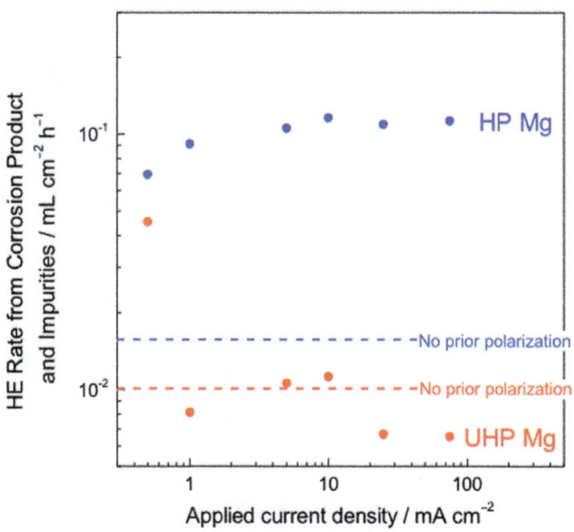

Fig. 11 HE rate from the corrosion product and surface enriched impurities for HP and UHP Mg determined by extrapolation as shown in Fig. 10.[5] Note that 1×10^{-2} ml cm^{-2} h^{-1} = 2.2×10^{-5} A cm^{-2}. Reprinted with permission from Elsevier.

small. The anodic HE rates associated with the corrosion film and accumulated impurity elements for the pretreated UHP Mg are close to that for untreated UHP Mg at the OCP. The value for 0.5 mA cm^{-2} is anomalous and a discussion of this point can be found in the original paper.[5]

As noted above, the rate of HE might increase on a dissolving surface because of an increase in the catalytic activity of the dissolving area. It can be assumed that the total rate of anodic HE is the sum of the HER occurring on the previously corroded surface, which itself is the sum of the HE on the corrosion film and that associated with accumulated impurities, and the HER occurring at the actively dissolving anode sites:

$$HE_{total} = HE_{anode} + HE_{film} + HE_{impurities} \tag{4}$$

A schematic for this process is shown in Fig. 12. It differs from the mechanism in Fig. 4 by noting the HE occurring on the film and impurities separately, and including HE_{anode}. The corrosion attack is shown in this figure as progressing from left to right, leaving behind a corroded area. Knowing the total rate of anodic HE from volumetric measurements and using the rate of anodic HE associated with the film and impurities, HE_{film} + $HE_{impurities}$, determined by the extrapolation method described above, it is possible to assess the anodic HE rate associated with the actively dissolving anodic regions, HE_{anode}. Fig. 13 shows the different components of the anodic HE. The dashed lines represent HE_{total} measured volumetrically. The symbols represent the HE_{anode} values, which were determined by subtracting the values of HE_{film} + $HE_{impurities}$ at each applied current density given in Fig. 11 from HE_{total}. It is clear that anodic HE rates from the film and impurities are small and that HE_{anode} is close to HE_{total}, especially for the higher applied current densities.

A mechanism focused on the actively dissolving anodic regions as the primary site for anodic HE is not inconsistent with the observations described above. Enhanced HER during cathodic polarization for a sample that was previously

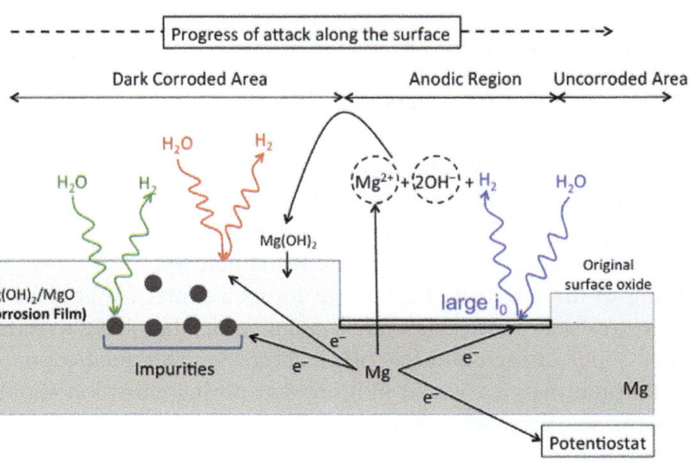

Fig. 12 A schematic representation of the anodic HE process that distinguishes the contribution of corrosion film, accumulated impurities, and anodic area.

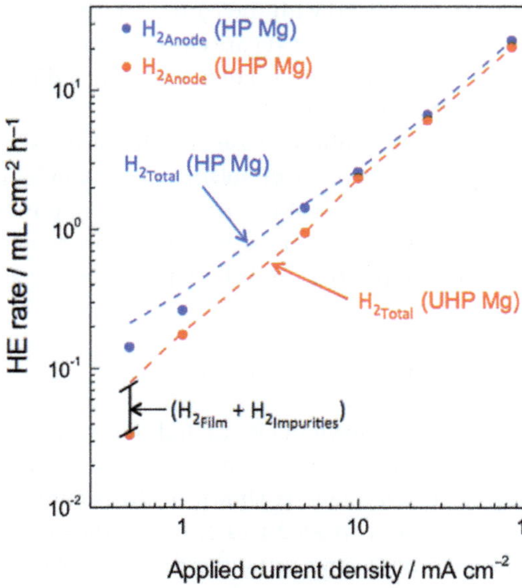

Fig. 13 Anodic HE rate as a function of anodic applied current density.[5] The dashed lines are HE_{tot} measured volumetrically and the symbols are the HE rates from the dissolving anode sites, determined by subtracting the values of HE from the film and impurities shown in Fig. 11. Reprinted with permission from Elsevier.

polarized anodically is strong evidence for the effects of the corrosion film and surface-enriched impurities. These factors might play a significant role during cathodic polarization and OCP dissolution, but their effects appear to be small under the conditions of anodic polarization. The SVET observations of distinct cathodic regions are also not inconsistent because the SVET only measures net current from spatially separated regions and so is not sensitive to a cathodic reaction co-located with a larger anodic reaction, which would be sensed as a net anodic region.

Anodic HE on Al

It was shown above that HE occurs in actively growing pits in Al at high rates. Owing to gradients in potential and chemistry as well as changing active area, factors that are inherent in all pits, it is not possible to assess the kinetics of HE in a controlled fashion as has been done for Mg as long as the corrosion morphology includes pitting. As a result, work was undertaken to find conditions of nominally uniform corrosion of Al and then measure the anodic HE.

According to the Pourbaix diagram, Al forms a protective layer of Al oxide (Al_2O_3) in neutral solutions. This oxide is amphoteric, dissolving to form aluminate and Al^{3+} ions at high and low pH, respectively. This prediction, which is based on thermodynamics, would indicate that pitting corrosion should occur only in neutral solutions because a stable passive film is required for the establishment of pitting. The first approach taken to address anodic HE in Al was to study the behavior in alkaline solutions, which are commonly used for etching of Al. Anodic HE was not observed in very alkaline solutions; the HE rate decreased

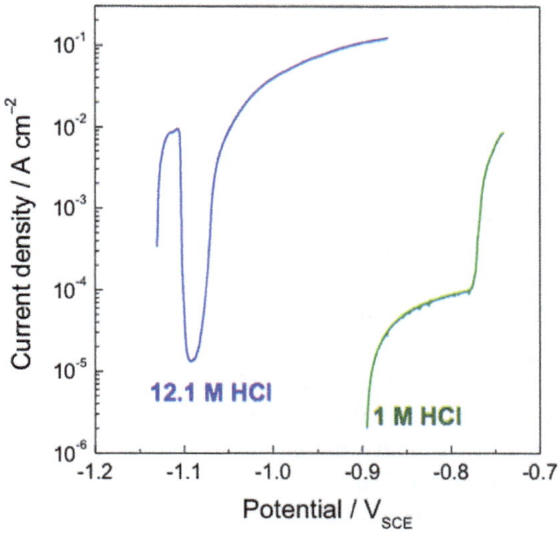

Fig. 14 Potentiodynamic polarization curves for 99.99% Al in HCl solutions measured at 0.1 mV s^{-1}.

with increasing applied anodic current density as expected from the Tafel Law, eqn (3).

Passivity should also be absent in low pH solutions. Furthermore, the pH of pit environments in Al has been measured to be 3,[24] so film-free dissolution should be possible in HCl solutions of this pH and lower. However, some retained passivity and pit formation can be observed on surfaces of pure Al samples polarized to high potentials in HCl solutions with concentrations as high as 4 M. Fig. 14 shows polarization curves for 99.99% Al in 1 M and 12.1 M HCl. The polarization curve in 1 M HCl shows a clear passive region and breakdown potential. The surface following polarization exhibits a high density of crystallographic pits but also clear evidence of unattacked passive surface, Fig. 15a. This shows stability of the passive film to very low pH values and suggests that the pit environment has a pH much lower than 3. Beck reported that the pH drops precipitously as the $AlCl_3$ concentration approaches saturation, reaching as low as −0.25.[25] However, the presence of crystallographic pits

Fig. 15 Optical microscope images of the surfaces of pure Al following measurements of the polarization curves. (a) 1 M HCl, (b) 12.1 M HCl.

rather than polished pits suggests that the pit solution was not saturated, and the copious hydrogen bubble evolution should effectively mix the pit environment with the bulk environment.

The polarization curve measured in 12.1 M HCl, which was taken directly from the acid bottle without dilution, exhibited active dissolution followed by an apparent active/passive transition, Fig. 14. The cause of this behavior is unknown, but it was reproducible and the current decrease was accompanied by an observable cessation of HE. The passive behavior was short-lived as the current increased again to very high values within about 100 mV. Following polarization, the surface was roughened, with more attack at grain boundaries, but no evidence of pitting, Fig. 15b.

The absence of pitting in 12.1 M HCl allowed the HE rate to be measured during anodic galvanostatic polarization, Fig. 16. The HE rate decreased as the applied anodic current density increased to 5 mA cm^{-2}. However, as the current density was increased further, the rate of HE increased substantially, providing clear evidence for anodic HE in Al, even in the absence of pitting corrosion. The amount of anodic HE was approximately 20% of the net applied current, which is similar to the values determined in the Al thin film pitting experiments mentioned above.[4] However, this amount is less than that found on Mg, which has been found to be about 50% of the net applied anodic current.[15]

As was done for Mg, cathodic potentiodynamic polarization measurements were performed on the Al samples following the anodic galvanostatic polarization. In contrast to Mg, these samples exhibited decreased cathodic reactivity compared to a sample that was not prepolarized, Fig. 17, indicating that robust catalytic enhancement of the surface is not required for anodic HE. It should be noted that precipitation of an Al corrosion product film is also not expected for this solution.

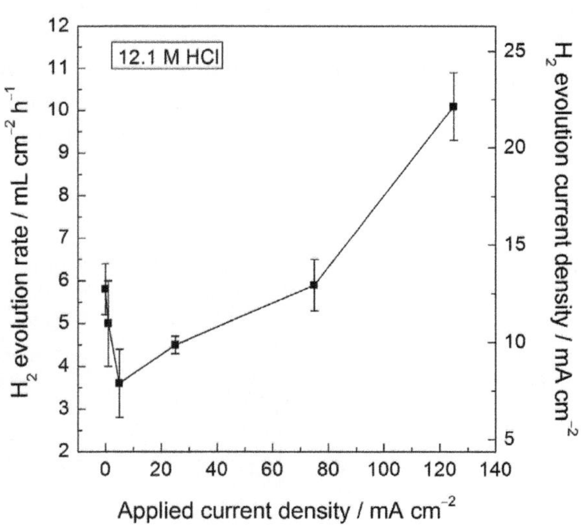

Fig. 16 HE rate measured on pure Al in 12.1 M HCl solution at different applied anodic current densities. The polarization time was 300 s for each case.

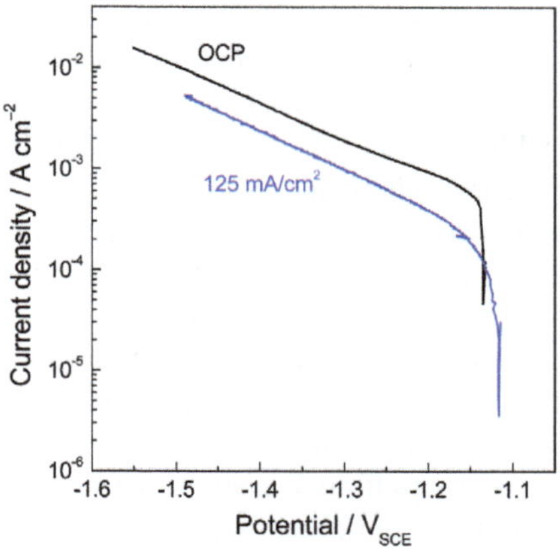

Fig. 17 Cathodic polarization curves measured on a freshly prepared surface and on a sample that was previously polarized at 125 mA cm^{-2} for 300 s.

Mechanism of anodic HE

As described above, the increasing rate of HE with increasing applied anodic current density or potential can be explained by an increase in the exchange current density that overcomes the decreasing exponential term containing the potential. The question is of course why this would happen. One approach to providing an explanation is from the definition of the exchange current density that falls out of the development of the Butler Volmer equation. The formalism presented by Bockris and Reddy[26] will be used here. The current density for an activated process can be assessed from the activation energy using the Arrhenius equation:

$$i = \frac{FkT}{h} c_{\mathrm{i}} \, \mathrm{e}^{-\Delta G^{0\neq}/RT} \tag{5}$$

where F is the Faraday constant, k is the Boltzmann constant, T is the temperature, h is the Planck constant, c_{i} is the ionic concentration, $\Delta G^{0\neq}$ is the activation energy and R is the ideal gas constant. The work associated with moving a charged species through a portion of the potential drop across the double layer (the portion to the activated state) is added to the chemical activation energy:

$$\Delta G^{0\neq} = \Delta G^{0\neq}_{\mathrm{chem}} + \beta e_0 \Delta\phi \tag{6}$$

where $\Delta G^{0\neq}_{\mathrm{chem}}$ is the chemical component of the activation energy, β is the symmetry factor, e_0 is the elemental charge, and $\Delta\phi$ is the potential drop across the double layer. Substituting eqn (6) into eqn (5), one arrives at the Tafel equation, eqn (3), with the exchange current density defined as:

$$i_0 = \frac{FkT}{h} c_{\mathrm{i}} \, \mathrm{e}^{-\Delta G^{0\neq}_{\mathrm{chem}}/RT} \tag{7}$$

It is therefore reasonable to consider that the exchange current density is determined primarily by the chemical activation energy of the electron transfer process, $\Delta G^{0\neq}_{\mathrm{chem}}$, and that a decrease in this value would lead to an increase in i_0. Again, if this decrease in $\Delta G^{0\neq}_{\mathrm{chem}}$ is larger than the increase in $\beta e_0 E$, then the total rate of the cathodic reaction would increase with increasing E.

One approach to addressing $\Delta G^{0\neq}_{\mathrm{chem}}$ is through density functional theory (DFT). DFT is a powerful approach for addressing issues in corrosion science, as is described by several papers in this meeting and volume. The approach has been used to probe the likely pathways for the HER on an Mg(0001) surface.[27] Considering elemental steps of the process separately, the activation energies for different transition states were calculated. Those with the lowest activation energy were considered most likely. This approach might be used to address changes in i_0 that might occur during dissolution. For example, it is possible that the dissolving Mg or Al surface is always filmed, even during high rate dissolution. This is not unreasonable given the extreme reactivity of these metals in the unoxidized state. If the chemical activation energy decreases as dissolution increases owing to a change in the nature of the surface such as a surface film, then the rate of the HER would increase with increasing E.

Another explanation for the increase in anodic HE with potential involves an increase in the number of active sites with increasing dissolution rate. For this to explain anodic HE, the rate of increase of the sites must be greater than the decreasing driving force for the reaction caused by the increasing potential. It can be shown that the validity of this explanation will depend on the relative kinetics of the anodic dissolution reaction and the HER.[28]

Possible implications for anodic HE in Al

There are several implications for the high rate of anodic HE in pits and other forms of localized corrosion in Al and Al alloys. One aspect is the moderating effect that the local cathodic reaction would have on the local pH. As mentioned above, high concentrations in the pit solution should be prevented by the strong evolution of hydrogen bubbles, but the co-location of the HER at a sizeable rate will also prevent the development of low pH values at the site of the anodic reaction.

Another role that high rates of HE in pits and cracks might play is related to environmental cracking. Although for Al alloys it is difficult to make a clear distinction between anodic effects on cracking and the role of hydrogen, which is always present during localized corrosion of Al, a strong case has been made for the embrittlement by hydrogen playing a significant role.[29,30] An interesting example is the situation of AA5083, an Al–Mg alloy that can form a β phase, Al_3Mg_2, along grain boundaries during prolonged exposure at intermediate temperatures that can be experienced, for instance, in service in certain maritime applications.[31–33] The β phase is anodic to the matrix and preferentially dissolves. When sufficient grain boundary β phase is present, the alloy is considered to be sensitized and will exhibit intergranular corrosion (IGC) and intergranular stress corrosion cracking (IGSCC) in chloride-containing environments. This explanation for the IGSCC susceptibility of sensitized AA5083 by the anodic dissolution of grain boundary β phase makes perfect sense, except that transmission electron microscopy imaging of even highly sensitized material does not reveal a continuous β phase along the grain boundaries.[33] In fact, intergranular fracturing of highly sensitized AA5083 by Ga embrittlement, which does not cause corrosion and thus preserves the details of the grain boundary fracture surface, reveals distinctly discontinuous β phase particles along the grain boundary.[34] The sustained intergranular fracture of the grain boundary ligaments between the β particles has been explained by a hydrogen embrittlement mechanism, such that the total fracture mechanism involves coupled dissolution of the β phase particles and hydrogen embrittlement.[35] HE at the crack tip in association with the localized anodic dissolution creates atomic H, which can be absorbed by the Al, transported to the fracture process zone preceding the crack tip, and result in embrittlement of the β-free grain boundary regions. Acidification caused by aluminum ion hydrolysis was described as being critical in promoting hydrogen evolution.[35] However, the high rate of anodic HE that likely occurs at the crack tip probably allows H to play a role independent of the crack solution acidification by hydrolysis. This possibility of high rate anodic HE influencing SCC has already been suggested.[36]

It is also possible that anodic HE promotes an embrittlement effect even in the absence of an applied stress, resulting in a hydrogen-related mechanism for nominally unstressed IGC in high strength Al alloys. The application of a strain causing a stress of less than 10% of the yield stress along the through thickness direction of a plate of AA2024-T3 resulted in a sizeable increase in the rate of IGC under aggressive conditions, essentially causing IGSCC.[37] Given this low threshold for the acceleration of intergranular attack, it seems that the corrosion product could provide enough stress to drive the attack, and that the concomitant anodic HE within the IGC site/crack might play a role in driving the IG separation.

Concluding remarks

This introductory paper has focused on a particular topic, the increase in the rate of the HER with increasing potential on actively dissolving Mg and Al surfaces, which is sometimes referred to as the negative difference effect. This topic is only a small part of the vast field of corrosion chemistry, but it embodies many of the important aspects of corrosion that were listed in the introduction: strong potential and compositional gradients, transport, metal composition and microstructure, stress effects, advanced analytical methods and modeling. It is also a somewhat controversial issue that has attracted a lot of attention recently, so it is a good topic to set the stage for discussions on corrosion chemistry.

The recent literature in the area of anodic HE on Mg has focused on the role of the corrosion product layer, both accumulated impurity atoms in or below the layer, as well as the nature of the Mg oxy-hydroxide itself. A careful analysis indicates that this product layer and accumulated impurities cannot play a large role in the high levels of HE found during anodic polarization of Mg at high currents or potentials. Eliminating these aspects leaves the site of the anodic reaction as the location for most of the anodic HE. A convenient way to consider this effect is through changes in the exchange current density for the HER, which might be controlled by the chemical activation energy associated with details in the surface characteristics.

Analogous HE on dissolving Al was found in concentrated HCl, an environment in which pitting corrosion does not occur. The rate of anodic HE on Al is less than on Mg, but that is not unexpected as Mg is a more reactive element. The exact mechanism for the increase in the rate of HE with increasing potential on

Mg and Al is not clear at this point. However, several testable hypotheses have been suggested, as listed in Table 1, and it is expected that a verified and agreed upon understanding will develop in the coming years.

References

1 G. S. Frankel, *J. Electrochem. Soc.*, 1998, **145**, 2186–2198.

2 S. Glanvill, A. Du Plessis, S. Street, M. Basham, T. Rayment and A. Davenport, *In situ* tomographic investigations of initiation of localised corrosion of aluminium alloy AA2024, poster presentation at Faraday Discussion: Corrosion Chemistry, London, 2015.

3 D. M. Drazic and J. P. Popic, *J. Appl. Electrochem.*, 1999, **29**, 43–50.

4 G. S. Frankel, *Corros. Sci.*, 1990, **30**, 1203–1218.

5 S. Fajardo and G. S. Frankel, *Electrochim. Acta*, 2015, **165**, 255–267.

6 S. Thomas, N. V. Medhekar, G. S. Frankel and N. Birbilis, *Curr. Opin. Solid State Mater. Sci.*, 2015, **19**, 85–94.

7 A. Atrens and W. Dietzel, *Adv. Eng. Mater.*, 2007, **9**, 292–297.

8 G. Song, A. Atrens, D. Stjohn, J. Nairn and Y. Li, *Corros. Sci.*, 1997, **39**, 855–875.

9 R. L. Petty, A. W. Davidson and J. Kleinberg, *J. Am. Chem. Soc.*, 1954, **76**, 363–366.

10 A. Samaniego, B. L. Hurley and G. S. Frankel, *J. Electroanal. Chem.*, 2015, **737**, 123–128.

11 Z. M. Shi, F. Y. Cao, G. L. Song and A. Atrens, *Corros. Sci.*, 2014, **88**, 434–443.

12 T. Cain, S. B. Madden, N. Birbilis and J. R. Scully, *J. Electrochem. Soc.*, 2015, **162**, C228–C237.

13 L. Rossrucker, K. J. J. Mayrhofer, G. S. Frankel and N. Birbilis, *J. Electrochem. Soc.*, 2014, **161**, C115–C119.

14 J. Swiatowska, P. Volovitch and K. Ogle, *Corros. Sci.*, 2010, **52**, 2372–2378.

15 G. S. Frankel, A. Samaniego and N. Birbilis, *Corros. Sci.*, 2013, **70**, 104–111.

16 G. Williams, N. Birbilis and H. N. McMurray, *Electrochem. Commun.*, 2013, **36**, 1–5.

17 G. Williams, personal communication, 2015.

18 N. Birbilis, A. D. King, S. Thomas, G. S. Frankel and J. R. Scully, *Electrochim. Acta*, 2014, **132**, 277–283.

19 M. Taheri, J. R. Kish, N. Birbilis, M. Danaie, E. A. McNally and J. R. McDermid, *Electrochim. Acta*, 2014, **116**, 396–403.

20 D. Lysne, S. Thomas, M. F. Hurley and N. Birbilis, *J. Electrochem. Soc.*, 2015, **162**, C396–C402.

21 A. Samaniego, N. Birbilis, X. Xia and G. S. Frankel, *Corrosion*, 2015, **71**, 224–233.

22 S. H. Salleh, S. Thomas, J. A. Yuwono, K. Venkatesan and N. Birbilis, *Electrochim. Acta*, 2015, **161**, 144–152.

23 L. Rossrucker, A. Samaniego, J.-P. Grote, A. M. Mingers, C. A. Laska, N. Birbilis, G. S. Frankel and K. J. J. Mayrhofer, *J. Electrochem. Soc.*, 2015, **162**, C333–C339.

24 K. P. Wong and R. C. Alkire, *J. Electrochem. Soc.*, 1990, **137**, 3010–3015.

25 T. R. Beck, *Electrochim. Acta*, 1984, **29**, 485–491.

26 J. O. M. Bockris, A. K. N. Reddy and M. Gamboa-Aldeco, *Modern Electrochemistry 2A*, Kluwer Academic, New York, 2000.

27 K. Williams, personal communication, 2015.

28 R. C. Newman, personal communication, 2015.
29 R. P. Gangloff, in *Enginering Materials Advisory Services*, ed. A. Blom, West Midlands, UK, 2002, pp. 3401–3433.
30 J. R. Scully, G. A. Young and S. W. Smith, *Gaseous Hydrogen Embrittlement of Materials in Energy Technologies, Vol. 1: The Problem, Its Characterisation and Effects on Particular Alloy Classes*, 2012, pp. 707–768.
31 D. R. Baer, C. F. Windisch, M. H. Engelhard, M. J. Danielson, R. H. Jones and J. S. Vetrano, *J. Vac. Sci. Technol., A*, 2000, **18**, 131–136.
32 J. R. Pickens, J. R. Gordon and J. A. S. Green, *Metall. Trans. A*, 1983, **14**, 925–930.
33 J. L. Searles, P. I. Gouma and R. G. Buchheit, *Metall. Mater. Trans. A*, 2001, **32**, 2859–2867.
34 N. Birbilis, R. Zhang, M. L. C. Lim, R. K. Gupta, C. H. J. Davies, S. P. Lynch, R. G. Kelly and J. R. Scully, *Corrosion*, 2013, **69**, 396–402.
35 C. B. Crane, Validation of the Coupled Dissolution-Hydrogen Embrittlement Mechanism of IGSCC in Low Temperature Sensitized AA5083-H131, PhD Thesis, University of Virginia, 2013.
36 D. M. Drazic and J. P. Popic, The negative difference effect and stress corrosion cracking, *5th European Symposium on Electrochemical Engineering*, Institution of Chemical Engineers Symposium Series, 1999, 145, pp. 111–120
37 X. D. Liu, G. S. Frankel, B. Zoofan and S. I. Rokhlin, *Corros. Sci.*, 2004, **46**, 405–425.

Faraday Discussions

PAPER

Aqueous solution/metal interfaces investigated *in operando* by photoelectron spectroscopy

O. Karslıoğlu,[a] S. Nemšák,[b] I. Zegkinoglou,†[a] A. Shavorskiy,‡[a] M. Hartl,[ac] F. Salmassi,[d] E. M. Gullikson,[d] M. L. Ng,[e] Ch. Rameshan,§[a] B. Rude,[a] D. Bianculli,[i] A. A. Cordones,[a] S. Axnanda,¶[f] E. J. Crumlin,[f] P. N. Ross,[d] C. M. Schneider,[b] Z. Hussain,[f] Z. Liu,*[fg] C. S. Fadley*[dh] and H. Bluhm*[a]

Received 7th January 2015, Accepted 16th January 2015
DOI: 10.1039/c5fd00003c

We describe a new *in operando* approach for the investigation of heterogeneous processes at solid/liquid interfaces with elemental and chemical specificity which combines the preparation of thin liquid films using the meniscus method with standing wave ambient pressure X-ray photoelectron spectroscopy [Nemšák *et al.*, *Nat. Commun.*, **5**, 5441 (2014)]. This technique provides information about the chemical composition across liquid/solid interfaces with sub-nanometer depth resolution and under realistic conditions of solution composition and concentration, pH, as well as electrical bias. In this article, we discuss the basics of the technique and present the first results of measurements on KOH/Ni interfaces.

[a]*Chemical Sciences Division, Lawrence Berkeley National Laboratory, Berkeley, CA 94720, USA. E-mail: HBluhm@lbl.gov*

[b]*Peter-Grünberg-Institut PGI-6, Forschungszentrum Jülich GmbH, 52425 Jülich, Germany*

[c]*Julius -Maximilians-Universität Würzburg, Sanderring 2, 97070 Würzburg, Germany*

[d]*Materials Sciences Division, Lawrence Berkeley National Laboratory, Berkeley, CA 94720, USA. E-mail: CSFadley@ucdavis.edu*

[e]*SLAC National Accelerator Laboratory, 2575 Sand Hill Road, Menlo Park, CA 94025, USA*

[f]*Advanced Light Source, Lawrence Berkeley National Laboratory, Berkeley, CA 94720, USA. E-mail: ZLiu2@lbl.gov*

[g]*State Key Laboratory of Functional Materials for Informatics, Shanghai Institute of Microsystem and Information Technology, Chinese Academy of Sciences, Shanghai 200050, China*

[h]*Department of Physics, University of California Davis, Davis, CA 95616, USA*

[i]*Engineering Division, Lawrence Berkeley National Laboratory, Berkeley, CA 94720, USA*

† Present address: Lehrstuhl für Experimentalphysik IV, Ruhr-Universität Bochum, D-44780 Bochum, Germany.

‡ Present address: MAX IV Laboratory, Lund University, P.O. Box 118, SE-221 00 Lund, Sweden.

§ Present address: Vienna University of Technology, Institute of Materials Chemistry, 1060 Vienna, Austria.

¶ Present address: BASF Catalysts LLC, 25 Middlesex Avenue, Iselin, NJ 08830, USA.

Introduction

Liquid/solid and gas/solid interfaces govern the majority of reactions in corrosion science. Metals undergo changes to their surface under atmospheric conditions, when thin aqueous layers are condensed on their surface, or when they are immersed in a bulk solution. Corrosion is inherently an electrochemical process and is characterized by a concurrent transfer of mass and charge across the metal/solution interface in a coupled electrochemical reaction, where anodic and cathodic processes occur simultaneously either at different sites on the same material or on different parts of the system.[1] The transfer of metal atoms into solution through oxidation in the anodic reaction leads to the structural weakening of the material and can ultimately result in mechanical failure, with potentially dire consequences.

A detailed understanding of the processes governing corrosion requires the investigation of the metal surface in interaction with its surrounding gas-phase or liquid-phase environment on the atomic scale. Surface science investigations into corrosion processes can be performed in principle in two ways, either by comparing the state of the surface before and after the corrosion process has taken place (*i.e.*, *ex situ*), or *in operando*, *i.e.* monitoring the surface during the process. While the whole arsenal of surface science methods can be used to perform *ex situ* investigations, the use of *in operando* methods often poses experimental challenges. These include the need to operate under elevated pressure conditions, since the partial pressure of water and aqueous solutions at environmental temperatures ranges from some tenths of Torr to tens of Torr. Another challenge is enhancing the signal from the interface region over that originating from the bulk liquid and solid. Over the past decades, considerable progress has been made in the adaptation of vacuum-based surface science techniques to investigations of surfaces under realistic relative humidity and even liquid/solid interfaces. Surface sensitive spectroscopies that can be used under these conditions include optical methods, such as infrared spectroscopy (IR)[2,3] and vibrational sum-frequency generation (VSFG);[4,5] X-ray emission spectroscopy (XES);[6] near edge X-ray absorption fine structure (NEXAFS) with fluorescence detection[7] as well as surface X-ray diffraction (SXRD).[8] Among the imaging techniques that can be used to study surfaces in the presence of gases and liquids are scanning force microscopy (SFM) in both contact[9] and non-contact[10] modes, as well as scanning tunneling microscopy (STM),[11] and in recent years also transmission electron microscopy (TEM)[12] and scanning electron microscopy (SEM).[13] A recent overview of surface science techniques used for measuring liquid/solid interfaces is given in ref. 14.

X-ray photoelectron spectroscopy (XPS) is an excellent method for monitoring corrosion processes, since it is exquisitely surface sensitive due to the short mean free path of electrons in matter at the typical kinetic energies (KE) used in XPS (a few 100 to a few 1000 eV). XPS permits quantitatively determining the elemental composition of the surface and, through detection of so called "chemical shifts", also the oxidation states of the different components of the surface. Since XPS uses electrons as probes, changes in local electrical potentials can also be monitored as they will lead to rigid kinetic energy shifts experienced by all core level and valence peaks and can thus be distinguished from shifts due to chemical changes.[15] XPS is thus ideally suited to study corrosion processes since it permits monitoring both changes in the chemistry as well as electronic properties of the surface simultaneously and in non-contact.

However, corrosion processes commonly take place under ambient conditions around room temperature and with an equilibrium water vapor pressure. At higher pressures, the scattering of electrons by gas molecules and the risk of arcing in the electrostatic lens or the detector pose serious problems. These limitations have been overcome through ambient pressure X-ray photoelectron spectroscopy (APXPS), which combines all advantages of vacuum-based XPS – quantitative elemental and chemical information at high surface sensitivity – with the possibility to measure samples at elevated pressures.[16–18] The technique was first introduced by the Siegbahn group in Uppsala, Sweden in the early 1970s. These spectrometers used several differential pumping stages between the sample cell and the hemispherical electron analyzer, with a small entrance aperture to the first pumping stage brought into close vicinity of the sample surface, to both minimize the path length of the electrons through the high pressure region and the leakage of gas into the spectrometer. The inelastic mean free path (IMFP) of electrons with a kinetic energy of 100 eV (*i.e.*, those with the highest surface sensitivity) is about 1 mm in 1 Torr of water vapor. Typical distances between the sample and the entrance aperture are ~1 mm. APXPS was further developed in the late 1990s by introducing electrostatic lenses in the differential pumping stages which overcame the trade-off between collection angle for electron detection and maximum pumping efficiency. A schematic layout of such a system is shown in Fig. 1. These instruments can now operate at pressures above 20 Torr, thus passing the important threshold of the equilibrium vapor pressures of water at room temperature. As an illustrative example of the

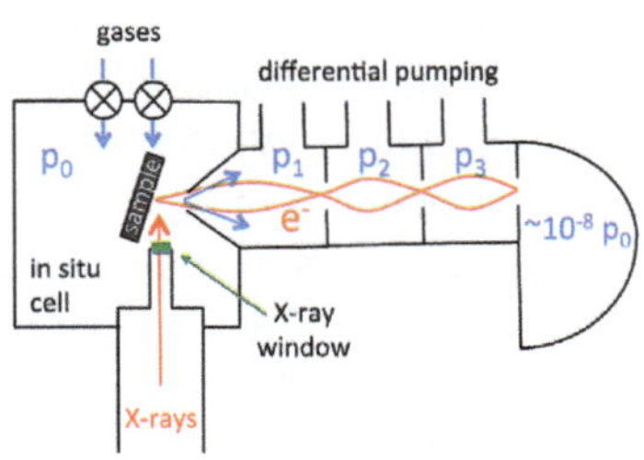

Fig. 1 Principal layout of a typical APXPS experiment. The sample is located inside the *in situ* cell, which is backfilled with a gas or gas mixture to a pressure p_0. X-rays from a laboratory source (X-ray anode) or synchrotron beamline are admitted to the *in situ* cell through an X-ray transparent window, most often silicon nitride with a thickness of ~100 nm and an active area of ~1 mm^2. The sample is placed in close proximity of the entrance aperture of the differentially pumped lens system of a hemispherical electron analyzer. The typical aperture size is ~0.1 mm^2, and the typical sample–aperture distance about 1 mm. Electrons and gas molecules escape through the entrance aperture into the first differential pumping stage, which is typically at a pressure of $p_1 < 10^{-3} p_0$, *i.e.* scattering of electrons by gas molecules in the first differential pumping stage is greatly reduced. Many APXPS systems feature several differential pumping stages, which successively decrease the pressure by several orders of magnitude. Electrostatic lenses in the differential pumping stages focus the electrons onto the apertures that separate the stages, thus maintaining a similar solid angle of electron detection as in a vacuum-based XPS instrument.

importance of pressure, recent APXPS studies of Cu surfaces have shown the importance of the surface crystallography for the wettability of metals: while Cu(110) is covered by a mixed hydroxyl/water layer at a relative humidity (RH) of 5%, Cu(111) does not react with water vapor at the same RH, due to the higher dissociation barrier of water on the less reactive (111) surface.[19,20] From these data and other studies of surfaces under ambient RH[21,22] a picture emerges in which most oxide and metal surfaces are hydroxylated and show adsorbed molecular water at relative humidities far below those present under normal ambient conditions, with significant consequences for atmospheric corrosion processes on metals.

While the investigation of solid/vapor interfaces at relevant relative humidities with high chemical and surface sensitivity using APXPS has become a standard method over the past decade, the investigation of solid/liquid interfaces using photoelectron spectroscopy is still a challenging task. Kolb *et al.* developed a quasi-*in situ* method for determining the composition and charge state of electrochemical interfaces using UHV XPS combined with an electrochemical cell in which the sample surface was dipped into an aqueous solution under potential, then removed, dried and transferred to the XPS measurement chamber.[23] Even though most of the water evaporated from the interface, the electrical double layer (EDL) was shown to be preserved so that measurements of the potential drop in the EDL became possible.[24,25] While this method has been proven to be a viable strategy for the investigation of the static properties of the EDL, it is not suitable for studies of corrosion processes *in operando* due to the lack of charge and mass transport during the XPS measurements. Artifacts can also result from the drying of the surface. True *in operando* measurements require the presence of a thicker solution layer in contact with a solution reservoir at the interface.

Since electrons are strongly scattered by ions and molecules in a liquid, there is a limit to the thickness of the solution layer in normal XPS measurements. Fig. 2 presents calculated and measured data of the inelastic mean free path of electrons in liquid water.[26] For maximum achievable kinetic energies in experiments using an Al Kα lab source (1486.7 eV) the IMFP is only about 6 nm, or 20 water monolayers. Recent developments of high KE APXPS instruments have opened the opportunity to detect electrons with much higher KE.[27] For a Cu Kα source (8047.8 eV) or a hard X-ray beamline at a synchrotron with 8 keV incident photon energy, 8 keV electrons with an IMFP of $\sim$20 nm in liquid water can be detected. One can argue that at these water film thicknesses the properties of the liquid/solid interface approach or are that of a true bulk liquid/solid interface.

For the study of solid/liquid interfaces with stable liquid layers of a thickness between several nm and several 10 nm, XPS measurements of all core levels of interest with sufficient signal-to-noise ratios are complicated by the expected low signal from the interface due to the attenuation of the electrons in the gas as well as liquid phase. One intriguing approach to preparing liquid layers with thicknesses in the nm range is the adaptation of the well-established liquid jet method, which has been successfully used to measure the composition of liquid/vapor interfaces.[28] For the investigation of liquid/solid interfaces, liquid jets are prepared from solutions containing suspended nanoparticles, where some of the nanoparticles reside sufficiently close to the surface of the liquid jet so that their interface with the solution can be probed using XPS.[29] This method has been successfully used to investigate, *e.g.*, the surface charge density of silica particles

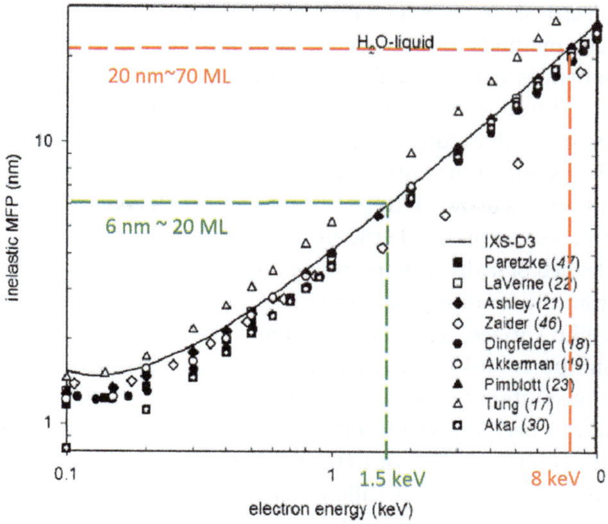

Fig. 2 Electron inelastic mean free path in liquid water as a function of electron kinetic energy. For the excitation energy of an Al Kα lab source (1486.7 eV) the maximum IMPF is about 6 nm, which corresponds to a water film thickness of about 20 monolayers. At higher excitation energies, *e.g.* 8 keV, which is a realistic value for high KE APXPS instruments, the mean free path increases to ~20 nm, or ~70 monolayers of water. The drawback of higher photon energies is the lower photoemission cross section and thus the decrease in the photoemission intensity. Adapted and reproduced with permission from ref. 26.

as a function of particle size and pH of the solution, but has not yet been tested for its applicability to corrosion studies.[29]

Another recently explored method to measure liquid/solid interfaces is the detection of photoelectrons through a thin membrane (*e.g.*, multilayer graphene), with the photoelectron analyzer located on the vacuum side of the membrane.[30] The advantage of this approach is that it overcomes the need for a thin solution layer in XPS investigations of the liquid/solid interface. The challenge here is to prepare ultrathin solid films of the material of interest so that photoelectrons can penetrate the solid on their way to the detector, as well as the enhancement of the interface signal over that from the bulk, which is in fact closer to the spectrometer. Also, this method does not lend itself to a standing wave approach as described below.

A further possibility to prepare thin liquid films is to expose the sample to precisely controlled relative humidities just below saturation (RH 100%), since the water film thickness increases rapidly as the RH approaches 100%. The challenges of this approach are three-fold: (I) the precise control of temperature and water vapor pressure in the experiment, since any small deviation will lead to rapid liquid film growth or desorption. (II) The control of the concentration and pH of the solution. While alkali halide solution films have successfully been formed by re-hydrating either drop-cast or evaporated alkali halide films inside a humidity-controlled vacuum chamber,[31,32] the exact control of the concentration and pH of the film is challenging. (III) Finally, the connection of the liquid film to an external electrical circuit, which might be achieved using an approach akin to scanning probe methods to immerse an electrode into the thin solution film.

These challenges can be overcome using the so-called meniscus method, which was first introduced many decades ago by Bockris and colleagues for the determination of the dissolution of gases in porous electrodes,[33] and has recently been applied by some of the authors of this article for the first time to studies of liquid/solid interfaces using APXPS.[27] The principle is shown in Fig. 3. The sample is immersed in a solution inside the measurement cell and can then be electrochemically cleaned of hydrocarbons and other contamination. When it is partially withdrawn from the solution, a meniscus will extend to a height z_1 (for neat water on a hydrophilic surface, a few mm)[34] above the bulk solution surface. Under favorable conditions and in the presence of a vapor pressure p_{vap} (which is a function of the solution temperature T_{sol}) close to saturation, a thin liquid film can be stabilized over the whole range over which the sample was originally dipped into the solution (z_2). Close to the top of the liquid film (at $\sim z_2$) parts of the liquid film are sufficiently thin to allow collection of photoelectron spectra from the liquid/solution interface. Liquid films formed in that manner have been shown to be stable for many hours.[27] This method was used in the present experiments and is described in more detail in the Experimental section.

The last challenge that needs to be overcome in the study of liquid/solid interfaces using APXPS is enhancing the signal from the narrow interfacial region

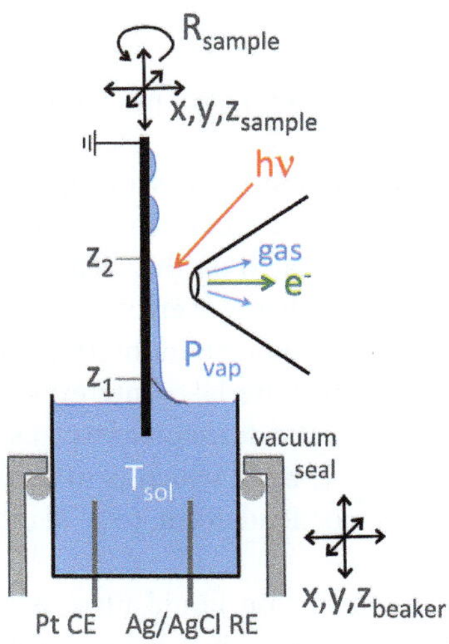

Fig. 3 Schematic of the meniscus method for the preparation of thin liquid films. For hydrophilic surfaces the meniscus will extend to a height z_1 (for neat water a few mm) above the bulk solution surface (for simplicity the formation of the liquid film is shown for only one side of the sample). Under favorable conditions and in the presence of water vapor pressures close to saturation a thin liquid film can be stabilized over the whole range that the sample was originally dipped into the solution (z_2). This part of the film is in electrical contact with the bulk solution. The front aperture of the differentially pumped lens system of the APXPS system is also shown. For full electrochemical control, a counter and reference electrode are placed into the solution, while the sample (working electrode) is held on ground potential, in case of a metal with its Fermi level aligned to that of the photoelectron spectrometer.

over that originating from the bulk of the liquid layer and solid substrate. To that end, some of the authors of this paper have recently introduced standing wave ambient pressure photoemission spectroscopy (SWAPPS), which is based on well-established standing wave photoemission spectroscopy, where the exciting X-ray wave field is tailored into a standing wave that is scanned through the interface of interest and thus provides much increased depth-resolution in XPS experiments (see Fig. 4).[35,36] The advantage of the standing wave approach is that it provides a built-in ruler, *i.e.*, the period of the standing wave, to determine absolute depth distributions of chemical species throughout the probed volume. The combination of SWAPPS with the meniscus method for the preparation of thin liquid layers thus has the potential to provide detailed chemical information about all components of the system (bulk liquid, liquid/solid interface, bulk solid) with sub-nm resolution in the direction perpendicular to the interface, with full control over the concentration and pH of the solution, as well as the applied potentials. In the following we describe the experimental details and present a proof-of-principle investigation for the reaction of Ni immersed in a 0.1 M KOH solution under oxidizing conditions.

Experimental

Preparation of thin liquid films

Fig. 3 describes the experimental set-up used in the current investigations. A glass beaker (outer diameter 3.2 cm, height 5 cm) is mounted inside a vacuum chamber using an UltraTorr (O-ring seal) fitting. An Ag/AgCl reference electrode (eDAQ, leakless miniature) is glued into a hole in the bottom of the beaker. A Pt counter electrode is fused into the bottom of the glass beaker. The UltraTorr fitting is

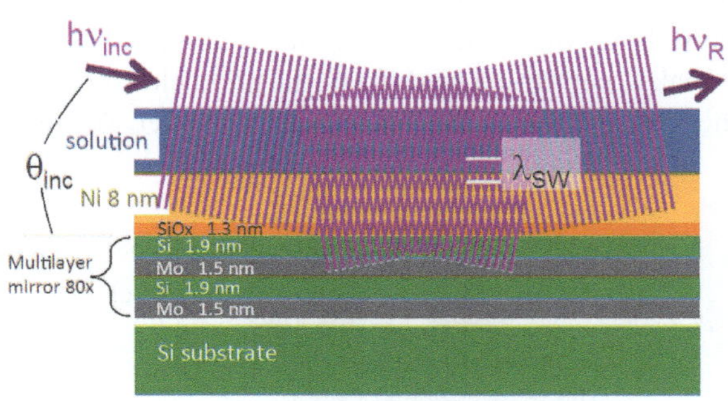

Fig. 4 Principal layout of a SWAPPS experiment, with the dimensions of the multilayer and sample indicated as used in the present experiments. The Si/Mo multilayer mirror with a periodicity of 3.4 nm is grown onto a Si substrate. Incident photons ($h\nu = 3.1$ keV) are reflected off the multilayer mirror under the first order Bragg angle ($\theta_{inc} \sim 3.4°$) and interfere with the reflected X-rays, leading to the formation of a standing wave with a periodicity close to that of the multilayer mirror in the direction perpendicular to the surface. When the incidence angle of the X-rays is changed around the first order Bragg condition (rocking curve), both the amplitude and the phase of the standing wave change. The Ni sample is grown on top of the SiO_2 cap layer of the multilayer mirror. The phase change during a rocking curve shifts the standing wave by a distance of half of its wavelength and thus allows variation of the amplitude of the standing wave across the metal/solution interface.

attached *via* an XYZ manipulation stage to the bottom flange of a high-vacuum chamber, which in turn is mounted to the front end of the differentially pumped electron lens system of the APXPS spectrometer. The sample is mounted on an XYZ manipulator and a rotational stage. A commercial KOH solution (Alfa Aesar, 30% w/v) is diluted to 0.1 M with ultra-pure water, and then degassed in a separate chamber by pumping before being introduced into the beaker inside the experimental cell. The solution temperature T_{sol} is allowed to equilibrate to the temperature of the surrounding parts of the experiment and not actively controlled. The experimental cell is slowly evacuated using an external roughing pump until the pressure in the cell, measured using a capacitance pressure gauge, is close to that of the expected vapor pressure of the solution at room temperature (p_{vap}), at which point the external pump is isolated. During the experiments the measurement cell is pumped through the front aperture (diameter 0.1 mm) of the differentially pumped electrostatic lens; the loss in gas molecules in the experimental cell is compensated for by evaporation from the solution in the beaker, and from a separate water reservoir (~5 cm^3; not shown) that acts as a buffer to reduce the decrease of the solution level in the beaker and limit the change in solution temperature due to evaporative cooling. With this procedure experiments can be performed up to 24 h without the necessity to refill the beaker with fresh solution, and with only minor changes in the solution concentration.

At the beginning of the experiment the sample is immersed as far as possible into the solution and its electrochemical properties are characterized (see Fig. 5c) using a Bio-Logic SP-150 potentiostat. By using appropriate parameters for cyclic voltammograms (CV), the sample surface can also be cleaned, *e.g.*, from carbon contamination. When the sample is then partially retracted from the solution, a meniscus forms, which – for a hydrophilic sample – extends to a height z_1 above the bulk liquid surface (see Fig. 3), with z_1 being close to the capillary length, which is of the order of a few mm for water.[34] A thin, continuous solution film extends beyond the meniscus up to a height z_2, the depth to which the sample was originally immersed into the solution. This film is in electrical contact with the solution in the beaker, confirmed by shifts in the photoelectron peaks of the thin film upon application of a bias to the solution in the beaker. The film is stabilized

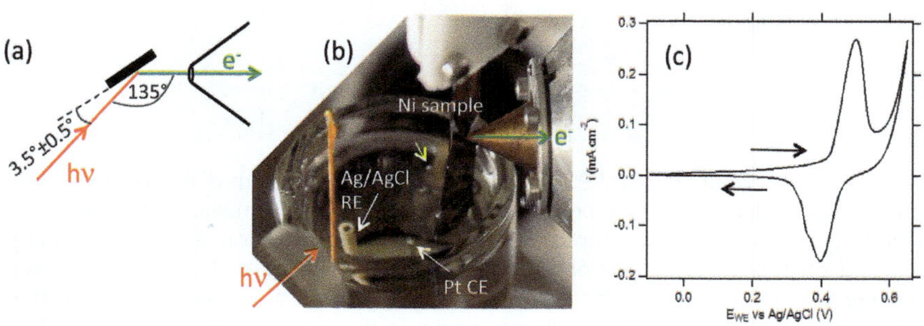

Fig. 5 (a) Top view of the measurement geometry in the SWAPPS experiments. (b) Photograph of the sample inside the measurement chamber, dipped into a 0.1 M KOH solution at +0.6 $V_{Ag/AgCl}$. The yellow arrow indicates a bubble that is formed in the solution. (c) Cyclic voltammogram of a nickel film on top of a multilayer structure in 0.1 M KOH taken inside the measurement chamber in the configuration shown in (b). The scan rate was 20 mV s^{-1}.

by the high RH (in case of an aqueous solution) close to 100% due to the presence of the vapor pressure above the solution. The initial, electrochemical removal of hydrophobic contaminations also supports the formation of a continuous liquid film. Above the height z_2 there is also water adsorption (from the gas phase), but these parts of the liquid film are usually not in electrical contact with the solution in the beaker, most likely due to hydrophobic contamination which prevents the formation of a continuous film. In some cases the thin water film between z_1 and z_2 breaks up and forms areas where patches of electrolyte remain with the thickness and ion concentration of the continuous film, but without electrical contact to the solution in the beaker. A photograph of the sample in the measurement position is shown in Fig. 5b.

We note that a similar approach for the preparation of a thin film has also been used to study the interface between an ionic liquid (vapor pressure < 10^{-9} Torr) and a metal electrode, where the sample was not in a vertical position but only slightly tilted out of the horizontal direction.[37]

Multilayer sample preparation

A Si/Mo multilayer was prepared on a Si(100) substrate at the Lawrence Berkeley National Laboratory Center for X-ray Optics. The multilayer mirror consists of 80 repeats of a Si 1.5 nm/Mo 1.9 nm bilayer, yielding a multilayer periodicity of 3.4 nm. The set-up of the sample is shown in Fig. 4. A cap layer of 1.3 nm of SiO_2 terminates the mirror. A Ni layer with a thickness of $\sim$8 nm was grown on the SiO_2 cap layer using magnetron sputtering in an ultrahigh vacuum chamber with a base pressure of better than 1×10^{-8} Torr.

Electrochemical characterization

Cyclic voltammetry (CV) was performed on the Ni sample using a Bio-Logic SP-150 potentiostat, with the sample as working electrode and connected to the common ground. Fig. 5c shows a cyclic voltammogram (-0.1 $V_{Ag/AgCl}$ to $+0.65$ $V_{Ag/AgCl}$) for the thin Ni layer immersed in a 0.1 M KOH solution, with characteristic oxidation and reduction peaks at $+0.5$ $V_{Ag/AgCl}$ and $+0.4$ $V_{Ag/AgCl}$, respectively, in good agreement with literature data.[38,39] The stability of the sample inside the solution was tested in repeated cyclic voltammograms over the course of several hours, during which the voltammograms showed little change. The measurements on the thin Ni film deposited onto the multilayer mirror were also compared to those on a bulk polycrystalline Ni foil and showed very good agreement.

APXPS and standing wave measurements

The basic principle of SWAPPS is shown in Fig. 4. Focused X-rays are incident on the sample at the first-order Bragg angle of the multilayer mirror, defined from $\lambda_x = 2d_{ML} \sin \theta_{inc}$, where λ_x is the X-ray wavelength, d_{ML} is the multilayer period, and θ_{inc} is the X-ray incidence angle, in the present case 3.4° for 3.1 keV X-rays and a multilayer periodicity of 3.4 nm. This leads to the formation of a standing wave above the mirror with a wavelength λ_{sw} close to the mirror periodicity.[40,41] The SW extends above the mirror (*i.e.*, through the deposited Ni film and the adsorbed solution layer) in the volume where the incident and reflected X-rays overlap, and is scanned vertically by changing the incidence angle (rocking curve), thus selectively probing the interface, the bulk solid or bulk liquid (*cf.* Fig. 4). Core photoelectron peak intensities from different chemical states in the sample are then monitored as a function of incidence angle, with the resultant variations being characteristic of the depth distribution of a given species. Beyond simple Bragg reflection from the mirror, interference due to X-rays reflected from the top of the sample and from other interfaces in the sample (*e.g.* the mirror surface) can lead to additional fine structure in the rocking curves that are often referred to as Kiessig fringes,[40,41] with these also containing structural information. These in general are described by $m\lambda_x = 2D \sin \theta_{inc}$, where m is the usually unknown order $= 1, 2, 3, \ldots$ and D is the distance between the surface of the sample and the reflecting interface, for the commonly very large m in experiment, the separation between these fringes can be used to estimate the distance involved from D (nm) $= 0.62/[h\nu(\text{keV})\Delta\theta(\text{rad})]$, where $h\nu$ is the photon energy. For more detailed discussions of the technique see ref. 32, 35, 36, 40 and 41. The combination of the standing wave approach and the intrinsic surface sensitivity of XPS is then used for the determination of the vertical positions and widths of all interfaces present in the system.

The measurements were performed at beamline 6.0.1 at the Advanced Light Source in Berkeley, CA, using a Phoibos NAP 150 electron energy analyzer (Specs Surface Nano Analysis GmbH, Berlin). The incident X-rays had a photon energy of 3100 eV. The incident photon flux at this energy is $\sim 10^{12}$ photons per s. The measurement geometry is shown in Fig. 5a. The sample–aperture distance is estimated to be 0.2 mm. The incident X-ray beam spot is about 0.1 mm, but is horizontally enlarged by about 16 times due to the shallow incidence angle on the sample of $\sim 3.4°$. Ni 2p, O 1s, and C 1s spectra were collected under all experimental conditions. Valence band spectra taken under dry conditions were used to calibrate the binding energy scale. The sample was connected to the common ground in all experiments. Due to the strong attenuation of the photoelectron signal by the water gas phase, at a pressure of *ca.* 17 Torr, acquisition times of up to 5 min for a single energy region were necessary, in particular for the Ni 2p and 3p core levels, where attenuation due to adsorbed water or the KOH solution led to additional attenuation of the signal. The Ni 3p and O 1s spectra for the rocking curves were taken in the "snapshot mode" using a 2D delayline detector. Pass energies of 200 eV and 150 eV were used for the Ni 3p and O 1s regions, respectively, giving binding energy windows of 19.9 eV and 14.9 eV, sufficient for the capture of all spectral components. The complete rocking curve measurement took approximately 5 h. An angle step size of 0.02° was used in the experiments. The precision of the manipulator's rotation was measured with a Möller-Wedel ELCOMAT 3000 autocollimator, and shown to be reliable at least down to a step size of 0.005°.

Data analysis

The X-ray optical and photoemission simulations were performed using the Yang X-Ray Optics (YXRO) software,[40,41] which simulated the depth distribution of the electric field strength as a function of beam incidence angle and also the photoemission rocking curves for different layers and elements. As a qualitative indicator of the anticipated sensitivity to vertical position in such rocking curve measurements, Fig. 6 shows a set of rocking curves as a "delta layer" is moved in 0.1 nm steps vertically through a hypothetical liquid layer on top of a multilayer of

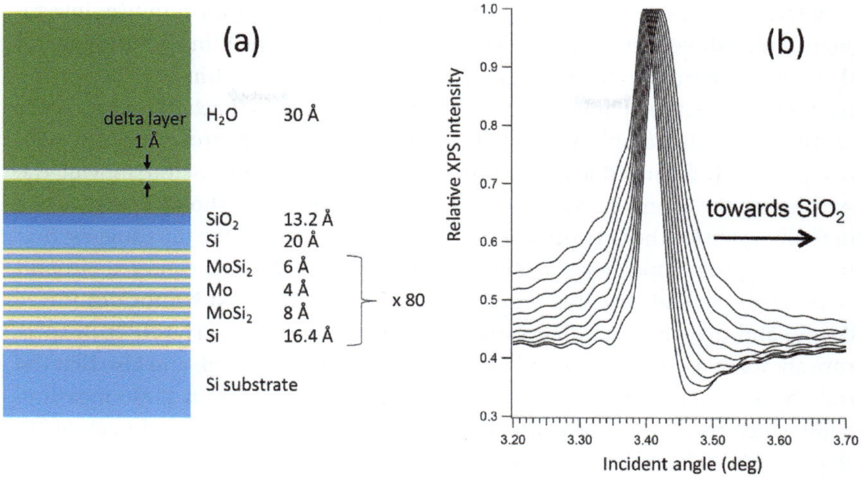

Fig. 6 Determination of the sensitivity in the present SWAPPS rocking curves to depth, using an incident photon energy of 3100 eV. (a) Layout of the model system used in the calculations. A water "delta layer" of 0.1 nm thickness is scanned through a 1 nm thick water layer on top of a SiO_2 cap layer on the multilayer mirror. (b) Calculated O 1s rocking curves, in steps of 0.1 nm changes of the height of the "delta layer" above the SiO_2/water interface. Both the peak position as well as the shape of the curves are changing as a function of the distance of the "delta layer" from the SiO_2 interface. A shift in the peak position of the rocking curve by 0.02° corresponds to a distance of 0.7 nm.

our configuration. These results indicate that, with the 3100 eV photon energy and multilayer mirror we have used, peaks separated by 0.02° originate from regions of the sample that are 0.7 nm apart. A key parameter in these calculations is the inelastic mean free paths for photoelectrons of the relevant core levels in liquid water; these are taken from Emfietzoglou and Nikjoo,[26] and can also be found in Table 1 (see below). The values for the IMFP in liquid water agree well with theoretical predictions by Tanuma, Powell, and Penn.[42]

Results and discussion

In the following we will describe proof-of-principle measurements of combining the meniscus method and SWAPPS measurements to study the interface of a 0.1 M KOH solution with polycrystalline Ni. We will first show results of static measurements before discussing the findings from rocking curve SWAPPS experiments.

Table 1 Data and parameters used for the calculation of liquid water layer thickness under "humid" and "immersed" conditions

Core level	Electron kinetic energy (eV)	IMFP (nm)	Dry	Humid	Immersed	Humid	Immersed
			Intensity (CPS eV)			d_w^L (nm)	
Ni 2p	~2240	7.5	48 000	2560	1100	13	18
Ni 3p	~3030	9.0	11 400	1510	300	11	21
C 1s	~2820	8.5	2300	860	2380	—	—

Before the SWAPPS measurements the state of the Ni thin film was investigated under three different conditions: as prepared ("dry"), in the presence of high relative humidity ("humid"), and after preparation of a thin KOH film using the meniscus method ("immersed"). Under dry conditions (background pressure 1×10^{-2} Torr), the sample was already covered by an oxide/hydroxide layer, which was most likely formed due to exposure to ambient conditions before moving the sample into the chamber. Ni $2p_{3/2}$, O 1s and C 1s spectra are shown in Fig. 7. The literature value for the O 1s peak of the oxide is 529–530 eV.[43] We thus suggest that the oxide peak appears as a shoulder to the broad peak centered around 531.5–532.0 eV, and that this broad peak is mainly due to Ni hydroxides (BE $\sim$ 531 eV) and oxygen containing carbonaceous contamination. The Ni $2p_{3/2}$ spectrum shows that not all of the Ni film is oxidized or hydroxylated, and also that the oxide/hydroxide layer is thin compared to the probing depth in these measurements since the Ni 2p spectrum shows mainly a sharp metallic Ni peak at BE $\sim$853 eV. The calculation of the oxide/hydroxide layer thickness for this dry state, using the standing-wave technique under UHV, shows a thickness of 1.8 nm. The carbon overlayer thickness is 0.8 nm from the same analysis. The C 1s signal consists of at least two peaks, which we assign to hydrophilic (oxygenated, high BE) and hydrophobic (non-oxygenated, low BE) carbon.

After the dry measurements the Ni sample was investigated under humid conditions at a water vapor pressure of 17 Torr, which corresponds to $\sim$85% RH at a sample temperature of 22 °C. The water vapor was supplied from degassed pure water; no KOH solution was present in the chamber in the experiments under humid conditions. There is a significant decrease in the intensity of all XPS signals due to scattering of electrons by water vapor. In addition, the Ni 2p, Ni 3p (not shown) and O 1s signals from Ni metal and Ni oxide/hydroxide are attenuated by the liquid water layer which is present at this high RH. A quantitative analysis of the decrease of the C 1s signal is consistent with a model where, under humid conditions, a similar amount of carbon as in the "dry" sample is present, with attenuation only by the water vapor. This indicates that most of the carbon contamination is located at the liquid/water vapor interface. Sources for the

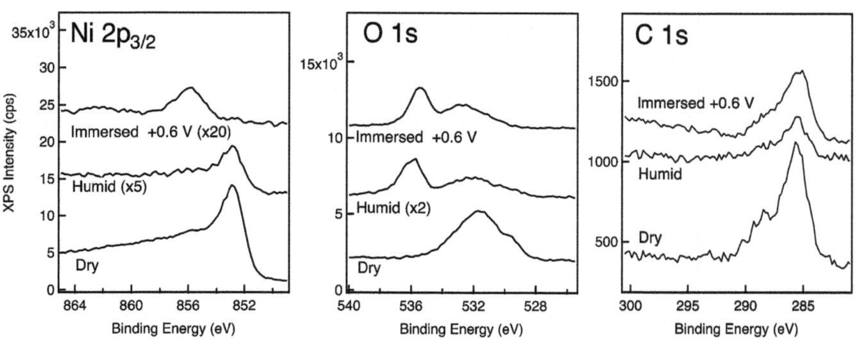

Fig. 7 Ni $2p_{3/2}$, O 1s and C 1s APXPS spectra of the Ni sample deposited onto the multilayer mirror taken under three different conditions: as is, in low vacuum (10^{-2} Torr, "Dry"); in the presence of 17 Torr water vapor ("humid"); and after forming a thin KOH solution film using the meniscus method, at a potential of +0.6 $V_{Ag/AgCl}$ ("immersed"). All measurements were performed at room temperature.

carbon contamination under humid conditions are the original contamination on the "dry" sample, but also additional contamination from the background vapor due to the limited pumping during the experiments at high humidity. There is no detectable change in the main Ni $2p_{3/2}$ peak position, which indicates that Ni is still mostly in the metallic state. The gas phase water peak is the most intense peak in the O 1s spectrum. The liquid water layer that is formed at high RH has an O 1s peak at 533–534 eV, which effectively broadens the Ni oxide/hydroxide signal centered around 532 eV towards higher binding energies. Thus, the broad O 1s signal around 532 eV is a combination of liquid water, oxide and hydroxides of Ni, and O containing carbonaceous contamination.

The thickness of the liquid water layer can be estimated from the Ni 2p and Ni 3p signals before and after the introduction of the vapor into the chamber using

$$\frac{I_{\mathrm{Ni}}}{I_{0,\mathrm{Ni}}} = \exp\left(-\frac{d_{\mathrm{w}}^{\mathrm{L}}}{\lambda_{\mathrm{w}}^{\mathrm{L}}\cos\theta}\right)\exp\left(-\frac{d_{\mathrm{w}}^{\mathrm{G}}}{\lambda_{\mathrm{w}}^{\mathrm{G}}}\right), \tag{1}$$

with $I_{0,\mathrm{Ni}}$ as the Ni signal intensity for the "dry" sample, I_{Ni} the Ni signal intensity for the "humid" or the "immersed" sample, $\lambda_{\mathrm{w}}^{\mathrm{L}}$ the inelastic mean free path of electron in liquid water, $\lambda_{\mathrm{w}}^{\mathrm{G}}$ the inelastic mean free path of electron in gaseous water, θ the angle between the surface normal and the analyzer axis (45°), $d_{\mathrm{w}}^{\mathrm{L}}$ the thickness of the liquid water layer, and $d_{\mathrm{w}}^{\mathrm{G}}$ the effective thickness of the water vapor that the electrons traverse before reaching low vacuum conditions in the first differential pumping stage (*i.e.*, approximately the sample–aperture distance of ~0.2 mm). The equation can be simplified combining the two terms on the right-hand side, under assumption of a hypothetical liquid layer (without the vapor) which gives the same I/I_0 value.

$$\frac{I_{\mathrm{Ni}}}{I_{0,\mathrm{Ni}}} = \exp\left(-\frac{d_{\mathrm{w}}^{\mathrm{L_{eq}}}}{\lambda_{\mathrm{w}}^{\mathrm{L}}\cos\theta}\right) \tag{2}$$

with

$$d_{\mathrm{w}}^{\mathrm{L_{eq}}} = d_{\mathrm{w}}^{\mathrm{L}} + d_{\mathrm{w}}^{\mathrm{G}}\,\frac{\lambda_{\mathrm{w}}^{\mathrm{L}}}{\lambda_{\mathrm{w}}^{\mathrm{G}}}\cos\theta. \tag{3}$$

The ratio of the inelastic mean free-paths in water vapor compared to liquid water is approximately the inverse of the densities of these media, such that:

$$\frac{\lambda_{\mathrm{w}}^{\mathrm{L}}}{\lambda_{\mathrm{w}}^{\mathrm{G}}} \approx \frac{\rho_{\mathrm{w}}^{\mathrm{G}}(17\ \mathrm{Torr},\ 20\ ^{\circ}\mathrm{C})}{\rho_{\mathrm{w}}^{\mathrm{L}}} = \frac{17\ \mathrm{g\ m^{-3}}}{10^{6}\ \mathrm{g\ m^{-3}}} = 1.7\times10^{-5}. \tag{4}$$

From combining eqn (3) and (4), and considering that $\cos\theta = \cos 45^{\circ} = 0.707$, it follows that

$$d_{\mathrm{w}}^{\mathrm{L_{eq}}} = d_{\mathrm{w}}^{\mathrm{L}} + d_{\mathrm{w}}^{\mathrm{G}}\,\frac{\lambda_{\mathrm{w}}^{\mathrm{L}}}{\lambda_{\mathrm{w}}^{\mathrm{G}}}\cos\theta = d_{\mathrm{w}}^{\mathrm{L}} + (200\ \mu\mathrm{m})\left(1.7\times10^{-5}\right)(0.707), \tag{5}$$

which yields $d_{\mathrm{w}}^{\mathrm{L_{eq}}} = d_{\mathrm{w}}^{\mathrm{L}} + 2.4$ nm. Based on this analysis, the thickness of the liquid water layer is estimated to be 11–13 nm. The parameters and data that are used in the calculation are summarized in Table 1.

After the sample was immersed in 0.1 M KOH solution, several CV cycles were run between +0.65 and −0.9 $V_{Ag/AgCl}$ before stopping at +0.6 $V_{Ag/AgCl}$ during the anodic sweep. Data were acquired at a constant potential of +0.6 $V_{Ag/AgCl}$ for the spectra shown in Fig. 7. From Pourbaix diagrams[44] of Ni, as well as from assignments of the CV features,[38] we expect the presence of hydroxides and oxyhydroxides at the interface under the pH and bias in our experiments. This is borne out by the Ni 2p spectrum, where the main peak BE has shifted to 856 eV, close to the literature value for oxide/hydroxide/oxyhydroxide. The O 1s spectrum shows a much stronger contribution from liquid water, as compared to the "humid" sample. The O 1s gas phase peak shifts to lower BE (higher KE) as expected from a negative potential applied to the solution; the gas phase peak shift confirms that the part of the liquid film under investigation is connected to the bulk solution reservoir. The water vapor peak shifts by ~0.4 eV, compared to the expected value of 0.6 eV for a full shift due to the applied bias. This can explained by the fact that gas phase peaks shifts less than surface peaks due to the pinning of the vacuum level of the gas phase to that of the sample and surrounding surfaces.[15,17] The C 1s spectra show the existence of at least 2 peaks (hydrophilic/hydrophobic), however, their intensity is now higher than the C 1s intensity for the "dry sample". This might be an indication that, during the retraction of the sample from the solution, a Langmuir–Blodgett-type film deposition of hydrocarbons took place due to hydrophobic molecules floating on the surface of the solution in the cell (*e.g.*, from trace contamination of the KOH solution or the water that was used for dilution). Another contribution to the extra carbon is probably residual CO_2 that is strongly absorbed into the liquid to form HCO_3^-. The K 2p signal which is expected at ~293 eV is not clearly visible. For a 0.1 M KOH solution, the K/H_2O ratio is ~1/550 and thus too low to observe K in the bulk of the solution using XPS. Using the previously explained analysis method, the thickness of the liquid water layer under immersed conditions is estimated to be 18–21 nm, *i.e.* ~50–100% thicker than the water layer under "humid" conditions.

We now proceed to discussing the results of the standing wave measurements using SWAPPS, where we compare measurements of the dry sample with those of the sample under immersed conditions. An initial experimental characterization of the dry sample using standing-wave photoemission was performed under UHV conditions at beamline 9.3.1 of the Advanced Light Source. The structure and composition of the sample was revealed through measuring Ni $2p_{3/2}$ and O 1s rocking curves and by an iterative simulation of the rocking curves, followed by a comparison with their experimental counterparts (Fig. 8, left panel) using the methods described in the experimental section. The measured rocking curves for the two core-level spectra (Ni $2p_{3/2}$, O 1s) are thus representative of the two primary layers: metallic Ni, and the oxidized Ni surface. All curves are normalized to their maxima. The metallic Ni layer was found to be 6.8 nm thick, capped by 1.8 nm of oxidized nickel, most likely in the form of NiO. The roughness of the interface between Ni and NiO and the top of the NiO layer is 1.0 nm, which is somewhat larger than the *ca.* 0.5 nm roughness measured by AFM for the Ni film just after growth. The carbon contamination layer at the surface is 0.8 nm thick, and we do not show any rocking curves from this contaminant element. This experimentally-determined model of the dry sample was used as a starting point for the analysis of the sample under immersed conditions.

In the immersed case, spectra of two regions were again collected: Ni 3p and O 1s. Ni 3p and O 1s spectra were fitted with 1 doublet and 3 peaks, respectively. The

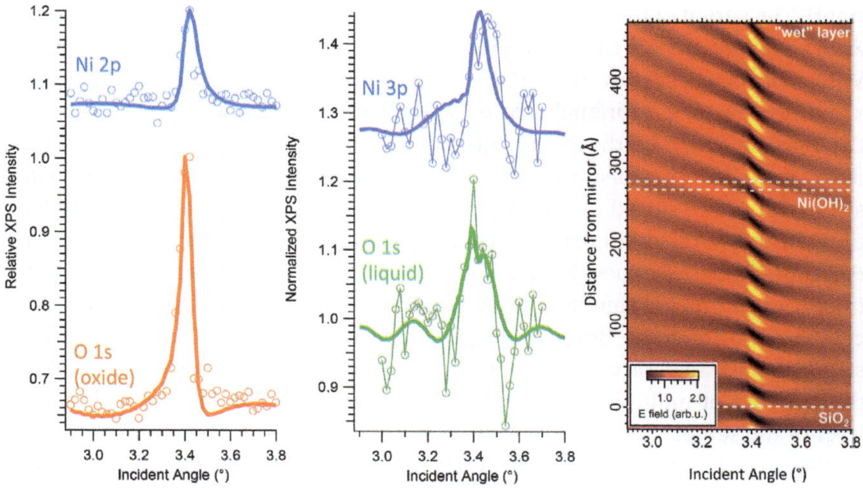

Fig. 8 SWAPPS rocking curves (dots) and iterative fits (solid lines) taken under different conditions: left panel: Ni $2p_{3/2}$ and O 1s (oxide component) for the as-prepared sample, in ultra-high vacuum; middle panel: Ni 3p and O 1s (liquid component) for a sample with a thin KOH solution film prepared using the meniscus method, at a potential of +0.6 $V_{Ag/AgCl}$; right panel: simulated electric field strength as a function of beam incidence angle and depth for the immersed sample (rocking curves in the middle panel). White dashed lines show the boundaries between the "wet" layer and $Ni(OH)_2$ and between $Ni(OH)_2$ and the multilayer mirror.

rocking curves in the middle panel of Fig. 8 show the peak intensities normalized to their mean value, as a function of incidence angle of the X-ray beam and the results of the simulation for the optimized sample structure as a solid line. The initial analysis of the nickel, which we assume is present in the form of hydroxide/oxyhydroxide, as discussed previously for the static data, and the solution layer thicknesses revealed that both of those layers are much thicker than a standing wave period (~3.4 nm) and therefore there are several nodes and antinodes in different depths buried within those layers (see the electric field plot – Fig. 8, right panel). The examination of such thick layers using standing wave photoemission, especially with harder X-rays (larger IMFPs), has an additional challenge. At these high kinetic energies (>2 keV) photoelectrons carry information from a larger depth and the intensity modulation at the Bragg angle is attenuated due to the signal integration along the axis perpendicular to the sample surface. This is the case in the present experiment and evidenced by the observation that the standing wave features in the rocking curves for the immersed sample are much broader and show smaller amplitudes than those from the thin layers of the dry sample.

Other relatively strong features adjacent to the Bragg peak and with a period of ~0.2° are observed in the rocking curves measured on the immersed sample. These are no doubt due to some kind of Kiessig interference, and we can estimate the *D* value involved from the prior equation to be about 40–60 nm, which is roughly the distance from the surface of the liquid film to the top of the multilayer mirror. In order to fully identify the origin of these oscillations, one has to examine the angular and depth dependence of the electric field strength (Fig. 8, right). The period of the standing wave is decreasing as the incidence angle of the X-rays is increasing, a trend expected from the two interfering waves involved. With increasing distance from the surface of the mirror, the difference of the standing wave phase for angles below and above the Bragg angle increases and the maxima enhance the photoemission signal from different depths at the different angles. These features, which in their period along the depth axis deviate slightly from the period of the mirror, are responsible for the observed oscillations in the O 1s rocking curves. They have proven to be especially helpful in determining the exact position of, as well as interdiffusion between, the different interfaces for thicker samples, as these features have different periods in depth compared to the main Bragg peaks generated by the mirror and therefore the standing-wave analysis can overcome the uncertainty in depth position caused by the presence of multiple nodes and antinodes of the standing-wave within one layer of the sample.

Using the iterative fitting of the X-ray optical and photoemission simulations to the experimental data, the thickness of the hydroxide/oxyhydroxide layer for the immersed sample was found to be 27 nm, with a water layer thickness of 20 nm. The surface of the wet layer is perfectly sharp within our depth resolution (<1 nm), while the interface between the hydroxide/oxyhydroxide layer and the KOH solution exhibits a roughness of 1 nm. The thickness of the hydroxide/oxyhydroxide layer also agrees well with the expected value for the expansion of the originally ~8 nm of metallic Ni transformed into a stoichiometric hydroxide by approximately a factor of 3.4, *i.e.* to ~27 nm. Under the given pH and bias conditions, no significant dissolution of Ni is expected. The standing wave analysis concurs with this expectation, as all of the Ni atoms are located in the solid layer according to the proposed model. A schematic of the proposed structural models for the Ni sample under dry and immersed conditions is shown in Fig. 9. The general model takes into consideration the expansion of the dry

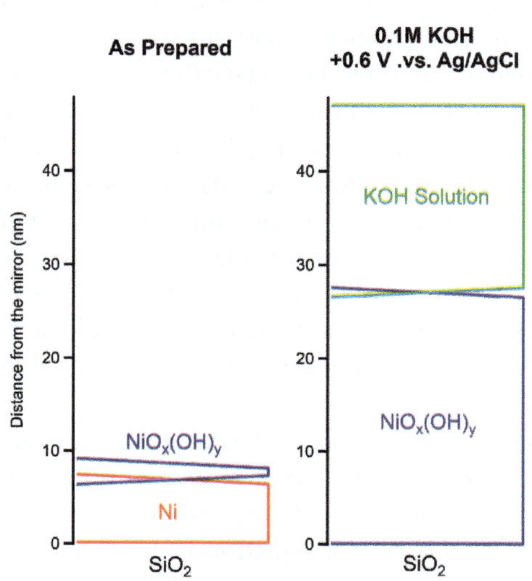

Fig. 9 Model based on the detailed analysis of the rocking curves in Fig. 8, showing the relative concentration profiles of the sample surface as prepared (left panel) and after immersion in 0.1 M KOH at +0.6 $V_{Ag/AgCl}$ (right panel).

nickel film into oxidized/hydroxylated nickel, as well as the quantitative analysis of the Ni signal attenuation for the immersed sample. The standing wave analysis enabled refinement of these rough estimates by determining the positions of the interfaces and surfaces with high precision (within one period of the standing wave) and also provides unique information on the interfacial and surface roughness.

Conclusions

The preceding data and discussion show that SWAPPS, combined with hard X-rays and with the preparation of thin solution layers using the meniscus method, is an exciting new approach to characterize all components of the liquid/solid interface, from bulk species in the liquid and solid to species located at the interface under potential control. In this paper, we determined the complete oxidation/hydroxylation of an 8 nm thick Ni film in the presence of a 20 nm thick 0.1 M KOH solution at a bias of +0.6 $V_{Ag/AgCl}$. This measurement would not have been possible in a conventional XPS experiment without the use of the standing wave method, which provides an intrinsic ruler in the direction perpendicular to the interface and extends the probing depth and the depth resolution in XPS. We anticipate that the resolution of the method can be further increased by using longer periods of the standing wave and by measuring more of the Kiessig fringes adjacent to the Bragg peak. The range of applications in corrosion science thus seems to be very large.

Acknowledgements

This work was supported by Laboratory Directed Research and Development projects through Lawrence Berkeley National Laboratory. The ALS and its beamlines 6.0.1 and 9.3.1 are supported by the Director, Office of Science, Office of Basic Energy Sciences of the US Department of Energy at the Lawrence Berkeley National Laboratory under Contract no. DE-AC02-05CH11231. HB, OK, AS, IZ, MH, and CR acknowledge support by the Director, Office of Science, Office of Basic Energy Sciences, and by the Division of Chemical Sciences, Geosciences, and Biosciences of the US Department of Energy at the Lawrence Berkeley National Laboratory under Contract no. DE-AC02-05CH11231. SN and CF also acknowledge support from the DOE Division of Materials Sciences, also under Contract DE-AC02-05CH11231 and ARO-MURI grant W911-NF-09-1-0398. Kai Liu, Peter K. Greene and Edward C. Burks (UC Davis) are acknowledged for the preparation of the Ni sample and AFM characterization of the as prepared samples. MH acknowledges support from the Deutscher Akademischer Austauschdienst (DAAD). M. L. Ng gratefully acknowledges the financial support from Wenner-Gren Foundations in Stockholm, Sweden. CF was also supported during this work by the French LabEx PALM, APTCOM Project, no. ANR-10-LABX-0039.

References

1 E. McCafferty, *Introduction to Corrosion Science*, Springer, New York, 2009.
2 S. J. Peters and G. E. Ewing, *J. Phys. Chem. B*, 1997, **101**, 10880.
3 V. Sadtchenko, G. E. Ewing, D. R. Nutt and A. J. Stone, *Langmuir*, 2002, **18**, 4632.

4 Y. R. Shen, *Nature*, 1989, **337**, 519.
5 G. Rupprechter, *MRS Bull.*, 2007, **32**, 1031.
6 J. Forsberg, L. C. Duda, A. Olsson, T. Schmitt, J. Andersson, J. Nordgren, J. Hedberg, C. Leygraf, C. Asstrup, D. Wallinder and J.-H. Guo, *Rev. Sci. Instrum.*, 2007, **78**, 083110.
7 J.-J. Velasco-Velez, C. H. Wu, T. A. Pascal, L. F. Wan, J. Guo, D. Prendergast and M. Salmeron, *Science*, 2014, **346**, 831.
8 S. Ferrer, M. D. Ackermann and E. Lundgren, *MRS Bull.*, 2007, **32**, 1010.
9 G. Binnig, C. F. Quate and C. Geber, *Phys. Rev. Lett.*, 1986, **56**, 930.
10 J. Hu, X.-d. Xiao, D. F. Ogletree and M. Salmeron, *Science*, 1995, **268**, 267.
11 B. L. M. Hendriksen and J. W. M. Frenken, *Phys. Rev. Lett.*, 2002, **89**, 046101.
12 M. J. Williamson, R. M. Tromp, P. M. Vereecken, R. Hull and F. M. Ross, *Nat. Mater.*, 2003, **2**, 532.
13 S. Thiberge, A. Nechushtan, D. Sprinzak, O. Gileadi, V. Behar, O. Zik, Y. Chowers, S. Michaeli, J. Schlessinger and E. Moses, *Proc. Natl. Acad. Sci. U. S. A.*, 2004, **101**, 3346.
14 F. Zaera, *Chem. Rev.*, 2012, **112**, 2920.
15 E. J. Crumlin, H. Bluhm and Z. Liu, *J. Electron Spectrosc. Relat. Phenom.*, 2013, **190**, 84.
16 D. E. Starr, Z. Liu, M. Hävecker, A. Knop-Gericke and H. Bluhm, *Chem. Soc. Rev.*, 2013, **42**, 5833.
17 H. Bluhm, *J. Electron Spectrosc. Relat. Phenom.*, 2010, **177**, 71.
18 M. Salmeron and R. Schlögl, *Surf. Sci. Rep.*, 2008, **63**, 169.
19 S. Yamamoto, K. Andersson, H. Bluhm, G. Ketteler, D. E. Starr, Th. Schiros, H. Ogasawara, L. G. M. Pettersson, M. Salmeron and A. Nilsson, *J. Phys. Chem. C*, 2007, **111**, 7848.
20 K. Andersson, G. Ketteler, H. Bluhm, S. Yamamoto, H. Ogasawara, L. G. M. Pettersson, M. Salmeron and A. Nilsson, *J. Am. Chem. Soc.*, 2008, **130**, 2793.
21 A. Verdaguer, G. M. Sacha, H. Bluhm and M. Salmeron, *Chem. Rev.*, 2006, **106**, 1478.
22 G. E. Ewing, *Chem. Rev.*, 2006, **106**, 1511.
23 D. M. Kolb, D. L. Rath, R. Wille and W. N. Hansen, *Ber. Bunsen-Ges. Phys. Chem.*, 1983, **87**, 1108.
24 E. R. Kötz, H. Neff and K. J. Müller, *J. Electroanal. Chem.*, 1986, **215**, 331.
25 W. Zhou and D. M. Kolb, *Surf. Sci.*, 2004, **573**, 176.
26 D. Emfietzoglou and H. Nikjoo, *Radiat. Res.*, 2007, **167**, 110.
27 S. Axnanda, E. J. Crumlin, B. Mao, S. Rani, R. Chang, W. Stolte, P. G. Karlsson, M. O. M. Edwards, M. Lundqvist, R. Moberg, P. N. Ross, Z. Hussain and Z. Liu, *Sci. Rep.*, 2015, **5**, 9788, DOI: 10.1038/srep09788.
28 M. Faubel, B. Steiner and J. P. Toennies, *J. Chem. Phys.*, 1997, **106**, 9013.
29 M. A. Brown, N. Duyckaerts, A. Beloqui Redondo, I. Jordan, F. Nolting, A. Kleibert, M. Ammann, H. J. Wörner, J. A. van Bokhoven and Z. Abbas, *Langmuir*, 2013, **29**, 5023.
30 A. Kolmakov, D. A. Dikin, L. J. Cote, J. Huang, M. K. Abyaneh, M. Amati, L. Gregoratti, S. Günther and M. Kiskinova, *Nat. Nanotechnol.*, 2011, **6**, 651.
31 K. Arima, P. Jiang, X. Deng, H. Bluhm and M. Salmeron, *J. Phys. Chem. C*, 2010, **114**, 14900.

32 S. Nemšák, A. Shavorskiy, O. Karslioglu, I. Zegkinoglou, P. K. Greene, E. C. Burks, K. Liu, A. Rattanachata, C. S. Conlon, A. Keqi, F. Salmassi, E. M. Gullikson, S.-H. Yang, H. Bluhm and C. S. Fadley, *Nat. Commun.*, 2014, **5**, 5441, DOI: 10.1038/ncomms6441.
33 J. O. 'M. Bockris and B. D. Cahan, *J. Chem. Phys.*, 1969, **50**, 1307.
34 P.-G. de Gennes, F. Brochard-Wyart and D. Quéré, *Capillarity and Wetting Phenomena*, Springer, New York, 2004.
35 C. S. Fadley, *J. Electron Spectrosc. Relat. Phenom.*, 2013, **190**, 165.
36 A. X. Gray, *J. Electron Spectrosc. Relat. Phenom.*, 2014, **195**, 399.
37 D. Weingarth, A. Foelske-Schmitz, A. Wokaun and R. Kötz, *Electrochem. Commun.*, 2011, **13**, 619.
38 K. Wang, G. S. Chottiner and D. A. Scherson, *J. Phys. Chem.*, 1993, **97**, 10108.
39 P. F. Luo, T. Kuwana, D. K. Paul and P. M. A. Sherwood, *Anal. Chem.*, 1996, **68**, 3330.
40 S.-H. Yang, A. X. Gray, A. M. Kaiser, B. S. Mun, B. C. Sell, J. B. Kortright and C. S. Fadley, *J. Appl. Phys.*, 2013, **113**, 073513.
41 S.-H. Yang, *X-Ray Optics (YXRO) program for simulating standing-wave photoemission above multilayers*, 2011, https://sites.google.com/a/lbl.gov/yxro/home.
42 S. Tanuma, C. J. Powell and D. R. Penn, *Surf. Interface Anal.*, 1994, **21**, 165.
43 M. C. Biesinger, B. P. Payne, L. W. M. Lau, A. Gerson and R. St. C. Smart, *Surf. Interface Anal.*, 2009, **41**, 324.
44 B. Beverskog and I. Puigdomenech, *Corros. Sci.*, 1997, **39**, 969.

Faraday Discussions

PAPER

Structure–charge relationship – the case of hematite (001)†

Johannes Lützenkirchen,*[a] Frank Heberling,[a] Filip Supljika,[b] Tajana Preocanin,[b] Nikola Kallay,[b] Florian Johann,[c] Ludger Weisser[c] and Peter J. Eng[d]

Received 12th December 2014, Accepted 16th January 2015

DOI: 10.1039/c4fd00260a

We present a multidisciplinary study on the hematite (001)–aqueous solution interface, in particular the relationship between surface structure (studied *via* surface diffraction in a humid atmosphere) and the macroscopic charging (studied *via* surface- and zeta-potential measurements in electrolyte solutions as a function of pH). Upon aging in water changes in the surface structure are observed, that are accompanied by drastic changes in the zeta-potential. Surprisingly the surface potential is not accordingly affected. We interpret our results by increasing hydration of the surface with time and enhanced reactivity of singly-coordinated hydroxyl groups that cause the isoelectric point of the surface to shift to values that are reminiscent of those typically reported for hematite particles. In its initial stages after preparation the hematite surface is very flat and only weakly hydrated. Our model links the entailing weak water structure with the observed low isoelectric point reminiscent of hydrophobic surfaces. The absence of an aging effect on the surface potential *vs.* pH curves is interpreted as domination of the surface potential by the doubly coordinated hydroxyls, which are present on both surfaces.

Introduction

The solid–aqueous solution interface is of interest in a wide range of scientific disciplines, including corrosion and corrosion inhibition,[1–3] catalyst preparation,[4–6]

[a]*Institut für Nukleare Entsorgung, Karlsruhe Institut für Technologie, P.O. Box 3640, 76021 Karlsruhe, Germany. E-mail: Johannes.luetzenkirchen@kit.edu; frank.heberling@kit.edu; Fax: +49 721 608 2 3927; Tel: +49 721 608 2 4023*

[b]*Laboratory of Physical Chemistry, Department of Chemistry, Faculty of Science, University of Zagreb, Horvatovac 102a, 10001 Zagreb, Croatia. E-mail: nkallay@chem.pmf.hr; tajana@chem.pmf.hr; fsupljika@chem.pmf.hr*

[c]*Oxford Instruments GmbH, c/o Asylum Research, Hauptstrasse 161, DE-68259 Mannheim, Germany. E-mail: Florian.Johann@oxinst.com; Ludger.Weisser@oxinst.com*

[d]*GeoSoilEnviroCars, University of Chicago, 5640 South Ellis, Chicago, IL 60637, USA. E-mail: eng@cars.uchicago.edu*

† Electronic supplementary information (ESI) available: ESI includes details on the surface diffraction model and surface complexation modelling for both fresh and aged hematite-001. See DOI: 10.1039/c4fd00260a

colloid chemistry,[7–9] nanotechnology,[10–12] geochemistry,[13–15] soil chemistry,[16–18] clay and clay mineral chemistry,[19–21] environmental chemistry[22–24] *etc.*, and overlaps thereof.[25–27] As a consequence many experimental and theoretical studies exist and highly detailed knowledge has been acquired from various points of view. Modern experimental techniques probe adsorbate structure,[28] surface structure[14] or interfacial water structure.[29,30] Theoretical work attempts to yield more insight into the sometimes difficult interpretation of such data.[31] Yet, several aspects still hamper the full understanding of the interfacial systems. One important aspect in that context is that different scientific disciplines sometimes develop independently from each other. Correspondingly, discoveries or observations made in one discipline are not necessarily taken up in another. A further drawback is related to the fact that many studies include no more than one technique to gather experimental or computational data. While the comparison between experiment and theory is often included, the precise relation of various experimental results originating from different techniques is rather sparse. It is therefore important to establish more accurate links between experiments and models emerging from different techniques (macroscopic and structural) and simulations.

Some time ago we reconciled the interfacial behaviour of α-Al_2O_3 (sapphire) single crystals with (001) orientation[32] in order to find out whether and which of the available data could be understood in one phenomenological picture. This attempt required a review of the available literature including a particular focus on the origin and treatment of the samples. It was finally postulated that the remaining data along with our own could be understood by a dual charging mechanism. The reconciliation included data from non-linear optic techniques (sum frequency and second harmonic generation), streaming potential and colloid adhesion studies, contact angle experiments and surface complexation models all as a function of pH, which is a master variable for the oxide–aqueous solution interface. The dual charging mechanism that allows an explanation of the experimental data involves classical protonation and deprotonation of doubly coordinated aluminol groups which are the only surface hydroxyl groups on the ideal surface, plus a rather unconventional contribution from ion adsorption in the electrochemical double layer. The latter (besides being unconventional for the oxide–electrolyte interface community) has been under hot debate for nearly a decade by now. It involves the physical adsorption of hydroxide ions and protons at interfacial water layers with a preference for the hydroxide ion. Strong adsorption of the hydroxide ion is assumed mostly by experimentalists.[33–38] However, others find that the proton is preferentially adsorbed,[39] that the charge is negligible,[40] or that there is no spectroscopic evidence for hydroxide ions at the surface for very high NaOH concentrations or in the absence of an inert background electrolyte.[41,42]

An accepted generic feature is the occurrence of the isoelectric point (IEP) in aqueous solutions at pH 2 to 4 for surfaces as different as diamond, PTFE, oil, gases, gold, *etc.*[43–45] Unlike oxide surfaces, all these systems have in common the absence of surface functional groups that could cause pH dependence. Without the occurrence of shifts in the IEP with changes in electrolyte concentrations and composition, experimentalists are left with only one solution to the problem that can explain the charge and the pH dependence at the same time. Even for ice surfaces the low IEP has been observed.[46,47] Especially in the case of ice this is surprising, as one would expect the IEP to be where bulk water is neutral. The simplest interpretation of the low IEPs involves the hydroxide ion to be more strongly attracted to the interface than the proton or in other words that dissociation of water at interfaces is very different from dissociation of bulk water. On a similar note the negative charge of gas bubbles in pure water at pH 7 is argued to be due to the presence of hydroxide ions at the interface in absence of a better explanation.[48]

On oxide minerals the protonation and deprotonation of surface hydroxyls has been traditionally invoked to explain pH-dependent charging. The corresponding p*K* values are successfully predicted from the surface structure for well-defined particles.[49–52] For various single crystal samples, the predicted p*K* values fail to reproduce the experimentally observed IEP, which finally led us to postulate dual charging, involving the traditional picture of charged surface hydroxyls plus the unconventional charging of the adjacent water layer. According to the traditional picture, points of zero charge for most oxide particles should coincide in the absence of specific adsorption of background electrolytes and the corresponding IEPs should be directly related to the p*K* values of the surfaces, *i.e.* at pH about 6 in the case of hematite (001) and sapphire (001).[30,53,54] Instead, the measured IEPs are at pH about 4, in the range of those observed for the "inert" surfaces discussed above. On sapphire (001) the predicted point of zero charge is confirmed in sum frequency generation studies[30] or contact angle investigations,[32] while the IEP of 4 is found in other experiments.[53,55] Importantly, some of the seemingly contradictory results have been obtained on identical samples.[32] The difference can therefore safely be assumed to be real and can phenomenologically be explained by a dual charging mechanism.

On sapphire (001) both contributions to the dual charging mechanism (surface hydroxyls and water layer) arise from protons and hydroxide ions. The adsorption of protons to or the release of protons from the surface aluminols would classify them as surface charge/potential determining ions, whereby their contribution to charge accumulation is restricted to the plane of the surface hydroxyls. The second contribution which arises from uptake of protons on or release of protons from adjacent water layers in turn affects the zeta-potential. In particular on surfaces like sapphire (001) the surface functional groups (involving only doubly coordinated oxygens) are rather stable and do not protonate or deprotonate within the usual pH-range.

Surfaces like silver halides on the other hand do not have the disadvantage that the same ions affect both surface and zeta-potential, and in particular protons and hydroxide ions do not affect the surface potential, which is controlled by silver and halide ions. In such systems simultaneous measurements of surface and zeta-potential clearly illustrate the difference between the action of the surface potential/charge determining ions and the action of protons and hydroxide ions, which were found to influence the zeta-potential.[56]

Furthermore, studies of both flat single crystals and colloidal particles showed, that hydroxide ion adsorption at interfacial water layers was more pronounced on the flat surfaces[57] in agreement with MD-simulations.[58] Knowing both, surface and zeta-potentials, is therefore key to the understanding of flat interfaces with defined crystallographic orientation, which are often taken as model substrates to investigate contaminant uptake, dissolution, growth and other phenomena. In the case of sapphire,[32,59] we relied on surface potentials estimated from surface complexation modelling. For the silver halides both quantities were measured[57] thus avoiding model inherent assumptions. A dual charging model was shown to reproduce all experimental observations qualitatively.[57]

The present study tries to push the approach further by applying various experimental techniques to identical samples and investigating the effect of aging on the hematite (001) surface. There is some indication from previous second harmonic generation and atomic force microscopy (AFM) experiments which report relatively low points of zero charge. This suggests that this surface has unexpected properties similar to sapphire (001).[60] In those studies it was reported that annealing of the surface led to changes in the experimental results, but no conclusions about the origin of the changes were possible.[60] Direct zeta-potential measurements for hematite (001) are not available to our knowledge but the hematite (001) surface has seen recent interest from various groups.

Surface diffraction is the major technique to study the structure of the surface and/or the adjacent water layer and to deduce which surface functional groups are present[61,62] and what kind of water structure forms at the interface.[29] An important conclusion has been that the water on annealed hematite (001) is weakly structured.[29] Another conclusion was that a bi-domain structure persists at the surface where the contributions from both domains varied in the different studies.[61,62] In the remainder of the text we will discuss this bi-domain structure which has been observed by other methods as well.[63,64] The two distinct structures that emerged from these studies are O_3–Fe–Fe–R and O_3–Fe–O_3–R. How these terminations relate to the bulk hematite structure is displayed in Fig. 1. For the oxygen terminated surface (O_3–Fe–Fe–R, Fig. 1, left) the surface is composed of doubly coordinated oxygen atoms. By removing the top oxygen layer the double iron termination (Fe–Fe–O_3–R, Fig. 1, 2nd structure) is obtained. Further removal of an iron layer yields the single iron termination (Fe–O_3–Fe–R, Fig. 1, 3rd structure). The fourth option (noted O_3–Fe–O_3–R, Fig. 1, right) is terminated by singly coordinated oxygen atoms. It may be understood as the Fe-terminated surface with water/hydroxide adsorbed to the top Fe atoms. This termination may form upon addition of Fe-ad atoms onto the oxygen termination,[65] with subsequent rapid hydroxylation.[66] The O_3–Fe–O_3–R-termination may also be understood as an oxygen termination with iron vacancies in one of the top iron layers (indicated by transparent iron atoms in Fig. 1, upper panel, right). It is important to note, that due to the low scattering cross-section of protons, surface diffraction does not allow to distinguish whether the top oxygen atoms are oxygen (O^{2-}), hydroxide (OH^-), or adsorbed water molecules (H_2O).

Another series of studies involved surface potential measurements.[67–71] The bulk part of the data suggests a broad plateau-like region in surface potential *vs.* pH curves with an increase at sufficiently low and a decrease at sufficiently high pH. Various interpretations of the data have been proposed. One interpretation was based on a shielding mechanism,[71] by which reactive groups could not be protonated or deprotonated, but this mainly explains the kinetics of the reaction and does not apply to our investigation. Two major difficulties appear in interpreting the experimental data: the determination of the point of zero surface potential and the absolute value of the surface potentials. The points of zero potential on two different hematite (001) samples have been previously fixed at the mid-pH of the plateau (suggesting a value of about 6)[70] or inferred from numerical procedures (resulting in a value of about pH 8.3).[71]

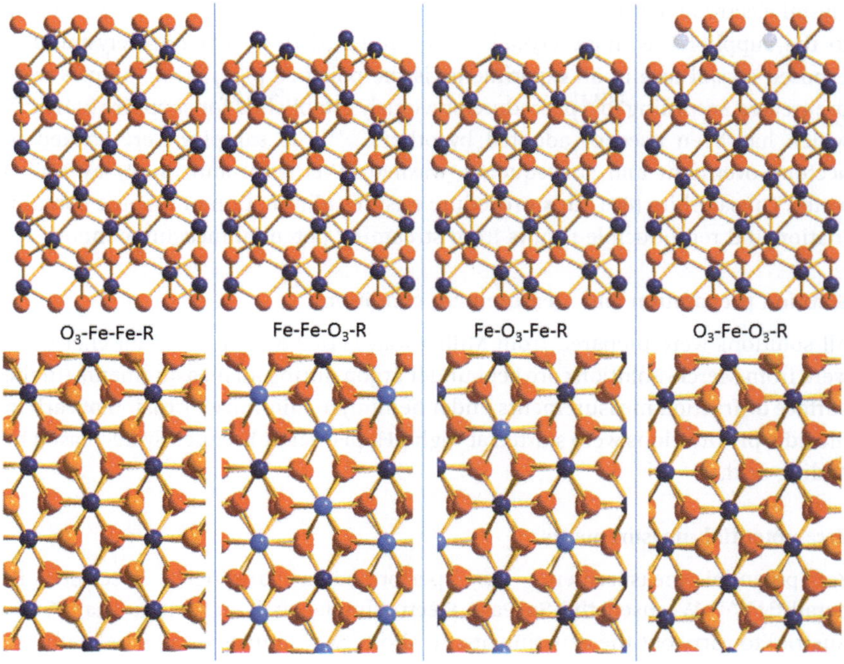

Fig. 1 A selection of chemically distinct surface terminations of hematite (001) (O: red, Fe: blue). The top panel shows side views of four possible terminations, the lower panel shows the corresponding top views. In the top views the surface atoms are highlighted in orange (oxygen) or light blue (iron).

In our present work we combine CTR and surface potential data and include zeta-potential measurements. Thus we probe the surface structure and the concomitant surface potential as well as the water structure and the concomitant interfacial potential, which results from the net charge within the electrokinetic shear plane.

A second aspect addresses the aging of the hematite surface. To this end we have exposed the crystal to solutions for extended periods of times, and followed the changes with zeta-potential measurements. The initial (referred to as "fresh") and final (referred to as "aged") states were analysed by surface diffraction to generate a consistent picture between the structure of the interface and the concomitant charging properties. The measurements were all performed on samples from the same source, treated in the same way and subject to similar time dependence. High resolution atomic force microscopy (AFM) on a fresh sample confirms various features of the surface diffraction study.

Experimental

Sample origin and preparation

The samples were purchased from "Surfacenet" (Münster, Germany). They are polished on one side and have dimensions of 10 × 20 mm^2. These samples fit exactly into the set-up used for zeta-potential measurements. Samples for surface potential measurements are smaller so that they allow the construction of a single crystal electrode. For the AFM study a 10×10 mm^2 sample was used. According to the supplier the single crystals were produced from a natural crystal. The exposed crystal face in all experiments is the polished (001) face. Samples were pre-treated according to a procedure previously established for sapphire (001),[32,59] which has been used or adopted by others.[72,73] The samples were soaked in acetone overnight and subsequently washed by ethanol and finally by MilliQ water. On sapphire (001) this procedure ensures elimination of organic contamination and reproducible results from streaming potential measurements.

Solution preparation and measurement conditions

All solutions were prepared from MilliQ water (18.2 MΩ cm) and all chemicals were from Merck. Solutions are kept under argon during all measurements (in the surface diffraction measurements under helium) to minimize intrusion of carbon dioxide. pH titrations were started at high pH. pH was subsequently decreased by adding acid.

Zeta-potential measurements

Zeta-potential measurements were performed as previously described in detail.[32,57,59,74] We used the SurPass Electrokinetic Analyzer (Anton Paar, Graz, Austria) to carry out the experiments. In short, two samples were glued with the unpolished side to two stamps. The exposed polished side (cleaned as described above) was rinsed by ethanol and MilliQ water and then the stamps were inserted into a cell to create a rectangular flow channel. The channel width was adjusted to about 100 μm. An experiment was initiated from a solution of a given electrolyte concentration (NaCl), in which the pH was increased by adding NaOH (50 mM) and then a titration was started by adding HCl (50 mM). Equilibration times were at least 10 minutes per point and at least 4 replicate measurements were performed at each pH. pH measurements in the SurPass set-up were performed using the built-in device, which was calibrated against three buffer solutions. All measurements were performed at room temperature.

After the first series of measurements the samples were kept in an aqueous environment (MilliQ water). Subsequent measurements surprisingly led to strong shifts in the zeta-potential. This is unlike observations with sapphire (001) where the charging behaviour can be reproduced over extended time periods with a given set of *unannealed* samples.

Surface potential measurements

Surface potential measurements were performed as previously described.[70] The potential of the Single Crystal (SCr)-hematite electrode was measured by a Metrohm 713 pH-meter. A silver/silver chloride electrode, filled with 3 mol dm^{-3} KCl solution, with a salt bridge (Metrohm, 6.0729.100) filled with sodium chloride ($c = 10^{-3}$ mol dm^{-3}), was used as a reference. pH was measured using a combined Metrosensor glass electrode (Metrohm, 6.0259.100). The pH-meter was operated by batteries (Metrohm, 826) and used for the measurements of pH. Systems were thermostatted at 25.0 °C. The experiments were performed under an argon atmosphere. The measuring system (thermostatted vessel, electrodes, magnetic stirrer) was placed in a Faraday cage. The surface of the SCr-hematite electrode was cleaned with ethanol and rinsed with water before all measurement runs (*i.e.*

beyond the standard cleaning procedure). Sodium chloride ($c = 10^{-3}$ or 10^{-2} mol dm^{-3}) was used as background electrolyte. Potentiometric titrations were performed by adding hydrochloric acid ($c = 10^{-1}$ mol dm^{-3}) or sodium hydroxide ($c = 10^{-1}$ mol dm^{-3}). Time intervals between additions were between 10 and 15 minutes. Potentiometric titrations were entirely automated. Data were collected every 4 s.

CTR-data collection

Crystal truncation rod (CTR)-data were recorded at the Advanced Photon Source (APS at Argonne National Laboratory) at Sector 13 (GSECARS) at the 13 BMC station.

The samples used for collection of CTR-data include a fresh sample and an aged sample that had been retained from the last series of zeta-potential measurements. Beamline optics contain a focusing SI (111) double crystal monochromator. The photon energy was fixed to 15 keV. The sample was oriented relative to the beam on a Newport 2 + 2 + kappa diffractometer. Measurement of the specular CTR was performed with the surface normal of the sample in a horizontal position, while for the measurements of the off specular CTRs the incident angle of the beam relative to the surface was maintained at a constant value of 2°. The diffraction signal was recorded by a PILATUS 2D pixel array detector with 195 × 487 pixels (vertical × horizontal). The size of the beam was limited by horizontal and vertical slits to 500 μm × 500 μm (vertical × horizontal).

The sample was mounted onto the diffractometer in an environmental sample cell, which was constantly flushed with humid He-gas (close to 100% relative humidity) in order to maintain the sample surface fully hydrated during the measurements.

CTR-data integration

Integration of diffracted intensities from the PILATUS images was performed using the tdl software package (https://github.com/xraypy/tdl).

A total of 692 and 654 structure factors were recorded for the fresh and the aged hematite samples, respectively. The measured structure factors were averaged in the $p3m$ plane group, resulting in 524 and 439 unique structure factors, respectively. Note that a single hematite (001) termination exhibits $p3$ symmetry only. However, in the hematite unit cell 6 chemically equivalent terminations exist, which are crystallographically distinct, but equally likely to constitute the actual surface. These terminations are linked by the threefold symmetry axis along hematite [001] and the glide plane symmetry along the same axis (space group $R\bar{3}c$). As the surface structure is modelled assuming an equal abundance of these symmetry related domains at the surface, as in previous studies,[61,62] the total symmetry considered for hematite (001) CTR data reduction may be increased from $p3$ to $p3m$ without any loss of information. This reduces the number of unique structure factors and, correspondingly, reduces the computational effort during modelling.

CTR-data modelling approach

Data modelling is performed using a modified version of the python interface structure refinement package (https://github.com/xraypy/tdl). The modified version was expanded to allow for the treatment of multiple surface terminations and symmetry related surface domains, following an approach described earlier.[75] The abundance of chemically distinct surface terminations is thereby adjusted as a fitting parameter during data modelling, while an equally weighted average is taken over the symmetry related domains. Structure factors of the two chemically distinct terminations are summed in a coherent fashion. In agreement with previous studies,[61,62] incoherent summation did not yield acceptable fits to the data.

Preliminary modelling approaches, using a single surface termination, failed to describe the data for both samples by a simple stoichiometric surface termination. Occupancies of the surface iron sites rather indicated a surface termination consisting of a mixture of O_3–Fe–Fe–R and O_3–Fe–O_3–R domains very similar to previous results.[61,62] In our modelling approach we therefore started with a linked two domain model as reported by Trainor *et al.*,[62] where the O_3–Fe–Fe–R and O_3–Fe–O_3–R domains share common structural parameters in layers that are not directly exposed to the surface or mimicking adsorbed water molecules (modelled as oxygen atoms, because of the low X-ray scattering cross-section of hydrogen). In contrast to Trainor *et al.*[62] and Tanwar *et al.*[61] this "linked model" did not yield a satisfactory description of the CTR data ($\chi^2 > 5$) for both data sets. Therefore, after the initial adjustment of the linked model, the structural parameter space was limited to a narrow range around the common value (± 0.005 in fractional coordinates) and the model was readjusted with individual parameters for both domains.

All atom displacements are modelled such that the threefold symmetry of the surface structure is maintained. The iron atoms have fixed lateral positions as they sit on the threefold symmetry axes. The vertical displacement of the oxygen atoms in one layer is described by one parameter fixing the oxygen atoms at equal vertical positions. Lateral positions of the three oxygen atoms in one layer are described by two parameters [*e.g.*: O_1: (x, y), O_2: $(-x + y, -x)$, O_3: $(-y, x - y)$] to keep the threefold symmetry.

During the model adjustment the bond valence sums of the iron atoms in the surface structure according to Brown and Altermatt[76] are constrained to stay within 5% of the nominal iron valence of +III. The bond valence calculations on the surface oxygen atoms simultaneously yield information, which can be used to draw conclusions about their protonation states.

AFM measurements

High resolution AFM measurements were carried out using a Cypher ES apparatus (Asylum Research). Both imaging and force spectroscopy were performed in AC mode with an "Arrow UHFAuD" cantilever from NanoWorld. The cantilever was excited using blueDrive photothermal excitation with oscillation amplitudes <1 nm in liquid environments and <5 nm in air. The fresh sample was studied to gain insight into the roughness of the sample and to try to support results from the CTR study concerning surface structure of and interfacial water structure on the fresh sample.

Surface complexation modelling

Surface complexation modelling was performed with a modified version of FITEQL2. The surface and zeta-potential data were fitted to a set of surface chemical equations involving a model for the electrical double layer (see ESI†).

Results and discussion

Zeta-potential measurements

Fig. 2 shows the pH function of the zeta-potential as it changes from the fresh sample to a steady state (aged sample). The initial state of the surface produced repeatedly an IEP of about 4. Later, the IEP shifted gradually to higher pH until it reached a steady state value of about pH 9.

The initial IEP obtained for the fresh sample was identical to the ones routinely obtained for "inert" surfaces,[77] but also for isostructural sapphire (001) samples[78] or even flat samples of TiO_2 (ref. 79) or ZnO.[80] The occurrence of this IEP for so many different samples could be challenged in the sense that it is triggered by the method or by systematic errors or artefacts. For sapphire (001) the IEP of about 4 (ref. 32) has been interpreted in terms of charging of the interfacial water layer adjacent to the surface. Weak ordering of water at both sapphire- and hematite (001) surfaces[29] supports this interpretation. The observation of an IEP at about pH 9 for the aged hematite sample, *i.e.* identical to the bulk of IEPs for hematite particles, suggests that the method is not at the origin of those low IEPs. The transient change could be due to structural changes at the surface of the sample and/or changes related to the interfacial water structure.

Surface potential measurements

The surface potential *vs.* pH response was not prone to significant changes even for an extended period of time. We always observed the same overall features with a broad range of constant potential at pH values between 4 and 9. This plateau-behaviour could be explained by the MUSIC model as discussed previously.[67,70] Briefly, the doubly coordinated surface hydroxyls were rather inactive towards protonation and deprotonation since their p*K* values are about 0 and 12.[54] The

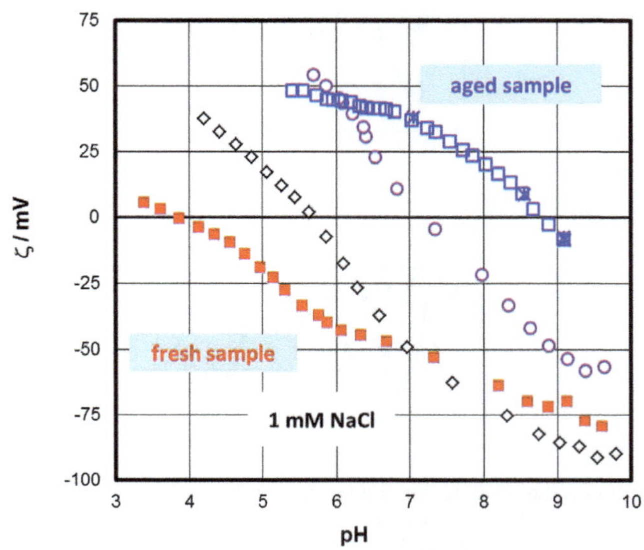

Fig. 2 The effect of aging in aqueous solution on the zeta-potential of the hematite samples measured in 1 mM NaCl solution.

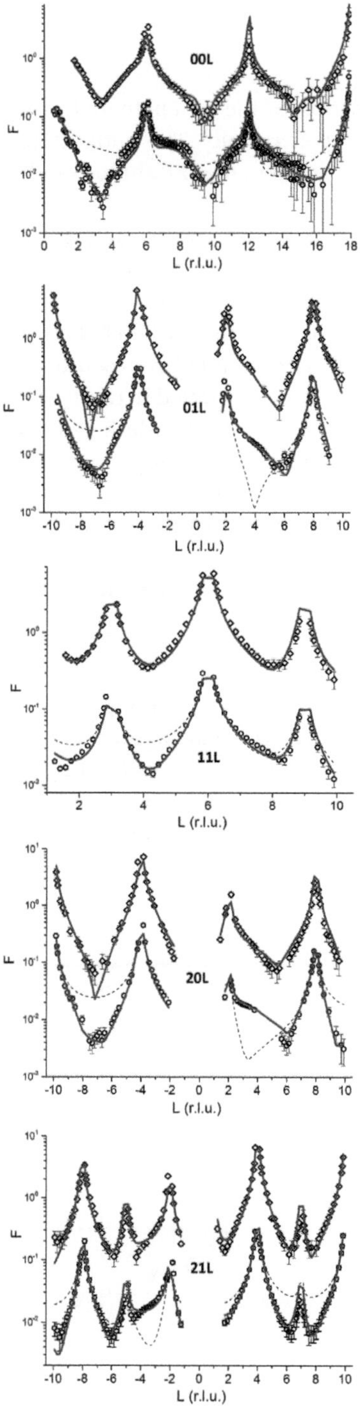

Fig. 3 CTR data for the aged (upper graphs, diamonds) and fresh (lower graphs, circles) are shown along with calculated structure factors (solid lines). The thin dashed lines in the lower graphs indicate the structure factors expected for bulk terminated hematite. The data for fresh and aged hematite are offset for clarity.

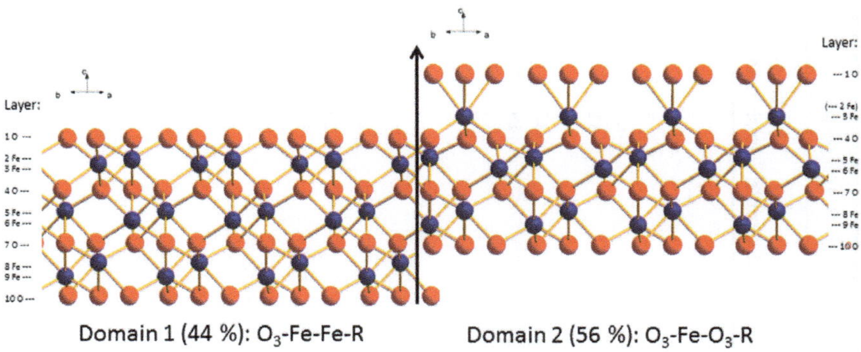

Fig. 4 Best fit structural model for the surface of the fresh hematite sample (oxygen: red, iron: blue). Indicated are the domain and layer nomenclature as used in Table SI1.†

insensitivity of the surface potential *vs.* pH curves to aging confirms our previous claim that a surface with plateau-like behaviour is thermodynamically stable.[70]

Surface diffraction data

CTR-data and best fit model calculations are shown in Fig. 3. As an example the 00*L* rod (top panel in Fig. 3) indicates that the surface diffraction data change with time. Clearly between $6 < L < 9$ a shoulder is present on the fresh sample that disappears on the aged sample.

The comparison with the bulk terminated hematite (dashed lines) shows that the real surfaces are different from the ideal surface. The full lines correspond to the structural models for the two samples obtained in the present study. The agreement between experiment and model is rather good ($\chi^2 = 3.97$ and $\chi^2 = 2.97$ for the fresh and aged samples, respectively). Both surfaces are bi-domain, with different contributions from the two terminations.

The structural parameters corresponding to the best fit models are reported in ESI (Table SI1† for the fresh hematite sample and in Table SI2† for the aged hematite sample). For comparison, the structural coordinates of the bulk

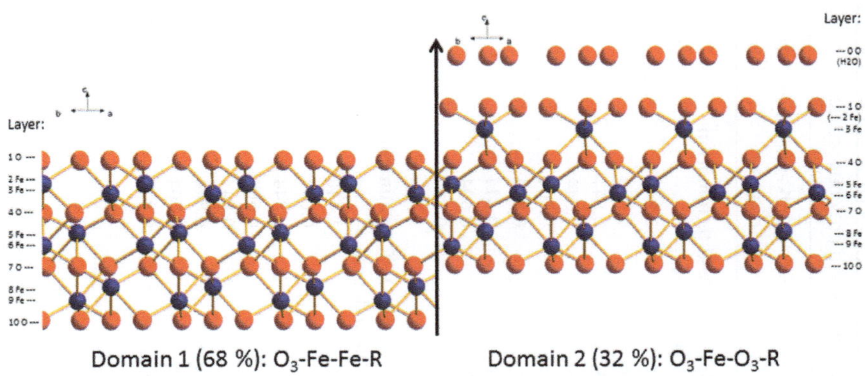

Fig. 5 Best fit structural model for the surface of the aged hematite sample (oxygen: red, iron: blue). Indicated are the domain and layer nomenclature as used in Table SI2.†

Table 1 Comparison of structure relaxations (percent difference between interlayer distances of the relaxed and the bulk structure) at the hematite (001)–water interface in this and previous studies

Layer	Bulk–interlayer distance	This study – CTR aged hematite	Trainor[62]-CTR	Trainor[62]-DFT	Catalano[29]-XR	Tanwar[61]-CTR unreacted sample	This study CTR fresh hematite
Panel 1: O_3–Fe–Fe–R termination							
Domain abundance		68%	62%	—	100%	54%	44%
$1O_3$–2Fe	0.844	16	23	35	34	1	5
2Fe–3Fe	0.604	**−28**	**−48**	**−41**	**−17**	**−18**	**−67**
3Fe–$4O_3$	0.844	−1	13	12	7	4	31
$4O_3$–5Fe	0.844	−3	2	2	1	1	8
5Fe–6Fe	0.604	**−12**	**−6**	6	**−22**	0	**−35**
6Fe–$7O_3$	0.844	9	5	−4	7	3	19
$7O_3$–8Fe	0.844	2	0	0	7	−1	−3
8Fe–9Fe	0.604	0			−6	0	0
9Fe–$10O_3$	0.844	0			1	−1	0
Panel 2: O_3–Fe–O_3–R termination							
Domain abundance		32%	38%	—	0%	46%	56%
$1O_3$–3Fe	1.448	−38	−24	−28	—	2	20
3Fe–$4O_3$	0.844	58	19	20	—	1	14
$4O_3$–5Fe	0.844	26	24	26	—	1	−11
5Fe–6Fe	0.604	**−38**	**−48**	**−69**	—	**−18**	**−23**
6Fe–$7O_3$	0.844	−7	12	15	—	4	10
$7O_3$–8Fe	0.844	7	0	0	—	1	−3
8Fe–9Fe	0.604	4			—	0	−4
9Fe–$10O_3$	0.844	−6			—	3	2

terminated hematite structure are given in Table SI3.† Ball and stick representations of the best fit structures for both surfaces are shown in Fig. 4 and 5. The z-offsets between the two domains in the structural pictures are artificial (only for illustration). The patchwise distribution is probably real, but the dimensions of the patches cannot be quantified from the CTR-data. The fact that only coherent addition of the contributions from the two terminations yield reasonable fits to the data, points, however, towards a small scale mixing of the domains. Interestingly, AFM images of the fresh surface (to be discussed below) support the patchwise structure and even a recent theoretical study suggests that the patchwise organisation is energetically favourable.[81]

From Fig. 4 and 5, it is clear that the contribution of domain 2 has decreased upon aging. Furthermore, the distances between the top oxygen atoms and the first iron layer on domain 2 is dramatically different. Finally on the aged sample a water layer is found, which in the sketch is shown over domain 2 for illustrative purposes only, since such information cannot be obtained from the CTR model. The occupancy in the water layer is relatively low (~20% of a mono layer), suggesting stronger ordering than for the fresh sample, but clearly not the very strong ordering found on other cuts of hematite.[29]

Table 1 shows a compilation of the structure relaxations observed at the hematite (001)–water interface in this and in previous CTR[61,62] and X-ray reflectivity[29] studies. DFT results by Trainor *et al.*[62] are listed for comparison.

From a qualitative comparison some remarkable similarities exist. For the O_3–Fe–Fe–R termination (domain 1) the topmost oxygen layer relaxes away from

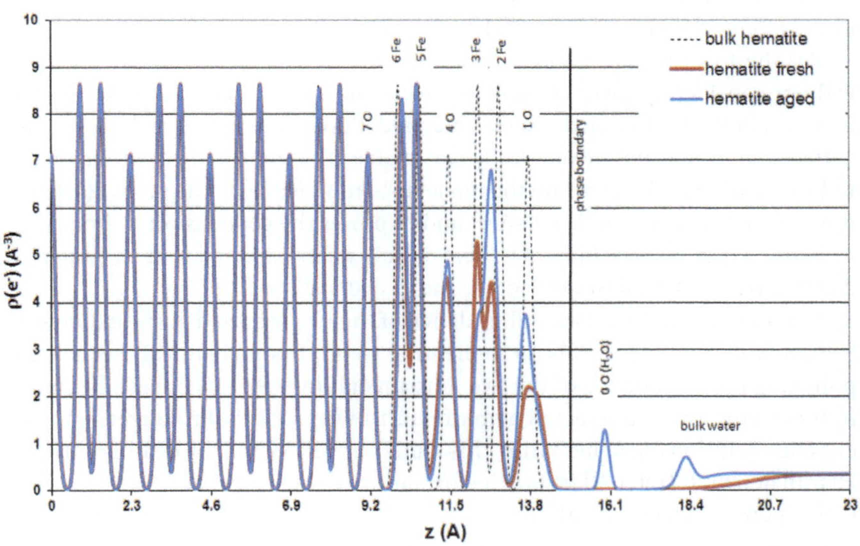

Fig. 6 Electron density distribution according to the hematite (001) surface structures of bulk hematite (dashed line), the fresh hematite sample (thick red line), and the aged hematite sample (blue line), projected onto the surface normal. The origin marks the arbitrary transition between undisturbed bulk and surface structures. The thick black line labeled "phase boundary" is placed between the terminating oxygen atoms (O^{2-}, OH^-, or H_2O) and the adsorbed water. It is no more than a guide to the eye, since the position is somewhat arbitrary.

the underlying iron layer ($1O_3$–2Fe). The two iron layers 2Fe and 3Fe contract. This contraction extends in most experimental observations (except Tanwar *et al.*[61]) to the iron layers underneath (5Fe–6Fe). The iron layers are marked in boldface in Table 1. The Fe–Fe contraction also emerges for the 5Fe and 6Fe layers in the O_3–Fe–O_3–R termination. For the O_3–Fe–O_3–R termination two different trends concerning the relaxations of the topmost layers occur. For the aged hematite in this study, the annealed hematite by Trainor *et al.*,[62] and in nice agreement with DFT results by Trainor *et al.*,[62] the top oxygen layer (1 O) contracts towards the 3Fe layer, while the 3Fe–$4O_3$ and $4O_3$–5Fe layers expand.

Deviating from this trend, the structure reported by Tanwar *et al.*[61] seems to be an intermediate case, while for the fresh hematite sample in the present study we observe an opposite relaxation of $1O_3$ away from 3Fe.

Based on the CTR results in this study an explanation for the relaxation of $1O_3$ away from 3Fe on the fresh hematite sample, might be a significantly weaker hydration of the surface of the fresh hematite sample compared to the hydration of the aged hematite sample. The surface hydration of the two samples is best compared by regarding the electron density distribution of the surface structures projected onto the surface normal (Fig. 6).

For the aged hematite sample Fig. 6 shows a peak for the partially occupied adsorbed water layer (~20% of a monolayer) at $z = 16$ Å and a relatively well structured first bulk water layer (z (water) = 18.25 Å, $U_0 = 0.039$ Å^2). The interlayer distance between consecutive bulk water layers is, however, small (d (water) = 0.81 Å) and the increase in the vibrational distribution from layer to layer is relatively large, $\bar{U} = 0.45$ Å^2, indicating that bulk water layers beyond the first layer are not structurally resolved. The bulk water model used here is identical to the one presented by Fenter and Sturchio (2004).[14]

For the fresh hematite sample good fits to the data could only be obtained without considering distinct adsorbed water molecules. The bulk water profile included in the best fit model is unstructured. Furthermore, it is so far above the surface that it may not be physically reasonable. The improvement of the model fit by including bulk water into the model is hardly visible on the plots. However, it has a numerically significant effect and improves the goodness of fit from $\chi^2 = 4.29$ ($p = 37$) to $\chi^2 = 3.97$ ($p = 41$). Best fit bulk water parameters are: z (water) = 20.62 Å, $U_0 = 1.3$ Å^2, d (water) = 1 Å, and $\bar{U} = 0.4$ Å^2.

Comparison of the relative abundance of the two reported surface domains (reported in Table 1) reveals an interesting trend. Upon chemical mechanical polishing the abundance of the two domains is similar (44/56: fresh hematite in this study, 54/46: unreacted hematite in Tanwar *et al.*[61]). Prolonged aging in aqueous solution (aged hematite, this study) as well as mild annealing [8 times 30 min at 600 °C (ref. 62)] favors the formation of the O_3–Fe–Fe–R termination. Only upon extensive annealing (12 h at 1100 °C) a fully O_3–Fe–Fe–R terminated hematite was inferred.[29] This agrees with the stability trends reported by Trainor *et al.*[62] based on DFT calculations. These indicate that the O_3–Fe–Fe–R termination is more stable than O_3–Fe–O_3–R at elevated temperatures in dry environments *e.g.* during annealing. Furthermore, the stability of the O_3–Fe–Fe–R termination is slightly favorable compared to O_3–Fe–O_3–R at room temperature in aqueous environments. This agrees well with the

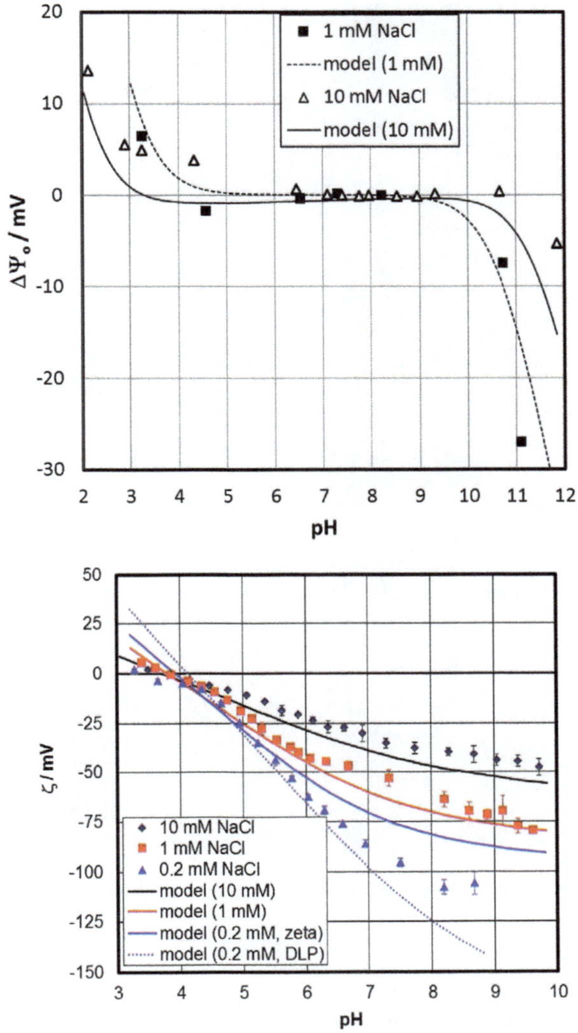

Fig. 7 Surface (top)- and zeta (bottom)-potentials for the fresh hematite (001) sample. Symbols are experimental data and lines are calculated by the proposed surface complexation model. In the lower graph, two model lines are plotted for the 0.2 mM NaCl data. The one noted "zeta" points to the model-inherent slip plane distance, while the one noted "DLP" pertains to the potential at the onset of the diffuse layer.

observed evolution of the surface termination of hematite from 44% O_3–Fe–Fe–R on fresh hematite to 68% O_3–Fe–Fe–R on the aged hematite sample.

The evaluation of the surface diffraction data does not include hydrogen atoms. However, the surface is expected to be rapidly hydroxylated.[66] Cutting the surface during its manufacturing will result in some termination exposing iron and/or oxygen. To include hydrogen atoms and the concomitant role of interfacial water, we used a bond valence analysis based on the CTR results. The bond valence analyses of the fresh and aged surfaces show some differences. The bond valence for the various doubly coordinated oxygen atoms

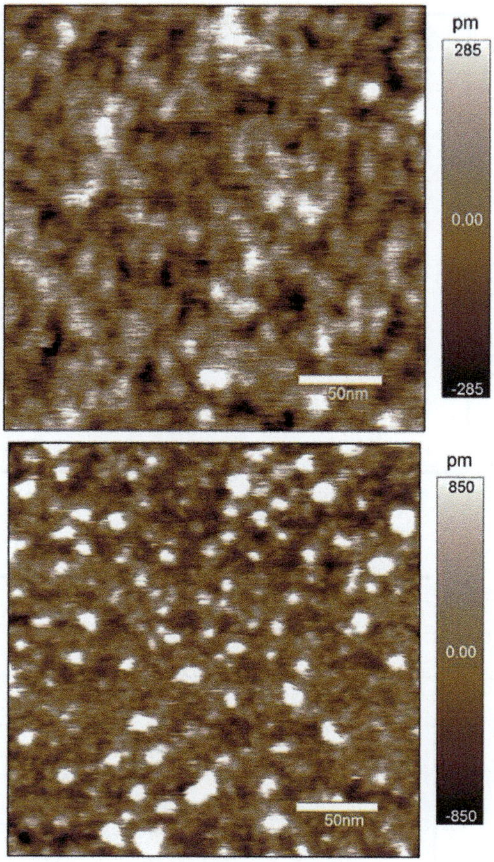

Fig. 8 High resolution AFM image of the fresh sample in air (top) and in water (bottom).

differs by no more than about 0.2 valence units, *i.e.* there is no significant difference. The undersaturation of the doubly coordinated oxygens is about one valence unit and addition of one proton would satisfy the oxygen valence for both fresh and aged hematite. This supports the view that these oxygen atoms are neither protonated nor deprotonated at the conditions of the CTR measurements. However, concerning the singly coordinated oxygen atoms, the fresh hematite sample exhibits an undersaturation by 1.75 valence units. The simplest way to reduce this deficit is to assume that the group is doubly protonated and has originated from adsorption of a water molecule at the iron termination.[66] This would result in a stable, aquo-group terminated configuration and might remain non-reactive (as is concluded later to explain the zeta-potential data). In turn, on the aged sample the undersaturation of the singly coordinated oxygen is about 1.45 valence units. In that case two adsorbed protons would not allow a stable configuration. A more likely explanation is therefore interaction with water to yield a hydroxyl group and involve hydrogen bonding between neighboring hydroxyl groups or with adjacent water molecules. This would explain the presence of structured water on the aged surface.

Surface complexation modelling

For the following it has to be mentioned that CTR data have not been collected at various pH conditions or even in the presence of salt. Also the amount of water in our study is much lower than in the work by Catalano.[29] A future comprehensive study should cover such aspects in order to test some of the model assumptions that are used in the following.

Fresh hematite (001)

Fig. 7 shows the surface- and zeta-potentials of the fresh sample. The surface potential exhibits the broad flat response in the mid pH-range that is believed to be due to the p*K* values of the doubly coordinated hydroxyl groups as discussed above. The point of zero charge of the sample would therefore be at pH about 6.5, close to the value reported for sapphire (001) based on sum frequency generation data.[30] The surface-potential curves appear to exhibit an ionic strength influence similar to observations by Boily *et al.*[67] The zeta-potentials in turn exhibit an IEP at about pH 4, which appears to decrease with decreasing sodium content. The dual charging mechanism therefore needs to be involved to explain the data that have been obtained on identical samples.

Based on the CTR data, the iron-terminated surface is dominant on the fresh sample. Fig. 3 shows schematically the presence of two domains. These are not separated by large distances. High resolution AFM (Fig. 8) indicates that the surface should rather be imagined as a mix of small 10 to 30 nm patches of the two domains. The CTR data do not show significant water structuring. This could explain behavior similar to hydrophobic/inert surfaces. This agrees with previous studies by Catalano who found weak water ordering on hematite (001).[29] Catalano involved bulk water above the surface studied in his work so that his results cannot be directly compared to our study, but in a qualitative way between the fresh and the aged sample the fresh sample surface clearly exhibits weaker water structuring.

In contrast to the macroscopic data, there is no salt present in the CTR study. Since the measurement, new evidence has appeared that related faces (the gibbsite basal plane) cause adsorption and layering of background electrolyte ions in $CaCl_2$-solutions at pH 6.[82] This was clearly shown by both AFM force curves and their interpretation in terms or electrostatic potentials and direct imaging for a 2 : 1 electrolyte. Even with increasing KCl and NaCl an increasing positive potential was reported for the gibbsite basal plane. Our own, unpublished surface potential data for sapphire (001) support the adsorption of KCl. At a constant pH of about 6 the surface potential increases with increasing KCl concentration. Finally, the presence of salt hydrates has been previously postulated based on XPS studies on hematite particles, wherein particularly with increasing contributions from basal planes an enhanced presence of salt hydrates was reported.[83] Based on the above, our present model approach for the fresh surface includes the following features:

– protonation/deprotonation of the surface hydroxyls (based on MUSIC type calculations[52,54]) to account for the conventional picture of oxide–aqueous solution interfaces (no attempt was made to relate p*K* values directly to the CTR results, the values are in the expected and previously reported range)

– adsorption of NaCl within the Stern plane to account for the experimental evidence on this phenomenon at gibbsite basal planes and hematite particles

– adsorption of protons and hydroxide ions beyond the NaCl plane to account for the pH dependence of the zeta-potential and the observed isoelectric point that is typically found for inert surfaces

– singly coordinated groups are assumed to be inert due to weak interaction with water as inferred from the CTR data.

A sketch of the interfacial model is given in the ESI (Fig. SI1†) and the model parameters are given in Table SI4.† Interestingly, simple versions of the model, including only one or two of the above mechanisms failed to reproduce the experimental data.

Lines in Fig. 7 show the fit to the surface- and zeta-potential based on the model described above. The surface potential is well described and the inclusion of the salt interactions allow shifts in the surface potential curves with salt content to be described. The traditional picture does not involve a significant effect of background electrolytes on the surface potential[56,57] and conventional surface complexation models agree with this.[57]

The surface potential cannot be measured on an absolute scale and is therefore plotted as the difference ($\Delta\Psi_0$) to an arbitrary reference potential. The model inherent surface potential is affected by the salt level assuming adsorption of NaCl within the Stern plane. Diffuse layer potential data on gibbsite (positive in the absence of salt) show that addition of NaCl or KCl increases the potential.[82] Macroscopic data on gibbsite have been interpreted by adsorption of sodium and chloride on the basal plane, involving (unlike our model) protons.[84] Spectroscopic data for gibbsite agrees with the adsorption of both sodium and chloride at low pH.[85] Even for hematite particles, such evidence has been presented.[83] The relative change of the surface potential as a function of pH is well described by the model. For the zeta-potential, the slip-plane separation is the most important parameter involved. The good description of the zeta-potentials is to a large extent due to its adjustment. At the lowest salt content, the experimental data at high pH are between the potential with the assumed slip-plane separation (full line) and the one at the onset of the diffuse layer (dashed line). We involve a slip plane distance parameter that is related to the ionic strength (see ESI† for more detailed information). An individual adjustment of the slip plane distance would yield an even better fit to the data.

Whatever option is chosen, the model is in the correct range of potentials. At low pH, at the smallest salt content, the misfit between experimental and simulated zeta-potentials is largest. Under these conditions the ionic strength is not constant, since the addition of acid starts affecting the overall value. Therefore, we have carried out calculations with the correct concentrations of all components. This shifts the model for nominal 0.2 mM concentrations towards the one for 1 mM thereby decreasing the misfit; while the shift of the IEP is not reproduced. Overall, the results warrant additional CTR-investigations in the presence of NaCl.

High resolution AFM does not show clear terraces on the fresh sample surface in air. However, the root mean square roughness is estimated to be 1.2 Å under these conditions (Fig. 8, top), comparable to the result from the CTR-study (1.1 Å), *i.e.* very low. The various small areas/patches might correspond to the two domains that are discussed in the context of the CTR-data. Addition of

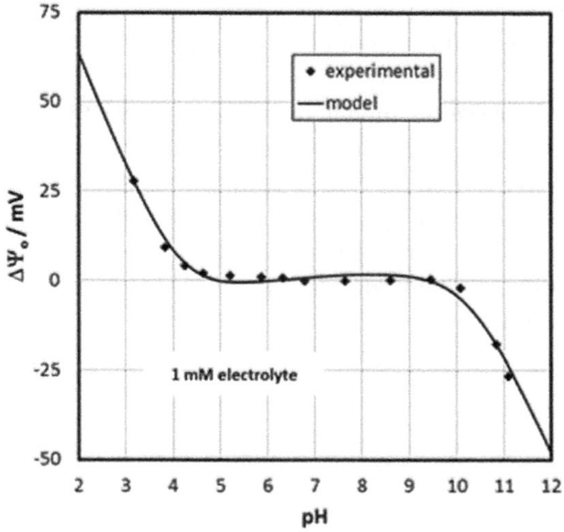

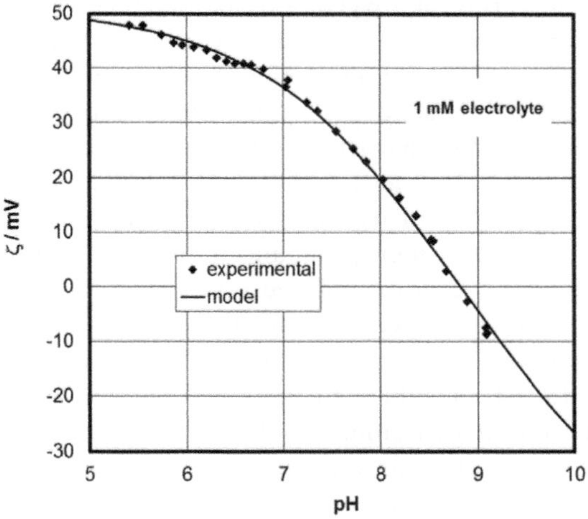

Fig. 9 Surface (top) and zeta (bottom)-potential for the aged hematite-sample, and model based on the concept and parameters given in ESI.†

water increases the roughness to between 3.9 and 4.4 Å, addition of 1 mM KCl solution further increases it to 5 to 7 Å. The addition of aqueous solutions at pH about 6 thus has a significant effect on the roughness (Fig. 8, bottom) but this effect has not been observed in our CTR experiments. For resolution of hydration water layers[86] or adsorbed ions or ion layers our sample is probably too rough. However, we believe that sufficient circumstantial evidence[82–84] exists by now that salt layers are present in related, as well as in very different systems. Correspondingly, the assumption of salt in the Stern layer on hematite (001) is considered reasonable.

Force curves with a negative tip probe at pH 6 (not shown) were always repulsive in water and 1 mM KCl supporting the zeta-potential measurements for the fresh sample. Additional measurements in 1 mM HCl also produced repulsive interactions, in line with the shift of the IEP that occurs with decreasing sodium concentration (Fig. 7). In 1 mM HCl potassium/sodium ions are absent and cannot contribute to the charging of the surface so that a "salt" layer cannot form.

Our surface complexation model does not need to resort to iron-surface species similar to acidic groups invoked for sapphire (001) to explain the low isoelectric point.[55] We rely on the CTR-data in this respect and our model is able to describe the system very well on that basis. We note that many parameters are involved, but most of them have been taken from associated systems (such as hydroxide adsorption on inert surfaces or expected protonation/deprotonation constants of the surface hydroxyls).

Aged hematite (001)

Based on the CTR data, the oxygen-terminated surface (O_3–Fe–Fe–R) is dominant on the aged sample. This supports a previous study where an annealed sample had been left to age and developed a surface potential response that was interpreted in terms of oxygen-termination.[70] The measured surface potentials of the present sample indeed show the same response (for both fresh and aged states), while the zeta-potentials evolve with time (Fig. 2). The IEP of the aged sample is at about pH 9 at the steady state condition, which is similar to that measured on particles.[87] Compared to the data for the fresh sample, the CTR data for the aged sample show adsorbed water molecules and the presence of a water film. This might suggest that the surface rearrangement is related to an increasingly pronounced interaction with water. In this interpretation the singly-coordinated hydroxyl groups would interact more strongly with the interfacial water and would be able to affect the electrokinetic potentials substantially. Surprisingly, the experimental data suggest that these reactive groups do not contribute to the measured surface potential. In a model the singly-coordinated groups could be treated as adsorbates.[65] This would be in line with a previous study on sapphire (001),[88] where the intentional addition of dissolved aluminum caused a strong shift of the isoelectric point, similar to the present case. The action of the water can be held responsible for the difference between the fresh and aged samples. Another issue could be how the ad-atoms are distributed over the full surface. But this cannot be inferred from our data. No high resolution AFM images are available for the aged samples.

The results in Fig. 9 show that the proposed model is capable of describing the surface (Fig. 9, top) and zeta (Fig. 9, bottom)-potentials in an excellent way.

The data for the aged hematite provide the first experimental observation *via* streaming potential measurements of an IEP around 9 for a flat oxide surface that would be expected based on the available data for particles that are established with far more popular methods, such as electrophoresis or accoustophoresis. The low IEPs previously reported for many oxide single crystals[79,80] have raised suspicion,[89] since they do not obey "surface chemistry" in terms of the MUSIC model and because they are rather generic similar to hydrophobic/inert surfaces. It is interesting to note that the model for the aged hematite (001) surface produces the slight increase of the surface potential with increasing pH within the plateau region, which has been observed previously,[71] and which so far lacked an explanation.

Summary of the surface complexation models

Despite the surface characterization of the two samples, a substantial number of assumptions is involved in the model design. The following points address weak aspects of our modelling:

- The assumption about the reactivity of the singly coordinated groups is ultimately linked to the outcome of the zeta-potential measurements. In particular on the fresh surface inclusion of the singly coordinated groups would not allow a low isoelectric point to be modelled. The main argument is the lack of water interaction in the sense that a strongly ordered water film is not observed.
- The salt layering is unconventional in the sense that it does not involve proton co-adsorption (for chloride adsorption) or proton release (for sodium adsorption) unlike the model proposed by Rosenqvist *et al.*[84]
- The major parameter for describing the zeta-potential is the slip-plane separation. We currently have to consider it as a pure fitting parameter, but we expect that in future with the advent of new techniques and enhanced computational power this parameter can be eliminated.

On the other hand, the model is quite successful in describing experimental results.

- Our model results offer for the first time an explanation for an observation previously reported by Chatman *et al.*[68,71] which was not discussed in detail at the time. The relatively small increase of measured surface potential with increasing pH within the plateau region is at odds with all the expectations for common oxide–water interface models. Apparently the model designed here for the aged hematite (001) (Fig. 9, top panel) is able to cover this behavior. Actually, the increase of surface potential with increasing pH reported by Chatman *et al.*[71] could be seen as independent support for our model concept.
- The model also has the potential to describe the experimentally observed offset of the protonation/deprotonation with variable salt content.[67] Usually a surface complexation would treat such experimental observations with one consistent set of parameters (extrapolated to infinite dilution). However, in the cited case it was necessary to apply different p*K* values for the data measured at the different salt contents.

Conclusions

The present study investigates transient changes at the hematite (001) surface upon aging in aqueous solutions. We observe changes with time in the zeta-potential and the surface diffraction data, while the surface potential response remains unaffected by the structural changes. AFM investigations show that the presence of water and electrolyte solutions affects the surface roughness of our sample. The initially very smooth surface becomes rougher in the presence of water. The surface diffraction data suggest that the fresh surface is not well hydrated. In particular the singly coordinated groups appear neither to affect the surface potential–pH response nor to influence the zeta-potentials of the fresh surface. A bond-valence analysis of the surface diffraction data supports the lack of hydration in that these aquo-groups are well satisfied in terms of bond valence. With time these reactive groups become more reactive, probably due to a change in bond distances that triggers or is triggered by interactions with water. In the case of the fresh hematite, the adjacent water layer, which is expected to be weakly structured similar to the one on hydrophobic surfaces, dominates the observed zeta-potentials *via* preferential hydroxide adsorption in that layer. The inner layer becomes affected by protonation (low pH) and deprotonation (high pH) of the doubly coordinated groups, and salt ion adsorption, which allows explanation of the measured surface potentials, while the singly coordinated groups remain inactive, similar to previous suggestions.[82,83]

The surface diffraction data for the aged surface show distinct water adsorption. Consequently, the model for the aged surface involves both groups and the action of the singly-coordinated groups explains the high IEP observed. Salt ion adsorption and doubly coordinated surface groups account for the surface potential response, which remains unchanged upon aging. The parameters used in the models are in the range expected based on previous investigations on related systems.

Acknowledgements

CTR measurements were performed at GeoSoilEnviroCARS (Sector 13), Advanced Photon Source (APS), Argonne National Laboratory. GeoSoilEnviroCARS is supported by the National Science Foundation – Earth Sciences (EAR-1128799) and Department of Energy-GeoSciences (DE-FG02-94ER14466). This research used resources of the Advanced Photon Source, a U.S. Department of Energy (DOE) Office of Science User Facility operated for the DOE Office of Science by Argonne National Laboratory under Contract no. DE-AC02-06CH11357. During the proofreading period of this article, one of the co-authors, Prof. Nikola Kallay, passed away unexpectedly to us all. We dedicate this paper to him.

Notes and references

1 C. Bataillon, F. Bouchon, C. Chainais-Hillairet, C. Desgranges, E. Hoarau, F. Martin, S. Perrin, M. Tupin and J. Talandier, *Electrochim. Acta*, 2010, **55**, 4451–4467.

2 M. Finšgar and J. Jackson, *Corros. Sci.*, 2014, **86**, 17–41.

3 F. F. Eliyan and A. Alfantazi, *Corrosion*, 2014, **70**, 880–898.

4 J. Brunelle, *Pure Appl. Chem.*, 1978, **50**, 1211–1229.

5 J. A. Schwarz, C. Contescu and A. Contescu, *Chem. Rev.*, 1995, **95**, 477–510.

6 T. Mang, B. Breitscheidel, P. Polanek and H. Knozinger, *Appl. Catal., A*, 1993, **106**, 239–258.

7 W. Stumm, *Colloids Surf., A*, 1997, **120**, 143–166.

8 T. Hiemstra, *Langmuir*, 2012, **28**, 15614–15623.

9 Z. Zhang, P. Fenter, L. Cheng, N. C. Sturchio, M. J. Bedzyk, M. Predota, A. Bandura, J. D. Kubicki, S. N. Lvov, P. T. Cummings, A. A. Chialvo, M. K. Ridley, P. Benezeth, L. Anovitz, D. A. Palmer, M. L. Machesky and D. J. Wesolowski, *Langmuir*, 2004, **20**, 4954–4969.

10 G. E. Schaumann, A. Philippe, M. Bundschuh, G. Metreveli, S. Klitzke, D. Rakcheev, A. Grün, S. K. Kumahor, M. Kühn, T. Baumann, F. Lang, W. Manz, R. Schulz and H.-J. Vogel, *Sci. Total Environ.*, 2015, DOI: 10.1016/j.scitotenv.2014.10.035.

11 A. Massarsky, V. L. Trudeau and T. W. Moon, *Environ. Toxicol. Pharmacol.*, 2014, **38**, 861–873.
12 S. Wagner, A. Gondikas, E. Neubauer, T. Hofmann and F. von der Kammer, *Angew. Chem., Int. Ed.*, 2014, **53**, 12398–12419.
13 J. A. Davis and K. F. Hayes, *Geochemical processes at mineral surfaces*, American Chemical Society, Washington, DC, 1986.
14 P. Fenter and N. C. Sturchio, *Prog. Surf. Sci.*, 2004, **77**, 171–258.
15 A. F. White and S. L. Brantley, *Chemical Weathering Rates of Silicate Minerals*, 1995, vol. 31, pp. 1–22.
16 J. Bachmann and R. R. van der Ploeg, *J. Plant Nutr. Soil Sci.*, 2002, **165**, 468.
17 H. B. Bradl, *J. Colloid Interface Sci.*, 2004, **277**, 1–18.
18 G. Sposito, *The chemistry of soils*, Oxford University Press, 2008.
19 C. E. Weaver and L. D. Pollard, *The chemistry of clay minerals*, Elsevier, 2011.
20 S. L. Swartzen-Allen and E. Matijevic, *Chem. Rev.*, 1974, **74**, 385–400.
21 G. Lagaly and S. Ziesmer, *Adv. Colloid Interface Sci.*, 2003, **100–102**, 105–128.
22 J. Tolls, *Environ. Sci. Technol.*, 2001, **35**, 3397–3406.
23 W. Stumm and J. J. Morgan, *Aquatic chemistry: chemical equilibria and rates in natural waters*, John Wiley & Sons, 2012.
24 R. P. Schwarzenbach, P. M. Gschwend and D. M. Imboden, *Environmental organic chemistry*, John Wiley & Sons, 2005.
25 Y. Wang, *Chem. Geol.*, 2014, **378–379**, 1–23.
26 D. Langmuir, P. Hall and J. Drever, *Environmental Geochemistry*, Prentice Hall, New Jersey, 1997.
27 Z. Navrátilová and P. Kula, *Electroanalysis*, 2003, **15**, 837–846.
28 L. Charlet and A. Manceau, *Int. J. Environ. Anal. Chem.*, 1992, **46**, 97–108.
29 J. G. Catalano, *Geochim. Cosmochim. Acta*, 2011, **75**, 2062–2071.
30 L. Zhang, C. Tian, G. A. Waychunas and Y. R. Shen, *J. Am. Chem. Soc.*, 2008, **130**, 7686–7694.
31 M. Sulpizi, M. Salanne, M. Sprik and M. P. Gaigeot, *J. Phys. Chem. Lett.*, 2013, **4**, 83–87.
32 J. Luetzenkirchen, R. Zimmermann, T. Preocanin, A. Filby, T. Kupcik, D. Kuettner, A. Abdelmonem, D. Schild, T. Rabung, M. Plaschke, F. Brandenstein, C. Werner and H. Geckeis, *Adv. Colloid Interface Sci.*, 2010, **157**, 61–74.
33 J. K. Beattie, A. M. Djerdjev, A. Gray-Weale, N. Kallay, J. Luetzenkirchen, T. Preocanin and A. Selmani, *J. Colloid Interface Sci.*, 2014, **422**, 54–57.
34 P. Creux, J. Lachaise, A. Graciaa, J. K. Beattie and A. M. Djerdjev, *J. Phys. Chem. B*, 2009, **113**, 14146–14150.
35 R. Zimmermann, N. Rein and C. Werner, *Phys. Chem. Chem. Phys.*, 2009, **11**, 4360–4364.
36 M. Corti, M. Bonomo and A. Raudino, *Langmuir*, 2012, **28**, 6060–6066.
37 V. Kuznetsov and G. Papastavrou, *J. Phys. Chem. C*, 2014, **118**, 2673–2685.
38 T. Cecchi, *J. Phys. Chem. C*, 2013, **117**, 25579–25585.
39 P. Jungwirth, *Faraday Discuss.*, 2009, **141**, 9–30.
40 V. L. Shapovalov, H. Mohwald, O. V. Konovalov and V. Knecht, *Phys. Chem. Chem. Phys.*, 2013, **15**, 13991–13998.
41 B. Winter, M. Faubel, R. Vácha and P. Jungwirth, *Chem. Phys. Lett.*, 2009, **474**, 241–247.

42 J.-S. Samson, R. Scheu, N. Smolentsev, S. W. Rick and S. Roke, *Chem. Phys. Lett.*, 2014, **615**, 124–131.
43 J. Luetzenkirchen, T. Preocanin and N. Kallay, *Phys. Chem. Chem. Phys.*, 2008, **10**, 4946–4955.
44 R. Zimmermann, U. Freudenberg, R. Schweiss, D. Kuettner and C. Werner, *Curr. Opin. Colloid Interface Sci.*, 2010, **15**, 196–202.
45 T. W. Healy and D. W. Fuerstenau, *J. Colloid Interface Sci.*, 2007, **309**, 183–188.
46 N. Kallay and D. Cakara, *J. Colloid Interface Sci.*, 2000, **232**, 81–85.
47 N. Kallay, A. Cop, E. Chibowski and L. Holysz, *J. Colloid Interface Sci.*, 2003, **259**, 89–96.
48 M. Takahashi, *J. Phys. Chem. B*, 2005, **109**, 21858–21864.
49 T. Hiemstra, J. C. M. Dewit and W. H. Vanriemsdijk, *J. Colloid Interface Sci.*, 1989, **133**, 105–117.
50 T. Hiemstra, W. H. Vanriemsdijk and G. H. Bolt, *J. Colloid Interface Sci.*, 1989, **133**, 91–104.
51 T. Hiemstra, P. Venema and W. H. VanRiemsdijk, *J. Colloid Interface Sci.*, 1996, **184**, 680–692.
52 P. Venema, T. Hiemstra, P. G. Weidler and W. H. van Riemsdijk, *J. Colloid Interface Sci.*, 1998, **198**, 282–295.
53 J. P. Fitts, X. M. Shang, G. W. Flynn, T. F. Heinz and K. B. Eisenthal, *J. Phys. Chem. B*, 2005, **109**, 7981–7986.
54 T. Hiemstra and W. H. Van Riemsdijk, *Langmuir*, 1999, **15**, 8045–8051.
55 D. Yang, M. Krasowska, R. Sedev and J. Ralston, *Phys. Chem. Chem. Phys.*, 2010, **12**, 13724–13729.
56 N. Kallay, T. Preocanin, F. Supljika, J. Luetzenkirchen and M. Lovkovic, *J. Colloid Interface Sci.*, 2012, **375**, 167–171.
57 A. Selmani, J. Luetzenkirchen, N. Kallay and T. Preocanin, *J. Phys.: Condens. Matter*, 2014, **26**(24), 244104.
58 R. Vácha, R. Zangi, J. Engberts and P. Jungwirth, *J. Phys. Chem. C*, 2008, **112**, 7689–7692.
59 J. Luetzenkirchen, *Langmuir*, 2013, **29**, 7726–7734.
60 A. G. Stack, S. R. Higgins and C. M. Eggleston, *Geochim. Cosmochim. Acta*, 2001, **65**, 3055–3063.
61 K. S. Tanwar, S. C. Petitto, S. K. Ghose, P. J. Eng and T. P. Trainor, *Geochim. Cosmochim. Acta*, 2009, **73**, 4346–4365.
62 T. P. Trainor, A. M. Chaka, P. J. Eng, M. Newville, G. A. Waychunas, J. G. Catalano and G. E. Brown, *Surf. Sci.*, 2004, **573**, 204–224.
63 X. G. Wang, W. Weiss, S. K. Shaikhutdinov, M. Ritter, M. Petersen, F. Wagner, R. Schlogl and M. Scheffler, *Phys. Rev. Lett.*, 1998, **81**, 1038–1041.
64 C. M. Eggleston, A. G. Stack, K. M. Rosso, S. R. Higgins, A. M. Bice, S. W. Boese, R. D. Pribyl and J. J. Nichols, *Geochim. Cosmochim. Acta*, 2003, **67**, 985–1000.
65 C. M. Eggleston, A. G. Stack, K. M. Rosso and A. M. Bice, *Geochem. Trans.*, 2004, **5**, 33–40.
66 P. Liu, T. Kendelewicz, G. E. Brown Jr, E. J. Nelson and S. A. Chambers, *Surf. Sci.*, 1998, **417**, 53–65.
67 J.-F. Boily, S. Chatman and K. M. Rosso, *Geochim. Cosmochim. Acta*, 2011, **75**, 4113–4124.
68 S. Chatman, P. Zarzycki and K. M. Rosso, *Phys. Chem. Chem. Phys.*, 2013, **15**, 13911–13921.

69 N. Kallay and T. Preocanin, *J. Colloid Interface Sci.*, 2008, **318**, 290–295.
70 J. Luetzenkirchen, T. Preocanin, F. Stipic, F. Heberling, J. Rosenqvist and N. Kallay, *Geochim. Cosmochim. Acta*, 2013, **120**, 479–486.
71 S. Chatman, P. Zarzycki, T. Preoanin and K. M. Rosso, *J. Colloid Interface Sci.*, 2013, **391**, 125–134.
72 M. M. Gentleman and J. A. Ruud, *Langmuir*, 2010, **26**, 1408–1411.
73 E. Anim-Danso, Y. Zhang and A. Dhinojwala, *J. Am. Chem. Soc.*, 2013, **135**, 8496–8499.
74 B. M. Cabanas, J. Luetzenkirchen, S. Leclercq, P. Barboux and G. Lefevre, *J. Nucl. Mater.*, 2012, **430**, 150–155.
75 S. A. de Vries, P. Goedtkindt, W. J. Huisman, M. J. Zwanenburg, R. Feidenhans'l, S. L. Bennett, D. M. Smilgies, A. Stierle, J. J. De Yoreo, W. J. P. van Enckevort, P. Bennema and E. Vlieg, *J. Cryst. Growth*, 1999, **205**, 202–214.
76 I. D. Brown and D. Altermatt, *Acta Crystallogr., Sect. B: Struct. Sci.*, 1985, **41**, 244–247.
77 R. Zimmermann, S. Dukhin and C. Werner, *J. Phys. Chem. B*, 2001, **105**, 8544–8549.
78 G. V. Franks and Y. Gan, *J. Am. Ceram. Soc.*, 2007, **90**, 3373–3388.
79 S. Roessler, R. Zimmermann, D. Scharnweber, C. Werner and H. Worch, *Colloids Surf., B*, 2002, **26**, 387–395.
80 D. Rothenstein, B. Claasen, B. Omiecienski, P. Lammel and J. Bill, *J. Am. Chem. Soc.*, 2012, **134**, 12547–12556.
81 A. Kiejna and T. Pabisiak, *J. Phys. Chem. C*, 2013, **117**, 24339–24344.
82 I. Siretanu, D. Ebeling, M. P. Andersson, S. L. S. Stipp, A. Philipse, M. C. Stuart, D. van den Ende and F. Mugele, *Sci. Rep.*, 2014, **4**, 4956.
83 A. Shchukarev, J.-F. Boily and A. R. Felmy, *J. Phys. Chem. C*, 2007, **111**, 18307–18316.
84 J. Rosenqvist, P. Persson and S. Sjöberg, *Langmuir*, 2002, **18**, 4598–4604.
85 A. Shchukarev and S. Sjöberg, *Surf. Sci.*, 2005, **584**, 106–112.
86 F. Takeshi, *Sci. Technol. Adv. Mater.*, 2010, **11**, 033003.
87 M. Kosmulski, *Chemical Properties of Material Surfaces*, CRC Press, 2001.
88 J. Lutzenkirchen, A. Abdelmonem, R. Weerasooriya, F. Heberling, V. Metz and R. Marsac, *Geochem. Trans.*, 2014, **15**, 9.
89 E. Mączka, J. Luetzenkirchen and M. Kosmulski, *J. Colloid Interface Sci.*, 2013, **393**, 228–233.

DISCUSSIONS

Solid/fluid interface: general discussion

Janine Mauzeroll, Geoffrey Thornton, Trevor Rayment, Vincent Maurice, David Williams, Frank Heberling, Philippe Marcus, Clara Wren, Kirsi Yliniemi, Rob Lindsay, Stephen Lyth, Tom Majchrowski, Hadeel Hussain, Gregory Hunt, Frank Renner, Geraint Williams, Roger Newman, Gerald Frankel, Johannes Lützenkirchen and Hendrik Bluhm

DOI: 10.1039/c5fd90044a

Stephen Lyth opened a general discussion of the paper by G. Frankel by asking: The corrosion chemistry community has made interesting insights into the role of hydrogen evolution as a degradation mechanism in metals. Can these findings be turned around to design more efficient catalysts for electrochemical hydrogen production?

Gerald Frankel replied: Thank you for this interesting question. There is no doubt that the combination of a Mg anode and any other cathode forms a potent hydrogen production engine. Mg alloys such as Mg–Fe generate even more H_2 than pure Mg.[1] However, a catalyst should be a non-consumable material. The energy that goes into making Mg or another active metal like Al is lost during the process, so it is an expensive way to create hydrogen. I doubt that understanding of the situation at dissolving metal anodes can be translated into non-consumable catalysts.

1 A. Samaniego, N. Birbilis, X. Xia and G. S. Frankel, Hydrogen Evolution During Anodic Polarization of Mg Alloyed with Li, Ca, or Fe, *Corrosion*, 2015, **71**, 224–233.

Roger Newman asked: Regarding the enhanced evolution of hydrogen during anodic polarization of magnesium, the relevance, if any, of the chloride ion was not mentioned. This seems to be universally present in the testing solutions. One can imagine that a Mg atom with an adsorbed anion might be more able to manifest its high reactivity, including for hydrogen evolution, than if it has an adsorbed OH. Does the effect exist over a wide range of anions? If it does, then indeed the anion adsorption idea would not be correct and the most likely explanation is simply that a fast-dissolving metal has more (albeit momentarily) bare metal sites which are the sites of hydrogen evolution (I believe this was suggested by Professor Williams).

As a footnote to the latter suggestion, this is somewhat related to a concept for activation of aluminum by low melting point (LMP) alloying elements that I developed with a student in the 90s.[1,2] There too we appealed to the momentary presence of bare Al sites, which in that case were occupied by diffusing LMP.

1 F. Sato and R. C. Newman, Mechanism of Activation of Aluminum by Low Melting Point Elements: Part 1 – Effect of Zinc on Activation of Aluminum in Metastable Pitting, *Corrosion*, 1998, **54**, 955–963.
2 F. Sato and R. C. Newman, Mechanism of Activation of Aluminum by Low-Melting Point Elements: Part 2 – Effect of Zinc on Activation of Aluminum in Pitting Corrosion, *Corrosion*, 1999, **55**, 3–9.

Gerald Frankel responded: In a recent paper, anodic HE was collected on HP Mg in chloride-free buffer solutions over a range of pH values.[1] Citrate (pH 3), borate (pH 7) and carbonate (pH 10.5) solutions were used. Anodic HE was observed in all of the solutions. However, in the citrate solution the HE rate increased only at higher applied current densities when the buffer capacity was seemingly overwhelmed by local pH changes. The bottom line is that chloride is not needed, but pH change is important.

Please see my response to Prof. Williams regarding the issues associated with explaining the effect solely by an increase in the number of active sites.

1 L. Rossrucker, A. Samaniego, J.-P. Grote, A. M. Mingers, C. A. Laska, N. Birbilis, G. S. Frankel and K. J. J. Mayrhofer, The pH Dependence of Magnesium Dissolution and Hydrogen Evolution During Anodic Polarization, *J. Electrochem. Soc.*, 2015, **162** C333–C339.

David Williams asked: Is it possible that, under the extreme dissolution conditions prevailing (which could be considered as kink sites moving extremely rapidly across the surface) that Al atoms at the kink sites could react with water either to form the aluminium hydroxide anodically or to split water to form hydrogen and aluminium hydroxide in a chemical process – in which case there would not be any current flowing through the external circuit. The balance between the two reactions could depend on the surface density and orientation of water molecules.

Gerald Frankel answered: This is an interesting notion that might be rephrased as follows: Assuming that Mg or Al dissolution occurs at active sites on the surface such as kink sites, it is possible that the number of active sites per unit exposed area increases as the dissolution rate increases, and these active sites can also reduce water. The result would be more water reduction with more dissolution.

This perspective describes how the active sites for both the anodic dissolution and hydrogen evolution reactions would be linked. However, as the applied anodic current or potential increases, the rate of the HER should decrease exponentially. Therefore, an explanation would still be needed to explain why the hydrogen evolution rate remains about the same percentage of the applied anodic current. One would have to assume that the rate of HER on these sites is independent of potential.

Janine Mauzeroll asked: Would it not be a good idea to also consider the complex equilibria that occurs in the magnesium system to account for some of the negative difference effect observed?

Gerald Frankel answered: I suppose you are referring to solution equilibria in the electrolyte near the dissolving surface. It is a complex situation owing to the rapid dissolution and the copious hydrogen evolution. It is generally understood that the pH will increase because of weak hydrolysis and strong water reduction. However, there is evidence of initial pH decrease because of a short dominance of hydrolysis.[1] And pH changes seem to be critical for the anodic HE effect on Mg. So, yes we need to understand details of both the surface and the near solution phase.

1 L. Rossrucker, A. Samaniego, J.-P. Grote, A. M. Mingers, C. A. Laska, N. Birbilis, G. S. Frankel and K. J. J. Mayrhofer, *J. Electrochem. Soc.*, 2015, **162**, C333.

Geraint Williams queried: Are there any local effects observed during the anodic polarisation of the Mg specimens, especially at high imposed current density values? For example, is the entirety of the "black" surface strongly evolving hydrogen, or are HE sites limited to a few regions on the exposed Mg. If the latter best represents observed behaviour, is there evidence that the number of sites strongly evolving hydrogen increases with potential or applied current density? Maybe the increase in "anodic" HE with anodic current density is an area effect, where the rise in current density progressively produces more regions of "nascent", totally film-free Mg which are the principal sites of both anodic dissolution and cathodic HE?

Gerald Frankel answered: Thanks for your question. This is a good point as simple optical observation should be able to help sort out the location of the HER. Unfortunately, this location is difficult to assess, particularly at high applied anodic currents when the HE is also large. Because of the copious HE at the surface, the entire surface is masked so it is impossible to determine the exact origin of the bubbles. This was especially difficult in our case because the sample surface was vertically oriented to allow for easy viewing of the surface through the

Fig. 1 Image of the surface during polarization at 25 mA cm^{-2}

cell wall. The accompanying figure (Fig. 1) shows an image of the surface during polarization at 25 mA cm^{-2}.

Images of the surfaces of electrodes after passage of a fixed charge but at different current densities exhibit increased coverage of dark corrosion product with increasing current density. This might indicate the important role of the dark regions. However, the variations are not orders of magnitude as are the increases in HE rate. Regarding the increase in active sites, please see my response to David Williams.

Clara Wren opened a general discussion of the papers by Hendrik Bluhm: What happens to the electrons ejected from the solid? The energy of the electrons would be the energy of the photoelectrons minus the binding energy and work function and, hence, relatively low, on the order of ~100 eV. This is not sufficient to penetrate the adsorbed water layer but is sufficient to ionize water molecules close to the metal or metal hydroxide surface. Could the ionization by the secondary electrons affect the water chemistry?

Hendrik Bluhm answered: The kinetic energy of the photoelectrons is the energy of the incident X-rays, minus the electron binding energy and work function of the electron analyzer. The incident photon energy in these experiments is 3100 eV, which leads to photoelectron kinetic energies of ~2570 eV (O 1s), ~2240 eV (Ni 2p), and 2820 eV (C 1s), sufficient to penetrate solution layers with thicknesses in the 10 nm range (see Table 1 in our paper). The influence of secondary electrons on the water chemistry has to be evaluated for each experiment to avoid artifacts in the experiments. This can be done by monitoring the spectral features as a function of sample irradiation and time. We have no indication of secondary electron-induced chemistry in the present experiments.

Geoffrey Thornton asked: In the paper by Springell *et al.*[1] it is claimed that X-rays of energy 17 keV cause radiolysis of water. Do you have any evidence one way or another for radiolysis at the lower energies that you employ?

1 Springell *et al.*, *Faraday Discuss.*, 2015, DOI:10.1039/c4fd00254g.

Hendrik Bluhm answered: We do not have evidence for X-ray induced radiolysis in the present experiments and neither in past experiments carried out resonantly at the O K-edge (incident photon energy ~530 eV) of bulk ice samples,[1] where radiolysis would manifest itself in the appearance of a OH-related peak in oxygen K-edge NEXAFS data, which was not observed.

1 D. E. Starr, D. Pan, J. T. Newberg, M. Ammann, E. G. Wang, A. Michaelidis, H. Bluhm, *Phys. Chem. Chem. Phys.*, 2011, **13**, 19988.

Philippe Marcus said: I have two questions:

(1) Looking at your Ni 2p spectra, it appears that the nickel surface is already oxidized prior to exposure to water. Do you have a way to prepare clean surfaces?

(2) A problem encountered with ultrathin water layer is the dissolution of the metal, leading to saturation and redeposition on the surface. Is this an issue in your experiments?

Hendrik Bluhm answered:

(1) Upon introduction into the measurement chamber, a Ni sample will always be oxidized since it was exposed to air during transfer into the measurement chamber. Specialized experimental set-ups with *in situ* Ni film preparation and introduction into the measurement chamber without exposure to air or water vapor before the start of the experiments are conceivable. The oxide can also be reduced electrochemically inside the solution. Another option for the preparation of a clean Ni film could be the direct (galvanic) deposition of Ni onto the multilayer substrate.

(2) To the second question: We did not see any evidence for dissolution of the Ni sample in these experiments. The solubility is very low under basic conditions, *e.g.* for NiO of the order of 10^{-7} mole L^{-1} (at pH = 13), with dissolution rates of 10^{-12} to 10^{-15} moles/(cm^2 s).[1] The solubility of $NiOH_2$ in neutral conditions is in the range of 10^{-5} moles L^{-1} and will be even smaller at pH = 13.[2]

1 P. R. Tremaine and J. C. LeBlanc , *J. Chem. Thermodynamics*, 1980, **12**, 521.
2 W. M. Haynes (ed), Physical Constants of Chemical Compounds, in *CRC Handbook of Chemistry and Physics*, 95th Edn, 2014.

Vincent Maurice asked: The thickness of 27 nm found for the hydroxide/oxy-hydroxide layer appears quite large. This is about one order of magnitude larger than for the passive film produced in similar conditions of passivation at pH 13 (from ~1.5 to ~4.5 nm with potential).[1,2] How would you explain such a difference?

1 H.-W. Hoppe and H.-H. Strehblow, *Surf. Interface Anal.*, 1989, **14**, 121.
2 A. Seyeux *et al.*, *J. Solid State Electrochem.*, 2005, **9**, 337–346.

Hendrik Bluhm replied: We believe that the complete oxidation of the Ni film in the present case as compared to partial oxidation in the case of a passive film at the same pH is due to the fact that the present measurements are performed under active oxidation conditions, *i.e.* at a sustained potential of +0.6 V.

Geoffrey Thornton asked: The signal-to-noise level of the standing wave data are rather limited. What are the prospects for significant improvement based on instrument development?

Hendrik Bluhm replied: The main causes for the signal attenuation in these measurements are the scattering of electrons by the liquid film and the gas phase surrounding the sample. These scattering processes depend on the given experimental conditions; for instance, for measurements at lower background vapor pressures and thinner liquid films the signal-to-noise ratio is greatly improved, while in the present case, for thicker liquid films at a RH close to 100% at room temperature scattering is stronger. The signal-to-noise in the measurement can be improved through either a better detection efficiency of the electron spectrometer or the use of incident X-rays with higher brilliance, such as will become available in future ultimate storage ring sources. In the experiments presented here we used a state-of-the-art ambient pressure photoelectron spectrometer with a theoretical maximum electron collection angle of $\pm 22°$. The greatest room for improvement lies in a stronger horizontal focusing of the incident X-ray beam. Due to the small entrance aperture of the differentially-pumped photoelectron spectrometer, and shallow incident angle in standing wave experiments, the illuminated area of the sample is in general much larger than the detected area in SW-APXPS (here by a factor of ~8). Thus, improved horizontal focusing will lead to a higher signal-to-noise-ratio in the experiments.

Trevor Rayment enquired: Does the presence of an organic contaminant layer at the surface of the electrolyte have any effect upon the wetting properties of the electrolyte since this might affect the thickness of the wetting layer which is a critical component of the system?

Hendrik Bluhm replied: The presence of the organic contaminant layer at the liquid/vapor interface may have an influence on the condensation/evaporation rates of the liquid,[1] or on the concentration of ions at the liquid/vapor interface.[2] We do not expect a direct effect of the contamination layer at the liquid/vapor interface on the wetting properties of the liquid/solid interface since the substrate surface is electrochemically cleaned inside the bulk solution before the sample is pulled out of the solution (always staying in contact with the bulk solution in the beaker).

1 J. F. Davies, R. E. H. Miles, A. E. Haddrell and J. P. Reid, *PNAS*, 2013, **110**, 8807.
2 M. J. Krisch, R. D'Auria, M. A. Brown, D. J. Tobias, J. C. Hemminger, M. Ammann, D. E. Starr and H. Bluhm, *J. Phys. Chem. C*, 2007, **111**, 13497.

Trevor Rayment remarked: It would be helpful to have more information about the opportunities and limitations which result from the use of a standing wave method for the use of ambient pressure XPS in the study of corrosion. For example how does the method constrain the thickness and smoothness of the electrode?

Hendrik Bluhm replied: The roughness of the electrode interfaces should not be more than a fraction of the standing wave period for an accurate characterization of the liquid/solid interface properties. However, the period of the multilayer can be increased beyond the few nm in this study so as to look at greater roughness. The total thickness of the electrode is limited by the effective attenuation of the X-ray standing wave field by the electrode material, as the X-rays must penetrate significantly into the substrate multilayer mirror; *e.g.* an upper value for the case of Ni and incident X-rays with an energy of 3.1 keV is about 100 nm at an incident angle of 3.5°.

Frank Heberling commented: The sample surface investigated in your study, deposited as a thin film on a multilayer mirror, appears very special. What kinds of surface can be or have already been investigated by the standing-wave *in situ* XPS method you describe in your paper?

Would it be feasible to investigate the surface speciation on single crystal samples if a total external reflection standing wave (TERXSW) field would be used for the excitation of the photoelectrons?

Hendrik Bluhm answered: So far only measurements on Ni (present case) and iron oxide films[1] have been performed using SW-APXPS, but we expect that the method is applicable to a wide range of substrate types due to the progress that has been made in the preparation of a wide variety of thin film systems using, *e.g.*, PLD or MBE. Near-total-reflection (NTR equivalent to TERXSW measurements have recently been demonstrated for solid-solid oxide interfaces.[2] However, measurements with harder X-rays may be difficult due to the very large beam spot along the beam propagation direction for the extremely shallow angles of incidence needed in these experiments. This is due to the very small effective sample area imposed by the entrance aperture of the ambient pressure photoelectron spectrometer. At soft X-ray energies, this should be less of a problem, however.

1 S. Nemšák, A. Shavorskiy, O. Karslioglu, I. Zegkinoglou, P. K. Greene, E. C. Burks, K. Liu, A. Rattanachata, C. S. Conlon, A. Keqi, F. Salmassi, E. M. Gullikson, S.-H. Yang, H. Bluhm and C. S. Fadley, *Nat. Commun.* 2014, **5**, 5441.
2 M. Marinova, *et al.*, *Nano Lett.*, 2015, **15**(4), 2533.

Clara Wren asked: Could you specify conditions in the solution? Was it aerated? What was the reason for the 0.1 M KOH? Nickel could irreversibly oxidize in an aerated basic solution and cathodic cleaning may not be able to reduce the resultant nickel oxide/hydroxide back to the metallic state. A surface of nickel oxide immersed in water is always covered with a nickel hydroxide. How do you ensure that the hydroxide being monitored is the real-time layer as claimed?

Hendrik Bluhm replied: The solution was degassed before use through evacuation in a separate vacuum chamber. Our past experiments have shown that hydroxide solutions are less prone to beam-induced changes (compared to solutions containing, *e.g.*, halide anions), hence the choice of KOH. All experiments in this study were performed at a potential of +0.6 eV, due to the limited amount of beamtime in this proof-of-principle study. We will explore the reduction of NiO_x under cathodic conditions in a next set of experiments that will help to answer these questions.

Roger Newman asked: How low does the current in the electrochemical cell have to be in order for adequate potential control to be maintained in the thin liquid layer? It seems that a fairly easy calculation would give the answer, which I imagine will be very low indeed. In that case the method would only seem to work for reactions that have gone to completion, like UPD. For growth of an oxide film there will be a residual current.

Hendrik Bluhm replied: The potentials measured by XPS at a location of the thin film ~2 cm above the bulk liquid surface in the beaker are in agreement with the applied potentials to the bulk liquid, *i.e.* the currents along the thin liquid film are sufficiently low under our experimental conditions. A simple estimate for a 20 nm thick 0.1 M KOH film with a specific conductance of 20 mS cm^{-1} yields an upper limit for the current of around 20 nA. At this current a full potential drop occurs over the whole film length. We stress, however, that, since at least part of the sample is immersed into the electrolyte, most of the current measured during cyclic voltammetry originates from the part of the sample in contact with the bulk solution. The current passing through the thin film is limited by ion diffusion, and therefore could be significantly lower than the total current. In other measurements we found that, as the current in cyclic voltammetry approaches 10 μA, one observes potential drops across the thin film length.

Frank Renner said: You described the thin liquid film as very stable. Can you also pull the specimen surface out of the solution and study the subsequent changes of the dry surface *versus* the wetted one?

Hendrik Bluhm replied: Yes, that is possible and would simulate a more commonly used *ex-situ* characterization of an electrochemically-treated sample, provided that the liquid reservoir is also removed from the vacuum chamber to provide a vacuum environment.

Tom Majchrowski stated: In Figure 7 of your paper, XP spectra, there is a significant shift in the position of the O 1s signal under humid conditions, however there is no shift in the position of the corresponding Ni $2p_{3/2}$ signal, does that suggest that Ni is not oxidised and oxygen reacts with C to form carbonate species?

Hendrik Bluhm replied: The apparent shift in the O 1s signal when going from dry to humid conditions is due to the appearance of the adsorbed water peak at around 534 eV and the gas phase water peak at 536 eV. The Ni 2p signal is dominated by the metallic component due to the thin oxide/hydroxide layer at the Ni surface, combined with the reduced surface sensitivity when using high kinetic energy photoelectrons. There is no evidence for an increase in the carbonate species coverage under humid conditions, based on the C 1s spectrum, where carbonates would appear at a binding energy of ~290 eV.

Gregory Hunt asked: Have you considered how this method could be applied to non-aqueous systems? What would be the challenges that need overcoming to do this with an oil and an organic matrix?

Hendrik Bluhm responded: This method is applicable to various types of liquid/solid interfaces and is not limited to aqueous solutions. The challenge that needs to be overcome is the preparation of a sufficiently thin and stable liquid film at the solid/liquid interface. In fact, systems like those you mention would be more amenable, in the sense that their vapor pressures are lower, such that the entrance aperture to the ambient pressure photoelectron spectrometer could be somewhat larger to increase intensities.

Rob Lindsay opened a general discussion of the paper by Johannes Lützenkirchen: Inspecting the AFM images in Fig. 8 of your paper, I was somewhat surprised by the apparent quality of your diffraction data, *i.e.* small error bars on structure factors. Given the small apparent terrace sizes, I would have expected that it would have been difficult to obtain significant intensity away from Bragg peaks. Could you comment on the width of surface reflections in the region of an anti-Bragg position, and perhaps show examples of reflections from the datasets acquired.

Frank Heberling and Johannes Lützenkirchen replied: The total height scale of the relevant AFM image (Fig. 8 upper image, the height scale is depicted on the right) is 570 pm. The surface heterogeneity is obvious, but the 10 nm to 20 nm wide islands (by which the surface is covered) have a height of no more than about 2 Å. Therefore, even though it is obvious that the surface is showing a distinct heterogeneity, it may be considered extremely flat. To show this more clearly we

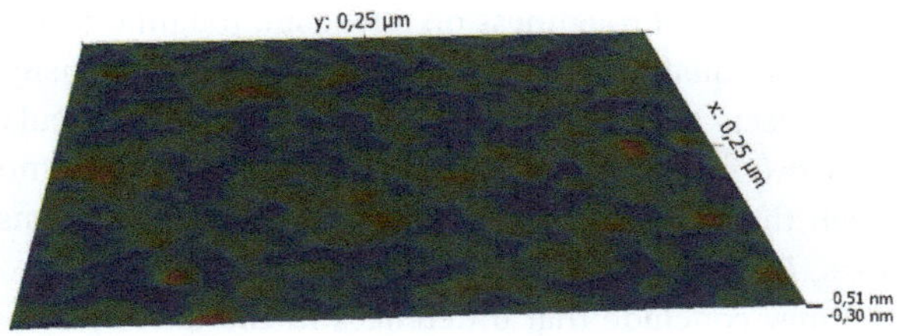

Fig. 2 3D projection of the AFM data presented in the upper image in Fig. 8 of our paper. Despite the impression that might be generated by the strong contrast chosen in Fig. 8, this projection shows that the surface is actually very flat. In this projection the roughness would hardly be visible if it was not highlighted by the color scheme.

have generated a new representation of the AFM data of the fresh hematite sample in air, which is shown in Fig. 2.

The different colors in Fig. 1 correspond roughly to the expected terrace heights on hematite (001). This indicates that the roughness of the sample corresponds quite nicely with the concept of the "β model" used to account for roughness during CTR modelling.

Fig. 3 presents a detail of an image measured by the PILATUS pixel array detector at the −1, −1, L CTR at L = 7.53. The shape of the X-ray reflection is largely determined by the beam footprint on the sample surface. The minor reflections at the lower left, and the two fold splitting of the main reflection, are due to the fact that the hematite crystal is not a perfect single crystal as often observed for oxide minerals.

Geoffrey Thornton commented: The main difference between CTR's obtained from the as-prepared substrate and that exposed to solution is the specular rod, *i.e.* 0,0,L. This is very sensitive to roughness. Have you analysed the data without including the specular rod to test for consistency.

Frank Heberling and Johannes Lützenkirchen responded: No, we have not performed such an analysis. However, according to the "β" model[1] which is widely

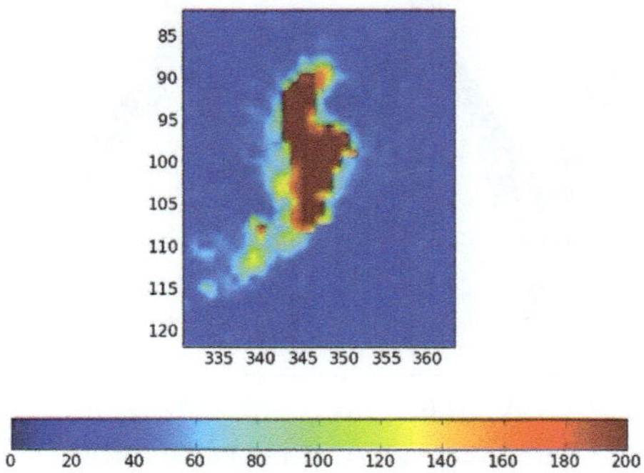

Fig. 3 X-ray reflection at the Anti-Bragg position at L = 7.53 on the −1, −1, L CTR measured at the fresh hematite sample.

used to describe the effect of roughness on CTR-data, roughness affects specular and off-specular CTRs equally, *i.e.* it decreases the scattered intensity in the anti-Bragg regions, between the Bragg peaks. Comparing the specular CTR-data obtained with our two samples (Fig. 4) it is obvious that there is no systematic difference between the two datasets concerning the scattered intensities in the anti-Bragg regions, *i.e.* around L = 3, 9, or 15.

Therefore, we may conclude that differences in the data must be related to changes in the structure and not to changes in roughness. These differences are certainly most pronounced on the specular data, but they do occur throughout the whole data set. Other CTRs with obvious significant differences are for example the negative side of the 21L CTR and the positive side of the 20L CTR.

From the CTR data analysis we obtain β values of 0.17 and 0.18 for the fresh and aged hematite samples, respectively. These values are relatively small, compared to previous studies on the hematite (001) surface. Thus Tanwar *et al.*[2] obtained $\beta = 0.18 - 0.23$ and Catalano[3] $\beta = 0.23$ (reported as $\sigma(\text{rms}) = 1.4$ Å).

For the fresh hematite sample we had the possibility to compare the roughness obtained from the CTR analysis with the roughness measured by high-resolution AFM and found an extraordinarily good agreement, *i.e.* $\sigma(\text{rms, CTR}) = 1.1$ Å and $\sigma(\text{rms, AFM}) = 1.2$ Å. One should not be misled by a first impression of the high-resolution AFM image, which appears quite heterogeneous. The total height range of the image at comparable conditions to the CTR measurements covers no more than 6 Å, *i.e.* the surface is heterogeneous, but still very flat.

In summary roughness should not be a major issue in the presented CTR study, and we see no reason for a special treatment of the specular CTR or for its exclusion from the analysis.

1 I. K. Robinson, *Phys. Rev. B*, 1986, **33**, 3830–3836.
2 K. S. Tanwar, S. C. Petitto, S. K. Ghose, P. J. Eng and T. P. Trainor, *Geochim. Cosmochim. Acta*, 2009, **73**, 4346–4365.
3 J. G. Catalano, *Geochim. Cosmochim. Acta*, 2011, **75**, 2062–2071.

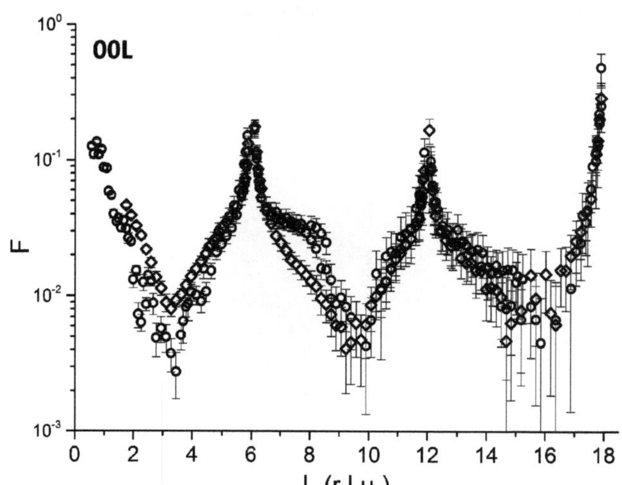

Fig. 4 Direct comparison of specular CTR data measured on the fresh (circles) and aged (diamonds) hematite (001) samples.

Kirsi Yliniemi commented: When modelling the surface potentials of hematite samples (for example in Fig. 7 in your paper) have you taken into account the interaction term between the neighbouring surface sites? If not, could you comment on how much the interaction between the neighbouring sites may influence the protonation/de-protonation behaviour.

Johannes Lützenkirchen and Frank Heberling answered: The models used in our paper do not consider next-neighbor interaction. Coupling between protonation and deprotonation reactions on the surface sites within the model occurs *via* mean-field electrostatic terms. Ising models consider those kinds of next-neighbor effects which you refer to and have been used by Borkovec and Koper to describe charging of polyelectrolytes.[1] Borkovec has also shown for oxides, that the mean-field models we are using are equivalent to the Ising models.[2] The relation between the two approaches involves some interesting aspects. For one thing, unlike polyelectrolytes, it is almost impossible to completely protonate or deprotonate an oxide surface within the normal pH range although some publications seem to show saturation, *i.e.* a plateau which has been interpreted as site density, *e.g.* for goethite.[3] Subsequent work has shown that the saturation data are doubtful.[4] The maximum degree of protonation/deprotonation is expected to be much smaller on oxides than on polyelectrolytes (like polyacrylic acid, which can be completely deprotonated). From this point of view next-neighbor effects on oxides might be expected to be less pronounced than on polyelectrolytes. Also, when dealing with next neighbors the situation is likely to become highly complex on oxides which is shown in Fig. 5 for the (110) face of goethite.

On this plane various sites occur. They have different proton affinities. These can be predicted within the MUSIC model framework.[5]

The sites have a given structural arrangement, which requires in an Ising model the appropriate interaction parameters. The MUSIC approach probably requires a smaller number of parameters.

Fig. 6 shows the individual contributions of the different sites to the overall charging of this plane based on the MUSIC model.

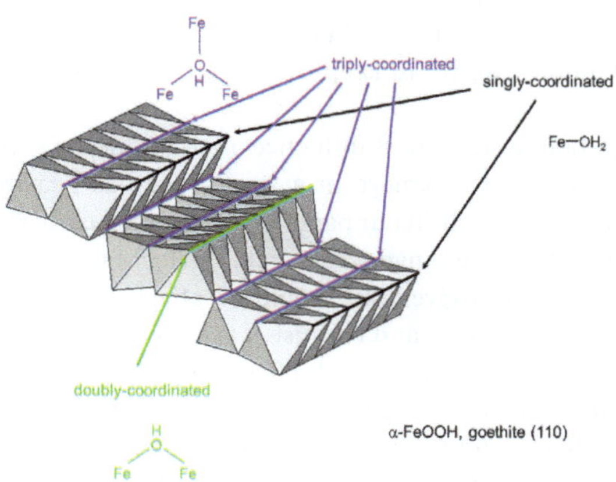

Fig. 5 Surface structure of the ideal (110) face of goethite.

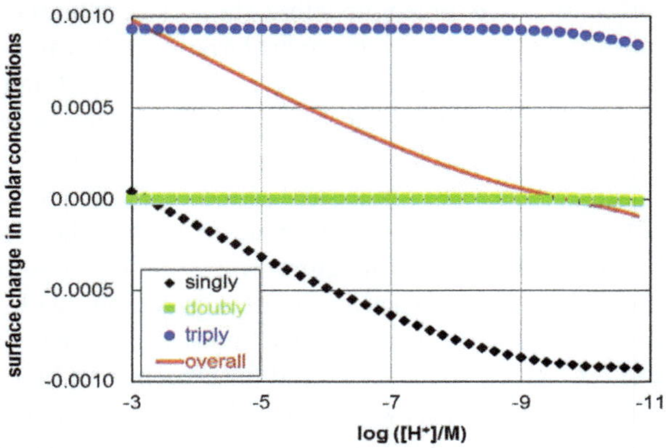

Fig. 6 Contributions of the different sites to the overall charging of goethite (110) as a function of the pH (here we plot the logarithm of the proton concentration, which for a given ionic strength is proportional to the pH).

The overall point of zero charge occurs at pH 9.2. The pH-dependent charge mainly arises from the reaction of the singly co-ordinated group. As in our paper the doubly co-ordinated groups within this model do not show significant charging within the normal pH range. The triply co-ordinated groups have a constant charge which hardly changes with pH in the pH range of interest. Actually the model can be further complicated by involving two types of triply co-ordinated groups (one with high and one with low proton affinity) resulting in a structurally even more complex situation. The model can also be simplified by excluding the doubly co-ordinated groups. The influence of the different sites on the overall protonation is best discussed by considering the singly co-ordinated sites. Their intrinsic pK_a value is 7.7. The conditional pK_a is between 3 and 4, which means a shift of about 4 pH units caused by the triply co-ordinated sites. In other words here the interaction among the different sites occurs *via* the model inherent electrostatic term(s). It would probably be possible to design an Ising model for this surface as well. As pointed out before one would need to consider the interactions between the different sites under the given structural constraints. Addition of site-interaction parameters to the present model would increase the number of parameters. However, with the present number of adjustables within the bare MUSIC model we already obtain a good fit to the data. Following the principle of parsimony, we try to keep the model to a minimum of adjustable parameters.

The next-neighbour issue and its influence on pK_a values of surface groups has been raised previously by Bickmore in a sense.[6] He has heavily criticized the MUSIC model (which we apply in our paper) for neglecting close range effects. His argument was that the close-range effects would affect next-neighbor pK values. Since bond distances on a given site would be affected due to protonation or deprotonation of adjacent sites and because it is known that the MUSIC model is highly sensitive to bond-distances[7] such effects would be substantial. From our point of view again such close-range effects are covered in the common model by the smeared-out charge and the resulting electrostatic factors supplemented by the restricted degree of protonation/deprotonation. Whether the close-range features should be combined with larger scale features is an interesting question.

The electrostatic factors are sometimes also interpreted as the ratios of activity coefficients of the surface species involved in a given reaction.

Overall, many aspects of the surface reactions could be improved based on detailed knowledge, but it usually involves new parameters that have to be adjusted or assumed. As pointed out previously we apply the principle of parsimony.

1 M. Borkovec and G. J. M. Koper, *J. Phys. Chem.*, 1994, **98**, 6038–6045.
2 M. Borkovec, *Langmuir*, 1997, **13**, 2608–2613.
3 L. Lovgren, S. Sjoberg and P. W. Schindler, *Geochim. Cosmochim. Acta*, 1990, **54**, 1301–1306.
4 J. Lützenkirchen, J.-F. Boily, L. Lovgren and S. Sjoberg, *Geochim. Cosmochim. Acta*, 2002, **66**, 3389–3396.
5 W. H. v. Riemsdijk and T. Hiemstra, in *Interface Science and Technology*, ed. L. Johannes, Elsevier, 2006, vol. 11, pp. 251–268.
6 B. R. Bickmore, K. M. Rosso and S. C. Mitchell, in *Interface Science and Technology*, ed. L. Johannes, Elsevier, 2006, vol. 11, pp. 269–283.
7 J.-F. Boily, J. Lützenkirchen, O. Balmes, J. Beattie and S. Sjoberg, *Colloid Surf. A-Physicochem. Eng. Asp.*, 2001, **179**, 11–27.

Rob Lindsay said: Regarding your approach to fitting the SXRD data, I was wondering if you had tried to model the experimental data by adopting partial site occupancies?

Frank Heberling and Johannes Lützenkirchen responded: As described in the paper, a preliminary approach to fit the data included one domain and partial occupancy of the top most iron site. The two domain model is to some extent equivalent to such an approach. The difference is that the consideration of two distinct surface domains allows us to address different local structural relaxations in the two domains. Thus regions where the topmost iron site is occupied are modeled separately from regions where the site is unoccupied. These additional degrees of freedom in the model turned out to be crucial to obtain a satisfactory fit to the data.

Philippe Marcus said: Oxides are extremely important in the field of corrosion chemistry because several metals and alloys are protected against corrosion by a thin (or ultra-thin) oxide layer.

However thin oxide layers on metals may differ in many ways from bulk oxides that are not supported by metals (for example the work function can be different).

According to you, can we take the values of the zeta-potential for bulk oxides and use them for oxide films on metals?

Johannes Lützenkirchen and Frank Heberling replied: This is an intriguing question. We believe there are several aspects to this question. One concerns how the chemistry of the oxide film on a metal relates to the surface of a bulk oxide. If one tends to believe that the reactivity is quite similar for both, the difficult part will be to choose the correct data from the literature. The major part of the available charging data is for oxide-particles. Data for such oxide particles can vary a lot (*i.e.* IEPs can cover ranges of several pH units). So the question would be which data to choose. The answer would be largely arbitrary.

A second more interesting aspect concerns the actual structure of the oxide film on a metal. If it can be characterized in terms of bond distances and nature and number of surface functional groups, then the available models[1] can in principle predict at least the point of zero charge and maybe more than that. This would involve the assumption that the supporting metal does not affect the solid–liquid interface additionally (*i.e.* beyond constraining bond distances in the film).

If bond distances in the oxide film (as affected by the underlying metal) are different from a nominally identical bulk oxide surface, the charging properties between the two samples will probably differ. The proton affinity constants are actually very sensitive to the bond-distances.[2]

A final aspect has been discussed by McCafferty.[3] According to him the relation between IEP and the potential of zero charge is quite complex. In the models as we use them and for the systems we are interested in we are not considering space charge issues.

Experience has shown that one should measure the points of zero charge and the zeta-potentials for unknown samples or samples that are not sufficiently characterized for application of MUSIC type models or where any other uncertainties exist under the conditions of interest. This is also true for bulk oxide samples.

1 W. H. v. Riemsdijk and T. Hiemstra, in *Interface Science and Technology*, ed. L. Johannes, Elsevier, 2006, vol. 11, pp. 251–268.
2 J.-F. Boily, J. Lützenkirchen, O. Balmes, J. Beattie and S. Sjoberg, *Colloid Surf. A-Physicochem. Eng. Asp.*, 2001, **179**, 11–27.
3 E. McCafferty, *Electrochim. Acta*, 2010, **55**, 1630–1637.

David Williams asked: Could you describe the surface changes as a slow, hydration-driven surface reconstruction that is spatially heterogeneous (and thus perhaps also susceptible to spatially heterogeneous surface contamination)?

Johannes Lützenkirchen and Frank Heberling replied: The question about the process behind the changes is crucial. Unfortunately we know very little about the mechanisms. We are confident that surface contamination does not play a major role because the cleaning procedures we apply have been previously elaborated very carefully for single crystals.[1] Cleaning substrates in this way also led to simultaneous observation of the very same features on the companion surface sapphire (001) which we report here for hematite (001). On sapphire (001) we observed them on one kind of sample, *i.e.* flat surface potential curve and steep zeta potential feature. What makes us confident is that those two features had been independently observed in a substantial number of studies.[2]

Furthermore, the cleaning procedure we use has been applied by others in a paper that came out simultaneously as ours[3] and in subsequent work.[4,5] In particular these latter studies involve highly surface specific techniques that would detect minor amounts of adventitious carbon compounds.

As for the major difference in the surface properties with the changes over time, the outcome namely just a change in the extent of hydration was certainly disappointing. A simple and far more appealing cause would have been the change from a surface that is initially oxygen terminated (for which we can explain the low isoelectric point, IEP) to one that involves singly co-ordinated groups (*i.e.* which has the bi-domain properties) and allows a straightforward explanation for the high IEP.

It would also agree with independent work, where we have found that the presence of dissolved aluminum can increase the IEP of sapphire (001)[6] to the same extent as we report here hematite (001) on simple aging. In that paper[6] we also cite previous work on hematite (001) where the formation of adatoms had been reported. So there is some consistency there. We are also currently re-evaluating the CTR data with new software in the hope of obtaining more satisfactory relationships between the surface structure and the charging.

We were very happy to see the high IEP appearing on the aged sample, because in the previous work on single crystals it has been very rare that IEPs of bare oxide surfaces above say pH 5 were measured (either by streaming potential or in force distance measurements). Even on aged steel samples usually low IEPs were obtained,[7,8] while one would tend to expect the IEP of iron oxides (particles), which are usually around pH 9.

We agree that the study of the actual transformation would be highly interesting. Furthermore, we believe that future studies should involve a number of different techniques over the experimentation period. Such multi-technique studies have the potential to shed light on the processes.

1 T. Rabung, D. Schild, H. Geckeis, R. Klenze and T. Fanghanel, *J. Phys. Chem. B*, 2004, **108**, 17160–17165.
2 J. Lützenkirchen, R. Zimmermann, T. Preocanin, A. Filby, T. Kupcik, D. Kuttner, A. Abdelmonem, D. Schild, T. Rabung, M. Plaschke, F. Brandenstein, C. Werner and H. Geckeis, *Adv. Colloid Interface Sci.*, 2010, **157**, 61–74.
3 M. M. Gentleman and J. A. Ruud, *Langmuir*, 2010, **26**, 1408–1411.
4 E. Anim-Danso, Y. Zhang, A. Alizadeh and A. Dhinojwala, *J. Am. Chem. Soc.*, 2013, **135**, 2734–2740.
5 E. Anim-Danso, Y. Zhang and A. Dhinojwala, *J. Am. Chem. Soc.*, 2013, **135**, 8496–8499.
6 J. Lützenkirchen, A. Abdelmonem, R. Weerasooriya, F. Heberling, V. Metz and R. Marsac, *Geochem. Trans.*, 2014, **15**, 9.
7 N. Kallay, D. Kovacevic, I. Dedic and V. Tomasic, *Corrosion*, 1994, **50**, 598–602.
8 B. M. Cabanas, J. Lützenkirchen, S. Leclercq, P. Barboux and G. Lefevre, *J. Nucl. Mater.*, 2012, **430**, 150–155.

Hadeel Hussain asked: In your paper it was not mentioned which model you have used to simulate surface roughness...I assume it is the beta roughness model? Have you tried using different models, as I'm sure you know, it is a relatively simple model and so may not accurately represent the rough surface you have as evident from the AFM images.

Frank Heberling and Johannes Lützenkirchen replied: In the electronic supplementary information (ESI) we mention that we use the beta model to account for the influence of roughness on the model. We did not test other models. As already mentioned in the previous discussion for the fresh hematite sample we were able to compare the roughness derived from the beta model (β = 0.17 is equivalent to σ_{rms} = 1.1 Å) to the rms roughness measured by AFM, which is σ_{rms} = 1.2 Å. Such agreement, and the type of roughness as mentioned in the answer to the earlier question from Robert Lindsay, supports the applicability of the beta model to our data. Concerning the "rough surface as evident from the AFM image", as already mentioned in the previously, we think it is important to point out that the total height scale of the relevant AFM image (as depicted on the right of Fig. 8 in our paper) is 570 pm. The surface heterogeneity is obvious, but the 10 nm to 20 nm wide islands (by which the surface is covered) have a height of only about 2 Å. Therefore, even though it is obvious that the surface is showing a distinct heterogeneity, it may be considered as extremely flat.

Hadeel Hussain queried: To me it seems your χ^2 is quite high (usually with CTR data an acceptable model has a χ^2 less than 2 and in many cases less than 1.5) and could possibly be due to the error bars you have on your CTR data. How have you calculated these?

Frank Heberling and Johannes Lützenkirchen replied: For a model refinement on a full set of CTRs including specular and off-specular data, and considering the constraints on the degrees of freedom within the model due to symmetry constraints at hematite (001), it is our notion that a $\chi^2 < 4$ is an acceptable fit. We do admit that there is some space for improvement of our structural models. As pointed out previously we are currently re-evaluating the data for other reasons. As for the use of the goodness-of-fit parameter, we prefer not to put too much weight on absolute values of χ^2. In our opinion χ^2 is mainly useful to compare various fits to one specific dataset. Discussing the goodness-of-fit parameter concerns, however, an important and critical issue regarding the development of structural models based on CTR data. The determination of the error bars on the structure factor amplitudes have a direct impact on χ^2. Therefore the uncertainties associated with structure factor amplitudes are a key issue when deciding how many degrees of freedom should be allowed in a structural model.

Our standard approach to estimate error bars on CTR data is based on counting statistics. As such the error of the integrated intensity, I_{err}, is estimated as the square root of the total intensity, I_{tot}, plus the background intensity, I_{bgr}: $I_{err} = (I_{tot} + I_{bgr})^{0.5}$. To calculate structure factor amplitudes, $|F|$, the integrated intensity, I, is multiplied by a correction factor, c, that depends on geometry and normalized by the beam intensity, I_0. Subsequently the square root of the resulting expression is taken. This leads to the following equation for the error, F_{err}, associated with $|F|$: $F_{err} = 1/2 \times (c/I_0 \times I_{err}/I)^{0.5}$.

For the two datasets presented in the manuscript we were in the comfortable situation that several symmetry equivalent CTRs were available. By comparing these equivalent CTRs we can calculate standard deviations for the structure factors and compare them to error estimates based on counting statistics. Doing so we found excellent agreement. For the fresh hematite sample the average uncertainty based on counting statistics is 19% and the average standard deviation based on symmetry equivalent CTRs is 17%. For the aged sample the agreement is even better and we found 18% average uncertainty with both methods.

Consequently, for the datasets presented here we are confident that the errors associated with the structure factors are reasonable.

Faraday Discussions

PAPER

Identification of bulk oxide defects in an electrochemical environment

Mira Todorova* and Jörg Neugebauer

Received 28th November 2014, Accepted 8th January 2015
DOI: 10.1039/c4fd00238e

We discuss how electronic-structure calculations can be used to identify the dominant point defects that control the growth and dissolution of the oxide barrier layer formed if a metal comes into contact with a corrosive environment. Using the example of the Zn/ZnO/H_2O interface we develop and apply a theoretical approach that is firmly based on *ab initio* computed defect formation energies and that unifies concepts of semiconductor defect chemistry with electrochemical concepts. Employing this approach we find that the commonly invoked and chemically intuitive defects such as the doubly negatively charged oxygen vacancy in electrochemically formed ZnO films may not be present. Rather, hitherto not discussed defects such as the oxygen interstitial or unexpected charge states, such as the neutral oxygen vacancy, are found. These new defect types will be shown to critically impact our understanding of fundamental corrosion mechanisms and to provide new insight into strategies to develop alloys with higher corrosion resistance.

I. Introduction

Considering the omnipresent oxidising environment it is hard to believe that metals play such an important role in our society. What makes the application of these materials possible is the formation of thin oxide layers at the metal surface, which dramatically enhance their corrosion resistance. Thus, it is not surprising that passivity of metals is a central topic in corrosion science and that the modelling of oxide growth is a major issue. The growth of the oxide layer, the dissolution of the oxide film and the precipitation of dissolving species from solution have been identified as the key phenomena involved in the formation of oxide films.[1]

Understanding the stability and growth kinetics of oxide films requires a detailed knowledge of conditions, with which the metal/oxide/water-system is confronted, *i.e.*, of the involved species and of their transport mechanisms. Since the appearance of Wagner's theory[2] on oxide growth considerable effort has been made to understand the elementary processes and the role of point defects (both vacancies and interstitials) in the growth of oxide films. The main models are

Max-Planck-Institut für Eisenforschung GmbH, Department for Computational Materials Science, Max-Planck-Str. 1, D-40237 Düsseldorf, Germany. E-mail: m.todorova@mpie.de

based on this theory[2] and apply to different modes of oxide film growth: the Cabrera–Mott[3] and the Fehlner–Mott[4] models were developed to treat oxide films formed thermally in air, while the Point Defect Model[5] (PDM) was developed for electrochemically formed oxide films. Newer models involve explicit atomistic kinetic Monte Carlo simulations of the oxide growth[6] or try to improve on the description of the interfacial potential drops.[1] These models are critically reviewed, for example, in ref. 1.

A central input quantity, essential for all of these models, is the thermodynamic and kinetic behaviour of point defects. To be more specific, knowing what kind of defects are present in the oxide in appreciable concentrations for a given electrochemical environment is critical to understand the growth/dissolution of the protective oxide film. In semiconductor defect chemistry, where essentially identical point defect properties are considered and studied, quantum-mechanical calculations based on density functional theory (DFT) have evolved into an indispensable tool to accurately predict the thermodynamic stability and diffusion barriers of neutral and charged point defects in a wide range of semiconductors, including oxides.

Despite addressing principally identical questions the two research fields of semiconductor defect physics and electrochemistry/corrosion are almost disconnected. In this work we discuss how the information about charged point defects in oxides, obtained by concepts originally developed by semiconductor physicists, can be used to obtain a detailed insight into the type and concentration of defects present in an oxide under corrosive conditions. Our approach is based on a grand canonical description of all elementary building blocks involved in the reactions, such as various charged point defects, complexes with extrinsic defects, electrons and holes, and unifies concepts in semiconductor physics and electrochemistry.[7] Some of its aspects, relevant in the context of the present work, will be briefly recapped in the next section of this manuscript. Focusing on the Zn/ZnO/H_2O system we use these concepts to construct a novel type of phase diagram that directly show the dominant defects as well as the stability of the oxide against dissolution. In Section III we discuss how such defect stability phase diagrams can be utilised to identify which point defects will be important in the context of corrosion and oxide film growth.

II. Theory

A. Point defects in oxide growth models

Passive films are considered to consist of an inner compact oxide layer and an outer precipitated layer.[8] The compact layer is supposed to be a native oxide of the material, while the outer layer is assumed to be a mixture of oxide and hydroxide, as discussed for the case of Zn in ref. 8. In this work we focus on the native oxide of Zn, which is ZnO, and its native defects.

Before discussing which point defects are commonly assumed to control oxide growth, we would like to comment on our notation, which resembles that used in semiconductor defect chemistry, but is distinct from the Kröger–Vink[9] notation typically used in discussions on oxide films. Our notation resembles the Kröger–Vink notation in as far as possible point defects are represented in both of them in the following way: a Zn vacancy is V_{Zn}, a Zn interstitial is Zn_i and an antiside defect, for example a zinc on an oxygen lattice site, is Zn_O. Each of these defects D can occur in a charge state q. In the historical Kröger–Vink[9] notation a neutral charge state is indicated by a superscript $\times$, a negative charge state by a superscript $'$ and a positive charge state by a superscript $\bullet$. In this work we denote the charge state q of the defect D^q as $q = 0$ for a neutral defect; as $q = +1$, if one electron is removed; and as $q = -1$, if one electron is added, following the nomenclature developed in semiconductor defect chemistry.[10]

Traditionally, defects in oxides are expressed in terms of full or half reactions that individually obey conservation laws with respect to the atomic sites (in solids), numbers and charges. From a historic perspective this approach is convenient to describe and analyse experimental electrochemical set-ups. However, using quantum chemical approaches to describe such reactions is computationally expensive: large cells and time scales are needed, while the number of possible reactions scales approximately as the square of the involved reaction educts and products. We will therefore use a grand canonical approach,[10,11] as in semiconductor physics. The approach is based on the calculation of individual defects and the conservation laws with respect to electron and particle numbers are realised not by specific reactions, but by abstract thermodynamic reservoirs for the electrons and the relevant chemical species. These reservoirs are described by chemical potentials for the atomic species and the Fermi energy for the electrons.

The above approach allows an efficient decomposition of the realistic system we are eventually interested in, and which typically contains a variety of defect species in different charge states that interact electrostatically, *via* charge transfer and by mass transport, into the most fundamental building blocks that can be straightforwardly handled by modern *ab initio* techniques. This decomposition allows us to focus on the properties of individual defects and only put them into relation with each other *a posteriori* within the context of a grand canonical formulation, as will be shown later. The main advantage is that we do not have to formulate all possible *individual reactions* that obey the conservation laws but only the reaction of each *individual defect* with its thermodynamic reservoirs. The conservation laws need then only be enforced on the total system containing all the defects by modifying/controlling the thermodynamic reservoirs for the various chemical species and the electrons. Due to this the number of calculations scales only linearly with the number of involved defects, rather than with the square we would need if we describe it in terms of charge neutral reactions. The full approach will be outlined in detail in the next section.

Before introducing and applying the above formalism to the $Zn/ZnO/H_2O$ system let us briefly recap our present understanding about which point defects participate in the oxide growth.[5,8] The assumed relevant defects are summarized in Fig. 1 (left). Following the conventional picture a doubly positively charged oxygen vacancy and a zinc interstitial defect, which may assume different charge states, are dominant at the Zn/ZnO interface, while a negatively charged zinc vacancy is prevalent at the ZnO/H_2O interface. Under oxidizing conditions, the interface is far away from thermodynamic equilibrium implying that there will be a substantial gradient in the O and Zn chemical potentials. The schematic behaviour of the two chemical potentials, which are Zn-rich at the Zn/ZnO interface and more O-rich at the ZnO/H_2O interface, are included in Fig. 1 (left). Due to the gradient in the chemical potentials defects with a net Zn (O) excess will move towards more oxygen (zinc) rich conditions, *i.e.* towards the ZnO/H_2O

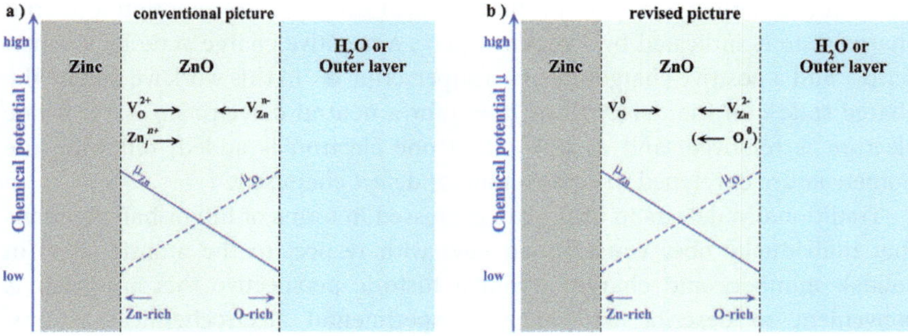

Fig. 1 Schematic representation of the $Zn/ZnO/H_2O$ system and the defects commonly assumed to be involved in the growth of a compact oxide layer during corrosion. Also schematically shown are the chemical potential (solid and dashed blue lines). (a) Conventional picture: positively charged oxygen vacancies and Zn interstitials are assumed to be created in the vicinity of the Zn/ZnO interface and transported towards the $ZnO/(H_2O$/outer layer) interface; negatively charged Zn vacancies are generated in the vicinity of the $ZnO/(H_2O$/outer layer) interface and transported through the compact oxide layer towards the Zn/ZnO interface. (b) Revised picture derived in Sec. III based on DFT-calculated defect formation energies: neutral oxygen vacancies are generated in the vicinity of the Zn/ZnO interface and transported towards the $ZnO/(H_2O$/outer layer) interface; doubly negatively charged Zn vacancies and neutral O interstitials are generated in the vicinity of the $ZnO/(H_2O$/outer layer) interface and transported through the compact oxide layer towards the Zn/ZnO interface.

(Zn/ZnO) interface. The arrows attached to each of the defects indicate the corresponding migration direction.

Fig. 1 (right) is based on the insight we gain from the grand canonical approach and will be discussed in Section III.C.

B. Defect formation energies and chemical potentials

The formal equivalency between defects in semiconductors and ions in solution allows us to extend concepts typically used to study charged point defects in semiconductors to electrochemical systems.[7] As discussed in Section II.A, the derived grand-canonical approach is particularly well suited to be used in conjunction with *ab initio* calculations, since it needs as input only the properties of the individual charged defects within a system and couples them only afterwards *via* the conditions of charge neutrality. Besides providing information on the stability and concentration of ions in solution, the approach we are going to describe in the following provides a natural link between density-functional theory calculations and experimentally measurable quantities, which determine and characterize the state of electrochemical systems, such as pH and electrode potential. In the following we recap some of the concepts, in particular the ones pertaining to defects in oxides, which become important in the context of the later discussion.

A key quantity to describe oxide growth is the local concentration of the various point defects. Assuming local thermodynamic equilibrium, *i.e.*, the defect concentration is in thermodynamic equilibrium with its local chemical potential, and the dilute limit where defect–defect interactions are negligible,[12] the concentration of a defect D in charge state q is given by:

$$c_{D^q} = c_0 e^{-\Delta_f G(D^q)/k_B T} \tag{1}$$

Quantities that enter eqn (1) are c_0, the highest possible concentration which can be realised within the system by occupying every available site where the defect can be formed, the Boltzmann constant k_B, the temperature T, and the free energy of formation of a charged defect in a semiconductor or an ion in solution $\Delta_f G(D^q)$. The above equation implies that defects with a high free energy of formation will be present only in small concentrations, while defects with low formation energies will occur in high(er) concentrations. The free energy of formation itself is given by the change in enthalpy $\Delta_f H$ and entropy $\Delta_f S$. Since the formation of the defect/ion requires an exchange of Δn_i atoms and q electrons with their respective thermodynamic reservoirs, the formation energy becomes a function of chemical potentials μ_i and Fermi energy μ_e:[7,11]

$$\Delta_f G(D^q) = \Delta_f H - T\Delta_f S - \sum_i \Delta n_i \mu_i + q\mu_e \tag{2}$$

Often, this general form for the formation energy of a defect in a condensed phase is simplified when applied to defects in semiconductors, by approximating the free energy of defect formation $\Delta_f G(D^q)$ by the formation energy of a defect $E^f(D^q)$. For a targeted accuracy in the defect formation energies of 0.1 eV (9.65 kJ mol^{-1}) this is a well justified approximation, since pressure induced changes in the enthalpic $\Delta_f H$ and entropic $\Delta_f S$ contributions, other than the configurational entropy $\Delta_f S_{conf}$ which is already included in eqn (1), are usually significantly smaller[10,11] and can thus be neglected. The formation energy of a defect in a semiconductor[10,11] is then obtained from the difference in *ab initio* calculated total energies for a super-cell containing the defect $E_{tot}(D^q$: host) and one for the unperturbed system E_{tot}(host),

$$E^f(D^q) = E_{tot}(D^q : \text{host}) - E_{tot}(\text{host}) - \sum_i \Delta n_i \mu_i + qE_F. \tag{3}$$

We note, that the reservoir for electron exchange is denoted by E_F, the Fermi energy, which is the convention in the field of semiconductor physics. Formally, μ_e and E_F, are equivalent and both describe the same quantity, but for a constant material specific offset.

In a first step, the chemical potentials of the electrons μ_e (or Fermi energy E_F) and of the chemical species μ_i are treated as variables. They can be used to analyse the influence of the electrochemical environment (chemical potentials, over-potential, pH) on the defect formation energy and concentrations. Due to thermodynamic/electronic constraints the electron and chemical potentials are restricted by upper and lower bounds.

For example, the electron chemical potential μ_e can assume within the oxide only values within the electronic band gap, *i.e.* the lower (upper) boundary for the electron potential μ_e is the valence band maximum (VBM) [conduction band minimum (CBM)]. For the chemical potentials of the involved species boundaries exist, if the considered oxide becomes unstable against the formation of another phase. A natural upper boundary for any species is the formation of its elemental state: for ZnO, consisting of Zn and O, this would be bulk fcc Zn and O_2 molecules, respectively. Since potentials are invariant against a constant shift for convenience the energy zero of the chemical potentials is commonly chosen such that the chemical potential of the elementary phase becomes zero. Thus, a general upper limit is $\mu_i \leq 0$.

While the formation of the elementary phases is an upper limit it is not always the lowest and thus critical one. Often the formation of (parasitic) phases sets in before the elemental phase can be formed. For the system discussed here, where the three chemical elements Zn, O, H are present, a possible parasitic phase is $Zn(OH)_2$. Then the resulting boundary conditions is, for example for Zn, $\mu_{Zn} \leq \Delta_f H(Zn(OH)_2) - 2\mu_O - 2\mu_H$.

Finally, the chemical potentials are not independent but linearly dependent on each other in a way that ensures thermodynamic stability of the involved phases. To be more specific we consider ZnO, where the two relevant chemical potentials are μ_{Zn} and μ_O. In thermodynamic equilibrium these two chemical potentials cannot be varied independently from each other. They are bound by the condition that ZnO must be a stable phase, which means that the stoichiometric sum of μ_{Zn} and μ_O must be equal to the enthalpy of formation of the compound,

$$\mu_O + \mu_{Zn} = \Delta_f H(ZnO). \tag{4}$$

For any thermodynamically stable compound the formation energy is negative. The above equation defines *lower boundaries* for the variation of μ_O and μ_{Zn} in the following way: the lowest chemical potential Zn can assume within the ZnO system is defined by the upper limit of the oxygen chemical potential as $\mu_{Zn}^{min} = \Delta_f H(ZnO) - \mu_{O[O_2]}$ and *vice versa*, the upper boundary for μ_{Zn} defines the lower boundary for the variation of μ_O, *i.e.* $\mu_O^{min} = \Delta_f H(ZnO) - \mu_{Zn[Zn\text{-}bulk]}$.

C. Electron potential

While the definition and calculation of an absolute alignment (energy zero) for the chemical potentials is straightforward and easily achieved, *e.g.*, by referencing all chemical potentials with respect to that of the corresponding elementary phases (see discussion above) the definition of an absolute potential for the electron reservoir is conceptually much less straightforward. In the semiconductor community the Fermi energy describing this potential is commonly referenced to the top of the valence band. While this is sufficient to describe, *e.g.*, optical transitions or doping within a given bulk system it does not allow to compare the formation energy between two or more separate bulk systems as *e.g.* ZnO and water. A common absolute reference is the vacuum level in semiconductor physics or the standard hydrogen electrode (SHE) employed in electrochemistry. How the energy of these two references can be related has been discussed in detail by Trasatti[13] and in a recent publication by us.[7] In the following we use the notation μ_e whenever we want to indicate alignment on an absolute energy scale and E_F when we utilize formation energies computed by the semiconductor community where the alignment is with respect to the VBM.

D. Computational details

In order to utilize the above described formalism the needed key input quantities are the defect formation energies defined by eqn (3). These require calculations of the total energies for the perfect bulk system (the host) and of the possible defects in all relevant charge states. Additionally, to obtain the alignment in the chemical potentials described above the total energy $E_{\rm tot}^{{\rm phase},i}$ of the elementary phases of each involved species i is needed to get the chemical potentials entering eqn (3):

$$\mu_i = \mu_i^{\rm aligned} - E_{\rm tot}^{{\rm phase},i} \tag{5}$$

Since the focus of the present paper is to show how the techniques and data produced in the semiconductor community can be used in the context of wet corrosion and oxide film formation we refrain from re-computing defect formation energies whenever such data exist in the literature. Thus, in the following we show and discuss how existing data on defect formation energies can be used in the context of corrosion and oxide growth.

Over the last few years ZnO has received a lot of attention in the semiconductor community due to its potential as a direct wide bandgap semiconductor for optoelectronic, as well as, electronic devices. Since for such devices doping and compensation by native defects is a fundamental aspect a large number of very detailed *ab initio* studies on the energetics and electronic structure of point defects have been reported in the literature. The actual calculation of accurate defect formation energies is challenging due to the bandgap problem encountered by the conventionally employed semilocal DFT xc functionals, finite size effects of the employed periodic supercells (particularly for charged defects) or band alignment, to name only a few.[11] Recently developed hybrid functionals such as HSE,[14] which mix in a certain amount of exact-exchanged, allow the bandgap problem to be overcome and provide at the same time total energy capabilities important to compute the equilibrium configurations of the defects, including atomic relaxation. Various correction schemes to treat charged defects[11,15–17] allow to reduce largely finite size effects and obtain the accurate dilute limit even with rather modest supercell sizes in the order of 100 atoms.

Specifically, for ZnO we will use the formation energies computed by Oba *et al.*[18] which have been obtained using HSE and charge corrections. Since the oxygen interstitial defect has not been considered in ref. 18 we also use formation energies computed by Janotti *et al.*[19]

As shown by several studies, employing HSE leads to a dramatic improvement in the description of metal oxide compound properties.[20] To improve the accuracy of semiconductor defect calculations in many HSE studies the mixing parameter between the use of non-local Fock-exchange and the semilocal DFT functional is not 0.25 as suggested in the original formulation[14] but adapted to provide an optimal (but material specific) value. For ZnO, values significantly larger than 0.25 have been used: in the calculations of Oba *et al.*[18] $\alpha = 0.375$ is used, while the calculations of Janotti *et al.*[19,21] are performed using $\alpha = 0.36$. Both choices improve the description of the position of the 3d Zn states, the band gap, the lattice parameters and the heat of formation $\Delta_f H$ compared to experiment, resulting in an equally satisfying description, as can be seen in Table 1.

Taking the defect formation energies from ref. [18, 19 and 21] and plugging them into eqn (3) we obtain for any defect and charge state its formation energy as a function of the Fermi energy and the chemical potentials. As outlined in the previous subsection we have to consider only one chemical potential (we chose here the oxygen one). Following the conventions in the semiconductor community these energies are plotted as function of the Fermi energy and for a fixed

Table 1 Calculated and experimental lattice parameters, formation energy, and band gap for ZnO in the wurtzite structure. The calculated values are from DFT calculations with the HSE exchange–correlation functional,[14] in which the fraction of non-local Fock-exchange α has been increased compared to the standard value (see text). Values for $\alpha = 0.375$ are from ref. 18 and for $\alpha = 0.36$ from ref. 21. The experimental values are taken from ref. 22

	$\alpha = 0.375$	$\alpha = 0.36$	Experiment
a (Å)	3.25	3.25	3.25
c (Å)	5.20	5.26	5.21
$\Delta_f H$ (eV)	−3.13	−3.43	−3.63
E_{gap} (eV)	3.43	3.29	3.44

chemical potential. The results for the two extreme conditions of the chemical potentials, *i.e.*, O-rich and Zn-rich, are shown in Fig. 2.

Analyzing these diagrams provides important insight into which defects dominate, into their electronic behaviour, *i.e.*, do they accept or donate electrons, as well as their possible charge states. The defect with the lowest formation energy for a given set of chemical potentials and Fermi level will be the dominant one and occur in much higher concentrations than all other defects. For O-rich conditions Fig. 2a shows that under p-type conditions (*i.e.* for Fermi energies close to the valence band) a Zn_O antisite in a 4+ charge state is dominant. Going towards higher Fermi levels (above 0.6 eV) the oxygen vacancy in a 2+ charge state, and above $E_F = 2.2$ eV the neutral oxygen vacancy, become the dominant defects. Under Zn-rich conditions and going from p to n-type conditions first the oxygen 2+ vacancy, then a neutral oxygen (split) interstitial and a 2+ Zn vacancy are stable.

While the 2+ oxygen and 2− zinc vacancy are chemically intuitive and invoked in the conventional picture (see Fig. 1a) the previously postulated Zn interstitial is absent, while neutral defects (such as the oxygen vacancy or the oxygen split

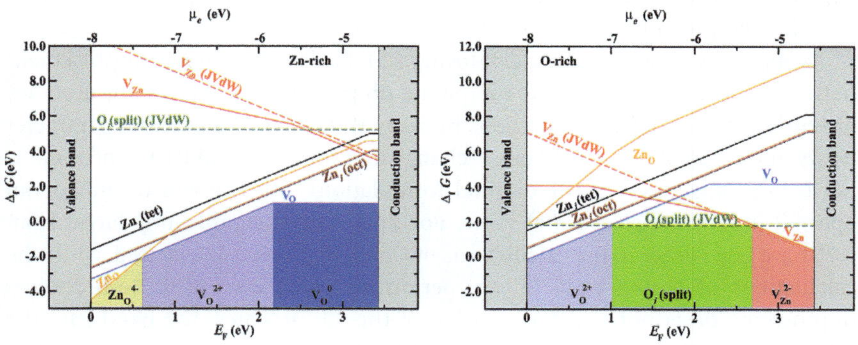

Fig. 2 Formation energies of native point defects of ZnO shown as a function of Fermi-level position for Zn-rich (right) and O-rich (left) conditions. Only segments corresponding to the lowest energy charge states for each defect are shown. The coloured areas depict the regions where the given defect species dominates. The zero on the Fermi-level E_F-axis corresponds to the VBM (valence band maximum). The slope of each line corresponds to the respective charged state. The *ab initio* calculated defect formation energies used to construct these diagrams were taken from ref. 18. The dashed-lines marked "(JVdW)" [V_{Zn}^{2-} and O_i (split)] are constructed using HSE-data obtained by Janotti and Van de Walle.[19,21]

interstitial) or a charged defect, such as the Zn_O antisite, have never been considered. Whether these defects become important under realistic electrochemical conditions will be discussed in Sec. III.

E. Defect transition levels and charge neutrality

An important benefit of representing defect formation energies in a diagram such as Fig. 2 is that the diagram provides direct insight into which defects are the thermodynamically most stable ones, thus allowing us to derive phase diagrams. It also provides direct insight into the electronic behaviour of any given defect by giving the possible charge states (oxidation states), as well as, the position of the Fermi level where the transition from one state into the other occurs, *i.e.*, the transition (redox) levels in the electronic bandgap.

To derive this information we note that the slope of the defect formation energies in Fig. 2 is according to eqn (3) the charge q of the defect: the position of the Fermi energy E_F enters only the last term qE_F. Consequently, the slope of the formation energy gives the charge (oxidation) state the defect has at a given Fermi level while the kinks in the formation energy imply a change in the charge state, thus giving the position of the transition level. Therefore, diagrams such as Fig. 2, which have been computed for a large number of bulk semiconductor/oxide materials and have been reported in the literature, can be used directly to extract all relevant defect properties for a given system.

We note that the actual numbers depend sensitively on the chosen xc functional as well as on the level of convergence. In fact, for well-studied semiconductor systems such as, *e.g.*, GaAs several sets of *ab initio* calculations exist that give differences of several eV(!) for the same defect. It is therefore of paramount importance when using existing literature data to carefully check that they are based on state-of-the-art hybrid xc functionals, include charge corrections and are well converged (*e.g.* with respect to supercell size, *k*-point sampling, *etc.*).

Fig. 2 shows nicely that state-of-the-art defect calculations have reached a maturity giving, if properly applied, essentially identical results. This can be seen for the example of the Zn vacancy for which an independent calculation by Janotti *et al.*[21] has been reported in the literature and marked as JVdW (dashed red line) in Fig. 2. Since in this study the authors focused on the technologically relevant case of n-type conditions where the 2+ charge state is the most stable one, the comparison is possible only for this charge state. An almost perfect match with a deviation of less than 0.1 eV is found. This level of agreement allows us to include in our considerations the O interstitial computed by the same group and which has not been considered in the original study of Oba *et al.*[18]

Having the energetic and electronic properties of all *individual* defects as a function of chemical potentials and Fermi level allows us to construct the full picture of a realistic system consisting of a variety of defects. For the following discussions we assume that all defects are in the dilute limit, *i.e.*, their concentration is so low that direct defect–defect interaction can be neglected. For the system considered here, ZnO and water, defect/ion concentrations are expected to be well below 1%, so that this assumption should be well justified.

While in the dilute limit a direct interaction between the defects can be safely excluded, an interaction with the chemical and electron reservoir occurs and has to be included. To be more specific let us first focus on the interaction with the electron reservoir. To realize a specific charge state q the defect has to exchange electrons with the electron reservoir. For example, to create the 2+ oxygen vacancy the defect has to transfer two electrons to the Fermi reservoir while creating a Zn 2− vacancy requires the transfer of two electrons from the reservoir to the defect. An important condition for any system is that it must be neutrally charged. In the grand canonical concept of an electron reservoir this simply means that the number of electrons transferred to the reservoir must be identical to the number of electrons taken out, *i.e.*, in the charge neutral case the reservoir will be completely empty and all transferred electrons will be distributed over the defects.

Formally, the above condition of charge neutrality can be expressed as follows:[7,11]

$$q_{\text{tot}}(\mu_i, E_{\text{F}}) = \sum_i q_i c_i(\mu_i, E_{\text{F}}) + c_{\text{h}} - c_{\text{e}} = 0 \tag{6}$$

This equation states that the sum of all charges q present in the system (associated with a charged defect/ion D^q, a free hole or a free electron) multiplied with their respective concentration c_i, as obtained from eqn (3) for the defects, or c_{h}, c_{e}, for the electrons/holes must become zero. In practice, one plugs into eqn (6) an initial value for the Fermi energy. If the total sum is positive/negative, the Fermi level is increased/decreased until one gets the Fermi level E_{F}^0 at which the total charge vanishes.

An important advantage of this approach is that we do not need to start from individual charge neutral reactions as commonly performed when describing electrochemical reactions. Rather, all conceivable reactions are intrinsically included in eqn (6). This equation thus provides the formal justification (i) why complex electrochemical systems can be decomposed into isolated non-charge neutral non-stoichiometric defects, that can be treated separately and (ii) that interact/couple only *via* their respective chemical/electronic reservoirs.

III. Discussion

A. Point defect stability phase diagrams

As discussed in the last section, in the field of semiconductor defect chemistry point defect formation energies are usually evaluated and discussed within plots such as the ones shown in Fig. 2. A disadvantage of these plots is that for each set of chemical potentials a separate plot is needed. For example, the phase diagrams shown in Fig. 2a and b strictly apply only for extreme Zn-rich and O-rich conditions implying that under intermediate conditions other defects may become most stable. Extending the concept of Pourbaix-diagrams which are a well-established concept in electrochemistry, showing the most stable bulk and defect (*i.e.* ions in water or other solvents) phases as function of pH and overpotential, we recently proposed a schema to construct defect phase diagrams that span the full configuration space as defined by the relevant chemical potentials and the chemical potential of the electron.[7,23] The key idea is to replace the pH value, which is a function of the oxygen chemical potential and the overpotential, by the oxygen chemical potential *via* a Legendre transformation. Using eqn (3) we can identify for any set of chemical potential(s) and overpotential the defect (*i.e.* type and charge state) with the lowest formation energy and thus directly construct such a phase diagram. For ZnO such a defect phase diagram is shown in Fig. 3 and yields for any thermodynamically allowed combination of oxygen chemical potential and overpotential, the dominant defect.

As mentioned in Section II.C, in electrochemistry the absolute value of the Fermi level μ_e rather than the relative level E_F with respect to the top of the valence band is the relevant one. To align the two levels we utilise the universal alignment of the H^+/H^- transition level observed in semiconductors, insulators and water.[24] The absolute level is shown on the upper *x*-axis both in Fig. 2 and 3.

As seen in Fig. 3, the set of defects that can become stable in ZnO is identical to the one identified from Fig. 2 considering only extreme Zn and O rich conditions. However, as becomes obvious also from Fig. 3 the region of Fermi levels where the respective defect becomes the dominant one strongly depends on the specific environment (oxygen chemical potential μ_O).

Since Fig. 3 is formally equivalent to a Pourbaix diagram we can immediately superimpose the well-known region of water stability, bounded in Fig. 3 by the two dashed black lines. The upper-right/lower-left line gives the boundary towards water becoming thermodynamically unstable against H_2/O_2 formation.

An important insight we gain from Fig. 3 is that in a substantial region defect formation energies become negative, *i.e.*, the formation of defects is exothermic implying that the oxide becomes unstable at these conditions. It is interesting to note that this instability of an oxide against exothermic formation of point defects is equivalent to the well-known water instability where the formation of the "intrinsic point defects" (solvated H^+ and OH^-) becomes exothermic. The regions, where ZnO becomes unstable, as a result of the negative defect formation energies of $Zn_O^{4+}(V_O^{2+})$ are marked in Fig. 3 as hashed yellow (blue) areas. The fact that the system always becomes unstable against defect formation under extreme p-type conditions (*i.e.* independent of the choice of the chemical potentials) may explain why p-type conductivity remains elusive in ZnO.

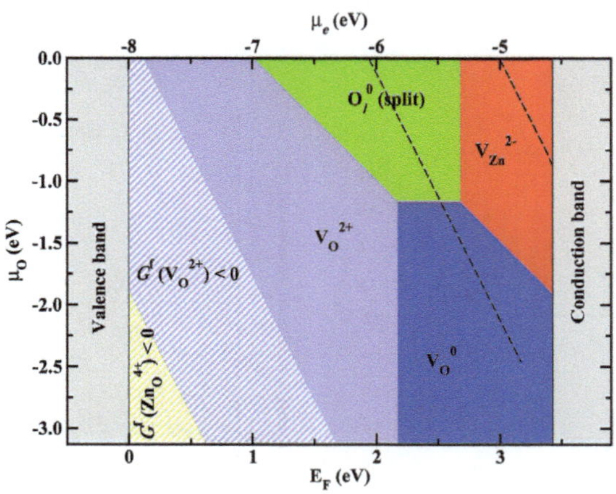

Fig. 3 Defect stability phase diagram for ZnO showing the regions where the respective native point defects dominate. Striped areas correspond to regions where ZnO becomes thermodynamically unstable against the formation of the respective point defects having negative formation energy in this region. The zero energy on the E_F-axis corresponds to the VBM. The μ_e-axis shows the electrode potential on an absolute scale. The black dashed-lines mark the boundaries of the electrochemical stability window of water.

B. Fermi energy position within the zinc/ZnO/water system

To connect the defect phase diagram (Fig. 3) with the oxide passivation layer bounded by Zn-bulk and water we need to know the position of the Fermi level as well as of the oxygen chemical potentials as a function of distance to the respective interfaces. To estimate the position of the Fermi level within the oxide film we first align the band structures of Zn, ZnO and H_2O with respect to the vacuum level. This is shown in Fig. 4, where we use the work function of zinc [Φ = 4.3 eV (ref. 25)] (which is by definition the Fermi energy position with respect to the vacuum level). For ZnO and water we use the universal alignment approach[24] mentioned previously.[7,26] For ZnO we assume n-type conditions to be consistent with experimental observations.[8,27,28]

Fig. 4(a) shows the alignment before the systems come into contact, *i.e.*, before electronic charge transfer occurs. Fig. 4(b) shows schematically the alignment after electronic charge transfer which results in a constant and identical Fermi level throughout the system. Since both the metal and water have an infinitely higher propensity to screen charges compared to a semiconductor, the potential drop at the Zn/ZnO and at the ZnO/H_2O interfaces will occur on the ZnO side of the respective interfaces,[29,30] leading to the formation of a space-charge layer and band bending. Since the Fermi energy of zinc lies above the CBM of ZnO there will be a flow of electrons from the metal to the conduction band of the oxide. As a consequence in this region the Fermi level will be above the conduction band, resulting in excess electrons forming a 2-dimensional electron gas (2DEG) and driving the system to extreme n-type conditions.

The Fermi level at the Zn/ZnO interface is pinned by the metal and cannot be changed. The opposite interface is formed by water for which the position of the Fermi level is not a constant but varies depending on the environment (chemical potentials, pH, overpotential). As shown in Fig. 3 the Fermi level cannot be chosen freely but is bounded by the condition of water stability, *i.e.*, that the native defects in water must have an endothermic formation energy. This region of allowed Fermi levels is marked in Fig. 4 by hashed boxes for extreme H-rich and

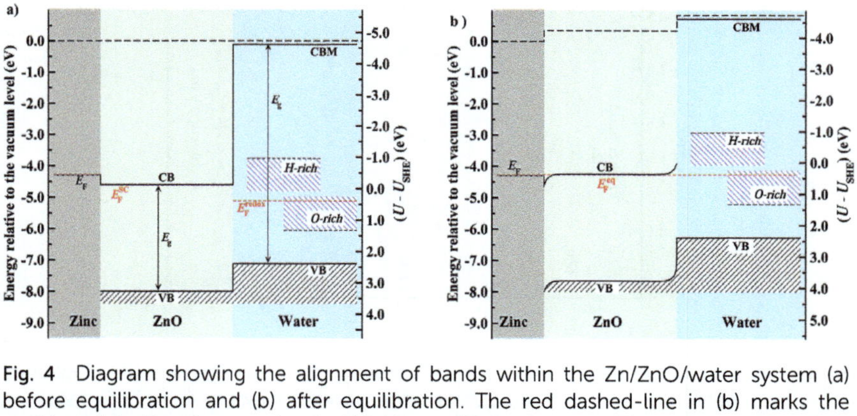

Fig. 4 Diagram showing the alignment of bands within the Zn/ZnO/water system (a) before equilibration and (b) after equilibration. The red dashed-line in (b) marks the position of the Fermi-energy level within the equilibrated system. The energy region within which the water Fermi-energy can be varied is marked by blue-striped regions depicting the extremes of possible chemical potential conditions which are H-rich and O-rich. For further details see text.

O-rich conditions. As can be seen the region of permitted Fermi levels is restricted to a rather small region in the water band gap. Since the Fermi level in water is variable, the exact shape of the band bending will depend on it. In Fig. 4 we therefore chose an intermediate value for E_F to sketch the band bending. Depending on the specific set-up band bending can be positive or negative, *i.e.*, extreme n-type conditions up to semi-insulating conditions may exist at this interface (see Fig. 3).

C. Point defects in the context of barrier oxide growth

We now have all the ingredients needed to identify the defect species that drive oxide growth. Since the two interfaces are chemically very different we start with the Zn/ZnO interface. This interface turns out to be conceptionally simpler since both Fermi level and chemical potential are fixed and well defined. Since the oxide layer is in direct contact with the Zn bulk and the Fermi level is slightly above the top of the conduction band (see Fig. 4b) we identify from the defect phase diagram (Fig. 3) the neutral oxygen vacancy as the most stable defect.

This may be surprising in two aspects: first, the oxygen vacancy in the context of oxide growth/corrosion has always been assumed to be 2+ which, however, based on Fig. 3 can be safely excluded. Second, to enable oxide growth defects with excess Zn should move from the Zn-bulk interface to the water interface while O excess defects should move in the opposite direction. Defects with Zn excess are Zn interstitials, Zn antisites and O vacancies and should be preferentially formed at the Zn interface while defects with O excess are O interstitials, O antisites and Zn vacancies that should form at the water interface. Indeed, Fig. 3 shows that at the Zn interface the defect that can be formed most easily, *i.e.*, the one which has the lowest formation energy is the O vacancy. However, Fig. 3 also indicates that when going from Zn-rich towards O-rich conditions the formation energy becomes higher and above a critical value becomes even higher in energy than defects with O excess such as the O interstitial and the Zn vacancy. In other words it looks like the defect would have to go in a direction where its formal formation energy becomes higher, *i.e.*, where the defect becomes energetically less favourable. This apparent discrepancy can be easily resolved by noting that the driving force for defect migration is the gradient in the chemical potential which will only disappear once thermodynamic equilibrium is achieved, *i.e.*, if the gradient vanishes.

Based on Fig. 3 together with the alignment diagram (Fig. 4) we can also identify the dominant defects for the interface of the ZnO with water. Under more oxygen-rich conditions the 2+ Zn vacancy [under n-conditions ($\mu_e > -5.3$ eV)] and the charge neutral O interstitial [under more electronegative conditions ($\mu_e < -5.3$ eV)] become the dominant defects. Both are defects with O excess, consistent with our expectations to what defects should be formed at this interface to promote the growth of the oxide film. Under less oxidizing conditions ($\mu_O \leq -1.2$ eV) the neutral O vacancy becomes again stable. Since this is an oxygen deficient defect growth of the oxide passivation layer growth is expected to slow down.

The approach outlined in this study provides thus an efficient and physically transparent approach to identify the relevant defects in the passivation layer under realistic electrochemical conditions. The defects identified here, the neutral O vacancy on the Zn interface and the Zn vacancy and O interstitial on the water interface (see Fig. 1b) are rather different compared to the ones assumed in the conventional picture (Fig. 1a). While the expected V_{Zn}^{2-} is indeed found at the water side our formalism shows that this is true only if the water overpotential is above 0.6 V *vs.* SHE, otherwise a hitherto not considered O interstitial in a neutral charge state will prevail. For the interface with Zn bulk the expected O vacancy is found to be dominant, however in the neutral not the 2+ charge state.

The identification of neutral rather than of charged defects that drive oxide growth has important consequences. In the conventional picture where transport by a 2+ O vacancy is assumed two criteria apply to activate this mechanism: (i) formation energy and diffusion barrier for the O vacancy have to be sufficiently low to enable diffusion at operating temperatures and (ii) electrons must be sufficiently mobile to ensure local charge neutrality. If electron transport for a given system is inefficient or negligible (*e.g.* for highly resistive oxides where the Fermi level is deep in the bandgap) the moving charged defects would quickly built up an energetically costly space charge which would quickly suppress further migration. This well-known concept to improve the corrosion resistance of a material is however counteracted if the materials transport inside the oxide is realized by charge neutral defects. In this case the defect carries its electron with itself. Consequently, making the material more intrinsic by reducing the Fermi level (*e.g.* by doping) would not slow down the transport. Only if the Fermi level comes into a region of the defect phase diagram where a charged defect becomes dominant the conventional picture would apply again. For the oxide considered here, ZnO, this would be the case if the Fermi level drops by more than 1.5 eV below the conduction band (see Fig. 3). Under realistic conditions it may be difficult/impossible to achieve such extremely intrinsic (resistive) conditions. The methods outlined here would be ideally suited to perform a systematic search for suitable dopants that are able to bring the Fermi level into the desired region and thus to design materials with a higher corrosion resistance.

IV. Summary and conclusions

In the present work we outlined a formalism and strategies that allow us to connect the vast knowledge of intrinsic point defects in semiconductors and particular in oxides acquired over the years with issues relevant in wet corrosion. To demonstrate and discuss this strategy for a specific and technologically relevant example we focus on the growth and stability of the barrier oxide that is formed once Zn comes into contact with a water film. The resulting oxide film, ZnO, is instrumental to *e.g.* realize the corrosion protection in galvanized steels. At the same time, ZnO has also been a well-studied semiconductor material providing ample defect data. The construction of defect phase diagrams (Fig. 3) turned out to be key in connecting these two very different worlds. These diagrams use as input only defect energies that are often readily available from *ab initio* calculations reported in the literature and an absolute alignment in the electronic levels as described in ref. 7. Using these phase diagrams allows us to make a direct connection to electrochemistry conditions, *e.g.*, to directly include conditions of water stability. This, together with a careful consideration of band alignments (Fig. 4) makes it possible to identify the relevant defects.

Applying this formalism revealed a number of surprises for an, at first glance, extremely well studied oxide such as ZnO: compared to the standard and chemically intuitive picture we find that neutral rather than 2+ charged O vacancies become prevalent or that neutral O interstitials (which have not been considered so far in the context of Zn corrosion) can become the dominant defect species. These findings are not only important because new/different defects compared to the commonly expected ones are observed. Probably even more important is that charge neutral defects become the dominant species under electrochemically relevant conditions. Such defects behave qualitatively different compared to charged ones as outlined in Sec. III.C. As also discussed this insight may be used to identify new strategies towards improving corrosion resistance.

As a last remark we note, that the approach derived and discussed here for the Zn/ZnO/H_2O system is general and can be directly employed to other metal/oxide/electrolyte systems as well as for Zn/ZnO/gas systems relevant for high temperature corrosion. We believe, that the insights gained from the strategies and concepts derived in this paper may significantly enhance our understanding on formation and evolution of oxide passive layers on metal surfaces in corrosive environments and help to develop systematic approaches to search for alloys with improved corrosion resistance.

Acknowledgements

We thank A. Janotti and C. G. Van de Walle for providing us with the HSE-calculated formation energy of the O_i^0 (split) interstitial point defect and stimulating discussions. The funding by the European Research Council under the EU's 7th Framework Programme (FP7/2007–2013)/ERC Grant agreement 290998 is gratefully acknowledged. This work is supported by the Cluster of Excellence RESOLV (EXC 1069) funded by the Deutsche Forschungsgemeinschaft.

References

1 A. Seyeux, V. Maurice and P. Marcus, *J. Electrochem. Soc.*, 2013, **160**, C189.
2 C. Wagner, *Hoppe-Seyler's Z. Physiol. Chem.*, 1933, **222**, 21.
3 N. Cabrera and N. F. Mott, *Rep. Prog. Phys.*, 1949, **12**, 163.
4 F. P. Fehlner and N. F. Mott, *Oxid. Met.*, 1970, **2**, 59.
5 D. D. Macdonald, *Electrochim. Acta*, 2011, **56**, 1761.
6 B. Diawara, Y. A. Beh and P. Marcus, *J. Phys. Chem. C*, 2010, **114**, 19299.
7 M. Todorova and J. Neugebauer, *Phys. Rev. Appl.*, 2014, **1**, 014001.
8 S. Thomas, I. S. Cole, M. Sridhar and N. Birbilis, *Electrochim. Acta*, 2013, **97**, 192.
9 F. A. Kröger and H. J. Vink, *Solid State Phys.*, 1956, **3**, 307; F. A. Kröger, *The Chemistry of Imperfect Crystals*, North Holland Publishing Co., Amsterdam, 1974.
10 C. G. Van de Walle and J. Neugebauer, *J. Appl. Phys.*, 2004, **95**, 3851.
11 C. Freysoldt, B. Grabowski, T. Hickel, J. Neugebauer, G. Kresse, A. Janotti and C. G. Van de Walle, *Rev. Mod. Phys.*, 2014, **86**, 253.
12 There is no principle difficulty in obtaining concentration dependent free energies of defect formation and an extension of the formalism to such cases is straight forward 7.

13 S. Trasatti, *Pure Appl. Chem.*, 1986, **58**, 955.
14 J. Heyd, G. E. Scuseria and M. Ernzerhof, *J. Chem. Phys.*, 2003, **118**, 8207; J. Heyd, G. E. Scuseria and M. Ernzerhof, *J. Chem. Phys.*, 2006, **124**, 219906.
15 C. Freysoldt, J. Neugebauer and C. G. van de Walle, *Phys. Rev. Lett.*, 2009, **102**, 016402; *Phys. Status Solidi B*, 2011, **248**, 1067.
16 G. Makov, R. Shah and M. C. Payne, *Phys. Rev. B: Condens. Matter*, 1996, **53**, 15513.
17 S. Lany and A. Zunger, *Phys. Rev. B: Condens. Matter Mater. Phys.*, 2008, **78**, 235104; *Modell. Simul. Mater. Sci. Eng.*, 2009, **17**, 084002.
18 F. Oba, A. Togo, I. Tanaka, J. Paier and G. Kresse, *Phys. Rev. B: Condens. Matter Mater. Phys.*, 2008, **77**, 245202.
19 A. Janotti, Private communication, The HSE-basis set is described in ref. 21.
20 A. Janotti, D. Segev and C. G. Van de Walle, *Phys. Rev. B: Condens. Matter Mater. Phys.*, 2006, **74**, 045202.
21 D. Steiauf, J. L. Lyons, A. Janotti and C. G. Van de Walle, *APL Mater.*, 2014, **2**, 096101.
22 W. M. Hayness, *CRC Handbook of Chemistry and Physics*, CRC Press Inc., Boca Raton FL, 95th edn, 2014.
23 M. Todorova and J. Neugebauer, *Surf. Sci.*, 2015, **631**, 190–195.
24 C. G. van de Walle and J. Neugebauer, *Nature*, 2003, **423**, 626.
25 R. A. Serway and J. W. Jewett, *Physics for Scientists and Engineers with Modern Physics*, Brooks and Cole, Boston, MA, 2010.
26 J. V. Coe, *Int. Rev. Phys. Chem.*, 2001, **20**, 33; J. V. Coe, A. D. Earhart, M. H. Cohen, G. J. Hoffman, H. W. Sarkas and K. H. Bowen, *J. Chem. Phys.*, 1997, **107**, 6023.
27 A. Janotti and C. G. Van de Walle, *Rep. Prog. Phys.*, 2009, **72**, 126501.
28 G. Trabanelli, F. Zucchi, G. Brunoro and G. Gilli, *Electrodeposition Surf. Treat.*, 1975, **3**, 129.
29 H. Gerischer, *Electrochim. Acta*, 1990, **35**, 1677.
30 S. Al-Halli and M. Willander, *Sensors*, 2009, **9**, 7445.

Faraday Discussions

PAPER

A mechanistic model for oxide growth and dissolution during corrosion of Cr-containing alloys†

M. Momeni and J. C. Wren*

Received 1st December 2014, Accepted 16th February 2015
DOI: 10.1039/c4fd00244j

We have developed a corrosion model that can predict metal oxide growth and dissolution rates as a function of time for a range of solution conditions. Our model considers electrochemical reactions at the metal/oxide and oxide/solution interfaces, and the metal cation flux from the metal to the solution phase through a growing oxide layer, and formulates the key processes using classical chemical reaction rate or flux equations. The model imposes mass and charge balance and hence, is labeled as the Mass Charge Balance (MCB) model. Mass and charge balance dictate that at any given time the oxidation (or metal cation) flux must be equal to the sum of the oxide growth flux and the dissolution flux. For each redox reaction leading to the formation of a specific oxide, the metal oxidation flux is formulated using a modified Butler–Volmer equation with an oxide-thickness-dependent effective overpotential. The oxide growth and dissolution fluxes have a first-order dependence on the metal cation flux. The rate constant for oxide formation also follows an Arrhenius dependence on the potential drop across the oxide layer and hence decreases exponentially with oxide thickness. This model is able to predict the time-dependent potentiostatic corrosion behaviour of both pure iron, and Co–Cr and Fe–Ni–Cr alloys.

1. Introduction

Several corrosion models have been developed for predicting the rate of metal dissolution in the presence of an oxide film.[1–13] To obtain the rate of corrosion under a given driving force (V), many of these models focus on solving the transport rate equations for individual charge carriers (interstitial cations and anions, cation and anion vacancies, and electrons and holes) across the oxide film, in addition to the rates of their creation at respective interfaces (*i.e.* electrochemical redox reaction rates). Since the rate of charge transport (charge flux) depends on the electric field gradient (electric potential), the electrochemical

Department of Chemistry, The University of Western Ontario, London, N6A 5B7, Ontario, Canada. E-mail: jcwren@uwo.ca

† Electronic supplementary information (ESI) available. See DOI: 10.1039/c4fd00244j

potentials of the metal, the oxide and the solution phases are important parameters in determining the corrosion rate. These potentials may change with time as corrosion progresses. However, these models do not specifically define the driving force for corrosion as a function of quantifiable potentials such as the equilibrium potential of a redox pair involved in corrosion ($E_{\rm rdx}^{\rm eq}$), or the electrode potential (corrosion potential ($E_{\rm corr}$) on an open circuit, or the applied potential ($E_{\rm app}$) during polarization).

In these models, the distribution of the driving force for corrosion on a corroding surface is often presented schematically as shown in Fig. 1. The implicit assumptions in this schematic are that the driving force for corrosion (V) is the potential difference between the Fermi-levels in the metal and the solution phases, that the driving force is distributed between the metal/oxide (m|ox) and oxide/solution (ox|sol) interfaces and the oxide film present, and that the potential may not be constant across the oxide film. The models differ in their assumptions on how the driving force is distributed and on how the potential distribution changes as the oxide film grows. For example, the potential drop across an oxide film is assumed to be independent of oxide thickness in the Cabrera–Mott model,[4] while it increases with oxide thickness in other models.[9–13] Alternatively, the potential difference at the ox|sol interface may be assumed to be constant as the oxide grows (the Point Defect Model (PDM))[9] while the potential difference at the m|ox interface is assumed to be constant in the Generalized Model for Oxide Film Growth.[12,13]

In addition, these models do not explicitly express the potentials that control the charge transport rates as a function of quantifiable potentials, such as $E_{\rm corr}$ or $E_{\rm rdx}^{\rm eq}$. The models assign different rates for the transport of different charge carriers across the solid oxide phase (ions, ion vacancies, electrons and holes). The individual transport rate parameters are difficult to verify, and this limits the predictive capabilities and the application ranges of these models. Oddly, mass and charge balance for the overall corrosion process, clear physical requirements, are not generally invoked in these models.

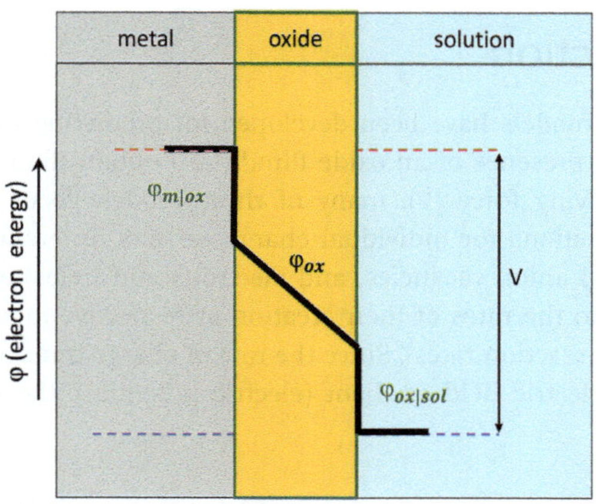

Fig. 1 Commonly accepted scheme for the distribution of the potential difference between the metal and solution phases, V, in a metal/oxide/solution system.

We have developed a corrosion kinetic model that can simulate both oxide film growth and metal dissolution as a function of time for a range of potentials, pHs and temperatures. Our model considers many of the elementary processes that are included in other models: electrochemical redox reactions at the m|ox and ox|sol interfaces, the transport of charged species across the oxide film, metal oxide formation and growth, and metal ion dissolution. The rates of the individual elementary reactions/processes are formulated using classical chemical reaction rate, and mass and charge flux equations. However, our model imposes mass and charge balance requirements on these rates, and reaction thermodynamic and kinetic constraints on electrochemical redox reactions. Hence, we have labeled our model the Mass Charge Balance (MCB) model.

In this paper, we describe the fundamental physical and chemical processes that underlie the MCB model and show how it can predict both oxide growth and dissolution during corrosion of an alloy. The rationales for the MCB model assumptions and the rate or charge flux equations used in the model are presented. In particular, we establish the driving force for corrosion as a function of equilibrium potential and how the driving force is distributed among the m|ox and ox|sol interfaces and across the oxide layer. The model includes relationships between the potential drop across the oxide film and the film thickness, and between the potential drop and the activation energy for oxide formation, and the rate of oxide growth as a function of the oxide thickness. We present a few comparisons of model simulations of the time-dependent corrosion current and oxide growth during potentiostatic polarization with data obtained for pure iron (using data from Sato *et al.*[14]) and for Cr-containing alloys: a Co–Cr alloy, Stellite 6,[15] and an Fe–Ni–Cr alloy, Alloy 800.

2. The MCB model

2.1 Overview of the MCB model

The MCB model considers corrosion to consist of four elements: electrochemical redox reactions at the m|ox and ox|sol interfaces, the transport of charged species across the oxide film, metal oxide formation and growth, and metal ion dissolution. The rates of the individual elementary reactions/processes in the model are formulated using classical chemical reaction rate, and mass and charge flux equations. The MCB model imposes mass and charge balance requirements on these rates, and reaction thermodynamic and kinetic constraints on electrochemical redox reactions. The mass and charge balance requirements invoked in the MCB model dictate that the rate of metal oxidation must equal the rate of its coupled solution species reduction, and the rate of metal oxidation must equal the sum of the rates of oxide formation and metal dissolution. This allows us to avoid the need for detailed modeling of charge transport across the oxide film. Instead, the MCB model takes into account the dependence of the potential drop across the oxide film on the type and thickness of the oxide(s) that grow with time.

Metal alloys may contain more than one active element that may form an oxide or hydroxide. This oxide can contain only a single metal element, or it can be a mixed oxide/hydroxide with more than one metal element. In addition, transition metals have many stable oxidation states and this, combined with the possibility of forming several different stable oxides and hydroxides, leads to the possible formation of many different layers of oxides/hydroxides on a metal surface. The

MCB model recognizes that different metal oxides can form and that the oxide composition and structure may change as corrosion progresses.[15,16] The result can be a complex and shifting set of oxides that form and grow as a function of time, even at a fixed potential.[15] The different types of metal oxides can have different oxide layer resistances and this will affect the potential drop across the oxide, and evolution in the nature of the oxide layer with time will change the potential drop as a function of time.

Irrespective of the type of oxide that forms and the rate of its formation, oxide formation is an electrochemical reaction and constrained by reaction thermodynamics. The thermodynamic constraints invoked in the MCB model dictate that metal oxidation (coupled with solution reduction) leads to the formation of a certain type of oxide with a driving force given by the difference in the equilibrium potentials of the two coupled redox half-reactions for that process. Energy pathway minimization prevents an oxide that requires a higher free energy of reaction from forming in competition. The MCB model assumes that the thermodynamic driving force is distributed between the m|ox and ox|sol interfaces and the oxide layer, in a manner somewhat similar to that shown in Fig. 1. Due to the potential distribution, the effective driving force for metal oxidation decreases as the oxide grows. In the MCB model, the distribution of the driving force at the m|ox and ox|sol interfaces and across the oxide layer is dictated by the mass and charge balance requirements. That is, the potential is distributed such that the rate of metal oxidation that produces metal cations must be the same as the rate of the metal cations moving across the oxide film, and these rates must be the same as the sum of the rates of metal oxide formation and metal ion dissolution.

The MCB model assumes that for a given type of oxide there is a charge distribution across the oxide layer (there is a higher metal cation concentration near the m|ox interface and a higher oxygen anion concentration near the ox|sol interface). In this case the oxide film on a corroding surface resembles a p–n junction in a solid-state diode device and is not a uniform semiconductor. Consequently, the potential drop across an oxide layer ($\Delta V_{\text{oxide}} = -\Delta\varphi_{\text{oxide}}$) increases linearly with oxide thickness. An increase in ΔV_{oxide} decreases the effective overpotentials for the redox half-reactions at the two interfaces. The result is that the metal oxidation rate can be formulated using a modified Butler–Volmer equation with an effective overpotential, provided that one can define the linear rate of oxide growth with time.

The MCB model assumes that the rate of oxide growth has a first order dependence on the flux of metal cations and that the oxide growth rate constant has an Arrhenius dependence on the activation energy for the metal oxide formation. The activation energy increases with an increase in ΔV_{oxide}, and hence, the rate constant for metal oxide formation decreases exponentially with an increase in oxide thickness.

The last key component of the MCB model takes into account the competition between oxide formation and dissolution for the metal cations produced by metal oxidation. Due to the mass and charge balance requirements the rate of metal oxidation must be the same as the sum of the rates of metal oxide formation and dissolution. In contrast to the changing oxide growth rate with oxide thickness, the rate constant for metal dissolution at the ox|sol interface is generally assumed to be independent of oxide thickness, but dependent on the type of dissolving oxide and the metal cation dissolution properties of the contacting solution (pH, temperature, *etc.*).

The principles behind the MCB model assumptions and the formulation of the rate equations are described next.

2.2 Elementary electrochemical and transport processes

The elementary physical processes considered in the MCB model are schematically presented in Fig. 2.[15] Metal oxidation occurs at the m|ox interface (Process 1a) and the reduction of aqueous species occurs on a counter electrode (Process 1b). On a naturally corroding surface (an open circuit) both the metal oxidation and the aqueous species reduction occur on the same (on a macroscopic scale) surface. When an alloy electrode is polarized in an electrochemical cell the two redox reactions occur on separated surfaces. The redox half-reactions are coupled *via* a flux of metal cations from the m|ox interface to the ox|sol interface (Process 2). The metal cations at the ox|sol interface can be hydrated and dissolve into the solution (Process 3), or combine with oxygen anions in the solution (O^{2-} or OH^-) to form a solid metal oxide that attaches to the ox|sol interface (Process 4). In this schematic, the metal cations are depicted as moving from the m|ox interface to the ox|sol interface. This does not mean that the individual metal cations physically move through the oxide layer, but rather that there is relative movement of the interfaces with respect to each other. Oxygen anions moving from the ox|sol to the m|ox interface result in the same transport rate equation. The net result is the transfer of metal species from the metal phase to the solution phase.

The oxyhydroxides of transition metals typically exhibit semiconducting properties.[17] For a chemically inert semiconductor, charge transport through the semiconductor is normally accomplished by movement of electrons (for an n-type) and holes (for a p-type). On a corroding metal surface, transfer of more massive charged species (metal cations and/or oxygen anions) also occurs. Movement of relatively massive ions through a solid oxide phase is not easy. To account for the charge flux through a solid oxide lattice, many mechanisms, such as transport of metal cations (or oxygen anions) *via* interstitials, or cation and anion vacancies, and electron hopping (or ion exchanges), have been proposed.[4,5,9,11–13] Irrespective of the ion transport mechanism, the charge flux

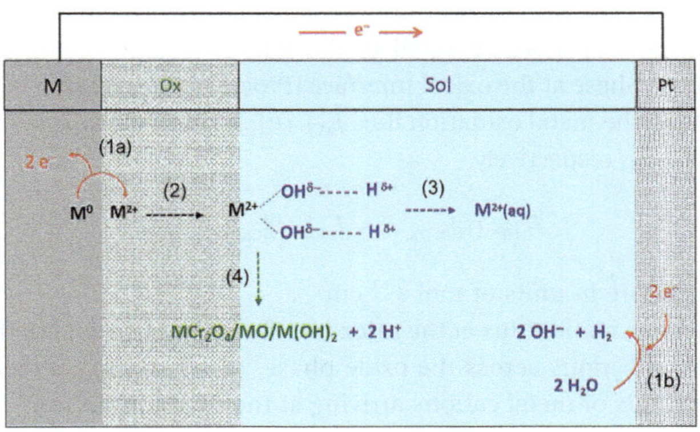

Fig. 2 Schematic of the elementary processes considered in the MCB model.

through a corroding surface can be modeled as the net flux of metal cations from the m|ox interface to the ox|sol interface.

For simplicity, only one active metal element, M, and its oxidation to one oxidation state, M^{2+}, are shown in Fig. 2. Similarly, only water is reduced. However, for an alloy, the set of elementary reactions will be much larger, taking into account all of the metal components of the alloy and their possible stable oxidation states, and the solution redox conditions. For example, in a highly oxidizing solution (*e.g.*, containing H_2O_2) the oxide growth process (Process 4) on a Ni–Fe–Cr alloy may consist of (1) oxidative conversion of a pre-existing layer of defective Cr_2O_3 to $FeCr_2O_4$, followed by (2) formation and growth of Fe_3O_4 and $NiFe_2O_4$ and then (3) formation and growth of $NiO/Ni(OH)_2$.[16] On a Co–Cr alloy, Process 4 may consist of (1) conversion of pre-existing Cr_2O_3 to $CoCr_2O_4$, followed by (2) the formation and growth of $CoO/Co(OH)_2$ and then (3) the formation and growth of CoOOH and Co_3O_4.[15]

2.3 Mass and charge balance

In the MCB model, the rates of individual reactions/processes are formulated using classical chemical reaction rate and mass and charge flux equations. The rates of the individual processes shown in Fig. 2 cannot vary independently. The mass and charge balance requirements dictate that at any given time, the rate of metal oxidation must satisfied.

$$\text{Oxidation rate} = \text{rate (1a)} = \text{rate (1b)} = \text{rate (2)} = \text{rate (3)} + \text{rate (4)} \tag{1}$$

The mass balance dictates that the rates of the reactions that occur in series must be the same. Also, the total rate of reactions in parallel is the sum of the individual reaction rates. Hence, the slowest in a series of reactions dictates the oxidation rate while dissolution (3) and oxide formation (4) in parallel compete for the metal cations.

For processes occurring at an interface, the rates are better expressed in terms of fluxes than in term of the change in concentration of a species (although the flux may depend on the concentration gradient of a species at the interface). Furthermore, in electrochemical studies of corrosion, the current (charge flux) is the measured quantity. The mass and charge balance requirements in terms of charge flux are:

(a) The flux of positive charges from the metal to the oxide phase at the m|ox interface (Process 1a) must be equal to the flux of negative charges from the solution to oxide phase at the ox|sol interface (Process 1b). Hereafter, these fluxes are referred to as the metal oxidation flux, $J_{M\#^{n+}}(t)|_{m|ox}$, and the solution reduction flux, $-J_{red\#}(t)|_{ox|sol}$, respectively.

$$J_{M\#^{n+}}(t)|_{m|ox} = -J_{red\#}(t)|_{ox|sol} \tag{2}$$

where the fluxes are in units of mol s^{-1} cm^{-2}.

(b) The metal oxidation flux at the m|ox interface must be equal to the average flux of the metal cations across the oxide phase, $\langle J_{M^{n+}}(z, t)\rangle_{oxide}$, (Process 2) and hence the total flux of metal cations arriving at the ox|sol interface:

$$J_{M\#^{n+}}(t)|_{m|ox} = \langle J_{M\#^{n+}}(z, t)\rangle_{oxide} = J_{M\#^{n+}}(t)|_{ox|sol} \tag{3}$$

The flux of the metal cations may vary along the oxide layer but the MCB model does not formulate this in detail. The average flux is assumed to be inversely proportional to the thickness of the oxide layer, $L_{\text{oxide}}(t)$, that may be initially present or growing.

$$\langle J_{\text{M\#}^{n+}}(z,t)\rangle_{\text{oxide}} = \frac{1}{L_{\text{MO}}(t)}\left(\int_0^{L_{\text{MO}}(t)} J_{\text{M\#}^{n+}}(z,t)\text{d}z\right) \tag{4}$$

Again it should be emphasized that a flux of the metal cations from the m|ox to ox|sol interface does not mean the physical movement of individual cations through the solid oxide phase but rather the relative movement of the interfaces with respect to each other. The flux of the oxygen anions from the ox|sol to m|ox interface in the opposite direction yields the same flux equation for positive charges.

(c) The charge flux must be equal to the sum of the fluxes of metal cations that dissolve into the solution phase (Process 3) and those that are used for growing an oxide film (Process 4) and these are referred to as the dissolution flux, $J_{\text{diss\#}}(t)|_{\text{sol}}$, and the oxide growth flux, $J_{\text{MO\#}}(t)|_{\text{oxide}}$, respectively.

$$J_{\text{M\#}^{n+}}(t)|_{\text{ox|sol}} = J_{\text{diss\#}}(t)|_{\text{sol}} + J_{\text{MO\#}}(t)|_{\text{oxide}} \tag{5}$$

The condition of equal fluxes for metal oxidation and solution reduction dictates the potential on a naturally corroding surface (E_{corr}) with the net current at E_{corr} being zero. Under polarization, the rate of oxidation (or reduction) occurring on the working electrode must equal the rate of reduction (or oxidation) occurring on the counter electrode, and this rate depends on the polarization potential (E_{app}).

In the MCB model we formulate the metal oxidation flux, $J_{\text{M\#}^{n+}}(t)|_{\text{m|ox}}$, and the oxide growth flux, $J_{\text{MO\#}}(t)|_{\text{oxide}}$, as functions of corrosion parameters (potentials, pH, T, *etc.*) and the other fluxes are determined using the mass and charge balance equations (eqn (1)–(5)).

In formulating $J_{\text{M\#}^{n+}}(t)|_{\text{m|ox}}$ and $J_{\text{MO\#}}(t)|_{\text{oxide}}$, the MCB model takes the free energy of reaction for the redox reaction, forming a specific oxide MO as the driving force (or reaction potential) for the reaction. The driving force is then distributed between the m|ox and ox|sol interfaces and the oxide film present on the surface. How the potential is distributed between the three components is discussed in Section 2.5.

Equal rates for metal oxidation and the sum of the metal oxide formation and metal ion dissolution rates then dictate the rate of oxide growth and its dependence on pH and temperature. Since metal oxidation results in both metal cation dissolution and oxide formation, the competing kinetics of these two pathways affects the rate of oxide growth.[15,16] These fluxes are discussed in Section 2.6.

2.4 Formulation of the metal oxidation flux, $J_{\text{M\#}^{n+}}(t)|_{\text{m|ox}}$

The overall redox reaction of M during corrosion can be expressed as,

$$\text{M} + \text{Ox} \leftrightarrows \text{M}^{n+} + \text{red} \tag{6}$$

where Ox represents the solution oxidant and red represents its reduced species. Knowing the nature of the metal and the solution redox species, we can calculate the Gibbs free energy of this reaction. The driving force for the overall reaction (the free energy of reaction, $-\Delta_r G(t)$) is the difference in electrochemical potential of the reacting system at time t ($E_{rdx}(t)$) and at equilibrium (E_{rdx}^{eq}):

$$-\Delta_r G(t) = -nF\Delta_r E(t) \tag{7a}$$

$$-\Delta_r G(t) = nF(E_{rdx}(t) - E_{rdx}^{eq}) \tag{7b}$$

By convention the electrochemical potential scale uses the reduction potential with respect to the standard hydrogen electrode potential (SHE), but the scale zero point is not important. A chemical reaction depends on the difference in potential and not the absolute values of the potentials.

The overall redox reaction is often expressed using two half-reactions:

$$\text{Ox: M} \leftrightarrows \text{M}^{n+} + ne^- \tag{8a}$$

$$\text{Red: Ox} + ne^- \leftrightarrows \text{red} \tag{8b}$$

This division is used for convenience in evaluating reaction thermodynamics. The electrochemical equilibrium potential for the overall redox reaction (6) can then be expressed using the equilibrium potentials of the two half-reactions,

$$\Delta E_{rdx}^{eq} = E_{red}^{eq} - E_{ox}^{eq} \tag{9a}$$

On the potential scale with respect to V_{SHE} this becomes

$$\Delta E_{rdx}^{eq} = E_{rdx}^{eq}\ (V_{SHE}) \tag{9b}$$

For an electrochemical reaction of a specific redox pair, rdx#, the net rate of the reaction (or the net flux of charges) can be defined by the Butler–Volmer equation. In terms of current:

$$i_{rdx\#}(t) = i_{rdx\#}^{eq}\left(\exp\left(\frac{nF}{RT}\alpha_{rdx\#}\eta_{rdx\#}(t)\right) - \exp\left(-\frac{nF}{RT}(1-\alpha_{rdx\#})\eta_{rdx\#}(t)\right)\right) \tag{10a}$$

$$\eta_{rdx\#}(t) = E_{elec}(t) - E_{rdx\#}^{eq} \tag{10b}$$

where $i_{rdx\#}^{eq}$ is the exchange current, or the anodic or cathodic current at equilibrium, n is the number of electrons involved in the reaction, $\alpha_{rdx\#}$ is the transfer coefficient (typically with a value of 0.5), F is Faraday's constant, R is universal gas constant (8.314 J mol^{-1} K^{-1}), T is absolute temperature (K), $\eta_{rdx\#}(t)$ is the overpotential at the reaction interface, and $E_{elec}(t)$ is the electrode potential or the potential at the reaction interface at time t. The electrode potential is the potential that we measure as the corrosion potential, E_{corr}, on an open circuit or the applied potential, E_{app}, on polarization. When $E_{elec}(t)$ is sufficiently more positive or more negative than the equilibrium potential, $E_{rdx\#}^{eq}$, the Butler–Volmer equation can be approximated to:

$$i_{\text{rdx}\#}(t) \approx i_{\text{ox}\#}(t) = i_{\text{ox}\#}^{\text{eq}}\left(\exp\left(\frac{0.5nF}{RT}\eta_{\text{ox}\#}(t)\right)\right) \quad \text{for } \eta_{\text{rdx}\#}(t) > 0 \tag{11a}$$

$$i_{\text{rdx}\#}(t) \approx i_{\text{red}\#}(t) = i_{\text{red}\#}^{\text{eq}}\left(-\exp\left(-\frac{0.5nF}{RT}\eta_{\text{red}\#}(t)\right)\right) \quad \text{for } \eta_{\text{rdx}\#}(t) < 0 \tag{11b}$$

where

$$i_{\text{rdx}\#}^{\text{eq}} = i_{\text{ox}\#}^{\text{eq}} = -i_{\text{red}\#}^{\text{eq}} \tag{11c}$$

$$\eta_{\text{rdx}\#}(t) = \eta_{\text{ox}\#}(t) = -\eta_{\text{red}\#}(t) \tag{11d}$$

The transfer coefficient is often empirically determined for a particular corrosion process.[18] In the MCB model the transfer coefficient for each elementary redox reaction, rdx#, is fixed at 0.5 and is not an adjustable parameter.

On a bare metal surface, there is only one reaction interface, the metal/solution interface, and the overpotential at the interface is defined as in eqn (10b). In the presence of an oxide film, the corrosion process involves reactions between three phases and at two different interfaces. The metal oxidation half-reaction (8a) occurs at the m|ox interface and the aqueous reduction half-reaction (8b) occurs at the ox|sol interface and/or on a counter electrode. The rate of each redox half-reaction can still be expressed using the Butler–Volmer equation (eqn (11)). However, not all of the free energy of reaction is available due to the potential barrier of the oxide film, $V_{\text{oxide}}(t)$:

$$\eta_{\text{rdx}\#}(t) = \eta_{\text{ox}\#}(t) = -\eta_{\text{red}\#}(t) = (E_{\text{elec}}(t) - E_{\text{rdx}\#}^{\text{eq}}) - \Delta V_{\text{oxide}}(t) \tag{12}$$

Thus, the rate of metal oxidation depends strongly on how the thermodynamic driving force is distributed between the two interfaces and across the oxide film.

For each possible metal/metal cation oxidation reaction, rdx#, in a specific solution environment the metal oxidation flux is formulated by a modified Butler–Volmer equation with the overpotential for the metal oxidation as defined in eqn (12):

$$J_{\text{M}\#^{n+}}(t)\big|_{\text{m|ox}} = J_{\text{rdx}\#}^{\text{eq}}\left(\exp\left(\frac{0.5nF}{RT}\eta_{\text{rdx}\#}(t)\right)\right) \tag{13}$$

2.5 Potential distribution

The potential energy for an interfacial charge transfer process is often described using the Fermi-levels (the total chemical potential of electrons (φ)) in the reacting phases. For a given interfacial redox reaction, the change in the chemical potential of electrons and the change in the chemical potential of the redox species must be the same. Although the electron potential energy scale uses a different reference point and is opposite in sign to that of the hydrogen reduction potential scale, the relative values are the same in both scales, $\Delta\varphi = -\Delta E$.

At phase equilibrium, the Fermi-levels of the two reacting phases at the interface must be the same. On a bare metal surface the Fermi-levels of the metal and solution phases at the m|sol interface must be the same, $\varphi_{\text{f}}^{\text{eq}} = \varphi_{\text{m}}^{\text{eq}} = \varphi_{\text{sol}}^{\text{eq}}$. Thus,

$$\eta_{\text{ox\#}}(t) = \varphi_{\text{m}}(t) - \varphi_{\text{m}}^{\text{eq}} \tag{14a}$$

$$-\eta_{\text{red\#}}(t) = \varphi_{\text{sol}}^{\text{eq}} - \varphi_{\text{sol}}(t) \tag{14b}$$

$$\varphi_{\text{m}}(t) - \varphi_{\text{sol}}(t) = E_{\text{red}}^{\text{eq}} - E_{\text{ox}}^{\text{eq}} = \Delta E_{\text{rdx}}^{\text{eq}} = E_{\text{rdx}}^{\text{eq}}\ (\text{V}_{\text{SHE}}) \tag{14c}$$

These relationships are schematically presented in Fig. 3. The driving force for corrosion on a bare metal surface is equivalent to the difference between the Fermi-levels of the metal and solution phases at time t, and this is the same as the difference in the equilibrium potentials of the two half-reactions. If there is no change in the electrochemical potential of the solution as corrosion progresses, the corrosion rate on a bare metal surface does not change.

The Fermi-level in the solution at equilibrium is the electron energy level where the density of unoccupied electron energy states ($\varphi_{\text{E(Ox)}}$) and the density of occupied electron energy states ($\varphi_{\text{E(red)}}$) are the same. In the presence of an oxide, the Fermi-levels of the solution and the oxide at the ox|sol interface must be the same. The Fermi-level in the metal phase at equilibrium is the electron energy level where the density of unoccupied electron energy states ($\varphi_{\text{E(M}^{n+})}$) and the density of occupied electron energy states ($\varphi_{\text{E(M)}}$) are the same. In the presence of an oxide film, the Fermi-levels of the metal and the oxide at the m|ox interface must be the same. The questions are then, "What is the Fermi gap across the oxide layer?" and "Is this gap constant across the oxide layer during corrosion?".

Most of the transition metal oxides present on corroding surfaces exhibit semiconducting properties.[17] The Fermi-level of a pure n-type semiconductor lies closer to the lowest energy of the conduction band, φ_{CB}, whereas the Fermi-level of a pure p-type semiconductor lies closer to the highest energy of the valence band, φ_{VB}. In order for corrosion to progress at any appreciable rate, the Fermi-level of the metal at the interface must lie above the φ_{CB} of the semiconducting oxide, while the Fermi-level of the solution phase must lie below the φ_{VB} of the oxide:

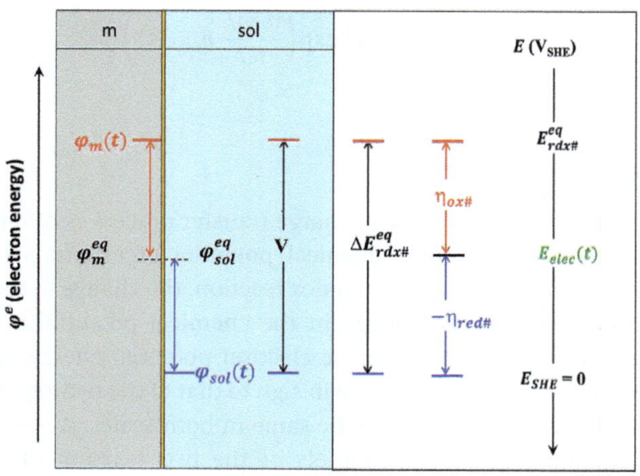

Fig. 3 Relative positions of the redox reaction potentials at time t during corrosion on a bare metal surface. The potential drops across the double layer and diffusion layer are not considered for simplicity.

$$\varphi_{\rm m}(t) > \varphi_{\rm CB} \text{ and } \varphi_{\rm sol}(t) < \varphi_{\rm VB} \tag{15}$$

These conditions reduce the overpotential that is available for metal oxidation or solution reduction:

$$\eta_{\rm ox}(t) = \varphi_{\rm m}(t) - \varphi_{\rm m}^{\rm eq} - \Delta\varphi_{\rm oxide}(t) \text{ for a p-type semiconductor} \tag{16a}$$

$$\eta_{\rm red}(t) = \varphi_{\rm sol}(t) - \varphi_{\rm sol}^{\rm eq} + \Delta\varphi_{\rm oxide}(t) \text{ for an n-type semiconductor} \tag{16b}$$

The mass and charge balance conditions further dictate that the effective overpotentials for the metal oxidation and its coupled solution reduction are related as given in eqn (13). Thus, for both n-type and p-type semiconductors,

$$\eta_{\rm rdx}(t) = \eta_{\rm ox}(t) + (-\eta_{\rm red}(t)) = \varphi_{\rm m}(t) - \varphi_{\rm sol}(t)\ \Delta\varphi_{\rm oxide}(t) = E_{\rm rdx}^{\rm eq} - \Delta V_{\rm oxide}(t) \tag{17}$$

These relationships are schematically presented in Fig. 4 (here the potential drops across the space charge layers, such as Mott–Schottky and double layers, are not considered for simplicity, as on a corroding surface these barriers should be negligible compared to the potential barrier of the oxide film).

On a pure semiconductor, the potential drop, $\Delta\varphi_{\rm oxide}(t)$, is the band gap ($V_{\rm bg}$). If the oxide film is a pure phase, the band gap does not change with an increase in oxide thickness and the growth of that oxide should not affect the interfacial charge transfer rate. This can explain some of the observations where the potential drop across an oxide film is independent of oxide thickness and justifies this assumption in the Cabrera–Mott model.[4] However, typically the oxide film composition on a corroding surface will not be uniform. Instead, there will be a charge distribution within the oxide lattice; the M^{n+} concentration will be higher nearer the m|ox interface and the O^{2-} concentration will be higher nearer the ox|sol interface. Therefore, the oxide near the m|ox interface will behave more like a p-type semiconductor (due to doping of the positive charges) while near the ox|sol interface the oxide will behave more like an n-type semiconductor (due to doping of the negative charges). The oxide film present on a corroding surface will then behave like a p–n junction in a solid-state diode device.

In the absence of any external potential the Fermi-levels of two reacting phases at the reaction interface must be the same. At equilibrium (no external force) the $\varphi_{\rm CB}$ of the p-type semiconductor is higher than the $\varphi_{\rm CB}$ of the n-type semiconductor. This results in a potential barrier to the flow of electrons (the majority of charge carriers). Similarly, holes cannot flow forward (from p-type to n-type regions) unless a positive external potential ($V_{\rm ext}$) is applied to overcome the potential barrier across the junction, and this potential barrier ($V_{\rm j}$, the junction potential) is not the band gap ($V_{\rm bg}$). Only when $V_{\rm ext}$ is larger than $V_{\rm j}$ and the Fermi-level in the n-type semiconductor region is raised above the $\varphi_{\rm CB}$ of the p-type semiconductor region can electrons flow from the n-type to the p-type regions.

We can envision the potential distribution across an oxide film on a corroding surface as being similar to that in a p–n junction. In the presence of an oxide layer, the corrosion redox reaction can occur only when the potential is sufficient to overcome the oxide potential barrier, $\Delta\varphi_{\rm oxide}$ (equivalent to $V_{\rm j}$), so that electrons can migrate from the ox|sol interface to the m|ox interface. If the potential at the m|ox interface is the same as the aqueous redox potential, there will not be any

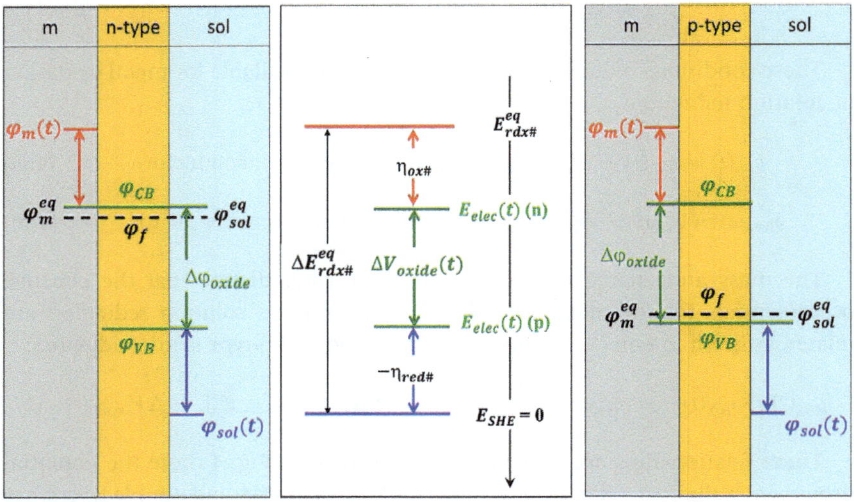

Fig. 4 Relative positions of the reaction potentials at a time t during corrosion: (middle panel) on the reduction potential scale (E (V_{SHE})), (left panel) for an n-type film on the electron energy scale (φ^e), and (right panel) for a p-type film on the electron energy scale (φ^e).

current – *i.e.*, no metal oxidation. The potential distribution on such a corroding system is schematically shown in Fig. 5.

Fig. 5 presents the potential energy distribution on a corroding surface at a specific time under specific solution redox conditions. If the oxide grows as corrosion progresses the potential barrier across the oxide film increases. For a system where a specific metal oxidation coupled with a specific aqueous reduction reaction, that leads to growth of a specific oxide film, MO#, it is reasonable to assume that $\Delta V_{\text{oxide}}(t)$ $(=-\Delta\varphi_{\text{oxide}}(t))$ is proportional to the changing oxide thickness:

$$\Delta V_{\text{oxide}}(t) = \Delta V_{\text{oxide}}(0) + \Delta V_{\text{MO\#}}(t) \tag{18a}$$

$$\Delta V_{\text{MO\#}}(t) = \varepsilon_{\text{MO\#}} L_{\text{MO\#}}(t) \tag{18b}$$

where $\Delta V_{\text{oxide}}(0)$ is the potential drop over, if present, the pre-existing oxide layer, $\varepsilon_{\text{MO\#}}$ is the proportionality constant or the specific potential gradient (potential drop per unit length) of oxide MO# and $L_{\text{MO\#}}(t)$ is the thickness of oxide MO# grown over time t.

Since the reaction potential energy is distributed such that it will satisfy the mass and charge balance requirements, the effective overpotential for metal oxidation will change according to eqn (17). For a given solution redox condition where $E^{\text{eq}}_{\text{rdx\#}}$ is constant, the potential gaps at the two interfaces and across the oxide film at two different times are schematically shown in Fig. 6. For simplicity, $\Delta V_{\text{oxide}}(0)$ is assumed to be zero. This schematic also illustrates that even with a constant solution redox environment, as the oxide layer thickens the corrosion potential, E_{corr}, which is the Fermi-level at pseudo equilibrium (or steady state), increases. This occurs even though the effective overpotential and, hence, the rate

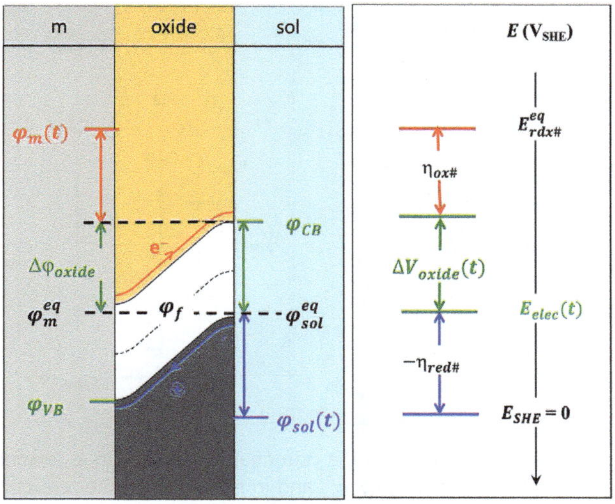

Fig. 5 Relative positions of the reaction potentials at a time t during corrosion in the presence of a n–p type oxide film: (left panel) on the electron energy scale (φ^e) and (right panel) on the reduction potential scale (E (V_{SHE})).

of the metal oxidation decrease. If the electrode potential, E_{app}, instead of E_{rdx}^{eq}, is maintained constant, as is the case for potentiostatic polarization, the effective overpotential for metal oxidation decreases as the oxide grows, as schematically shown in Fig. 7.

The potential energy diagrams presented in Fig. 5–7 address a system where there is only one redox reaction occurring. In a real system for an alloy with multiple elements, there may be multiple redox reactions that occur in parallel or in series. Nevertheless, the same principles apply to each individual redox reaction (rdx#) with its own electrochemical equilibrium potential, $E_{rdx\#}^{eq}$. The existence of multiple redox reactions is also the reason that E_{corr} depends on the aqueous redox environment and the type(s) of oxide that can be formed.

2.6 Formulation of the oxide growth and the dissolution fluxes

The electric potential barrier across the oxide film is an important rate controlling parameter. As discussed above, for a specific redox reaction that leads to a specific oxide film, it is reasonable to assume that $\Delta V_{MO\#}(t)$ is proportional to the oxide thickness, $L_{MO\#}(t)$ (eqn (18)). The oxide growth flux and the time dependence of $L_{MO\#}(t)$ can be established as follows. Oxide formation will compete with dissolution for the metal cations. Assuming that both processes have a first order dependence on $[M\#^{n+}]$ with rate constants, $k_{MO\#}$ and $k_{diss\#}$, respectively, the mass balance requirement results in:

$$J_{MO\#}(t)|_{oxide} = f_{k\text{-}MO\#}(t) J_{M\#^{n+}}(t)|_{m|ox} \quad (19)$$

$$J_{diss\#}(t)|_{sol} = (1 - f_{k\text{-}MO\#}(t)) J_{M\#^{n+}}(t)|_{m|ox} \quad (20)$$

$$f_{k\text{-}MO\#}(t) = \left(\frac{k_{MO\#}(t)}{k_{MO\#}(t) + k_{diss\#}} \right) \quad (21)$$

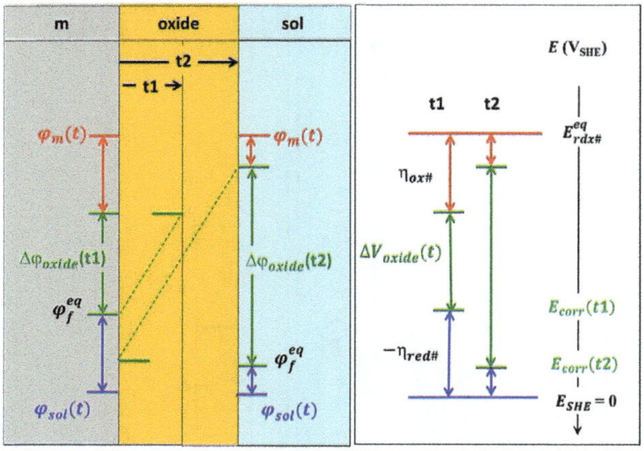

Fig. 6 Effect of linear oxide growth on the potential distribution in a corroding system: (left panel) on the electron energy scale (φ^e) and (right panel) on the reduction potential scale (E (V_{SHE})).

The ratio of the rate constants, $f_{k\text{-MO\#}}(t)$, depends strongly on pH and temperature. An increase in temperature will increase both the oxide formation and the dissolution rates, whereas a change in pH will primarily affect the dissolution rate.

Oxide formation is a chemical reaction and its rate constant can be assumed to have a normal Arrhenius dependence on the activation energy for the reaction. The electric potential energy gap across the oxide layer ($\Delta V_{MO\#}(t)$) contributes to the activation energy for the formation of an oxide (MO#) ($\Delta E_{a_{MO\#}}(t)$):

$$\Delta E_{a_{MO\#}}(t) = \Delta E_{a_{MO\#}}(0) + c'\Delta V_{MO\#}(t) = \Delta E_{a_{MO\#}}(0) + c_{MO\#}L_{MO\#}(t) \tag{22}$$

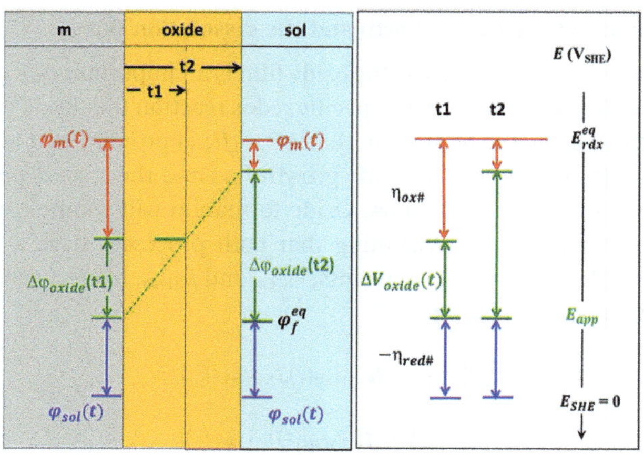

Fig. 7 Effect of linear oxide growth on the potential distribution during potentiostatic polarization: (left panel) on the electron energy scale (φ^e) and (right panel) on the reduction potential scale (E (V_{SHE})).

where c' is the dependency of activation energy of oxide growth of the potential drop across a layer of MO#, and $c_{MO\#}$ is the specific activation energy gradient of the oxide. The activation energy for the oxide formation increases as the oxide grows and the rate constant for the oxide formation decreases accordingly:

$$k_{MO\#}(t) = k_{MO\#}(0)\exp\left(-\frac{c_{MO\#}L_{MO\#}(t)}{RT}\right) \quad (23a)$$

where

$$k_{MO\#}(0) = k_{0\text{-}MO\#}\exp\left(-\frac{(\Delta E_{a_{MO\#}}(0))}{RT}\right) \quad (23b)$$

(Note that the other contributors to the reaction activation energy are included in $\Delta E_{a_{MO\#}}(0)$ and the value of $k_{0\text{-}MO\#}$, which is the pre-exponential factor for the oxide formation, and they are assumed to be constant with time).

The fluxes calculated in the model are related to physical parameters that can be measured. For example, the metal oxidation flux, $J_{M\#^{n+}}(t)|_{m|ox}$, can be measured as anodic current if metal oxidation current can be effectively separated from the water reduction current in an electrochemical cell:

$$i_{ox\#}(t) = nFJ_{M\#^{n+}}(t)|_{m|ox} \quad (24)$$

The dissolution flux, $J_{diss\#}(t)|_{sol}$, is related to the amount of dissolved metal:

$$A_{sol}(J_{diss\#}(t)|_{sol})dt = dm_{diss\#}(t) \quad (25)$$

where $dm_{diss\#}(t)$ represents the amount of dissolved metal over time dt (mol), and A_{sol} is the surface area exposed to the solution (cm^2). The oxide growth flux, $J_{MO\#}(t)|_{oxide}$, is related to the thickness of the oxide:

$$\upsilon_{MO\#}(J_{MO\#}(t)|_{oxide})dt = dL_{MO\#}(t) \quad (26)$$

where $\upsilon_{MO\#}$ is the molar volume of MO# (cm^3 mol^{-1}).

Eqn (23) shows that the rate constant for oxide formation decreases exponentially with increase in oxide thickness. If the metal oxidation flux does not depend on oxide thickness, the oxide grows at a progressively slower rate (exponentially slower with time). However, since the metal oxidation flux also decreases exponentially with oxide thickness (eqn (11c), (13) and (18)), the slower rate of oxide growth with time also slows down the rate of decrease in metal oxidation flux with time (eqn (22)). As derived in more detail in the ESI,† the net effect is that the oxide thickness can be approximated to increase logarithmically with time under constant electrode potential ($E_{elec}(t)$) conditions (*i.e.*, constant E_{corr} or E_{app}):

$$L_{MO\#}(t) \approx \frac{1}{\lambda_{MO\#}}\left(\ln\left(\lambda_{MO\#}J''_{MO\#}\right) + \ln t\right) \quad (27a)$$

$$\lambda_{MO\#} = \frac{0.5nF}{RT}\varepsilon_{MO\#} \quad (27b)$$

$$J''_{MO\#} = \upsilon_{MO\#}f_{k\text{-}MO\#}(0)J^{eq}_{rdx\#}\left(\exp\left(\frac{0.5nF}{RT}\left(E_{elec}(t) - E^{eq}_{ox\#} - \Delta V_{oxide}(0)\right)\right)\right) \quad (27c)$$

$$f_{k\text{-MO\#}}(0) = \left(\frac{k_{\text{MO\#}}(0)}{k_{\text{MO\#}}(0) + k_{\text{diss\#}}}\right) \tag{27d}$$

where $\lambda_{\text{MO\#}}$ represents a constant related to the potential drop across a unit length of the layer of oxide in inverse-length equivalent units, and $J''_{\text{MO\#}}$ represents the constant component of the metal cation flux.

The approximated analytical solution of the flux equations in the MCB model (eqn (27a)) has the form that we recognize for logarithmic film growth as previously reported by McDonald in his PDM model.[6] They expressed the rate law for film growth (for $L_{\text{MO\#}}(t) > 5$ Å) as

$$L_{\text{MO}}(t) \approx \frac{1}{2K}(\ln(2KA(B-1)) + \ln t) \tag{28}$$

where K, A and B are constants. That their derivation for the oxide growth rate is based on very different physical and chemical descriptions, but results in the same rate expression, suggests that the simpler approach used in the MCB model is sound.

2.7 Summary of the mathematical formulation of the model and the model parameters

The MCB model is summarized in Table 1. It consists of three key flux equations: metal oxidation flux, $J_{\text{M\#}^{n+}}(t)|_{\text{m|ox}}$, metal oxide growth flux, $J_{\text{MO\#}}(t)|_{\text{oxide}}$, and metal cation dissolution flux, $J_{\text{diss\#}}(t)|_{\text{sol}}$. Due to mass and charge balance requirements the oxide growth and dissolution fluxes cannot vary independently, and their sum must be the same as the metal oxidation flux. Thus, the MCB model consists of really only two independent flux equations. These flux equations are applied to each redox pair (designated with # in the flux equation) of metal oxidation and solution reduction.

In the MCB model, these flux equations are formulated based on well-established classical rate equations. The metal oxidation flux is formulated using a modified Butler–Volmer equation with an effective overpotential, where the effective overpotential takes into account the decrease in the driving force due to the potential drop across the oxide layer that is present or growing during corrosion. The oxide growth flux is formulated based on a first-order dependence of the oxide formation on the metal cation flux, and on an Arrhenius dependence of the rate constant on activation energy, which increases with an increase in the potential drop across the oxide layer. The metal ion dissolution flux is simply the difference between the metal oxidation flux and the metal oxide growth flux. The ratio of the oxide growth flux to the dissolution flux is determined by their rate constants. The rate constant of the oxide growth changes with time as the oxide grows while the rate constant for dissolution from a given oxide surface is constant with time.

In the MCB model, the model parameters are: (1) the equilibrium potentials of the two coupled half-redox reactions ($E^{\text{eq}}_{\text{red\#}}$ and $E^{\text{eq}}_{\text{ox\#}}$), (2) the potential drop over the initially present oxide layer ($\Delta V_{\text{oxide}}(0)$) and the specific potential drop over the MO# oxide that is growing ($\varepsilon_{\text{MO\#}}$), and (3) the rate constant for MO# oxide formation without an oxide barrier ($k_{\text{MO\#}}(0)$). The last term can be further divided into two more fundamental parameters, a pre-exponential factor and an activation energy ($k_{0\text{-MO\#}}$ and $\Delta E_{\text{a}_{\text{MO\#}}}(0)$). These model parameters for a given alloy

Table 1 Mathematical formulation of the model

Flux equations used in the model	Model output
1 $J_{M\#^{n+}}(t)\|_{m\|ox} = J^{eq}_{rdx\#}\left(\exp\left(\frac{0.5nF}{RT}\eta_{ox\#}(t)\right)\right)$	Current (t) $i_{ox\#}(t) = nFJ_{M\#^{n+}}(t)\|_{m\|ox}$
2 $J_{MO\#}(t)\|_{oxide} = \left(\frac{k_{MO\#}(t)}{k_{MO\#}(t) + k_{diss\#}}\right)J_{M\#^{n+}}(t)\|_{m\|ox}$	Oxide thickness (t) $L_{MO\#}(t) = \upsilon_{MO\#}\int_{t=0}^{t}(J_{MO\#}(t)\|_{oxide})dt$
3 $J_{diss\#}(t)\|_{sol} = J_{M\#^{n+}}(t)\|_{m\|ox} - J_{MO\#}(t)\|_{oxide}$	Dissolved amount (t) $m_{diss\#}(t) = A_{sol}\int_{t=0}^{t}(J_{diss\#}(t)\|_{sol})dt$

	Time dependent terms in the flux equations	Model parameters
1	$\eta_{rdx\#}(t) = E^{eq}_{red\#} - E^{eq}_{ox\#} - \Delta V_{oxide}(t)$ $\eta_{ox\#}(t) = E_{elec}(t) - E^{eq}_{ox\#} - \Delta V_{oxide}(t)$	$J^{eq}_{rdx\#}$, $E^{eq}_{red\#}$, $E^{eq}_{ox\#}$
2	$\Delta V_{oxide}(t) = \Delta V_{oxide}(0) + \varepsilon_{MO\#}L_{MO\#}(t)$	$\Delta V_{oxide}(0)$, $\varepsilon_{MO\#}$, $\upsilon_{MO\#}$
3	$\Delta E_{a_{MO\#}}(t) = \Delta E_{a_{MO\#}}(0) + c_{MO\#}L_{MO\#}(t)$	$\Delta E_{a_{MO\#}}(0)$, $c_{MO\#}$
4	$k_{MO\#}(t) = k_{MO\#}(0)\exp\left(-\frac{c_{MO\#}L_{MO\#}(t)}{RT}\right)$	$k_{MO\#}(0)$, $k_{diss\#}$

depend on the corrosion environment (which includes the type and concentration of aqueous redox species present, pH and temperature). The effects of the environmental parameters on the overall corrosion kinetics are thus modeled through their effects on the model parameters.

The flux equations can be numerically solved using any standard computer software differential equation solver. The results presented below were obtained using MATLAB.

3. Examples of model simulation results

The fluxes that the MCB model calculates correspond to measurable quantities, the current, the oxide thickness and the amount of dissolved metal as a function of corrosion time. These are all independently measurable quantities and the model's capability for predicting corrosion kinetics over a wide range of environmental conditions can be verified experimentally. We have applied the MCB model to simulate the potentiostatic polarization of a number of alloys including carbon steel, stainless steel, Co–Cr alloy Stellite 6, and Fe–Ni–Cr alloys Inconel 600 and Alloy 800. The preliminary results are very promising in all cases modelled to date and a few examples are presented here.

3.1 Oxide thickness on pure iron

We have applied the MCB model to predict the thickness of iron oxide grown as a function of E_{app} and compared the results to experimental data obtained from potentiostatic polarization of pure iron in mildly basic solutions by Sato *et al.*[14] The model simulation results and the experimental data are compared in Fig. 8, showing an excellent agreement. For this simulation, following parameters were

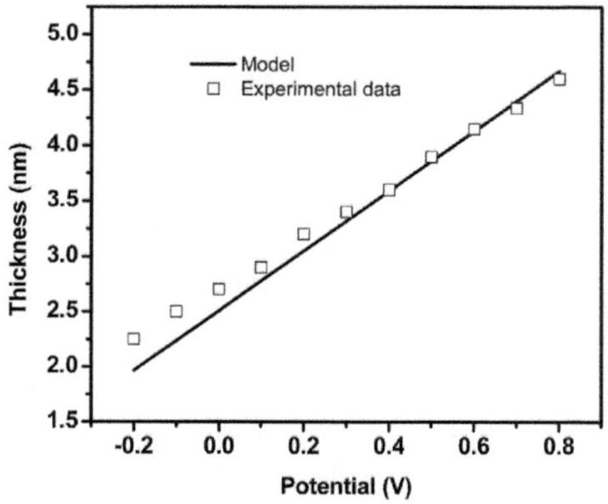

Fig. 8 Measured average oxide thickness on pure iron after 1 h potentiostatic polarization at 25 °C in a range of pHs from 7.45 to 10.45 in 0.15 N boric-borate solution (the symbols are data from Table 2 of ref. 14). The straight line is the prediction of the MCB model at pH = 10.45 (note that Sato's work showed no dependence of oxide thickness on pH over the range studied).

used: $i_{Fe_3O_4}{}^{eq} = 6 \times 10^{-6}$, $f_{k\text{-}MO\#}(0) = 0.9$, $\upsilon_{MO\#} = 30$, $\varepsilon_{Fe_3O_4} = 3.7 \times 10^7$ and $E^{eq}_{ox\#}$ was calculated based on the eqn (5) of Sato's work.[19]

3.2 Corrosion of Cr-containing alloys

Model simulations and the results of polarization tests (at $-0.6\ V_{SCE}$) of two alloys, Co–Cr (Stellite 6) and Fe–Ni–Cr (Alloy 800), are shown in Fig. 9. The tests were performed at two different pHs, 10.6 and 8.4, in 0.01 M borate buffer solutions at 25 °C. The polarization potentials modelled are near the corrosion potentials measured on these alloys in deaerated solutions.[15] The model predictions of the current behaviour are in very good agreement with the data.

In these simulations, the surface was assumed to be initially covered with a 2 nm thick layer of chromium oxide (Cr_2O_3).[20] With this Cr_2O_3 layer present, the only oxidation pathway that is thermodynamically possible at $-0.6\ V_{SCE}$ is the conversion of the chromium oxide to chromite ($CoCr_2O_4$ or $FeCr_2O_4$).[15] When this conversion is complete, Fe_3O_4 may grow over the chromite layer for the Fe–Ni–Cr alloy. For simplicity this process is not considered in the following discussion as it did not occur under the test conditions. Since the solubility of chromium is much lower than that of cobalt (for the Co–Cr alloy) and iron and nickel (for the Fe–Ni–Cr alloy) under the test conditions,[21] we assumed that only cobalt dissolution occurred (from the Co–Cr alloy), or that only iron dissolution occurred (from the Fe–Ni–Cr alloy). The solubility of nickel is also lower than that of iron and hence nickel dissolution from the Fe–Ni–Cr alloy at this low potential was not modeled. Nevertheless, the oxidative conversion of Cr_2O_3 to chromite requires additional modeling considerations.

Alloys which contain more than about 10% Cr typically display corrosion resistance because of the presence of a thin protective layer of air-formed Cr_2O_3 on the surface. This is the case for the Fe–Cr–Ni and Co–Cr alloys that we have studied. Mott[3] has shown that there can be a 5 nm maximum thickness of

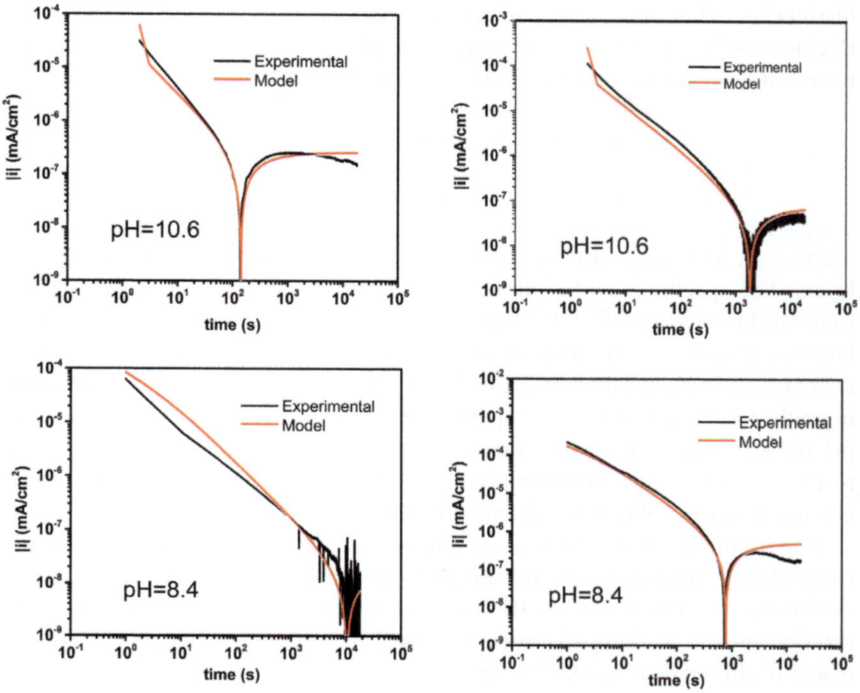

Fig. 9 Current observed during polarization at -0.6 V_{SCE} of (left panel) Co–Cr alloy Stellite 6 and (right panel) Fe–Ni–Cr Alloy 800 at (top row) pH 10.6 and (bottom row) pH 8.4. Experimental results are in black and modelling results are in red.

chromium oxide formed after long time of exposure to room temperature air. Even in deaerated solutions ($E_{corr} \approx -0.48$ V_{SCE} and -0.59 V_{SCE} on Stellite 6 in deaerated solutions at pH 10.6 and 8.4, respectively[21]), this chromium oxide is converted to a mixed element chromate layer ($CoCr_2O_4$ for cobalt alloys and $FeCr_2O_4$ for Fe–Cr–Ni alloys):

$$M + Cr_2O_3 + 2\ OH^- \rightarrow MCr_2O_4 + H_2O + 2e^- \text{ on WE} \tag{29a}$$

$$Ox + ne^- \rightarrow \text{red on CE} \tag{29b}$$

while this occurs there are two types of oxide in the oxide film: a more chromite-like layer and a more chromium oxide-like layer.

During conversion of chromium oxide to chromite, the thickness of the chromium oxide layer decreases while that of the chromite layer increases correspondingly:[20]

$$L_{Cr_2O_3}(t) = L_0 - f_l L_{MCr_2O_4}(t) \tag{30}$$

where $L_{Cr_2O_3}(t)$ and $L_{MCr_2O_4}(t)$ are the thicknesses of the chromium oxide and chromite layers, and L_0 is the initial air-formed chromium oxide thickness. The factor f_l is used to adjust for differences in the unit cell lengths of Cr_2O_3 and chromite oxide. In actuality there may not be a sharp division of the oxide into two distinct layers but a gradation between the two oxide types. In the model the oxide thicknesses in eqn (31) are for pure-oxide-phase equivalent thicknesses. The potential drop across the film can then be expressed as:

$$\Delta V_{\text{oxide}}(t) = \varepsilon_{\text{Cr}_2\text{O}_3} L_{\text{Cr}_2\text{O}_3}(t) + \varepsilon_{\text{MCr}_2\text{O}_4} L_{\text{MCr}_2\text{O}_4}(t) \tag{31}$$

$$\Delta V_{\text{oxide}}(t) = \varepsilon_{\text{Cr}_2\text{O}_3} L_0 + (\varepsilon_{\text{MCr}_2\text{O}_4} - f_l \varepsilon_{\text{Cr}_2\text{O}_3}) L_{\text{MCr}_2\text{O}_4}(t) \tag{32}$$

where $\varepsilon_{\text{Cr}_2\text{O}_3}$ and $\varepsilon_{\text{MCr}_2\text{O}_4}$ are the specific potential drops across the different oxide layers. The specific potential drop, $\varepsilon_{\text{MO}\#}$, is characteristic of the oxide[9] (with a value in the range of 10^5 to 10^7 V cm^{-1}). The values for $\varepsilon_{\text{Cr}_2\text{O}_3}$ and $\varepsilon_{\text{MCr}_2\text{O}_4}$ are not known and the values used in the simulations were those that yielded best fits of data on a given alloy. The values used in the simulations shown in Fig. 8 are listed in Table 2. Ideally, at a given temperature these values are fixed, independent of pH and E_{app}. The best-fit values of these parameters in the MCB model are indeed nearly the same at two different pHs. The values of the other model parameters are also listed in Table 2 and discussed below.

Under potentiostatic polarization, the aqueous reduction reaction that is not coupled with metal oxidation, but coupled with aqueous oxidation on the counter electrode, can also occur on the working electrode. This aqueous reduction reaction on the working electrode is treated as a separate independent redox reaction with its own equilibrium potential (or the difference of the equilibrium potentials of the two half-reactions of aqueous reduction and aqueous oxidation). For example, reduction of H_2O or dissolved O_2 (at an impurity level) can occur on the working electrode coupled with oxidation of H_2 or H_2O on the counter electrode:

$$2H_2O + 2e^- \rightarrow H_2 + 2OH^- \text{ on WE} \tag{33}$$

$$H_2 + 2OH^- \rightarrow 2H_2O + 2e^- \text{ on CE} \tag{34}$$

and/or

$$O_2 + 4H^+ + 4e^- \rightarrow 2H_2O \text{ on WE} \tag{35}$$

$$2H_2O \rightarrow O_2 + 4H^+ + 4e^- \text{ on CE} \tag{36}$$

The aqueous redox reactions do not contribute to the metal oxidation flux or oxide growth flux, but only to the net current. Nevertheless, the aqueous reduction flux on the working electrode is expressed in a manner similar to that used for the

Table 2 Fitting parameters for Cr-alloy potentiostatic simulations

Alloy system	pH	$i_{\text{MCr}_2\text{O}_4}^{\text{eq}}$ (mA cm^{-2})	$i_{\text{aq-rdx}}^{\text{eq}}$ (mA cm^{-2})	$\varepsilon_{\text{Cr}_2\text{O}_3}$ (V cm^{-1})	$\varepsilon_{\text{MCr}_2\text{O}_4}$ (V cm^{-1})	$c_{\text{MCr}_2\text{O}_4}$	$f_{k\text{-MCr}_2\text{O}_4}(0)$
Co–Cr	10.6	1.0×10^{-7}	-3.5×10^{-19}	1.7×10^6	5.5×10^6	1×10^7	0.91
	8.4	5.0×10^{-7}	-3.0×10^{-20}	1.2×10^6	5.5×10^6	1×10^7	0.15
Fe–Ni–Cr	10.6	1.0×10^{-6}	-1.0×10^{-19}	1.9×10^6	3.2×10^6	1×10^7	0.88
	8.4	1.0×10^{-6}	-1.0×10^{-19}	1.2×10^6	3.2×10^6	1×10^7	0.26

metal oxidation flux, using an effective cathodic overpotential as a function of potential drop across the oxide layer.

The values of the model parameters, the exchange current density ($i_{MCr_2O_4}{}^{eq}$) for the metal oxidation (29) and the exchange current ($i^{eq}_{aq\text{-}rdx\#}$) for the solution redox reaction ((30) and (31)) and the initial rate constants ratio, $f_{k\text{-}MO\#}(0)$, are also listed in Table 2. Although these values were obtained from best fits to the data, these values are nearly the same for a given alloy and all within acceptable ranges. The exchange current densities depend on the solution conditions and factors related to the surface characteristics. The larger variation in the exchange currents on Stellite 6 due to pH change can be attributed to the presence of two alloy phases, Cr-rich and Co-rich phases, on Stellite 6.

The ratio, $f_{k\text{-}MO\#}(0)$, depends on the solution environment, the type of growing oxide, and other factors, but this ratio is always less than one, and the ratio should be higher at a pH where the oxide solubility is lower. The solubilities of Fe^{2+} and Co^{2+} species are lower at pH 10.6 than at pH 8.4, and the best-fit values for the ratio, $f_{k\text{-}MO\#}(0)$, do indeed reflect this pH dependence.

At a higher potential, more oxidation pathways are available. In the MCB model the individual oxidation reactions are modelled separately and their contributions are added to obtain the overall corrosion kinetics.

4. Summary

In this study, a new classical model for oxide growth and metal dissolution is presented. This Mass Charge Balance (MCB) model is based on mass and charge balance and consists of three key flux equations: metal oxidation, oxide growth, and dissolution flux. The mass and charge balance requirements dictate that the oxide growth and dissolution fluxes cannot vary independently, but their sum must be the same as the metal oxidation flux. The metal oxidation flux is formulated using a modified Butler–Volmer equation with an oxide-thickness-dependent effective overpotential. The oxide growth and dissolution fluxes have a first-order dependence on the metal oxidation flux. Mass balance dictates that the ratio of the oxide growth and the dissolution fluxes is determined by their respective first-order rate constants. The rate constant for oxide growth is assumed to have a normal Arrhenius dependence on the activation energy for the reaction where the potential drop across the growing oxide layer contributes to the activation energy. Thus, the rate constant for oxide growth decreases exponentially with oxide thickness while the rate constant for dissolution remains constant. The analytical solution of this model results in a logarithmic dependence of the thickness of oxide on time.

The MCB model is able to predict the time dependent potentiostatic corrosion behaviour of both pure iron, and Co–Cr and Fe–Ni–Cr alloys.

List of acronyms and symbols

V	Driving force for corrosion (V)
$E^{eq}_{rdx\#}$	Equilibrium potential of a redox pair # involved in corrosion (V)
E_{corr}	Corrosion potential (V)
E_{app}	Applied potential during polarization (V)

$E_{rdx\#}(t)$	Electrochemical potential of the reacting system at time t (V)
$E^{eq}_{ox\#}$	Equilibrium potential for oxidation half-reaction (V)
$E^{eq}_{red\#}$	Equilibrium reduction half reaction potential (V)
$E_{elec}(t)$	Electrode potential at time t. It is E_{corr} in an open circuit and E_{app} in potentiostatic polarization (V)
$\eta_{rdx\#}(t)$	Overpotential at the reaction interface (V)
$\eta_{ox\#}(t)$	Anodic overpotential (V)
$\eta_{red\#}(t)$	Cathodic overpotential (V)
$i^{eq}_{rdx\#}$	Exchange current density (A cm^{-2})
$i_{rdx\#}(t)$	Current density at time t (A cm^{-2})
n	Number of electrons involved in the reaction.
$\alpha_{rdx\#}$	Transfer coefficient or symmetry factor, normally equal to 0.5
F	Faraday's constant (96 485 C mol^{-1})
R	Universal gas constant (8.314 J mol^{-1} K^{-1})
T	Absolute temperature (K)
m|ox	Metal/oxide interface
ox|sol	Oxide/solution interface
$\varphi_{m\|ox}$	Fermi-level at the metal oxide interface (V)
$\varphi_{ox\|sol}$	Fermi-level at the oxide solution interface (V)
$\Delta\varphi_{oxide}(t)$	Potential drop across the oxide layer at time t (V)
$J_{M\#^{n+}}(t)\|_{m\|ox}$	Metal oxidation flux at the metal/oxide interface (mol s^{-1} cm^{-2})
$J_{red\#}(t)\|_{ox\|sol}$	Solution reduction flux (mol s^{-1} cm^{-2})
$\langle J_{M^{n+}}(z, t)\rangle_{oxide}$	Average flux of metal cations across the oxide layer (mol s^{-1} cm^{-2})
$J_{M\#^{n+}}(t)\|_{ox\|sol}$	Total flux of metal cations arriving at the ox|sol interface (mol s^{-1} cm^{-2})
$J_{MO\#}(t)\|_{oxide}$	Oxide growth flux (mol s^{-1} cm^{-2})
$J_{diss\#}(t)\|_{sol}$	Dissolution flux (mol s^{-1} cm^{-2})
$L_{MO\#}(t)$	Thickness of the MO# oxide layer (cm)
$-\Delta_rG(t)$	Free energy of reaction (J mol^{-1})
φ^{eq}_{m}	Fermi-level of metal at equilibrium
φ^{eq}_{sol}	Fermi-level of solution at equilibrium
$\varphi_{E(Ox)}$	Density of unoccupied electron energy states of oxidants
$\varphi_{E(red)}$	Density of occupied electron energy states of reduction reaction products
$\varphi_{E(M^{n+})}$	Density of unoccupied electron energy states of metal cation
$\varphi_{E(M)}$	Density of occupied electron energy states of metal atom
φ_{CB}	Lowest energy of conduction band
φ_{VB}	Highest energy of valance band
$\varepsilon_{MO\#}$	Specific potential gradient of oxide (V cm^{-1})
$\Delta E_{a_{MO\#}}(t)$	Activation energy barrier for oxide growth at time t (J mol^{-1})
$\Delta E_{a_{MO\#}}(0)$	Activation energy barrier for oxide growth at time $t = 0$ (no oxide on the surface) (J mol^{-1})
$\Delta E_{a_{oxide}}(t)$	Activation energy barrier for oxide growth across the oxide present on the surface at time t (J mol^{-1})
$c_{MO\#}$	Specific activation energy gradient of oxide (J mol^{-1} cm^{-1})
A_{sol}	Surface area exposed to solution (cm^2)
$m_{diss\#}$	Dissolved amount of metal cations (mol)
$\Delta V_{oxide}(t)$	Potential drop across an oxide layer at time t (V)
$\Delta V_{oxide}(0)$	Potential drop across an oxide layer at time zero (V)
$\Delta V_{MO\#}(t)$	Potential drop across the layer of MO# at time t (V)

$\upsilon_{MO\#}$	Molar volume of oxide ($cm^3\ mol^{-1}$)
$L_{Cr_2O_3}$	Thickness of air-formed chromium oxide (cm)
$L_{MCr_2O_4}$	Thickness of growing chromite (cm)
f_l	Relative monolayer length of Cr_2O_3 to chromite
$f_{k\text{-}MO\#}$	Relative ratio of the oxide formation and dissolution constants
$\lambda_{MO\#}$	Constant related to the potential drop in the oxide (cm^{-1})
J''_{MO}	Constant component of metal cation flux

Acknowledgements

This research was funded under the Natural Science and Engineering Council of Canada (NSERC) and Atomic Energy of Canada Limited (AECL) Industrial Research Chair Agreement. The electrochemical analysis equipment was purchased using a grant from the Canada Foundation for Innovation.

References

1 N. F. Mott, *Trans. Faraday Soc.*, 1939, **35**, 1175–1177.
2 N. F. Mott, *Trans. Faraday Soc.*, 1940, **35**, 472–483.
3 N. F. Mott, *Trans. Faraday Soc.*, 1947, **43**, 429–434.
4 N. Cabrera and N. F. Mott, *Rep. Prog. Phys.*, 1949, **12**, 163.
5 F. P. Fehlner and N. F. Mott, *Oxid. Met.*, 1970, **2**, 59–99.
6 C. Y. Chao, L. F. Lin and D. D. Macdonald, *J. Electrochem. Soc.*, 1981, **128**, 1187–1194.
7 L. F. Lin, C. Y. Chao and D. D. Macdonald, *J. Electrochem. Soc.*, 1981, **128**, 1194–1198.
8 G. T. Burstein and A. J. Davenport, *J. Electrochem. Soc.*, 1989, **136**, 936–941.
9 D. D. Macdonald, *J. Electrochem. Soc.*, 1992, **139**, 3434–3449.
10 V. Battaglia and J. Newman, *J. Electrochem. Soc.*, 1995, **142**, 1423–1430.
11 M. Bojinov, G. Fabricius, T. Laitinen, K. Mäkelä, T. Saario and G. Sundholm, *Electrochim. Acta*, 2000, **45**, 2029–2048.
12 K. Leistner, C. Toulemonde, B. Diawara, A. Seyeux and P. Marcus, *J. Electrochem. Soc.*, 2013, **160**, C197–C205.
13 A. Seyeux, V. Maurice and P. Marcus, *J. Electrochem. Soc.*, 2013, **160**, C189–C196.
14 N. Sato, K. Kudo and T. Noda, *Electrochim. Acta*, 1971, **16**, 1909–1921.
15 M. Behazin, J. J. Noël and J. C. Wren, *Electrochim. Acta*, 2014, **134**, 399–410.
16 Q. W. Knapp and J. C. Wren, *Electrochim. Acta*, 2012, **80**, 90–99.
17 S. Fujimoto, in *Passivation of Metals and Semiconductors, and Properties of Thin Oxide Layers*, ed. P. Marcus and V. Maurice, Elsevier Science, Amsterdam, 2006, pp. 285–290.
18 W. A. Mueller, *Can. J. Chem.*, 1960, **38**, 576–587.
19 N. Sato, T. Noda and K. Kudo, *Electrochim. Acta*, 1974, **19**, 471–475.
20 K. J. Vetter, *Electrochim. Acta*, 1971, **16**, 1923–1937.
21 A. Y. Musa, M. Behazin and J. C. Wren, *Electrochim. Acta*, 2015, **162**, 185–197.

Faraday Discussions

PAPER

Mechano-electrochemistry effects due to deformation of copper oxide films

Denis Kramer,[a] Yikun Wang[b] and Julian Wharton*[b]

Received 3rd December 2014, Accepted 10th February 2015
DOI: 10.1039/c4fd00247d

In an attempt to elucidate the relationship and underlying processes of metal oxidation under stress, we combined the electrochemical characterisation with Density-Functional-Theory (DFT) calculations to interrogate the (100) surface of copper. The oxidised (100) surface shows a missing-row reconstruction, which is believed to be driven by surface stress. Hence, additional mechanical stimuli might have a significant impact on the onset of Cu oxidation. We find that different surface sites respond differently to strain. O_{ads} at the thermodynamically favoured high-coordination hollow site (O coordinated to four Cu) is stabilised by up to 130 meV by imposing 2% tensile strain onto the surface, while the low-coordination top site (O coordinated to one Cu) shows a markedly different sensitivity. By cramping into the hollow site, O_{ads} induces compressive stress into the (100) surface, an effect that is largely absent for the adsorption at the top site. We also find that the thermodynamic advantage of reconstructive underpotential surface oxidation is diminished under tensile strain. Hence, imposing tensile stress counter-balances the oxygen induced surface stress, which might have an implication on the onset of bulk copper oxidation. Studying Cu(100) single crystal surfaces in perchloric acid using cyclic voltammetry, we were able to confirm sensitivity of the electrochemical response towards the elastic strain.

1 Introduction

Bulk copper and surface oxidation have attracted immense interest over the years for the technical relevance as well as physical richness involved.[1–7] Copper remains the material of choice for electric wires and is the industrial standard pipe material for various chemicals (*e.g.*, refrigerants). In both cases, oxidation greatly threatens service performance by either reducing electrical conductivity or ductility. Copper oxidises to Cu_2O (cuprite) in neutral and alkaline media. In addition to being the oldest material for semiconductor electronics, Cu_2O finds new applications in diverse fields, including

[a]*Engineering Materials, Engineering Sciences, University of Southampton, Hants, SO17 1BJ, UK*
[b]*National Centre for Advanced Tribology at Southampton (nCATS), Engineering Sciences, University of Southampton, Hants, SO17 1BJ, UK. E-mail: j.a.wharton@soton.ac.uk*

nanoelectronics, spintronics, and photovoltaics. We, however, are mostly motivated by an interest to understand the corrosion behaviour of metallic copper under mechanical stress.

As a first step towards this goal, we focus here on the underpotential oxidation of the copper(100) surface, because formed oxide structures might act as precursors for bulk oxide film formation.[8] Numerous studies have elucidated the interaction of O_{ads} with copper surfaces and unveiled an intriguingly complex picture: (100) as well as (110) are fairly reactive and reconstruct in the presence of O_{ads}, while (111) is relatively inert. From the gas phase, O_2 dissociatively adsorbs at four-fold hollow sites (*cf.* the inset of Fig. 4 for an illustration of the adsorption mode), but a continuous $c(2 \times 2)$-O structure does not form. Instead, nano-structured $c(2 \times 2)$-O domains, separated by oxygen-deficient anti-phase boundaries, prevail up to a coverage of $\Theta \simeq 0.3$ ML. Larger $c(2 \times 2)$-O domains are unstable due to the accumulation of high stresses. The induced surface stress is the driving force for a $(2\sqrt{2} \times \sqrt{2})$R45°-O reconstruction in the gas-phase,[9] where O_{ads} orders in a $c(2 \times 2)$-O superstructure, but every forth surface Cu row along [110] is missing. The ejected surface Cu forms nanometer-sized terraces onto which oxygen adsorbs as well.[10]

The reconstruction of Cu(100) has also been observed in alkaline environments. At potentials above −810 mV *vs.* NHE, zig-zag chains are observed, which are attributed to OH pulling Cu atoms out of the top layer and into the hollow site.[11] Two such Cu atoms are then linked by an OH group. The $c(2 \times 2)$ superstructure is reported to prevail in dilute acid.[12]

2 Methodology

2.1 Theory

Total energy calculations. The use of Density-Functional-Theory (DFT) to investigate bulk phase stability,[13] nano-particles,[14] and surfaces[15,16] in aqueous environments has gained popularity over the last years. We use DFT in the generalised gradient approximation (GGA) as parameterised by Perdew, Burke, and Ernzerhof[17] (PBE) and implemented *via* the projector augmented wave function method in the Vienna *ab initio* Simulation Package (VASP)[18] to study the energetics of oxygen adsorption on Cu(100). Slabs were cut from the fully relaxed bulk with a lattice constant of 3.635 Å. We found that slabs consisting of eleven atomic Cu layers were sufficiently thick to mimic bulk behaviour in the interior of the slab. Atomic coordinates were allowed to relax, while the unit cell was fixed at dimensions equivalent to the fully relaxed bulk. The Brillion zone of the $c(1 \times 1)$ unit cell was sampled with a 9×9 Monkhorst mesh parallel to the surface. The mesh density was reduced appropriately for larger supercells. Only a single *k*-point is sampled perpendicular to the surface and a vacuum of at least 15 Å was maintained between slab images to minimize spurious interactions between slabs. The slabs were strained or compressed by adjusting the lattice constants appropriately to ensure isotropic strain parallel to the surface. Note that slab surfaces remain unstrained normal to the surface, because atomic coordinates were allowed to relax. We belief this to be in good agreement with experiments, where a stress normal to the surface would imply a compensating backpressure of the aqueous phase.

Adsorption energies. To probe the energetics of oxygen adsorption, a simple difference scheme is used, where the total energy of a clean or reduced slab $E^{\mathrm{red}}_{\mathrm{slab}}$ is subtracted from the total energy of an oxidised slab $E^{\mathrm{ox}}_{\mathrm{slab}}$ with adatoms:

$$E_{\mathrm{ads,O}} = (E^{\mathrm{ox}}_{\mathrm{slab}} - E^{\mathrm{red}}_{\mathrm{slab}})/N_{\mathrm{O}} - (\mu_{\mathrm{O}} - \mu^{\mathrm{o}}_{\mathrm{O}}) \tag{1}$$

Note that the oxygen absorption energy depends on the oxygen chemical potential, μ_{O}, reflecting the oxidative strength of the environment. For consistency with experiments, we choose to reference the oxygen chemical potential, $\mu^{\mathrm{o}}_{\mathrm{O}}$, to the oxygen dimer including a correction due to Wang *et al.* to compensate for inaccurate self-energy in the GGA of the O_2 dimer.[19]

Surface Pourbaix diagrams. The thermodynamically favoured surface termination in aqueous environments is controlled by pH and potential. Surface Pourbaix diagrams, therefore, are a convenient tool. These can be obtained from DFT slab calculations to first approximation by reflecting the aqueous environment in a thermodynamic framework. Note that this is an approximation, because it implicitly assumes that the interface (free) energies Metal–Aq. and Metal–Vac. have the same difference across terminations and therefore cancel each other out when comparing terminations. The surface energy of a slab, γ, can be expressed as

$$(2A)\gamma = G_{\mathrm{slab}} - \sum_i N_i\mu_i. \tag{2}$$

Note that the factor $2A$ is a consequence of the symmetry of the slab, which has two surfaces. The chemical potentials of the constituents, μ_i, are given by the equilibrium between the surface and bulk phases, namely the metal bulk and the aqueous phase. The surface energy can therefore be expressed as

$$(2A)\gamma = G_{\mathrm{slab}} - N_{\mathrm{Cu}}G_{\mathrm{Cu}} - N_{\mathrm{O}}\mu_{\mathrm{O}} - N_{\mathrm{H}}\mu_{\mathrm{H}}, \tag{3}$$

if only oxygen- and hydrogen-containing species adsorb. The free energy of bulk copper is denoted by G_{Cu}. Neither μ_{O} nor μ_{H} are convenient variables, because they cannot be controlled independently in aqueous environments. Because water is a reactive species, the chemical potentials of oxygen and hydrogen are not independent variables and are linked by

$$\mu_{\mathrm{H_2O}} = 2\mu_{\mathrm{H}} + \mu_{\mathrm{O}}. \tag{4}$$

Finally, μ_{H} can be expressed as a linear combination of pH and voltage by

$$\mu_{\mathrm{H}} = \mu_{\mathrm{H^+}} - F\varphi. \tag{5}$$

By rewriting the chemical potentials in eqn (3) in terms of eqn (4) and (5), a convenient expression for the surface energy as a function of pH and voltage is obtained:

$$(2A)\gamma = G_{\mathrm{slab}} - N_{\mathrm{Cu}}G_{\mathrm{Cu}} - N_{\mathrm{O}}\mu_{\mathrm{H_2O}} - (N_{\mathrm{H}} - 2N_{\mathrm{O}})(\mu_{\mathrm{H^+}} - F\varphi) \tag{6}$$

At room temperature, the proton chemical potential is simply given by $\mu_{H^+} = -0.0594pH$, if eV is used as unit.

Eqn (6) is linear in pH and voltage. Hence, the surface energy of each slab termination can be expressed as a plane in (γ, pH, φ)-space. The intersection of all planes representing possible terminations forms a convex hull, which after projection onto the (pH, φ)-plane can be visualised as a Pourbaix diagram. Finally, we neglect entropic contributions to the surface energy, which are likely to largely cancel each other out between terminations, by calculating γ based on slab energies, E_{slab}, rather than slab free energies, G_{slab}.

2.2 Experimental

Electrodes and electrochemical cell. The experimental setup consisted of a single compartment three-electrode cell as sketched in Fig. 1. Cyclic voltammetry (CV) was carried out with graphite rod and silver/silver chloride (Ag/AgCl, 3.5 M KCl, +0.205 V *vs.* NHE) counter and reference electrodes, respectively, using a Gamry Reference 600 potentiostat, Framework 6.25 and PHE200 Physical Electrochemical software. The working electrode disc was a Cu(100) single crystal (99.999% Cu supplied by Goodfellow Cambridge Limited) with an exposed surface area of 0.24 cm^2. Prior to electrochemical testing, the Cu electrode was mechanically polished to 1 μm, cleaned with acetone and then finally air dried. CVs were carried out in a naturally aerated solution at a scan rate of 5 mV s^{-1} with a step size of 2 mV, over a potential range of −741 mV to −241 mV *vs.* Ag/AgCl (−536 mV to −36 mV *vs.* NHE), in order to study the electrochemical response. Five CV scans were performed for each condition, and it is the data from the last cyclic scan that are reported here. CVs were carried out with and without elastic deformation to induce a tensile strain at the exposed electrode surface. Tensile stress was generated by positioning a steel ball bearing at the rear surface of the disc within the electrode assembly.

Chemicals. All the solutions were prepared with deionised water (≥18.2 MΩ cm, Millipore) using analytical grade chemicals (supplied by Sigma Aldrich). The test solution was 0.1 M $KClO_4$ (potassium perchlorate) adjusted to pH 5 using perchloric acid ($HClO_4$).

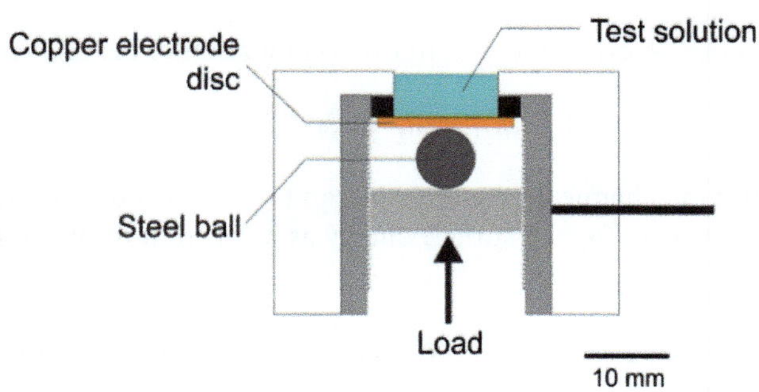

Fig. 1 Cross-sectional view of the experimental cell assembly detailing the setup for generating elastic stress at the copper surface.

3 Results

3.1 Energetics of oxygen adsorption on unreconstructed Cu(100)

Adsorption energies of oxygen at different sites of the unreconstructed Cu(100) surface are given in Table 1; relative changes are visualised in Fig. 2. We used a strain of 2% to ensure that the calculated effects are not obscured by numerical noise. Note that this is well outside the strain that can be sustained elastically by polycrystalline bulk copper, but is broadly in line with the large experimentally observed inward relaxation of Cu(100) of 1.0–2.1% (ref. 20–22) and surface strain in excess of 1% that can be induced chemically.[23] Similarly a large surface strain is observed at the beginning of epitaxial oxide growth on Cu(111), which was reported to exhibit 1.7% compressive strain.[24] No plastic deformation, however, was observed in the slabs used in the DFT calculations, due to the defect-free single crystal nature of the computational domain. Adsorption energies under full coverage and 50% coverage in the $c(2 \times 2)$ configuration were calculated. Coulomb repulsion between the surface oxygen atoms is the main reason for the generally lower adsorption energies under half coverage. Adsorption energies show a clear trend with oxygen coordination. The four-fold coordinated hollow site provides the most stable absorption site in both cases, followed by the two-fold coordinated bridge site. The least stable adsorption site is the top site.

A single oxygen–copper bond is so unfavoured that the presence of oxygen triggers a substantial rumbling of the top copper layer for half coverage, as can be seen from the insets of Fig. 2. Rumbling refers to the outward relaxation of every other surface copper atom leading to a zig-zag configuration of the top copper layer as highlighted in the left inset of Fig. 2. This substantial rumbling might also be the reason why the top site does not follow the trend of all other sites with respect to strain. It is the only site where tensile strain results in a destabilisation of adsorbed oxygen at 1/2 ML coverage. For all other sites, tensile strain results in a stabilisation of up to 130 meV. As we will show later, stabilising a full monolayer of oxygen on Cu(100) requires more oxidating conditions than bulk copper can sustain. However, even at half coverage, a stabilisation effect of about 50 meV is achieved for the important hollow site. The bridge site shows a similar behaviour although this adsorption mode is extremely unlikely to be experimentally active due to the energetic cost of almost 1 eV relative to the hollow site.

It is worth noting that Cu–O bond distances are insensitive to strain. This can be rationalised by a scissor-like effect where oxygen relaxes inward if the top layer

Table 1 Oxygen adsorption energies and Cu–O bond lengths on unreconstructed Cu(100) at different strains; the negative strain corresponds to compression

			d(Cu–O) [Å]			$E_{ads,O}$ [eV]		
Coverage	Ordering	Site	−2%	0%	2%	−2%	0%	2%
1 ML	$c(1 \times 1)$	Hollow	1.88	1.88	1.88	−1.66	−1.80	−1.92
		Bridge	1.81	1.80	1.80	−1.08	−1.13	−1.17
		Top	1.77	1.76	1.76	−0.19	−0.18	−0.18
1/2 ML	$c(2 \times 2)$	Hollow	1.94	1.95	1.96	−2.87	−2.92	−2.95
		Bridge	1.80	1.80	1.80	−1.90	−1.94	−1.98
		Top	1.78	1.73	1.72	−0.99	−0.92	−0.89

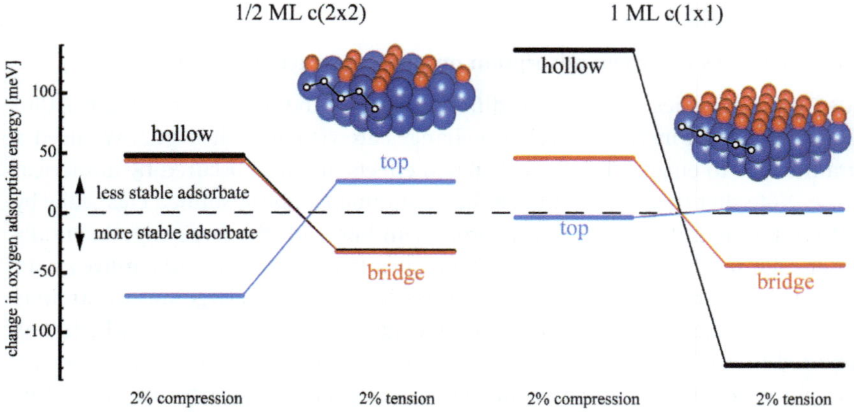

Fig. 2 Relative change of oxygen adsorption energies on unreconstructed Cu(100) under strain showing atypical behaviour of the top site; the negative values correspond to a stronger adsorption.

Table 2 Considered surface terminations of Cu(100) for construction of the unstrained surface Pourbaix diagram shown in Fig. 3

				Instability[a] [meV Å^{-2}]		
Species	Coverage	Unit cell	Reconst.	−2%	0%	2%
None	0	$c(1 \times 1)$	No	—	—	—
OH	1/6	$c(6 \times 2)$	Yes		88	
	1/2	$c(2 \times 2)$	No	120	123	130
	1	$c(1 \times 1)$	No	360	343	320
O	1/2	$(2\sqrt{2} \times 2)$R45°	Yes	—	—	—
	1/2	$c(2 \times 2)$	No	8	4	<1
	1	$c(1 \times 1)$	No	—	—	—

[a] Distance from the convex hull is used as a measure of instability.

Cu–Cu distance increases and outward if the Cu–Cu distance decreases for the hollow and bridge site. Because of the oxygen coordinates with a single Cu atom at the top site at full coverage, we did not expect a sensitivity of the bond lengths in this case; but it is remarkable that the bond length at the top site remains roughly 1.75 Å at half coverage where the top Cu layer rumbles.

In summary, we find oxygen adsorption on the unreconstructed Cu(100) surface to be most favourable at the hollow site at half and full coverage. Putting the surface under tensile strain results in a stabilisation of adsorbed oxygen. The stabilisation effect increases with coverage from about 45 meV at 1/2 ML to about 130 meV at full coverage for 2% tensile strain.

3.2 Thermodynamically favoured surface terminations

Unstrained Cu(100). The terminations given in Table 2 were considered to establish the thermodynamically most favourable surface termination as a function of pH and voltage. These were selected based on literature reports of

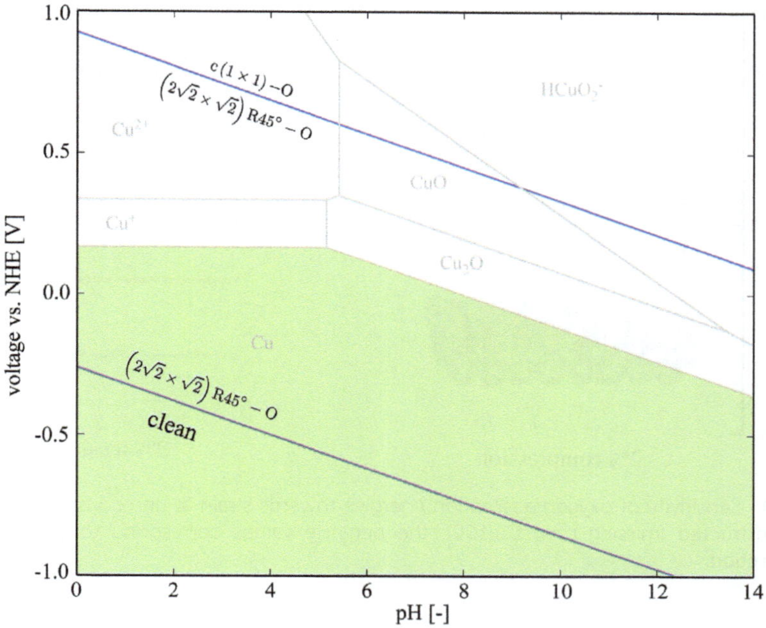

Fig. 3 DFT-derived surface Pourbaix diagram of thermodynamically favoured terminations of unstrained Cu(100) with a long-range order. See Table 2 for the considered terminations.

Cu(100) surface studies in aqueous and gaseous environments as well as common structures. The resulting Pourbaix diagram is shown in Fig. 3.

We find three terminations to be thermodynamically favoured and only two to be feasible within the bounds of bulk stability, which is indicated as the shaded area in Fig. 3. The unreconstructed clean surface is favoured at low voltages and pH. But oxygen adsorption becomes feasible well before bulk oxidation or dissolution occurs. We predict the missing row reconstruction to form above −550 mV *vs.* NHE at pH 5, which is some 700 mV before bulk stability is compromised. We also predict that a full monolayer of oxygen would be thermodynamically feasible under extremely oxidising conditions, but it is unlikely that this termination can be experimentally realised due to limited bulk stability of metallic copper.

It is also noteworthy that the $c(2 \times 2)$-O termination is very close in energy to become thermodynamically favoured. It is unstable by a mere 4 meV Å^{-2}. Thermal energy is in the order of 25 meV at room temperature, making thermal activation of the $c(2 \times 2)$-O termination likely. We, therefore, consider it likely that oxygen adsorbs at the hollow site up to 0.5 ML with increasing voltage and remains in close competition with the missing row reconstruction of the $(2\sqrt{2} \times 2)$R45° termination until bulk stability is lost.

Finally, our calculations indicate that neither of the considered OH orderings yields a termination low enough in energy to compete with the oxygen-terminated surfaces. In particular, our calculations indicate that the $c(6 \times 2)$-OH model might not be the correct explanation of the zig-zag superstructure observed by Kunze *et al.* in alkaline media.[11]

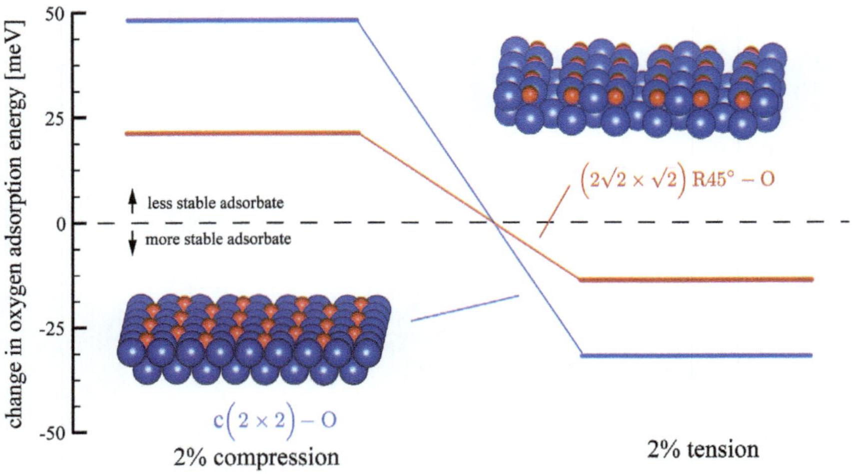

Fig. 4 Sensitivity of oxygen adsorption energies towards strain at unreconstructed and reconstructed (missing row) Cu(100); the negative values correspond to a stronger adsorption.

Cu(100) under tensile strain. We considered the same terminations given in Table 2 to investigate the surface stability under tensile strain. The resulting Pourbaix diagram is very similar to the one shown in Fig. 3 for the unstrained surface.

It should, however, be highlighted that the $c(2 \times 2)$-O termination gains in stability relative to the missing row $(2\sqrt{2} \times 2)$R45°-O configuration. In fact, the competition is so close that we consider both terminations to be degenerate within the precision of our calculations. The energetics of oxygen adsorption at the hollow site are more sensitive towards strain than the three-fold coordinated adsorption site in the missing-row reconstruction. As shown in Fig. 4, oxygen is stabilised by 30 meV in the four-fold coordinated hollow site under 2% strain, while the three-fold coordinated adsorption site in the missing row reconstruction only gains about half that energy.

This is in agreement with oxygen induced compressive surface stress. As the missing row reconstruction relieves at least part of the induced surface stress, it stands to reason that this termination benefits less from superimposing a tensile strain than the unreconstructed surface. That there is nonetheless an energetic benefit might indicate that the missing row reconstruction does not fully compensate for the induced surface stress.

3.3 Electrochemical response of Cu(100)

Fig. 5 shows a pair of cyclic voltammograms for Cu(100) in 0.1 M $KClO_4$ (pH 5) spanning a potential region from −536 mV to −36 mV and return. The potential window was chosen such that H_2 evolution and Cu dissolution (bulk oxidation) is avoided. Two distinct reductive features (C1 and C2) are visible, while no clearly defined cathodic activity is seen up to −150 mV, where the onset of an oxidative process, possibly the beginning of bulk Cu dissolution, is detected. The C1 feature is at potentials of −300 mV and −318 mV *vs.* NHE for a Cu(100) single crystal

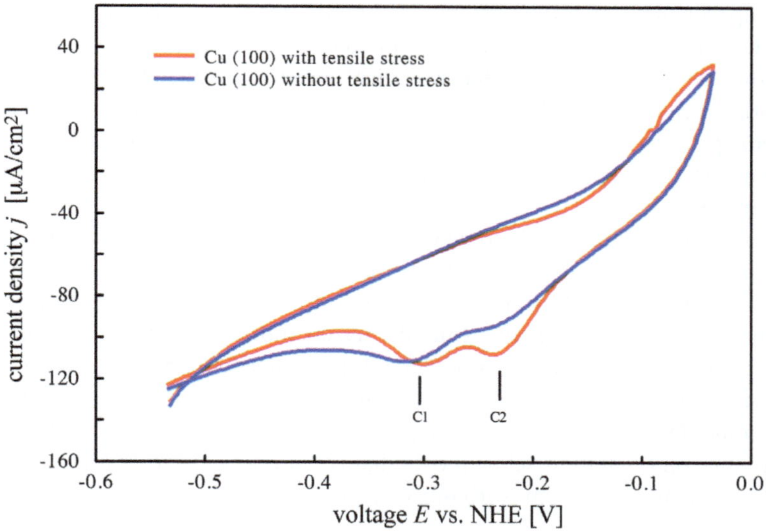

Fig. 5 Cyclic voltammogram of unstrained and elastically strained Cu(100) in 0.1 M $KClO_4$ adjusted to pH 5 using $HClO_4$; taken with a scan rate of 5 mV s^{-1} and a step size of 2 mV.

surface with and without imposed tensile stress, respectively. There is a positive shift in the order of 18 mV for this reductive surface process. The C2 feature at approximately 236 mV *vs.* NHE is most pronounced for the Cu(100) surface subject to tensile stress. Inspection by eye indicates that the C2 feature becomes somewhat more pronounced with subjecting the surface to tensile stress. The large double layer capacity and general skew of the CVs make it hard to define a reliable base line for quantitative analysis of the charge associated with the two reductive processes.

4 Discussion

4.1 Underpotential oxidation of Cu(100)

The surface state and structure of copper in aqueous solutions are fundamentally important to understanding its corrosion performance, in addition to being of technologically importance in catalysis, electrodeposition and electronics. The characterisation of underpotential oxidation at the Cu(100) interface, however, is often sparsely discussed in the literature. Previous studies have suggested a surface reconstruction taking place in the underpotential range during the initial stages of oxide formation.[11,25,26] Underpotential oxidation involves the electrosorption of oxygen species to form either very thin films or monolayer thick Cu_2O at potentials negative to the reversible potential of the Cu_2O formation. This oxide structure is often considered as a precursor to Cu_2O growth.

Our DFT calculations and experimental results indicate a substantial underpotential activity of Cu(100). While we calculate a specific adsorption of oxygen to happen above −500 mV *vs.* NHE at pH 5, we measure a surface activity at around −300 mV *vs.* NHE. We consider the agreement between the DFT prediction and the experiment remarkable, in spite of the discrepancy of 200 mV, given the rather crude approximations of the *First Principles* model: (1) no explicit treatment of the aqueous phases and (2) heuristic corrections of the reference states.

Kunze *et al.* also found electrochemical activity of Cu(100) in the underpotential region.[11] They, however, were able to detect an anodic as well as a cathodic peak around −800 mV *vs.* NHE in 0.1 M NaOH. The activity at around −300 mV *vs.* NHE reported here is in good agreement after compensating for the different pH values.

Kunze *et al.* were also able to detect substantial surface reconstruction in the underpotential region using scanning-tunneling-microscopy (STM), which they attributed to an interesting adsorption mode of OH leading to a zig-zag feature within a $c(6 \times 2)$ superstructure. Our DFT results indicate that this model of the observed reconstruction is thermodynamically unlikely with the missing row reconstruction and the $c(2 \times 2)$ termination being thermodynamically significantly more favourable. Cruickshank *et al.* observed a $(\sqrt{2} \times \sqrt{2})$R45° adsorbate superlattice on Cu(100) in acidic media that could not be detected on Cu(111),[12] which is in better agreement with our DFT predictions than the model of Kunze *et al.*

We, therefore, belief that Cu(100) reconstructs in perchloric acid environments similarly to the gas phase, and even if the resulting reconstruction differs somewhat from the missing row reconstruction, it is thermodynamically in close competition to it, which is likely to involve the removal of Cu atoms from the top layer, and might coexist with oxygen adsorption at the hollow site on unreconstructed Cu(100). Furthermore, surface-enhanced Raman spectroscopy (SERS) data from Chan *et al.* indicate that the adsorbing species is indeed oxygen rather than hydroxide.[27] They detected a broad SERS band at 625 cm^{-1} which prevailed between −700 and −100 mV *vs.* SCE in 0.1 M $HClO_4$. This feature was attributed to O_{ads} after deuteration of the electrolyte failed to produce the redshift expected from hydrogenated adsorbates.

4.2 Double peak and asymmetry of cathodic and anodic characteristics

The CVs in Fig. 5 have two original features that we were unable to link to previous literature: (1) the reductive double peak (C1 and C2) and (2) the apparently missing cathodic analogues.

It is noteworthy that the C2 feature is more sensitive to mechanical stimuli than the C1 feature. Given the close competition between the oxygen adsorption at the hollow site of unreconstructed Cu(100) and in the missing row reconstruction predicted by our DFT calculations, it is tempting to attribute the C1 feature to one and the C2 feature to the other. Recalling that the three-fold adsorption site provided by the missing row reconstruction is slightly more stable, it is logical to attribute C1 to oxygen desorption from this site and C2 to desorption from the four-fold hollow site. This assignment is also in agreement with the relative growth of the C2 feature under tensile strain. As our DFT calculations indicate, adsorption on the hollow site benefits more from tensile strain, because of the higher compressive surface stress associated with this adsorption mode. Hence, it stands to reason that this mode gains in importance under tensile strain.

Our DFT calculations do not allow us to shed more light on the observed asymmetry between oxidative and reductive behaviour in the underpotential region, because a potential electrokinetic origin is not captured. However, it is not uncommon for oxidative surface processes to yield a substantially different response from their reductive counterparts, especially if a rearrangement of metal atoms is involved. The archetypal case is Pt, where surface oxidation is characterised by a plateau in CVs with some fine structure and the reduction back to a bare Pt surface by a marked peak similar to the ones seen in Fig. 5. Ongoing research, therefore, focuses on linking the asymmetric behaviour to the reconstruction tendencies of Cu(100).

In addition, very little is known about the structural sensitivity of the cathodic processes on Cu single crystal surfaces.[28] It is generally well-established that the state of the Cu surface sites can influence the oxygen reduction reaction (ORR) reactivity, with the most reactive site towards oxygen redaction involving a Cu(I) species and the least reactive the metal itself, Cu(0).[29,30] It is, therefore, possible that the stabilisation of the unreconstructed $c(2 \times 2)$-O termination under tensile strain might influence the kinetics of the ORR.

5 Conclusion

Our work allows for new understanding of the influence of mechanical induced strains on initial oxidation and reduction processes at a Cu(100) surface. We investigated the underpotential oxidation of Cu(100) as a function of mechanical stimuli. We found that the surface stress induced by the adsorption of O_{ads} plays a crucial role in understanding the surface oxidation of Cu(100) and the electrochemical response in the underpotential region to the strain. While oxidative surface stress is likely relieved *via* a reconstruction of the surface if no mechanical stimulus is applied, subjecting the surface to tensile strain diminishes the underlying driving force for reconstruction and brings oxygen adsorption at the unreconstructed surface (hollow site) on par with reconstructive oxygen adsorption. The stabilisation of a four-fold coordinated adsorption site under tensile stress might impact ORR kinetics, and we belief the close competition between the two adsorption modes to be the reason for the double peak in the CV data shown in Fig. 5.

Appendix

A static finite element (FE) model was constructed to quantify the experimentally induced strain using ANSYS 14.0. The model consists of a copper disc (14 mm in diameter and 1 mm in thickness) in contact with a rigid ball bearing (7 mm in diameter). The copper disc was parameterised with a Young's modulus of 110 GPa, a Poisson's ratio of 0.343, a yield stress of 33.3 MPa and a tangent modulus of 0 GPa. The model considers material and geometric nonlinearities. Element SOLID185 with simplified enhanced strain formulation (keyopt (2) = 3) and the contact element couple TARGE170 and CONTA175 were adopted to construct the 3D model. The behaviour of the contact surfaces between the disc and the ball were set to be standard. The disc is restrained on the 1 mm wide circumference. The rigid ball was set to have an upward displacement of 0.1 mm.

The von Mises strain (effective strain), ε_e, was determined using the following expression:

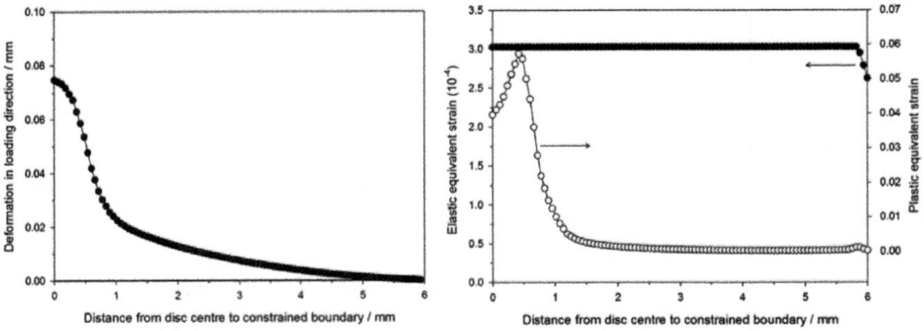

Fig. 6 Normal deformation and strain of the loaded Cu disc as estimated by finite-element analysis.

$$\varepsilon_e = \frac{1}{1+\nu}\sqrt{\frac{1}{2}\left[(\varepsilon_1-\varepsilon_2)^2+(\varepsilon_2-\varepsilon_3)^2+(\varepsilon_3-\varepsilon_1)^2\right]},$$

where ν is the effective Poisson's ratio (0.5 for plastic strain) and $\varepsilon_{1\text{-}3}$ are the principal strains.

The resultant deformation and strain distribution for 0.1 mm indention of the ball is shown in Fig. 6. The plastic strain field induced by the ball is relatively localised in the centre of the disc with a peak value in excess of 5% that falls off rather rapidly. Outside of an inner radius of $r_i = 1.5$ mm, the strain is mostly elastic. Given the relatively small fraction of surface where appreciable plastic deformation is expected, we consider it reasonable to assume that the electrochemical response discussed in Section 3.3 is largely dominated by elastic effects.

References

1 C. Ammon, A. Bayer, H. P. Steinrück and G. Held, *Chem. Phys. Lett.*, 2003, **377**, 163–169.
2 K. Bange, D. Grider, T. Madey and J. Sass, *Surf. Sci. Lett.*, 1984, **137**, A44.
3 A. B. Belonoshko and A. Rosengren, *Langmuir*, 2010, **102**, 8942–8952.
4 T. E. Eriksen, P. Ndalamba and I. Grenthe, *Corros. Sci.*, 1989, **29**, 1241–1250.
5 G. Hultquist, P. Szakálos, M. J. Graham, A. B. Belonoshko, G. I. Sproule, L. Gråsjö, P. Dorogokupets, B. Danilov, T. Aastrup, G. Wikmark, G. K. Chuah, J. C. Eriksson and A. Rosengren, *Catal. Lett.*, 2009, **132**, 311–316.
6 A. Spitzer and H. Lüth, *Surf. Sci. Lett.*, 1982, **120**, A337–A338.
7 Q. L. Tang and Z. X. Chen, *Surf. Sci.*, 2007, **601**, 954–964.
8 A. Soon, M. Todorova, B. Delley and C. Stampfl, *Surf. Sci.*, 2007, **601**, 5809–5813.
9 F. Jensen, F. Besenbacher, E. Laegsgaard and I. Stensgaard, *Phys. Rev. B: Condens. Matter Mater. Phys.*, 1990, **42**, 9206–9209.
10 K.-i. Tanaka, T. Fujita and Y. Okawa, *Surf. Sci.*, 1998, **401**, L407–L412.
11 J. Kunze, V. Maurice, L. H. Klein, H.-H. Strehblow and P. Marcus, *J. Electroanal. Chem.*, 2003, **554–555**, 113–125.
12 B. J. Cruickshank, D. D. Sneddon and A. A. Gewirth, *Surf. Sci.*, 1993, **281**, L308–L314.
13 K. A. Persson, B. Waldwick, P. Lazic and G. Ceder, *Phys. Rev. B: Condens. Matter Mater. Phys.*, 2012, **85**, 235438.

14 B. C. Han, C. R. Miranda and G. ceder, *Phys. Rev. B: Condens. Matter Mater. Phys.*, 2008, **77**, 1–9.
15 R. Benedek, M. M. Thackeray and A. van de Walle, *J. Mater. Chem.*, 2009, **20**, 369.
16 H. A. Hansen, J. Rossmeisl and J. K. Norskov, *Phys. Chem. Chem. Phys.*, 2008, **10**, 3722–3730.
17 J. P. Perdew, K. Burke and M. Ernzerhof, *Phys. Rev. Lett.*, 1996, **77**, 3865–3868.
18 G. Kresse and D. Joubert, *Phys. Rev. B: Condens. Matter Mater. Phys.*, 1999, **59**, 1758–1775.
19 L. Wang, T. Maxisch and G. ceder, *Phys. Rev. B: Condens. Matter Mater. Phys.*, 2006, **73**, 195107.
20 J. Wan, Y. L. Fan, D. W. Gong, S. G. Shen and X. Q. Fan, *Modell. Simul. Mater. Sci. Eng.*, 1999, **7**, 189.
21 H. Davis and J. Noonan, *Surf. Sci.*, 1983, **126**, 245–252.
22 D. M. Lind, F. B. Dunning, G. K. Walters and H. L. Davis, *Phys. Rev. B: Condens. Matter Mater. Phys.*, 1987, **35**, 9037–9044.
23 D. Sekiba, Y. Yoshimoto, K. Nakatsuji, Y. Takagi, T. Iimori, S. Doi and F. Komori, *Phys. Rev. B: Condens. Matter Mater. Phys.*, 2007, **75**, 115404.
24 Y. S. Chu, I. K. Robinson and A. A. Gewirth, *J. Chem. Phys.*, 1999, **110**, 5952–5959.
25 V. Jovi and B. Jovi, *J. Electroanal. Chem.*, 2003, **541**, 13–21.
26 V. Maurice, H.-H. Strehblow and P. Marcus, *Surf. Sci.*, 2000, **458**, 185–194.
27 H. Y. H. Chan, C. G. Takoudis and M. J. Weaver, *J. Phys. Chem. B*, 1999, **103**, 357–365.
28 T. Jiang and G. Brisard, *Electrochim. Acta*, 2007, **52**, 4487–4496.
29 F. King, M. Quinn and C. Litke, *J. Electroanal. Chem.*, 1995, **385**, 45–55.
30 G. Kear, B. Barker and F. Walsh, *Corros. Sci.*, 2004, **46**, 109–135.

Faraday Discussions

ROYAL SOCIETY OF CHEMISTRY

PAPER

Effects of molybdenum on the composition and nanoscale morphology of passivated austenitic stainless steel surfaces

Vincent Maurice,* Hao Peng, Lorena H. Klein, Antoine Seyeux, Sandrine Zanna and Philippe Marcus*

Received 24th November 2014, Accepted 12th January 2015
DOI: 10.1039/c4fd00231h

Surface analysis by time-of-flight secondary ion mass spectrometry, X-ray photoelectron spectroscopy and scanning tunnelling microscopy has been applied to provide new insight on Mo effects on the composition and nanostructure of the passive films grown in sulfuric acid on well-controlled Fe–17Cr–14.5Ni–2.3Mo(100) austenitic stainless steel single crystal surfaces. A duplex hydroxylated oxide matrix, 1.8–1.9 nm thick, is formed with a strong partition between Cr(III) and Fe(III) in the inner and outer layers, respectively. Cr(III) is increasingly enriched by preferential iron oxide dissolution upon passivation and ageing. Ni, only present as oxide traces in the film, is enriched in the alloy underneath. Mo, mostly present as Mo(IV) in the Cr-rich inner layer prior to anodic polarisation, becomes increasingly enriched (up to 16% of cations) mostly as Mo(VI) in the Fe-rich outer layer of the passive film, with ageing promoting this effect. Metallic Mo is not significantly enriched below the passive film produced from the native oxide covered surface. Mo does not markedly impact the nanogranular morphology of the native oxide film nor its local thickness variations assigned to substrate site effects on Cr(III) enrichment. Site specific preferential passivation still takes place at the (native) oxide-covered step edges of the alloy surface, and transient dissolution remains preferentially located on the terraces. Nanostructures, possibly Mo-containing, and healing local depressions formed by transient dissolution during passivation, appear as a specific effect of the Mo presence. Another Mo effect, observed even after 20 h of passivation, is to prevent crystallisation at least in the Fe-rich outer part of the passive film where it is concentrated mostly as Mo(VI) (*i.e.* molybdate) species.

1. Introduction

Stainless steels (SS) owe their high technical importance as structure materials to the formation of a continuous and protective surface oxide layer, the passive film,

PSL Research University, Chimie ParisTech – CNRS, Institut de Recherche de Chimie Paris, 75005 Paris, France. E-mail: vincent.maurice@chimie-paristech.fr; philippe.marcus@chimie-paristech.fr

only a few nanometres thick at ambient temperature that provides high corrosion resistance in chemically aggressive aqueous environment. Surface analysis, mostly using X-ray photoelectron spectroscopy (XPS), has demonstrated strong chromium enrichment for passive films on ferritic Fe–Cr,[1–17] and Fe–Cr–Mo[6,18–20] alloys and on austenitic Fe–Cr–Ni[21–26] and Fe–Cr–Ni–Mo[21,27–44] alloys. In acidic media, the strong chromium enrichment in the passive range is a consequence of the extremely small dissolution rate of Cr(III) oxide compared to Fe(II)/Fe(III) oxides. It is reduced by the dissolution of chromium as Cr(VI) in the transpassive range, or in alkaline solutions, by the decreasing solubility of Fe(II)/Fe(III) oxides. For Fe–Cr–Ni alloys, no or very little oxidized nickel enters the composition of the passive film and the alloy region underneath the film is enriched in metallic nickel.[21–26] The film thickness is in the range 1–3 nm depending on alloy and passivation conditions.[6,8,9,11,13,23,25,31,38]

Alloying a few percents of molybdenum improves markedly the resistance to localized corrosion by pitting and Mo bearing stainless steels, including austenitic AISI 316L, offer a better corrosion protection engineering solution in chloride-containing environments where passive film breakdown can be followed by pit initiation. Molybdenum enters the composition of the Cr-enriched passive film as Mo(IV) or Mo(VI) according to various studies, remains at few atomic percent level and has no marked effect on the film thickness.[21,23,29,31,34,36,38,43] The beneficial role of molybdenum on corrosion resistance has long been discussed and several effects have been proposed. They can be sorted into two main classes depending on whether Mo is proposed to mitigate passive film breakdown,[20,21,27,29,32–37,40,44] or to promote passive film repair.[19,21,28–30] However, which effect prevails remains subject to debate or several effects may combine to improve passivity and pitting corrosion resistance.

The atomic structure of passive films on stainless steel has been clarified mostly using scanning tunnelling microscopy (STM) and shown to depend on the Cr content of the alloy and on polarization conditions. For short polarization times in acid solution ($\leq$2 h), long range structural ordering (*i.e.* crystallinity) decreases with increasing Cr content of the alloy.[45–48] However, ageing under anodic polarization promotes crystallisation[13,25] in parallel to the growth of the inner barrier oxide layer, increasingly enriched in Cr(III) as a result of the dehydroxylation of the Cr(III) hydroxide layer preferentially formed in the early stages of passivation.[3,11,13,25] The observed atomic lattice is consistent with a (111) orientation of the oxygen sublattice (fcc) of the main oxide constituent (Cr_2O_3) of the barrier oxide layer.[13,25,45,46]

The nanostructure of passive films on stainless steel is much less documented despite the fact that intergranular sites joining crystalline grains of the passive film have been shown to play a key role in passivity breakdown and the initiation of localized corrosion.[49–54] It is only very recently that STM has been applied to study the morphological modifications brought at the nanometre scale by passivation in acid solution of a Fe–18Cr–13Ni stainless steel surface.[55] The data could be rationalised in terms of local dissolution of iron-rich oxide species and growth of insoluble Cr(III)-enriched species competing during the passivation process. To our knowledge, no data have been reported on the nanostructure nor the atomic structure of passive films grown on Mo bearing austenitic Fe–Cr–Ni alloys.

In the present work, time-of-flight secondary ion mass spectrometry (ToF-SIMS), XPS and STM have been combined to study the composition and the nanostructure of the passive film formed in sulfuric acid on a Fe–Cr–Ni–Mo alloy of composition close to that of AISI 316L with the objective to gain new insight on Mo effects. Well-oriented single crystal surfaces were used since a well-controlled topography is a prerequisite to resolve the nanoscale morphology of a passivated surface. A complementary *in situ* STM study will be reported separately. To our knowledge, the present work is the first of its kind combining ToF-SIMS, XPS and STM analysis on Mo bearing passivated austenitic stainless steel single crystal surfaces.

2. Experimental

A Fe–17Cr–14.5Ni–2.3Mo (wt%) single crystal oriented parallel to (100) within ±1° by Laue back-diffraction was used. The *ex situ* surface preparation was the same as that previously adopted for a Fe–18Cr–13Ni(100) single crystal.[25,55] It included mechanical polishing, electrochemical polishing and high-temperature annealing in hydrogen reducing atmosphere as previously described. For transfer to the electrochemical cell, the sample was exposed to air (~5 min) leading to the growth of an air-formed native oxide film also characterized by ToF-SIMS, XPS and STM.

The electrochemical cell used for passivation was that of the STM set up. Details on the cell and its cleaning can be found elsewhere.[56–60] All potential values given hereafter are with respect to the standard hydrogen electrode (V/SHE). The electrolyte was 0.05 M H_2SO_4(aq) prepared from ultra pure chemicals (Merck, Suprapur®) and Millipore® water (resistivity > 18 MΩ cm). It was not deaerated. Since cathodic reduction of the Cr(III)-containing oxide films irreversibly destroys the terrace topography of the substrate required for high resolution STM as shown previously on chromium,[61,62] it was avoided. Thus the present work addresses the modifications induced by transformation of the air-formed native oxide into a passive film. The surface was passivated by a potential step from the $E_{OCP} = -0.17$ V to $E_{pass} = 0.5$ V, beyond the active peak measured at −0.21 V and in the passive domain ranging up to 0.90 V. The passivation time (t_{pass}) was 2 or 20 h. After passivation, the samples were rinsed with Millipore® water, dried with filtered compressed air and transferred for analysis by ToF-SIMS, XPS or STM.

ToF-SIMS analysis was performed using a ToF-SIMS 5 spectrometer (Ion Tof – Munster, Germany) operating at about 10^{-9} mbar. A pulsed 25 keV Bi^+ primary ion source delivering 1.3 pA current over a 100 × 100 μm^2 area was employed for analysis in static SIMS conditions. Depth profiling was carried out by interlacing static analysis with sputtering using a 0.5 keV Cs^+ sputter beam giving a 12.5 nA target current over a 500 × 500 μm^2 area. Analysis was centred inside the eroded crater to avoid edge effects. The profiles were recorded with negative secondary ions which have higher yield for oxide matrices than for metallic matrices. The Ion-Spec software was used for data acquisition and processing.

A VG ESCALAB 250 spectrometer operating at about 10^{-9} mbar was used for XPS analysis. The X-ray source was an Al Kα monochromatised radiation ($h\nu$ = 1486.6 eV). Survey spectra were recorded with a pass energy of 100 eV at a step size of 1 eV and high resolution spectra of the Fe 2p, Cr 2p, Ni 2p, Mo 3d, O 1s, and C 1s core level regions were recorded with a pass energy of 20 eV at a step size of

0.1 eV. Take-off angle of the analysed photoelectrons was 90°. Data processing (curve fitting) was performed using the Avantage software provided by Thermo Electron Corporation. An iterative Shirley-type background and Lorentzian/Gaussian peak shape with a fixed ratio of 30/70 and a tail mix of 100% were used except for the Fe^0 $2p_{3/2}$ (91/8 and 11.5%), Cr^0 $2p_{3/2}$ (35/65 and 0.66%) and Mo^0 $3d_{5/2-3/2}$ (35/65 and 64.7%) metallic components. The binding energies were calibrated by setting the C 1s signal corresponding to olefinic bonds ($-CH_2-CH_2-$) at 285.0 eV.

The STM setup consisted of a PicoSPM base, a STM S scanner, a PicoScan 2100 controller, and the Picoscan software (Agilent Technologies). W tips electro eroded from 0.25 mm wire in 3 M NaOH(aq) were used. All STM images were acquired in air in the constant current topographic mode and were corrected from the background plane. No filtering was used. Complementary AFM analysis was performed with an Agilent 5500 AFM/SPM microscope operated in intermittent contact (tapping®) mode. In this mode, the cantilever supporting the tip oscillates at its resonant frequency and the feedback loop maintains the oscillation amplitude constant (constant tip–sample interaction) by adjusting the vertical position z of the scanner at each (x,y) data point of the scan. The tip used was made of n-doped silicon and had a nominal radius of curvature less than 10 nm.

3. Results and discussion

3.1. Composition studied by ToF-SIMS and XPS

Fig. 1 shows the ToF-SIMS negative ion depth profiles for the air-formed native oxide film and the passive films formed at $E_{pass} = 0.5$ V for 2 and 20 h on the Mo bearing stainless steel surface. The intensity is plotted in logarithmic scale *versus* sputtering time for the following selected secondary ions: $^{17}OH^-$, $^{18}O^-$, $^{104}Cr_2^-$, $^{112}Fe_2^-$ and $^{116}Ni_2^-$ ($^{16}O^-$ secondary ions saturate the detector). The $^{18}O^-$ ion profile is used to define the oxide film region (ending at 84% of the intensity on the plateau[63]) and the Fe_2^-, Cr_2^- and Ni_2^- ion profiles are used to define the metallic substrate region (starting at the beginning of the intensity plateaux). The intermediate region is assigned to the interfacial region between oxide film and metallic substrate. An intense OH^- signal is measured throughout the region of the oxide films, indicating in all cases a hydroxylated oxide matrix confirmed by XPS.

In Fig. 2, the profiles of secondary ions ($^{88}FeO_2^-$, $^{84}CrO_2^-$, $^{91}NiO_2^-$ and $^{128}MoO_2^-$) characteristic for oxide species present in the films are overlapped using a linear intensity scale. This allows the evolution after passivation and ageing to be seem more clearly. The colour-coded dashed lines mark the limits between oxide film and interface regions. Their respective positions do not reflect actual thickness variations of the oxide films as revealed by the thickness values obtained by XPS analysis and discussed later, possibly owing to variation of the sputtering yield. It is shown that the FeO_2^- profiles all have their maximum in the outer parts of the oxide film region. Oxidized iron is therefore preferentially located in the outer part of the films regardless of the conditions of formation of the oxide. Its intensity decreases after passivation and ageing, which is linked to the preferential dissolution of the iron oxide generated by passivation in acid solution as expected. The CrO_2^- profiles also exhibit an intensity maximum but are located deeper in the films compared to the FeO_2^- profiles. Oxidized

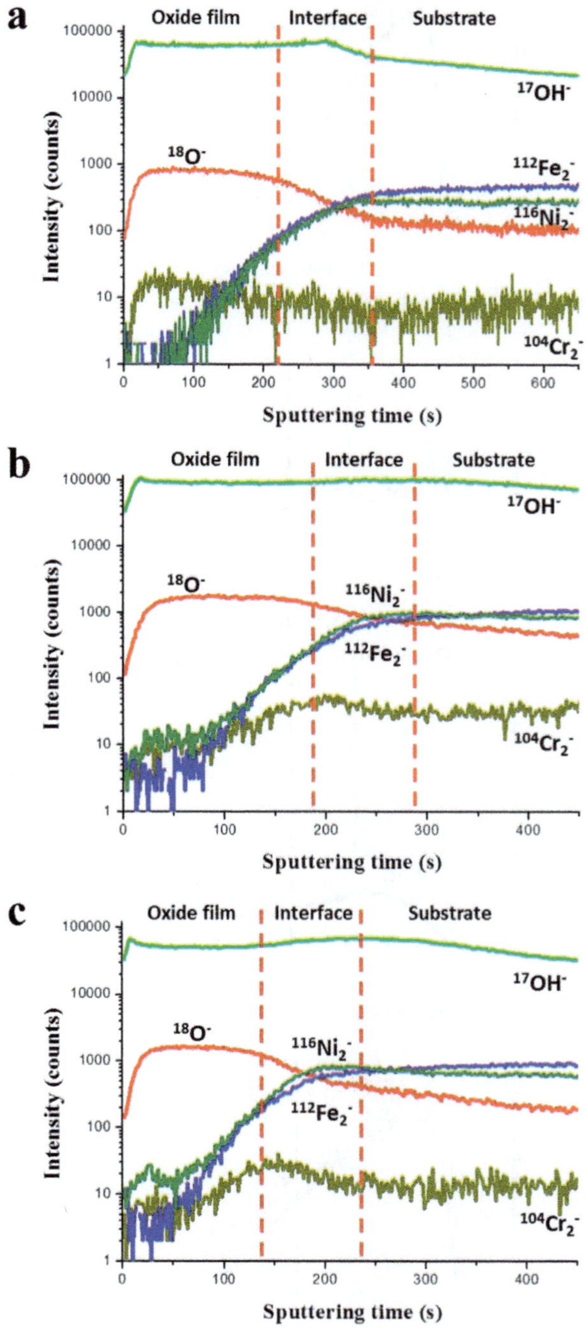

Fig. 1 ToF-SIMS negative ion depth profiles of the air-formed native oxide film (a) and passive films formed on the Fe–17Cr–14.5Ni–2.3Mo(100) single crystal stainless steel surface in 0.05 M H_2SO_4(aq) at 0.5 V/SHE for 2 h (b) and 20 h (c). Dashed lines mark the limits between oxide film, interface and substrate regions.

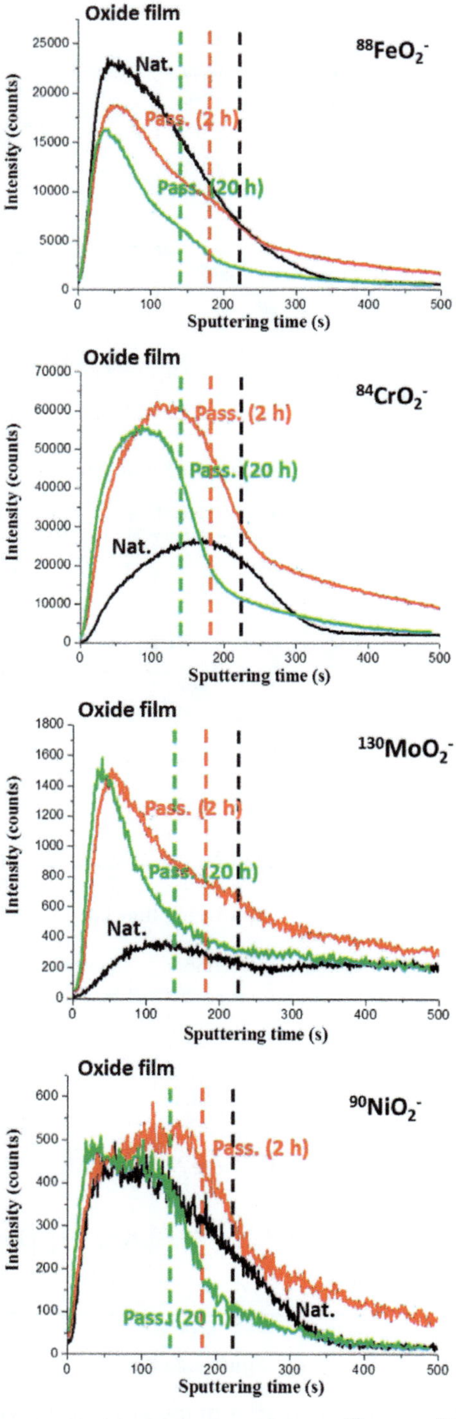

Fig. 2 Comparison of the ToF-SIMS depth profiles of $^{88}FeO_2^-$, $^{84}CrO_2^-$, $^{91}NiO_2^-$ and $^{128}MoO_2^-$ ions for the air-formed native oxide film (Nat.) and passive films formed on Fe–17Cr–14.5Ni–2.3Mo(100) single crystal stainless steel surface in 0.05 M H_2SO_4(aq) at 0.5 V/SHE for 2 h (Pass. 2 h) and 20 h (Pass. 20 h). The colour-coded dashed lines mark the limits between oxide film and interface regions.

chromium is thus preferentially concentrated in the inner part of the films. Its intensity increases significantly after passivation, confirming enrichment of chromium in the passive films.

Throughout the native oxide film, the MoO_2^- profile coincides better with the CrO_2^- profile with a maximum in the inner part of the film. However, after passivation, the MoO_2^- profile peaks in the outer part of the film and coincides with the FeO_2^- profile. It also increases significantly in intensity. This shows that Mo is incorporated in the inner part of the oxide film before passivation and in the outer part, and in larger amounts, after passivation, where oxidized iron is more concentrated.

The NiO_2^- profiles show plateaux of very low intensity throughout the region of the oxide films, showing a homogeneous distribution. The intensity of the NiO_2^- ions, much lower than those of CrO_2^-, FeO_2^- and even MoO_2^- ions, shows that Ni is present only as oxide traces in the films. Passivation and ageing do not seem to generate significant changes of concentration and distribution.

Fig. 3 shows the XP Fe 2p, Cr 2p, Ni 2p, Mo 3d and O 1s core level spectra for these samples. Binding energies (BE), full widths at half-maximum (FWHM) and relative intensities of the component peaks obtained by curve fitting are compiled in Table 1. For all samples, nickel is only detected in the metallic state since a single peak component, assigned to Ni^0 $2p_{3/2}$ [42,64] of the alloy matrix underneath the films, is observed. The traces of oxidized nickel observed by ToF-SIMS are then below the detection limit of XPS (<1 at%). In contrast, several peak components, assigned to different oxidation states, are observed for the other elements constitutive of the alloy: Fe^0 $2p_{3/2}$,[6,33,35,42] Cr^0 $2p_{3/2}$ [6,31] and the doublet Mo^0 $3d_{5/2}$–Mo^0 $3d_{3/2}$ [6,31,33,35,38,42] for the metallic matrix of the alloy below the films, and Fe^{3+} $2p_{3/2}$,[6,33,35,42] Cr^{3+} $2p_{3/2}$,[6,31,33,35] and the doublets Mo^{4+} $3d_{5/2}$–Mo^{4+} $3d_{3/2}$,[6,31,33,35,38,42] and Mo^{6+} $3d_{5/2}$–Mo^{6+} $3d_{3/2}$ [6,31,33,35,38,42] for the hydroxylated oxide matrix of the films. Satellite peaks [Fe^{sat} $2p_{3/2}$,[6,33,35,42] Cr^{sat} $2p_{3/2}$ [6,31,33,35]] are also observed for the more intense oxidized iron and chromium. The O 1s core level reveals three components, O^{2-}, OH^- and H_2O, assigned to the oxide, hydroxide and water ligands of the films, respectively.[31,33,42]

The intensity ratio of the hydroxide to oxide components ($I[OH^-]/I[O^{2-}]$) markedly increases from 0.36 before passivation to 1.03 and 0.91 after passivation for 2 and 20 h, respectively, showing that the oxide matrix of the passive films is markedly more hydroxylated than that of the native oxide film and with no in-depth significant variations as shown by the ToF-SIMS profiles (Fig. 1). For the passivated samples, an additional $Cr^{3+(hyd)}$ $2p_{3/2}$ component assigned to Cr(III) ions bonded to an hydroxylated oxide matrix[33,35,42] was used for curve fitting the Cr 2p spectra. In contrast, adding such a component was not necessary for peak fitting the Fe 2p and Mo 3d spectra. Still, it appears reasonable to consider that the fitted Fe^{3+} $2p_{3/2}$, Mo^{4+} 3d and Mo^{6+} 3d components also include cations bonded to hydroxide ligands. For all samples, no Fe^{2+} $2p_{3/2}$ component at 708.3 ± 0.2 eV [65–67] was necessary to fit Fe 2p spectra, showing that iron cations are essentially in the 3+ oxidation state. However, it cannot be excluded that a small fraction of them are in the 2+ oxidation state. The slight decrease of the $I[OH^-]/I[O^{2-}]$ ratio for the 20 h passivated sample reflects the process of dehydroxylation generated by ageing in the passive state, in agreement with previous data.[11,13,25]

The intensity ratio $I[Cr^{3+(ox)(hyd)}]/I[Fe^{3+(ox)}]$ increases from 0.55 for the native oxide film to 1.13 (2 h sample) and 1.63 (20 h sample) for the passive films,

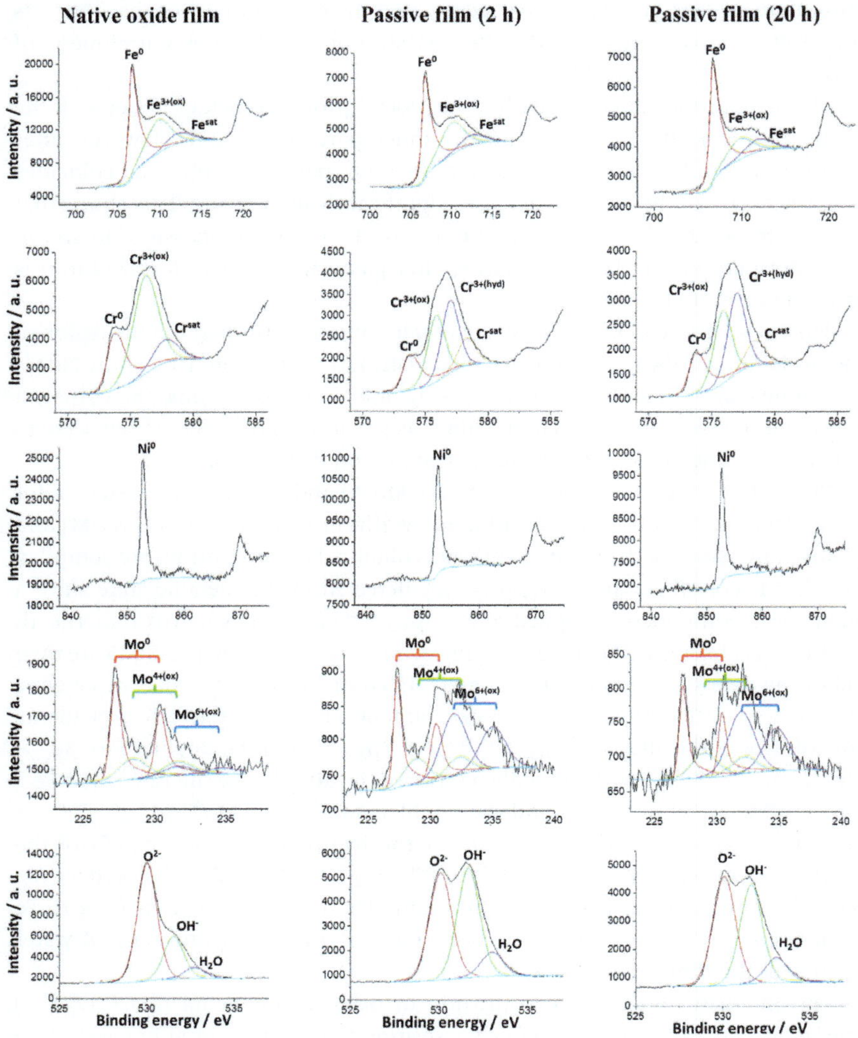

Fig. 3 XP core level Fe 2p, Cr 2p, Ni 2p, Mo 3d and O 1s spectra for the air-formed native oxide film and passive films formed on Fe–17Cr–14.5Ni–2.3Mo(100) single crystal stainless steel surface in 0.05 M H_2SO_4(aq) at 0.5 V/SHE for 2 and 20 h.

highlighting the enrichment in oxidized chromium in the films after passivation and ageing in the passive state due to preferential dissolution of the iron oxide in acid medium. This confirms the trend measured by ToF-SIMS and is in agreement with previous data.[11,13,25] The intensity ratio $I[Mo^{(ox)}]/I[Mo^0]$, increasing from 0.65 before passivation to 1.95 and 2.32 after passivation for 2 and 20 h, respectively, is indicative of the Mo enrichment in the oxide film after passivation. Mo enrichment is also obtained by ageing from 2 to 20 h, but it is much less pronounced. The intensity ratio $I[Mo^{6+(ox)}]/I[Mo^{4+(ox)}]$ also increases, from 0.44 before passivation to 2.62 and 2.60 after passivation for 2 and 20 h, respectively. Molybdenum is thus preferentially present in the 4+ oxidation state in the native oxide film and in the 6+ oxidation state after passivation, in agreement with previous work.[29,31,42]

Table 1 Binding energy (BE), full-width at half-maximum (FWHM) and relative intensity (RI) of the XP core level components obtained by peak fitting for the native and passive oxide films formed on the Fe–17Cr–14.5Ni–2.3Mo(100) single crystal stainless steel surface. The relative intensity is normalised to the most intense component of the core level

	Native oxide film			Passive film (2 h)			Passive film (20 h)		
	BE (eV)	FWHM (eV)	RI	BE (eV)	FWHM (eV)	RI	BE (eV)	FWHM (eV)	RI
Fe^0 $2p_{3/2}$	706.7	0.8	1.00	706.7	0.8	1.00	706.7	0.8	1.00
$Fe^{3+(ox)}$ $2p_{3/2}$	709.8	3.5	0.67	710.0	3.5	0.69	709.8	3.4	0.38
Fe^{sat} $2p_{3/2}$	712.3	3.5	0.20	712.5	3.5	0.21	711.8	3.4	0.22
Cr^0 $2p_{3/2}$	573.7	1.3	0.46	573.7	1.3	0.55	573.7	1.3	0.56
$Cr^{3+(ox)}$ $2p_{3/2}$	576.2	2.4	1.00	575.9	1.7	0.92	575.9	1.7	0.90
$Cr^{3+(hyd)}$ $2p_{3/2}$	—	—	—	577.0	1.7	1.00	577.0	1.7	1.00
Cr^{sat} $2p_{3/2}$	577.8	2.4	0.23	578.3	1.7	0.37	578.4	1.7	0.30
Mo^0 $3d_{5/2}$	227.2	0.8	1.00	227.3	0.7	0.71	227.3	0.7	0.60
Mo^0 $3d_{3/2}$	230.4	0.8	0.67	230.4	1.0	0.47	230.4	0.7	0.40
$Mo^{4+(ox)}$ $3d_{5/2}$	228.5	2.5	0.45	228.9	2.3	0.38	229.0	2.5	0.39
$Mo^{4+(ox)}$ $3d_{3/2}$	231.6	2.5	0.30	232.2	2.3	0.26	232.2	2.5	0.26
$Mo^{6+(ox)}$ $3d_{5/2}$	231.6	2.5	0.20	231.8	2.5	1.00	232.0	2.5	1.00
$Mo^{6+(ox)}$ $3d_{3/2}$	234.7	2.5	0.13	235.0	2.5	0.67	234.8	2.5	0.67
Ni^0 $2p_{3/2}$	852.8	1.0	1.00	852.7	1.1	1.00	852.7	1.1	1.00
O^{2-} 1s	530.0	1.4	1.00	530.1	1.5	0.98	530.1	1.5	1.00
OH^- 1s	531.5	1.4	0.36	531.7	1.5	1.00	531.6	1.5	0.91
H_2O 1s	532.7	1.4	0.09	533.0	1.5	0.21	533.1	1.5	0.22
C–C 1s	285.0	1.3	1.00	284.8	1.5	1.00	284.7	1.5	1.00

This ratio remains constant during ageing. These results show that the passivation in acid medium has a marked effect on the Mo content in the film and the fractions of cations Mo^{6+} and $Mo^{4+(ox)}$. Ageing produces a weaker effect.

Based on these ToF-SIMS and XPS data, the duplex model presented in Fig. 4 was used to calculate the composition and thickness of the oxide films and the composition of the alloy underneath. For all films, the model assumes an essentially oxide/hydroxide matrix with partition of iron and chromium in the outer and inner layers, respectively. Molybdenum is considered to be homogeneously distributed in the inner layer for the native oxide film and in the outer layer of the passive films. This duplex model differs from those previously proposed for Fe–Cr–Ni–Mo alloys passivated in acid and neutral solutions for

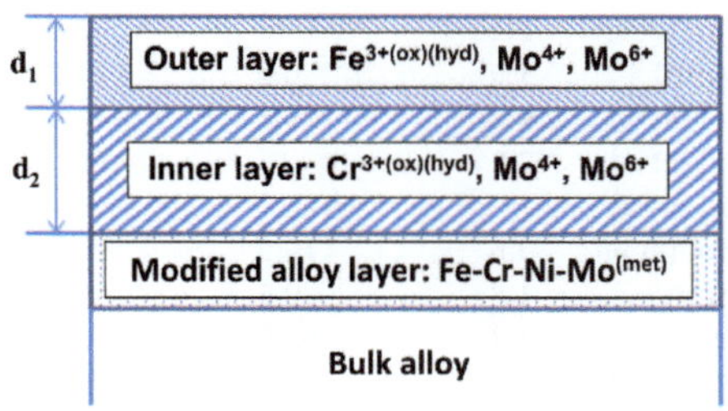

Fig. 4 Duplex model for calculating the chemical composition of the oxide films formed on the Fe–17Cr–14.5Ni–2.3Mo(100) single crystal stainless steel surface.

which the outer layer consisted of chromium hydroxide and the inner layer of mixed iron/chromium oxide.[21,34,38,42] It is however consistent with previous Auger electron spectroscopy depth profiling data showing maximum Cr enrichment in the inner part of the film.[6,18,28] Clearly, our ToF-SIMS data do not support a hydroxide/oxide partition as no peak is observed in the outer region of the film for the OH^-, CrO_2^- and also CrO^- (not shown) ion profiles. The low OH^- peak at the beginning of the ToF-SIMS profiles and the low intensity H_2O 1s XP component suggest a slight surface hydration. It is neglected in the model because of the low $I[H_2O]/I[OH^-] + I[O^{2-}]$ XP ratio (<10%). Also, an abrupt interface between the outer and inner layers of the films appears as a simplification according to the ToF-SIMS data, and decreasing and increasing concentrations of iron and chromium oxides, respectively, could be a more realistic description.

The thickness of each layer of the oxide films and the concentration in Fe, Cr, Ni and Mo of the outer and inner layer of the films and of the modified alloy underneath the films can be calculated from the intensities of the Fe^{3+} $2p_{3/2}$, Cr^{3+} $2p_{3/2}$ ($I[Cr^{3+(ox)}\ 2p_{3/2}] + I[Cr^{3+(hyd)}\ 2p_{3/2}]$), Mo^{4+} $3d_{3/2}$, Mo^{6+} $3d_{3/2}$, Fe^0 $2p_{3/2}$, Cr^0 $2p_{3/2}$, Ni^0 $2p_{3/2}$ and Mo^0 $3d_{3/2}$ component peaks based on the exponential attenuation of the photoelectron intensity with increasing escape depth from the surface. The procedure uses intensity ratios and is described in detail elsewhere.[68] The results are compiled in Table 2.

The total thickness is found to be 1.8–1.9 nm for the passive films, in agreement with previous data (1.4–1.7 nm) for Mo implanted stainless steel after passivation in sulfuric acid.[42] The inner layer is thicker than the outer layer. The total film thickness does not vary markedly after passivation, however with slight thickening of the inner layer and thinning of the outer layer. The ageing process has no significant effect on the thickness. Chromium is enriched in all studied films and this enrichment is more pronounced after passivation, due to the preferential dissolution of iron oxide in acid solutions.[29,38,42,43,69] Cr(III) ions

Table 2 Thickness and chemical composition of the duplex native and passive oxide films on the Fe–17Cr–14.5Ni–2.3Mo(100) single crystal stainless steel surface as obtained from XPS analysis

		Global film	Outer layer	Inner layer	Modified alloy
Native oxide film	*d* (nm)	1.7	0.7	1.0	—
	[Fe] (at%)	41.2	100	—	60.8
	[Cr] (at%)	56.7	—	96.4	13.1
	[Ni] (at%)	—	—	—	24.6
	[Mo] (at%)	2.1	—	3.6	1.5
Passive film (2 h)	*d* (nm)	1.9	0.6	1.3	—
	[Fe] (at%)	29.2	89.1	—	51.0
	[Cr] (at%)	67.2	—	100	14.7
	[Ni] (at%)	—	—	—	32.8
	[Mo] (at%)	3.6	10.9	—	1.5
Passive film (20 h)	*d* (nm)	1.8	0.5	1.3	—
	[Fe] (at%)	23.3	83.9	—	50.5
	[Cr] (at%)	72.2	—	100	15.1
	[Ni] (at%)	—	—	—	33.1
	[Mo] (at%)	4.5	16.1	—	1.3

represent 67% and 72% of the metal cations for the passive films formed for 2 and 20 h, respectively. Nickel oxide, observed by ToF-SIMS, is present only as traces not detected by XPS in this study, in agreement with previous work.[29,42] As a consequence, the alloy underneath the passive film is slightly depleted in metallic Cr (~15 at% for the passivated surfaces as compared to ~18 at% in the bulk alloy) and strongly enriched in metallic Ni (~33 at% for the passivated surfaces as compared to ~14 at% in the bulk alloy), also in agreement with previous work.[29,42] This enrichment in Ni is more pronounced for the passive films than for the native oxide film, consistently with the preferential dissolution of iron upon passivation. Ageing in the passive state has little effect on the Cr enrichment in the passive film and no significant effect on the enrichment in Ni in the modified alloy region below the films.

Concerning molybdenum, although the concentration is high, about 11 and 16% of metal cations in the outer layer of the passive films after 2 and 20 h, respectively, it is only about 4% if referred to the whole film. For the native oxide film in which Mo is concentrated in the inner layer, this value is 2.1%. In the modified alloy layers underneath the films, the Mo concentration (1.3–1.5 at%) is not significantly different from that in the bulk alloy (1.3 at%). We can therefore conclude that passivation causes Mo enrichment, concentrated in the outer part of the film and mostly in the 6+ oxidation state as discussed earlier, in agreement with previous work in the presence of molybdates.[29,31,42] Ageing in the passive state causes further enrichment of the outer layer in Mo, most likely due to preferential dissolution of iron oxide in the passive state. However, our data show that the passivation-induced Mo enrichment in the oxide film does not modify significantly the Mo concentration in the metallic layers underneath. This is in contrast with previous studies[21,31,38] that reported Mo enrichment below the passive film in a polycrystalline alloy (on which more dissolution may have taken place during passivation).

3.2. Nanoscale morphology studied by STM and AFM

Fig. 5 shows STM topographic data obtained at the nanometre scale on the Fe–17Cr–14.5Ni–2.3Mo(100) single crystal stainless steel surface covered by an air-formed native oxide film. The observed topography is characterized by a succession of steps and terraces. Terraces, up to hundreds of nanometres in width, alternate with series of narrower terraces from 30 to 54 nm in width. This terrace topography is that of the substrate. The steps separating the terraces result from the residual disorientation of the surface from the nominal crystallographic orientation (100) obtained only on the terraces. The step height ranges from 0.9 to 1.3 nm. Given the reticular distance of two successive (100) planes of the alloy (0.18 nm from bulk values), the measured steps are multi-atomic [from 5 to 7 (100) planes high].

Both on the wider (Fig. 5b) and narrower (Fig. 5c) terraces of the alloy, the presence of three main surface levels is clearly evidenced. It is confirmed by AFM images (not shown). The corresponding top (T), medium (M) and bottom (B) topographic levels are marked in red, green and blue, respectively, in Fig. 5b and c and in the extracted line profile in Fig. 5d. Table 3 compiles the height difference (ΔZ) values measured between these three main surface levels by STM. It also includes values obtained by AFM. The very good agreement between the STM and

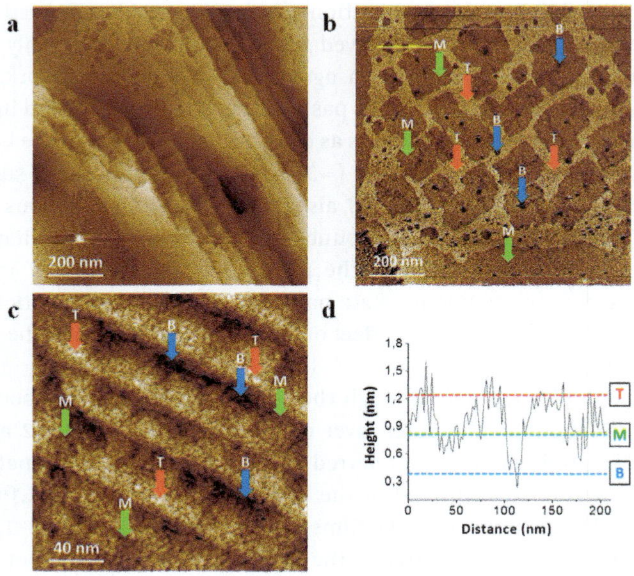

Fig. 5 STM topographic data for the Fe–17Cr–14.5Ni–2.3Mo(100) stainless steel surface covered by the air-formed native oxide film (a) top view image: Z range $\Delta Z = 8$ nm, bias voltage $V_t = -1.6$ V, tunnel current $I = 0.5$ nA. (b) Top view image: $\Delta Z = 2$ nm, $V_t = -1.7$ V, $I = 0.5$ nA. (c) Top view image: $\Delta Z = 2$ nm, $V_t = -1.6$ V, $I = 0.5$ nA. (d) Height profile along the segment in (b). The three main topographic levels are marked B (bottom), M (medium) and T (top) symbols.

AFM values confirms the topographic origin of the surface levels and allows ruling out any significant artefact in the STM measurement caused by electronic effect. The height difference between the topmost (T) and lowest (B) surface levels ($\Delta Z_{T-B} = 1.02 \pm 0.20$ nm) is significantly lower than the equivalent thickness of the native oxide film measured by XPS (1.7 nm). It appears thus consistent with a native oxide film fully covering the alloy substrate but presenting local deviations (of up to about $\pm$ 0.5 nm as inferred from the ΔZ_{T-B} value of about 1 nm) from an average thickness value of 1.7 nm. The formation of three main topographic levels was also observed on a Fe–18Cr–13Ni(100)[55] surface prepared under the same conditions, and ΔZ values are listed in Table 3 for comparison.

Fig. 5c evidences the nanogranular morphology of the air-formed oxide film. Grains, better resolved by STM than by AFM, are observed on each of the three main surface levels of the oxide film, *i.e.* independently of the local thickness. Their lateral grain dimension is 5.29 ± 0.86 nm and the apparent depth measured in the boundaries separating the grains is 0.35 ± 0.05 nm as measured by STM (Table 3, the AFM resolution does not allow extraction of reliable values of these parameters). The apparent depth measured between the grains is consistent with full coverage of the substrate by the oxide, even in the thinnest ($\sim$1.2 nm) regions of the films. This nanogranular morphology is assigned to the formation of the film by a nucleation and growth mechanism, with the small dimensions of the grains being a direct consequence of the high density of oxide nuclei formed by exposure at ambient pressure and room temperature. Table 3 shows no significant variation of the dimensions of the grains owing to the presence of Mo in the

Table 3 Height differences between bottom (B), medium (M) and top (T) topographic surface levels (ΔZ_{M-T}, ΔZ_{B-M} and ΔZ_{B-T}), grain lateral dimension (L_G) and intergranular depth (ΔZ_{IG}) for the native oxide film and passive films grown on the Fe–17Cr–14.5Ni–2.3Mo(100) single crystal stainless steel surface, and comparison with data for the Fe–18Cr–13Ni(100) surface

Sample	Technique	ΔZ_{T-M} (nm)	ΔZ_{M-B} (nm)	ΔZ_{T-B} (nm)	L_G (nm)	ΔZ_{IG} (nm)	Ref.
Native oxide on FeCrNiMo	STM	0.49 ± 0.06	0.55 ± 0.12	1.02 ± 0.20	5.29 ± 0.86	0.35 ± 0.05	This work
	AFM	0.61 ± 0.07	0.60 ± 0.08	1.49 ± 0.15	—	—	This work
Passive film on FeCrNiMo ($E_{pass} = 0.50$ V; $t_{pass} = 2$ h)	STM	0.97 ± 0.20	1.22 ± 0.17	2.27 ± 0.25	11.49 ± 2.56	0.54 ± 0.15	This work
	AFM	0.92 ± 0.10	0.96 ± 0.13	2.00 ± 0.30	—	—	This work
Passive film on FeCrNiMo ($E_{pass} = 0.50$ V; $t_{pass} = 20$ h)	STM	1.17 ± 0.16	1.30 ± 0.21	2.14 ± 0.22	8.73 ± 2.86	0.51 ± 0.18	This work
	AFM	0.90 ± 0.13	0.97 ± 0.17	2.06 ± 0.26	—	—	This work
Native oxide on FeCrNi	STM	0.36 ± 0.05	0.72 ± 0.09	1.21 ± 0.09	5.43 ± 0.71	0.25 ± 0.06	55
Passive film on FeCrNi ($E_{pass} = 0.50$ V; $t_{pass} = 3$ h)	STM	1.09 ± 0.23	0.64 ± 0.15	1.97 ± 0.20	9.57 ± 1.73	0.60 ± 0.23	55
Passive film on FeCrNi ($E_{pass} = 0.50$ V; $t_{pass} = 20$ h)	STM	1.03 ± 0.20	1.06 ± 0.19	1.95 ± 0.26	8.24 ± 2.00	0.49 ± 0.17	This work

alloy, showing again that its presence in small amounts does not affect the nanoscopic morphological characteristics of the native oxide film.

Quite interestingly, one also notices in Fig. 5c that the step edges are nearly saturated with grains associated with the topmost surface level of the oxide film and that the terraces are covered with grains preferentially associated with the medium and lowest surface levels. This site differentiation is indicative of a relation between the nucleation and growth mechanism of the oxide film and its local thickness. The XPS data in Table 2 show that the native oxide is enriched in Cr(III) as a result of the preferential oxidation of chromium of the prepared alloy surface. It is then likely that the oxide formed at step edges and in the T areas is locally not only thicker, as directly observed by STM, but also more enriched in Cr(III). For balance, the locally thinner oxide formed on the terraces in the M and B areas would be then less enriched in Cr(III). Crystallographic defect sites of the alloy surface, such as step edges, would thus promote nucleation of the oxide film, as a result of a decreased energy barrier for oxidation of both iron and chromium. They would also promote local thickening and Cr(III) enrichment in the oxide as a result of faster growth preferentially consuming chromium by surface diffusion also from the neighbouring areas. As a result, the neighbouring areas would be less Cr(III)-enriched and thinner. The B areas of the oxide film would correspond to the thinnest and least enriched in Cr(III) since they have provided chromium for the enrichment of both the M and T areas. On the terraces, such promoting effects on local thickening and Cr(III) enrichment would be less pronounced however with other crystallographic features playing a role as suggested by the nearly square arrangements adopted by the T areas (Fig. 5b). Comparison with the data for the non Mo bearing alloy shows no significant effect of molybdenum on the formation of the three topographic surface levels (separated by similar ΔZ values if one considers dispersion, Table 3) and their site differentiation. Thus, it appears that the nucleation and growth mechanism and resulting local thickness variation of the native oxide film are not significantly modified by the presence of a small amount (3.6% cation fraction) of Mo in the inner Cr-rich layer of the native oxide film.

No crystalline lattice could be measured on this surface at higher resolution, suggesting the growth of an amorphous air-formed oxide film, as on the Fe–18Cr–13Ni(100) surface.[55]

Fig. 6 shows STM topographic data for the Fe–17Cr–14.5Ni–2.3Mo(100) passivated for 2 h. It is evidenced that anodic polarization in the passive range does not destroy the typical terrace topography of the substrate produced by the surface preparation, and modifications must be appreciated more locally. Three main surface levels are still present (as marked in Fig. 6a and shown by the line profile in Fig. 6c), however with significant increases of the ΔZ values separating them (Table 3) also measured by AFM. Two phenomena compete during passivation in acid solution: (i) transient dissolution, predominantly of soluble Fe(II) species, either pre-existing on the surface oxide film or produced by anodic oxidation and consuming the alloy, which contributes to lower the apparent surface level, and (ii) growth, predominantly of insoluble Cr(III) and Mo(VI) species produced by anodic oxidation, which contributes to raise the apparent surface level. Both contribute to an increase in the ΔZ values between the local surface levels. XPS shows no significant variation of the total thickness of the surface oxide film after passivation and thus, the observed variations must be dominated

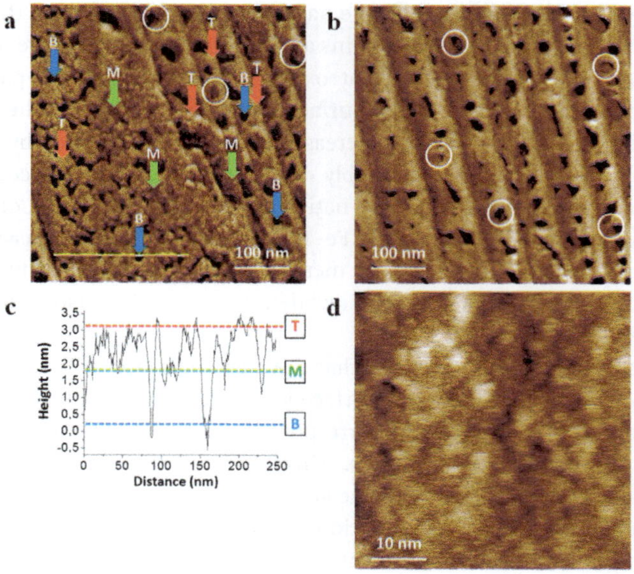

Fig. 6 STM topographic data for the Fe–17Cr–14.5Ni–2.3Mo(100) stainless steel surface passivated in 0.05 M H_2SO_4(aq) at 0.5 V/SHE for 2 h (a) top view image: Z range $\Delta Z = 5$ nm, bias voltage $V_t = -1.7$ V, tunnel current $I = 0.5$ nA. (b) Top view image: $\Delta Z = 5$ nm, $V_t = -1.7$ V, $I = 0.5$ nA. (c) Height profile along the segment in (a). (d) Top view image: $\Delta Z = 4$ nm, $V_t = -1.7$ V, $I = 0.5$ nA. The three main topographic levels are marked B (bottom), M (medium) and T (top) symbols. Circles mark depressions partially filled by nanostructures.

by passivation-induced transient dissolution. It is expected that transient dissolution will vary depending on the local thickness and composition of the pre-formed native oxide film. In the lowest surface level areas, dissolution of the alloy should be promoted since the film is thinner and dissolution of the oxide should also be promoted since the film is locally less Cr(III) enriched as discussed above. Conversely, in the topmost surface level areas where the film is the thickest and possibly the most Cr(III)-rich, dissolution of both the alloy and the oxide should be less pronounced. This mechanism would account for the observed increase of the difference between the surface topographic levels after passivation. The role of Mo on these passivation-induced local modifications of the stainless steel surface is difficult to assess. Table 3 compiles similar data for Mo-free Fe–18Cr–13Ni(100) surface passivated in similar conditions (3 h at 0.5 V) in 0.5 M H_2SO_4. The ΔZ values also increase after passivation and are similar to those measured for the Mo bearing surface, showing no significant effect of the presence of Mo. Only the ΔZ_{M-B} value varies differently, which remains unexplained at present. Mo is known for decreasing dissolution in the active state. This effect is not confirmed by the ΔZ values that are the same after 2–3 h of passivation. However, the present data do not strictly address the active–passive transition since they are obtained on surfaces initially covered by the air-formed native oxide. Data obtained after shorter polarisation time and/or at lower potential (*e.g.* in the pre-passive region) could be more conclusive on the effect of Mo on the passivation-induced local modifications of the stainless steel surface associated with the competing mechanisms of transient dissolution and oxide growth.

The image in Fig. 6c also reveals the nanogranular morphology of the passive film. The lateral dimensions of the grains are larger than for the native oxide film (Table 3). This increase suggests that oxide growth generated by passivation occurs both perpendicularly to the surface so as to compensate the effect of transient dissolution (no thickness increase of the film is measured by XPS) and laterally. The lateral growth is probably due to a phenomenon of coalescence accompanying the transformation of native oxide to passive film. Comparison with data obtained on the Mo-free Fe–18Cr–13Ni(100) surface passivated in similar conditions, that also show an increase of the grain lateral dimension,[55] points to no significant effect of molybdenum given the dispersion of the measurements.

One also notices in Fig. 6a and b that the step edges are still saturated with grains associated with the topmost surface level of the passive film and that, on the narrower terraces, depressions corresponding to the lowest surface level are located in the centre of the terraces. This confirms site differentiation after passivation-induced modifications. The absence of depressions at the step edges suggests that local transient dissolution, accompanying the formation of the passive film as discussed above, was attenuated, or even blocked, at these sites. This is consistent with the native oxide-covered alloy surface being more enriched in chromium along the step edges and less in the terraces in the vicinity of step edges as discussed above. A similar site differentiation was observed for the passivated Mo-free Fe–18Cr–13Ni(100) surface,[55] and it appears not affected by the presence of Mo.

Fig. 6 also reveals an interesting feature of the passivated surface. The depressions (some are marked by circles in Fig. 6a and b) observed on the narrower terraces appear partially filled by nanostructures. In some cases, the depressions even appear fully filled and only their contours remain identified. The presence of such nanostructures suggests late growth of the passive film in the sites where transient dissolution was stronger or redeposition of dissolved corrosion products. The formation of these structures tends to homogenize the topography of the passivated surface, healing nanostructural heterogeneities (*i.e.* local depressions) created by transient dissolution. Such nanostructures were not observed on the passivated Mo-free Fe–18Cr–13Ni(100) surface,[55] and it is suggested that their formation is promoted by the presence of Mo.

Fig. 7 shows STM topographic data for the Fe–17Cr–14.5Ni–2.3Mo(100) passivated for 20 h. All nanoscopic features of the passivated surface are confirmed: subsisting terrace topography of the substrate (Fig. 7a); three main topographic levels on the terraces (Fig. 7b), site-specific formation of depressions (Fig. 7a) and nanogranular morphology of the passive film (Fig. 7c). The values of the height differences between topographic surface levels and grain dimensions compiled in Table 3 show no significant variation caused by ageing in the passive state from 2 to 20 h. Also, comparison with data for passivated Mo-free Fe–18Cr–13Ni(100) surface[55] shows no marked effect of the presence of Mo on these nanoscopic features. The presence of nanostructures (marked by circles in Fig. 7a) filling the depressions created by passivation-induced transient dissolution is confirmed for the Mo bearing alloy surface.

No crystalline lattice could be measured on these passivated surfaces at higher resolution, either after 2 or 20 h of passivation. Fig. 6d and 7c show two examples of higher resolution images that display no sign of crystallisation of the oxide films,

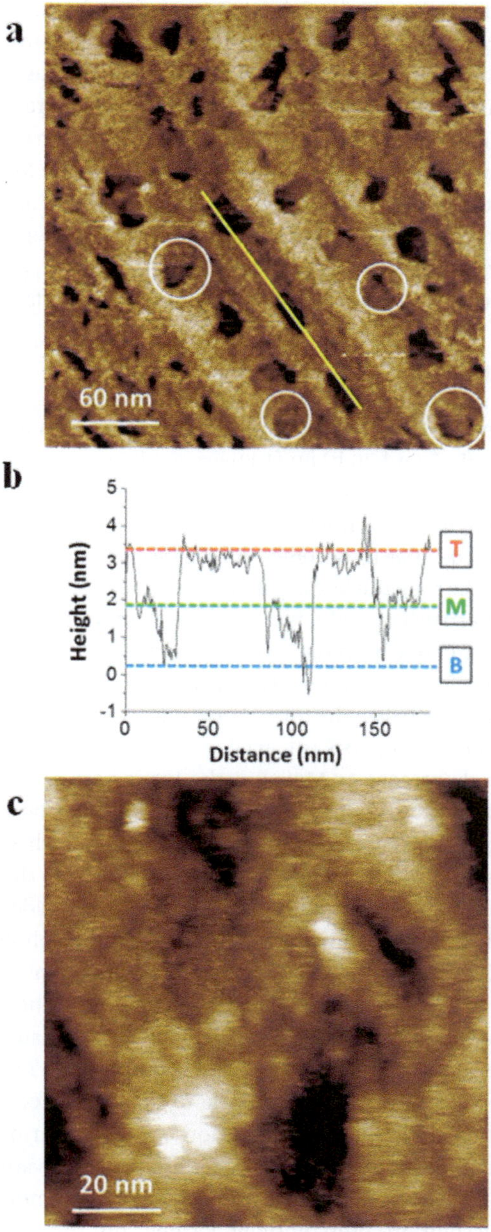

Fig. 7 STM topographic data for the Fe–17Cr–14.5Ni–2.3Mo(100) stainless steel surface passivated in 0.05 M H_2SO_4(aq) at 0.5 V/SHE for 20 hours (a) top view image: Z range $\Delta Z =$ 4 nm, bias voltage $V_t = -1.9$ V, tunnel current $I = 0.5$ nA. (b) Height profile along the segment in (a). (c) Top view image: $\Delta Z = 4$ nm, $V_t = -3.0$ V, $I = 0.5$ nA. The three main topographic levels are marked B (bottom), M (medium) and T (top) symbols. Circles mark depressions partially filled by nanostructures.

even after ageing for 20 h. Typically, when a crystalline oxide film forms, images obtained at this magnification display the terrace topography associated with the termination of the oxide lattice.[13,25,48,55–58,62] This is not observed in the presented images and could not be obtained in the present study, in contrast with data obtained for Mo-free austenitic stainless steel surfaces studied in similar conditions of passivation and measurement.[13,25,55] This suggests that the passive films produced in these conditions on the Mo bearing stainless steel surface retain an amorphous structure, at least in their outer parts, in the conditions tested in this work, including after 20 h ageing under anodic polarisation. It is possible that Mo, concentrated in the outer parts of the passive films, stabilizes the formation of an amorphous outer layer of molybdate-containing Fe-rich oxide. This amorphous outer layer could prevent the observation by STM of the inner Cr-rich oxide layer, possibly crystallized as usually observed after ageing on Mo-free stainless steel surfaces.[13,25,55]

4. Conclusion

The oxide films grown in air or by anodic polarisation in sulfuric acid (0.05 M H_2SO_4(aq); 0.5 V/SHE; 2 and 20 h) on (100)-oriented Fe–17Cr–14.5Ni–2.3Mo single crystal stainless steel surfaces can be described by a duplex hydroxylated oxide matrix with a strongly Cr(III)-enriched inner layer and a strongly Fe(III)-enriched outer layer. The oxide matrix becomes markedly hydroxylated after passivation ($[OH^-]/[O^{2-}] \sim 1$). The total thickness (1.8–1.9 nm) does not markedly increase after passivation, however with a slight increase of the inner part compensated by a decrease of the outer part. Chromium is increasingly enriched, up to 67 and 72% of metal cations, by preferential iron oxide dissolution upon passivation and ageing, and depleted in the alloy below the film. Nickel, present in the films as oxide traces below the detection limit of XPS (<1 at%), is enriched in the alloy underneath. Molybdenum, mostly present as Mo(IV) in the inner part of the air-formed film, becomes concentrated and increasingly (up to 16% of metal cations) enriched mostly as Mo(VI) in the outer part of the passive film with ageing promoting enrichment. No significant Mo enrichment is observed in the alloy below the passive film produced by anodic polarization of the native oxide covered surface.

The air-formed native oxide film has a nanogranular morphology and develops 3 main topographic surface levels reflecting local variations ($\pm$0.5 nm) of the thickness (1.7 nm in average), assigned to local variations of the Cr(III) enrichment promoted by substrate site effects. Passivation increases the lateral dimensions of the nanograins, as for a Mo-free stainless steel surface. It also increases the height difference between surface levels, also owing to the local effect of thickness and Cr(III) enrichment on the transient dissolution occurring during passivation but with no marked effect of Mo. Site specific preferential passivation takes place at the oxide-covered step edges of the alloy surface, most likely more Cr(III)-enriched, with transient dissolution preferentially located on the less Cr(III)-enriched terraces, similar to that observed in the absence of Mo. Local nanostructures, possibly Mo-containing, growing in the depression formed by local transient dissolution and healing them, appear as a specific effect observed in the presence of molybdenum. Another Mo effect, observed even after 20 h of passivation, is to prevent crystallisation at least in the outer part of the passive film where it is concentrated mostly as Mo(VI) (*i.e.* molybdate) species.

References

1 M. Seo and N. Sato, *Surf. Sci.*, 1979, **86**, 601.

2 I. Olefjord and B. Brox, in *Passivity of Metals and Semiconductors*, ed. M. Froment, Elsevier, Amsterdam, 1983, p. 561.
3 G. Hultquist, M. Seo, T. Leitner, C. Leygraf and N. Sato, *Corros. Sci.*, 1987, **27**, 937.
4 D. F. Mitchell and M. Graham, *Surf. Interface Anal.*, 1987, **10**, 259.
5 S. Mischler, H. J. Mathieu and D. Landolt, *Surf. Interface Anal.*, 1988, **11**, 182.
6 P. Marcus and I. Olefjord, *Corros. Sci.*, 1988, **28**, 589.
7 C. Calinski and H.-H. Strehblow, *J. Electrochem. Soc.*, 1989, **136**, 1328.
8 R. Kirchheim, B. Heine, H. Fischmeister, S. Hofmann, H. Knote and U. Stolz, *Corros. Sci.*, 1989, **29**, 899.
9 J. E. Castle and J. H. Qiu, *Corros. Sci.*, 1989, **29**, 591.
10 C. Hubschmid, H. J. Mathieu and D. Landolt, *Proceedings of the 12th International Corrosion Congress*, NACE, Houston, 1993, p. 3913.
11 W. P. Yang, D. Costa and P. Marcus, *J. Electrochem. Soc.*, 1994, **141**, 2669.
12 S. Haupt and H.-H. Strehblow, *Corros. Sci.*, 1995, **37**, 43.
13 V. Maurice, W. Yang and P. Marcus, *J. Electrochem. Soc.*, 1996, **143**, 1182.
14 L. J. Oblonsky, M. P. Ryan and H. S. Isaacs, *J. Electrochem. Soc.*, 1998, **145**, 1922.
15 D. Hamm, K. Ogle, C.-O. A. Olsson, S. Weber and D. Landolt, *Corros. Sci.*, 2002, **44**, 1443.
16 P. Keller and H.-H. Strehblow, *Corros. Sci.*, 2004, **46**, 1939.
17 C.-O. A. Olsson and D. Landolt, *Electrochim. Acta*, 2003, **48**, 1093.
18 S. Mischler, A. Vogel, H. J. Mathieu and D. Landolt, *Corros. Sci.*, 1991, **32**, 925–944.
19 K. Hashimoto, K. Asami and K. Teramoto, *Corros. Sci.*, 1979, **19**, 3–14.
20 M.-W. Tan, E. Akiyama, A. Kawashima, K. Asami and K. Hashimoto, *Corros. Sci.*, 1995, **37**, 1289–1301.
21 I. Olefjord and B.-O. Elfström, *Corrosion*, 1982, **38**, 46.
22 B. Brox and I. Olefjord, *Surf. Interface Anal.*, 1988, **13**, 3.
23 E. De Vito and P. Marcus, *Surf. Interface Anal.*, 1992, **19**, 403.
24 C.-O. A. Olsson and S. E. Hörnström, *Corros. Sci.*, 1994, **36**, 141.
25 V. Maurice, W. Yang and P. Marcus, *J. Electrochem. Soc.*, 1998, **145**, 909.
26 T. Yamamoto, K. Fushimi, M. Seo, S. Tsuri, T. Adachi and H. Habazaki, *Corros. Sci.*, 2009, **51**, 1545.
27 K. Sugimoto and Y. Sawada, *Corros. Sci.*, 1977, **17**, 425–445.
28 H. Ogawa, H. Omata, I. Itoh and H. Okada, *Corrosion*, 1978, **34**, 52–60.
29 I. Olefjord, *Mater. Sci. Eng.*, 1980, **42**, 161–171.
30 W. Yang, R.-C. Ni, H.-Z. Hua and A. Pourbaix, *Corros. Sci.*, 1984, **24**, 691–707.
31 I. Olefjord, B. Brox and U. Jelvestam, *J. Electrochem. Soc.*, 1985, **132**, 2854–2861.
32 A. R. Brooks, C. R. Clayton, K. Doss and Y. C. Lu, *J. Electrochem. Soc.*, 1986, **133**, 2459–2464.
33 C. R. Clayton and Y. C. Lu, *J. Electrochem. Soc.*, 1986, **133**, 2465–2473.
34 Y. C. Lu, C. R. Clayton and A. R. Brooks, *Corros. Sci.*, 1989, **29**, 863–880.
35 C. R. Clayton and Y. C. Lu, *Corros. Sci.*, 1989, **29**, 881–898.
36 J. E. Castle and J. H. Qiu, *Corros. Sci.*, 1989, **29**, 591–603.
37 J. E. Castle and J. H. Qiu, *Corros. Sci.*, 1989, **29**, 605–616.
38 I. Olefjord and L. Wegrelius, *Corros. Sci.*, 1990, **31**, 89–98.
39 H. Habazaki, A. Kawashima, K. Asami and K. Hashimoto, *Corros. Sci.*, 1992, **33**, 225–236.
40 D. D. Macdonald, *J. Electrochem. Soc.*, 1992, **139**, 3434–3449.

41 A. Elbiache and P. Marcus, *Corros. Sci.*, 1992, **33**, 261–269.
42 E. De Vito and P. Marcus, *Surf. Interface Anal.*, 1992, **19**, 403–408.
43 N. E. Hakiki, M. D. C. Belo, A. M. P. Simões and M. G. S. Ferreira, *J. Electrochem. Soc.*, 1998, **145**, 3821–3829.
44 M. Bojinov, G. Fabricius, T. Laitinen, K. Mäkelä, T. Saario and G. Sundholm, *Electrochim. Acta*, 2001, **46**, 1339–1358.
45 M. P. Ryan, R. C. Newman and G. E. Thompson, *Philos. Mag. B*, 1994, **70**, 241.
46 M. P. Ryan, R. C. Newman and G. E. Thompson, *J. Electrochem. Soc.*, 1994, **141**, L164.
47 C. L. McBee and J. Kruger, *Electrochim. Acta*, 1972, **17**, 1337.
48 H. Nanjo, R. C. Newman and N. Sanada, *Appl. Surf. Sci.*, 1997, **121–122**, 253.
49 V. Maurice, L. H. Klein and P. Marcus, *Electrochem. Solid-State Lett.*, 2001, **4**, B1.
50 V. Maurice, T. Nakamura, L. Klein and P. Marcus, in *Local Probe Techniques for Corrosion Research*, ed. R. Oltra, V. Maurice, R. Akid and P. Marcus, EFC Publications No. 45, Woodhead Publishing Ltd, Cambridge, 2007, p. 71.
51 P. Marcus, H.-H. Strehblow and V. Maurice, *Corros. Sci.*, 2008, **50**, 2698.
52 A. Seyeux, V. Maurice and P. Marcus, *Electrochem. Solid-State Lett.*, 2009, **12**, C25.
53 T. Massoud, V. Maurice, F. Wiame, L. H. Klein, A. Seyeux and P. Marcus, *J. Electrochem. Soc.*, 2012, **159**, C351.
54 T. Massoud, V. Maurice, F. Wiame, L. H. Klein, A. Seyeux and P. Marcus, *Corros. Sci.*, 2013, **69**, 245.
55 T. Massoud, V. Maurice, L. H. Klein and P. Marcus, *J. Electrochem. Soc.*, 2013, **160**, C232–C238.
56 D. Zuili, V. Maurice and P. Marcus, *J. Electrochem. Soc.*, 2000, **147**, 1393.
57 A. Seyeux, V. Maurice, L. H. Klein and P. Marcus, *J. Solid State Electrochem.*, 2005, **9**, 337.
58 A. Seyeux, V. Maurice, L. H. Klein and P. Marcus, *J. Electrochem. Soc.*, 2006, **153**, B453.
59 V. Maurice, L. H. Klein, H.-H. Strehblow and P. Marcus, *J. Phys. Chem. C*, 2007, **111**, 16351.
60 N. Li, V. Maurice, L. H. Klein and P. Marcus, *J. Phys. Chem. C*, 2012, **116**, 7062.
61 V. Maurice, W.-P. Yang and P. Marcus, *J. Electrochem. Soc.*, 1994, **141**, 3016–3027.
62 D. Zuili, V. Maurice and P. Marcus, *J. Phys. Chem. B*, 1999, **103**, 7896.
63 ASTM E 1438-91 (1996), Standard Guide for Measuring Widths of Interfaces in Sputter Depth Profiling Using SIMS.
64 P. Marcus and J. M. Grimal, *Corros. Sci.*, 1992, **33**, 805–814.
65 N. S. McIntyre and D. G. Zetaruk, *Anal. Chem.*, 1977, **49**, 1521–1529.
66 V. Stambouli, C. Palacio, H. J. Mathieu and D. Landolt, *Appl. Surf. Sci.*, 1993, **70–71**, 240–244.
67 G. Bhargava, I. Gouzman, C. M. Chun, T. A. Ramanarayanan and S. L. Bernasek, *Appl. Surf. Sci.*, 2007, **253**, 4322–4329.
68 H. Peng, PhD Thesis, Thèse de Doctorat, Université Pierre et Marie Curie, 2014.
69 J. B. Lumsden and R. W. Staehle, *Scr. Metall.*, 1972, **6**, 1205–1208.

Faraday Discussions

PAPER

First stages of siderite crystallisation during CO_2 corrosion of steel evaluated using *in situ* synchrotron small- and wide-angle X-ray scattering

Bridget Ingham,[ab] Monika Ko,[bcd] Nick Laycock,[e] Nigel M. Kirby[f] and David E. Williams*[bd]

Received 11th November 2014, Accepted 17th December 2014
DOI: 10.1039/c4fd00218k

We use *in situ* synchrotron small- and wide-angle X-ray scattering (SAXS/WAXS) to demonstrate that the formation of crystalline siderite ($FeCO_3$) during the corrosion of steel in CO_2-saturated brine – a problem of practical interest relating to the growth of protective scales on the interior surface of oil and gas production pipelines – is preceded by the formation of a colloidal precipitate in the solution and an amorphous surface layer, both assumed to be amorphous ferrous carbonate. Grazing incidence SAXS shows instantaneous film formation upon the application of an anodic potential, followed by development of a separate population of particles at later times, then by the formation of crystalline species, observed by WAXS. These observations can be interpreted in terms of crystal nucleation within the amorphous surface layer. Traces of Cr^{3+} in the solution significantly accelerate the precipitation rate of the colloidal precursor and accelerate the appearance of the crystalline scale. We speculate on the significance of these observations for the nucleation, growth and morphology of the corrosion scale and hence its protectiveness.

Introduction

The dependence of nucleation and subsequent growth of crystalline surface scales is of particular importance in relation to the formation of corrosion scales that may be protective or not, according to their growth morphology. It connects

[a]*Callaghan Innovation, P.O. Box 31310, Lower Hutt 5040, New Zealand. E-mail: bridget.ingham@callaghaninnovation.govt.nz*
[b]*MacDiarmid Institute for Advanced Materials and Nanotechnology, Victoria University of Wellington, P.O. Box 600, Wellington 6140, New Zealand*
[c]*Quest Integrity Group, P.O. Box 38096, Lower Hutt 5040, New Zealand*
[d]*School of Chemical Sciences, University of Auckland, Private Bag 92029, Auckland 1022, New Zealand. E-mail: david.williams@auckland.ac.nz*
[e]*College of Engineering and Physical Sciences, University of Birmingham, Edgbaston, Birmingham B15 2TT, UK*
[f]*Australian Synchrotron, 800 Blackburn Road, Clayton, Vic. 3168, Australia*

to general questions about the precursor states for crystal nucleation onto surfaces from solution, and the dependence of crystal nucleation on supersaturation and on the presence of other solution species. Addressing this interesting and experimentally challenging question requires methods both for control of supersaturation and for measurement of crystal and precursor composition and size distribution through the stages of nucleation and growth.[1,2] Small-angle X-ray scattering (SAXS) is a powerful technique for measuring the size distribution of particles (1–100 nm in size) in solution within a defined region of space. The high flux of synchrotron radiation allows the temporal evolution of the size distribution of nanoparticles to be followed[3] and the formation of amorphous precursor states for crystalline species to be studied.[4] A combination of small- and wide-angle X-ray scattering (SAXS/WAXS) allows the process of nucleation and growth of amorphous phases and their subsequent evolution into crystalline phases to be followed in great detail.[5] Grazing-incidence SAXS (GI-SAXS)[6] provides information about the development of surface films[7,8] and nucleation on surfaces.[9]

The particular system investigated here concerns the formation of corrosion scales which can be protective or otherwise depending on the growth morphology. In the case of corrosion of pipeline steels in hot brine saturated with carbon dioxide, the scale formed is siderite, $FeCO_3$. A highly crystalline, protective scale is often formed.[10–12] The magnitude of the corrosion rate and the occurrence or otherwise of localized corrosion ('mesa' corrosion) is determined by the nucleation, growth, morphology and stability of this scale.[13–15] In previous work using grazing incidence X-ray diffraction (XRD), we have demonstrated that there is a significant induction time from the start of anodic polarisation for the observation of crystalline siderite. The induction time is dependent on the solution and metal alloy composition, metal microstructure and surface roughness and the presence or otherwise of scale inhibitors.[16–20] Chukanovite, $Fe_2(OH)_2CO_3$ can also be formed,[21–24] dependent on temperature and pH. It was noted that the critical supersaturation, S_{crit}, for the nucleation of crystalline siderite appears to be very high. The strong temperature dependence of the corrosion rate[11,25,26] can be related to the temperature dependence of S_{crit} and of the crystallisation rate,[27] determining whether or not a protective scale is formed.[28] The development of predictive models for the corrosion rate requires estimated values for the rate of formation and growth of the scale, and methods for codifying the development of scale morphology. This in turn requires estimation of S_{crit} and crystallite growth rate. As noted by Remita *et al.* values given in the literature for the precipitation rate constant of siderite vary very widely, by a factor of 10^5.[29] Literature on the value of S_{crit} for precipitation of siderite, and studies on the subsequent growth rate of the crystals formed, is limited.[27,30–32] Sun and Nešić[30] measured the corrosion layer accumulation rates of siderite for values of supersaturation between 10 and 350. By adding an excess of iron to a CO_2-saturated solution and monitoring the system until precipitation ceased, then measuring the concentration of iron, the implication was that S_{crit} should be around 10 for a temperature of 80 °C. Tomson *et al.*[32] measured concentration as a function of time when Fe^{2+} was slowly added to a CO_2-saturated solution. From the time of the onset of precipitation, they estimated S_{crit} around 3 at 55 °C, in a solution that also contained Ca^{2+}. In the absence of Ca^{2+} at 55 °C, the precipitation rate appeared to become very low at supersaturation less than about 10^2:[27] in this work, the solutions were rigorously de-oxygenated to avoid the possible presence of Fe^{3+}. Remita *et al.*[29] discuss the importance of slow kinetics on siderite precipitation, and quote a supersaturation factor ($[Fe^{2+}]/[Fe^{2+}]_{eq}$) of around 7. Some of the discrepancies in the literature could be because the demonstrated effect of high-valent metal ions[18] could mean that the presence of traces of Fe^{3+} might be important. The apparent supersaturation for the formation of crystalline siderite on the electrode surface can be estimated from our previous work (see Discussion) and is unfeasibly large ($\approx 10^7$). The literature on the crystallisation of other carbonates from amorphous precursors offers a possible resolution.

Precipitation of other carbonates is known to involve polymer like pre-nucleation clusters and the formation of amorphous phases in solution as precursors to crystalline formation. In particular, amorphous calcium carbonate has been extensively characterised as a precursor to its crystalline forms.[33–36] Clarkson *et al.*[34] studied the spontaneous precipitation of calcium carbonate from aqueous solution under various conditions of supersaturation, temperature and presence/absence of inhibitor. They reported the formation of intermediate phases prior to detection of crystalline phases: homogenous precipitation of amorphous calcium carbonate (ACC) was observed under high initial supersaturation while heterogeneous precipitation of hydrated calcium carbonate colloids was observed under low initial supersaturation. Decomposition of these intermediate phases could then result in the formation of the crystalline phase. Using synchrotron small angle X-ray scattering, Bolze *et al.*[4] demonstrated the formation of colloidal ACC particles in solution as a precursor to stable crystalline calcite. They noted a relatively constant mass density (or number of growing particles) following the nucleation period. Using a similar technique, Pontoni *et al.*[5] observed the transformation of these amorphous particles to crystals by dissolution and subsequent heterogeneous nucleation on the walls of their test container. Fernandez-Martinez *et al.* studied heterogeneous nucleation of $CaCO_3$ on mica.[9] In this case, ACC was not observed. Subsequent studies using levitated droplet techniques so that the only interface was a liquid–gas interface showed the formation of a stable, nanosized, liquid-like amorphous $CaCO_3$ phase that underwent a 'solution assisted transformation' to calcite.[37] A liquid-like phase was deduced as the form of the amorphous carbonate formed at high supersaturation, from which the crystals formed.[38–40] Such colloid particles have also been proposed as precursors in the precipitation of magnesium carbonate.[41] Similarly, amorphous ferrous carbonate (AFC) has been detected as a precursor to siderite.[41,42] Sel *et al.*[41] measured the crystallisation kinetics of AFC in solution and reported a slow crystallisation process (the amorphous phase was persistent for at least 9 to 10 hours at 25 °C following synthesis). These authors emphasised the need for rigorous removal of Fe^{3+} species. Hence it seems reasonable to propose that amorphous iron carbonate species in solution may act as precursors for the nucleation of both crystalline siderite and chukanovite.

In the present work, we have combined synchrotron (GI)-SAXS/WAXS with electrochemistry and demonstrate the delayed nucleation and growth of crystalline siderite following formation of a precursor amorphous film on a carbon steel electrode surface in CO_2-saturated brine at elevated temperature. We also demonstrate the formation of amorphous colloidal species in solution near a steel electrode and show the important effect of trace Cr^{3+} in solution on the precipitation rate of the amorphous species and on the transformation to crystalline siderite.

Experimental

The SAXS/WAXS beamline of the Australian Synchrotron[43] was used to conduct the experiments, which used two different set-ups. For the first, a purpose-built electrochemical cell with heating capability was used, which has been described previously for *in situ* X-ray diffraction experiments.[16–18] The cell has a reservoir at the top (volume 50 mL) into which the counter (Pt wire) and reference (Ag/AgCl) electrodes are placed, with a tapering neck at the base down to a narrow slot. The working electrode (a polished 1.5 mm diameter rod of carbon steel, embedded in epoxy resin) was mounted in the base of the slot. The slot depth above the working electrode was 1 mm. The beam [100 μm (h) × 240 μm (w) in size] intersected the electrode at grazing incidence but the incidence angle was ill-defined because the small size of the electrode meant that the surface was slightly rounded by polishing. The cell was periodically translated so that the beam either intersected the electrode surface or passed through the solution in the slot approximately 200 μm above the surface of the working electrode, in order to monitor the scattering signal from the growing film and solution respectively. The positioning of the WAXS detector restricted the range of q_z observable in GI-SAXS. Here, $q = 4\pi \sin \theta/\lambda$ denotes the scattering vector: θ is half of the scattering angle with respect to the incident beam (the z direction is perpendicular to and xy parallel to the surface) and λ is the X-ray wavelength. The X-ray path length through the solution was approximately 2 mm. The electrochemical test solution was 0.5 M NaCl, and was saturated with CO_2 by bubbling a stream of CO_2 gas at 1 bar. $CrCl_3$ was added to some solutions to a concentration of 10 μM. The pH was adjusted to a calculated value of 6.8 at 80 °C by adding 2 M NaOH in a 1 : 50 volume ratio. The solution was heated to 80 ± 1 °C. CO_2 was bubbled in the reservoir continuously throughout the experiment, and the cell was sealed to prevent oxygen ingress. The sample was first polarised at −1 V (*vs.* Ag/AgCl) for 5 minutes to reduce any residual surface oxide, then allowed to reach a steady state at open-circuit (5 minutes), then a potentiostatic or galvanostatic anodic current programme was applied. The exposure time was 10 s, with a 10 s pause between each position (*i.e.* the time between two measurements at the same position was 40 s). SAXS images were recorded using a Pilatus 1 M detector, and WAXS images were collected simultaneously using a Pilatus 200k detector. The scans collected before the initiation of the anodic reaction were averaged and used as the background scattering.

A second experimental set-up was used for direct injection of Fe^{2+} solution into CO_2-saturated 0.5 M NaCl. A syringe was connected to a stepper motor to allow remote control of the injection while data were being collected. $FeSO_4$ solution (50 mM in 0.5 M NaCl) freshly prepared in nitrogen-degassed MilliQ® water was injected at 1.7 μL s^{-1} through a 20 nm filter† into 200 mL of 0.5 M NaCl inside a temperature-controlled vessel with CO_2 continuously bubbling. A peristaltic pump was used to continuously extract solution to be passed through a 2 mm

† Ferrous sulfate solutions are not stable and if left standing will develop a precipitate detectable by SAXS at a rate dependent on concentration: slow at 5 mM but relatively rapid at 50 mM.

diameter quartz capillary for sampling by the X-ray beam, after which it returned to the vessel. SAXS images were continuously recorded with an exposure time of 1 s. Scans were averaged together in groups of 10. A series of static solutions of different Fe^{2+} concentrations were measured in a similar manner to determine the effect of Fe^{2+} ions on the solution scattering.

For all experiments, the X-ray energy was 15 keV ($\lambda = 0.8266$ Å) and the sample-detector distance was 7280 mm, calibrated using silver behenate. Data were processed using the ScatterBrain software (Australian Synchrotron), normalized to the intensity recorded at the beamstop and calibrated to the scattering of water. The background scattering from the solution and the cell windows was subtracted, then fitting was performed over the range $0.004 < q < 0.1$ Å^{-1} using the Irena 'Modeling II' macros.[44]

Results

1. Concurrent SAXS and WAXS measurement shows that a colloidal species is formed above the electrode in advance of a crystalline deposit on the electrode, with characteristics altered by the presence of Cr(III) in solution

Fig. 1 shows SAXS from the X-ray beam passing through the solution above the steel electrode surface, and WAXS when the beam intersected the electrode surface, for a potentiostatic experiment ($E = -500$ mV Ag/AgCl) in 0.5 M NaCl, pH = 6.8 at 80 °C with continuous CO_2 bubbling. The beam passing above the electrode detected a species with size scale 10 nm (observable as a broad peak

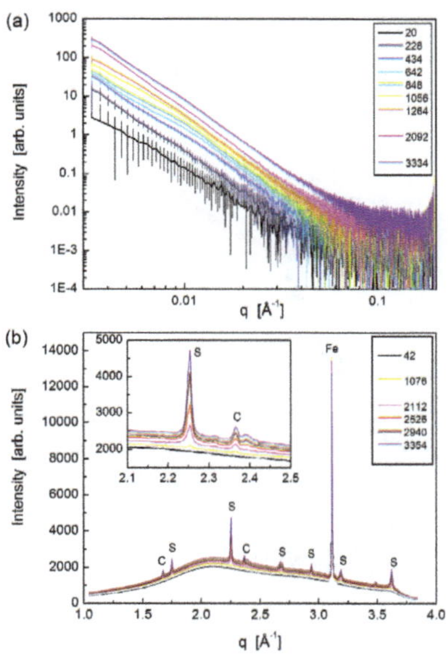

Fig. 1 *In situ* snapshots at selected times, as indicated (in seconds), of raw data collected during an electrochemical experiment with CO_2-saturated 0.5 M NaCl at pH = 6.8 at 80 °C under potentiostatic control at −500 mV (*vs.* Ag/AgCl). (a) SAXS with the beam passing above the electrode and (b) WAXS with the beam intersecting the electrode. The solution signal before applying the potential was subtracted from the SAXS data as a background. For the WAXS data, peaks corresponding to siderite ($FeCO_3$) and chukanovite ($Fe_2(OH)_2CO_3$) are labelled as 'S' and 'C' respectively, along with that from the Fe electrode.

centred around $q = 0.01\ \text{Å}^{-1}$) within 20 s from when the electrode potential stepped from the rest potential. The formation of crystalline siderite and chukanovite on the surface occurred at significantly later times. The SAXS and WAXS data were fitted and the results are plotted on the same time scale axis as the electrochemical current density in Fig. 2. The SAXS data in Fig. 1a were fitted using two populations of scatterers having log-normal distributions. The first had the total volume, mean radius, and standard deviation as fitted parameters (the standard deviation was constrained to be ≤ 0.8), while the second had the mean radius fixed at 135 nm and the standard deviation fixed at 0.5 – only the total volume was fitted. Attempts to fit the data using a single population failed. The first population of particles detected above the electrode surface, which may exist either in the solution or on the cell windows, formed with an average radius 15 nm which remained approximately constant during the period before the crystalline phase was observed, then started to rise (Fig. 2b). The total volume of particles detected increased monotonically with time. The main peak of each phase in the WAXS patterns in Fig. 1b was fitted as in our previous studies[16–18] to obtain the peak area, which is proportional to the volume of crystalline material (Fig. 2a).

The results are consistent with earlier work[16–18,45] given that the acceleration and deceleration of the cell during its translation might induce some convection, to which the crystalline scale is sensitive.[45] The current density was initially high and relatively constant for an 'induction period' of around 30 minutes before it started to increase. During this induction period, the electrode surface is being roughened.[16,45] Crystalline siderite was observed after a significant delay and its

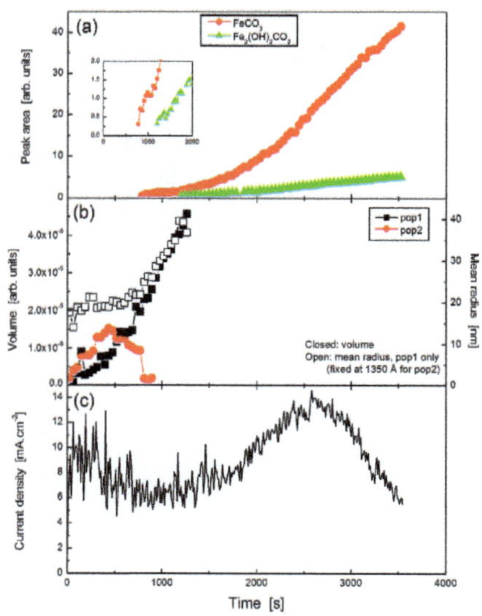

Fig. 2 Derived results from Fig. 1 for an electrochemical experiment with CO_2-saturated 0.5 M NaCl at pH = 6.8 at 80 °C under potentiostatic control at −500 mV (*vs.* Ag/AgCl): (a) WAXS peak area for the most intense peak of each phase. The inset shows the development at the early stages, locating the initiation time for the crystalline phases; (b) SAXS fitting parameters (the mean radius for pop2 is fixed and so not plotted); (c) electrochemical current density.

formation rate increased as the current rose. Chukanovite was observed as a minority phase that formed after the siderite film began to form. The current peaked and then dropped to a lower value than the initial current density, indicating an approach to passivation.[16] The siderite growth curve passed through a point of inflection at the current peak, indicating the approach to the plateau shown previously.[16]

Fig. 3 and 4 show the effect of addition of 10 μM Cr^{3+} to the solution. Fig. 3 shows that, in contrast to the result of Fig. 1, the onset of scattering from above the electrode was delayed after the potential step that initiated the experiment. Crystalline species were observed on the electrode earlier than in the absence of Cr^{3+}. The SAXS data in Fig. 3a were fitted using two populations of scatterers having log-normal distribution. The standard deviations were fixed at 0.3 for the larger scatterers, which emerged first, and 0.8 for the smaller scatterers, which became dominant at later times. The total volume of particles and the mean radii of these distributions were fitted parameters. Attempts to fit the data using a single population failed. These, together with the peak area of the main peak of each phase in the WAXS patterns (Fig. 3b) and the electrochemical current density, are plotted on the same time axis in Fig. 4. The scattering species, though of similar or smaller size to those observed in the absence of Cr^{3+}, grew steadily even before the onset of crystallisation.

Again, the results are consistent with previous work.[18] The current rise corresponded to the nucleation of crystalline siderite onto the electrode surface. The induction time before observation of the current rise and the formation of siderite

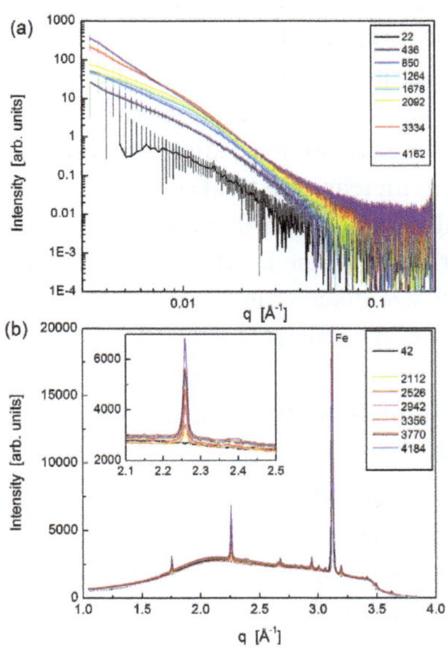

Fig. 3 *In situ* snapshots at selected times, as indicated (in seconds), of (a) SAXS and (b) WAXS raw data collected during an electrochemical experiment with CO_2-saturated 0.5 M NaCl with 10 μM $CrCl_3$ at pH = 6.8 at 80 °C under potentiostatic control at −500 mV (*vs.* Ag/AgCl). The solution signal prior to applying the potential was subtracted from the SAXS data as a background. For the WAXS data, peaks corresponding to siderite ($FeCO_3$) and chukanovite ($Fe_2(OH)_2CO_3$) are labelled as 'S' and 'C' respectively, along with the Fe electrode.

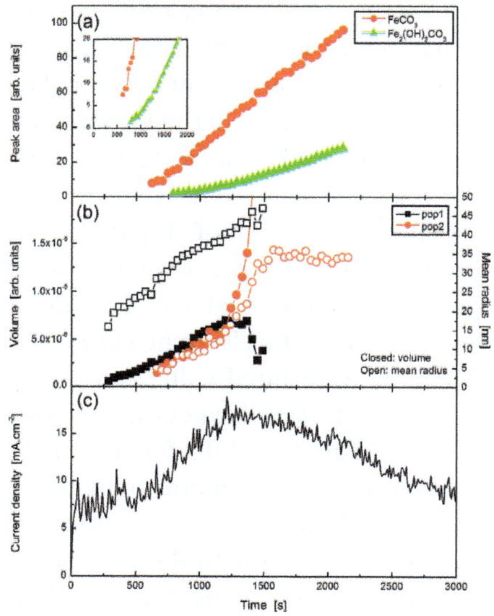

Fig. 4 Derived results from Fig. 3 for an electrochemical experiment with CO_2-saturated 0.5 M NaCl with 10 μM $CrCl_3$ at pH = 6.8 at 80 °C under potentiostatic control at −500 mV (*vs.* Ag/AgCl): (a) WAXS peak area for the most intense peak of each phase. The inset shows the development at the early stages, locating the initiation time for the crystalline phases; (b) SAXS fitting parameters for the two log normal distribution populations; (c) electrochemical current density.

was significantly shortened in the presence of Cr^{3+}. Chukanovite was observed as a minority phase nucleating after siderite. In contrast to the shortened timescale for the observation of siderite on the electrode surface, there was a delay to the observation of particles above the electrode. Again these particles had an initial size of around 15 nm. They were observed shortly before the detectable nucleation of crystalline siderite, and increased in total volume and radius steadily over time. The volume of the smaller population (having a broad size distribution) rapidly increased after 1500 s (the peak time for the current density) but this may simply indicate that the overall size distribution was becoming increasingly broad, not that a separate population of particles was developing.

2. GI-SAXS shows the formation of an amorphous thin film on the electrode preceding crystallisation

Fig. 5 shows grazing incidence (GI-SAXS) patterns measured with the X-ray beam intersecting the electrode surface, for the electrochemical experiment with CO_2-saturated 0.5 M NaCl at pH = 6.8 at 80 °C under potentiostatic control at −500 mV (*vs.* Ag/AgCl). The specular reflection is the vertical streak in the image above the beamstop. Fig. 6a and b show the time variation of the intensity along a cut through the pattern at fixed q_z (indicated by the dotted lines in Fig. 5), and compares the development of the pattern with that observed in the presence of 10 μM Cr^{3+} in the solution (Fig. 6c and d). The GISAXS intensity at a fixed q_z corresponds to scattering from objects of a particular size in the direction perpendicular to the electrode surface, *e.g.* particle height, cavity depth, or film thickness. The lateral scattering corresponds to the size of the scattering objects

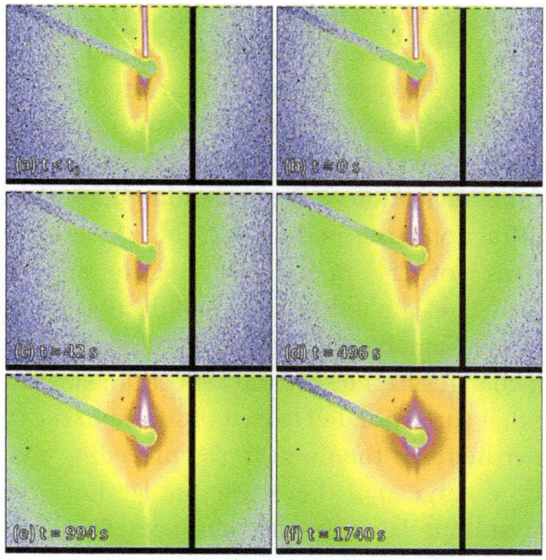

Fig. 5 SAXS scans collected when the X-ray beam was striking the electrode, during an electrochemical experiment with CO_2-saturated 0.5 M NaCl at 80 °C under potentiostatic control at −500 mV (*vs.* Ag/AgCl). The intense signal running vertically from the beam stop in the centre of the image is the specular reflection from the sample. The dotted line shows the cut at constant q_z along which intensity is plotted in Fig. 6.

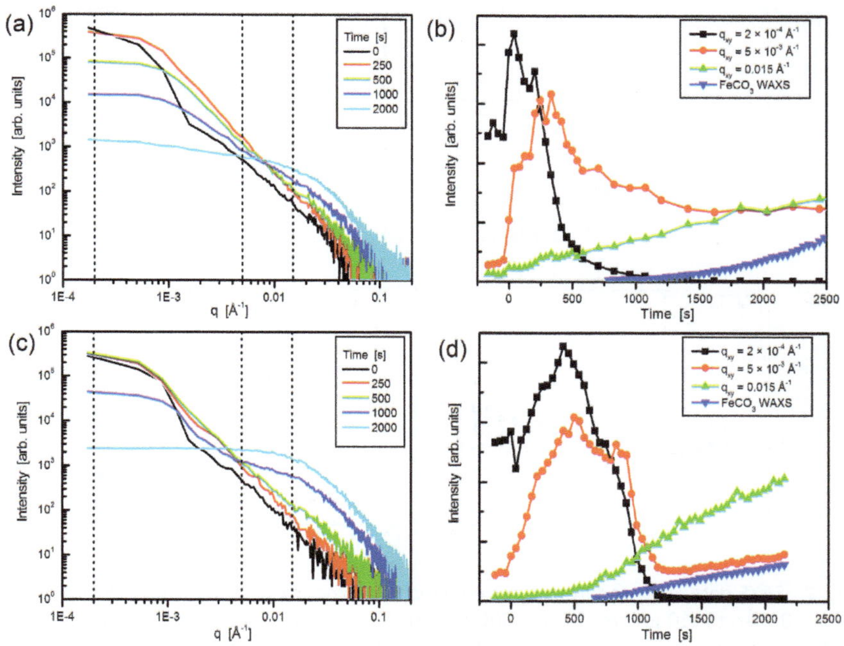

Fig. 6 Analysis of GI-SAXS series from which selected scans are shown in Fig. 5, showing (a and c) intensity in q_{xy} along a cut at $q_z = 0.0248$ Å^{-1} and (b and d). Intensity as a function of time for three fixed (q_{xy}, q_z) points is indicated by the black dotted lines in (a and c) respectively. The corresponding solution SAXS recorded when the beam was positioned above the electrode was subtracted for each time point. The behaviour in the absence (a and b) and presence (c and d) of 10 μM Cr^{3+} in the solution is compared.

in plane, parallel to the electrode surface. The initial intense scattering in Fig. 6a and c up to $q_{xy} \sim 10^{-3}$ Å^{-1} (black squares in Fig. 6b and d) corresponds to the specular reflectance. This decreases in intensity at later times. At large q_{xy} (green triangles in Fig. 6b and d), the data resemble a Guinier scattering pattern at later times corresponding to scattering objects approximately 7 nm in radius. This shows the effect of the nucleation of the crystalline film, which is evidenced by the emergence of diffraction peaks in the WAXS (blue inverted triangles in Fig. 6b and d). In the absence of Cr^{3+} in the solution, the specular reflection first increased, approximately to a plateau, and then decreased. At intermediate q_{xy} (red circles in Fig. 6b, the smallest q_{xy} at which lateral scattering can reliably be distinguished from changes in the specular reflection; size scale, $q^{-1} \approx 20$ nm) there was a plateau that corresponded to that observed in the specular reflection, followed by a further increase then a decrease preceding the formation of the crystalline deposit. At large q_{xy}, the scattering at longer time tracked the formation of the crystalline layer. In the presence of Cr^{3+} in the solution the temporal variation was different (Fig. 6d). Both the specular reflection and the scattering at intermediate q_{xy} rose on a similar timescale, to a maximum, before falling after the crystalline deposit was detected. Thus the temporal development of structures of a particular lateral size scale was significantly altered by the presence of Cr^{3+} in the solution.

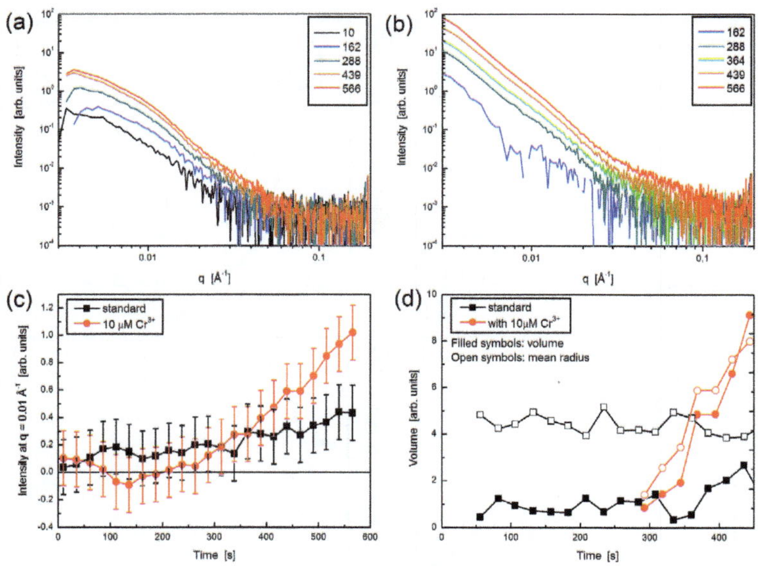

Fig. 7 SAXS data collected from a current stepping experiment, from zero to 55 μA cm^{-2} at $t = 0$ then to 110 μA cm^{-2} at $t = 300$ s, with CO_2-saturated 0.5 M NaCl at 80 °C in the absence (a) and presence (b) of Cr^{3+} in the solution, with the beam passing approximately 200 μm above the surface of the working electrode. The solution scan collected at open circuit before initiation of the current was subtracted as a background. The legend shows the time per s following the application of the current. (c) Location of the time of appearance of scattering species after the current step from zero at $t = 0$, by plotting scattering at fixed q; and (d) derived mean particle size and total volume of scattering species, comparing the behaviour in the absence and in the presence of 10 μM Cr^{3+} in the solution.

3. Galvanostatic step experiments show that both the precipitation rate and the growth rate of the colloidal species is increased in the presence of Cr(III) in the solution

Fig. 7 shows the background-subtracted SAXS data and derived parameters with the beam position kept fixed passing above the electrode when the electrode was subject to a galvanostatic current stepping progressively to higher values. The behaviour in the absence and presence of Cr^{3+} in solution is compared. In the absence of Cr^{3+}, scattering corresponding to a mean particle radius of 25 nm (log-normal distribution with standard deviation 0.25 increasing to 0.8 with increasing time) was observed at the lowest current applied (55 μA cm^{-2}) after a delay of approximately 30–60 s. Approximately 70 s after the step to 110 μA cm^{-2} at $t = 300$ s, the total volume of particles started to increase steadily, but with little change in the particle size. In contrast, in the presence of 10 μM Cr^{3+}, no additional scattering above the background was detected before 250 s. Immediately after the step to 110 μA cm^{-2} at 300 s, the mean particle size increased rapidly before stabilising at ≈60 nm. We develop in the Discussion the demonstration that the delay in appearance of particles above the electrode can be related to the critical supersaturation for nucleation of a precipitate and to the precipitation rate near the electrode, with a longer delay implying a greater precipitation rate or smaller critical supersaturation.

4. A solution injection experiment demonstrates $S_{crit} \approx 2.5$ for formation of a colloid

Fig. 8 shows the increase of SAXS with time when a 50 mM $FeSO_4$ solution was slowly injected through a 20 nm filter into rapidly stirred, CO_2-saturated 0.5 M

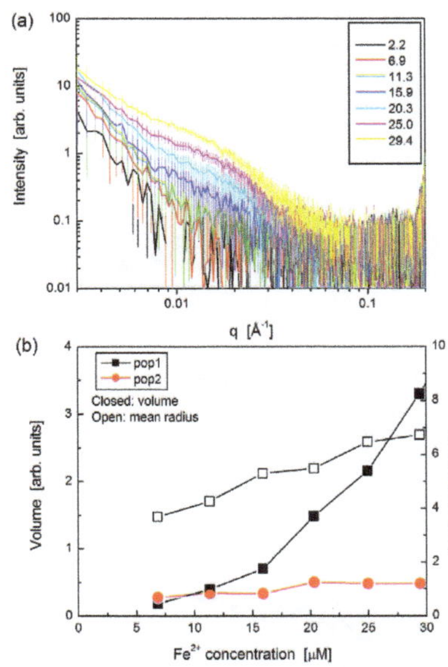

Fig. 8 (a) SAXS data collected at various times during injection of a 50 mM $FeSO_4$ solution into 200 mL of a CO_2-saturated 0.5 M NaCl solution at 80 °C at 0.1 mL min^{-1}. The legend shows the corresponding Fe^{2+} concentration in solution. The initial solution scan before injection commenced was subtracted as a background. (b) Derived mean particle radius and total volume of scattering species as a function of the Fe^{2+} concentration in solution. The mean radius of pop2 was fixed at 150 nm.

NaCl at pH 6.8 at 80 °C. Above a threshold concentration of approximately 5 μM Fe^{2+}, located by extrapolating the concentration-dependence of the derived total volume of scattering species, two populations of scattering species were observed: one with a mean radius ≈4 nm, that increased with increase of Fe^{2+} concentration and the other of much larger particles (≈150 nm). SAXS of $FeSO_4$ solutions freshly prepared in deoxygenated water showed detectable particles of size scale >10 nm for solutions of concentration 5 mM and above. From the onset Fe^{2+} concentration to observing particles, and knowing the solubility product K_{sp} and CO_3^{2-} concentration, we can estimate S_{crit} for the formation of a colloid:

$$S_{crit} = \frac{[Fe^{2+}][CO_3^{2-}]}{K_{sp}}$$

K_{sp} under these conditions is around 1.37×10^{-10} mol^2 L^{-2} at 80 °C.[46] The CO_3^{2-} concentration is estimated to be around 7×10^{-5} mol L^{-1}.[20] If the Fe^{2+} concentration at the onset of precipitation is around 5×10^{-6} mol L^{-1}, this yields a value for S_{crit} of around 2.5.

5. An interrupted galvanostatic measurement demonstrates heterogeneous nucleation of the colloid on the cell windows

A current density of 0.56 mA cm^{-2} was applied for 1 minute, then turned off. After a further 10 minutes, the current was turned on again. SAXS data were collected continuously through the experiment and are shown in Fig. 9. A single population of scatterers having a log-normal distribution was used to fit the data. The mean radius and total volume of particles were independent fitted parameters and are

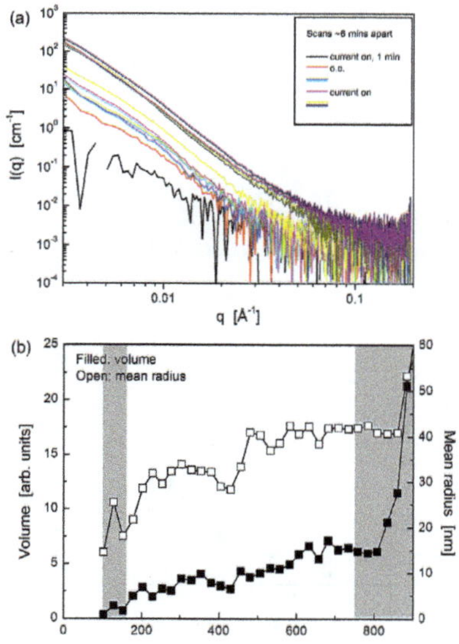

Fig. 9 (a) SAXS data collected from an experiment with CO_2-saturated 0.5 M NaCl at 80 °C where a galvanostatic current density of 0.56 mA cm^{-2} was applied for 1 minute, turned off, and turned on again after 10 minutes. The initial solution scan before the current was first applied was subtracted as a background. (b) SAXS fitting parameters for the log normal distribution populations; the shaded areas show where the current was applied.

shown in Fig. 9. Particles formed during the initial 1 min application of the current, having a mean radius of around 20 nm. During the period that the current was off, the total volume increased by a factor of three and the mean size increased by a factor of two. If the particles observed were forming in solution, one would expect the signal to decay during the 10 minutes that the current was turned off, as the particles either settled or diffused away. However, if they were forming on or adhering to the windows in the path of the X-ray beam, the signal would not change significantly (and may increase as Fe^{2+} concentration relaxed, or as an effect of X-ray exposure). The latter situation is what was observed here.

Discussion

The data are consistent with the formation and transformation of an amorphous, perhaps gel-like surface layer, which then gives way to crystalline species with kinetics significantly altered by the presence of Cr^{3+}. The ideas are consistent with Ostwald's 'rule of stages'.[47,48] The data are consistent with the dissolution of the surface 'gel' layer leading to amorphous colloidal species both in solution and nucleating on other surfaces, with precipitation rate also significantly altered by the presence of Cr^{3+}. The discussion can be developed first from a consideration of the apparent supersaturation required for the nucleation of a crystalline surface film, then from analysis of the results of SAXS from the solution above the electrode and comparison with the supersaturation required to form amorphous species by injection of ferrous sulfate into the CO_2-saturated NaCl solution, and finally from consideration of the GI-SAXS results.

For the purpose of discussing the apparent supersaturation achieved in a galvanostatic experiment, and the detection of precipitating species, Fig. 10 illustrates results from an idealized diffusion geometry for the slot above the electrode: for the purpose of modeling, the experimental geometry (Fig. 10a, inset, b) is approximated by the simple linear geometry (Fig. 10a, inset, a). A constant

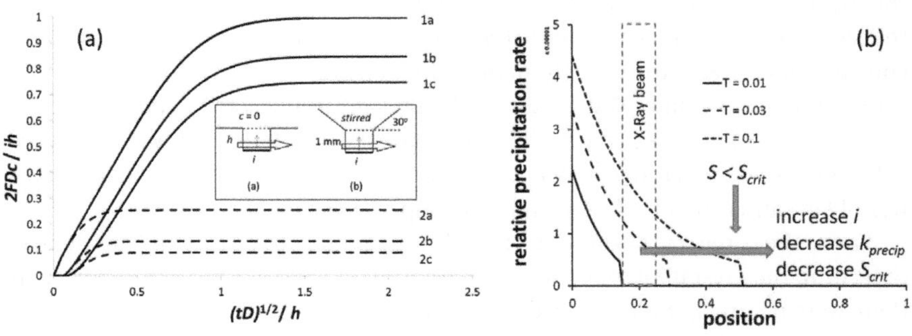

Fig. 10 Results of finite difference solution of the diffusion problem for Fe^{2+} and CO_2 with siderite precipitation. (a) Time evolution of Fe^{2+} concentration with $K_p = 0$ (curve 1a at the electrode surface and 1b, 1c at positions corresponding approximately to the bottom and top of the X-ray beam) and $K_p = 20$ (curves 2, at the electrode surface, bottom and top of the X-ray beam). The inset illustrates the experimental geometry, inset b, and the idealized calculation geometry, inset a. (b) Relative precipitation rate at different time $T = tD/h^2$ as a function of position above the electrode, calculated for $K_p = 20$ with flux $2Fic_0/(hD) = 10^{-5}$ and critical supersaturation $S^*_{scaled} = 10^{-3}$, corresponding approximately to the experimental conditions of Fig. 7 with $S_{crit} \approx 5$.

current density is applied to cause dissolution of an electrode at the base of the slot. The precipitating reagent (CO_2) is present outside the slot at uniform concentration. The X-ray beam detects the formation of precipitate within a section of the slot above the electrode.

Ignoring the effects of hydrolysis of the metal ions, the reaction scheme for the precipitation of siderite can be written:

$$CO_2 + H_2O \underset{k_{-1}}{\overset{k_1}{\rightleftharpoons}} H_2CO_3$$

$$H_2CO_3 \overset{K_1}{\rightleftharpoons} H^+ + HCO_3^- \overset{K_2}{\rightleftharpoons} H^+ + CO_3^{2-}$$

$$Fe^{2+} + CO_3^{2-} \xrightarrow{k_2} FeCO_3$$

The discussion focuses on the attainment of the critical condition for precipitation, and hence considers the case when the CO_2 concentration is not greatly perturbed. In that case, the discussion is simplified by noting that $[CO_3^{2-}] \ll [HCO_3^-] + [H_2CO_3]$ and writing $[CO_3^{2-}] = \beta[CO_2]$ with β constant ($\approx 5.6 \times 10^{-3}$ at 80 °C, pH 6.8). Then, ignoring migration in view of the excess of NaCl, the diffusion equation for Fe^{2+} and CO_2 can be written:

$$\frac{\partial C_{Fe^{2+}}}{\partial T} = \frac{\partial^2 C_{Fe^{2+}}}{\partial X^2} - K_p C_{Fe^{2+}} C_{CO_2}$$

$$\frac{\partial C_{CO_2}}{\partial T} = \frac{\partial^2 C_{CO_2}}{\partial X^2} - K_p C_{Fe^{2+}} C_{CO_2}$$

where the scaled time, $T = tD/h^2$, position, $X = x/h$ and concentration $C = c/c_0$. Here, D denotes the diffusion coefficient of Fe^{2+} (2.1×10^{-5} cm^2 s^{-1} at 80 °C), h is the depth of the slot (≈ 1 mm), c_0 is the saturation concentration of CO_2 (outside the slot; ≈ 0.013 M at 80 °C and pH 6.8) and t is the time. The rate constant for consumption of Fe^{2+} through formation of siderite, $K_p = \beta k_2 K_2 h^2/D$ if the supersaturation, $S \geq S^*$, and $=0$ if $S < S^*$. In the scaled variables, the supersaturation, $S_{scaled} = C_{Fe^{2+}} C_{CO_2} = SK_{sp}/\beta c_0^2$. Fig. 10 shows the results of a finite-difference solution of these equations with boundary conditions at the electrode of constant flux of Fe^{2+} and zero flux of CO_2, and at the top of the slot of $C_{Fe^{2+}} = 0$ and $C_{CO_2} = 1$.

From the time to observe a crystalline product and the current density preceding the crystalline scale nucleation (~1500 s, and ~6 mA cm^{-2}, Fig. 1) if there were no precipitation preceding the crystalline scale then, from Fig. 10a the approximate surface concentration of Fe^{2+} can be calculated (≈ 0.1 M) and hence the apparent supersaturation ($\approx 10^7$) which is physically unreasonable. The SAXS (Fig. 1 and 2) detects the colloidal, amorphous precipitate which is indeed present and which precedes the crystallisation. The model is therefore of an induction time for crystallization from a preceding amorphous material. This induction time for crystallization from an amorphous precursor is evidently decreased in the presence of a small concentration of Cr^{3+} in the solution (Fig. 3 and 4).

When precipitation is occurring, Fig. 10b illustrates that a precipitation front would move out from the electrode surface, with a delay before the front moved through the position of the X-ray beam, as indeed was observed (Fig. 7). We verified that the delay time for a given (low) current density was consistent with the critical supersaturation determined in the solution injection experiment (Fig. 8). Fig. 10b shows that, the higher the precipitation rate, the longer it takes for the precipitation front to reach the position of the X-ray beam and hence to give rise to detectable scattering. This calculation illustrates that the significant additional delay in appearance of precipitate in the X-ray beam caused by addition of trace Cr^{3+} to the solution is due to an increase in precipitation rate or a decrease in critical supersaturation. The SAXS experiment (Fig. 4 and 7) also showed that the presence of trace Cr^{3+} in solution caused a significant increase in the growth rate of the amorphous colloidal particles initially formed.

Although the experimental set-up was not ideal for GI-SAXS, the results are informative. The interpretation must take into account the development of surface roughness, previously demonstrated to occur during the induction period before the appearance of crystalline material.[16,45] The fractal dimension of the diffraction ring (a useful measure of 'spottiness'[49]) increases over time and has been interpreted as due to the removal by dissolution of smaller grains of the metal, resulting in an increase in surface roughening over time.[45,49] While the fractal dimension of the diffraction ring does not necessarily relate to a fractal dimension of the surface morphology, we note that the SAXS of a rough interface gives a power law dependence on q of diffuse scattering intensity, with exponent decreasing as fractal dimension increases. This effect will be superimposed upon effects associated with the development of surface layers. The results in the absence of added Cr^{3+} (Fig. 6) show a clear temporal separation of three stages in the development of the surface layer, corresponding to different lateral size scales. The first stage is the immediate increase of specular reflection to a plateau, followed by a steep drop in signal, Fig. 6b. A surface layer of amorphous $FeCO_3$ would have a significantly smaller density than Fe and thus a significantly smaller X-ray reflectivity. Roughening of the surface will also cause a decrease in reflectivity. The steep drop in specular reflectivity therefore can be interpreted as due to the coverage of the surface by a continuous film of amorphous $FeCO_3$. The increase prior to this is clearly a consequence of the electrochemical polarisation. Given the magnitude of the initial current density, the critical supersaturation for amorphous $FeCO_3$ would be exceeded almost immediately. We therefore interpret the increase in specular reflection to be a consequence of multiple reflection and refraction of the X-rays caused by the presence of a discontinuous layer comprising 'blobs' of amorphous $FeCO_3$, which subsequently grow to spread over the surface and eventually cover it. The next phase is the appearance of a new set of scatterers with lateral size scale ≈ 20 nm, indicated by the change of scattering intensity at intermediate q_{xy}. That result could be interpreted as due to aggregation within the surface layer. Thus, the results indicate that the initial stage is the formation of a thin layer of amorphous ferrous carbonate on the electrode, within which crystalline siderite or chukanovite subsequently forms. Speculatively, these stages could be the spreading of patches of an amorphous gel-like layer until it covers the surface, concurrent with the development of lateral inhomogeneity and clusters within this layer, analogous to the prenucleation clusters[50] noted for example for solution precipitation of $CaCO_3$ (ref. 51) then finally the nucleation and growth of crystals, which is signalled by the appearance of a crystalline diffraction pattern.

A gel-like AFC layer covering the steel surface would form a barrier that would control the initial current density, depending on the thickness (primarily) of this layer and also on its density, which would control its ionic conductivity. The effect of the presence of trace Cr^{3+} in the solution could be interpreted as a significant acceleration and overlapping of all three stages. A different morphology of the amorphous 'gel' as it nucleates and spreads over the surface could account for the different temporal development of the specular reflectance. The development of lateral inhomogeneity and clusters within the gel layer would seem to have been significantly affected, such that the development of lateral inhomogeneity preceding crystal nucleation overlapped in time with the spreading and thickening of the gel layer that is presumed to be present.

Fig. 11 gives a schematic model of the evolution of the surface state deduced from both the present and previous work. Important effects of surface roughness on the time scale for nucleation of crystalline phases have been observed, including in synchrotron experiments where the dominant effect on mass-transport to the surface should be the depth of the narrow slot above the electrode.[45] The idea of a secondary boundary layer, defined by the gel layer, helps understand these effects in that crystallisation would be controlled by the conditions specifically pertaining within the gel and thus sensitive to the local geometry and environment. The current rise upon nucleation of crystalline siderite, interpreted previously in terms of a local pH decrease,[17] can with the new observations be interpreted in part as due to the development of holes in the amorphous surface layer as it transforms into a crystalline layer. By analogy with the transformation of amorphous to crystalline calcium carbonate,[52] the transformation of hydrated AFC to crystalline siderite should result in a significant decrease in volume which would cause the AFC layer locally to become thinner or be punctured. This model offers an alternative explanation for the current rise upon crystallisation, whereby the presence of voids in the otherwise continuous AFC layer would decrease the potential drop across the surface layer and result in a rise in current density. Once the crystalline phases are formed, rapid growth of these materials may cause further acidification of the local solution due to the consumption of carbonate species, as previously proposed.[17]

The transformation of AFC to crystalline phases is known to be strongly time dependent and slow.[41] The crystallisation process may be dependent on a number of factors, including the concentration of Fe^{2+} within the AFC layer as indicated in Scheme 1 below. Chukanovite is also a crystallisation product, and the relative amounts of chukanovite and siderite have proven in previous work to be somewhat variable.[45] The observation of an amorphous surface gel layer and the hypothesis that crystalline species nucleate within the gel layer can aid interpretation. Since in the electrochemical experiments iron is dissolving through the surface layer, there would be an iron excess present within the layer. Then, there would be the possibility of formation of a hydrated, iron-excess gel that could resemble amorphous chukanovite:

$$FeCO_3 \text{ (amorphous)} + Fe^{2+} + 2H_2O = Fe_2(OH)_2CO_3 \text{ (amorphous)} + 2H^+$$

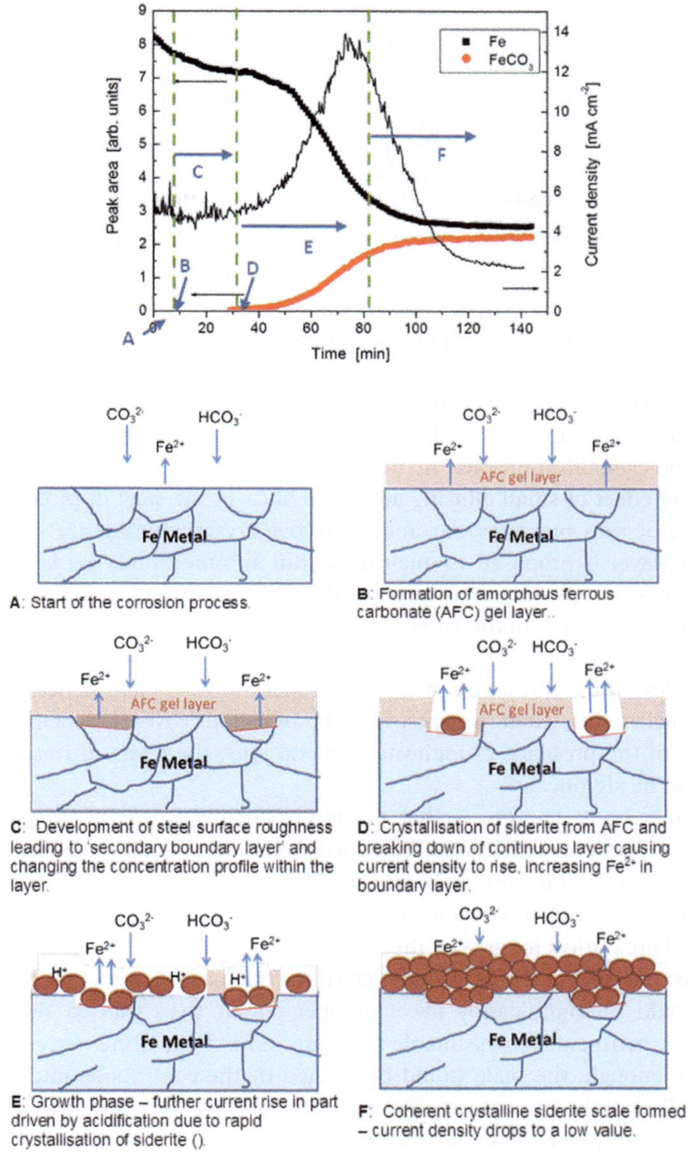

Fig. 11 Schematic scale precipitation model during CO_2 corrosion of carbon steel, relating the different stages to the evolution of the current density (light line) and the diffraction pattern (heavy lines) observed from species on the surface (top panel from ref. 17).

Given that, there is the possibility of multiple transformation pathways as indicated in Scheme 1.

In summary, this work taken with previous studies has shown that the nucleation and growth characteristics of crystalline siderite scales are unusual and that there is a complex inter-dependent relationship between the precipitation rate, the composition and the morphology of the formed scale. The important fact seems to be that direct formation of crystalline siderite from solution

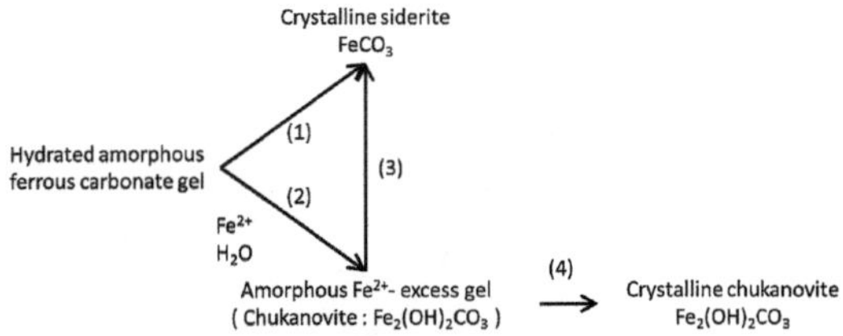

Scheme 1 Possible transformation paths to different crystalline products.

requires a very large critical supersaturation or a long induction time for nucleation. The formation of crystalline siderite is preceded by the formation of an amorphous colloidal ferrous carbonate (AFC) precipitate. The presence of Cr^{3+} in the solution has a significant effect on the precipitation and growth rate of AFC and on the crystallisation rate. As noted before, this phenomenon explains the very strong effect of small alloying additions of Cr in the steel in promoting the formation of thin but adherent and protective crystalline siderite scales. The crystalline layer is proposed to nucleate within an amorphous gel layer, whose thickness and composition may be dependent on the current density and surface boundary layer. In view of the known influence of a wide range of solution species on the nucleation and crystallisation of other carbonates, important effects of such species on the formation of crystalline siderite scales and hence corrosion rate of steel in CO_2 environments, is to be expected. Given the demonstrated influence of the presence of high-valent metal ions, the effect of traces of Fe^{3+} might also be significant.

One final point to note is that the electrochemical experiments generally resulted in formation of friable scales with relatively small crystallite size. In contrast, open circuit corrosion studies show scales comprising large, well-formed crystals. The discussion above implies that the conditions within a surface gel are critical for at least the initial development of a crystalline scale. In particular, the current density would determine the local supersaturation of Fe^{2+}, which would be significantly lower at open-circuit than that in the electrochemical experiments. A reasonable presumption is that, if the current density were high enough, the scale would be formed by the continuous nucleation of new, small crystals. In contrast, if the current density were low, the growth of crystallites might be expected to dominate.

Conclusions

The present work supports a model for CO_2 corrosion of pipeline steels in which corrosion-protective siderite scales nucleate within a thin surface 'gel' of amorphous iron carbonate. The precipitation rate of the amorphous carbonate is strongly increased as a consequence of the presence of Cr^{3+} in solution. The crystallisation rate of siderite from the amorphous precursor is also strongly increased in the presence of Cr^{3+}. We suggest that the conditions within a secondary boundary layer defined by the surface roughness and the current density determine the thickness of the surface gel and the subsequent crystallisation.

Acknowledgements

This work was funded by the New Zealand Ministry of Business, Innovation and Employment (MBIE) under contract C08X1003. Portions of this research were undertaken on the SAXS/WAXS beamline at the Australian Synchrotron, Victoria, Australia. We thank the New Zealand Synchrotron Group Ltd. for a grant to construct the electrochemical cell. We also thank Nick Birbilis (Monash University) for his generous advice and loan of equipment.

References

1 O. Devos, C. Gabrielli, M. Tlili and B. Tribollet, *J. Electrochem. Soc.*, 2003, **150**, C494–C501.

2 C. Gabrielli, G. Maurin, G. Poindessous and R. Rosset, *J. Cryst. Growth*, 1999, **200**, 236–250.

3 B. Ingham, T. H. Lim, C. J. Dotzler, A. Henning, M. F. Toney and R. D. Tilley, *Chem. Mater.*, 2011, **23**, 3312–3317.

4 J. Bolze, B. Peng, N. Dingenouts, P. Panine, T. Narayanan and M. Ballauff, *Langmuir*, 2002, **18**, 8364–8369.

5 D. Pontoni, J. Bolze, N. Dingenouts, T. Narayanan and M. Ballauff, *J. Phys. Chem. B*, 2003, **107**, 5123–5125.

6 G. Renaud, R. Lazzari and F. Leroy, *Surf. Sci. Rep.*, 2009, **64**, 255–380.

7 G. Kaune, M. A. Ruderer, E. Metwalli, W. Wang, S. Couet, K. Schlage, R. Röhlsberger, S. V. Roth and P. Müller-Buschbaum, *ACS Appl. Mater. Interfaces*, 2008, **1**, 353–360.

8 P. Müller-Buschbaum, in *Applications of Synchrotron Light to Scattering and Diffraction in Materials and Life Sciences*, Springer, 2009, pp. 61–89.

9 A. Fernandez-Martinez, Y. D. Hu, B. Lee, Y. S. Jun and G. A. Waychunas, *Environ. Sci. Technol.*, 2013, **47**, 102–109.

10 M. H. Nazari, S. R. Allahkaram and M. B. Kermani, *Mater. Des.*, 2010, **31**, 3559–3563.

11 M. B. Kermani and A. Morshed, *Corrosion*, 2003, **59**, 659–683.

12 A. Morshed, PhD thesis, University College London, 2002.

13 R. Nyborg and A. Dugstad, in *Paper no. 3642, CORROSION/2003*, NACE International, 2003.

14 R. Nyborg, in *Paper No. 48, CORROSION/98*, NACE International, 1998.

15 W. Sun, S. Nesic and R. C. Woollam, *Corros. Sci.*, 2009, **51**, 1273–1276.

16 B. Ingham, M. Ko, G. Kear, P. Kappen, N. Laycock, J. A. Kimpton and D. E. Williams, *Corros. Sci.*, 2010, **52**, 3052–3061.

17 B. Ingham, M. Ko, N. Laycock, J. Burnell, P. Kappen, J. A. Kimpton and D. E. Williams, *Corros. Sci.*, 2012, **56**, 96–104.

18 M. Ko, B. Ingham, N. Laycock and D. E. Williams, *Corros. Sci.*, 2014, **80**, 237–246.

19 M. Ko, B. Ingham, N. Laycock and D. E. Williams, *Corros. Sci.*, 2015, **90**, 192–201.

20 M. Ko, N. J. Laycock, B. Ingham and D. E. Williams, *Corrosion*, 2012, **68**, 1085–1093.

21 R. De Marco, Z.-T. Jiang, B. Pejcic and E. Poinen, *J. Electrochem. Soc.*, 2005, **152**, B389.

22 I. Azoulay, C. Remazeilles and P. Refait, *Corros. Sci.*, 2012, **58**, 229–236.
23 P. Refait, J. A. Bourdoiseau, M. Jeannin, D. D. Nguyen, A. Romaine and S. Sabot, *Electrochim. Acta*, 2012, **79**, 210–217.
24 C. Rémazeilles and P. Refait, *Polyhedron*, 2009, **28**, 749–756.
25 M. Honarvar Nazari, S. R. Allahkaram and M. B. Kermani, *Mater. Des.*, 2010, **31**, 3559–3563.
26 C. D. Waard and U. Lotz, in *A Working Party Report on Predicting CO_2 Corrosion in the Oil and Gas Industry*, The Institute of Materials, UK, 1994, p. 30.
27 S. Yean, H. A. Alsaiari, A. T. Kan and M. B. Tomson, *SPE International Oilfield Scale Conference*, 2008.
28 A. Dugtad, in *Paper No. 14, CORROSION/92*, NACE International, 1992.
29 E. Remita, B. Tribollet, E. Sutter, F. Ropital, X. Longaygue, J. Kittel, C. Taravel-Condat and N. Desamaise, *J. Electrochem. Soc.*, 2008, **155**, C41–C45.
30 W. Sun and S. Nesic, *Corrosion*, 2008, **64**, 334–346.
31 M. L. Johnson and M. B. Tomson, in *Paper No. 268, CORROSION/91*, NACE International, 1991.
32 H. A. Alsaiari, A. Kan and M. Tomson, *SPE J.*, 2010, **15**, 294–300.
33 D. Gebauer, A. Völkel and H. Cölfen, *Science*, 2008, **322**, 1819–1822.
34 J. R. Clarkson, T. J. Price and C. J. Adams, *J. Chem. Soc., Faraday Trans.*, 1992, **88**, 243–249.
35 H. Elfil and H. Roques, *AIChE J.*, 2004, **50**, 1908–1916.
36 F. C. Meldrum and R. P. Sear, *Science*, 2008, **322**, 1802–1803.
37 S. E. Wolf, J. Leiterer, M. Kappl, F. Emmerling and W. Tremel, *J. Am. Chem. Soc.*, 2008, **130**, 12342–12347.
38 A. F. Wallace, L. O. Hedges, A. Fernandez-Martinez, P. Raiteri, J. D. Gale, G. A. Waychunas, S. Whitelam, J. F. Banfield and J. J. D. Yoreo, *Science*, 2013, **341**, 885.
39 A. Myerson and B. Trout, *Science*, 2013, **341**, 855.
40 A. Accardo, M. Burghammer, E. D. Cola, M. Reynolds, E. D. Fabrizio and C. Riekel, *Langmuir*, 2011, **27**, 8216–8222.
41 O. Sel, A. V. Radha, K. Dideriksen and A. Navrotsky, *Geochim. Cosmochim. Acta*, 2012, **87**, 61–68.
42 C. Jimenez-Lopez and C. S. Romanek, *Geochim. Cosmochim. Acta*, 2004, **68**, 557–571.
43 N. M. Kirby, S. T. Mudie, A. M. Hawley, D. J. Cookson, H. D. T. Mertens, N. Cowieson and V. Samardzic-Boban, *J. Appl. Crystallogr.*, 2013, **46**, 1670–1680.
44 J. Ilavsky and P. R. Jemian, *J. Appl. Crystallogr.*, 2009, **42**, 347–353.
45 M. Ko, B. Ingham, N. Laycock and D. Williams, *Corros. Sci.*, 2015, **90**, 192–201.
46 W. Sun, S. Nešić and R. C. Woollam, *Corros. Sci.*, 2009, **51**, 1273–1276.
47 J. Nyvlt, *Cryst. Res. Technol.*, 1995, **30**, 443–449.
48 R. Van Santen, *J. Phys. Chem.*, 1984, **88**, 5768–5769.
49 B. Ingham, *J. Appl. Crystallogr.*, 2014, **47**, 166–172.
50 D. Gebauer, M. Kellermeier, J. D. Gale, L. Bergstrom and H. Coelfen, *Chem. Soc. Rev.*, 2014, **43**, 2348–2371.
51 D. Gebauer, A. Volkel and H. Colfen, *Science*, 2008, **322**, 1819–1822.
52 J. Ihli, W. C. Wong, E. H. Noel, Y.-Y. Kim, A. N. Kulak, H. K. Christenson, M. J. Duer and F. C. Meldrum, *Nat. Commun.*, 2014, **5**, 3169.

Faraday Discussions

PAPER

Localized dealloying corrosion mediated by self-assembled monolayers used as an inhibitor system

B. R. Shrestha,[a] A. Bashir,[a] G. N. Ankah,[a] M. Valtiner[a] and F. U. Renner*[ab]

Received 10th December 2014, Accepted 23rd January 2015
DOI: 10.1039/c4fd00256c

The structure and chemistry of thiol or selenol self-assembled organic monolayers have been frequently addressed due to the unique opportunities in functionalization of materials. Such organic films can also act as effective inhibition layers to mitigate oxidation or corrosion. Cu–Au alloy substrates covered by self-assembled monolayers show a different dealloying mechanism compared to bare surfaces. The organic surface layer inhibits dealloying of noble metal alloys by a suppression of surface diffusion at lower potentials but at higher applied potentials dealloying proceeds in localized regions due to passivity breakdown. We present an *in situ* atomic force microscopy study of a patterned thiol layer applied on Cu–Au alloy surfaces and further explore approaches to change the local composition of the surface layers by exchange of molecules. The pattern for the *in situ* experiment has been applied by micro-contact printing. This allows the study of corrosion protection with its dependence on different molecule densities at different sites. Low-density thiol areas surrounding the high-density patterns are completely protected and initiation of dealloying proceeds only along the areas with the lowest inhibitor concentration. Dealloying patterns are highly influenced and controlled by molecular thiol to selenol exchange and are also affected by introducing structural defects such as scratches or polishing defects.

Introduction

Corrosion is a major problem in various technical fields from the oil, gas, and chemical industry[1] to biomedical implants[2,3] or infrastructures.[4] In particular, premature catastrophic failure of structural components can be initiated by corrosion and poses a considerable risk for the environment and human society. Several studies report on the huge economical loss caused by corrosion each year. Considerable effort has therefore been made to design and develop complex alloys which deliver good mechanical properties as well as excellent corrosion

[a]*Max-Planck-Institut für Eisenforschung GmbH, Max-Planck-Straße 1, D-40237 Düsseldorf, Germany. E-mail: frank.renner@uhasselt.be*
[b]*Institute of Materials Research (IMO), Hasselt University, Wetenschapspark 1, B-3590 Diepenbeek, Belgium*

resistance. Unfortunately, such corrosion resistant alloys often contain large quantities of expensive elements such as Cr, Ni, or Mo. One further way to mitigate corrosion of materials in applications is the use of inhibitors, and organic systems are today widespread. Especially the oil and gas industry is a large consumer of organic inhibitors.[1] At the same time there is also a large research effort to find environmentally friendly inhibitors.[5] The effect of inhibitors varies often depending on (even slight) changes in the environment, temperature, or protected materials, even "from one oil well to another".[1] A better understanding of the underlying mechanisms of corrosion protection by organic inhibitors is therefore needed. A true mechanistic understanding on the atomic and molecular level is so far widely lacking. Thiol and selenol based films have been employed for oxidation and corrosion protection of copper.[6–10] The formation of self-assembled organic monolayer films (SAMs) based on various thiols and selenols has been widely studied due to the unique opportunities in functionalization of materials. In particular, with the wide and detailed knowledge of the structure and chemistry of such often highly ordered films on the molecular level, SAMs provide a promising model system for fundamental corrosion studies.

Molecular self-assembly is controlled through a strong covalent bonding to a substrate and non-covalent interactions between the molecules such as van der Waals forces, or π–π interactions.[11,12] Common examples include the formation of liquid crystal phases, micelles, vesicles, and Langmuir films.[13] Many SAMs are based on thiol or lately selenol molecules on Au, Ag, Pd, or Cu substrates.[11–14] The large possible variety (Fig. 1a) to combine the S- or Se-head group[15–17] with different organic ends or chains enables many possibilities to control the structure of SAMs films.[18] Recent research has concentrated on increasing the stability of the films by offering structural changes *via* the exchange of molecules. For Au substrates, selenols were found to be more stable than thiol-based molecules.[18] In general the stability of SAMs depends on the substrate symmetry and atomic distances in the surface plane as well as on the interplay with the chain interactions. The ability to form a well-ordered surface film thus depends often on the chosen chain length, an odd or even number of carbon atoms[18] in the chain, or the existence of aromatic building blocks.[6,19] The presence of such SAMs plays an important role in dealloying.

Dealloying is one form of corrosion where less noble element/s (one or more) are selectively dissolved. The nobler element(s) accumulates or diffuses on the surface eventually leading to passivation. At lower overpotentials above the equilibrium potential of Cu, bare Cu_3Au (111) surfaces are covered by an ultrathin film of Au of about 3 monolayers (ML), which protects from further selective dissolution of Cu, as was described earlier.[20,21] At intermediate overpotentials,

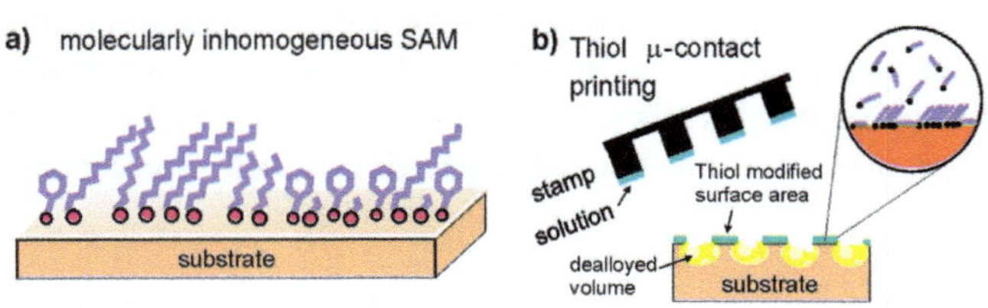

Fig. 1 (a) Schematic showing the formation of an organic SAMs layer on a metal substrate. (b) Schematic showing the formation of SAMs patches during stamping and dealloying.

10–15 ML thick Au islands are formed, before the protective effect of the Au barrier layer breaks down at the so-called critical potential (E_c) and bulk nanoporous Au is formed[22,23] due to uniform dealloying. But the uniform dealloying can be overruled by the application of SAMs.[23–26] We have recently started research on the effects of SAMs on alloys[23–25] based on noble metal binary systems.[27] Cu–Au alloy substrates covered by SAMs show a different dealloying mechanism compared to bare surfaces. The organic surface layer inhibits dealloying corrosion of noble metal alloys by a suppression of surface diffusion. After passivity breakdown at higher applied potentials, dealloying proceeds in localized regions. With progressing dealloying, the process results in the appearance of surface cracks which form due to an accumulation of local stress related to a volume shrinking of nanoporous gold during dealloying.[28] Together with the high level of control over the structural quality of self-assembled organic films, the system is a promising inhibitor model for understanding localized dealloying corrosion on the molecular scale.

Moreover, laterally structured SAMs can be applied on surfaces employing micro-contact printing (μCP) using simple polydimethylsiloxane (PDMS) stamps.[29–33] The process is sketched in Fig. 1b. This approach includes local chemical reactions[31–33] and possibly also a local exchange of molecules in the monolayer.

Here, we report initial results of a larger research effort where we want to employ the great opportunities offered by molecularly homogeneous and heterogeneous self-assembled organic films used as model inhibitor films towards dealloying and in particular localized corrosion. The use of patterned SAMs and their effect on dealloying is well suited for *in situ* atomic force microscopy to reveal molecular and atomic scale mechanisms of corrosion and corrosion protection. We will further discuss first results of a controlled exchange of thiol and selenol molecules to further manipulate the internal structure of the self-assembled inhibitor monolayers and the respective effect on dealloying corrosion.

Experimental

Standard H_2SO_4 was bought from Sigma Aldrich and diluted to get the required 0.1 M solution. Thiophenol (BT), benzeneselenol (BSe) were purchased from Aldrich and used without further purification. Self-assembled monolayers (SAMs) of target molecules were prepared by immersing the Au(111) substrates into a 1 mM (BT) and 4mM (BSe) ethanolic solution at room temperature. After the allotted immersion time, the samples were removed from the respective solutions, thoroughly rinsed with copious amounts of the solvent (ethanol) and finally dried in a stream of nitrogen. All solutions were freshly prepared before each experiment.

We report the results of the characterization by different techniques including *in situ* electrochemical dealloying by Atomic Force Microscopy (AFM), the exchange of organic molecules (BSe, BT) by Quartz Crystal Microbalance Dissipation (QCMD) and ex-situ Scanning Electron Microscopy after electrochemical dealloying.

(i) *In situ* electrochemical dealloying by atomic force microscopy

Cu_3Au (111) single-crystals with an orientation misfit less than 0.1° and polycrystalline alloys were obtained from MaTecK GmbH (Germany). The surfaces were prepared by repeating cycles of sputtering and annealing in an UHV chamber. The self-assembled monolayers (SAMs) were printed on well-prepared Cu_3Au crystals by using a polydimethylsiloxane (PDMS) stamp with 3 μm × 3 μm square patterns. Cu_3Au crystals were patterned by micro-contact printing (μCP), to yield a surface consisting of target molecules. The desired SAMs were fabricated from the stamp after exposing them to the 1 mM solutions of target molecules. *In situ* AFM experiments on Cu_3Au surfaces were carried out under potentiostatic conditions in 0.1 M H_2SO_4. A platinum wire was used as a counter electrode and a silver/silver-chloride (Ag/AgCl) micro-reference electrode was used as reference electrode. The reference electrode was homemade according to the protocol published by Hassel *et al.*[34] All potentials in this paper are referred to Ag/AgCl. Prior to the AFM experiments, the tip was cleaned in conc. sulfuric acid for 2 minutes, thoroughly rinsed with deionized water and ethanol and dried in a stream of nitrogen. *In situ*-AFM experiments were performed using a JPK Nano-Wizard I AFM (JPK Instruments AG, Berlin, Germany). A syringe pump (Micro-mechatronics AG, Germany) was used for pumping the electrolyte. For electrochemical control, a VoltaLab 40 potentiostat/galvanostat (Radiometer Analytical, France) was used. IR-drop compensation was performed using the current-interrupt method compensating for 95% of the measured IR-drop.

(ii) Adsorption and or replacement of organic molecule experiment by QCMD

Gold quartz crystals used in the QCMD experiments were purchased from Q-sense. 4 mM solutions of BT and BSe were used for this experiment. Firstly, ethanol was passed into the QCMD system (Q-Sense E1 from Biolin Scientific, Finland) to minimize the drift of the system. The data obtained in ethanol are used as the base line of zero, as all the SAMs solutions were prepared in ethanol. The drift in the experiments is less than 0.5 Hz h^{-1}. Once the drift is satisfactory, the adsorption of BT, BSe and the exchange of BT with BSe, or *vice versa*, on polycrystalline gold quartz crystals were measured during the flow of the respective solution.

(iii) Scanning electron microscopy after electrochemical dealloying

Polycrystalline Cu_3Au alloys were obtained from MaTecK GmbH (Germany) in polished form. The samples were first dipped in 0.1 M H_2SO_4 for 1 hour for chemical cleaning. Then BT SAMs was deposited onto the surface by exposing the sample in 4 mM SAMs solution, and an eventual exchange of the BT SAMs was done by BSe by exposing the surface onto the respective solution. One cycle of CV was performed from −0.2 V to 1.0 V *vs.* Ag/AgCl in 0.1 M H_2SO_4 at 10 mV s^{-1} and finally SEM images were taken using a Zeiss Leo scanning electron microscope.

Results and discussion

Localized and inhomogeneous corrosion, the likely case in any realistic application, is much less understood than the homogeneous case. The application of structured thiol SAMs by a pre-formed PDMS stamp is a well established simple lithography method. We have fabricated hexadecanethiol (HDT) SAMs patterns of 3 μm × 3 μm, separated by a gap of 2 μm, as an inhomogeneous film of a model inhibitor. Fig. 2 shows the sequence of AFM images obtained by following the

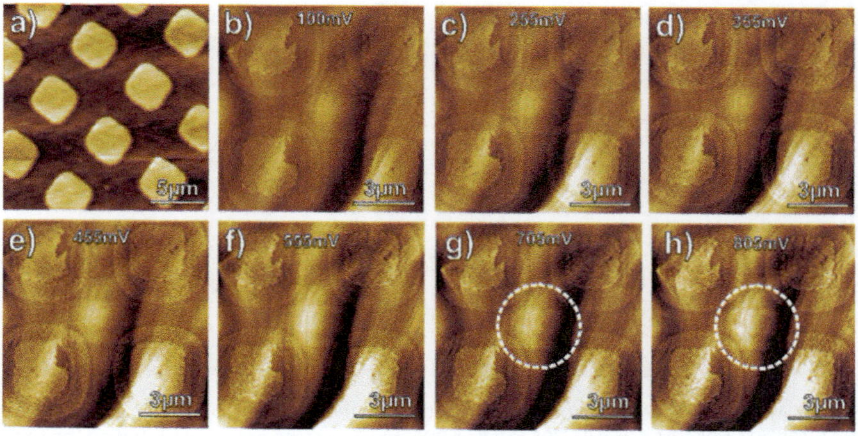

Fig. 2 Alloy corrosion studied with patterned inhibitor films by *in situ* AFM. We show the initial surface roughening and the evolution of the structure up to +805 mV. (a) Typical pattern on pure Au without halo effect (ideal press). The used pattern has been pressed harder to obtain a low-density halo around the central patch at (b) 100 mV, (c) 255 mV, (d) 355 mV, (e) 455 mV, (f) 555 mV, (g) 705 mV, and (h) 805 mV.

potential-dependent initiation of the dealloying process on the surfaces protected by the patterned thiol inhibitor film by *in situ* atomic force microscopy. Fig. 2a represents the used stamp with a typical regular SAMs pattern on pure gold. For our experiment we have intentionally pressed the stamp slightly harder on the Cu_3Au (111) surface in order to obtain a halo or additional ring of a less dense thiol phase around the central patch (Fig. 2b). For the experiments, we have chosen to image an area including 4 patches of HDT SAMs. Two of these show a macroscopic defect, which stems from a presumed defect in the used stamp. The defects allow us to better correlate the structural changes and changes in the surface morphology around the patterns.

The sequence of *in situ* images (Fig. 2b–h) from initially 100 mV to 805 mV does not show the typical Au islands, which form on bare Cu_3Au (111).[20,21] The mentioned (flat) islands indicate a regime of slightly higher initial Cu dissolution below the critical potential. At 805 mV, which is close to the critical potential of bare Cu_3Au surfaces, dealloying clearly occurs in the central part between the applied thiol patches (area indicated by a white circle). Yet, the part of the surface where no thiol film or islands (in the AFM images) can be seen is also still completely stable as revealed by the flat morphology showing the typical crystalline steps of intact well-prepared single-crystalline surfaces.

Fig. 3 shows a selection of enlarged sections focusing on the thiol SAMs around the two macroscopic defects of the sequence shown in Fig. 2. The central part, representing a dense thiol SAMs, does not show considerable changes of the local film morphology although it appears slightly rougher at higher applied potentials. In contrast, the halo shows major changes. At 100 mV the halo is formed by a smooth continuous film with a clear height contrast in comparison to the central dense SAMs. This directly points to a possible lying-down thiol phase within the halo film. The white arrows in the upper row of Fig. 3 follow a thiol island, which appears to be stable over the experimental run. The surrounding of

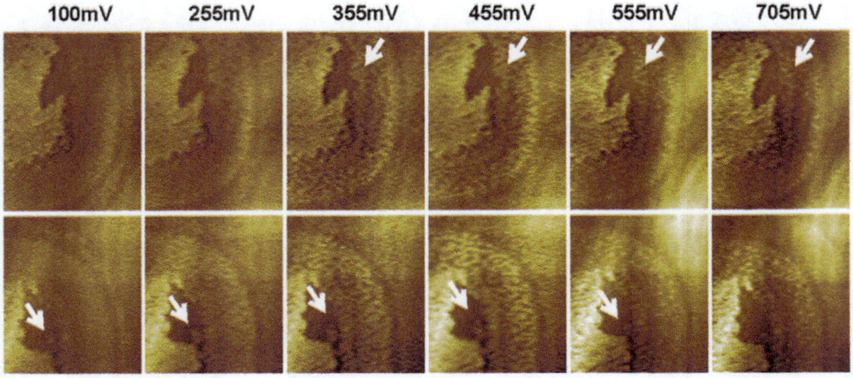

Fig. 3 Magnified view of the structure after stepping the potential to 100 mV, 255 mV, 355 mV, 455 mV, 505 mV, 605 mV, 705 mV and 805 mV. The same potential (stated on the top) was applied to the sample shown on the top and below.

this discernible island has nevertheless considerably changed at 705 mV. The initial dense and continuous lying-down phase has developed into islands covering the area of the initial halo. An interesting detail is that this area of the initial halo appears at the end with a clear height difference of the underlying substrate towards the outer surface which appears slightly lower. The same processes are mirrored in the lower row where the lower white arrow points to a thiol island which initially grows but finally vanishes completely. Such phenomena of evolution and disappearance of islands generated from organothiolate[35] SAMs/Au (111) have previously been observed. Fig. 4 focuses on the image obtained at 805 mV and reveals a height difference of the surface below the thiol islands of the halo and the outer (Au-enriched) surface of exactly 2 monoatomic steps (about 0.48 nm, a double step on Au (111)). On the outer surface mono-atomic steps are present (about 0.24 nm step height). The higher resolution image also reveals that the formed thiol islands within the halo area have the same height as the initial SAMs of the inner dense core of the applied pattern. Above the critical potential, at 805 mV, the area in between the thiol SAMs undergoes dealloying, leaving part of the surface in the vicinity of the halo region

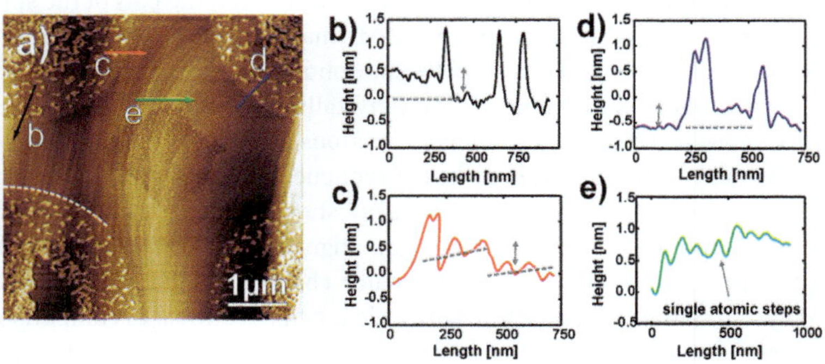

Fig. 4 High resolution atomic force micrograph showing the structure of the Cu_3Au alloy surface after stepping and holding the potential at 805 mV.

unaltered. Thus, an area in the form of a second ring becomes evident around the printed central dense SAMs where apparently thiol molecules again influence the stability of the surface and protect from a dealloying attack.

Fig. 5 depicts the AFM images of Fig. 2 in phase mode. The phase signals along the images indicate a reason for the observed second or outer halo area. At lower potentials, only one homogeneous halo ring around the central part is visible (Fig. 5a–c). But at higher potentials, another bright halo ring becomes visible (Fig. 5d–h) due to diffusion of thiol from the initial halo. Thus, we interpret this region as a lower-density thiol phase. This emerging second ring (see white arrow) extends right to the middle area of the image where dealloying commences (white circle in Fig. 5h).

Fig. 6 and 7 focus on the area where dealloying is clearly visible at potentials of 805 mV or higher. The first contrasting and important feature to our previous study on bare Cu_3Au surfaces[20,21] is the fact that no flat Au islands form below the onset of bulk dealloying at the critical potential. Without thiols 10–15 ML thick Au islands are formed on Cu–Au alloys at intermediate potentials well below the critical potential. Above the critical potential finally a bulk nanoporous Au network is formed due to passivity breakdown of the Au film. The suppression of the formation of Au islands was nevertheless observed in experiments where a continuous self-assembled thiol film was applied[23] and localized nanoporous etch pits appeared at the anodically shifted critical potential. The site of dealloying is given in this experiment by the applied pattern of SAMs and line profiles in the attacked regions clearly show the progression of dissolution (Fig. 6c and f). Initially at 805 mV, the observed bottom surface is about 1 nm lower than that of the surrounding area, and later at 905 mV, the pitted area is about 3 nm deep (Fig. 6c and f). To accumulate about 1 ML of Au, Cu from about 4 ML of Cu_3Au has to be removed by selective dissolution. In other words, a pit of 3 nm points to an accumulated Au thickness of about 1 nm, *i.e.* the dealloying process is still in an initial state. Full ligaments of a nanoporous Au structure are typically in the order of 10–40 nm.

The evolution of the morphology at the locally attacked sites during dealloying is also interesting to follow. Initial individual islands are first forming in low

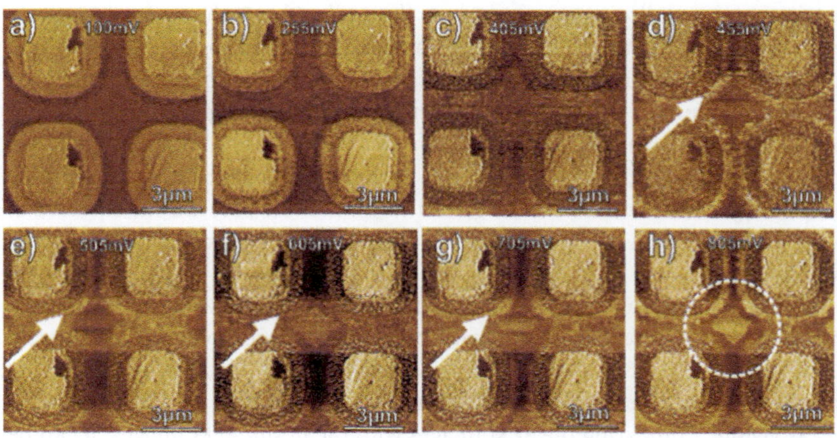

Fig. 5 Phase images of the structure after stepping the potential to 100 mV, 255 mV, 355 mV, 455 mV, 505 mV, 605 mV, 705 mV and 805 mV.

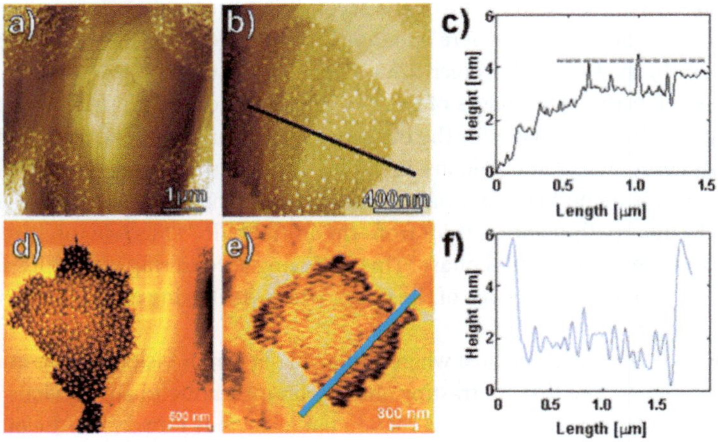

Fig. 6 High resolution AFM images and height profiles along the indicated lines at 805 mV (a–c) and 905 mV (d–f).

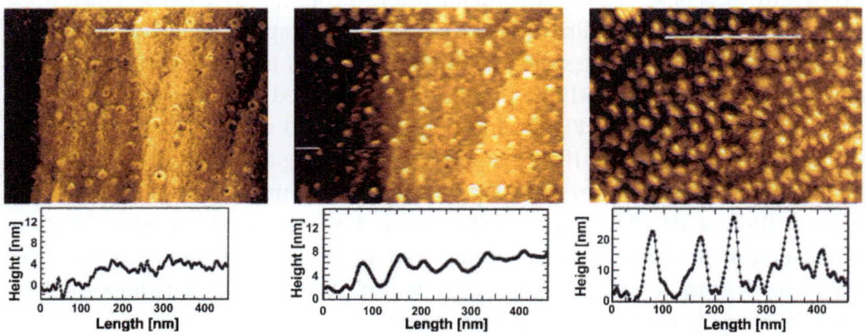

Fig. 7 High resolution atomic force micrographs showing the structure of the inner region with visible dealloying while holding the potential at 805 mV at different times.

density before dealloying proceeds into the bulk. Fig. 7 shows one sequence recorded from the inner part of an area undergoing dealloying. The receding bottom surface is initially following the original atomic steps of the well-prepared surface. With further dealloying, the interface develops a higher roughness and the islands cover the interface more densely. Here, the island formation is at potentials close to E_c, the first step in the evolution of nanoporosity. The surface morphology clearly resembles the island formation on bare Cu_3Au surfaces at much lower potentials and can be clearly distinguished from other alloys such as Cu–Pd.[36]

The *in situ* AFM imaging was stopped at 905 mV and SEM images, such as shown in Fig. 8a, were taken. Dealloying processes have visibly taken place over a large sample area and it always occurs at the farthest site from the printed SAMs patches. The dense central thiol SAMs as well as the first halo around this central area is visible. Comparing the SEM images with the *in situ* AFM images, it can be inferred that a large part of the area between the SAMs patches and also their halo is clearly free of dealloying and the actual dealloying has taken place only in a

Fig. 8 (a) SEM image of the Cu_3Au (111) surface after the *in situ* AFM experiment up to 905 mV. (b) Image after dealloying of a Cu_3Au polycrystalline surface modified with a continuous self-assembled mixed aminobenzenethiol monolayer. (c) SEM image of a grain surface partly free of thiols after a potential step to 1.3 V for 15 s and a CV.

limited area. Few of these potential dealloying sites are fully protected allowing no visible dealloying at all. Part of the thiol-free area was therefore apparently protected from massive Cu dissolution. We nevertheless note here that some limited number of spots have undergone a more severe dealloying attack. These originate at sites where visibly one of the central thiol patches is missing, *i.e.* the printing process has not properly succeeded and dealloying has locally spread out over a larger region. Due to such flaws and their associated inhomogeneous current distribution, it is not possible to correlate the measured total current with the observed changes in morphology, in particular, the local observation by AFM. Yet, since dissolution is a kinetically fast process, we can conclude that the applied potentials are homogeneous over the surface.

When we compare dealloying from patterned thiol-modified alloys (current study) with dealloying from homogeneously thiol-modified alloys (previous study[23]) a few interesting aspects can be shown. With the printed pattern of SAMs, we could control the site where dealloying corrosion initiates. In contrast, on homogeneously covered thiol-surfaces a stochastic distribution of dealloying sites has been described.[23] For these sites, the separation is much larger as natural structural inhomogeneity than for example the typical domain size as exemplified for a surface initially modified by a mixed aminobenzenethiol film, [24] shown in Fig. 8b. Still an open question therefore is the nature of the potential initiation site. Approaching this problem using pre-patterned inhomogeneous SAMs may well contribute to the molecular-level understanding of the inhibition effect. A further approach is represented in Fig. 8c where an initially continuous ethanethiol monolayer formed on Cu_3Au was partially removed by a short oxidative potential step. The effective removal of ethanethiol by such an oxidation step has been observed on Au by *in situ* electrochemical surface-sensitive X-ray diffraction, employing synchrotron light.[25] This process results in an inhomogeneous dealloying process (Fig. 8c). The driving force for these experiments has been to advance our understanding of the effect of structural inhomogeneity on the onset of dealloying for inhibitor-covered surfaces. Yet chemical or molecular inhomogeneity will potentially play a large role and provides further means to manipulate and control the stability of inhibitor-covered alloy surfaces.

Molecular inhomogeneity can be deliberately introduced by exchanging part of the applied SAMs on the molecular level and can be simultaneously monitored by

QCMD as the sensitivity of the technique is few ng cm^{-2}. Fig. 10 shows the areal mass change on gold quartz crystals during the adsorption and exchange of BT with BSe or *vice versa*. The base line shown until 300 seconds is the data obtained while ethanol was passed into the QCMD. Since all SAMs were assembled in ethanol, ethanol was referenced as the base line. As shown in Fig. 9, after 300

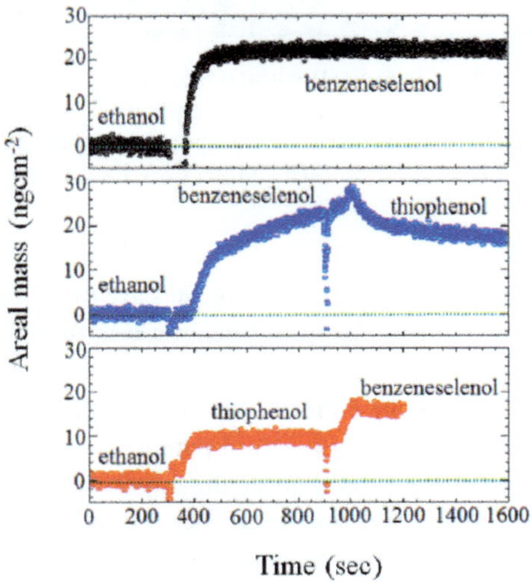

Fig. 9 Plot showing the areal mass change on Au with time during the adsorption or adsorption and exchange of respective molecules. The red line first shows the adsorption of the BT molecule and then the replacement by BSe. The black line represents the curve obtained during the adsorption of BSe and the blue line shows the adsorption of BSe and its potential exchange with BT.

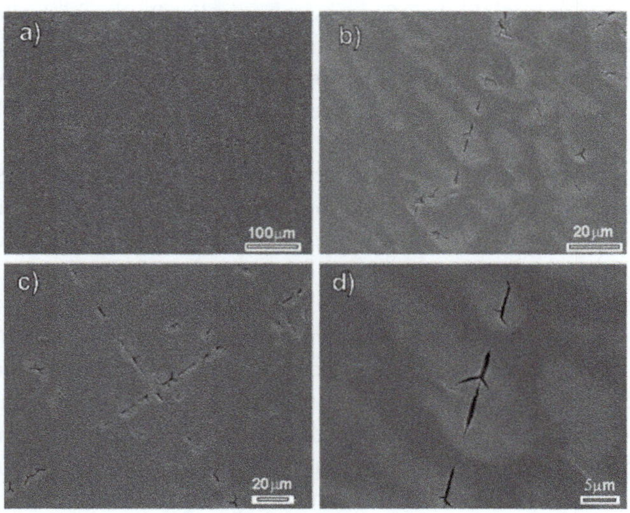

Fig. 10 SEM images after dealloying of a Cu_3Au polycrystalline surface modified with a self-assembled BT and exchanged with a BSe monolayer. (a)–(d) SEM images of the surface cracks after one CV from −0.2 to 1.0 V *vs.* Ag/AgCl at 10 mV s^{-1} in 0.1 M H_2SO_4.

seconds of passing ethanol, solutions containing SAMs were passed and their adsorption increases the areal mass in each case. In two other experiments, a different (BT or BSe) molecule was passed after 600 seconds, which either increased (see red line) or did not show any effect (see blue line) on the areal mass. Some limited disturbance of the measurement system is caused during the process of solution change.

Kaefer *et al.*[37] report molecular areas of 28.8 Å^2 and 38.5 Å^2 for BSe and BT, resulting in about 85 ng cm^{-2} and 43 ng cm^{-2} for a monolayer, respectively. We see in Fig. 9 that our measured values on a rough sample are lower but show the same ratio of about 2. Thus, a SAMs of BSe or BT is formed when the respective molecules are passed to the crystal (see black and red line) but when replacing the BT SAMs by BSe, we find that it is to about 80% exchanged. When BT was passed the areal mass decreased slightly, indicating some replacement of BT molecules as well. Interestingly, the exchange from a full BSe SAMs exhibits slower exchange kinetics, compared to exchanging BT by benezeneselenol. Yet, both ways converge to coverage of approximately 80% BSe and 20% BT.

Based on the exchange, we expect different thin-film properties, and as such a moderation of the dealloying process. These results on molecular exchange have been then applied to a polycrystalline Cu_3Au foil for a first test of respective changes in the behavior during dealloying corrosion. Fig. 10 shows the SEM recorded after the exchange of BT SAMs by BSe according to Fig. 9 (red curve). As seen in Fig. 9, the mass obtained during exchange of BT with BSe is not the same as when BSe alone is passed. So, we conclude that the exchange process did not replace BT by 100%. The polished surface was then dealloyed by a potential sweep from −200 mV to 1000 mV and back with a rate of 10 mV s^{-1}. As a result, the alloy is non-homogeneously dealloyed as can be seen in Fig. 10 where the brighter regions represent the dealloyed regions. Round as well as slightly elongated dealloyed areas with typical dimensions of between 1 μm and maximum 20 μm can be distinguished. Some polishing lines are also visible but, in general, the dealloyed regions are rather independent of these defect lines. The larger dealloyed regions, however, often show cracks along the crossing polishing lines. Independent of the cracking behavior we can conclude that the exchange process did indeed significantly change the dealloying initiation, as shown in Fig. 8b and 10b. With a homogeneous SAMs, dealloying occurs purely local, producing round areas in the beginning which later form cracks (Fig. 8b) but we see dealloying in more elongated and outspread forms indicating dealloying is not purely local anymore (Fig. 10b). Routes to control the growth and behavior of organic molecules on metal surfaces are also important for functionalized interfaces, for example in sensor applications. Inhibition of corrosion is another economically important process where metal–organic interfaces play a crucial role. Most of today's inhibitors are organic in nature.[1] Dealloying of Cu–Au alloys is due to clearly distinguished stages of surface evolution (ultrathin film – Au islands – nanoporosity) of the bare substrate and the localized nature of SAMs-modified (inhibited) surfaces, an in our view interesting model system. Here, we addressed different aspects of control over a localized corrosion process on Cu–Au alloys. Printed structured patterns produce a local variation of different inhibitor densities and enabled us, by varying the basic dimensions of the pattern, to address the important questions of inhibitors and the mechanistic understanding of material (surface) protection on the molecular level. Also, a combination of approaches with a molecular exchange, the second aspect is presented here, is possible. We presented the manipulation of the internal structure of a given SAMs by molecule exchange processes as a promising approach to address the molecular stability of an inhibitor monolayer film towards dealloying.

As reported for Cu and Au, the surface mobility may be orders of magnitude higher in solution (in particular when halides are present) than in vacuum or dry atmospheres.[38–41] Smaller structural defects should be removed by gentle chemical annealing in acidic electrolytes but larger polishing scratches will not be removed by this simple method (requires a number of sputter-annealing cycles in UHV). Using such a surface with still present scratches, however, allows in parallel to follow the effect of such macroscopic defects on the dealloying behavior. Although a slight increase of dealloying probability is visible, the density and size of the dealloyed sites are not largely influenced by these defects. But higher numbers of cracks are clearly visible along such polishing lines. Also, the cracks follow mostly the direction of the defect or polishing lines. Macroscopic structural defects thus do clearly influence the stability of our chosen model system, Cu_3Au, during dealloying.

Conclusion

The conclusions that are derived for the current study are as follows:

➢ *In situ* AFM studies of dealloying corrosion of surfaces modified with a micro-structured inhibitor pattern applied by μ-contact printing is a promising approach to study the effect of inhomogeneous organic molecular films.

➢ Even a small density of hexadecanethiol SAMs molecules can suppress the typical Au island formation on Cu_3Au (111) at intermediate potentials.

➢ This study also shows the potential dependent mobility of thiol islands. Diffusion of hexadecanethiol molecules can take place at elevated potentials.

➢ Only at higher potentials, dealloying initiates at the areas with the lowest concentration of thiol.

➢ The local dealloying pattern is highly influenced and controlled by a molecular exchange of thiols to selenols and is also affected by introducing structural defects such as scratches or polishing defects (lines).

References

1 M. Finsgar and J. Jackson, *Corros. Sci.*, 2014, **86**, 17–41.
2 Y. Chen, Z. Xu, C. Smith and J. Sankar, *Acta Biomater.*, 2014, **10**, 4561–4573.
3 R. A. Antunes and M. C. Lopes de Oliveira, *Acta Biomater.*, 2012, **8**, 937–962.
4 H. H. Uhlig, *Uhlig's Corrosion Handbook*, 3rd edn, ed. R. W. Revie, John Wiley & Sons, Inc., Hoboken, NJ, USA, 2011.
5 E. E. Oguzie, C. B. Adindu, C. K. Enenebeaku, C. E. Ogukwe, M. A. Chidiebere and K. L. Oguzie, *J. Phys. Chem. C*, 2012, **116**, 13603–13615.
6 J. Denayer, J. Delhalle and Z. Mekhalif, *J. Electroanal. Chem.*, 2009, **637**, 43–49.
7 G. Fonder, J. Delhalle and Z. Mekhalif, *Appl. Surf. Sci.*, 2010, **256**, 2968–2973.
8 P. E. Laibinis and G. M. Whitesides, *J. Am. Chem. Soc.*, 1992, **114**, 9022–9028.
9 G. K. Jennings, J. C. Munro, T. H. Yong and P. E. Laibinis, *Langmuir*, 1998, **14**, 6130–6139.

10 S. Hosseinpour, M. Gothelid, C. Leygraf and C. M. Johnson, *J. Electrochem. Soc.*, 2014, **161**, C50–C56.
11 J. C. Love, L. A. Estroff, J. K. Kriebel, R. G. Nuzzo and G. M. Whitesides, *Chem. Rev.*, 2005, **105**, 1103–1169.
12 F. Schreiber, *Prog. Surf. Sci.*, 2000, **65**, 151–256.
13 J. C. Love, D. B. Wolfe, R. Haasch, M. L. Chabinyc, K. E. Paul, G. M. Whitesides and R. G. Nuzzo, *J. Am. Chem. Soc.*, 2003, **125**, 2597–2609.
14 A. Bashir, D. Kaefer, J. Mueller, C. Woell, A. Terfort and G. Witte, *Angew. Chem., Int. Ed.*, 2008, **47**, 5250–5252.
15 J. V. B. J. A. Martínez, R. C. Milánc, J. Herrera, M. H. Farías and M. P. Hernández, *Appl. Phys. Lett.*, 2014, **320**, 287–293.
16 C. Vericat, M. E. Vela, G. A. Andreasen, R. C. Salvarezza, F. Borgatti, R. Felici, T. L. Lee, F. Renner, J. Zegenhagen and J. A. Martin-Gago, *Phys. Rev. Lett.*, 2003, **90**, 075505.
17 J. Jia, A. Bendounan, H. M. N. Kotresh, K. Chaouchi, F. Sirotti, S. Sampath and V. A. Esaulov, *J. Phys. Chem. C*, 2013, **117**, 9835–9842.
18 K. Szelagowska-Kunstman, P. Cyganik, B. Schuepbach and A. Terfort, *Phys. Chem. Chem. Phys.*, 2010, **12**, 4400–4406.
19 W. Azzam, A. Bashir, A. Terfort, T. Strunskus and C. Woll, *Langmuir*, 2006, **22**, 3647–3655.
20 F. U. Renner, A. Stierle, H. Dosch, D. M. Kolb, T. L. Lee and J. Zegenhagen, *Nature*, 2006, **439**, 707–710.
21 F. U. Renner, A. Stierle, H. Dosch, D. M. Kolb, T. L. Lee and J. Zegenhagen, *Phys. Rev. B: Condens. Matter Mater. Phys.*, 2008, **77**, 235433.
22 J. Erlebacher, M. J. Aziz, A. Karma, N. Dimitrov and K. Sieradzki, *Nature*, 2001, **410**, 450–453.
23 A. Pareek, S. Borodin, A. Bashir, G. N. Ankah, P. Keil, G. A. Eckstein, M. Rohwerder, M. Stratmann, Y. Gruender and F. U. Renner, *J. Am. Chem. Soc.*, 2011, **133**, 18264–18271.
24 A. Pareek, G. N. Ankah, S. Cherevko, P. Ebbinghaus, K. J. J. Mayrhofer, A. Erbe and F. U. Renner, *RSC Adv.*, 2013, **3**, 6586–6595.
25 G. N. Ankah, A. Pareek, S. Cherevko, J. Zegenhagen and F. U. Renner, *Electrochim. Acta*, 2014, **140**, 352–358.
26 F. U. Renner, A. Bashir, G. N. Ankah, B. R. Shrestha, D. Ma, P. Losada-Pérez, M. J. Duarte, M. Nellessen, A. Khorashadizadeh, P. U. Biedermann, D. Raabe and M. Valtiner, Star-shaped crystallographic cracking of localized nanoporous defects, submitted.
27 T. P. Moffat, F. R. F. Fan and A. J. Bard, *J. Electrochem. Soc.*, 1991, **138**, 3224–3235.
28 S. Parida, D. Kramer, C. A. Volkert, H. Roesner, J. Erlebacher and J. Weissmueller, *Phys. Rev. Lett.*, 2006, **97**, 035504.
29 A. Kumar and G. M. Whitesides, *Appl. Phys. Lett.*, 1993, **63**, 2002–2004.
30 J. L. Wilbur, A. Kumar, H. A. Biebuyck, E. Kim and G. M. Whitesides, *Nanotechnology*, 1996, **7**, 452–457.
31 J. Tien, A. Terfort and G. M. Whitesides, *Langmuir*, 1997, **13**, 5349–5355.
32 L. Yan, C. Marzolin, A. Terfort and G. M. Whitesides, *Langmuir*, 1997, **13**, 6704–6712.
33 L. Yan, X. M. Zhao and G. M. Whitesides, *J. Am. Chem. Soc.*, 1998, **120**, 6179–6180.

34 A. W. Hassel, K. Fushimi and M. Seo, *Electrochem. Commun.*, 1999, **1**, 180–183.
35 W. Azzam, A. Bashir, P. U. Biedermann and M. Rohwerder, *Langmuir*, 2012, **28**, 10192–10208.
36 G. N. Ankah, S. Meimandi and F. U. Renner, *J. Electrochem. Soc.*, 2013, **160**, C390–C395.
37 D. Kaefer, A. Bashir and G. Witte, *J. Phys. Chem. C*, 2007, **111**, 10546–10551.
38 J. Wiechers, T. Twomey, D. M. Kolb and R. J. Behm, *J. Electroanal. Chem.*, 1988, **248**, 451–460.
39 C. Alonso, R. C. Salvarezza, J. M. Vara and A. J. Arvia, *Electrochim. Acta*, 1990, **35**, 1331–1336.
40 M. Giesen and D. M. Kolb, *Surf. Sci.*, 2000, **468**, 149–164.
41 M. Giesen, *Prog. Surf. Sci.*, 2001, **68**, 1–153.

Faraday Discussions

ROYAL SOCIETY OF CHEMISTRY

DISCUSSIONS

Corrosion scales and passive films: general discussion

Gerald Frankel, Janine Mauzeroll, Geoffrey Thornton, Hendrik Bluhm, Jonathan Morrison, Vincent Maurice, Trevor Rayment, David Williams, Angus Cook, Gaurav Joshi, Alison Davenport, Simon Gibbon, Denis Kramer, Matthew Acres, Markus Tautschnig, Hiroki Habazaki, Philippe Marcus, David Shoesmith, Clara Wren, Tom Majchrowski, Rob Lindsay, Mary Wood, Mira Todorova, John Scully, Frank Renner, Anton Kokalj, Christopher Taylor, Sannakaisa Virtanen and Julian Wharton

DOI: 10.1039/c5fd90045j

Sannakaisa Virtanen opened a general discussion of the paper by Mira Todorova: In view of the fact that there is a lot of data available for ZnO (as a semiconducting oxide), how "easy or difficult" is it to use the same methodology for other oxides, such as for instance iron oxides or chromium oxide? What experimental data would be needed for some validation of the calculations?

Mira Todorova replied: There is no principle problem in applying the methodology to other oxides, *i.e.* iron oxides or chromium oxides. The question boils down to the availability of the relevant defect data. ZnO was a great oxide to consider as an example, because it is of interest to both the corrosion and the semiconducting community. Thus, there were already a lot of data available for ZnO. What we would now need to do is to expand on the available database of *ab initio* calculations for charged point defects in other oxides or materials of interest in the context of corrosion. Unfortunately, there is little, if any, such information available for most of the oxides which are interesting in the context of corrosion. Furthermore, some of the most interesting oxides in the context of corrosion, such as the iron or chromium oxides, are unfortunately very challenging for *ab initio* calculations, as they possess strong electronic correlations. This means that the required calculations are not just "routine calculations", which can be performed in a high-throughput fashion.

There are various experiments that can be performed to obtain information about the point defects present in a system, such as DLTS (deep level transient spectroscopy), Hall measurements, EPR, *etc.* A general discussion about the comparison of the discussed defect calculations to experiment can be found in ref. 1. A discussion focusing on the experimental identification of the discussed native point defects in ZnO can be found in ref. 2.

1 C. G. Van de Walle and J. Neugebauer, *J. Appl. Phys.*, 2004, **95**, 3851.
2 A. Janotti and C. G. Van de Walle, *Rep. Prog. Phys.*, 2009, **72**, 126501.

Philippe Marcus asked: If the defects are neutral, one would expect no effect of the electric field on oxide growth. Is this what you suggest?

Mira Todorova responded: This is a very interesting suggestion to experimentally verify the existence/presence of charge neutral defects. We have not considered this aspect in our paper but would like to call for corresponding experiments in the region of phase space where such defects are predicted to be thermodynamically stable.

Hendrik Bluhm questioned: Have you considered the presence of a hydroxide phase at the interface between ZnO and water? How would the presence of the hydroxide phase influence your model on the role of point defects in governing the growth/dissolution of the oxide formed on Zn?

Mira Todorova answered: A hydroxide phase at the interface was not considered. We chose ZnO for two reasons: (i) because of the availability of a vast amount of relevant data found for ZnO in the semiconductor literature and (ii) because ZnO is mentioned in the experimental literature on corrosion, as being the first compact oxide layer formed during corrosion on ZnO.[1] The methodology presented in our work can be equally well applied to a hydroxide phase. Unfortunately, to our knowledge the information about the point defects forming in a Zn-hydroxide, which is needed for our thermodynamic model, is not available in the literature yet. Once we have performed the necessary density-functional theory calculations for the point defects in the hydroxide phase, the presence of the hydroxide phase can be explicitly considered.

1 S. Thomas, *et al.*, *Electrochimica Acta*, 2013, **97**, 192.

Gerald Frankel asked: Regarding the neutral defects, are the electrons or holes kept localized to the vacancies by electrostatic interaction?

Mira Todorova replied: The neutral defects are true bonding states. A comprehensive study of the electronic and geometric structure of these defects can be found in ref. 1 (for the oxygen split interstitial defect O_i^0 (split)) and in ref. 2 (for the oxygen vacancy V_O^0). Following are briefly some of the main points, regarding the bonding nature of these defects, as found in the mentioned publications.

In the O_i^0 (split) configuration two oxygen atoms share a lattice site and form a bond of length 1.46 Å with each other. This leads to the appearance of two almost degenerate states in the band bap, which are completely filled. The electronic configuration of this defect resembles an antibonding $pp\pi^*$ molecular state found in an isolated O_2 molecule.

The oxygen vacancy defects involve 4 Zn dangling bonds (sp^3-hybrids) and two electrons. The Zn dangling bonds combine to a fully symmetric a_1 state located in the band gap and three almost degenerate states in the conduction band. For the neutral oxygen vacancy, the a_1 state is fully occupied by two electrons.

1 A. Janotti and C. G. Van de Walle, *Phys. Rev. B: Condens. Matter Mater. Phys.*, 2007, **76**, 165202.
2 A. Janotti and C. G. Van de Walle, *Rep. Prog. Phys.*, 2009, **72**, 126501.

Christopher Taylor enquired: The "Pourbaix diagram" for point-defects shows the regions of E_h–pH space subdivided, according to which point-defect has the lowest free energy of formation. However, I was wondering if it is not impossible for multiple types of point-defect to co-exist in the same domains, if they both have negative free energies?

Mira Todorova responded: Having a negative free energy of formation is not a prerequisite for the existence of a defect. Usually the reason for the presence of defects in a material is configurational entropy, not an exothermic formation enthalpy at $T = 0$ K. The defects shown in the defect stability phase diagram (Fig. 3 in our paper) are the dominant defect in the system, *i.e.* the defects which will be present in high concentration at the conditions at which they have the lowest formation energy. Thus, indeed the presence of other defects within the system is not excluded and various defects may coexist. Fig. 2 in our paper shows all the possible native defects within the ZnO system. There are various conditions, at which defects, other than the ones appearing in the defect stability phase diagram, have formation energies not much higher than the formation energies of the respective dominant defect. The likelihood and concentration in which such defects will be found in the system at the given conditions can be calculated using their formation energy at the conditions of interest and eqn (1).

Christopher Taylor asked: Is it possible to construct a point-defect phase diagram when the oxide film has more than one metallic element and/or the oxide film is formed over an alloy?

Mira Todorova answered: Yes, it is straight forward to construct defect phase diagrams for systems containing more than two elements. The methodology remains unchanged and allows to consider any number of elements. The only difference will be that such a diagram will no longer be 2-dimensional but must reflect the number of chemical species that can change independently.

John Scully commented: Thank you for your clarification that your map shows the most stable defect from the thermodynamic view and that there might be other "less stable" defects at a lower concentration which co-exists.

I have two questions. How do you rationalize experiments on oxide growth that suggest "transference number" or that many oxides grow by a combination of cations moving out and anions in? Marker studies indicate this and that experimental conditions can be changed to alter this behavior.

The second question regards the role of metastable point defects and the higher disordered state of passive films when they first begin to thicken. How should we think about these? Couldn't these metastable point defects be dominant factors early in growth?

Mira Todorova replied: Based on Fig. 2 in our paper I would conclude that, at the conditions relevant when ZnO comes into contact with water, all the metastable defects are rather high in energy to be expected to play any significant role. However, one should keep in mind that the model system we use here is still rather simple compared to the complexity of the oxide films formed in a corrosive environment. Therefore, a next step would be to expand this defect picture to the consideration of defects in the subsurface and surface region of the oxide as well as to include diffusion barriers.

Clara Wren remarked: Your calculations assume that ZnO is a relatively pure semi-conductor except for the small density of defects that you introduce. On a corroding surface, the oxide layer is not likely a uniform layer but will have a distribution of ions and ion vacancies. Thus, the electron energy at the metal/oxide interface is near the energy of the conduction band energy while that at the oxide/water interface is near the valence band energy. Since your model shows stability regions for different defects, could your model be used in predicting the kinetics of oxide growth?

Mira Todorova responded: You are, of course, correct that the oxide at the corroding surface will be much more complex than the rather perfect oxide assumed in our theoretical approach. Such a simple model will not be able to capture the whole complexity of the corrosion product. The consideration of the charged bulk point defects presents a first step, which provides some insight into the system and forms a solid basis which we can expand on in future. An obvious next step would be to explicitly include the presence of a surface/interface and investigate sub-surface and surface defects. While the model allows identifying trends, without knowledge of at least the migration barriers it is difficult to predict the kinetics of the oxide growth.

David Williams addressed Mira Todorova and Clara Wren: I'm interested to explore the connection of your defect stability diagrams to measurements of dependence of current density on applied electrode potential. The paper by Momeni and Wren is also relevant to this discussion, as they discuss that the potential difference between metal and solution is distributed across the two interfaces (metal–oxide and oxide–solution) and across the oxide film. The dependence on interfacial potential difference of the current flow across the interfaces can be described by a Butler–Volmer equation. The dependence of the current on the potential difference across the film can be described by a high-field growth law (Cabrera–Mott) or its linear low-field limit in which, of course, the charge on the migrating defect appears. The potential profile across the system can be drawn onto the diagram of μ_O against E_F where at least the end points (metal and solution) are known, and the path then drawn depends on assumptions about the potential differences across the interfaces and the electric field in the oxide film, and whether there are any composition variations (space charges) developed in the oxide near the interfaces. Where this path goes will determine the dependence of the current on the field in the oxide, through the charge on the migrating defects, and since the current is continuous it will then determine the magnitude of the potential differences across the interfaces. There ought to be a self-consistent solution which can be mapped onto experimental data, similar to the procedure used by Momeni and Wren. Could you comment?

Mira Todorova answered: This is an interesting suggestion, but requires additional conceptual development in order to include the potential profile and connect our defect stability diagrams with the Cabrera–Mott model. It is a point we will consider in the future.

Clara Wren commented: This question is primarily addressed to Dr. Mira Todorova. My comment on the discussion on defect stability regions is that Dr. Todorova's model, which shows stability regions for different defects, further confirms our assumptions regarding a corroding system. On a corroding surface, the oxide layer is not likely a pure semi-conductor, but will have a distribution of ions and ion vacancies, and the electron energy at the metal/oxide interface is near the energy of the conduction band energy while that at the oxide/water interface is near the valence band energy.

Philippe Marcus asked: The energy of defect formation is certainly an important factor, but dont't you think that the energy barrier (and hence activation energy) for the migration of defects is also very important for oxide growth?

Mira Todorova responded: Yes, indeed. The energy barrier for the migrations of defects is a very important parameter, which needs to be considered. The thermodynamic aspect we consider here should be regarded only as a first step towards modelling the oxide growth. Specifically, this step provides information about the (thermodynamically) dominant defects and about gradients. Whether these gradients can be realised will depend on the kinetics in the system.

In the semiconducting literature mainly the energetic position of the defect levels are of interest since this controls the electronic behaviour of the material. Consequently, there is a lot of data available about defect formation energies and different charged states (for materials of interest in this field), since these quantities yield information about the charge transition levels. Data on diffusion barriers is much sparser. Some information about the migration of oxygen and zinc defects in ZnO can be found in ref. 1, albeit at the LDA+U level.

The application and usefulness of such an *ab initio* derived methodology in corrosion science requires that the calculation of quantities like defect formation energies and migration barriers is pursued much more vigorously in the future, in particular for materials interesting in the context of corrosion.

1 A. Janotti and C. G. Van de Walle, *Phys. Rev. B: Condens. Matter Mater. Phys.*, 2007, **76**, 165202.

Denis Kramer queried: Do the supercell calculations of the neutral defects, especially the neutral oxygen vacancy, yield semi-conducting states with a band gap?

Mira Todorova said: Yes, the point defects yield semi-conducting states in the band-gap. The band structures for the perfect ZnO crystal, the neutral oxygen vacancy and 2+ oxygen vacancies, calculated with the HSE-functional, are shown in Fig. 3 of ref. 1. A discussion of the electronic structure of the oxygen vacancy (in its various charged states) can also be found in ref. 2.

1 F. Oba, A. Togo, I. Tanaka, J. Paier and G. Kresse, *Phys. Rev. B: Condens. Matter Mater. Phys.*, 2008, 77, 245202.
2 A. Janotti and C. G. Van de Walle, *Rep. Prog. Phys.*, 2009, 72, 126501.

Philippe Marcus asked: To what extent do the boundaries between stability domains of defects change if different hybrid functionals are used?

Mira Todorova responded: This is a very interesting and relevant question. Unfortunately, there are as yet no systematic investigations, which would allow us to give a general answer to this question. It is something that definitely needs to be investigated, since it will allow us to assess the accuracy we can achieve.

Simon Gibbon enquired: At the end of the penultimate paragraph of your paper you state that "this insight may be used to identify new strategies towards improving corrosion protection." Please could you explain what new strategies you envision?

Mira Todorova replied: The possibility to use a defect stability phase diagram to identify the defects relevant in the context of oxide growth and quantify the regions within which they are stable opens the door to perform systematic searches for alloying elements/dopants which may (i) make ZnO more intrinsic, *i.e.* shift the Fermi level towards p-type conditions, and (ii) prevent the formation of a defect, which has been identified as being harmful, or bind it by forming some complex, which prevents its migration.

Tom Majchrowski communicated: An interesting model is proposed, but how would it respond to the presence of alloying elements and/or impurities?

Mira Todorova communicated in reply: There is no fundamental difference between the treatment of charged point defects native to the considered oxide or charged point defects stemming from alloying elements and/or impurities within the proposed method. In the context of semiconductors, defects due to elements alien to the semiconductor are dopants and a subject of investigation. A brief glance at the ZnO literature revealed that there are, for example, first principles calculations for hydrogen,[1] nitrogen,[2] or Si and Ge[3] already available. It would be straight forward to use this information when constructing the defect stability phase diagram for ZnO. The configurational space which needs to be considered by the inclusion of dopants (impurities, alloying elements) increases and there is also always the question of availability of existent calculations for the impurity of interest.

1 A. Janotti and C. G. Van de Walle, *Rep. Prog. Phys.*, 2009, **72**, 126501.
2 J. L. Lyons, A. Janotti, and C. G. Van de Walle, *Appl. Phys. Lett.*, 2009, **95**, 252105.
3 J. L. Lyons, A. Janotti, and C. G. Van de Walle, *Phys. Rev. B: Condens. Matter Mater. Phys.*, 2009, **80**, 205113.

Rob Lindsay asked: Could you please clarify the charge removed/remaining upon creation of the various point defects (*e.g.* the neutral oxygen vacancy)?

Mira Todorova responded: Each Zn(O) atom in ZnO is tetrahedrally coordinated. The way in which chemists and semiconductor physicist think about the building blocks of ZnO is conceptually different, although both of course describe the same material system. In the chemical community ZnO is thought of as being built by Zn^{2+} and O^{2-} ions. In semiconductor physics each of the Zn(O) forms polarized covalent bonds with its 4 O(Zn) neighbours (sp^3-hybrides). A Zn atom contributes ½ electron to each of these bonds (the 2 Zn valence electrons are distributed onto 4 bonds) and an O atom contributes $^3/_2$ electrons to each bond (the 6 O valence electrons are distributed onto the 4 bonds). When an O atom is removed to form an oxygen vacancy 4 unpaired Zn dangling bonds remain, which can be filled with up to 2 electrons. In the case of a neutral oxygen vacancy, V_O^0, a neutral oxygen atom is removed from the lattice and the remaining 2 electrons of the Zn have to be distributed onto the Zn dangling bonds. The corresponding electronic structure was already touched upon in the answer to Prof. Frankel. When the oxygen vacancy is in the +1 charged state, V_O^{1+}, a O^{1-} ion has been removed from the lattice. One electron remains, which can be distributed onto the four Zn dangling bonds. This is 1 electron short of the maximum occupation possible within a covalent bond. When the oxygen vacancy is in the +2 charged state, V_O^{2+}, a O^{2-} ion has been removed from the lattice and all Zn dangling bonds are empty. The relaxation patterns found for each of these defects can be seen in Fig. 5 of ref. 27 mentioned in our paper, as well as a discussion of the electronic structure of these defects.

Anton Kokalj communicated: It is not clear to me how you align the levels (bands) in respect to the vacuum level. Namely, the position of the energy levels (in respect to the vacuum level) inside the material depends on the conditions at the surface (*e.g.*, type of surface termination, *etc.*). However, the "surface details" seem not to enter into your formalism.

Mira Todorova communicated in reply: The surface/interface dipole is indeed important in the context of alignment. Although this is not an aspect we mention explicitly in the paper, indirectly it is considered in Fig. 4(b) of our paper, where you may notice the two jumps in the vacuum level (at the positions of the interface) of the system after equilibration.

Specifically, we used the work function of Zinc to align the Fermi level of the metal on an absolute scale. For the purpose of this work differences in the work function, due to the orientation of the sample (which are typically small) as well as due to the formation of surface dipoles, were not considered, but can be easily included. To align the ZnO and water levels we used the universal alignment, which was found to exist for the hydrogen ($\pm$) transition level in semiconductors, insulators and water.[1] Based on the calculations we used, we would consider this alignment to be accurate within 0.2–0.3 eV. Our recent work allowed us to rationalize the standard hydrogen electrode on a microscopic level.[2] There, we also discuss the alignment of the $H^{+(aq)}$ and $OH^{-(aq)}$ defects and the hydrogen ($\pm$) level in water on an absolute scale.

1 C. G. Van de Walle and J. Neugebauer, *Nature*, 2003, **423**, 626.
2 M. Todorova and J. Neugebauer, *Phys. Rev. Applied*, 2014, **1**, 014001.

Jonathan Morrison opened the discussion of the paper by Clara Wren: Alloy 800 and Stellite are generally found in high temperature water environments;

have you compared your model output to literature data for oxide films formed under the expected service conditions? Has the model still proven to be successful?

Clara Wren answered: We have not yet applied our model to such high temperatures (in the range 250–320 °C), but we are planning to do so. We have used our model to simulate our corrosion experiments on a range of Fe–Ni–Cr and Co–Cr alloys performed at temperatures up to 150 °C. In these studies, we have performed both coupon corrosion tests and potentiostatic polarization experiments combined with post-test surface and solution analyses (ref. 15, 16, and 21 in our paper). The MCB model simulates the current as a function of time, the composition and layer structure of the oxides formed, and the amounts of dissolved metals well. More importantly, the best-fit model parameters have expected temperature dependences, indicating that the model can predict the effect of temperature.

David Shoesmith asked: Can your model predict the influence of temperature on film composition and thickness? Experimental observations show that the passive current does not change much with temperature (at least over the range 25 °C to 90 °C). Our experiments indicate the current is controlled by a combination of decreased Cr content (which should increase the current) which is counterbalanced by an increased thickness.

Clara Wren answered: Yes, our model can predict the effect of temperature on film composition and thickness. As noted in our response to the earlier question by Dr J. J. Morison, we have applied our model to temperatures of up to 150 °C. The MCB model simulations results are in good agreement with the experimental data (the current as a function of time, the composition and layer structure of the oxides formed, and dissolved amounts of different metals) obtained at all of the temperatures studied so far. More importantly, the best-fit model parameters show expected temperature dependences. In the MCB model, the rate constants for oxide formation and dissolution depend on temperature. The first-order rate constant for oxide formation will increase with temperature according to an Arrhenius dependence, while that for dissolution will have a more complicated temperature dependence (due to increasing rate constants for surface hydration and hydrolysis reactions and typically non-linear dependences of the solubilities of metal cation species on temperature). In our experimental studies we found that a Fe–Ni–Cr alloy forms an oxide with a mixed spinel oxide ($FeCr_2O_4/Fe_3O_4/NiFe_2O_4$) inner layer and a $NiO/Ni(OH)_2$ outer layer. The outer layer thickness increases with increasing temperature and this results in a decrease in Cr content of the oxide.

Janine Mauzeroll queried: Could you further explain the weighing that was applied for the water and oxygen reaction that concomitantly also occurs at the working electrode?

Clara Wren answered: The reduction reactions of H_2O or dissolved O_2 (at an impurity level) on the working electrode are coupled with oxidation of H_2 or H_2O on the counter electrode (reactions (1)–(4)):

$$2H_2O + 2e^- \rightarrow H_2 + 2OH^- \text{ on WE} \tag{1}$$

$$H_2 + 2OH^- \rightarrow 2H_2O + 2e^- \text{ on CE} \tag{2}$$

and/or

$$O_2 + 2H_2O + 4e^- \rightarrow 4OH^- \text{ on WE} \tag{3}$$

$$4OH^- \rightarrow O_2 + 2H_2O + 4e^- \text{ on CE} \tag{4}$$

These aqueous redox reactions do not contribute to the metal oxidation flux or oxide growth flux, but to the net current that is being monitored during polarization. Thus, in order to compare the model results with experimentally measured quantity current, we take into account the contribution of the solution reduction on the WE to the current. This solution reduction reaction is treated as a separate independent redox reaction with its own equilibrium potential (or the difference of the equilibrium potentials of the two half-reactions of aqueous reduction and aqueous oxidation).

In the simulation of the potentiostatic polarization experiments on Stellite-6 and Alloy 800, the solution reduction reaction on the WE is represented by one reaction:

$$\text{Ox(s)} + ne^- \rightarrow \text{Red(s)} + \text{on WE} \tag{5}$$

The solution reduction flux was modelled in the same way as the metal oxidation flux, *i.e.*, using a modified Butler–Volmer equation with an effective overpotential:

$$J_{sol-red}(t)\big|_{ox|sol} = -J^{eq}_{sol-rdx}\exp\left(\frac{0.5nF}{RT}\left(-\eta^{eff}_{sol-red}\right)\right) \quad when\ \eta^{eff}_{ox\#}(t) < 0 \tag{6}$$

$$\eta^{eff}_{ox\#}(t) = E^{eq}_{sol-red} - E^{elec} - f_{sol}\Delta V_{oxide} \tag{7}$$

where $E^{eq}_{sol-red}$ is the equilibrium potential for the solution redox reaction, and the factor f_{sol} is used to adjust for the difference in the potential barrier for the electron transfer for solution redox reactions and for ion and ion vacancy transfer for metal/metal oxide redox reactions.

Since the potential drop for the cathodic reaction is several orders of magnitude lower than that of the anodic reaction, and it is negligible in comparison to that of anodic reaction, f_{sol} is assumed to be zero. Due to a difficulty in accurately measuring the dissolved concentrations of H_2, O_2 and possibly other redox active impurities, $E^{eq}_{sol-red}$ was obtained by measuring the open circuit potential on a Pt working electrode in the same solution environment. In the Ar-purged solution at pH 10.6 the open circuit potential was measured to be 0.2 V_{SCE}.

Philippe Marcus enquired: Why do you consider only the flux of cations in the oxide?

What are the possible rate-determining steps and the values of the physical parameters needed to calculate real kinetics of oxide growth?

Clara Wren replied: We have considered the fluxes of different charged species, metal cations, oxygen anions and their vacancies and electrons and holes in the oxide. However, as stated in the paper, we have concluded that detailed modeling of the fluxes of individual charged species is not necessary because their fluxes cannot vary independently due to mass and charge conservation requirements. What is strictly observed in our MCB (mass and charge balance) model is that the flux of cations created by oxidation (referred to as the oxidation flux in the paper) must equal the sum of the fluxes of the cations leading to oxide formation (the oxide formation flux) and those leading to dissolution (the dissolution flux).

Our MCB model considers that the full corrosion system consists of four elementary steps, electrochemical reaction (metal oxidation coupled with reduction of aqueous species), transport of charged species through the oxide layer, dissolution of metal cations, and oxide formation, as schematically shown in Fig. 2 in our paper (due to mass and charge conservation, only three flux equations needed to be formulated). Since these steps are not independent but coupled processes, it is not necessary to define a rate-determining step. The mass and charge balanced rate equations will automatically adjust the overall corrosion rate to the rate dictated by the rate determining step, and the rate step can change depending on the corrosion conditions. The physical parameters needed to calculate the kinetics of oxide growth are listed in Table 1 of our paper. They are fundamental parameters or independently measurable parameters, including:

- equilibrium potentials for redox reactions of metal/metal oxide that can occur on a given alloy,
- specific (*i.e.*, linear) potential drop across an oxide,
- first-order rate constants for oxide formation and dissolution, and
- activation energy for oxide growth.

The values of some of these parameters used for our simulation results are also listed in Table 2 of our paper.

Markus Tautschnig said: It is assumed in the presented model that the electric potential drop across the layer increases linearly with the thickness of the layer. What is the physical explanation and reasoning leading to this assumption?

Clara Wren answered: The oxide layer that is growing on a corroding surface is not likely a pure semiconducting layer. During corrosion there will be continuous injection of metal cations at the metal/oxide interface and oxygen anions at the oxide/water interface. Thus, our model assumes that the oxide layer is more likely a p–n junction in a solid-state diode device as shown in Fig. 5 of our paper. The potential drop across the oxide layer resembles a junction potential rather than the semi-conductor band-gap. The junction potential arises from the resistance to the diffusion of electrons and holes and hence increases with the diffusion length.

John Scully asked: Oxidation and insertion of metal cations at the metal oxide interface was modeled using a Butler–Volmer type expression originally aimed at charge transfer controlled oxidation. Why not consider using another expression, perhaps more suitable to the metal/oxide interface, such as those developed by Cabrera and Mott, Macdonald, Young or Sato?

Clara Wren responded: The Butler–Volmer equation is a more fundamental flux equation for interfacial charge transfer, but it requires knowledge of the potential difference between the metal and oxide phases at the metal/oxide interface. This potential difference is not defined as a function of more fundamental parameters in the other models, and hence, the expression for the overall interfacial charge transfer in other models includes the effects of the solution redox environment and the oxide potential barrier semi-empirically. In the MCB model, the potential drop across the oxide layer is defined as a function of the type and thickness of the oxide that is growing and the driving force for the interfacial transfer is defined as a function of equilibrium potential for the metal oxidation coupled with solution reduction (see Table 1 in our paper). Thus, the MCB model can calculate the potential difference at the metal/oxide interface, (this is referred to as the effective overpotential for metal oxidation in our paper). Consequently the MCB model does not need to use a semi-empirical formula and uses the more fundamental rate equation for the interfacial charge transfer.

Christopher Taylor opened a general discussion of the paper by Denis Kramer: The strains described in the paper were 2D and equally applied in those two dimensions. Have you investigated the role of uniaxial stresses on the adsorption energy and geometry of oxygen on copper? A relevant paper might be the work by Francis and Curtin, "Mechanical work makes important contributions to surface chemistry at steps".[1]

1 M. F. Francis and W. A. Curtin, *Nat. Commun.*, 2015, **6**, 6261.

Denis Kramer answered: We have not yet considered an anisotropic strain, but agree that this is a very interesting topic.

Christopher Taylor questioned: Can you please further elaborate upon what the two peaks C1 and C2 in the signals for the cyclic voltammetry correspond to when strain is applied? Why does strain introduce a second peak?

Denis Kramer replied: The two peaks are assigned to two different adsorption modes of an oxygen species, one showing higher sensitivity towards strain than the other. Our modelling results suggest that C2 might be attributable to the unreconstructive adsorption mode as we find higher sensitivity of unreconstructed adsorption of oxygen towards strain than for reconstructive adsorption.

David Shoesmith asked: Is it possible that the introduction of sufficient strain into the Cu surface could lead to its susceptibility to corrosion in anoxic water?

Denis Kramer answered: Straining Cu will add mechanical energy to the Gibbs free energy of the solid, in principle shifting the stability windows of the metal. We, however, do not expect this thermodynamic effect to be very pronounced and think that secondary effects need to be considered. This includes the influence of the strain state on the adsorption energies of intermediates and related kinetic barriers of the corrosion reaction. Unfortunately, our results presently do not provide the opportunity to speculate on kinetic effects due to the stress state of the surface.

Vincent Maurice commented: The discussion of the CVs (Fig. 5 in your paper) appears not relevant for a (100) surface. As described, the surface preparation does not include the final stage that allows recrystallising the cold work layer formed by mechanical polishing. As a result, the topmost surface will not have exposed (100) planes but instead a highly defective structure.

Denis Kramer answered: A commercially available polished Cu single crystal oriented along [100] was used for the study. We agree that further pretreatment would have necessitated a reassessment of the surface orientation. This, however, was not the case and the single crystal was used as received. The good agreement with our *ab initio* modelling provides further corroborating evidence. Hence, we belief the interrogated surface to be reasonably representative of Cu(100).

Vincent Maurice further commented: The DFT-derived surface Pourbaix diagram for the unstrained Cu(001) only partially reproduces the available experimental data obtained at pH 13.[1] In this work, adsorption-induced surface reconstruction was indeed observed some 700 mV before 3D bulk oxide growth. However, the observed superstructure was *c*(2x6) and not $(2\sqrt{2}x\sqrt{2})R45°$ as modelled in the present paper for pH 13. Even if the atomic model proposed for the *c*(2x6) may not be fully correct, we note from Table 2 in your paper that its energy is below that of the unreconstructed adsorbed structure for OH adsorption that must be considered at alkaline pH.

1 J. Kunze, V. Maurice, L. H. Klein, H.-H. Strehblow and P. Marcus, *J. Electroanal. Chem.*, 2003, **554–555**, 113–125.

Denis Kramer replied: We agree that the (100) surface is likely to reconstruct in aqueous environments. Although we explicitly considered the reconstruction proposed by Kunze *et al.*,[1] we were unable to identify any OH super-structure using DFT that would be thermodynamically stable within reasonable bounds of pH and voltage. We do not exclude that an exhaustive search through all possible atomic arrangements within a *c*(2x6) supercell might be able to identify a thermodynamically stable OH-terminated configuration. Indeed, our results encourage the development of a refined model of the reconstructive OH adsorption mode as we predict reconstructive adsorption of O and OH to be energetically more favourable relative to their respective unreconstructed adsorption modes.

1 J. Kunze, V. Maurice, L. H. Klein, H.-H. Strehblow and P. Marcus, *J. Electroanal. Chem.*, 2003, **554–555**, 113–125.

Philippe Marcus asked: On Cu (111), combining *in situ* STM determination of the adsorbed overlayer structure, and electrochemical measurements of the charge passed during adsorption, it was shown that the adsorbed species (at pH 13) was OH, and not oxygen. Why would it be so different for Cu (100)?

Denis Kramer responded: Please see the answer to the previous question by Vincent Maurice. Further, our study focused on mildly acidic conditions (pH 5) where experimental evidence for adsorption of oxygen at Cu(100) within the missing row reconstruction was provided by Cruickshank *et al.* (*cf.* ref. 12 in our paper). Cruickshank *et al.* also pointed out that this is markedly different from Cu(111). The prevalence of O as opposed to OH as the adsorbing species in acidic media is further supported by surface-enhanced Raman spectroscopy (SERS) data from Chan *et al.* (*cf.* ref. 27 in our paper).

Hendrik Bluhm remarked: How well is the surface structure, crystallographic orientation and cleanliness controlled in the experiment? In particular, what is the miscut of the (001) orientation and how does this influence the terrace size/step density?

Denis Kramer answered: Please refer to the answer to the previous question asked by Philippe Marcus.

Rob Lindsay said: I was wondering if you have any strong evidence that the surface terminations you theoretically predict are present in your experimental measurements, *i.e.* have you conducted any characterisation beyond cyclic voltammograms. If not, it would be very interesting to do so, perhaps using surface diffraction, as it is possible that the surface terminations are more complex than suggested.

Denis Kramer replied: We fully agree that further sophisticated studies of the precise atomic arrangement at the interface under various conditions would be highly beneficial and likely to bring new insights.

Simon Gibbon addressed Denis Kramer: In your experiment the strain is induced by a spherical ball being pressed into the copper sheet. I would expect this to induce a radial strain field in the copper surface and that the surface reorganisation of the copper atoms may be different from surface reorganisation caused by a strain in the x-direction followed by a subsequent (or even simultaneous strain) in the y-direction. In your modeling work, how has the strain been applied?

Denis Kramer answered: We did not consider the radial nature of the strain within our atomistic modelling due to the difference in length scales between the atomistic domain and the radius of the electrode. Isotropic strain was introduced in the modelling work by scaling the lattice constants appropriately.

Clara Wren remarked: In your DFT calculations, what was the functional used? How did you introduce the stress in the DFT calculations? How do you model the interface where the periodicity of wave function ends, and how do you model the effect of the stress on the functional?

Denis Kramer replied: We used the Generalised-Gradient-Approximation (GGA) to DFT in a parameterisation due to Perdew, Burke, and Ernzerhof (PBE). Stress was introduced by straining the lattice constants. The wave functions were expanded in a plain-wave basis making the consideration of basis set superposition errors (BSSE) irrelevant. Further, we only sampled a single k-point perpendicular to the surface to minimise spurious interactions between slab images. We have not considered any influence of the stress on the functional itself. Note that only the exchange–correlation term of the functional is not precisely known and needs to be parameterised. While there are interesting approaches to improve the predictive power of PBE for adsorption phenomena (*e.g.*, the revised PBE out of the Norskov group), we are not aware of any work explicitly considering strain as part of the parameterisation of the exchange–correlation term.

Mira Todorova asked: There should be two in-equivalent adsorption geometries at 1/2 ML oxygen coverage on Cu when the system is under strain, due to the significant rumpling of the surface (*cf.* Fig. 2 of your paper). In one of them the coordination of O should be increased (structure shown in Fig. 2). Am I correct in assuming that this is the geometry which is affected when the structure is strained, while the energy of the other structure remains unchanged, regardless of the strain state?

Denis Kramer responded: The substantial rumbling was only seen for O adsorption at the top site. The rumbling is not a direct consequence of straining the surface. It is induced by the adsorption of oxygen. Straining the clean surface as well as a full ML of O did not lead to rumbling. We, therefore, report the adsorption energy of O at the recessed Cu atoms only.

Geoffrey Thornton commented: Scanning tunnelling microscopy does not have sufficient lateral resolution to detect tensile strain of a few percent on Cu(001). However, X-ray scattering does provide a potential route to measure the strain.

Julian Wharton replied to the comment: Yes we agree that the combination with more advanced refractive and/or scanning probe characterisation would be very beneficial to provide further experimental evidence.

Anton Kokalj communicated: The explanation that the $(2\sqrt{2}\times\sqrt{2})R45^\circ$ missing row (MR) reconstruction on Cu(100) is driven by compressive stress induced by chemisorbed oxygen is plausible, and calculations performed with imposing 2% tensile strain nicely confirm the argument. Nevertheless, I believe there is a deeper reason behind the MR reconstruction, which is actually confirmed and strengthened by the current stress argument. (We have investigated the MR reconstruction of O/Cu(100) and O/Ag(100) systems[1] and structurally related step-edge decoration of oxygen on Ag(410)[2] by DFT calculations; more details about the below given explanation for the stability of MR can be found in ref. 2.)

In the MR reconstruction oxygen atoms are fully decorating step-edges by forming the –O–Cu–O–Cu– chains. These oxygen atoms are sunk into the surface, *i.e.*, oxygen and Cu atoms are almost aligned, if looked at along the step-edge direction. Hence, they form the structure of stable electrostatic +–+–+–+– strings. Indeed, in this structure oxygen atoms are particularly stable. This reasoning is confirmed by calculation with 50% of O atoms removed, such that step-edge displays the –O–Cu–Cu–O–Cu–Cu– structure. The corresponding magnitude of adsorption energy is considerably reduced with respect to 100% decorated step-edge, because the above mentioned electrostatic stabilization is not operational in this case.

So what happens is this: in the parent structure, $c(2\times2)$–O/Cu(100), oxygen atoms would like to sink into the surface so that repulsive lateral interactions between negatively charged O adatoms would be screened. But they cannot do so, because they are too big (actually they are able to do it only slightly and this is the origin of the observed oxygen induced compressive stress). In contrast, in the MR structure, the O atoms have more space due to missing Cu atoms and they can sink into the surface to become electrostatically stabilized as explained above.

1 N. Bonini, A. Kokalj, A. Dal Corso, S. de Gironcoli, and S. Baroni, *Surf. Sci.*, 2006, **600**, 5074–5079.
2 N. Bonini, A. Kokalj, A. Dal Corso, S. de Gironcoli, and S. Baroni, *Phys. Rev. B: Condens. Matter Mater. Phys.*, 2004, **69**, 195401.

Denis Kramer communicated in reply: We agree that coulomb screening is a likely fundamental process influencing the observed phenomena, which is in full agreement with our findings and interpretation.

Mary Wood opened a general discussion of the paper by Vincent Maurice: Concerning the TOF-SIMS data presented in Figure 1 of your paper, how do you compensate for the different relative sensitivities of the different elements of interest to the technique? As far as I am aware, relative sensitivity factors (RSFs) have only been reported for trace elements within a matrix, such as silicon, making it difficult to apply to a system such as steel. However, these RSFs indicate that whilst nickel and iron show comparable sensitivity, chromium has a sensitivity that is generally around an order of magnitude higher, and oxygen has a sensitivity around three orders of magnitude lower. I was wondering how significant you expect these factors to be in this case?

Vincent Maurice replied: In Figure 1 of our paper, the relative intensities of the selected ions cannot be regarded as indicative of their concentration. O^- ions have a high ionisation yield when emitted from an oxide matrix. Here, the lower intensity $^{18}O^-$ isotope is reported since $^{16}O^-$ had an intensity close to saturation of the detector. The Fe^{2-}, Ni^{2-}, and Cr^{2-} negative ions have a relatively low intensity when emitted from a metallic matrix (positive ions have a higher ionisation yield for metallic ions but it is not possible to reverse the polarity during the depth profile). We have no calibration data of the RSFs for these ions in a metal matrix that would allow determining their relative concentrations. We emphasize that the quantification of the elemental concentration in the alloy was done using the XPS intensity data, not the ToF-SIMS data.

Rob Lindsay remarked: Your ToF-SIMS data suggest that O/OH species are present in the substrate. Could you comment on this result?

Vincent Maurice answered: We see this as a possible artefact effect of the surface analysis by the primary ion beam. Even though the extremely low primary ion dose used corresponds to static SIMS analytical conditions, the impinging ions create atomic plane mixing by pushing the target atoms below the topmost surface plane from which the secondary ions are mostly emitted. With continuous depth profiling, this effect continues even after entering the substrate region leading to the measurement of high intensity tails in the profiles.

Jonathan Morrison commented: In our own studies of the corrosion of 316L stainless steel under high temperature water conditions, we've seen that molybdenum has almost no effect on the oxide film, and is instead dissolved into solution. Have you performed a mass balance calculation to be certain that all of the molybdenum in your experiments has been retained in the oxide films, or to determine the total loss of molybdenum to the solution? How can you be certain that the molybdenum in your films represents the total amount present at the beginning of your test?

Vincent Maurice responded: The total amount of oxidized Mo present in our native oxide film is 2.1% of the total cation fraction in an oxide matrix 1.7 nm thick. After 2 h passivation, this amount increases to 3.6% of the total cation fraction in an oxide matrix 1.9 nm thick. This shows that not only the Mo initially present has been retained in the passive film but also that Mo enrichment took place, implying Mo consumption from the alloy substrate. Of course, one cannot exclude some Mo loss to the solution during passivation, implying an even higher consumption of Mo from the alloy. However, our data show no significant variation of the metallic Mo at the alloy surface.

Hiroki Habazaki questioned: Do you have some results on the potential dependent film composition and film structure? Mo(IV) is stable at low potentials close to the active dissolution region of stainless steel. So, the composition and film structure may change with potential.

Vincent Maurice answered: The potential dependence of the film composition and Mo(IV)/Mo(VI) balance has not been studied in the work reported here. From *in situ* STM data (to be published), the film nanostructure was found to be potentially dependent with the weaker sites continuously enlarged by dissolution for polarization in the pre-passive range before time-assisted curing promoted by the presence of Mo in the alloy.

Gerald Frankel asked: You have made several observations regarding the effect of Mo. Can you speculate as to which is the most important in conferring corrosion resistance?

Vincent Maurice replied: Two of the observed Mo effects, strong Mo(VI) enrichment after passivation and Mo-assisted curing of the weak sites, appear as linked. Mo(IV) present prior to passivation (2.5% of the cations in the inner layer 1 nm thick) is still present after passivation (3% of the cations in the outer layer 0.6 nm thick) and not enriched. In contrast, Mo(VI) gets markedly enriched (increasing from 1.0% initially to 7.9% after passivation). At the same time, Mo is observed to assist curing of the weak sites of the passive film revealed by transient dissolution and hence to likely concentrate as Mo(VI) in these weak sites. It does not exclude that Mo(VI) could also be present in the oxide matrix where no weak sites were revealed by transient dissolution but where a similar Mo(VI) enrichment/curing mechanism could operate but at a smaller scale, *e.g.* in the intergranular sites of the oxide matrix. Thus, this combined effect of Mo(VI) enrichment and curing of weak sites appears as most important for increasing the corrosion resistance in the passive state and in the initiation of localized corrosion for which the presented data are relevant.

Clara Wren commented: The passivation potential that you used in your experiments was 0.5 V, a very oxidising potential. The corrosion potentials on Ni–Fe–Cr alloys even in the presence of O_2 and H_2O_2 are not likely to reach that high a potential. Could you comment on whether the oxide structure that you see at such a high polarization potential can be extrapolated to the oxide structure that can be formed on naturally corroding surfaces? What are the equilibrium potentials for Mo to Mo(IV) and Mo(IV) to Mo(VI) conversion? Do you expect to form Mo(VI) oxides/hydroxides at the high polarization potential? What are the solubilities of the Mo(IV) and Mo(VI) hydroxides? Do their solubilities support your proposed oxide structure?

Vincent Maurice responded: A revisited Pourbaix diagram of the Mo-H_2O system at 25 °C, including the many forms of molybdates, has recently been proposed.[1] It shows that, in our passivation conditions (0.5 V/SHE, pH 1.3, RT), the Mo(VI) oxide phase (MoO_3) is the most stable expected, the Mo(IV)–Mo(VI) transition being in the 0.25–0.33 potential range at pH 1.3. Our results that show passivation-induced enrichment in Mo(VI) only (see also answer to the previous question asked by Gerald Frankel) are thus consistent with this prediction. However, this diagram also shows that the solubility limit for the formation of polymolybdates ($H_2Mo_8O_{26}{}^{2-}$) is pH = 1.9 +0.5 log a. Thus at pH 1.3, one can expect the transpassive dissolution of polymolybdates (0.07 M) and speculate their redeposition on the surface. A Mo enrichment of the passive film dominated by the formation of Mo(IV) species at lower potential has not been studied in the present work. Still, the experimental conditions of our model study are relevant for pitting initiation in the middle of the passive range as expected for a Mo-bearing alloy in chloride-containing environments.

1 P. A. Nikolaychuk and A. G. Tyurin, *Butlerov Communications*, 2011, **24**, 101–105.

Frank Renner remarked: My research group has recently addressed the corrosion and dissolution behavior of a stainless-type amorphous steel ($Fe_{50}Cr_{15}Mo_{14}C_{15}B_6$) and its respective partially and fully nanocrystalline derivates.[1–3] For the amorphous alloy a very flat and homogeneous Cr-rich passive oxide forms due to the ideal homogeneous distribution of the elements in the amorphous structure and the flat starting surface. With an increasing degree of (partial) crystallization this initially amorphous alloy forms Cr-depleted Mo-rich nanoscale crystallites which dissolve preferentially creating a nanoscale roughness of the forming Cr-rich passive film. This roughness (see Fig. 1) translates directly to respectively higher steady state dissolution. As a direct conclusion of this work we therefore realized the importance of the nanoscale morphology of the passive films on the relative stability of the alloy.

Do you have more information or can you speculate on the elemental composition of the different (lower and higher) patches visible on your surfaces? Does the surface morphology of your passive oxide change with time and do you observe a similar influence of changing surface morphology on the passive current, *i.e.* the steady-state dissolution rate?

1 M. J. Duarte, J. Klemm, S. O. Klemm, S. Borodin, K. Mayrhofer, M. Stratmann, A. H. Romero, D. Crespo, J. Serrano, P. Choi, D. Raabe and F. U. Renner, *Science*, 2013, **341**, 372–376.
2 M. J. Duarte, A. Kostka, J. A. Jimenez, P. Choi, J. Klemm, D. Crespo, D. Raabe and F. U. Renner, *Acta Mater.*, 2014, **71**, 20–30.
3 J. Klemm, M. J. Duarte, S. O. Klemm, P. Choi, K. J. J. Mayrhofer and F. U. Renner, *Corros. Sci.*, 2014, **89**, 59–68.

Vincent Maurice responded: It remains unknown from our present data for this system if the nanogranular morphology of the matrix of the oxide film revealed by STM presents variation of composition between grains or even in-between them (at the interface between grains). A recent conductive AFM study performed in our group on thicker passive films with larger grains formed on 316L stainless steel in simulated PWR water at 325 °C led to the conclusion that, indeed, there are variations of the oxide composition on different grains and in-between them.[1] As discussed previously (answer to the question asked by Gerald Frankel), one could speculate that Mo(VI) enrichment could also take place in the weaker (*i.e.* intergranular) sites of the observed matrix by a local dissolution-induced mechanism similar to that revealed at larger scale in the present study. Certainly, *in situ* monitoring of the time evolution of the granular morphology of the oxide matrix and its relation with the passive current appears relevant for future work in that prospect.

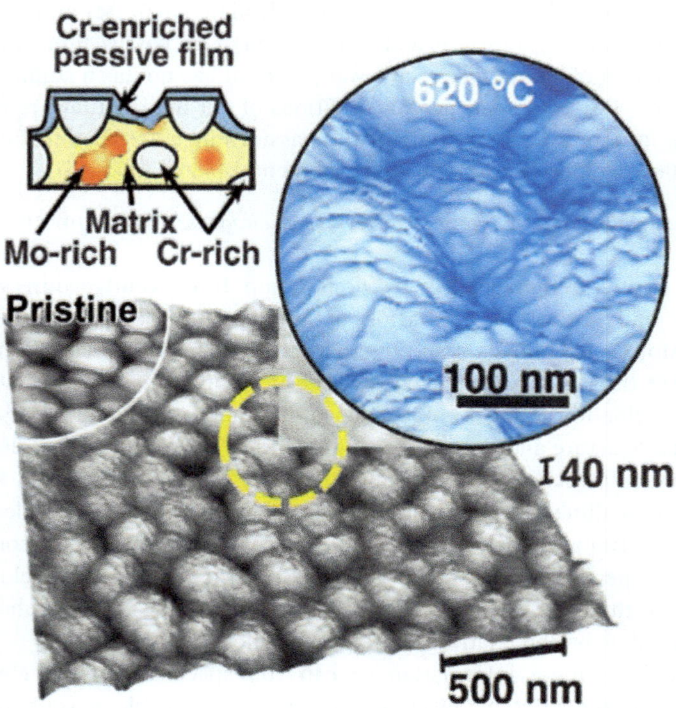

Fig. 1 AFM image of a partially nanocrystalline derivate of a stainless-type amorphous steel ($Fe_{50}Cr_{15}Mo_{14}C_{15}B_6$) after polarization in the Cr-dominated passive regime in 0.1 M H_2SO_4. Nanoscale roughness develops by selective dissolution of Cr-depleted Mo-enriched phases and results in a higher passive current. Adapted from ref. 1.

1 T. Massoud, V. Maurice, F. Wiame, L. H. Klein, A. Seyeux and P. Marcus, *Corros. Sci.*, 2014, **84**, 198–203.

Alison Davenport enquired: How do you think that Mo prevents localised corrosion of stainless steel? There is well-established evidence that it decreases the rate of propagation of pits by decreasing the rate of active dissolution (*e.g.* ref. 1). How might the effect of Mo on the passive film affect localised corrosion susceptibility?

1 R. C. Newman, *Corros. Sci.*, 1985, **25**, 331.

Vincent Maurice answered: It is clear from the presented data that Mo plays a role in the initiation stage of localized corrosion by pitting. Not only Mo is found strongly enriched (as Mo(VI)) in the passive film but it is found to act locally in the weak sites of the passive film where dissolution is preferential. It can be speculated that this beneficial effect will operate in the metastable pitting stage retarding pit initiation (see also my answer to the question previously asked by Gerald Frankel). This does not exclude that a Mo effect also operates in the pit propagation stage, by decreasing active dissolution.

Rob Lindsay remarked: Regarding the chemical nature of the Mo species incorporated into the outer part of the passive film, you suggest that it may be molybdate on the basis of its oxidation state (Mo(VI)). Under the acidic conditions of your experiments, would you expect polymolybdates to be present?

Vincent Maurice replied: The formation of polymolybdates by transpassive dissolution of Mo(VI) can be expected in our acidic solution as shown by a revised Pourbaix diagram for molybdenum (see answer to the previous question by Clara Wren). This implies a dissolution/redeposition rather than a solid state mechanism of formation, in which case the formation of polymolybdates bonded to Fe cations appears as a sound possibility since Mo(VI) is found in the Fe-rich outer part of the passive film.

David Shoesmith addressed Vincent Maurice and Philippe Marcus:Your results appear to confirm that Mo has no observable role on the properties of the passive film and that its primary function is to repassivate a very large number of metastable pitting events. This is consistent with our observations on Ni–Cr–Mo alloys (which contain considerably more Mo than the steel used in your experiments). Is the evidence you generate enough to confirm that the primary function of Mo is to form oxidized (Mo(VI) states that reseal the passive film? If so, do your results rule out the claim that the primary function of Mo is to suppress the anodic oxidation of the alloy?

Vincent Maurice and **Philippe Marcus** jointly responded: Our data indeed support the view that a primary role of Mo (present in our model alloy at similar concentration as in 316L stainless steel) is to promote a reseal of the passive film by Mo(VI) enrichment (see also Vincent Maurice's answer to the previous question by Gerald Frankel). However, we do not exclude that Mo also decreases anodic dissolution (see also Vincent Maurice's answer to the question previously asked by Alison Davenport).

Angus Cook asked: In Fig. 2 of your paper the ToF-SIMS data show that both the air passivated sample and the sample passivated potentiostatically for 20 h have similar intensity counts for the four secondary ion species surveyed when sputtering has reached the metal substrate (~400–500 s). The sample passivated potentiostatically for 2 h, however, shows a higher intensity reading at these sputtering times for all four secondary ions. Can you suggest a reason for this difference?

Vincent Maurice replied: This difference is also observed in Fig. 1 of our paper for all ions in the substrate region. A possible cause is that the low current (set to 1.3 pA) of the primary Bi^+ ions used for analysis may have varied in the different experiments. We emphasize that the quantification of the cation concentration was done using the XPS intensity data, not the ToF-SIMS data.

Rob Lindsay opened a general discussion of the paper by David Williams: Having begun some work in the area of CO_2 (sweet) oilfield corrosion, I have the impression that the literature is rather inconsistent, with many apparently simple questions outstanding. Do you agree with this opinion, and if so do you have any idea as to why this is so, beyond simply the complexity of the topic?

David Williams replied: I do agree with your opinion. As Professor Taylor's paper later in this Discussion will surely emphasise, corrosion is generally a complex, multi-faceted phenomenon, and it can be difficult to cut through to the basic processes that underly the behaviour. Often, to find these underlying processes needs new experimental techniques and we have been fortunate in that we have been able to deploy one such that turned out to be very appropriate for this study. We were also fortunate that in the course of developing this study we discovered the potentiostatic current transients that were correlated with the appearance of crystalline siderite. This was a complete surprise. We were then able to unravel a number of interacting effects.

John Scully asked: The notion of looking at the metal/solution interface and at the colloidal aspects of the solution at the same time *in situ* was particularly novel. What do you think was the effect of Cr on the crystallization and formation of Siderite?

I assume that the Cr that affected the process was in solution and came from the alloy. If this is the case, does that mean that the scaling processing depends critically on the Cr alloying content and the details of Cr release? This implies that every oil field alloy used pertinent applications will have subtle differences in behavior. Comments?

David Williams replied: I can only speculate as to the detailed mechanism of the Cr effect, which seems (on the limited evidence available) to be a general effect of highly-charged metal ions, which also includes Ca^{2+} and Mg^{2+} (references in our paper). The effects are on both the nucleation and growth of the amorphous colloid precursor and on its transformation to crystalline siderite. It might be that these ions adsorb at the boundary of pre-nucleation clusters (ref. 50 and 51 of our paper). I agree that the observations could imply that the scaling might depend critically on the alloy composition and how the species such as Cr are released. The GI-SAXS/WAXS experiment is an extremely powerful method for revealing details of the initial stages of surface scale formation and as such would be very appropriate for exploration of the effects of alloy composition. One aspect of our experiment that was not under good control was the transport of species to and from the surface. An important extension of the experiment would be to redesign the cell so that the effect of flow in the solution could be studied in detail, which I think would then allow your question to be better addressed.

Christopher Taylor said: Can you speculate upon the mechanism by which Cr(III) influences the formation of these amorphous phases? Do you think it may be acting as a catalytic agent in the nucleation of the siderite? Can you also briefly describe Ostwald's rule of stages and how it applies here?

David Williams answered: Ostwald's rule of stages states: "As a stepwise reaction proceeds, it is not the state with lowest Gibbs energy that is initially obtained, but the least stable one, lying nearest to the original state in Gibbs energy".[1] As applied to crystallisation, this rule has been used to underpin a mechanism that postulates a succession of steps, from pre-nucleation clusters, through amorphous states to the crystalline state, rather than a classical nucleation mechanism which postulates a critical nucleus size determined by the balance of interfacial and bulk (volume) Gibbs energy. In the case of calcium carbonate nucleation, highly dynamic, liquid and chain-like polymers of calcium carbonate ion pairs have been proposed.[2] I note that there may be an analogy to the effect of silica on the formation of prenucleation clusters of $CaCO_3$, where adsorption of colloidal silica on the periphery of the clusters is proposed as a mechanism both for stabilising the clusters and promoting their cross-linking into aggregates.[3] In the present case, Cr^{3+} or colloidal chromium oxyhydroxide-carbonate may play a similar role.

1 R. A. van Santen, *J. Phys. Chem.*, 1984, **88**, 5768–5769.
2 R. Demichelis, P. Raiteri, J. D. Gale, D. Quigley and D. Gebauer, *Nat. Commun.*, 2011, **2**, 590.
3 M. Kellermeier, D. Gebauer, E. Melero-García, M. Drechsler, Y. Talmon, L. Kienle, H. Cölfen, J. M. García-Ruiz and W. Kunz, *Adv. Funct. Mater.*, 2012, **22**, 4301–4311.

Gerald Frankel questioned: Do you have any evidence that the mechanism for the effect of Cr on siderite formation is the same for Cr as an alloying element as for Cr^{3+} in solution?

David Williams responded: We recently presented in detail a study of the effect of Cr additions to the steel and in solution.[1] Fig. 2 compares the evolution of current density and diffraction peak intensity for two steels of the same quenched and tempered microstructure where one steel contains 1 wt% Cr and the other does not. Cr^{3+} is added to the solution at concentrations from 0 to 100 μM in the case of the steel that does not contain Cr. Addition of Cr^{3+} to the solution in this case shortens the induction time for scale nucleation, as shown by the occurrence of the peak in the current density and the appearance of the diffraction peaks for siderite and chukanovite. In the case of the Cr-containing steel, the nucleation of a surface scale occurs almost instantaneously after initiation of the anodic reaction. The scale formed was very thin and adherent and there was some evidence for intergranular corrosion initiating.[1] Thus, there is a strong effect of Cr in the steel upon the nucleation and growth of the scale. It is an assumption that the mechanisms acting in the case of Cr in the steel and Cr in solution are similar. It is debatable whether Cr from the steel will be released into the solution or whether it forms a colloidal oxy-hydroxide immediately adjacent to the surface with this acting as seed for nucleation of the corrosion scale.

1 M. Ko, B. Ingham, N. Laycock and D. E. Williams, *Corros. Sci.*, 2014, **80**, 237–246.

David Shoesmith enquired: You observe the formation of chukanovite as well as siderite. Is this expected during pipeline corrosion? In groundwater systems whether siderite or chukanovite form tends to be determined by the relative concentrations of hydroxide and carbonate. Given that you see pH changes during film formation could this lead to chukanovite formation? Is it also possible that local changes in chemistry are a requirement for the Cr(111) effect. The high sensitivity to Cr(111) suggests this could be the case.

David Williams replied: I do not know whether chukanovite forms during pipeline corrosion. I suspect that, if it did, the scale morphology would be significantly affected and that the resultant scale would not be nearly as protective as siderite: that, however, is a guess. The formation of chukanovite in our

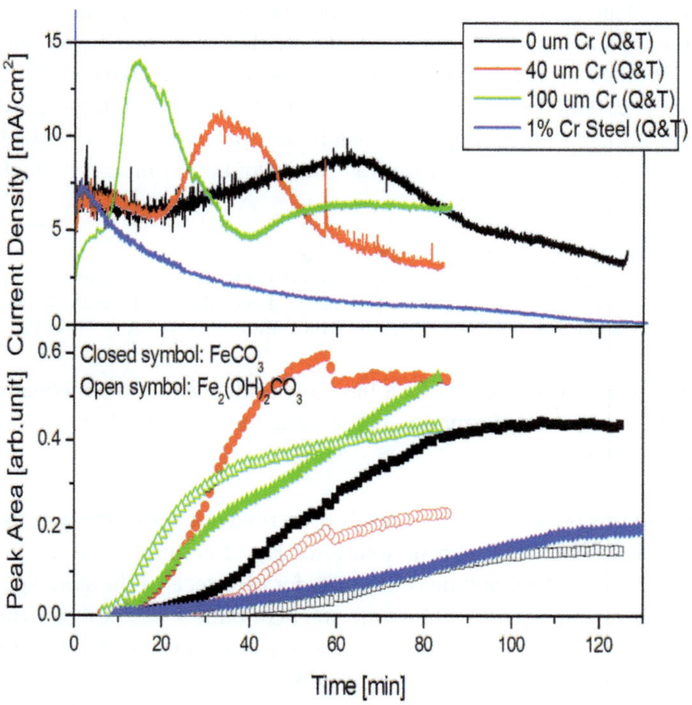

Fig. 2 Evolution of current density and diffraction peak intensity for two steels of the same quenched and tempered microstructure where one steel contains 1wt% Cr and the other does not.

experiments may indeed be a consequence of the pH changes that we deduce accompany siderite crystallisation, because usually the formation of chukanovite follows that of siderite and seems to be more pronounced the more rapidly siderite is formed. We also hypothesise in the paper that the transport of Fe^{2+} through the amorphous surface 'gel' could form an iron-rich gel that could be considered as an amorphous chukanovite. We hypothesise that the effect of Cr^{3+} is representative of a general effect of high-valent metal ions, based on our own observations of the effect of La^{3+} on the crystallisation rate[1] and on literature suggestions cited in our paper that the crystallisation of amorphous ferrous carbonate is sensitive to the presence of traces of Fe^{3+}.

1 M. Ko, B. Ingham, N. Laycock and D. E. Williams, *Corros. Sci.*, 2014, **80**, 237–246.

Gaurav Joshi commented: Having performed 6-day immersion corrosion experiments on polished pure Fe (macroelectrode (0.8 cm^2)), which was allowed to remain at OCP in $CO_{2\text{-sat}}$ water (across a range of [NaCl] @ 80 °C, pH 6.8 (comparable to the experimental conditions in your paper)), I present some data which might support the conclusions of your paper (amorphous siderite), in particular, to lines 40–41 on page 15, "presence of a discontinuous layer comprising 'blobs' of amorphous $FeCO_3$". Between regions of discrete crystalline siderite crystals, where the surface appears only to have suffered uniform corrosion, high resolution imaging in secondary electron and back-scatter electron mode in SEM indicate only a roughened surface. However, as the working distance and accelerating voltage are lowered, and the detector changed to one located within the electron beam column (annular), some darker string-like features are clearly elucidated. These have a lower electronic work function than the conducting Fe substrate and are suspected to be nanocrystalline strings/blobs of iron carbonate. *Ex situ* micro-Raman spectroscopy from such regions reproducibly provides a broad peak in the range 400–700 cm^{-1}, centered at 620 cm^{-1} and shoulders at 520 cm^{-1} and 690 cm^{-1}. These shoulders are interpreted to separate into contributions from Fe–O stretches from siderite (520 cm^{-1}) and magnetite/maghemite (690 cm^{-1}), but the central peak (620 cm^{-1}) has to remain unassigned at this point. Carbonate stretches may not have been detected due to the nano-crystalline structure (molecular vibration constraints, perhaps?) or high fluorescence of the substrate. Complementary work with FT-IR and higher resolution techniques (*e.g.* AFM-IR) might assist in resolving/revealing their nature. Please refer to ref. 1. In ref. 1, the idea of colloidal solution forming close to steel substrate has been made (in passing) on page 5.

1 A. Dugstad, *Mechanism of protective film formation during CO_2 corrosion of carbon steel*, NACE International, Corrosion 98, paper no 31, 1998.

Tom Majchrowski asked: In your paper you suggest formation of corrosion protective siderite scales that nucleate within a surface gel of amorphous carbonate. You also mentioned a positive effect of chromium ions on the film growth. Would you be able to elaborate in more detail on the nature of the bonding within the surface film? Is it covalent or ionic? Does the intermolecular interaction play an important role in maintaining the film's structure?

David Williams responded: I assume that the amorphous iron carbonate has similar bonding to amorphous calcium carbonate, and is a random hydrated ionic network. There is an extensive literature on amorphous calcium carbonate (references in our paper). Studying the bonding in these systems (*e.g.* by EXAFS) is difficult, because they are formed as a dilute colloid that is unstable to crystallisation.

Matthew Acres questioned: What chemical characterisation techniques, if any, were performed to distinguish the amorphous iron carbonate gel from a layer of atmospheric contamination?

David Williams answered: Chemical characterisation is extremely difficult for these thin surface layers that are also likely to transform if removed from the environment in which they are formed, and so we have not been able to do that. We can definitively state, however, that the species we detect form as a consequence of the anodic electrochemical reaction in the cell, because their appearance correlates with the current flowing.

Christopher Taylor queried: Can you determine if the nucleation of the siderite precipitate is heterogeneous (catalyzed by defects or otherwise on the surface of the material) or homogeneous (occurs in the aqueous phase, or in the near-surface aqueous phase)? How adherent are these films? If they are homogeneously nucleated, I would expect them to be less adherent since a direct bond between the surface and the film would not be anticipated.

David Williams responded: We have not detected the formation of crystalline siderite from the colloidal precursor in the solution or in the material deposited on the Kapton windows of the cell: Fig. 6 in our paper demonstrates that colloidal material was deposited or formed on and adhered to the cell windows, but there was no detectable WAXS signal from this material. Our GI-SAXS results imply, as we discussed, that crystalline siderite forms within an amorphous gel layer on the electrode surface, and we speculated on the possible importance of the gradient of Fe^{2+} concentration through the gel layer. This idea has elements of both homogeneous and heterogeneous nucleation. In some experiments, we replaced the Kapton window of the cell with a thin (1 μm thick) Fe foil in an attempt to see whether deposition of colloidal material from solution onto an iron surface was more efficient that on Kapton and whether it led to formation of crystalline siderite, but did not observe any crystalline material.

Rob Lindsay remarked: From reading your paper, I believe you have anodically polarised your sample to encourage scale growth on a shorter timescale. Have you attempted to undertake similar measurements at OCP? If so, what is the impact on scale formation, *e.g.* do you still see evidence for amorphous gel formation?

David Williams replied: We made one attempt to observe the surface film formation at the open circuit potential, by a transmission SAXS/WAXS experiment through a thin iron foil. Unfortunately, we did not vary the beam position during the 3 h run, so the X-ray beam carbonised the Kapton film that was backing the iron foil and the Kapton film that formed the opposing side of the cell. There was no detectable WAXS signal from the crystalline material. We could not distinguish any SAXS signal that might have been due to a surface layer from a signal due to the amorphous carbon and we did not scan the beam over other parts of the Fe foil after the exposure. We had not at that stage suspected or detected the surface amorphous film, which we only found after we analysed our data in detail as described in the paper. This sort of blurred decision making often accompanies long sleepless stretches of synchrotron time. We will have to make a new beamtime proposal.

Rob Lindsay opened a general discussion of the paper by Frank Renner: According to your paper, the Cu_3Au (111) sample is prepared in UHV and then exposed to the ambient laboratory atmosphere before being patterned. Could you comment on the surface composition/structure following this exposure, *i.e.* how is the surface modified relative to that formed in UHV?

Frank Renner replied: The UHV prepared Cu_3Au (111) is found to be quite stable in ambient air. The sputter-annealing cycles in UHV produce a flat surface which is ordered according to earlier X-ray crystal-truncation-rod measurements up to the topmost atomic layer.[1] After a (short) exposure to air, some preferential oxidation of the topmost Cu occurs, which leads to an initial partial Cu dissolution on a thiol-free alloy surface even below the Cu equilibrium potential with immersion into the electrolyte. Yet, this procedure only showed the formation of a partial, not a full monolayer of Au, and *ex situ* AFM images of the alloy surfaces in ambient air show well-ordered atomic terraces. The adsorption of thiols then leads to Au segregation to the surface. There are no pronounced differences found in the surface morphology, even the surface exposed to ambient air for longer time before the fabrication of patterned structures. As a further comment important for the frequently necessary re-polishing step, we subjected the surfaces in-between the UHV treatment after fresh polishing to a gentle electrochemical etching step below the critical potential which lead to much smoother surfaces.[2]

1 A. Stierle, A. Steinhäuser, A. Rühm, F. U. Renner, R. Weigel, N. Kasper and H. Dosch, *Rev. Sci. Instrum.*, 2004, **75**, 5302.
2 (G. N. Ankah, A. Pareek, S. Cherevko, A. A. Topalov, M. Rohwerder and F. U. Renner, *Electrochim. Acta*, 2012, **85**, 384).

Rob Lindsay further asked: Regarding substrate templating, you indicate that you have sometimes pressed the stamp slightly harder onto the surface. Could you clarify what is the result of this harder pressing (*e.g.* are you simply enlarging the surface area contacted by the liquid)?

Frank Renner responded: Indeed, by hard-pressing we enlarge the contact area. But we need to consider that the amount of solution or of dissolved thiol molecules at the side walls is apparently lower than at the direct contact area. At least the density of thiol molecules is lower within the additional area. Here, a lying-down thiol phase is observed.

Hendrik Bluhm requested: What can we potentially learn from other modes of scanning probe microscopy, such as friction force microscopy, on the early stages of corrosion, before the onset of detectable morphological changes?

Frank Renner answered: As a very short answer: we did not yet perform other modes of AFM such as friction force microscopy. But we do agree that the lateral surface friction behavior should reveal interesting insights on the stability of the different adsorbed thiol phases. In a currently started project, my research group will indeed follow such approaches and further include also force–distance curves.

Trevor Rayment asked: Have you given consideration to the electrochemical activity of the thiol layer, since in your experiment you need to start the experiments at low potentials where the alloy is stable but at low potentials the thiol monolayer will be reduced or made more mobile? There is a substantial literature on the electrochemical properties of thiols. Indeed there are studies of the UPD of Cu SAMs on gold which might be regarded as being in some ways the reverse of the dealloying that you are studying.

Frank Renner replied: The thiol layer is stable in the potential regime we applied here. Our initial potential is above the one required for a cathodic desorption of the thiol layer and the critical potential, which is roughly our anodic limit, is below the anodic oxidation of the adsorbed molecules. In fact we have studied the effect of a higher anodic pulse applied to a similar ethanethiol SAM-covered Cu_3Au surface[1] and did observe the full or partial removal of the thiol film which in consequence directly led to large changes in the observed nanoporous surface morphology. We did not address the influence of the UPD of Cu on the SAM in the process as we started in an initially Cu-free electrolyte and always increased the potential during the experiments.

1 G. N. Ankah, A. Pareek, S. Cherevko, J. Zegenhagen and F. U. Renner, *Electrochim. Acta*, 2014, **140**, 352.

David Williams addressed Frank Renner: I'm assuming that the thiol adsorbs on gold, and also, that the final nanostructure developed is sensitive to surface diffusion.[1] However, how does that translate into adsorption on the intermetallic compound Cu_3Au? Also, since thiol adsorbs on gold, as the dealloying proceeds, would there be some surface diffusion or desorption and readsorption of the thiol onto the developing gold nanostructure that would consequently impact the nanostructure development? Or does the thiol alter surface diffusion of gold and hence the coarsening of the nanostructure?

1 J. Erlebacher, M. J. Aziz, A. Karma, N. Dimitrov and K. Sieradzki, *Nature*, 2001, **410**, 450.

Frank Renner answered: In regard to the influence on the surface diffusion we do clearly see that while the dealloying or formation and coarsening of nanoporosity starts at nearly thiol-free parts of the surface, the thiol-covered parts show no sign of coarsening or nanoporosity. The thiols do thus indeed considerably alter the surface diffusion. On the other hand we have no sign of thiols diffusing into the forming pores which are producing fresh thiol-free surfaces below the initial surfaces. A change in the surface diffusion of thiol-covered Cu-Au surfaces was much earlier proposed by Moffat and Bard[1] and becomes in our experiments nicely visible. In regard of the initial state after adsorption the surfaces show a segregation of about 2–3 monolayers of Au, *i.e.* the thiols do indeed preferentially adsorb to Au. This effect can be clearly observed by *in situ* X-ray diffraction because already a single monolayer of Au on Cu_3Au (111) has a different in-plane lattice parameter compared to the substrate, and the ultrathin Au film approaches the expected Au lattice for slightly thicker films as we have shown in earlier work.[2–3] The adsorption of thiol molecules results in the formation of vacancy islands or ad-islands. These two features depend significantly on the various factors, *e.g.* stress on the Au surface due to thiolate molecules, molecular structures with or without a flexible alkyl spacer between the docking group and the aromatic backbone. The molecules used in this study are less flexible and introduce more stress on the substrate, which eventually leads to the formation of gold ad-islands. For the case of a less flexible system, these gold ad-islands are known to be quite mobile.[4]

1 T. P. Moffat, F. R. F. Fan and A. J. Bard, *J. Electrochem. Soc.*, 1991, **138**, 3224.
2 A. Pareek, G. N. Ankah, S. Cherevko, P. Ebbinghaus, K. J. J. Mayrhofer, A. Erbe and F. U. Renner, *RSC Adv.*, 2013, **3**, 6586.
3 A. Pareek, S. Borodin, A. Bashir, G. N. Ankah, G. A. Eckstein, M. Rohwerder, M. Stratmann, Y. Gründer and F. U. Renner, *J. Am. Chem. Soc.*, 2011, **133**, 18264.
4 W. Azzam, A. Bashir, P. U. Biedermann and M. Rohwerder, *Langmuir*, 2012, **28**, 10192.

Janine Mauzeroll asked two questions: 1. Did you consider using surface analytical IR methods to validate the thiol orientation giving the potential of non-negligeable interaction of the AFM tip in some of your results? 2. What is the achievable penetration depth with this dealloying method?

Frank Renner responded: We have so far indeed not used further techniques to address the different thiol structures. It is difficult to carry out the IR measurement on a patterned surface since the surface is not homogeneously covered with the target molecules. Also we see, as discussed in our paper, potential-dependent movements and formation of new standing-up thiol islands. However, we believe that in the patterned areas SAM molecules are densely packed and have no significant interaction with the AFM tip. We can further support the assumption of only weak interactions of the AFM tip with the thiol islands by SEM images of the same substrate or similarly treated surfaces which show no differences in the morphology between areas imaged by AFM and surrounding surfaces. The depth of homogeneous dealloying depends on the allied time and can reach from monolayers to complete bulk samples.

Rob Lindsay remarked: In Fig. 9 of your paper you demonstrate that adsorbed species can relatively easily interchange with those in solution. This result suggests that the adsorbed overlayers are rather dynamic in solution, *i.e.* adsorption/desorption events would be ongoing. On this basis, I am quite surprised that the templated substrates remain intact in solution. I would expect that the templating would become weaker as a function time through molecule desorption and adsorption at other locations. What is your expectation?

Frank Renner answered: We do see a slight out-diffusion of thiol molecules as we describe in our manuscript. Yet the effect is indeed not very strong which indicates in our view that the thiols are not frequently desorbed and re-adsorbed. The exchange mechanism then rather requires the displacement of one molecule by a molecule offering a more stable bond. Subsequently the same site triggers a growth of the more stable exchanged phase which may be called an intercalation process. It may here be illuminating to perform an exchange experiment on a patterned surface but we did so far only perform the patterning and the exchange in independent experiments as discussed in our present manuscript. The molecules in the early stage of adsorption or in the vicinity of the imprint are flat lying, poorly or less ordered and therefore may be more easily exchangeable.

Anton Kokalj communicated: How do the thiophenol and benzenoselenol molecules interact with the alloy surface, *i.e.*, only *via* the S and Se atoms or also with the benzene ring?

Frank Renner communicated in reply: Thiol and selenol moieties have a strong affinity to make a covalent bond to a gold surface and the benzene ring is then pointing upward also due to a low flexibility of the molecule (compare *e.g.* ref. 1).

1 M. Kind and C. Wöll, *Prog. Surf. Sci.*, 2009, **84**, 230.

Faraday Discussions

PAPER

Pitting of steam-generator tubing alloys in solutions containing thiosulfate and sulfate or chloride

William Zhang, Anatolie G. Carcea and Roger C. Newman*

Received 20th January 2015, Accepted 22nd February 2015
DOI: 10.1039/c5fd00008d

The pitting of nuclear steam generator tubing alloys 600, 690 and 800 was studied at 60 °C using dilute thiosulfate solutions containing excess sulfate or (for Alloy 600) chloride. A potentiostatic scratch method was used. In sulfate solutions, all alloys pitted at low potentials, reflecting their lack of protective Mo. The alloys demonstrated the most severe pitting at a sulfate : thiosulfate concentration ratio of ~40. Alloy 600 pitted worst at a chloride : thiosulfate ratio of ~2000. The results are interpreted through the mutual electromigration of differently charged anions into a pit nucleus, and differences in the major alloy component.

1 Introduction

Some corrosion phenomena in nuclear steam generator (SG) tubing have been postulated to occur at low temperature, due to the presence of both oxygen and 'reduced' sulfur species (*i.e.* oxidation state of the sulfur is less than 6). The reduced sulfur could be generated by reaction of sulfate impurity with hydrazine near heat transfer surfaces during SG operation. It is well known that reduced sulfur species are catalytic for the corrosion of iron and nickel, and their alloys; early work focused on pitting of simple stainless steels, such as type 304, in solutions with chloride and various reduced sulfur species.[1] Thiosulfate was found to be the most aggressive sulfur anion, much more so than sulfide, and the beneficial effect of Mo alloying was noted. Later, there were a number of studies of localized corrosion of alloys 600 and 800 in mixed solutions of chloride and thiosulfate, or other reduced sulfur species,[2–7] as well as many studies of stress corrosion cracking (SCC) of sensitized Alloy 600 in pure thiosulfate solutions. To our knowledge, there has been no systematic study of the pitting of annealed SG tubing alloys in chloride-free solutions (sulfate plus thiosulfate). This phenomenon was discovered in stainless steel in the early 1980s,[8,9] and independently by

University of Toronto, Department of Chemical Engineering and Applied Chemistry, 200 College Street, Toronto, Canada. E-mail: roger.newman@utoronto.ca; a.carcea@utoronto.ca; lianwei.zhang@utoronto.ca; Fax: +1-416-978-8605; Tel: +1-416-946-0604

Garner in the pulp and paper industry; a number of studies were published in the context of paper machines and microbial corrosion.[10,11] Thiosulfate was implicated in practical problems involving fouled fresh-water heat-exchangers,[12] although here some chloride was present.

In sulfate–thiosulfate pitting, sulfate plays the role of the strong-acid anion – just like chloride in ordinary pitting – that permits, within certain limitations, acidification of the pit cavity by cation hydrolysis. The limitations are the buffer character of sulfate at low pH ($pK_a \approx 1.92$; evidently this is not a dominant factor, otherwise pitting would not occur), and the tendency of thiosulfate to neutralize the acidity of the pit by disproportionation according to eqn (1):

$$S_2O_3^{--} + H^+ \rightarrow HSO_3^- + S \quad (1)$$

or by electroreduction as shown in eqn (2):

$$S_2O_3^{--} + 6H^+ + 4e \rightarrow 2S + 3H_2O \quad (2)$$

The catalytic effect of adsorbed sulfur ($S_{ads}{}^0$ or $S_{ads}{}^{\delta-}$) on anodic dissolution of Ni, Fe and NiFe(Cr) alloys is well known from the work of Marcus and others. Essentially, S adsorption in hollow sites weakens the metal–metal bonding in the surface, facilitating dissolution; it also hinders hydrogen adsorption, because the weakening of the metal–metal interaction lowers the hydrogen adsorption energy. On Fe or Ni, S adsorption occurs over a wide range of potential and pH,[13,14] and is even believed to occur on Cr, despite the instability of bulk Cr sulfides in aqueous solution,[15] although no kinetic information is available that shows an aggressive effect of S_{ads} on Cr. Stainless steel is less susceptible than pure Fe or Ni (or FeNi alloys), because the presence of Cr causes some desorption of S.[16] This effect is also demonstrated, to a much greater extent, by alloyed Mo, which strongly desorbs sulfur.[18] The first outcome of this, at least for temperatures <100 °C, is that chloride needs to be present for S-catalyzed pitting to occur in alloys with Mo. The activating effect of reduced sulfur is much less in alloys with Mo, as shown by direct kinetic measurements on artificial pits[19] and by studies in simulated solutions made from HCl with tetrathionate ($S_4O_6{}^{--}$).[20] SG tubing alloys are not alloyed with Mo, so they can be expected to have a generic susceptibility to reduced sulfur effects.

It was proposed in the early 1980s[1] that the remarkable aggressiveness of thiosulfate was associated with its combination of anionic nature, giving it the ability to enrich in a pit by electromigration (in contrast to sulfide), and its ability to deliver S_{ads} by electroreduction on the pit surface. This cathodic reaction was studied later by Marcus and Protopopoff.[17] But there must not be too much thiosulfate, relative to chloride or sulfate, otherwise the loss of acidity in the pit, by reactions (1) or (2), will be too great.[1]

Work of Ezuber[11] showed a strong influence of the sulfate to thiosulfate concentration ratio on sulfate–thiosulfate pitting of 304SS, as the ionic strength was varied over several orders of magnitude. This was rationalized by postulating that mutual electromigration of the anions into the pit preserved their concentration ratio locally. It was assumed – wrongly perhaps – that chloride would behave like sulfate in this respect, but the same level of detailed experimentation was never applied to chloride–thiosulfate pitting. Later, Laycock[24] obtained results that seemed to argue against a special significance of the anionic ratio for chloride–thiosulfate pitting. This topic will be further discussed later on. The ability of thiosulfate to catalyze pitting in extremely dilute – below millimolar – solutions[10] still requires a robust explanation. This is discussed later. Thiosulfate may also play some role in ordinary chloride pitting, as it has been detected near reactive manganese sulfide inclusions in stainless steel.[25]

The charge redistribution that occurs upon adsorption of S or H_2S on metal surfaces can now be studied using *ab initio* computational methods, simulating gas-phase adsorption and, increasingly, aqueous-phase 'electrosorption'.[21–23] It is clear from such studies that charge transfer from Fe or Ni to S is substantial, more perhaps than was envisaged in the work of the Marcus group. Yang and Whitten[23] carried out one of the earliest of such studies, and we quote the following: "For sulfur at the most stable three-fold site, our calculations show that the sulfur gains 0.61 electrons…mainly from the Ni 4s and 4p, and only 0.04e from the Ni 3d shells. The electrons are mainly transferred to the S 2p orbitals." They remark that the S_{ads}–Ni bond should not be considered ionic but covalent, in contrast to the O_{ads}–Ni bond.

The present study uses alloys 600, 800 and 690, using mainly the potentiostatic scratching technique to induce and measure pit propagation. The relative performance of these alloys, and the predictive significance if any of the anionic concentration ratio, are of interest, as is the lower limit of ionic strength at which such phenomena can be expected to occur.

2 Materials and methods

The 600, 800 and 690 alloys were obtained in the form of 1.5 mm plates from Metal Samples Inc., Munford, Alabama. Their compositions with respect to the main alloying elements are shown in Table 1. They were solution-annealed at 1050 °C for one hour in a 97.5% Ar/2.5% H_2 gas mixture and water quenched.

After mounting in epoxy resin, samples were used as working electrodes in a 3-electrode cell containing the sulfate–thiosulfate or chloride–thiosulfate solution. Tracepur® sodium sulfate was from EMD, assay >99%; sodium thiosulfate was from EM Science, an affiliate of Merck KGaA, Darmstadt, Germany, assay 99.5–100.5%. Sodium chloride was purchased from EMD, assay >99%. The resistivity of the DI water was 18.2 MΩ cm. A platinum counter electrode and an Hg/Hg_2SO_4/sat. K_2SO_4 reference electrode (MSE, 0.65 V *vs.* SHE at 25 °C) were also immersed for the sulfate–thiosulfate experiments; for chloride–thiosulfate, the reference electrode was Ag/AgCl/sat. KCl (0.199 V *vs.* SHE at 25 °C). All experiments reported here used an open beaker as the electrochemical cell, with freshly

Table 1 Compositions of the alloys studied, in wt%

Compositions	Alloy 600	Alloy 690	Alloy 800
Ni	Balance	Balance	31.2
Fe	9.53	10.1	Balance
Cr	16.1	29.6	20.0
C	0.06	0.03	0.07

poured solution – the oxidation rate of thiosulfate is slow enough to allow such a procedure, even in highly diluted solutions, which was confirmed by detailed ion chromatography (IC) measurements, shown in Table 2. IC tests were performed using the Dionex ion chronomatographic system (DX500) with a conductivity detector (ED40).

A diamond engraving pencil was used for scratching – the scratches were approximately 0.1 mm wide and 9.5 mm long. All the results shown here relate to a solution temperature of 60 °C – pitting occurred at 25 °C, but it was less stable and required much more effort for interpretation. A similar observation was made in the 1980s on stainless steel – the first reports of sulfate–thiosulfate pitting used high-purity alloys that were less resistant than ordinary type 304 stainless steel,[8,9] and gave the same kind of results at room temperature that were later obtained on commercial 304SS at 50 or 60 °C.[10,11]

The procedure for a sulfate–thiosulfate pitting test was as follows, exploiting the fact that sulfate–thiosulfate pitting ceases at a sufficiently positive potential, where thiosulfate can no longer be electroreduced in the pit.[8] First, the sample was passivated at 0 V (MSE) in 0.1 M Na_2SO_4 for half an hour. Then, it was transferred to the test cell and polarized at −0.91 V (MSE), below any possible pitting onset potential. The current was allowed to stabilize for 30 s. Then a scratch was made at this potential and the current transient (the scratch current) recorded for 180 s. Then, when no pitting occurred, the potential was stepped upwards by 100 or 25 mV, depending on the resolution required, and the process was repeated. If pitting occurred, as indicated by an increase in the current after initial repassivation (the scratch pitting current), the pits on the scratch were repassivated anodically by applying 0 V (MSE) for 10 s, then the potential was stepped back to the next scratching potential. All this is possible because sulfate–thiosulfate pitting is reluctant to initiate on undisturbed surfaces, although it does do so eventually.

Table 2 Ion chromatography measurements of thiosulfate solutions

Solution concentration	Conditions before IC test	Peak area, nS cm^{-1} min^{-1}	Retention time, min
10 μM	Freshly prepared	241 955	8.7
10 μM	Heated[a] and aerated 2 hours	235 573	8.8
100 μM	Freshly prepared	2 072 866	8.7
	Freshly prepared (repeat)	2 031 375	8.7
100 μM	Heated and aerated 2 hours	2 305 521	8.7
	Heated and aerated 2 hours	2 127 417	8.7
120 μM Na_2SO_4 + 10 μM $Na_2S_2O_3$	Freshly prepared	7 078 476(Na_2SO_4) 188 775($Na_2S_2O_3$)	4.2 8.8
120 μM Na_2SO_4 + 10 μM $Na_2S_2O_3$	Heated and aerated 2 hours	6 832 029(Na_2SO_4) 169 861($Na_2S_2O_3$)	4.3 8.7

[a] Heating and aeration for 2 hours at 60 °C.

Table 3 Lower/upper bounds of pitting with associated sulfate to thiosulfate ratio summarized from Fig. 4[a]

Ratio alloys	4000	2000	1000	**200**	100	40	20	10	3.3
A600	NP	LBP	P	P	P	P	P	UBP	NP
A800				NP	LBP	P	UBP	NP	
A690			NP	LBP	P	UBP	NP		

[a] NP-no pitting; LBP-lower bound of pitting; UBP-upper bound of pitting; P-pitting.

In order to define carefully the ranges of the potential and anionic concentration ratio for pitting, various interpolations and repeat scratches were made, and some sequences started at high potential and were worked down to a low potential. The final charts shown in this paper were the results of a large number of repeat tests.

For chloride–thiosulfate pitting, anodic potentials could not be applied to repassivate the pits, because ordinary chloride pitting would have occurred, but an anodic passivation in pure sulfate solution was still used, as before, to minimize the base currents and delay spontaneous pitting away from the scratches.

Pitted scratches were examined by SEM and EDX, usually after allowing a longer time for pit development than was used in the main electrochemical procedure.

3 Results and discussion

3.1 Electrochemical testing – sulfate–thiosulfate pitting

Fig. 1 shows typical results of scratching, for cases where pitting did, and did not, occur on Alloy 600. The increase in current due to pitting is obvious in this

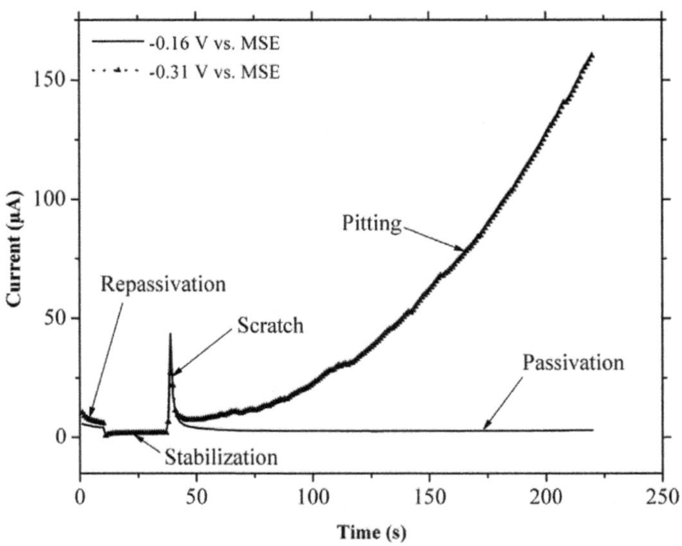

Fig. 1 Typical raw data for the scratch experiments: annealed Alloy 600 in 0.1 M Na_2SO_4 + 0.005 M $Na_2S_2O_3$ at 60 °C.

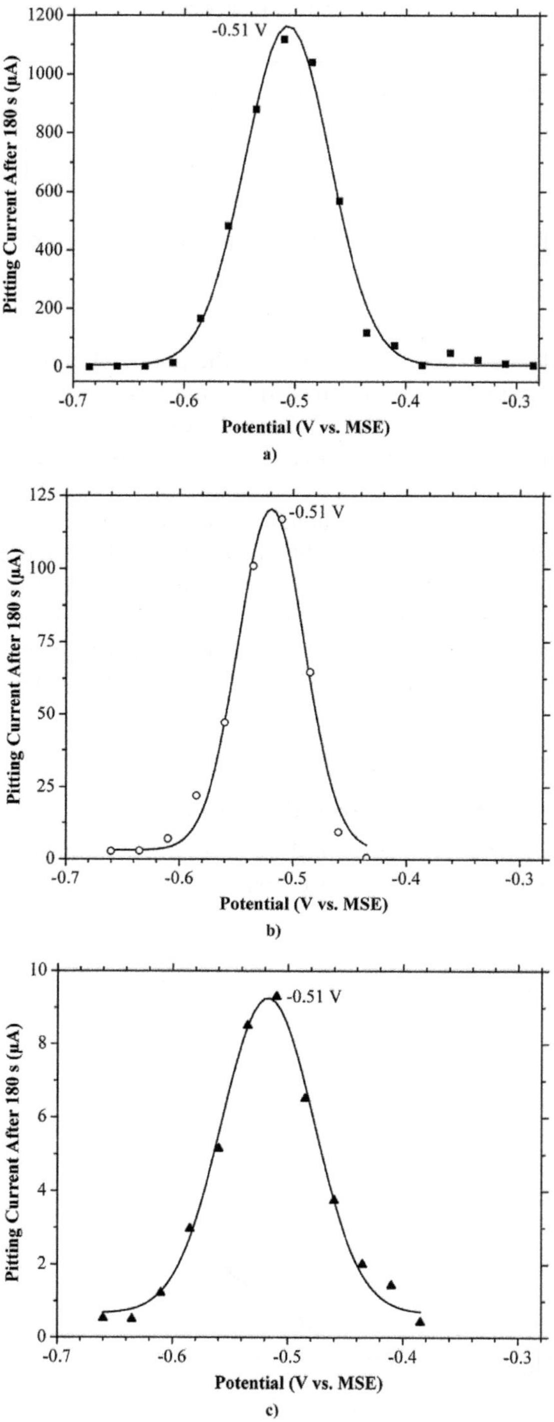

Fig. 2 Dependence of the scratch pitting current after 180 s potential, for tests done in 0.1 M Na_2SO_4 + 0.0025 M $Na_2S_2O_3$ on annealed alloys at 60 °C: (a) Alloy 600; (b) Alloy 800; (c) Alloy 690. The peak current was always observed at potential −0.51 V in this solution.

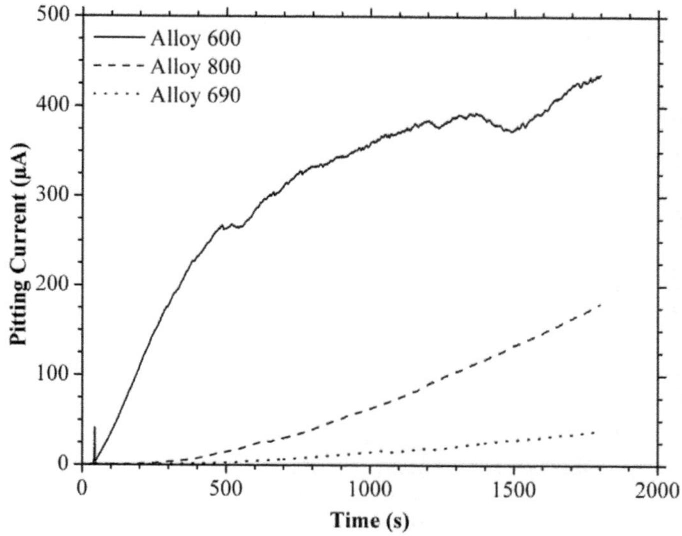

Fig. 3 Longer-term exposure of scratched samples at −0.61 V (MSE) showing the relative scratch currents; 0.1 M Na_2SO_4 + 0.001 M $Na_2S_2O_3$ at 60 °C.

example – but was more subtle for the other alloys, especially Alloy 690, as will be seen later. The current increase was always rather smooth, which is characteristic of thiosulfate-related pitting – the pits are strongly stabilized by sulfur adsorption and by the precipitation of a tight cap of corrosion product, to the point that it is quite easy to study the growth of single pits electrochemically.[26]

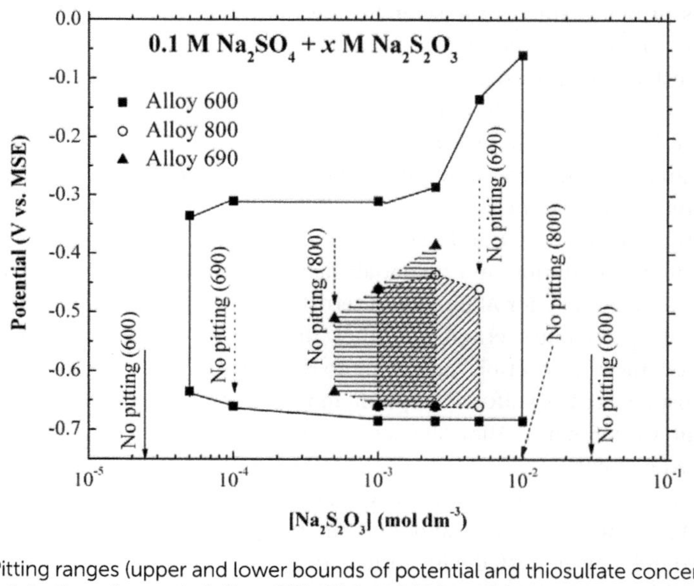

Fig. 4 Pitting ranges (upper and lower bounds of potential and thiosulfate concentration) for scratch testing of annealed alloys 600, 800 and 690 in 0.1 M Na_2SO_4 + x M $Na_2S_2O_3$ at 60 °C. Lower/upper bounds of pitting with associated sulfate to thiosulfate ratio are summarized in Table 3.

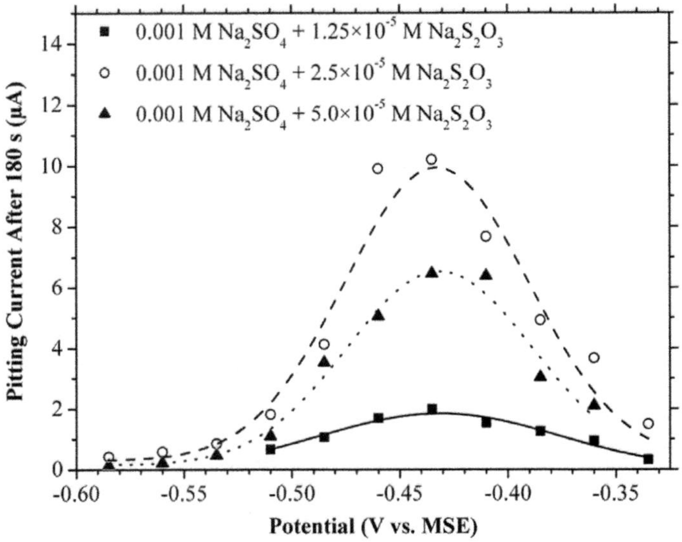

Fig. 5 The dependence of pitting rate on potential for annealed Alloy 600 in 1×10^{-3} M Na_2SO_4 + $Na_2S_2O_3$ with different concentrations at 60 °C, showing that the most severe anionic concentration ratio remains at 40.

Fig. 2 shows the scratch pitting current after 180 s as a function of potential, for each alloy. The initial increase with increasing potential is mainly due to a greater pit growth rate, while the decrease after the peak is mainly due to a decreased number of pit nuclei on each scratch. The much lower scratch currents for the 800 and 690 alloys are evident in this figure, although in the longer time-scale, the current continued to increase at a greater rate than for Alloy 600 (Fig. 3). This was primarily due to the scratches on Alloy 600 having completely pitted away and turned into grooves.

After exhaustive investigation of the pitting behaviour of the three alloys in relatively concentrated solution, the chart shown in Fig. 4 was obtained. This shows the following main features:

1 Pitting of annealed Alloy 600 occurs over a wide range of thiosulfate concentration and potential.

2 Pitting of Alloys 800 and 690 occurs over very similar, smaller ranges of thiosulfate concentration and potential; these ranges were close or identical to the most severe range for Alloy 600, in terms of scratch current after 180 s.

Depending on one's criterion, either the 800 or the 690 alloy could be considered 'more susceptible' according to these data. It may be that with nickel being most affected by sulfur adsorption, more so than iron, the higher (generally beneficial) Cr content of Alloy 690 is counteracted by its higher (generally detrimental) Ni content.

A maximum susceptibility at a sulfate–thiosulfate concentration ratio of around 40 is evident in Fig. 4, and was refined to 40 ± 10 in further work. Then, ratios around 40 were tested with various degrees of dilution of the solution. As shown in Fig. 5, the special ratio of 40 remained valid after a 100× dilution.

Finally, a solution with only 1 μM thiosulfate was tested, which gave easy pitting as shown in Fig. 6 – in this poorly conducting solution, the pits spaced themselves out along the scratch in a self-avoiding pattern.

3.2 Microscopy and analysis – sulfate–thiosulfate pitting

The different propensity of the three alloys to pitting is illustrated in Fig. 7. Alloy 600 has many and large pits. The 800 alloy has fewer pits, but of a similar size to Alloy 600. The 690 alloy clearly has a smaller pit size than the other alloys, for a given time of exposure after scratching. It could be that these pits have already repassivated spontaneously, but the current–time curves did not indicate any major noise or drops in current that might support such an interpretation. One

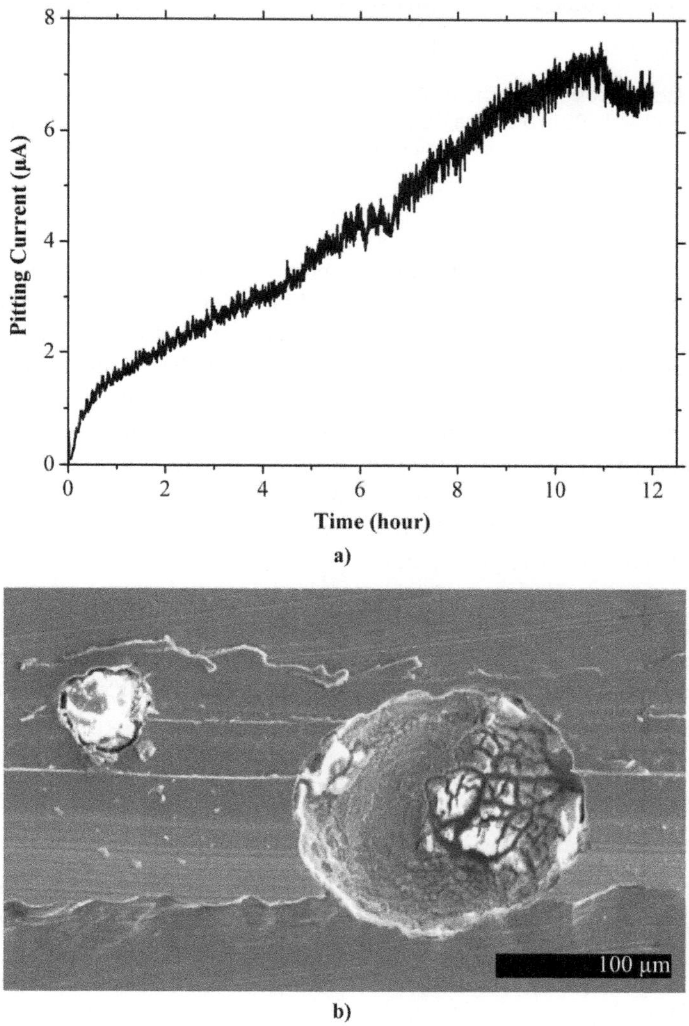

Fig. 6 Long-term exposure of scratched Alloy 600 in 4 × 10^{-5} M Na_2SO_4 + 1 × 10^{-6} M $Na_2S_2O_3$ for 12 hours at −0.435 V *vs.* MSE and 60 °C: (a) current–time curve; (b) pit morphology.

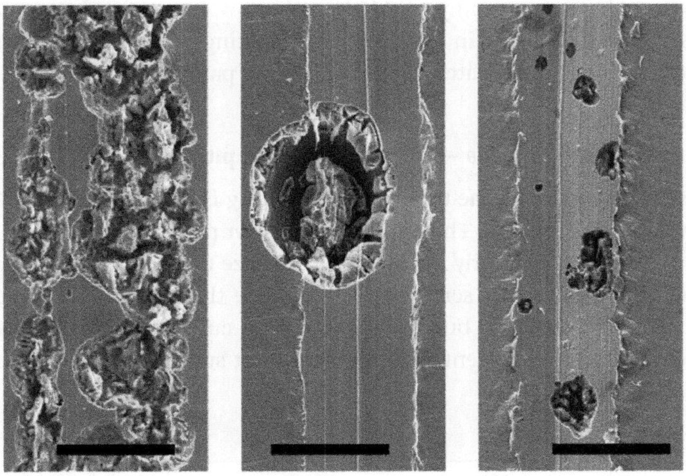

Fig. 7 Scratch corrosion morphologies of Alloy 600 in 0.1 M Na_2SO_4 + 0.001 M $Na_2S_2O_3$, and alloys 800 and 690 in 0.1 M Na_2SO_4 + 0.0025 M $Na_2S_2O_3$; all tests were done at 60 °C, −0.61 V (MSE) and exposed potentiostatically for 30 minutes after scratching. Marker bars are 100 μm.

factor to be considered is that nickel is more noble than iron, so for a given potential the 690 alloy would have slower dissolution kinetics than Alloy 800 (perhaps), but the 600 alloy is also mostly nickel. The most likely explanation is that the corrosion product cap is denser (less porous) on the 690 alloy, leading to a higher ohmic resistance across the cap (and/or a lower effective cation diffusivity, in the event that pit growth is diffusion controlled), and therefore to a slower pit growth rate. To the extent that the cap contains chromium [oxy]hydroxide, and the alloy has an exceptionally high Cr content, this may be expected.

In further work, the pitting currents could be rationalized in terms of area fraction of active surface, pit depth, *etc.* (or single pits could be created for the study).

All pits showed the enrichment of S and depletion of Fe and Ni, relative to Cr, in EDX analyses, as might be expected, see Fig. 8. So it is reasonable to characterize the pit cap as CrOOH containing some oxidized Fe and Ni, and unidentified S species, such as Ni_3S_2 perhaps.

3.3 Electrochemical testing – chloride–thiosulfate pitting

Initial scratch testing on Alloy 600, with no thiosulfate present, demonstrated some degree of pitting resistance up to at least −100 mV (Ag/AgCl) in 0.3 M NaCl at 60 °C, and when it did show persistent pitting at potentials around 0 V, the currents were low (sub-μA) and unstable. Given this result, there was no difficulty distinguishing ordinary chloride pitting from thiosulfate-catalyzed pitting, which invariably gave scratch currents of at least 0.5 μA and up to 150 μA in the most severe cases (after 240 s), with a smooth current increase. There was a subtle peak or shoulder in the rate of thiosulfate-catalyzed chloride pitting with potential, see

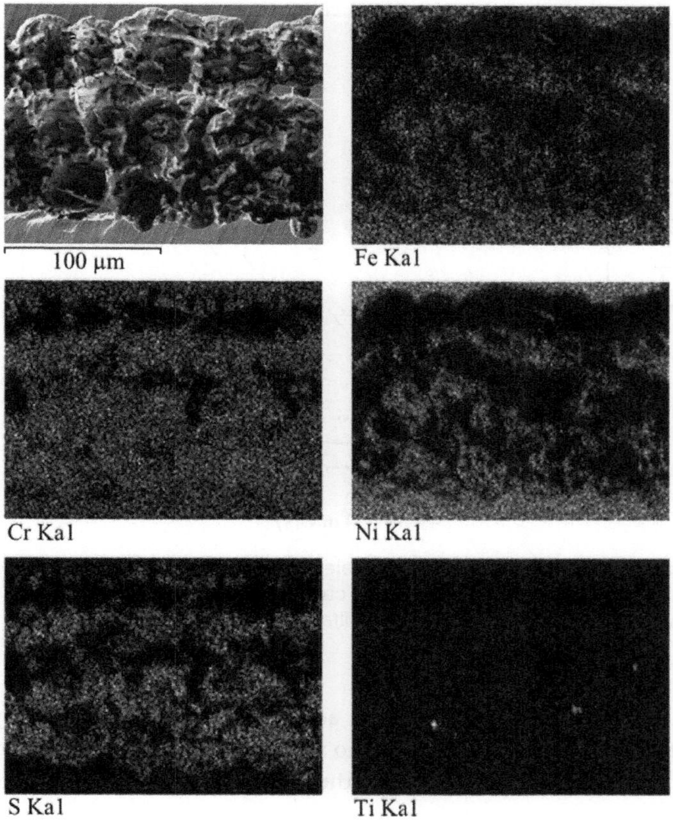

Fig. 8 EDX elemental mapping of a pitted scratch on Alloy 600, exposed to 0.1 M Na_2SO_4 + 0.001 M $Na_2S_2O_3$ for 30 minutes after scratching at −0.61 V (MSE) and 60 °C, showing S enrichment in the pits and relative depletion of the susceptible elements Ni and Fe compared to Cr.

Fig. 9, but this was not nearly as easy to discern as in the case of sulfate–thiosulfate pitting, where pitting ceased altogether at high potentials.

It was immediately apparent, see Fig. 9 and 10, that the amount of thiosulfate required for the pitting of Alloy 600 was much less in (strong) chloride than in a (strong) sulfate solution. More work is required to harmonize this result with the behaviour of type 304 stainless steel, where the original work of Newman, Isaacs and Alman[1] suggested an optimal chloride to thiosulfate ratio of no more than 25, while Franz[26] also used a ratio of 25 in her work on growth of single pits in 304SS. One possible explanation is that adsorbed S is more catalytic (activating) for a dissolution of Ni than Fe – perhaps due to the latter's greater affinity for oxygen, or to a difference in the degree of charge transfer on to the adsorbed S atoms, as there are several inter-related possibilities. In that case, a lower flux of thiosulfate will be required to maintain the activation of a Ni-based alloy compared to a Fe-based alloy, with other things being equal (the difference between sulfate and chloride will be discussed further below). Studies of chloride–thiosulfate pitting of Alloy 800, which is a high-alloy stainless steel, are in progress and should clarify this issue. Another possibility is that Alloy 600 has a different behaviour simply

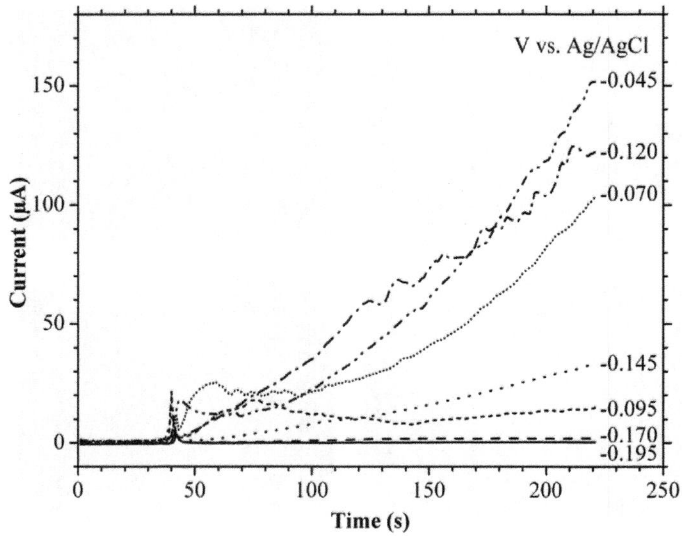

Fig. 9 Scratch current transients for annealed Alloy 600 in 0.1 M NaCl + 5 × 10^{-5} M $Na_2S_2O_3$, 60 °C, showing that the rate of current increase is not monotonic with the potential when pitting is catalyzed by thiosulfate.

because it is kinetically more noble in acids than type 304SS. Given that electroreduction of thiosulfate is required to stabilize the dissolution in the pit, the rate of this reduction will be lower, at the onset of pitting, in Alloy 600 than in

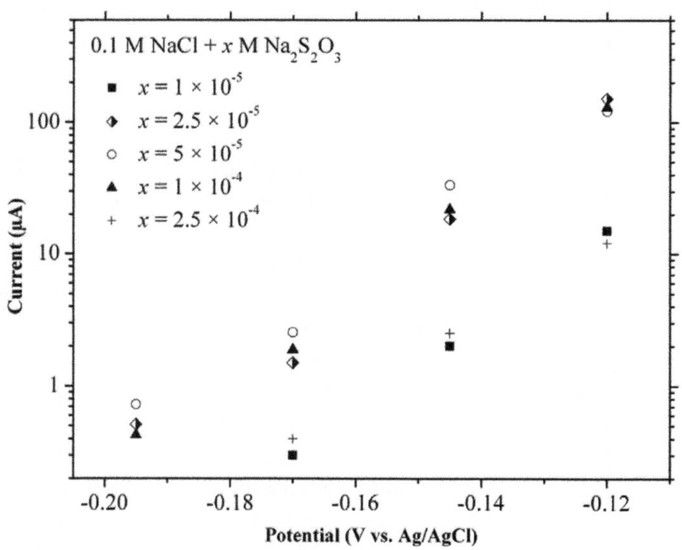

Fig. 10 Pitting corrosion of annealed Alloy 600 in 0.1 M NaCl + $Na_2S_2O_3$ with different concentrations at 60 °C, showing that pitting is rapid for chloride–thiosulfate concentration ratios in the range from 1000 to 4000. Little or no catalyzed pitting occurred at ratios of 400 or 10 000 – the behaviour reverted to the pure-chloride type (unstable pit growth and higher onset potential of pitting).

304SS, due to the displacement of the catalyzed pitting to higher potentials. So for Alloy 600, it is easy to swamp the pit solution with too much thiosulfate, leading to repassivation by the neutralizing effect of disproportionation. It may be (direct demonstration is difficult) that the dissolution kinetics in sulfate, at a given potential and in the absence of passivation, are faster than in chloride.

Experiments are in progress using strong HCl and H_2SO_4 solutions with tetrathionate addition, to model the effect of thiosulfate with a more acid-stable but similarly aggressive anion. The more noble character of Alloy 600, compared to 304SS or the other SG tubing alloys, is apparent in these experiments.

No S-catalyzed pitting was observed in solutions with chloride–thiosulfate concentration ratios of 400 or 10 000. Rapid S-catalyzed pitting was confined to the range of ratios from 1000 to 4000. Another difference from the sulfate based system was that pitting was much more difficult to stabilize in diluted solutions, because the 'optimal' thiosulfate concentration became absurdly low. So, there are two answers to the question – which is more severe in Alloy 600, sulfate–thiosulfate or chloride–thiosulfate pitting? From the perspective of relatively concentrated systems, chloride is more severe because less thiosulfate is required for pitting. But sulfate–thiosulfate pitting persists in solutions with much lower overall ionic strength, so it is a more plausible practical issue in dilute systems.

3.4 Considerations related to local chemistry and transport

In ordinary chloride pitting, the pitting potential, E_{pit}, varies with the bulk concentration of NaCl according to eqn (3):

$$E_{\text{pit}} = A - B\log[\text{Cl}^-] \quad (3)$$

where A is a kind of baseline pitting potential determined by the metal dissolution kinetics in the concentrated, acidic metal chloride solution in a pit. B is an IR drop in the pit nucleus. The value of B, for stainless steel at room temperature, is about 90–100 mV. Long ago, Galvele explained this behaviour in terms of reaction and transport in a pit nucleus[27,28] (others have speculated on adsorption and transport within the passive film as the underlying physics leading to this equation). In Galvele's treatment, the change in E_{pit} is just the change in the IR drop within a pit nucleus of particular geometry when the chemistry at the bottom of the pit (hydrolyzed cations; chloride ions) is at a critical value required to stabilize dissolution. Later, strong direct evidence for such an interpretation was provided by Laycock and Newman.[29] Galvele's result is obtained by treating Cl^- as a non-reacting ion in the cavity and equating its (outward) diffusion and (inward) migration fluxes at steady state. Now in theory, the value of B should be $-2.3RT/zF$, where z is the charge on the anion, *i.e.* -1. The departure from this value (59 mV at 298 K) is likely due to chloro-complexation of Cr in the pit solution, as pure iron does show a B value close to 60 mV. So, for sulfate–thiosulfate pitting, where sulfate plays the role of chloride, B should be close to 30 mV at 298 K, or 33 mV at 333 K, as shown in the present work. This assumes that SO_4^{--}, not HSO_4^-, is the appropriate form of sulfate to consider. We justify that by noting that pitting occurs at all – if the buffering effect of the HSO_4^-/SO_4^{--} equilibrium was important, pitting would likely not occur.

Galvele showed, amongst other things, that when two anions of different charge were present, the anion with the larger charge would accumulate more strongly at the bottom of the pit as the current density in the pit, or the potential, was increased. In other words, the idea that the ratio of anionic concentrations should be preserved in the enriched pit solution is not really correct unless the anions have the same charge (we assume that differences in mobility and diffusivity are small). This immediately explains why less thiosulfate, relative to the predominant anion, is required in a chloride–thiosulfate system than in a sulfate–thiosulfate system, and would also account for a lack of adherence of the former to a strict anionic concentration ratio criterion. When Ezuber[11] was studying thiosulfate-related pitting, he paid much more attention to sulfate–thiosulfate than chloride–thiosulfate. Pitting propensity in the sulfate–thiosulfate system showed a very good correlation with the anionic concentration ratio, and this led to the possibly incorrect assumption that the same would be true in chloride–thiosulfate solutions, or mixed solutions with both chloride and sulfate. In fact, the detailed measurements in chloride–thiosulfate were never done. It is possible that this omission can explain the relatively weak correlation of pitting with chloride to thiosulfate ratio observed by Laycock.[24]

Preliminary measurements of the B value for Alloy 600 in sulfate–thiosulfate solutions show a value of 40–50 mV. This can be seen by comparing the onset potentials for pitting in Fig. 2a and 5. Two orders of magnitude difference in ionic strength cause about a 90 mV shift in the kinetics. This requires more investigation, and is a little blurred by the gradual onset of the catalyzed pitting in these systems. It may be that this is the counterpart of the 90 or 100 mV value seen for stainless steels, *i.e.* a value that would be 33 mV if it were not for some complexation reaction. Alternatively, the discrepancy could be due to some protonation of sulfate in the pit, so that its charge, on average, is not exactly 2 – we know this must happen to some extent near the bottom of the pit. Anyway, we can certainly say that B is less than $2.3RT/F$, which is important.

Why is the B value significant? Because it helps to explain why pitting can occur in such extremely diluted sulfate–thiosulfate solutions. A low B value, such as 50 mV, implies that a 5 orders of magnitude decrease in ionic strength of the solution only elevates the pitting potential, E_{pit} (the lower potential limit of pitting in these data), by 250 mV. Not only does this still leave E_{pit} at a relatively low value by the standards of laboratory testing, it is still well within the range of natural open-circuit or corrosion potential that can be achieved on a passive alloy in a mildly aerated solution of near-neutral pH. Of course the exceptionally stable nature of thiosulfate-related pitting, including the dense pit cover, plays a role as well.

Finally we should consider the possibility of SCC in these solutions, as it occurs very easily in simple austenitic stainless steels with chloride instead of sulfate.[30] This is unlikely in the Ni-rich alloys but could feasibly occur in Alloy 800, by a de-alloying type of mechanism similar to that proposed for ordinary austenitic steels in chloride, or chloride with reduced sulfur.[31] This would likely require an elevated temperature.

4 Conclusions

• Alloys 600, 800 and 690 are susceptible to pitting in sulfate–thiosulfate solutions, including highly diluted versions. The potentials at which pitting occurs are easily attainable in practice with oxygen present. Galvele's *B* value for sulfate–thiosulfate pitting is low, facilitating pitting in highly diluted systems.

• Either Alloy 690 or Alloy 800 could be considered the least susceptible of the three alloys, depending on the criterion chosen.

• The pitting ceases at low or high potential, and at low or high sulfate to thiosulfate concentration ratio.

• The sulfate to thiosulfate concentration ratio has good predictive power across the range of ionic strengths examined, because sulfate and thiosulfate have the same charge and their concentration ratio should be preserved in the pit cavity. This is not the case for chloride–thiosulfate pitting.

• In Alloy 600, much less thiosulfate is required for chloride–thiosulfate pitting than for sulfate–thiosulfate pitting. This may not be a general observation across the range of NiCrFe systems. Ni seems to be more sensitive to S adsorption than Fe, so Ni-rich alloys require less thiosulfate. Another factor is that Alloy 600 is kinetically more noble, in the pit or pit-like environments, than the other alloys studied.

Acknowledgements

This research was supported by NSERC (Canada) and by UNENE, the University Network of Excellence in Nuclear Engineering. The support, interest and input of K. Sedman (Bruce Power) and P. J. King (B&W Canada) are greatly appreciated. Thanks also to N. Senior for assistance with the preparation of this document.

References

1 R. C. Newman, H. S. Isaacs and B. Alman, Effects of sulfur-compounds on the pitting behavior of type-304 stainless-steel in near-neutral chloride solutions, *Corrosion*, 1982, **38**, 261–265.

2 Z. Fang and R. W. Staehle, Effects of the valence of sulfur on passivation of alloys 600, 690, and 800 at 25 °C and 95 °C, *Corrosion*, 1999, **55**, 355–379.

3 J. T. Ho and G. P. Yu, Pitting corrosion of Inconel-600 in chloride and thiosulfate anion solutions at low-temperature, *Corrosion*, 1992, **48**, 147–158.

4 J. J. Park and S. I. Pyun, Stochastic approach to the pit growth kinetics of Inconel Alloy 600 in Cl- ion-containing thiosulphate solution at temperatures 25–150 °C by analysis of the potentiostatic current transients, *Corros. Sci.*, 2004, **46**, 285–296.

5 R. Roberge, Effect of the nickel content in the pitting of stainless-steels in low chloride and thiosulfate solutions, *Corrosion*, 1988, **44**, 274–280.

6 W. T. Tsai, Z. H. Lee, J. T. Lee, M. C. Tsai and P. H. Lo, Pitting and stress-corrosion cracking behavior of Inconel 600 alloy in thiosulfate solution, *Mater. Sci. Eng., A*, 1989, **118**, 121–129.

7 I. J. Yang, localized corrosion of Alloy 600 in thiosulfate solution at 25 °C, *Corrosion*, 1993, **49**, 576–584.

8 R. C. Newman, Pitting of stainless alloys in sulfate-solutions containing thiosulfate ions, *Corrosion*, 1985, **41**, 450–453.

9 R. C. Newman and K. Sieradzki, Electrochemical aspects of stress-corrosion cracking of sensitized stainless-steels, *Corros. Sci.*, 1983, **23**, 363–378.

10 R. C. Newman, W. P. Wong and A. Garner, A mechanism of microbial pitting in stainless-steel, *Corrosion*, 1986, **42**, 489–491.

11 R. C. Newman, W. P. Wong, H. Ezuber and A. Garner, Pitting of stainless-steels by thiosulfate ions, *Corrosion*, 1989, **45**, 282–287.

12 A. M. Brennenstuhl, T. S. Gendron and R. Cleland, Mechanisms of underdeposit corrosion in fresh-water cooled austenitic alloy heat-exchangers, *Corros. Sci.*, 1993, **35**, 699–711.

13 P. Marcus and E. Protopopoff, Potential-pH diagrams for adsorbed species – application to sulfur adsorbed on iron in water at 25 °C and 300 °C, *J. Electrochem. Soc.*, 1990, **137**, 2709–2712.

14 P. Marcus and E. Protopopoff, Potential-pH diagrams for sulfur and oxygen adsorbed on nickel in water at 25 °C and 300 °C, *J. Electrochem. Soc.*, 1993, **140**, 1571–1575.

15 P. Marcus and E. Protopopoff, Potential-pH diagrams for sulfur and oxygen adsorbed on chromium in water, *J. Electrochem. Soc.*, 1997, **144**, 1586–1590.

16 A. Elbiache and P. Marcus, The role of molybdenum in the dissolution and the passivation of stainless-steels with adsorbed sulfur, *Corros. Sci.*, 1992, **33**, 261–269.

17 P. Marcus and E. Protopopoff, Thermodynamics of thiosulfate reduction on surfaces of iron, nickel and chromium in water at 25 and 300 °C, *Corros. Sci.*, 1997, **39**, 1741–1752.

18 P. Marcus and M. Moscatelli, The role of alloyed molybdenum in the dissolution and the passivation of nickel-molybdenum alloys in the presence of adsorbed sulfur, *J. Electrochem. Soc.*, 1989, **136**, 1634–1637.

19 R. C. Newman, The dissolution and passivation kinetics of stainless alloys containing molybdenum .2. Dissolution kinetics in artificial pits, *Corros. Sci.*, 1985, **25**, 341–350.

20 A. J. Betts and R. C. Newman, The effect of alloyed molybdenum on the activation of anodic dissolution by reduced sulphur compounds, *Corros. Sci.*, 1993, **34**, 1551–1555.

21 J. Rossmeisl, E. Skulason, M. E. Bjorketun, V. Tripkovic and J. K. Norskov, Modeling the electrified solid–liquid interface, *Chem. Phys. Lett.*, 2008, **466**, 68–71.

22 C. D. Taylor and M. Neurock, Theoretical insights into the structure and reactivity of the aqueous/metal interface, *Curr. Opin. Solid State Mater. Sci.*, 2005, **9**, 49–65.

23 H. Yang and J. L. Whitten, Adsorption of SH and OH and coadsorption of S, O and H on Ni(111), *Surf.Sci.*, 1997, **370**, 136–154.

24 N. J. Laycock, Effects of temperature and thiosulfate on chloride pitting of austenitic stainless steels, *Corrosion*, 1999, **55**, 590–595.

25 E. G. Webb and R. C. Alkire, Pit initiation at single sulfide inclusions in stainless steel-II. Detection of local pH, sulfide, and thiosulfate, *J. Electrochem. Soc.*, 2002, **149**, B280–B285.

26 R. C. Newman and E. M. Franz, Growth and repassivation of single corrosion pits in stainless-steel, *Corrosion*, 1984, **40**, 325–330.

27 J. R. Galvele, Transport processes and mechanism of pitting of metals, *J. Electrochem. Soc.*, 1976, **123**, 464–474.

28 S. M. Gravano and J. R. Galvele, Transport processes in passivity breakdown .3. Full hydrolysis plus ion migration plus buffers, *Corros. Sci.*, 1984, **24**, 517–534.

29 N. J. Laycock and R. C. Newman, Localised dissolution kinetics, salt films and pitting potentials, *Corros. Sci.*, 1997, **39**, 1771–1790.
30 T. Haruna, T. Shibata and R. Toyota, Initiation and propagation of stress corrosion cracks for type 304L stainless steel in chloride solutions containing thiosulfate, *Corros. Sci.*, 1997, **39**, 1935–1947.
31 R. C. Newman, R. R. Corderman and K. Sieradzki, Evidence for de-alloying of austenitic stainless steels in simulated stress-corrosion crack environments, *Br. Corros. J.*, 1989, **24**, 143–148.

Faraday Discussions

PAPER

Atmospheric pitting corrosion of 304L stainless steel: the role of highly concentrated chloride solutions

Steven R. Street,*[a] Na Mi,[a] Angus J. M. C. Cook,[a] Haval B. Mohammed-Ali,[a] Liya Guo,[a] Trevor Rayment[b] and Alison J. Davenport[a]

Received 2nd December 2014, Accepted 16th January 2015
DOI: 10.1039/c4fd00246f

The morphology of atmospheric pitting corrosion in 304L stainless steel plate was analysed using $MgCl_2$ droplets in relation to changes in relative humidity (RH) and chloride deposition density (CDD). It was found that highly reproducible morphologies occur that are distinct at different RH. Pitting at higher concentrations, *i.e.* lower RH, resulted in satellite pits forming around the perimeter of wide shallow dish regions. At higher RH, these satellite pits did not form and instead spiral attack into the shallow region was observed. Increasing CDD at saturation resulted in a very broad-mouthed pitting attack within the shallow dish region. Large data sets were used to find trends in pit size and morphology in what is essentially a heterogeneous alloy. Electrochemical experiments on 304 stainless steel wires in highly saturated solutions showed that the passive current density increased significantly above 3 M $MgCl_2$ and the breakdown pitting potential dropped as the concentration increased. It is proposed that the shallow dish regions grow *via* enhanced dissolution of the passive film, whereas satellite pits and a spiral attack take place with active dissolution of bare metal surfaces.

Introduction

Atmospheric pitting corrosion occurs stochastically on stainless steels, with key factors influencing initiation and propagation not obvious on the scale that the phenomenon is typically observed.[1] As such, individual instances of pitting can only be considered as anecdotal not archetypal. To draw conclusions from observing pitting on industrial alloys, large data sets have become increasingly used to get a sense of the significant factors influencing atmospheric pitting corrosion.[2–4]

Key to the concept of atmospheric corrosion is the presence of small volumes of solution which are highly concentrated, with Cl-rich solutions being of particular interest.[5–7] The concentration of these solutions is governed by the

[a]*University of Birmingham, School of Metallurgy and Materials, Edgbaston, Birmingham, B15 2TT, UK. E-mail: SRS928@bham.ac.uk*
[b]*Diamond Light Source, Harwell Science and Innovation Campus, Didcot, Oxford, OX11 0DE, UK*

relative humidity (RH) to which the solutions are exposed.[8] A range of deposition methods have been used to study atmospheric corrosion, including inkjet printing,[9–11] manually-deposited salt crystals,[12] and dissolving in alcohol and drying.[13,14] Droplets have been widely used to analyse atmospheric corrosion phenomena in steels[15–20] and other alloys,[21–29] as their well-defined geometry allows claims to be made about mechanisms of corrosion, such as the available cathodic current[30] and chloride deposition density (CDD). The nuanced nature of droplet deposition allows the control of solution volumes and concentrations with relative ease.

Pit morphologies have been explored both analytically and schematically.[31,32] Vera Cruz *et al.*[33] described "colonies" of pits occurring in stainless steel with a single large pit surrounded by several smaller ones when under thin solute layers. The morphology is also influenced by the microstructure,[9,34] solution concentration at the pit mouth[32,35,36] and pit potential.[37] Instances have been recorded where these morphologies have had implications in failure under SCC and fatigue regimes.[38]

This work will explore the influence of concentrated $MgCl_2$ on the morphology of atmospheric pitting corrosion and full immersion current characteristics of 304L stainless steel.

Experimental

Atmospheric corrosion

Samples were studied using two protocols: the first samples were prepared and examined after 24 hours and for the second ones, the evolution of pitting was monitored at regular shorter intervals for 24 hours.

A cold-rolled and solution heat-treated stainless steel 304L plate (Table 1) of 3 mm thickness was used (Aperam, France). This plate is known to have up to 3% ferrite distributed along the rolling direction. The plate was cut into 75 mm × 25 mm samples and ground using 800 grit SiC paper, then cleaned in deionised (DI) water (18 MΩ cm) in an ultrasonic bath for 15 minutes. The samples were then rinsed with DI water, then dried by blowing air from an empty DI water bottle. The samples were then left for 24 hours under covered, ambient conditions before droplet deposition.

Arrays of 2 μl droplets of 0.27 M $MgCl_2$ were deposited, with a final CDD of approx. 750 μg cm^{-2}. The droplets were deposited using a Multiprobe II liquid handling system (Perkin-Elmer Life Sciences) at 5 mm intervals, in arrays of 4 × 14 or 4 × 15 (56–60 droplets). The droplets were deposited with the pipette tip at 0.91 mm from the plate surface at a rate of 10 μl s^{-1}. The full array was deposited within 8 minutes. The ambient temperature was between 20–23 °C and the humidity 25–44% RH during deposition. The experimental conditions used are reported in Table 2.

Table 1 Foundry specifications of the 304L plate used in the atmospheric corrosion experiments

Element (wt%)	Fe	Ni	Cr	Mn	S	C	Si	N	P
Foundry spec.	Balance	8–10.5	18–19.5	2	0.015	0.03	0.75	0.1	0.045

Table 2 Chloride deposition density (CDD) for the as-deposited droplets of $MgCl_2$ deposited on a 304L stainless steel plate. The samples sizes are 56–60, with the standard deviation given for the data set in each case. The droplets were deposited at 20–23 °C in ambient relative humidity (25–44% RH). The samples were put into controlled experimental humidity within 8 minutes of the first deposition

Humidity ± S.D. (% RH)	Average droplet CDD ± S.D. ($\mu g\ cm^{-2}$)			Droplet area on deposition (cm^2)		
	"80"	"750"	"5400"	"80"	"750"	"5400"
33 ± 2	80 ± 3	670 ± 40	5400 ± 300	0.048 ± 0.001	0.057 ± 0.003	0.051 ± 0.001
38 ± 2	—	820 ± 20	—	—	0.047 ± 0.001	—
43 ± 2	—	790 ± 40	—	—	0.048 ± 0.002	—
48 ± 2	—	670 ± 40	—	—	0.057 ± 0.003	—
56 ± 2	—	730 ± 40	—	—	0.052 ± 0.002	—

Once deposited, the plate was held in an ECO135 atmospheric chamber (TAS Ltd) at 30 ± 1 °C and at a controlled humidity between 33–56% RH for 24 hours (Fig. 1). This equates to droplet solution concentrations between 3.7 M and 5 M $MgCl_2$. After 24 hours, each droplet was photographed with a microscope and then the plate was washed with DI and methanol and dried using a hot hairdryer.

The samples of 304L stainless steel for the time-lapse experiments were prepared as above. 2 μl droplets of 0.27 M $MgCl_2$ were deposited on the sample and kept in a sealed transparent container in which humidity was controlled by saturated salt solutions.[40] The container was kept in an environmental chamber at 30 ± 1 °C. Once an hour, it was removed from the chamber and the droplet was photographed using an optical microscope. The time outside the environmental chamber was approx. 3 minutes per hour.

Electrochemical sweeps

The wires of the 304 stainless steel (250 μm diameter) were abraded along their length using 4000 grit SiC paper, and then rinsed with DI water. Each wire was then immersed in a $MgCl_2$ solution with the concentration ranging from 0.05–5 M. The wire electrode was bent into a J-shape so the cut ends were not left in solution. The surface area of the wire in solution was controlled to be 0.5 cm^2. A tape was placed 5 mm above the waterline on the wire to avoid this being a crevice corrosion site.

A SCE reference electrode and platinum auxiliary electrode were used in approx. 200 ml of solution. The solutions used were de-aerated by bubbling argon for 20 minutes and argon before the wire electrode was put into solution. Once the wire electrode was put into the solution, the flow of argon into the solution was stopped and a flow of argon into the air above the solution was started, to prevent oxygen diffusion into the solution during the experiment. A new wire electrode was prepared and used in each experiment.

Results

Effect of the chloride deposition density

Fig. 2 shows typical droplets at 33% RH for CDDs of 80 $\mu g\ cm^{-2}$, 750 $\mu g\ cm^{-2}$, and 5400 $\mu g\ cm^{-2}$. The droplets usually contained only one main pit, but in a few

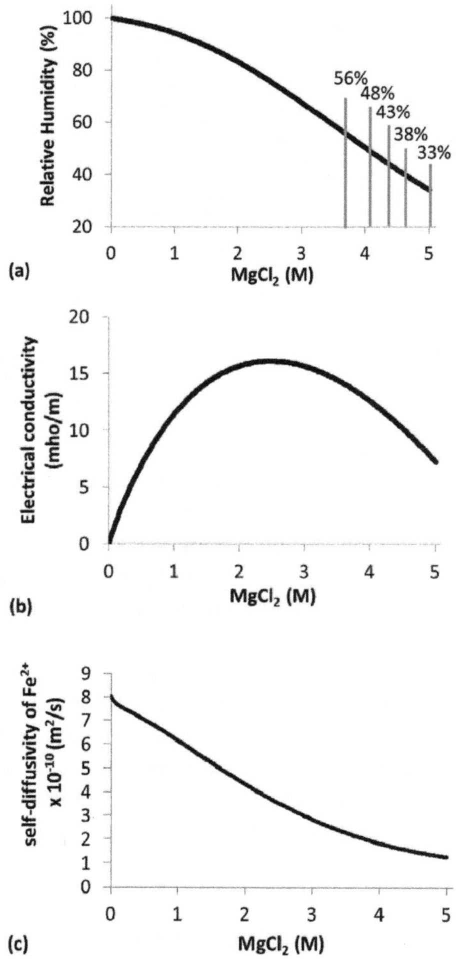

Fig. 1 Relationship between the solution concentration (M) of $MgCl_2$ and (a) relative humidity (%), (b) electrical conductivity (mho m^{-1}), and (c) self-diffusivity of Fe^{2+} ($m^2\ s^{-1}$) in $MgCl_2$ solutions at 30 °C, calculated using OLI Analyzer 9.[39] Humidities investigated in this work are marked in (a).

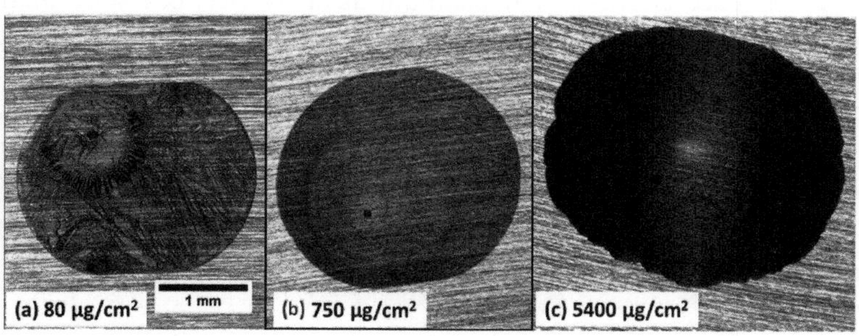

Fig. 2 Optical microscope images of droplets after 24 hours at 33% RH and 30 °C. (a) 80 µg cm^{-2}, (b) 750 µg per cm^2 CDD and (c) 5400 µg per cm^2 CDD. The dark spots in the droplets are pits.

cases two were observed. At a CDD of 80 μg cm^{-2}, droplets that contained pits often showed needle-like crystal growth along the surface of the droplet. Droplets with an initial CDD of 5400 μg cm^{-2} (Fig. 2c) increased in surface area during 24 hours with an average value of 0.078 ± 0.004 cm^2, an increase of 53%. Droplets at lower CDD did not change perimeter noticeably.

Most pits that occurred at a CDD of 80 μg cm^{-2} showed a shallow dish-shaped region, with one side significantly more attacked, and with several satellite pits occurring at the perimeter of the dish or nearby (Fig. 3a). Crystallographic etching can be seen in the shallow dish region. Out of 56 droplets, 43 pitted (77%) and of these, 36 showed this morphology. 18 of the 43 pitted droplets had needle-like crystals growing across the droplet surface, a phenomenon that did not occur in non-pitting droplets.

The same etched shallow dish region with satellite pits was observed for droplets with a CDD of 750 μg cm^{-2}, usually with one side showing more corrosion (Fig. 3b). The shallow dish region usually had a smoother circumference than at a CDD of 80 μg cm^{-2}. 52 out of 56 droplets pitted under these conditions, with 49 of those that pitted showed this morphology.

All 56 droplets with a CDD of 5400 μg cm^{-2} showed a single pit. 33 out of 56 droplets showed shallow dish regions with less pronounced etching than that observed at lower CDDs, with a deeper pit in the part of the shallow dish (Fig. 3c). Most of the remaining pits had very large pit mouth areas ≥100 μm in diameter, which had probably consumed an initial shallow dish that was no longer visible. One pit showed the satellite morphology seen at lower CDD.

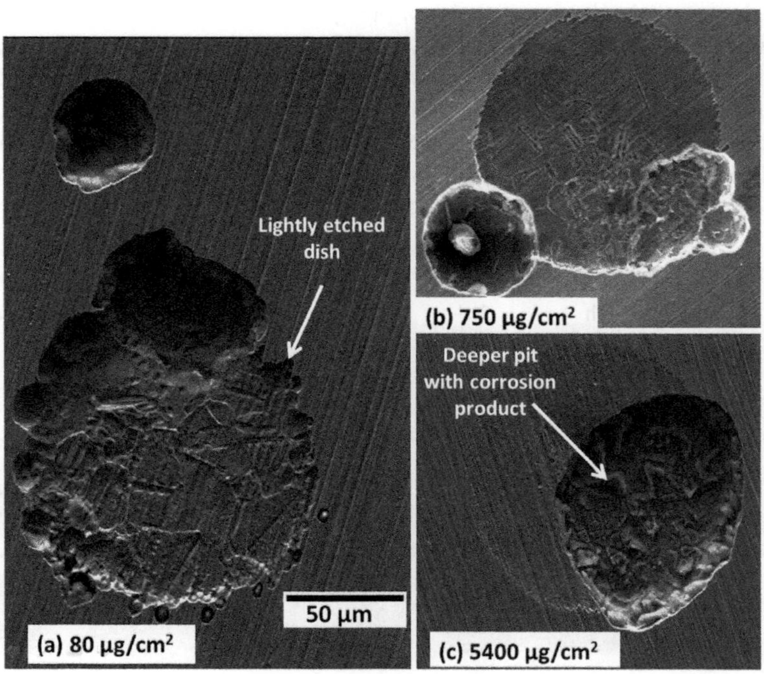

Fig. 3 SEM images of a pit grown in a $MgCl_2$ droplet with (a) 80 μg per cm^2 CDD, (b) 750 μg per cm^2 CDD and (c) 5400 μg per cm^2 CDD at 33% RH and 30 °C for 24 hours.

Time-lapse measurement of the pit growth

Fig. 4a shows images of a growing pit as a function of time under a droplet at a RH of 33% with a CDD of ~750 μg cm^{-2}. After 1 hour, the shallow dish region was still growing radially, but by 2 hours the diameter of the dish had stopped increasing, but further etching can be seen. In this same time period, small dark spots that developed into satellite pits had begun to appear around the shallow dish, becoming more prominent by 3 hours. These satellite pits grew concurrently with continued etching behaviour at the bottom of the shallow dish over the next 3 hours. By 24 hours, the shallow dish region had ceased to develop but the satellite pit mouths had each grown significantly in size.

At 43% RH (Fig. 4b), the initial pit growth was similar to that at 33% RH: after 1 hour, radial growth of the shallow dish region continued, but by 3 hours, radial growth had ceased but etching continued. No satellite pits formed; instead growth into the shallow region continued in a spiral fashion. These time-lapse images are on a smaller scale than those shown at 33% RH, and a smaller pit was formed in this instance. However, in both 33% and 43% RH high-throughput experiments, the shallow dish regions formed showed significant variation in size.

At an RH of 56%, only a small amount of radial growth occurred within the first hour, with a greater extent of attack into the metal (Fig. 4c). The radial growth of the pit mouth was completed by 4 hours, but the pit appeared to continue to grow deeper.

Pit morphologies

The characteristic pit morphologies of pits grown at a CDD of 750 μg cm^{-2} are shown in Fig. 5. At 33% RH, a shallow dish region that showed crystallographic etching with deeper attack on one side and satellite pits at or near the dish circumference was found for 49 out of 52 pits (Fig. 5a). Pit caps were seen on the top of the satellite pits,[41,42] and showed grinding marks that lined up with those on the bulk metal surface. EDX analysis of the pit caps gave typical compositions of ~40% Cr, ~45% O, ~5% S, and ~10% Cl (Fig. 6).

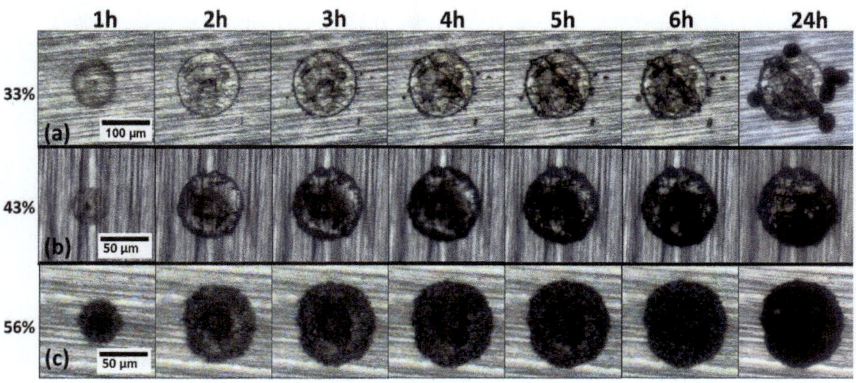

Fig. 4 Time-lapse images of pitting of 304L stainless steel under 750 μg per cm^2 $MgCl_2$ at 30 °C and (a) 33% RH, (b) 43% RH, and (c) 56% RH. All times are from deposition.

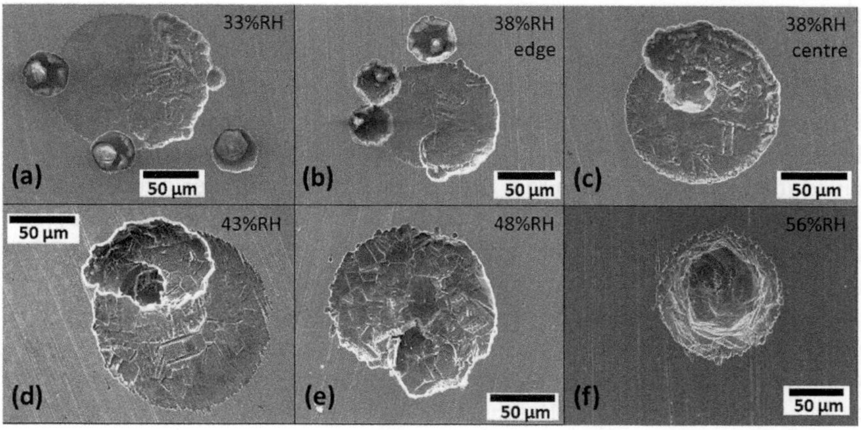

Fig. 5 Typical pit morphologies of 750 µg per cm^2 CDD $MgCl_2$ droplets at (a) 33% RH, (b) 38% RH near the droplet edge, (c) 38% RH near the droplet centre, (d) 43% RH, (e) 48% RH, and (f) 56% RH. The pits were grown for 24 hours at 30 °C.

At 38% RH, all 56 droplets pitted, with two dominant pit morphologies occurring. 37 pits showed the "satellite" morphology described above (Fig. 5b) and 16 pits showed a deeper "spiral" attack in the shallow dish region (Fig. 5c).

At 43% RH, 56 out of 60 droplets pitted, and 54 of these showed the "spiral" morphology, with an etched shallow dish region more strongly attacked at one side, with the attack continuing deeper into the metal (Fig. 5d). One pit showed the "satellite" pit morphology, and one pit had grown to propagate outside the droplet region and had developed an irregular morphology.

Pits with the "spiral" morphology were also found for 53/56 droplets at 48% RH (Fig. 5e). Two droplets contained pits that showed growth along the rolling direction of the steel, indicating a susceptibility to microstructural features and one droplet did not contain a pit.

All 56 droplets at 56% RH had pits that did not have a shallow dish region, but instead showed narrow circular mouths (Fig. 5f). 24 droplets (43%) showed at least two pits that were more than 500 µm apart. Analysis of the rust deposits around the pits suggested that these pits occurred consecutively not concurrently in each case.

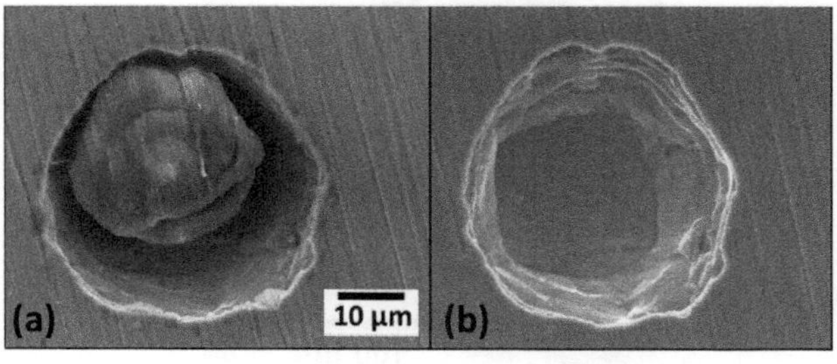

Fig. 6 Satellite pit from 33% RH (a) before and (b) after ultrasonic cleaning in DI water for 30 seconds. The pit caps of the type seen in (a) showed typical EDX compositions of ~40% Cr, ~45% O, ~5% S, and ~10% Cl.

Effect of the pit location on the size of the shallow dish region

Pits were observed to occur at all locations in the droplets under all conditions, with no clear trend in initiation site towards either edge or centre. However, close inspection suggested that the size of the shallow dish regions was affected by the location of the pit within the droplet. The diameter was determined by drawing a circle around the dish region that had remained after continued pitting (Fig. 7). This technique was based on the assumption that dish regions were circular; any pits that did not show this feature were not included. Plots of the diameter of the shallow dish as a function of distance from the droplet edge are shown in Fig. 8.

At 33% RH, the dish diameters showed a high degree of scatter, but despite the wide variation, a trend of a larger diameter dish towards the centre of the droplet can be observed in Fig. 8a. A similar trend existed at higher humidities, with less scatter at 38% RH and 48% RH. From 38% RH to 48% RH, droplets that occurred near the edge were not only smaller but had less variation in diameter compared to droplets formed towards the centre. At 38% RH, a transition occurred in the morphology regarding the pit location: satellite pits predominately formed near the edge of the droplets, and spiral pits were found towards the centre. At 43% and 48% RH, the majority of pits were spiral, and there were fewer dish regions smaller than 80 μm (Table 3). There are no data for pits formed at 56% RH as they showed no shallow dish region after 24 hours.

Electrochemistry

Fig. 9 shows polarisation curves of 304 wires in solutions of different concentrations of $MgCl_2$. In solutions between 0.05 M and 3 M, passive current densities are of similar magnitude. Each of these curves show metastable pitting peaks as the pitting potential is approached. In concentrations higher than 3 M (the solution in equilibrium with 68% RH), the passive current density increased with concentration (Fig. 9b). A small amount of metastable pitting was observed at 3.2 M (equivalent to 64% RH), but significantly less than at lower solution concentrations. Passive current densities are similar for solutions of 3.2, 3.5, and 4 M $MgCl_2$ (equivalent to 64% RH, 59% RH and 50% RH, respectively), but increase significantly at 5 M (33% RH).

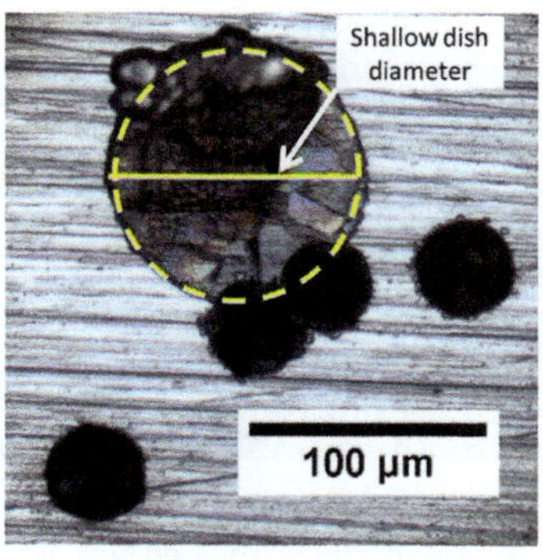

Fig. 7 Technique for measuring the shallow dish region seen during pitting.

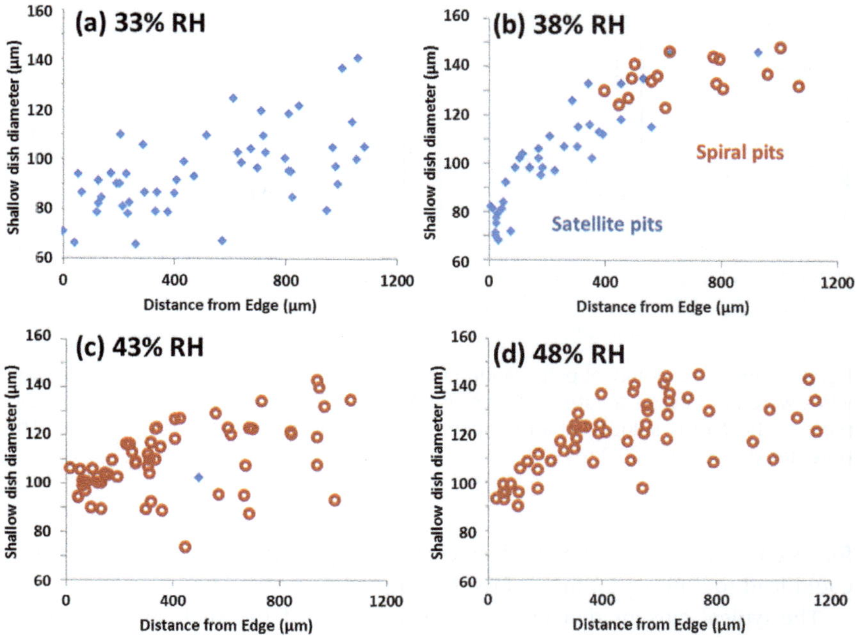

Fig. 8 Diameter of the shallow dish regions against the distance from the droplet edge in $MgCl_2$ droplets with a CDD of 750 µg cm^{-2} after 24 hours at 30 °C. Satellite pits are shown as blue diamonds and spiral pits are shown as red circles.

Table 3 Average diameters and standard deviations of the pits formed in $MgCl_2$ droplets with a CDD of 750 µg cm^{-2} after 24 hours at 30 °C

RH (%)	Diameter average (µm)
33	95 ± 17
38	110 ± 23
43	110 ± 14
48	120 ± 15

Discussion

Atmospheric corrosion is characterised by a limited electrolyte volume, differential aeration, and highly concentrated solutions in which a further increase in concentration leads to a decrease in ionic conductivity and diffusion coefficient of ions. These effects must be considered to understand mechanisms of growth in droplet experiments.

At RH values of 48% and below, localised corrosion starts with a shallow dish.[19,20,33] The surface of the dish is etched, suggesting that no salt layer is present, since dissolution under a salt layer leads to polished or roughened surfaces,[43,44] which can be seen in the deep pitting in Fig. 3. Crystallographic attack could take place on a bare metal surface, or on a surface with a passive film. It is known that the passive film on both iron and stainless steel is crystalline and

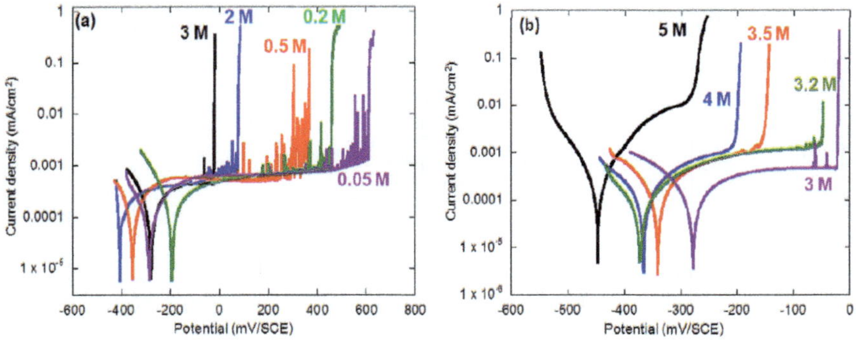

Fig. 9 Current *vs.* potential polarisation curves of 304 stainless steel. 250 μm diameter wires were put into de-aerated solutions immediately after abrasion using 4000 grit SiC papers. The $MgCl_2$ solutions had concentrations of (a) 0.05–3 M and (b) 3–5 M. The potential sweep rate is 0.2 mV s^{-1}.

forms epitaxially on the metal,[45,46] so crystallographic etching of the passive film could lead to differential rates of oxidation on different grains.

The typical current density for vertical growth of the shallow dish can be estimated from Faraday's law using a density of 7.9 g cm^{-3} and an average charge on the dissolving ions of 2.2 for 304L. The shallow dish regions have a typical depth of 10 μm, with the majority of growth occurring in the first two hours. This yields a current density of ~4 μA cm^{-2}, which is in the same order of magnitude as the passive current densities observed at a high chloride concentration in Fig. 9, indicating that dissolution of the shallow dish region may take place *via* concurrent passive oxide film growth and dissolution. Furthermore, it is clear from Fig. 9 that the passive current density increases significantly with solution concentration, suggesting that the dish shape may be influenced by the higher concentration of the solution in the vicinity of the initial breakdown site at the centre of the dish.

Maier and Frankel[19] have proposed the adsorption of S to the oxide layer around the pit, causing an attack in a radial fashion. Hastuty *et al.*[20] have also proposed a mechanism that causes radial growth by hydrolysis after going through a thiosulphate intermediate stage. However, the variation in dish diameter with distance from the droplet edge (Fig. 8) indicates that the controlling factor is not simply the presence of sulphur species.

Time-lapse imaging shows that for 33% and 43% RH at a CDD of 750 μg cm^{-2} (Fig. 4), the dish-shaped region stops growing radially and a deeper and more localised attack develops. At 33% RH, small satellite pits form around the shallow dish region that are protected by pit caps enriched in Cr, O, S, and Cl, suggesting that they form at the site of sulphide inclusions. Only limited attack continues into the shallow dish region. At 43% RH, these satellite pits are not observed and attack continues into the shallow dish, forming a spiral morphology. At 56% RH, no sign of the initial shallow dish region was seen as pits grew.

A similar change in morphology is seen at 33% RH on increasing the CDD from 80 μg cm^{-2}, where satellite pits are formed, to 5400 μg cm^{-2}, where part of the dish is attacked more deeply. In some cases the entire shallow dish region at 5400 μg cm^{-2} CDD was attacked, leaving no trace of it after 24 hours. For a fixed

RH, the solution concentration (and thus its resistivity) is constant, but the droplet height is greater for a higher CDD. This suggests that the increase in localised attack is associated with a lower resistance, and thus lower IR drop between the cathode and the anode, as well as a less constrained geometry that will allow faster diffusion of ions away from the pit.

The role of the IR drop and diffusion in controlling the extent of pitting attack is also valid for the transition from satellite to spiral pits between 33% and 43% RH, since for fixed CDD, the droplet height will be greater at 43% RH and the solution resistivity is lower, both leading to a lower IR drop, and the self-diffusion coefficient is higher. Furthermore, at 38% RH, satellite pits are seen near the edge of the droplet, where the droplet height is low, the resistance path is high to most of the droplet perimeter, and the diffusion rate of metal ions is lower, whereas spiral pits are observed when pits are located towards the centre of the droplets. Where the IR drop is lower, the interfacial potential within the pit will be higher, increasing the likelihood that it will be greater than the "breakdown potential" observed in the polarisation curves (Fig. 9). The lack of development of a shallow dish region in pitting at 56% RH may be linked to the lower passive current density at lower concentrations, but the higher solution conductivity and droplet height leads to a lower IR drop so that it is easier for the breakdown potential to be exceeded.

For the concentrated solutions and thin layer of electrolytes found at low RH, satellite pitting sites that initiate outside the shallow dish are able to overcome the transition from metastability[47] because pit caps form, and the escape of ions from the pits is low since diffusion is limited. The satellite pits draw a continually increasing share of the limited cathodic current[30] as they grow. This might be expected to the shallow dish of necessary cathodic current to continue, passivating this region.

At higher RH and with thicker solution layers, spiral growth is observed. Although the IR drop is reduced under these conditions compared to those showing satellite morphology, there remains enough to allow only partial attack of this region. Only a small area of the shallow dish region can sustain a sufficient current density to propagate, the rest is passivating. It is possible that the spiral geometry is favoured as it increases occlusion of the pit, but further investigations are necessary to confirm this. While satellite pitting sites are available around the shallow dish region, the diluted solution means that these are not able to be sustained, allowing growth into the shallow region to continue.

The considerable scatter in measurements of the shallow dish diameter and CDD show that there is a need for large data sets when dealing with commercial alloys and droplets as atmospheric corrosion models. Trends in these variables only make themselves apparent when high throughput techniques are used. It is also clear that the pitting behaviour is highly sensitive to local variations in conductivity and diffusivity with solution chemistry.

Conclusions

Atmospheric pitting corrosion of 304L stainless steel was analysed using droplets of $MgCl_2$ after 24 hours of pitting at 30 °C. It was found that the pit morphology is a sensitive function of RH and CDD.

Pits grown at RH values ≤48% grow initially as crystallographically-etched shallow dishes with a current density (estimated from the typical depth as a function of time) that is consistent with the passive current density of stainless steel in concentrated $MgCl_2$ solutions.

At low RH (33%), the shallow dishes develop small satellite pits that grow under caps rich in Cr, S, O and Cl. At higher RH (43% and 48%), attack into the shallow region continues, but only in a small region, developing a spiral attack with a roughened surface. At 56% RH, shallow dishes are not observed, and deeper and rougher pits are observed.

The change in pit morphology is attributed to changes in the IR drop between the anode and cathode, and changes in solution conductivity and diffusivity with RH. A deeper spiral attack is observed for higher droplets (higher CDD at fixed RH or higher RH at fixed CDD). For 38% RH, pits close to the droplet edge form satellites, whereas pits that form towards the centre of the droplet, where it is higher, show a spiral attack. Higher RH values not only increase the droplet height, but also increase the ionic conductivity and diffusivity of these highly concentrated solutions.

High throughput methods are particularly important for providing statistically significant data on corrosion behaviour of commercial alloys.

References

1 H. Strehblow, Mechanisms of Pitting Corrosion, in *Corrosion Mechanisms in Theory and Practice*, ed. P. Marcus, Marcel Dekker, Inc., New York, NY, 2nd edn, 2002.

2 N. S. Azmat, K. D. Ralston, T. H. Muster, B. C. Muddle and I. S. Cole, A High-Throughput Test Methodology for Atmospheric Corrosion Studies, *Electrochem. Solid-State Lett.*, 2011, **14**(6), C9–C11.

3 E. Schindelholz, B. E. Risteen and R. G. Kelly, Effect of Relative Humidity on Corrosion of Steel under Sea Salt Aerosol Proxies, *J. Electrochem. Soc.*, 2014, **161**(10), C460–C470.

4 E. Schindelholz, B. E. Risteen and R. G. Kelly, Effect of Relative Humidity on Corrosion of Steel under Sea Salt Aerosol Proxies, *J. Electrochem. Soc.*, 2014, **161**(10), C450–C459.

5 N. D. Tomashov, Development of Electrochemical Theory of Metallic Corrosion, *Corrosion*, 1964, **20**(1), 7t–14t.

6 T. Tsuru, A. Nishikata and J. Wang, Electrochemical Studies on Corrosion under a Water Film, *Mater. Sci. Eng., A*, 1995, **198**(1–2), 161–168.

7 A. Nishikata, Y. Yamashita, H. Katayama, T. Tsuru, A. Usami, K. Tanabe, *et al.* An Electrochemical Impedance Study on Atmospheric Corrosion of Steels in a Cyclic Wet-Dry Condition, *Corros. Sci.*, 1995, **37**(12), 2059–2069.

8 Y. Tsutsumi, A. Nishikata and T. Tsuru, Initial stage of pitting corrosion of type 304 stainless steel under thin electrolyte layers containing chloride ions, *J. Electrochem. Soc.*, 2005, **152**(9), B358–B363.

9 N. Mi, M. Ghahari, T. Rayment and A. J. Davenport, Use of inkjet printing to deposit magnesium chloride salt patterns for investigation of atmospheric corrosion of 304 stainless steel, *Corros. Sci.*, 2011, **53**(10), 3114–3121.

10 E. Schindelholz and R. G. Kelly, Application of Inkjet Printing for Depositing Salt Prior to Atmospheric Corrosion Testing, *Electrochem. Solid-State Lett.*, 2010, **13**(10), C29–C31.

11 N. S. Azmat, K. D. Ralston, B. C. Muddle and I. S. Cole, Corrosion of Zn under fine size aerosols and droplets using inkjet printer deposition and optical profilometry quantification, *Corros. Sci.*, 2011, **53**(11), 3534–3541.
12 L. H. Hihara and S. Li, Atmospheric corrosion initiation on steel from predeposited NaCl salt particles in high humidity atmospheres, *Corros. Eng., Sci. Technol.*, 2010, **45**(1), 49–56.
13 Z. Y. Chen, D. Persson, A. Nazarov, S. Zakipour, D. Thierry and C. Leygraf, *In situ* studies of the effect of CO_2 on the initial NaCl-induced atmospheric corrosion of copper, *J. Electrochem. Soc.*, 2005, **152**(9), B342–B351.
14 Z. Y. Chen, S. Zakipour, D. Persson and C. Leygraf, Effect of sodium chloride particles on the atmospheric corrosion of pure copper, *Corrosion*, 2004, **60**(5), 479–491.
15 Y. Tsutsumi, A. Nishikata and T. Tsuru, Pitting corrosion mechanism of type 304 stainless steel under a droplet of chloride solutions, *Corros. Sci.*, 2007, **49**(3), 1394–1407.
16 T. Tsuru, K. I. Tamiya and A. Nishikata, Formation and growth of micro-droplets during the initial stage of atmospheric corrosion, *Electrochim. Acta*, 2004, **49**(17–18), 2709–2715.
17 S. Li and L. H. Hihara, *In situ* Raman spectroscopic identification of rust formation in Evans' droplet experiments, *Electrochem. Commun.*, 2012, **18**, 48–50.
18 Y. H. Wang, Y. Y. Liu, W. Wang, L. Zhong and J. Wang, Influences of the three-phase boundary on the electrochemical corrosion characteristics of carbon steel under droplets, *Mater. Corros.*, 2013, **64**(4), 309–313.
19 B. Maier and G. S. Frankel, Pitting Corrosion of Bare Stainless Steel 304 under Chloride Solution Droplets, *J. Electrochem. Soc.*, 2010, **157**(10), C302–C312.
20 S. Hastuty, A. Nishikata and T. Tsuru, Pitting corrosion of type 430 stainless steel under chloride solution droplet, *Corros. Sci.*, 2010, **52**(6), 2035–2043.
21 S. Thomas, I. S. Cole and N. Birbilis, Compact Oxides Formed on Zinc during Exposure to a Single Sea-Water Droplet, *J. Electrochem. Soc.*, 2013, **160**(2), C59–C63.
22 P. C. King, I. S. Cole, P. A. Corrigan, A. E. Hughes, T. H. Muster and S. Thomas, FIB/SEM study of AA2024 corrosion under a seawater drop, part II, *Corros. Sci.*, 2012, **55**, 116–125.
23 N. S. Azmat, K. D. Ralston, B. C. Muddle and I. S. Cole, Corrosion of Zn under acidified marine droplets, *Corros. Sci.*, 2011, **53**(4), 1604–1615.
24 T. H. Muster, A. Bradbury, A. Trinchi, I. S. Cole, T. Markley, D. Lau, *et al.* The atmospheric corrosion of zinc: the effects of salt concentration, droplet size and droplet shape, *Electrochim. Acta*, 2011, **56**(4), 1866–1873.
25 J. F. Li, B. Maier and G. S. Frankel, Corrosion of an Al–Mg–Si alloy under $MgCl_2$ solution droplets, *Corros. Sci.*, 2011, **53**(6), 2142–2151.
26 I. S. Cole, T. H. Muster, D. Lau, N. Wright and N. S. Azmat, Products Formed during the Interaction of Seawater Droplets with Zinc Surfaces II. Results from Short Exposures, *J. Electrochem. Soc.*, 2010, **157**(6), C213–C222.
27 M. S. Venkatraman, I. S. Cole, D. R. Gunasegaram and B. Emmanuel, in *Modeling Corrosion of a Metal under an Aerosol Droplet*, ed. J. F. Nie and A. Morton, PRICM 7, 2010, Pts 1-3, 1650–1653.

28 Z. Y. Chen, D. Persson and C. Leygraf, *In situ* studies of the effect of SO_2 on the initial NaCl-induced atmospheric corrosion of copper, *J. Electrochem. Soc.*, 2005, **152**(12), B526–B533.

29 I. S. Cole, D. Lau and D. A. Paterson, Holistic model for atmospheric corrosion – Part 6-From wet aerosol to salt deposit, *Corros. Eng., Sci. Technol.*, 2004, **39**(3), 209–218.

30 Z. Y. Chen and R. G. Kelly, Computational Modeling of Bounding Conditions for Pit Size on Stainless Steel in Atmospheric Environments, *J. Electrochem. Soc.*, 2010, **157**(2), C69–C78.

31 Standard Guide for Examination and Evaluation of Pitting Corrosion, ASTM Standard G46-94, ASTM International, 2013.

32 N. J. Laycock and S. P. White, Computer simulation of single pit propagation in stainless steel under potentiostatic control, *J. Electrochem. Soc.*, 2001, **148**(7), B264–B275.

33 R. P. V. Cruz, A. Nishikata and T. Tsuru, Pitting corrosion mechanism of stainless steels under wet-dry exposure in chloride-containing environments, *Corros. Sci.*, 1998, **40**(1), 125–139.

34 A. J. Davenport, L. Guo, N. Mi, H. Mohammed-Ali, S. M. Ghahari, S. R. Street, *et al.* Mechanistic studies of atmospheric pitting corrosion of stainless steel for ILW containers, *Corros. Eng., Sci. Technol.*, 2014, **49**(6), 514–520.

35 S. M. Ghahari, D. P. Krouse, N. J. Laycock, T. Rayment, C. Padovani, T. Suter, *et al.* Pitting corrosion of stainless steel: measuring and modelling pit propagation in support of damage prediction for radioactive waste containers, *Corros. Eng., Sci. Technol.*, 2011, **46**(2), 205–211.

36 P. Ernst, N. J. Laycock, M. H. Moayed and R. C. Newman, The mechanism of lacy cover formation in pitting, *Corros. Sci.*, 1997, **39**(6), 1133–1136.

37 G. S. Frankel, The growth of 2-D pits in thin-film aluminium, *Corros. Sci.*, 1990, **30**(12), 1203–1218.

38 A. Turnbull, K. Mingard, J. D. Lord, B. Roebuck, D. R. Tice, K. J. Mottershead, *et al.* Sensitivity of stress corrosion cracking of stainless steel to surface machining and grinding procedure, *Corros. Sci.*, 2011, **53**(10), 3398–3415.

39 OLI Systems Inc., OLI Analyzer 9.0, 2013, http://www.olisystems.com/.

40 E104-02 Standard Practice for Maintaining Constant Relative Humidity by Means of Aqueous Solutions, West Conshohocken, PA: ASTM International, 2007.

41 G. Wranglen, PITTING AND SULFIDE INCLUSIONS IN STEEL, *Corros. Sci.*, 1974, **14**(5), 331–349.

42 M. P. Ryan, N. J. Laycock, R. C. Newman and H. S. Isaacs, The pitting behavior of iron–chromium thin film alloys in hydrochloric acid, *J. Electrochem. Soc.*, 1998, **145**(5), 1566–1571.

43 W. Xu, Synchrotron X-ray and Electrochemical Studies of Pitting Corrosion of Iron, PhD thesis, University of Birmingham, Birmingham, UK, 2014.

44 T. R. Beck, PITTING OF TITANIUM 2. ONE-DIMENSIONAL PIT EXPERIMENTS, *J. Electrochem. Soc.*, 1973, **120**(10), 1317–1324.

45 A. J. Davenport, L. J. Oblonsky, M. P. Ryan and M. F. Toney, The structure of the passive film that forms on iron in aqueous environments, *J. Electrochem. Soc.*, 2000, **147**(6), 2162–2173.

46 V. Maurice, W. P. Yang and P. Marcus, X-ray photoelectron spectroscopy and scanning tunneling microscopy study of passive films formed on (100) Fe–18Cr–13Ni single-crystal surfaces, *J. Electrochem. Soc.*, 1998, **145**(3), 909–920.
47 P. C. Pistorius and G. T. Burstein, Metastable pitting corrosion of stainless-steel and the transition to stability, *Philos. Trans. R. Soc., A*, 1992, **341**(1662), 531–559.

Faraday Discussions

PAPER

Initiation and propagation of a single pit on stainless steel using a local probe technique†

Stéphane Heurtault,[ab] Raphaël Robin,[a] Fabien Rouillard*[a] and Vincent Vivier*[b]

Received 8th December 2014, Accepted 2nd February 2015
DOI: 10.1039/c4fd00252k

Experiments about single pit initiation and propagation were performed on 316L stainless steel with the aim to determine the pitting corrosion behaviour for nuclear waste containers. The experimental setup permits to control the pit development at will and to create reproducible single pits in three dimensions. Radial and deep evolutions of a disc shaped pit were studied for propagation times of up to 10 hours. These evolutions were used to determine what limiting mechanism takes place during long-term pit propagation. Special attention has been paid to the chloride ion action on pitting. A minimum chloride concentration was found to be necessary in the bulk electrolyte to support pit propagation. The existence of a critical pit depth of 230 μm was also underlined.

1 Introduction

Pitting corrosion is an important issue for the long-term behaviour of nuclear waste containers. Some containers may contain chloride-bearing polymers, which result in the production of chloride hydrogen gas by radiolysis. With the presence of humid zones formed on the inner 316L stainless steel container surface, the HCl gas completely dissolves and may form highly concentrated hydrochloric acid zones. Under these conditions, pits can initiate and propagate.

Pitting corrosion has been widely studied in the last decades on stainless steel,[1,2] but since it is a stochastic process, it remains difficult to study experimentally. Indeed adding an aggressive anion, such as chloride ion, to the electrolyte leads to the formation of several pits with different stages of development on the material surface making any pitting analysis very difficult. Many statistical studies have been performed to determine the pitting susceptibility of a

[a]CEA, DEN, DPC, SCCME, Laboratoire d'Etude de la Corrosion Non Aqueuse, F-91191 Gif-sur-Yvette, France. E-mail: fabien.rouillard@cea.fr; Fax: +33-1-69081586; Tel: +33-1-69081614

[b]Université P. et M. Curie-CNRS, UMR8235, Laboratoire Interfaces et Systèmes Electrochimiques, 4 place Jussieu, F-75005 Paris, France. E-mail: vincent.vivier@upmc.fr; Fax: +33-1-44274074; Tel: +33-1-44274158

† Electronic supplementary information (ESI) available. See DOI: 10.1039/c4fd00252k

material.[3–6] These studies focused on the frequencies of pit initiation and pit repassivation on a material surface on which several pits can evolve simultaneously. The main results of this approach concern the probability to get a stable pit that cannot repassivate, the decrease of the pit generation rate with decreasing chloride concentration and the influence of potential on the pitting rates. However, these statistical studies do not permit to highlight the driving forces of pitting corrosion and need to be complemented with mechanistic investigations. In addition, usual electrochemical measurements result in an average value on the electrode surface. Thus, to determine the mechanisms governing pitting corrosion from its initiation to the propagation or its repassivation, it is necessary to be able to work on a single pit.

Many researchers have attempted to create single pits. Different methods have been used: an initial shape was given to the pit on the electrode surface using twin probe techniques,[7] one-dimensional shielded wires,[8] focus laser illumination,[9] or the electrode surface was partially covered to limit the number of pits.[10] These methods permit to study the early stage of pitting corrosion. However, they have some disadvantages. First of all, the methods are not entirely reproducible from one experiment to the other. In some case these methods also damage the material surface. Also they cannot be used to study long-term pit growth. Depth evolution was studied in the literature with a lead-in-pencil electrode.[11–14] It consisted of a wire, the diameter of which varies from 10 to 50 μm. This technique has the advantages to easily create single pits (such pits are called ‘artificial pits’) and to get simple pit geometry for further modelling. However, the lead-in-pencil electrode does not permit to have radius evolution. The radius evolution was independently investigated in the literature on aluminium,[15,16] using a horizontal plate of a few micrometers thick. This technique permitted to highlight the lateral evolution of the pit (but no work was performed on the depth evolution). Mixing the two last mentioned techniques with the introduction of a vertical foil in the lead-in-pencil technique, it was possible to study both lateral and vertical pit evolutions.[17,18] However, due to the low thickness of the foil (from 25 to 50 μm), which led to interactions between the pit and the resin, it was not possible to determine precise kinetic information about lateral evolution. In this article, an experimental setup allowing to locally inject chloride ions[19–21] has been used to analyze the long-term development of a single pit in three dimensions.

It is generally admitted in the literature that long-term experiments are missing for pit propagation.[1,10] That is the scope of this article with the creation and sustenance of a disc shaped single pit. Disc shaped pits are less studied than hemispherical pits despite the fact that they are more convenient for numerical modeling,[22–24] since they are obtained in a concentrated solution of chloride in which the number of pits is more important.

This paper aims at bringing new insight on the pit propagation description on 316L stainless steel. After the description of the experimental setup, the control of the initiation of a single pit will be detailed. Then, the propagation phase will be characterized and a particular attention will be paid to the description of both radial and deep evolutions. The last part of this article will focus on the pitting chemistry necessary to maintain the pit propagation.

2 Experimental

2.1 Instrumentation

Local electrochemical techniques have been widely used in corrosion investigation.[25,26] The instrumentation used in this work was based on a scanning electrochemical microscope (SECM), which has already been described elsewhere[19–21] and in which the amperometric probe was replaced by a glass capillary (*vide infra*). The setup, sketched in Fig. 1, was composed of a lab-made potentiostat coupled with a low-noise-current-to-voltage converter (Femto DLPCA200, BFI Optilas) with an adjustable gain (10^3 to 10^{11} V A^{-1}) and a three-axis positioning system (VP-25XA, Newport) driven by a 100 nm spatial-resolution motion encoder (ESP300, Newport). The entire setup was controlled by a software developed under Labview® environment. The glass microcapillary was used as a chloride ion dispenser to create a single pit on the stainless steel electrode. The top of the microcapillary was connected to a 300 μL Terumo® syringe filled with a NaCl in a 0.5 M H_2SO_4 solution. The effect of the NaCl concentration has been studied by varying its concentration in the syringe. A KD Scientific® syringe infusion pump was used to gradually release the solution at a fixed rate. The effect of the releasing rate was also studied as a relevant parameter for the pit initiation/propagation. The lifetime of a microcapillary was extended by washing it with a water–ethanol mixture after each experiment since the use of high concentration chloride solution may lead to crystal formation that obstruct its apex. All the experiments presented in this article were performed with the same glass microcapillary size with an inner diameter of 100 μm and an outer diameter of 1000 μm. The precise positioning of the microcapillary was performed by measuring the electrolyte resistance between the substrate and the reference electrode.[20] This technique was shown to be very sensitive to monitor the position of an insulating material in the close vicinity of a substrate.[27] In a preliminary work, different distances between the microcapillary and the sample have been tested. If the distance was too large, several pits developed on the stainless steel surface. Thus, distances

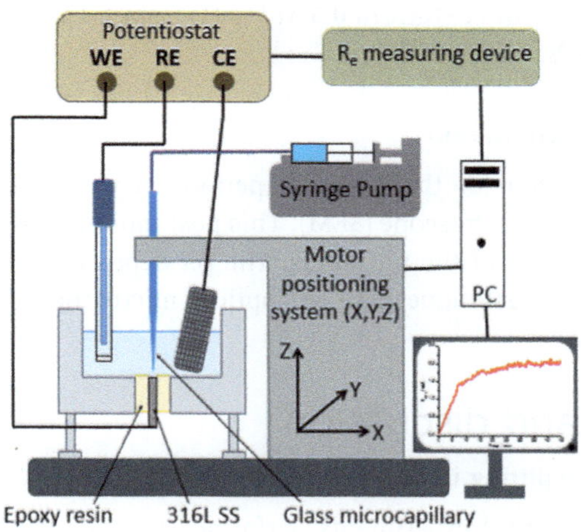

Fig. 1 Electrochemical device developed for generating a single pit and studying its propagation.

between 10 to 20 μm have been chosen because corresponding to the larger distance at which a single pit was always observed.

2.2 Electrode preparation

The electrochemical cell was a three-electrode cell with a platinum grid acting as counter electrode, a mercury saturated sulphate electrode (MSE) as reference electrode ($E = 650$ mV/NHE) and a 316L stainless steel disk as working electrode. The nominal composition of the 316L stainless steel (in wt%) was obtained from EDAX analysis: 65.9% Fe, 18.5% Cr, 10.6% Ni, 2.4% Mo, 1.7% Mn and 0.9% Si. The stainless steel surface and the thickness of the sample were large enough to study the 3D evolution of the pit, allowing the progress in depth and radius of the single pit for a time longer than 10 hours to be investigated.

Surface preparation was important because it determined the passive film nature, thickness, and its crystallization.[28,29] In this article, the same preparation was always used to have a reproducible passive film. The chosen surface preparation was previously used in the literature by Hisamatsu *et al.*[30] and consisted in several steps. First, the 316L stainless steel specimen was mechanically polished with 1200, 2400, and 4000 SiC paper, and then cleaned with a water and ethanol mixture in an ultrasonic bath for 10 min. After the polishing, the electrode was introduced in the electrolytic cell that will also be used for pitting corrosion experiments, and was then cathodically treated in 0.5 M H_2SO_4 at −0.9 V/MSE for 1 minute to reduce (at least partially) the native oxide film. In the next step, the formation of the passive film was achieved in the same electrolytic solution by performing a single potential sweep between −0.9 V/MSE and the working potential (*i.e.* the potential at which we wanted to initiate the pit) at a low scan rate (1 mV s^{-1}). Then, before releasing the chloride through the glass microcapillary, the working potential was held for 30 min to ensure that the passive film is thick enough. In this work, all the experiments were performed at 0 V/MSE, but in a previous work[21] the effect of the working potential on pitting was investigated. The thickness of the passive film at the end of the surface preparation was evaluated by electrochemical impedance spectroscopy to be about 1 nm. This result is in good agreement with the work of Strehblow and Haupt[29] who measured the passive film thickness of Fe–15Cr in 0.5 M H_2SO_4 from 1.1 nm at −0.4 V/MSE to 1.9 nm at 0.4 V/MSE.

2.3 Physical characterization

Ex situ characterization of the pits was performed with a Leica Stereoscan 440 scanning electron microscope (SEM). This post-mortem observation allowed accurate measurements of the pit radius. The pit depth was measured using a differential focusing technique with an optical microscope equipped with a micrometric screw.

3 Results and discussion

3.1 Control of the pitting initiation

Fig. 2 shows 11 identical experiments performed independently. The solution injected with the microcapillary was 3 M NaCl with a flow rate equal to 5.4 μL h^{-1}. The propagation time was varied between 3 and 40 minutes to allow *ex situ* pit

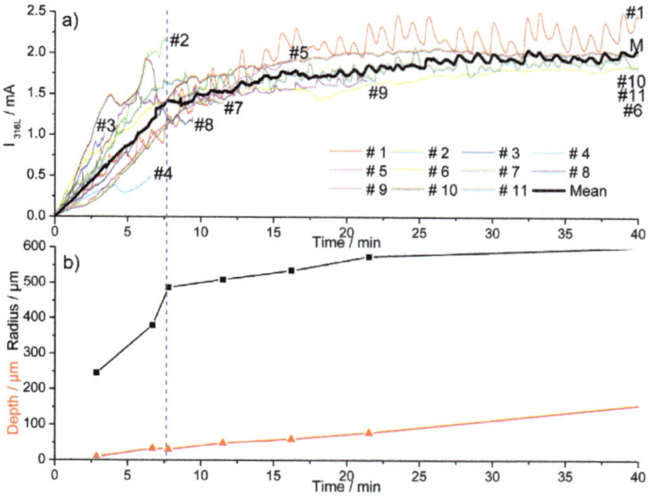

Fig. 2 Current recorded (a) and mean geometric parameters of the pits (b) during the initiation of the pitting corrosion for 11 different experiments at 0 V/MSE.

radius and depth characterizations at different propagation times. The current–time records for the 11 identical experiments plotted in Fig. 2a were similar and included different stages. First the initial time for the pitting in all of our experiments (*i.e.* $t = 0$ in Fig. 2) was defined as the time for which an increase in current due to pitting corrosion was monitored. The pit started to propagate after the breakdown of the passive layer (just beneath the microcapillary and independent of the metal microstructure) due to the aggressiveness of chloride ions. The first 10 to 15 minutes of the pit propagation were characterized by a sharp increase of current. Then the increases of the current slowed down. A similar current evolution between the 11 identical tests shows that the experimental setup permits to obtain reproducible single pits. Varying the working electrode disk diameter did not change the pitting reproducibility. Indeed, Fig. 3 shows that a similar current evolution was recorded for a 2 mm electrode diameter and a 4 mm electrode diameter. After 45 minutes of propagation, the pit radius and depth (determined with an optical microscope) were found to be 610 µm and 147 µm for a 2 mm in diameter working electrode, and 600 µm and 139 µm for a 4 mm electrode, respectively. It can be concluded that the electrode size did not modify the pitting evolution. However, the use of a smaller working electrode allowed us to investigate the early stage of the pitting since the background current (passive dissolution) is then minimized (data not shown), allowing a sharp evaluation of the onset of the current increase. The current evolution could be mainly explained by the increase of the pit radius while the pit depth seems to have a lower contribution, as shown in Fig. 2b. The initial sharp increase of current was linked with a huge increase of the pit radius while the pit depth did not exceed 25 µm. This was ascribed to the large concentration of chloride used for the local pit initiation. This initial stage of the pit formation is linked with the inner and outer diameters of the microcapillary used to generate the single pit. Then the increases of the current and the pit radius slowed down together. In the literature such decreases of growth in current and pit radius were already found

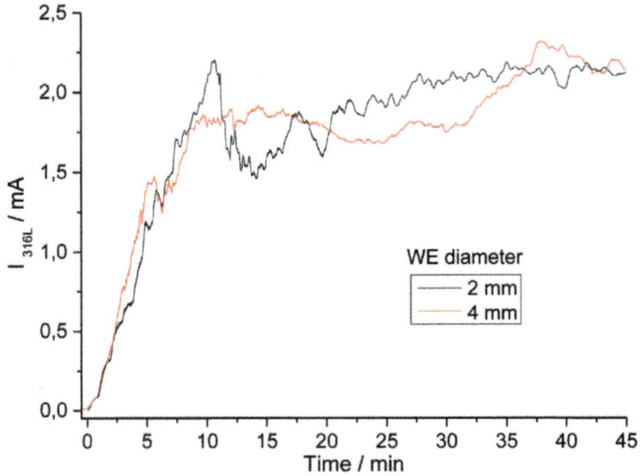

Fig. 3 Current recorded during the single pit propagation experiments varying the diameter of the stainless steel electrode at 0 V/MSE.

on iron under potentiostatic conditions and were explained by a salt film precipitation.[31]

The influence of the chloride amount brought by the microcapillary on the stainless steel disk was studied in two different ways: (i) by varying the flow rate of the syringe driver, and (ii) by varying the NaCl concentration in the syringe. With both of these methods, similar results were obtained. Increasing the NaCl concentration in the syringe modified the pitting current as shown in Fig. 4. The initial current slope increased as did the step value of the current. The amount of chloride released by the microcapillary significantly impacted the pit radius whereas the pit depth remained constant as shown in Fig. 5. A reproducible pit depth was found for experiments performed for 2 and 3 hours of propagation with variation of chloride introduction and the standard deviation was very low

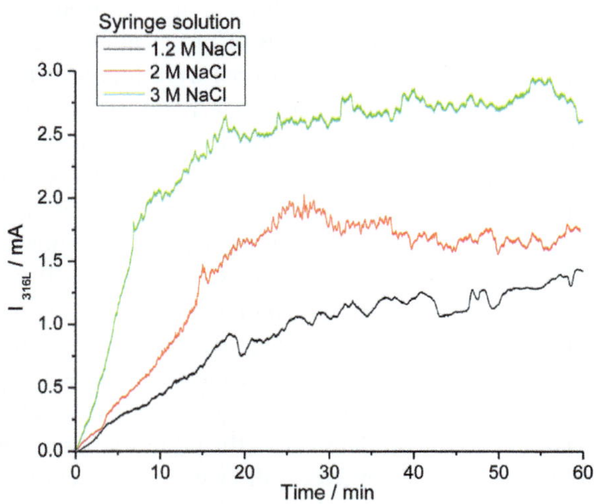

Fig. 4 Current–time curves for three single pits after 1 hour propagation at 0 V/MSE with the chloride concentration as a parameter.

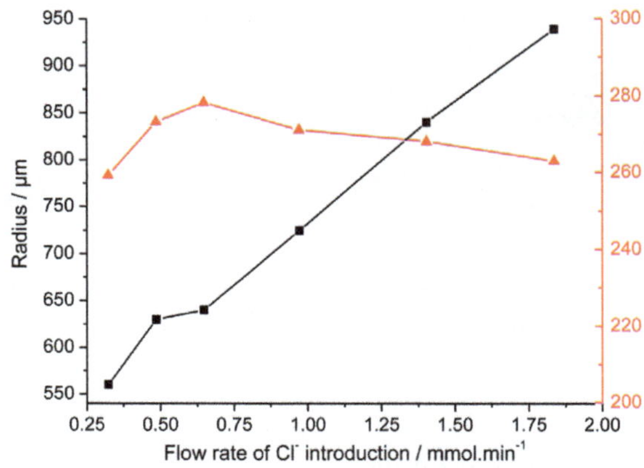

Fig. 5 Influence of the chloride amount injected with the microcapillary on the geometric parameters of the pit after 2 hours of propagation at 0 V/MSE.

between each test (Table 1). The microscopy measurements showed that it was the increasing radius that could explain the current increase when the pit solution is more concentrated in chloride. Besides, increasing the amount of chloride brought by the microcapillary reduced the probability to have a pit cover of residual metal above the pit (Fig. 6). This shows that the whole stainless steel surface below the glass microcapillary dissolves when the solution is concentrated enough in chloride. In addition, the small distance between the capillary and the steel electrode allowed for controlling the local environment in the close vicinity of the pit. Similar observations on the disappearance of the pit cover were reported in the literature but for increasing the potential[32] and temperature[33] instead of increasing the chloride concentration. In the following work, only pits without pit cover at the pit mouth were considered further. It should also be mentioned that the size of the microcapillary may play a non-negligible role on the pit initiation/propagation processes, especially the outer diameter of the glass part because the capillary and the steel substrate form a thin layer in which the electrolytic solution is confined.

3.2 Characterization of the pit propagation

The evolution of the pit geometry was investigated by performing 30 experiments for 11 different pit propagation times between 45 minutes and 10 hours. Chloride ions were introduced in the electrochemical cell at a constant flow rate of 5.4 µL h^{-1} using a 3 M NaCl and 0.5 M H_2SO_4 solution. The ratio pit depth/pit radius at

Table 1 Pit depth for two different propagation times with the variation of the chloride amount brought to the system between each test

Propagation duration	Number of performed experiments	Mean depth (µm)	Standard deviation for depth (µm)
2 hours	7	268	4
3 hours	10	360	8

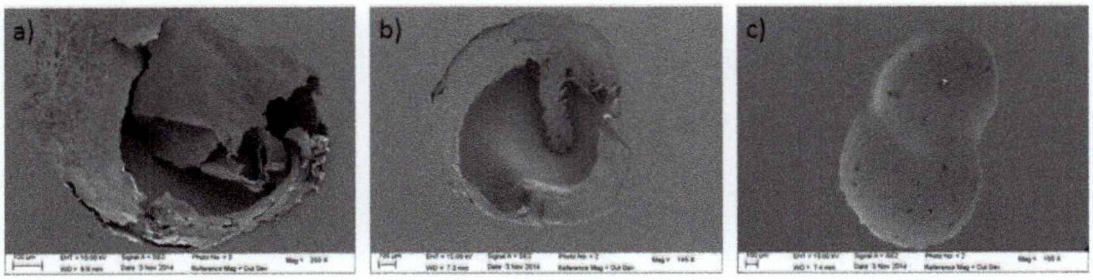

Fig. 6 SEM images after 1 hour of propagation for the pits generated with 1.2 M NaCl + 0.5 M H_2SO_4 (a), 2 M NaCl + 0.5 M H_2SO_4 (b) and 3 M NaCl + 0.5 M H_2SO_4 (c) in the syringe at 0 V/MSE.

the end of each experiment was calculated and plotted, shown in Fig. 7. For the whole tests, this ratio had approximately the same value of 0.4. As a consequence, the generated single pits are always disc shaped. It should also be mentioned that in the literature, most of the investigations are related to hemispherical pits.[1,7,30,32,34] Fig. 7 shows also the repartition of the 30 performed experiments between the 11 propagation times. Reproducible pits were obtained for the propagation phase of up to 10 hours, as shown in Fig. 8a, for which the chronoamperometry results between different experiments overlap. During the propagation phase, the current increases slightly. In our experiments, after 45 minutes of propagation, the current is proportional to $t^{0.09}$. The measurement of the pit depth and pit diameter with an optical microscope and SEM allowed the determination of experimental laws for the evolution of the pit and the radius (Fig. 8b). For the radius the 11 experiments presented in the previous section for the initiation phase were also used because it was shown that the pit radius significantly impacts the initiation phase whereas the pit depth evolution did not change between initiation and propagation. The following laws were found:

$$p = 126 + 1.57|t - 2700|^{0.54} \text{ for } t > 45 \text{ min} \tag{1}$$

$$r = 250 + 26.36|t - 180|^{0.34} \text{ for } t > 3 \text{ min} \tag{2}$$

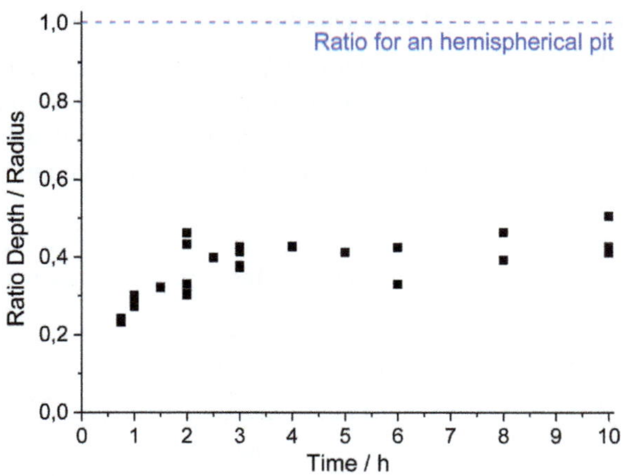

Fig. 7 Evolution of the pit depth/pit radius ratio measured with an optical microscope.

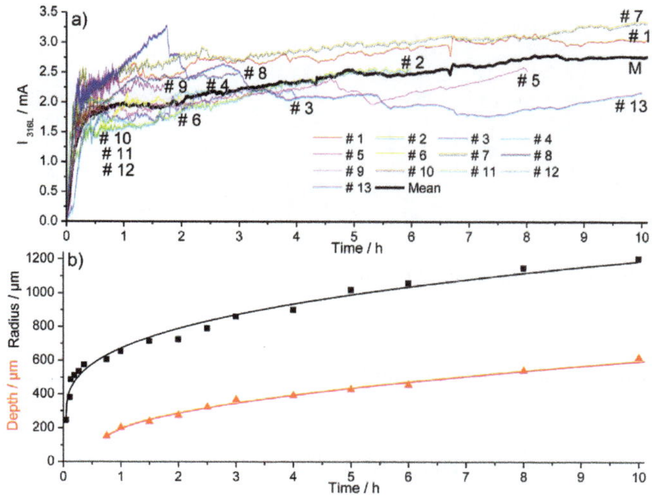

Fig. 8 Current recorded (a) and mean geometric parameters of the pits (b) during the propagation of pitting corrosion for 30 different experiments at 0 V/MSE.

where p is the pit depth (expressed in µm), r the pit radius (in µm) and t the propagation time (in s).

In the experimental laws (1) and (2), the value of the different parameters have a physical meaning. At 3 minutes (*i.e.* 180 s), we found a pit radius of 250 µm. After 45 minutes (*i.e.* 2700 s), a pit depth of 126 µm was found using eqn (1). These laws also revealed that the pit depth increases as a function of the square root of the time, whereas the radius grows as a function of $t^{0.34}$. An increase in square root of time is characteristic of a process limited by diffusion, suggesting that the pit bottom is rate determined by diffusion.

In the literature, for stainless steel Fe–18Cr–12Ni–2Mo–Ti in 0.5 M NaCl + 0.05 M H_2SO_4 solution, Szklarska-Smialowska *et al.*[35] found under potentiostatic conditions (at 0.860 V/NHE) that $p \sim t^{0.75}$ and $r \sim t^{0.94}$. The differences with our results can be explained by the fact that they worked on a large electrode with many pits. As a consequence they obtained an average value on an unknown value of pits and at different stages of the propagation process, whereas in this work, all the measurements are always performed on a single pit. In addition, fluctuations of the current that preceded the formation of stable pits[2] due to the true passivity breakdown[36] may have also vitiated the measurement.

A pit radius of 500 µm, corresponding to the glass microcapillary radius, was obtained after 15 minutes of propagation according to relation (2). For a pit radius greater than 500 µm, the pit growth occurs under the stainless steel surface where the glass microcapillary is not present. In Fig. 9a, a pit growth under the stainless steel surface is noticeable at the periphery of the pit for a disc shaped pit after more than 10 hours of propagation. This presence of a lacy cover is consistent with the potentiostatic investigations performed on 304 stainless steel foils in 1 M NaCl solution.[17,18] Fig. 9b did not show any crystallographic formation at the pit bottom, as it has been observed in the literature when the pit bottom potential is in the active state and when a salt film is not present.[37]

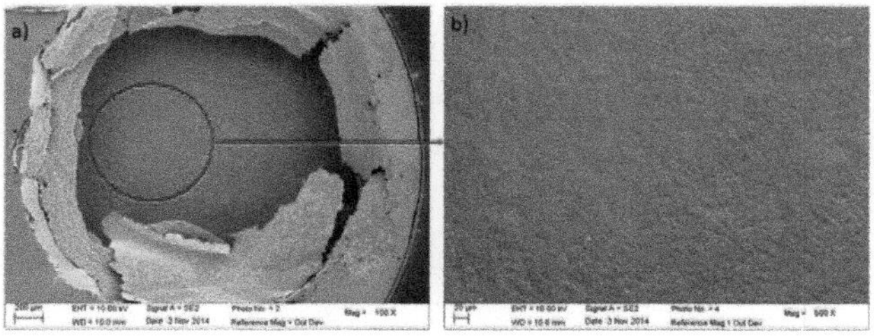

Fig. 9 SEM image of a disc shaped pit after 15 h of propagation at 0 V/MSE.

Taking the derivative of eqn (1) and (2) allows the vertical and lateral pit growth rates to be expressed as:

$$\frac{\mathrm{d}p}{\mathrm{d}t} = 0.86|t - 2700|^{-0.46} \quad \text{for } t > 45 \text{ min} \tag{3}$$

$$\frac{\mathrm{d}r}{\mathrm{d}t} = 8.98|t - 180|^{-0.66} \quad \text{for } t > 3 \text{ min} \tag{4}$$

For long-term propagation, the growth rate of the depth determined in relation (3) expressed in μm s^{-1} is lower than the growth rate of the radius written in relation (4) in μm s^{-1}. It is consistent with a disc shaped pit.

The Faraday law permits to link the electrochemical behavior with a mass loss proportional to the growth rates presented in relations (3) and (4). However, to use the Faraday law we need to know the oxidation number of species that dissolve. The hypothesis that there is no specific dissolution has been made and thus a mean oxidation number of 2.2 was used for 316L stainless steel. Fig. 10 validated the hypothesis of no specific dissolution. The pit volume corresponding to the metal dissolution was calculated by two different methods. The direct route uses the Faraday law, which allowed for obtaining the volume of dissolution from the measured current. In the second approach, the calculation of volume was made from optical microscopy measurements and the hypothesis of a cylindrical pit was made. During the whole propagation time, both methods gave the same dissolution volume, suggesting that the use of the Faraday law is a valid hypothesis. The observation that the fraction of the dissolved elements in the solution after pitting measured by ICP was equal to their fraction in the substrate confirms this result.[20,21]

From relations (3) and (4), using the Faraday law considering no specific dissolution and calculating the flux, the local current densities on the pit walls and at pit bottom were determined to be:

$$J_{\text{pit bottom}} = \frac{\rho n F}{M}\frac{\mathrm{d}p}{\mathrm{d}t} \quad \text{for } t > 45 \text{ min} \tag{5}$$

$$J_{\text{pit walls}} = \frac{\rho n F}{M}\frac{\mathrm{d}r}{\mathrm{d}t} \quad \text{for } t > 3 \text{ min} \tag{6}$$

With J_{local} being the local current density, ρ being the 316L stainless steel density (8 g cm^{-3}), n being the mean oxidation number involved in the corrosion

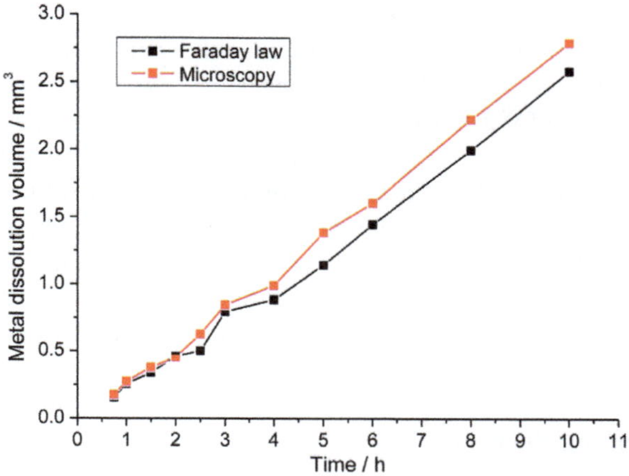

Fig. 10 Dissolution volume calculated by the Faraday law and by the optical measurements.

reaction (2.2), F being the Faraday constant (96 485 C mol^{-1}) and M being the 316L SS molar mass (56.1 g mol^{-1}).

These two relations are plotted in Fig. 11. The current density at the pit bottom decreases as a square root of time, revealing that the rate-limiting step at the pit bottom may be a diffusive process (see the ESI†). This might be the diffusion of metallic cations from the pit bottom to the pit mouth as defended several times in the literature.[10,11,38,39] Another explanation was to consider that the limiting diffusion is the transport of aggressive anions to the outside of the pit.[31]

Current density in a single pit had already been evaluated using artificial pit electrodes. Isaacs[11] determined the evolution of current density with time for propagation times of up to 45 minutes for a 308 stainless steel immerged in 1 M NaCl. The initial value of current density was about 1.2 A m^{-2}, which is in good agreement with other works from the literature.[37,40,41] Then the current density

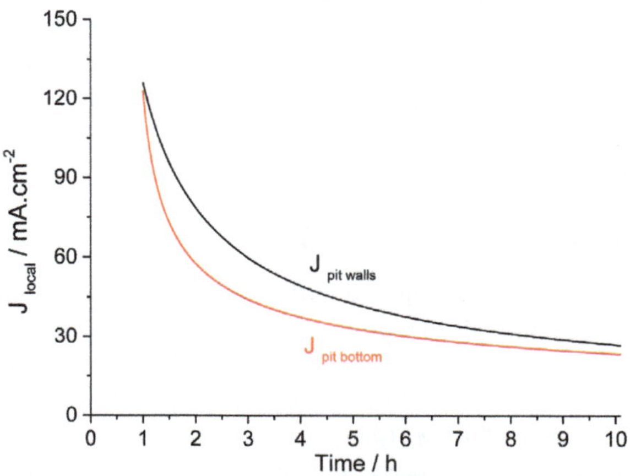

Fig. 11 Evolution of the current density on the pit walls and at the pit bottom for long-term pit propagation.

decreases in the square root of time to the value of 200 mA cm^{-2} after 45 minutes of propagation. This value could be consistent with a current density of 125 mA cm^{-2} that we observed after 1 hour of propagation. Newman[13] determined the current density value for a longer propagation time, performing a single pit of 1 mm depth on a Fe–17Cr–11Ni–2.7Mo alloy using 4 M KCl solution. After reaching this depth, he changed the solution to 1 M KCl solution and then measured the current density at different potentials. In our paper, using relation (1), a pit depth of 1 mm is obtained after 34 hours of propagation. At 34 hours of propagation, a pit density of 10 mA cm^{-2} is calculated using relation (5). This value is also in good agreement with the one obtained by Newman for a lower overpotential.

3.3 Critical pitting chemistry

3.3.1 Corrosion product concentration at the pit bottom

The current density at the pit bottom was found to be lower than the current density on the pit walls (Fig. 11). This can be explained by the local chemistry at the pit bottom, which slows the corrosion rate but also permits to sustain the pit propagation. Using the Fick first law, the concentration of metallic cations Me^{n+} at the pit bottom could be calculated from the following relationship:

$$\left[\mathrm{Me}^{n+}\right]_{\text{pit bottom}} = \frac{p\rho}{D_{\mathrm{Me}^{n+}} M} \frac{\mathrm{d}p}{\mathrm{d}t} \qquad (7)$$

With $D_{Me^{n+}}$ being the diffusion coefficient of the metallic cations which was taken equal to 0.5 10^{-5} cm^2 s^{-1}.[11] The bulk concentration of the metallic cations (far from the pit) was also assumed to be zero.

During the single pit propagation, a constant concentration of 1.4 mol L^{-1} was found for the metallic cation concentration at the pit bottom (Fig. 12). Such a value corresponds to 30% of the saturation concentration of the solubility of $FeCl_2$ (4.4 mol L^{-1}).[42] According to the literature,[43,44] the concentration of the corrosion products within the pit must be maintained at about 60% for allowing the pit to propagate. However, a lower corrosion product concentration had already been reported in the literature for 304 stainless steel. Hisamatsu[45] observed a pit propagation with a corrosion product concentration of 1.8 M at the pit bottom (*i.e.* 40% of the saturation). Moreover, more complex iron salt with a lower solubility could have formed. This could be the case in our experiment with the formation of sulphate iron salts.

3.3.2 Critical bulk chloride concentration

The concentration of chloride required in the bulk electrolyte to maintain the pit propagation was determined. 20 reproducible single pits were created at 0 V/MSE releasing a solution of 3 M NaCl + 0.5 M H_2SO_4 using the glass microcapillary with a constant flow rate of 5.4 μL h^{-1}. During the propagation, the bulk electrolyte was changed from a solution of 0.5 M H_2SO_4 to a solution of 0.5 M H_2SO_4 and variable NaCl concentrations (C_0) using peristaltic pumps working at constant flow rates. After 15 minutes, necessary to completely renew the bulk solution, the glass microcapillary was removed. Depending on the chloride concentration, C_0, of the

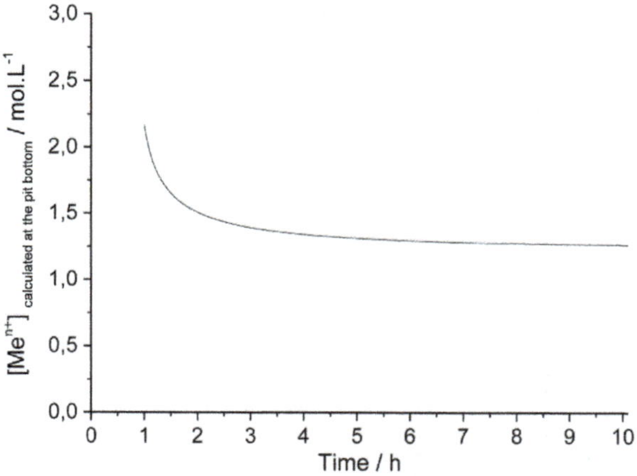

Fig. 12 Calculated concentration evolution at the pit bottom for long-term pit propagation.

bulk electrolyte, the pit propagation continued or stopped. As shown in Fig. 13, after removing the microcapillary at 2 hours and 3 hours, it is noticeable that the pit still propagated for a bulk chloride concentration of 250 mM for both propagation times and that the pit repassivated for a lower bulk chloride concentration of 235 mM.

The effect of the bulk chloride concentration was studied as a function of pit propagation time and pit depth (Fig. 14). Each point in Fig. 14 represents one of the 20 performed experiments. Two zones can be clearly seen. A pit propagation zone may be defined for the bulk chloride concentration when it was high enough to sustain pit propagation. A pit repassivation zone exists when the bulk chloride concentration was no longer sufficient to maintain the propagation process. Between both zones, a critical chloride concentration may be defined. From

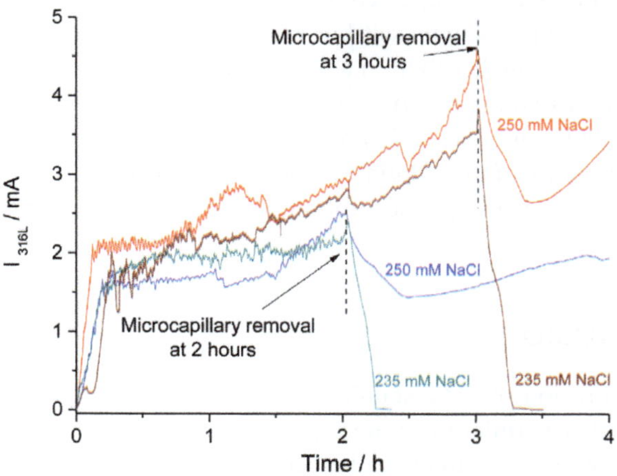

Fig. 13 Chronoamperometries for 2 pitting experiments in which the microcapillary was removed after 2 hours (blue curves) and after 3 hours (red curves) to see if a critical chemistry exists in the bulk.

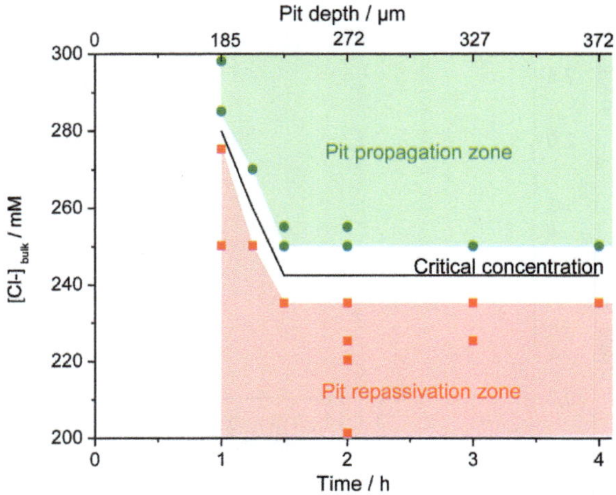

Fig. 14 Evolution of the critical chloride concentration necessary to continue the propagation of an opened pit with a radius of 800 µm as a function of propagation time and pit depth.

1 hour to 1 h 30 min of propagation, this critical concentration decreases. Then at 1 h 30 min of pit propagation, achieving a pit depth of 230 µm, it reaches a step value between 235 mM and 250 mM. Additional experiments were performed at 235 mM and 250 mM of bulk chloride concentration to confirm the reproducibility of the repassivation and propagation of the single pit after 1 h 30 min of propagation. Conversely to iron or nickel,[42] stainless steel repassivates at a high chloride concentration. The need of chloride ions to continue pit propagation was explained for the early stage of pit development by an increase of the salt film volume during pit propagation as the pit surface grows. As the salt film is composed of chloride ions, a supply of chloride ions is necessary to respond to the increased volume of the salt film.[42] In addition, it was shown that the sulfate concentration (data not shown) also plays a significant role, since an increase in sulfate concentration is accompanied by a current decrease. For instance, in a 6 M sulfate solution (adding sulfate salt but maintaining a constant pH) the pitting current was 8 times smaller than in a 0.5 M solution. These observations are in good agreement with those reported by Pistorius and Burstein[47] on a 304 stainless steel with the addition of SO_4^{2-} ion to a 1 M Cl^- solution. Indeed, these results indicate the influence of the chloride/sulphate ratio on the pitting behaviour and can be explained by a stronger adsorption of sulphate than chloride at the electrode interface.

4 Conclusions

The pitting corrosion of 316L stainless steel was investigated using an experimental setup that permits to create reproducible single pits. In the literature, pitting corrosion experiments usually focus on small hemispherical pits. Here, large disc shaped pits were analyzed with a propagation time of up to 10 hours. The experimental setup permitted to control at will the chloride ions brought into the pit solution. Experimental propagation laws for the geometrical parameters of the pit radius and depth were proposed. Calculations of local current densities were deducted to determine the limiting mechanism controlling the pit propagation. The pit growth at the pit bottom was limited by diffusion, probably due to the diffusion of the metallic cations from the bottom to the outside of the pit. A corrosion product concentration of 30% of the saturation was calculated at the pit bottom. This percentage, although lower than the one proposed in the literature, would be enough to prevent the pit repassivation. Experimentally, a minimum chloride concentration was determined in the bulk electrolyte to sustain the pit propagation. This concentration decreases with time until a critical pit depth of 230 μm is reached. Then, the bulk electrolyte concentration has to be at least 245 mM NaCl to sustain the pit propagation on 316L austenitic stainless steel. Further investigations are under progress to investigate the influence of the micro-capillary geometry and to determine geometrical laws for the pit radius and depth as a function of the chloride concentration.

Acknowledgements

The authors gratefully acknowledge AREVA NC for the funding of this project, and Françoise Pillier (UMR 8235) for the SEM observations.

References

1 Z. Szklarska-Smialowska, *Electrochem. Soc., Proc.*, 2002, **13**, 251.
2 J. Soltis, *Corros. Sci.*, 2015, **90**, 5.
3 T. Shibata and T. Takeyama, *Corrosion*, 1977, **33**, 243.
4 D. E. Williams, C. Wescott and M. Fleischmann, *Passivity of Metals and Semiconductors*, ed. M. Froment, 1983, p. 217.
5 D. E. Williams, J. Stewart and P. H. Balkwill, *Critical Factors in Localized Corrosion*, ed. G. S. Frankel and R. C. Newman, 1992, p. 36.
6 D. E. Williams, J. Stewart and P. H. Balkwill, *Corros. Sci.*, 1994, **36**, 1213.
7 I. L. Rosenfeld and I. S. Danilov, *Corros. Sci.*, 1967, **7**, 129.
8 T. Beck, *J. Electrochem. Soc.*, 1973, **120**, 1317.
9 R. K. Ulrich and R. C. Alkire, *Corros. Sci.*, 1983, **23**, 1153.
10 R. C. Alkire and K. P. Wong, *Corros. Sci.*, 1988, **28**, 411.
11 H. S. Isaacs, *J. Electrochem. Soc.*, 1973, **120**, 1456.
12 R. C. Newman and H. S. Isaacs, *Passivity of Metals and Semiconductors*, ed. M. Froment, 1983, p. 269.
13 R. C. Newman, *Corros. Sci.*, 1985, **25**, 341.
14 N. J. Laycock and R. C. Newman, *Mater. Sci. Forum*, 1995, **192–194**, 649.
15 G. S. Frankel, J. R. Scully and C. V. Jahnes, *J. Electrochem. Soc.*, 1996, **143**, 1834.
16 A. Sehgal, D. Lu and G. S. Frankel, *J. Electrochem. Soc.*, 1998, **145**, 2831.
17 P. Ernst and R. C. Newman, *Corros. Sci.*, 2002, **44**, 927.
18 S. M. Ghahari, D. P. Krouse, N. J. Laycock, T. Rayment, C. Padovani, T. Suter, R. Mokso, F. Marone, M. Stampanoni, M. Monir and A. J. Davenport, *Corros. Eng., Sci. Technol.*, 2011, **46**, 205.
19 N. Aouina, F. Balbaud-Célérier, F. Huet, S. Joiret, H. Perrot, F. Rouillard and V. Vivier, *Electrochim. Acta*, 2011, **56**, 8589.
20 N. Aouina, F. Balbaud-Célérier, F. Huet, S. Joiret, H. Perrot, F. Rouillard and V. Vivier, *Corros. Sci.*, 2012, **62**, 1.

21 N. Aouina, F. Balbaud-Célérier, F. Huet, S. Joiret, H. Perrot, F. Rouillard and V. Vivier, *Electrochim. Acta*, 2013, **104**, 274.
22 J. R. Galvele, *J. Electrochem. Soc.*, 1976, **123**, 464.
23 J. R. Galvele, *Corros. Sci.*, 1981, **21**, 551.
24 S. M. Gravano and J. R. Galvele, *Corros. Sci.*, 1984, **24**, 517.
25 L. Niu, Y. Yin, W. Guo, M. Lu, R. Qin and S. Chen, *J. Mater. Sci.*, 2009, **44**, 4511.
26 M. B. Jensen and D. E. Tallman, *Electroanalytical Chemistry A Series of Advances*, ed. C. Bard and A. J. Zoski, 2011, vol. 24, p. 171.
27 H. Bouazaze, J. Fransaer, F. Huet, P. Rousseau and V. Vivier, *Electrochim. Acta*, 2010, **55**, 1645.
28 A. J. Davenport, R. C. Newman and P. Ernst, *Passivity of Metals and Semiconductors*, ed. M. B. Ives, J. L. Luo and J. R. Rodda, 2001, Proceedings vol. 99–42, p. 65.
29 S. Haupt and H. H. Strehblow, *Corros. Sci.*, 1994, **37**, 43.
30 Y. Hisamatsu, T. Yoshii and Y. Matsumura, *Localized Corrosion*, ed. R. Staehle, NACE Houston, 1974, p. 427.
31 H. H. Strehblow and P. Marcus, *Corrosion Mechanisms in Theory and Practise*, ed. P. Marcus, 3rd edn, 2012, p. 349.
32 N. Sato, T. Nakagawa, K. Kudo and M. Sakashita, *Localized Corrosion*, ed. R. Staehle, NACE Houston, 1974, p. 447.
33 N. J. Laycock, M. H. Moayed and R. C. Newman, *J. Electrochem. Soc.*, 1998, **145**, 2622.
34 H. J. Engell and N. D. Stolica, *Z. Phys. Chem.*, 1959, **20**, 113.
35 Z. Szklarska-Smialowska and J. Mankowski, *Corros. Sci.*, 1972, **12**, 925.
36 G. T. Burstein and S. P. Mattin, *Critical Factors in Localized Corrosion 2*, ed. P. M. Natishan, R. G. Kelly, G. S. Frankel and R. C. Newman, 1995, p. 1.
37 N. Sato, *Corros. Sci.*, 1995, **37**, 1947.
38 P. C. Pistorius and G. T. Burstein, *Philos. Trans. R. Soc., A*, 1992, **341**, 531.
39 H. S. Isaacs and G. Kissel, *J. Electrochem. Soc.*, 1972, **119**, 1628.
40 W. Schwenk, *Corrosion*, 1964, **20**, 129t.
41 V. M. Novokovsky and A. N. Sorokina, *Proceedings Third International Congress Metallic Corrosion*, MIR Publishers, Moscou, 1969, vol. 1, p. 159.
42 K. J. Vetter and H. H. Strehblow, *Localized Corrosion*, ed. R. Staehle, NACE Houston, 1974, p. 240.
43 G. T. Gaudet, W. T. Mo, T. A. Hatton, J. W. Tester, J. Tilly, H. S. Isaacs and R. C. Newman, *AIChE J.*, 1986, **32**, 949.
44 H. S. Isaacs, *Corros. Sci.*, 1989, **29**, 313.
45 Y. Hisamatsu, *Passivity and its Breakdown on Iron and Iron Base Alloys*, ed. R. W. Staehle and H. Okada, NACE Houston, 1976, p. 99.
46 K. J. Vetter, *Electrochemical Kinetics*, 1967, p. 214.
47 P. C. Pistorius and G. T. Burstein, *Corros. Sci.*, 1992, **22**, 1885.

Faraday Discussions

PAPER

The influence of hydrogen peroxide and hydrogen on the corrosion of simulated spent nuclear fuel

Mayuri Razdan and David W. Shoesmith*

Received 15th November 2014, Accepted 16th January 2015
DOI: 10.1039/c4fd00221k

The synergistic influence between H_2O_2 and H_2 on the corrosion of SIMFUEL (simulated spent nuclear fuel) has been studied in solutions with and without added HCO_3^-/CO_3^{2-}. The response of the surface to increasing concentrations of added H_2O_2 was monitored by measuring the corrosion potential in either Ar or Ar/H_2-purged solutions. Using X-ray photoelectron spectroscopy it was shown that the extent of surface oxidation (U^V + U^{VI} content) was directly related to the corrosion potential. Variations in corrosion potential with time, redox conditions, HCO_3^-/CO_3^{2-} concentration, and convective conditions showed that surface oxidation induced by H_2O_2 could be reversed by reaction with H_2, the latter reaction occurring dominantly on the noble metal particles in the SIMFUEL. For sufficiently large H_2O_2 concentrations, the influence of H_2 was overwhelmed and irreversible oxidation of the surface to U^{VI} occurred. Subsequently, corrosion was controlled by the chemical dissolution rate of this U^{VI} layer.

1. Introduction

The Canadian concept to ensure the long term safe disposal of used nuclear fuel is based on multiple barriers: the used fuel bundles, a durable carbon steel container with an outer Cu shell, a clay buffer which seals around the container, and a deep stable geologic environment.[1] Based on this approach it can be calculated that long term containment should be achieved,[2] but it is judicious to assume container failure leading to the exposure of the fuel to groundwater will occur. On failure, the groundwater entering the container would be anoxic since oxidants, mainly dissolved O_2, trapped in the repository on sealing, will be rapidly consumed by minerals (*e.g.*, iron oxides) and biochemical reactions in the surrounding clays, and by corrosion of the copper container.[3] However, radiolysis of the groundwater caused by the radioactivity in the fuel can produce oxidizing conditions at the fuel surface, leading to an increase in solubility of the fuel by

Department of Chemistry and Surface Science Western, University of Western Ontario, London, ON, N6A 5B7, Canada. E-mail: dwshoesm@uwo.ca; Tel: +1-519-661-2111 ext. 86366

many orders of magnitude.[4,5] This could lead to fuel corrosion and the release of radionuclides would be expected.

The fuel corrosion process involves a sequence of oxidation, dissolution, and possibly, corrosion product deposition reactions,

$$U^{IV}O_2 \rightarrow U^{IV}_{1-2x}U^{V}_{2x}O_{2+x} \rightarrow U^{VI}O_2^{2+} \rightarrow U^{VI}O_3\cdot yH_2O \quad (1)$$

where $U^{IV}_{1-2x}U^{V}_{2x}O_{2+x}$ is a thin intermediate oxide layer[6] and $U^{VI}O_3\cdot yH_2O$ is an outer surface layer on the $U^{IV}O_2$ surface. Since, the anticipated ground water pH will be between 6 and 9 (when U^{VI} is at a solubility minimum), corrosion product films would be expected to form and to influence the corrosion rate.[7–9]

In Canadian groundwater, the key constituent likely to influence fuel dissolution is HCO_3^-/CO_3^{2-} (10^{-4} to 10^{-3} mol L^{-1}),[10] which increases the solubility of $U^{VI}O_2^{2+}$ by complexation[11] and buffers the pH. In the presence of HCO_3^-/CO_3^{2-}, the rate of oxidative dissolution is expected to increase, since the formation of $U^{VI}O_3\cdot yH_2O$ would be retarded, and the overall corrosion reaction would become limited by the rate of surface oxidation (to $U^{IV}_{1-2x}U^{V}_{2x}O_{2+x}$). For $[CO_3^{2-}] \geq 10^{-3}$ mol L^{-1}, both electrochemical[6,12,13] and chemical studies[14–16] show not only is $U^{VI}O_3\cdot yH_2O$ formation prevented but the formation of the underlying $U^{IV}_{1-2x}U^{V}_{2x}O_{2+x}$ layer is retarded[14,15] and oxidative dissolution becomes strongly promoted.[13]

In a reactor, irradiation leads to key changes in the composition and properties of the $U^{IV}O_2$ matrix.[17–19] Of these possible changes two are particularly important from a corrosion perspective: (i) rare earth (RE^{III}) doping of the $U^{IV}O_2$ matrix, which will significantly increase its conductivity, and (ii) the segregation of some fission products into noble metal (ε) particles.[20,21] This dispersion of noble metal particles throughout an electrically conducting matrix introduces the possibility of separated, and galvanically-coupled anodes and cathodes which would be expected to significantly influence fuel corrosion in the presence of radiolytic oxidants (particularly H_2O_2).[22,23]

Within a failed and groundwater-flooded container two corrosion fronts exist, Fig. 1, one on the fuel surface and a second on the inner surface of the carbon steel vessel where steel corrosion is sustained by reaction with water to produce Fe^{2+} and H_2:

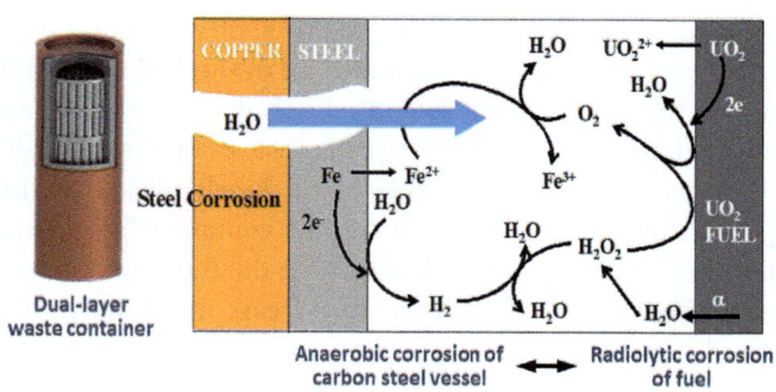

Fig. 1 Schematic showing a cut-away of a dual-walled waste container and the two corrosion fronts that would exist within the container after failure and flooding with groundwater.

$$Fe + 2H_2O \rightarrow Fe^{2+} + H_2 + 2OH^- \quad (2)$$

$$3Fe + 4H_2O \rightarrow Fe_3O_4 + 4H_2 \quad (3)$$

This introduces the possibility that the products of steel corrosion will scavenge the radiolytic oxidants (primarily H_2O_2) responsible for fuel corrosion. Many studies on the influence of Fe and Fe corrosion products on fuel corrosion have been published (ref. 24 and references therein) and inevitably show that the presence of Fe suppresses fuel corrosion and radionuclide release.

However, it is not possible to separate the redox controlling influences of Fe^{2+} and H_2 from steel corrosion and considerable effort has been expended in studying them separately. The influences of Fe^{2+} and Fe_3O_4 on fuel corrosion have been studied extensively[22] and Jonsson *et al.*[25] calculated that the consumption of H_2O_2 *via* the homogeneous Fenton reaction could suppress $U^{IV}O_2$ corrosion inside a failed container by a factor of >40.

$$Fe^{2+} + H_2O_2 \rightarrow Fe^{3+} + OH^{\bullet} + OH^- \quad (4)$$

$$Fe^{2+} + OH^{\bullet} \rightarrow Fe^{3+} + OH^- \quad (5)$$

There is now a considerable amount of evidence to show that H_2 has a more significant effect on the suppression of fuel corrosion than Fe^{2+}, primarily by H_2 oxidation on noble metal (ε) particles galvanically coupled to the UO_2 matrix.[24,26,27]

The overall rate of fuel corrosion will be determined by the relative rates of a series of possible reactions on the fuel surface. To determine the overall influence of these many reactions we have developed a model to determine the influence of steel corrosion products on the alpha (α) radiolytic corrosion of spent nuclear fuel within a failed waste container.[28–30] The key reactions incorporated in this model are illustrated in Fig. 2 and include: (1) a complete reaction set for the α-radiolysis

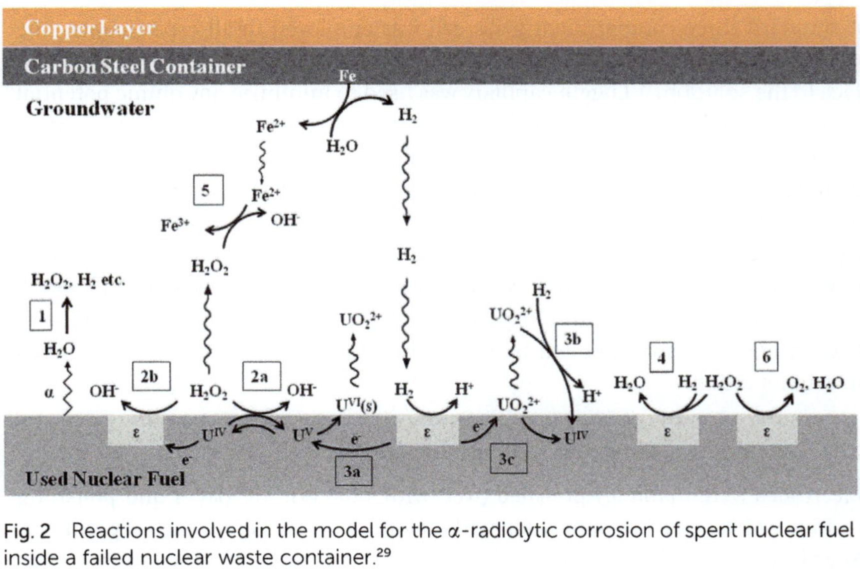

Fig. 2 Reactions involved in the model for the α-radiolytic corrosion of spent nuclear fuel inside a failed nuclear waste container.[29]

of H_2O including the generation of, and the interactions between, the radiolysis products; (2) the oxidative dissolution (corrosion) of $U^{IV}O_2$ supported by H_2O_2 reduction on both the $U^{IV}O_2$ surface (2a) and the noble metal (ε) particles (2b); (3) the reduction of oxidized surface species (U^V/U^{VI}) by H_2 oxidation on noble metal (ε) particles (3a) and of dissolved $U^{VI}O_2^{2+}$ either by reaction with H_2 in solution (3b) or with H_2 catalyzed on the fuel surface (3c); (4) the reaction of H_2O_2 with H_2 catalyzed by noble metal (ε) particles; (5) the scavenging of H_2O_2 by the Fenton reaction; and (6) the decomposition of H_2O_2 to O_2 and H_2O.

One key, but not comprehensively understood, feature of this model is the role the chemical state of the UO_2 surface plays in determining the fuel reactivity in H_2O_2/H_2-containing solutions; *i.e.*, reactions (2a) and (2b). Here, we investigate these reactions in solutions containing various relative concentrations of H_2O_2 and H_2 on 1.5 at% SIMFUEL in solutions with and without added HCO_3^-/CO_3^{2-}.

2. Experimental

2.1 Electrode materials

Experiments were conducted on a 1.5 at% SIMFUEL fabricated by Canadian Nuclear Laboratories Limited (Chalk River, Ontario). SIMFUEL is a simulated chemical fuel, produced by doping UO_2 with eleven stable elements (Y, La, Ce, Mo, Sr, Ba, Rh, Zr, Ru, Nd, and Pd) in the proportions required to simulate spent fuel with a specific in-reactor burn up. The noble metal dopants (Mo, Rh, Pd, and Ru) are unstable as oxides and separate during fabrication to form noble metal (ε) particles, uniformly distributed in the UO_2 matrix.[31] The microstructure of the material is typical of CANDU fuel with grains 8–15 μm in size and a density ~97% of the theoretical value. Secondary ion mass spectrometry confirms the presence of noble metal particles.[24] The electrodes used were ~2 mm thick and 1.2 cm in diameter and were prepared from pellets using a procedure described in detail elsewhere.[32]

2.2 Electrochemical cell and equipment

A standard three compartment glass cell was employed in all experiments. The working electrode was a 1.5 at% SIMFUEL disc, set in resin to expose only one flat face to the solution. A Luggin capillary was used to minimize any ohmic potential drop between the working and reference electrodes. The reference electrode was a commercially available saturated calomel electrode (SCE, Fischer Scientific). The counter electrode was a Pt foil, with a surface area of ~6 cm^2, spot-welded to a Pt wire (99.9% purity, Alfa Aesar). All potentials were measured, and are quoted, against the SCE scale. The cell was housed in a grounded Faraday cage to minimize interference from external sources of noise. A Solartron Model 1287 potentiostat was used to control experiments and record electrochemical data. Corrware™ software (supplied by Scribner and Associates) was used to control instrumentation.

2.3 Electrode polishing and solution preparation

Electrodes were manually polished (wet) with 1200 grit SiC paper and polishing residue was removed by sonication with deionized water before each experiment. Any air-formed oxides were reduced at a cathodic potential of −1500 mV (*vs.* SCE) applied for 5 min before each experiment. On removal from the electrochemical cell, the electrode was rinsed with deionized water and air dried prior to analysis by X-ray photoelectron spectroscopy (XPS).

Solutions were prepared using distilled deionized water purified using a NANOpure Diamond UV ultrapure water system from Barnstead International which removes organic and inorganic impurities ($\rho = 18.2$ MΩ cm). Experiments were carried out in a 0.1 mol L^{-1} NaCl (Caledon, >99%) solution purged with UHP Ar or 5% H_2/95% Ar ($[H_2] \sim 10^{-4}$ mol L^{-1}) gas (Praxair) with and without added HCO_3^-/CO_3^{2-}. Solutions were purged for a minimum of an hour prior to the start of an experiment to minimize dissolved O_2 levels. The 5×10^{-2} mol L^{-1} HCO_3^-/CO_3^{2-} solution was prepared with Na_2CO_3 and $NaHCO_3$ (Caledon, >99%) and the pH adjusted to ~9 (when required) using NaOH and an Orion Model 720A pH meter. Stock solutions of H_2O_2 were prepared using an appropriate amount of 3% w/v solution (Fisher Scientific) and the desired cell concentration (10^{-8} to 10^{-5} mol L^{-1}) was achieved by adding the required amount of this solution into the electrolyte. The H_2O_2 concentration used was determined by ultra-violet/visible spectrophotometry.

2.4 Experimental procedure

The corrosion potential (E_{CORR}) measurements were performed at ambient temperature. E_{CORR} was monitored continuously as the $[H_2O_2]$ was periodically increased. When required the solution was stirred using an external magnetic stirrer.

2.5 UV-vis spectrophotometry

Hydrogen peroxide concentrations were determined spectrophotometrically using a BioLogic Science Instruments MOS UV-vis spectrophotometer. The Ghormley tri-iodide method was employed with ammonium molybdate used to catalyze the oxidation of I^- to I_3^- by H_2O_2. The I_3^- produced has a maximum absorption at 350 nm with a molar extinction coefficient of 25 500 mol L^{-1} cm^{-1}.[33]

2.6 XPS surface analysis

A Kratos Axis Ultra spectrometer was used to record XPS spectra. Spectra were excited using an Al K_α (15 mA, 14 kV) monochromatic high energy ($h\nu = 1486.6$ eV) radiation source. The spectrometer work function was set to give a value of 83.96 eV for the binding energy (BE) of the Au $4f_{7/2}$ line of metallic Au. The instrument was calibrated to give a BE of 932.62 eV for the Cu $2p_{3/2}$ line of Cu metal. Survey spectra were collected over the energy range from 0 to 1100 eV with an X-ray spot size of ~300 to 700 μm at a pass energy of 160 eV. Charge neutralization was used on all specimens. The C 1s peak at 285.0 eV was used as a standard to correct for surface charging.

All spectra were analyzed using CasaXPS software (version 2.3.14) and involved a 50% Gaussian and 50% Lorentzian fitting routine with a Shirley background correction. The procedure used to deconvolute the U 4f spectra into contributions from U^{IV}, U^{V}, and U^{VI} has been described elsewhere.[33-36] The resolved components in both spin–orbit split peaks and the associated satellite structures were used to calculate the proportion of each oxidation state in the surface. The positions and shapes of the satellite structures were used to confirm the validity of the deconvolution of the U 4f peaks.

3. Results and discussion

Fig. 3 shows a series of measurements of E_{CORR} in stirred Ar-purged solutions with and without added HCO_3^-/CO_3^{2-}. Prior to the first H_2O_2 addition, E_{CORR} was allowed to stabilize, and each subsequent H_2O_2 addition was made only after E_{CORR} achieved a steady-state value at the existing $[H_2O_2]$.

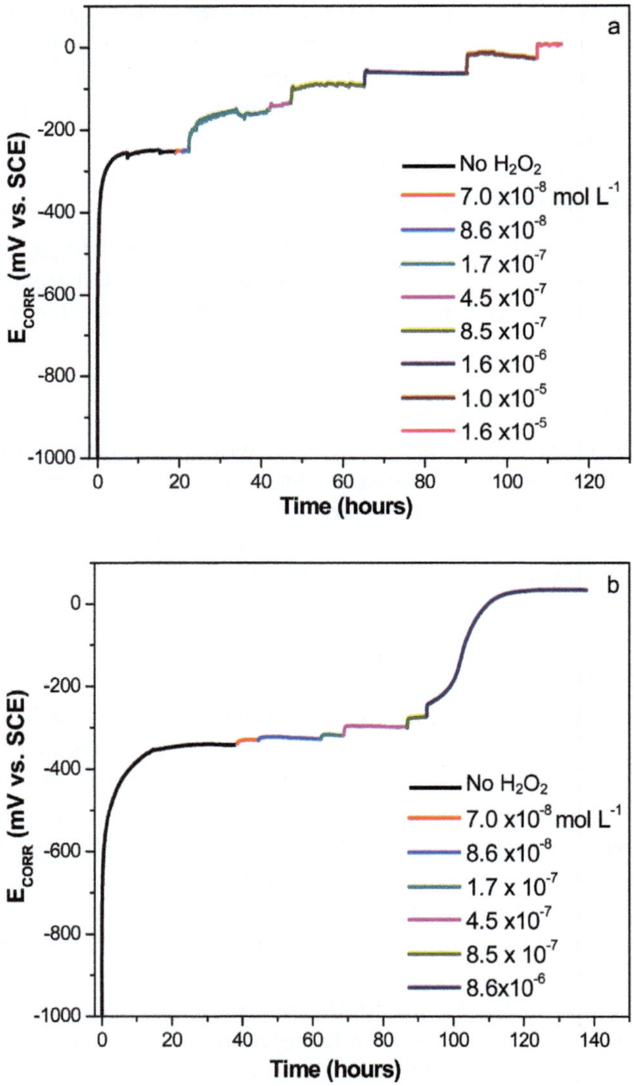

Fig. 3 E_{CORR} as a function of time for different $[H_2O_2]$ measured on a 1.5 at% SIMFUEL electrode in stirred Ar-purged 0.1 mol L^{-1} NaCl solution (pH ~ 9) (a) without and (b) with 0.05 mol L^{-1} HCO_3^-/CO_3^{2-}.

For the solution containing no HCO_3^-/CO_3^{2-}, the addition of $<10^{-7}$ mol L^{-1} H_2O_2 had no observable influence on E_{CORR}, Fig. 3(a). Subsequent additions to higher $[H_2O_2]$ led to increases in E_{CORR} to higher steady-state values, which depended on $[H_2O_2]$. This dependence of E_{CORR} on $[H_2O_2]$ suggests the $U^{IV}O_2$ surface becomes progressively more oxidized as the $[H_2O_2]$ increases. The lack of sensitivity of E_{CORR} to $[H_2O_2]$ below 10^{-7} mol L^{-1} is in contrast to previous experiments conducted at 60 °C,[27] in which E_{CORR} responded to H_2O_2 additions in the concentration range 10^{-11} to 10^{-10} mol L^{-1}.

Previous studies[6,27,37] suggest that the extent of oxidation of the surface (as defined by x in $U^{IV}_{1-2x}U^{V}_{2x}O_{2+x}$) is proportional to the E_{CORR} over the potential range −400 mV to ~0 mV in the absence of HCO_3^-/CO_3^{2-}.

For the experiments conducted in the solution containing 0.05 mol L^{-1} HCO_3^-/CO_3^{2-}, Fig. 3(b), E_{CORR} responded much less markedly to H_2O_2 additions over the concentration range 7×10^{-8} mol L^{-1} to 8.5×10^{-7} mol L^{-1}, changing only marginally. After the final addition of 8.6×10^{-6} mol L^{-1}, E_{CORR} underwent a major transition over a period of ~20 hours to a value >0 mV, effectively the same value achieved in the absence of HCO_3^-/CO_3^{2-} at higher $[H_2O_2]$, Fig. 3(a). This marked increase in E_{CORR} suggests a slow irreversible oxidation of the $U^{IV}O_2$ surface. The lack of response of E_{CORR} to $[H_2O_2]$ at the low concentrations suggests either the UO_2 is not oxidized or, more likely, oxidized to U^{VI} (as $U^{VI}O_2^{2+}$) and subsequently complexed by carbonate ($U^{VI}O_2(CO_3)_x^{(2-x)+}$)

$$U^{VI}O_3 \cdot yH_2O + 2HCO_3^- \rightarrow U^{VI}O_2(CO_3)_2^{2-} + (y+1)H_2O \quad (6)$$

and dissolved, thereby, maintaining the electrode surface unoxidized.[22]

Previous experiments showed that E_{CORR} became independent of H_2O_2 at $[H_2O_2] \geq 10^{-5}$ mol L^{-1}, irrespective of whether HCO_3^-/CO_3^{2-} was present or not.[38] Wren *et al.*[39] observed a similar trend from a dependence on $[H_2O_2]$ (produced by α-radiolysis) at low values of $[H_2O_2]$ to independence at high concentrations. This independence was attributed to the formation of an insulating U^{VI} surface layer (possibly $U^{VI}O_3 \cdot yH_2O$), which lead to a condition of redox buffering, with both $U^{IV}O_2$ corrosion and H_2O_2 decomposition occurring at rates determined by the rate of release of U^{VI} to solution to expose the underlying conductive $U^{IV}_{1-2x}U^{V}_{2x}O_{2+x}$. This mechanism has recently been confirmed[40] and the influence of HCO_3^-/CO_3^{2-} demonstrated.[41]

The results in Fig. 3 suggest the surface in both the present experiments is oxidized to U^{VI} at the final $[H_2O_2]$. Of more interest in the present study is the potential range over which the electrode surface is oxidized to $U^{IV}_{1-2x}U^{V}_{2x}O_{2+x}$ but not covered by a passive layer of insulating U^{VI} (as $U^{VI}O_3 \cdot yH_2O$). In the stoichiometric form, $U^{IV}O_2$ is a Mott–Hubbard insulator with a filled narrow U 5f band, located in the ~5 eV gap between the filled valence band and the empty conduction band. However, the introduction of surface non-stoichiometry, in the form of oxygen interstitial ions (O_I^{II}) accompanied by the creation of U^V atoms to maintain charge balance, creates holes in the 5f band and an increase in electrical conductivity. Under these conditions the surface should be able to sustain cathodic reactions, may be catalytic, and could be susceptible to corrosion.

Fig. 4 shows the response of E_{CORR} to H_2O_2 additions in stirred solutions purged with 5% H_2/Ar with and without added HCO_3^-/CO_3^{2-}. The initial E_{CORR} value recorded in the solution containing no HCO_3^-/CO_3^{2-} is significantly lower

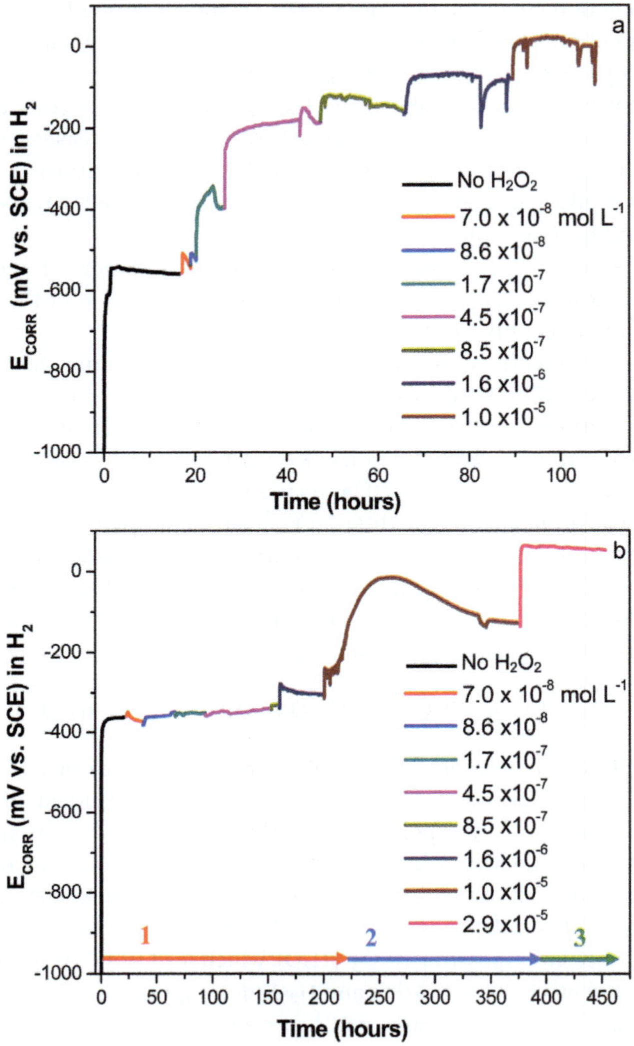

Fig. 4 E_{CORR} as a function of time for different $[H_2O_2]$ measured on a 1.5 at% SIMFUEL electrode in stirred 5% H_2/95% Ar-purged 0.1 mol L^{-1} NaCl (pH ~ 9) (a) without and (b) with 0.05 mol L^{-1} HCO_3^-/CO_3^{2-}.

than that observed in the Ar-purged solution (Fig. 3(a)). This is expected when dissolved H_2 is present,[24,27] since H_2 oxidation on the noble metal particles has been shown to retard oxidation of the galvanically-coupled RE^{III}-doped UO_2 matrix.[26,42] In the absence of HCO_3^-/CO_3^{2-}, Fig. 4(a), E_{CORR} responds in a similar manner to that observed in Ar-purged solutions, but steady-state values at individual $[H_2O_2]$ were not so readily achieved. For $[H_2O_2] \leq 10^{-7}$ mol L^{-1}, E_{CORR} initially increases on the addition of H_2O_2 before decreasing again. Previous experiments at a higher temperature (60 °C)[27] indicated that such a behavior in E_{CORR} can be attributed to the consumption of H_2O_2 by reaction with dissolved H_2.

A number of possible pathways for such a reaction were proposed. One possible pathway involves the production of $OH^{\bullet}$ radicals by H_2O_2 decomposition on noble metal (ε) particles

$$H_2O_2 \rightarrow 2(OH^{\bullet})_{\varepsilon} \quad (7)$$

followed by their scavenging by reaction with dissolved H_2

$$(OH^{\bullet})_{\varepsilon} + H_2 \rightarrow H^+ + H_2O \quad (8)$$

as illustrated in Fig. 5(a).

Alternatively, H_2 oxidation on noble metal (ε) particles

$$H_2 \rightarrow 2H^{\bullet} \quad (9)$$

$$2H^{\bullet} \rightarrow 2H^+ + 2e^- \quad (10)$$

could galvanically couple to H_2O_2 oxidation on the fuel surface

$$H_2O_2 \rightarrow 2(OH^{\bullet})_{UO_2} \quad (11)$$

$$2(OH^{\bullet})_{UO_2} + 2e^- \rightarrow 2OH^- \quad (12)$$

preventing the reaction of the $OH^{\bullet}$ radicals with UO_2 to cause corrosion, as illustrated schematically in Fig. 5(b). Since the number density of noble metal particles is low in the SIMFUEL used in our experiment, the second reaction pathway is intuitively more likely.[27]

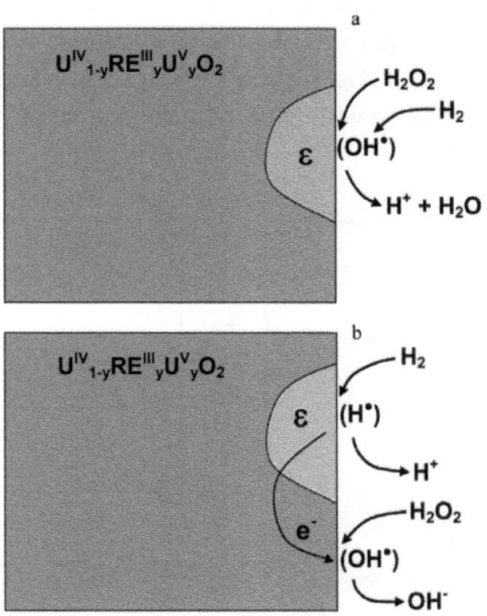

Fig. 5 Schematic illustrating the possible reaction pathways for the consumption of H_2O_2 by reaction with H_2 on a galvanically-coupled 1.5 at% SIMFUEL surface in the presence of H_2O_2 and H_2: (a) on noble metal (ε) particles; (b) by H_2 oxidation on ε-particles galvanically coupled to the oxide matrix.

At 60 °C, it was demonstrated that a concentration ratio, $[H_2]/[H_2O_2]$ of $\geq 10^6$ was required to completely inhibit oxidation of the UO_2 surface by H_2O_2. In the present experiments, E_{CORR} was not followed for a sufficient period of time at these low $[H_2O_2]$ to determine whether the value measured prior to H_2O_2 addition would eventually be re-established; *i.e.*, whether a dissolved $[H_2] \sim 10^{-4}$ mol L^{-1} was sufficient to inhibit surface oxidation for $[H_2O_2] \leq 10^{-7}$ mol L^{-1}.

Further increases in $[H_2O_2]$ in the absence of HCO_3^-/CO_3^{2-}, Fig. 4(a), lead to increased E_{CORR} values, which are maintained for substantially longer periods, although even at $[H_2O_2] \sim 8.5 \times 10^{-7}$ mol L^{-1} there is some indication that E_{CORR} is slowly decreasing, suggesting the catalyzed consumption of H_2O_2 by reaction with H_2 and a decrease in extent of oxidation of the UO_2 surface. At a $[H_2O_2]$ in the range of 10^{-6} to 10^{-5} mol L^{-1}, E_{CORR} apparently achieves a steady-state but is sensitive to convection. Short pauses in stirring (seen in Fig. 4(a) as negative potential excursions) lead to a decrease, and its re-establishment to an increase, in E_{CORR} to the original value. Two examples of such transients due to the interruption of convection are shown in Fig. 6.

The decrease in E_{CORR} occurs on the time scale of a few minutes consistent with the transport controlled depletion of H_2O_2 at the UO_2 surface. The recovery in E_{CORR} on restablishing a convective flow is considerably slower. Additionally, the comparison of the two transients shows the rate of recovery may depend on the value to which E_{CORR} is allowed to decay prior to the re-establishment of the convective flow conditions. The re-establishment of convective conditions would be expected to rapidly restore the surface $[H_2O_2]$, suggesting that the slow response in E_{CORR} must be related to slow changes in the chemical state of the $U^{IV}O_2$ surface.

A similar sensitivity to the local $[H_2O_2]$ was observed previously in thin-layer experiments, in which the local concentration of H_2O_2 was generated by α-radiolysis.[39] Since these previous experiments were conducted on UO_2, not SIMFUEL, this sensitivity to local $[H_2O_2]$ can be considered a feature of the UO_2 surface and not dependent on the presence of noble metal particles.

In the presence of HCO_3^-/CO_3^{2-}, as observed with Ar-purging (Fig. 3(b)), E_{CORR} does not increase significantly with H_2O_2 additions up to $[H_2O_2] \leq 10^{-6}$ mol L^{-1},

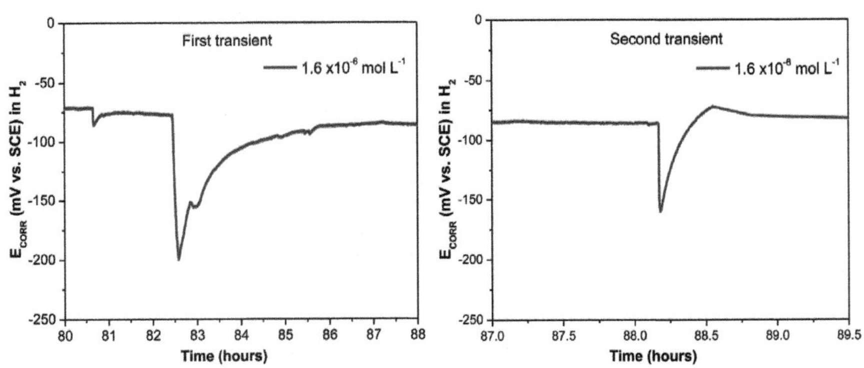

Fig. 6 E_{CORR} as a function of time for $[H_2O_2] \sim 1.6 \times 10^{-6}$ mol L^{-1} obtained on a 1.5 at% SIMFUEL electrode in 5% H_2/95% Ar-purged 0.1 mol L^{-1} NaCl (pH ~ 9) without HCO_3^-/CO_3^{2-}, showing the influence of short periods when stirring was ceased (E_{CORR} decreases) and then re-established (E_{CORR} increases).

Fig. 4(b). At higher $[H_2O_2]$, E_{CORR} begins to increase, indicating irreversible oxidation of the $U^{IV}O_2$ surface. However, the time-dependent behaviour at $[H_2O_2] = 10^{-5}$ mol L^{-1} can be contrasted to that observed in Ar-purged solution at approximately the same $[H_2O_2]$ (8.6×10^{-6} mol L^{-1}), Fig. 3(b). In the Ar-purged solution, E_{CORR} rises steadily to a steady-state value >0 mV, indicating irreversible oxidation of the $U^{IV}O_2$ surface, possibly to the U^{VI} state. By contrast, in H_2/Ar-purged solution, E_{CORR} initially increases before eventually decreasing towards a much lower steady-state value (<−100 mV). This transient behaviour suggests that while the added H_2O_2 may initially oxidize the surface, its consumption by reaction with H_2 eventually leads to a decrease in the extent of the surface oxidation.

The very slow nature of the E_{CORR} response suggests a competition between the H_2O_2 driven oxidation of the surface and its reversal by galvanically-coupled (on ε-particles) or direct (on the UO_{2+x} surface) H_2 oxidation. Since the inventory of H_2O_2 is limited by the amount added while the $[H_2]$ is maintained by continuous purging, the reversal in E_{CORR} from positive-going to negative-going may reflect the decrease in $[H_2O_2]$ as it is consumed by these reactions. Such transient behavior was much more marked at 60 °C even at considerably lower $[H_2O_2]$.[27] Similar transient behavior was observed at 60 °C on a SIMFUEL not containing noble metal (ε) particles, consistent with the α-radiolysis experiments,[39] indicating that the reversible oxidation of the $U^{IV}O_2$ surface is involved and not simply the consumption of H_2O_2 by H_2 catalyzed by noble metal particles.

Fig. 7 summarizes the final steady-state E_{CORR} values as a function of $[H_2O_2]$ for these two sets of experiments. In the absence of HCO_3^-/CO_3^{2-} an influence of H_2 on E_{CORR} becomes significant for $[H_2O_2] \leq 10^{-6}$ mol L^{-1}. In the solution containing HCO_3^-/CO_3^{2-}, the influence of HCO_3^-/CO_3^{2-} is clear by comparison to the values when no carbonate is present. The influence of H_2 is also clear, as E_{CORR} values are consistently lower when H_2 is present.

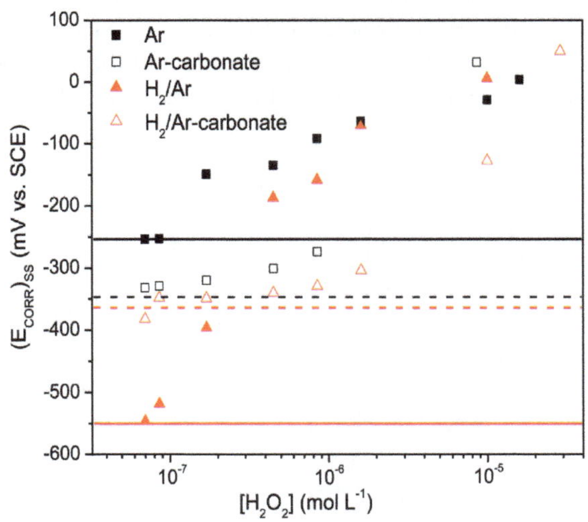

Fig. 7 Comparison of steady-state E_{CORR} values for various $[H_2O_2]$ obtained from Fig. 3 and 4 in Ar and 5% H_2/95% Ar-purged with and without HCO_3^-/CO_3^{2-} in solution. The full and dashed horizontal lines show the E_{CORR} values recorded in the individual experiments prior to the addition of H_2O_2.

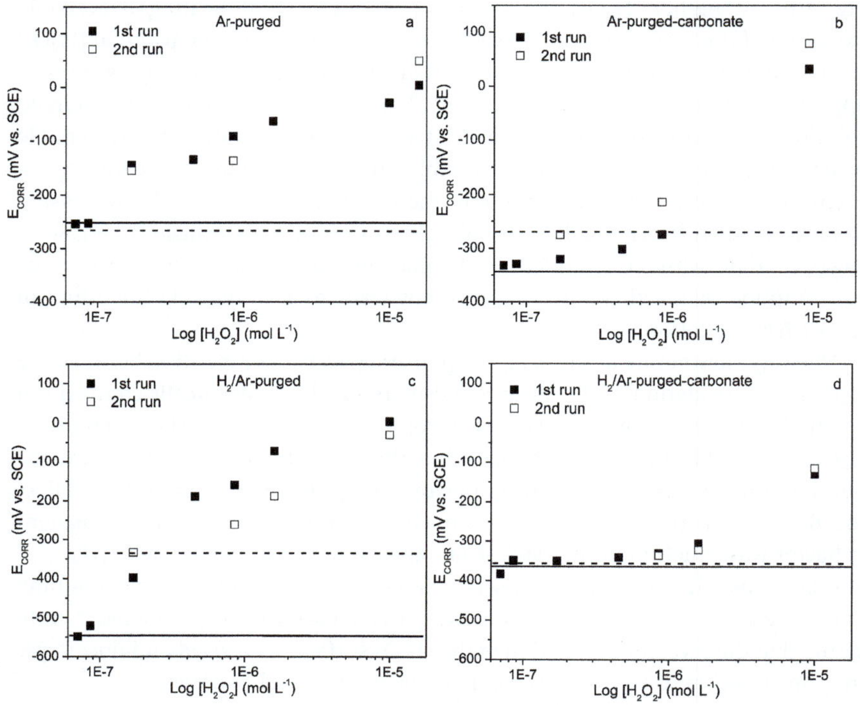

Fig. 8 Comparison of steady-state E_{CORR} values measured on a 1.5 at% SIMFUEL electrode for various $[H_2O_2]$ in (a and b) Ar and (c and d) 5% H_2/95% Ar-purged with and without HCO_3^-/CO_3^{2-}: solid data points from Fig. 3 and 4; open data points measured prior to the XPS analyses. The full and dashed horizontal lines show the E_{CORR} values recorded in the individual run without H_2O_2.

To confirm the extent of oxidation of the electrode surface, a series of experiments was conducted in all four solutions followed by XPS analyses. In these experiments only a single H_2O_2 addition was made and the E_{CORR} was followed until it established a steady-state value. The electrode was then removed and

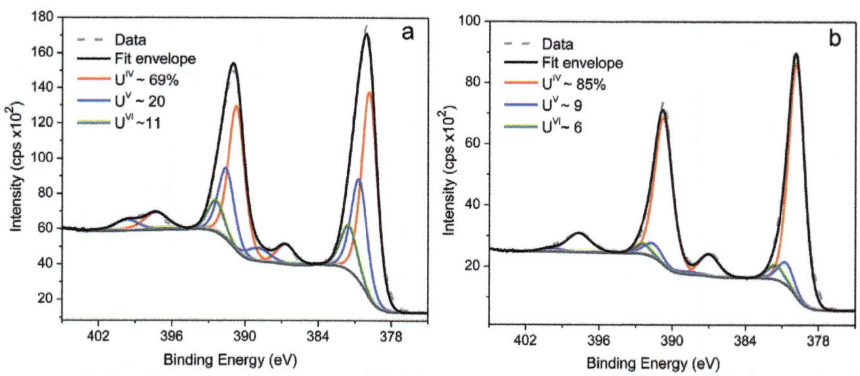

Fig. 9 XPS spectra recorded after corrosion until a steady-state E_{CORR} was achieved in a 0.1 mol L^{-1} NaCl + 5 × 10^{-2} mol L^{-1} HCO_3^-/CO_3^{2-} solution containing $[H_2O_2]$ ~ 8.5 × 10^{-7} mol L^{-1}; (a) Ar-purged; (b) H_2/Ar-purged.

analyzed by XPS. A series of experiments was performed in each solution for different $[H_2O_2]$. Steady-state E_{CORR} values were measured twice for each solution condition to determine reproducibility, but XPS measurements were only performed on one set of measurements for each solution. These E_{CORR} values are compared to the values plotted in Fig. 7 for each experimental condition, Fig. 8, showing that, while absolute values may vary somewhat, the overall trends, as a function of $[H_2O_2]$, are consistent.

Fig. 9 shows two examples of fitted and deconvoluted XPS spectra recorded in solutions with added HCO_3^-/CO_3^{2-} containing the same $[H_2O_2]$. The less oxidized state of the surface after exposure to the H_2/Ar-purged solution (Fig. 9(b)) compared to the Ar-purged solution (Fig. 9(a)) is clear from the fractions of the individual oxidation states present in the surface.

Fig. 10 compares the atomic fractions of the oxidized states (*i.e.*, $(U^V + U^{VI})/U_{total}$) (U_{total} also contains the U^{IV} fraction) as a function of $[H_2O_2]$ for all four conditions. In the Ar-purged solutions, the fractions measured when HCO_3^-/CO_3^{2-} is present are considerably lower than those measured when no HCO_3^-/CO_3^{2-} is present. When H_2 is present the extent of surface oxidation is further suppressed, especially if HCO_3^-/CO_3^{2} is present, when significant oxidation of the surface is only observed for $[H_2O_2] \geq 10^{-5}$ mol L^{-1}.

Fig. 11 shows a plot of the atomic fraction of the oxidized states as a function of E_{CORR}, confirming that the composition of the surface dictates the value of E_{CORR} as previously proposed for similar experiments performed at higher temperatures. These observations support the claim that, when dissolved H_2 is present, the response to convection and the E_{CORR} transients, such as those shown in Fig. 4(b), can be attributed to H_2O_2-stimulated oxidation/corrosion of the $U^{IV}O_2$ surface, which can be reversed by the reaction of the surface with dissolved H_2.

This is best illustrated by revisiting the influence of added H_2O_2 on the SIMFUEL surface in an H_2/Ar-purged solution containing HCO_3^-/CO_3^{2-}, Fig. 4(b).

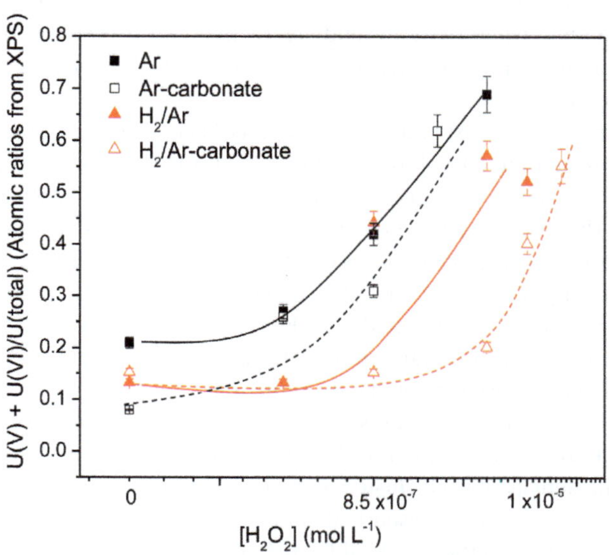

Fig. 10 The $(U^V + U^{VI})/U_{total}$ ratio as a function of $[H_2O_2]$ measured on a 1.5 at% SIMFUEL surface in Ar and 5% H_2/95% Ar-purged with and without HCO_3^-/CO_3^{2-}.

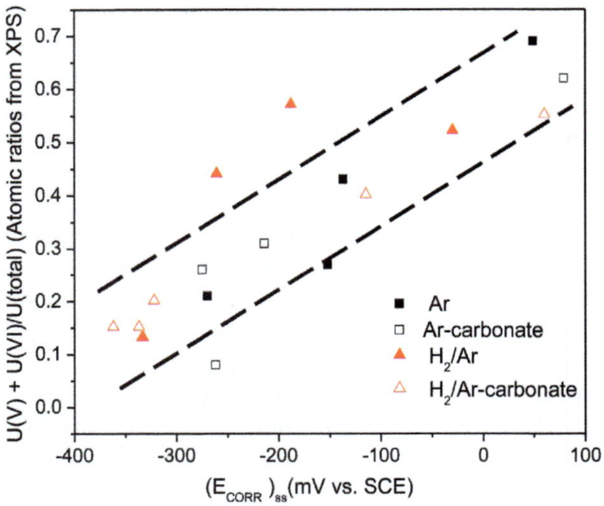

Fig. 11 Comparison of the $(U^V + U^{VI})/U_{total}$ ratio as a function of steady-state E_{CORR} values measured on a 1.5 at% SIMFUEL electrode at various $[H_2O_2]$ in Ar and 5% H_2/95% Ar-purged solutions with and without HCO_3^-/CO_3^{2-}.

In region 1 ($[H_2O_2] < 10^{-5}$ mol L^{-1}), oxidation of the $U^{IV}O_2$ surface is slow and, in the presence of a substantial HCO_3^-/CO_3^{2-}concentration, dissolution proceeds rapidly *via* reaction (6) with the surface remaining free of U^{VI}. In region 2, oxidation of the surface is accelerated at the higher $[H_2O_2]$, leading to the accumulation of oxidized states (U^V, U^{VI}) in the surface and an increase in E_{CORR}. Eventually, as depletion of the H_2O_2 proceeds, H_2 oxidation leading to matrix reduction, most likely on the galvanically-coupled ε-particles, overwhelms the matrix oxidation process. This reduction, coupled with the continuing dissolution of U^{VI} by complexation with HCO_3^-/CO_3^{2-} (reaction (6)) leads to a decrease in E_{CORR}. In region 3, when $[H_2O_2]$ is sufficiently high, an irreversible oxidation of the surface occurs, and, although not shown here, further increases in $[H_2O_2]$ have no influence on E_{CORR}.

The electrochemical oxidation of $U^{IV}O_2$ has been shown to proceed by the injection of oxygen interstitial ions (O_I^{II}) into vacant lattice positions available in the $U^{IV}O_2$ fluorite lattice, a process accompanied by, for a sufficiently low O_I^{II}, the creation of adjacent U^V states. Previously, we have shown, using scanning electrochemical microscopy, that this process is reversible for intermediate degrees of non-stoichiometry; *i.e.*, $x \sim 0.1$ in $U^{IV}_{1-2x}U^{V}_{2x}O_{2+x}$,[43] which adds credibility to the argument for reversibility presented here. For such a degree of non-stoichiometry, O_I^{II} can be considered to be randomly distributed within the top ~5 nm of the oxide surface, and hence extractable by reduction.

However, as the degree of non-stoichiometry increases and the composition approaches $UO_{2.25}$, cuboctahedral clusters begin to form and reversibility is lost.[43,44] As the density of such clusters increases, the lattice undergoes tetragonal distortions.[45] Although not presently demonstrated, it seems likely these distortions are a precursor to the irreversible surface oxidation, which occurs at high enough $[H_2O_2]$, see Fig. 3 and 4.

4. Summary

In terms of the overall mechanism illustrated in Fig. 2, our results confirm that reaction (2a) is reversible and that oxidation of UO_2 by H_2O_2 can be suppressed, in the presence of a sufficient concentration of dissolved H_2, by the oxidation of H_2, predominantly on ε-particles (reaction (3a)). The oxidation/dissolution (corrosion) reaction can be accelerated by the presence of HCO_3^-/CO_3^{2-} since the deposition of insulating U^{VI} corrosion products is prevented.

Under these conditions the $U^{IV}_{1-2x}U^{V}_{2x}O_{2+x}$ surface layer remains exposed to the solution, allowing the combination of reactions (2a) and (3b) to regulate the corrosion process. When exposed, this layer controls the surface redox conditions by catalyzing the decomposition of H_2O_2 to O_2 and H_2O.

For a sufficiently high $[H_2O_2]$, the UO_2 surface becomes irreversibly oxidized to the U^{VI} state. In the absence of HCO_3^-/CO_3^{2-}, this surface layer is most likely $U^{VI}O_3 \cdot yH_2O$, while, when HCO_3^-/CO_3^{2-} is present, corrosion is controlled by the rate of chemical dissolution of this U^{VI} layer.

Acknowledgements

This research was funded under the Industrial Research Chair agreement between the Natural Sciences and Engineering Research Council (NSERC, Ottawa, Canada) and the Nuclear Waste Management Organization (NWMO, Toronto, Canada).

References

1 J. McMurry, D. A. Dixon, J. D. Garroni, B. M. Ikeda, S. Stroes-Gascoyne, P. Baumgartner and T. W. Melnyk, *Evolution of a Canadian Deep Geologic Repository: Base Scenario*, Ontario Power Generation Report 06819-REP-01200-10092-R00, 2003.
2 F. King and M. Kolar, *The Copper Container Corrosion Model Used in AECL's Second Case Study,* Ontario Power Generation Report 06819-REP-01200-10041-R00, 2000
3 F. King, L. Ahonen, C. Taxen, U. Vuorinen and L. Werme, *Copper Corrosion under Expected Conditions in a Deep Geologic Repository*, Swedish Nuclear Fuel and Waste Management Co, Stockholm, Sweden, TR-01-23, 2001.
4 I. Grenthe, J. Fuger, R. J. Konings, R. J. Lemire, A. B. Muller, C. Nguyen-Trung, and H. Wanne, in *Chemical Thermodynamics of Uranium*, Elsevier, Amsterdam, 1992.
5 R. Guillamont, R. T. Fanghanel, I. Grenthe, V. Neck, D. Palmer and M. H. Rand, in *Chemical Thermodynamics*, Elsevier, Amsterdam, 2003, vol. 5.
6 D. W. Shoesmith, *J. Nucl. Mater.*, 2000, **282**, 1.
7 B. G. Santos, J. J. Noel and D. W. Shoesmith, *Corros. Sci.*, 2006, **48**, 3852.
8 B. G. Santos, J. J. Noel and D. W. Shoesmith, *J. Nucl. Mater.*, 2006, **350**, 320.
9 B. G. Santos, J. J. Noël and D. W. Shoesmith, *J. Electroanal. Chem.*, 2006, **586**, 1.
10 P. G. Keech, J. S. Goldik, Z. Qin and D. W. Shoesmith, *Electrochim. Acta*, 2011, **56**, 7923.
11 I. Grenthe, F. Diego, F. Salvatore and G. Riccio, *J. Chem. Soc., Dalton Trans.*, 1984, 2439.

12 D. W. Shoesmith and F. King, Ontario Power Generation Report: 06819-REP-01200-0038R00, 1998.
13 J. S. Goldik, J. J. Noël and D. W. Shoesmith, *Electrochim. Acta*, 2006, **51**, 3278.
14 M. M. Hossain, E. Ekeroth and M. Jonsson, *J. Nucl. Mater.*, 2006, **358**, 202.
15 J. de Pablo, I. Casas, J. Gimenez, V. Marti and M. E. Torrero, *J. Nucl. Mater.*, 1996, **232**, 138.
16 J. de Pablo, I. Casas, J. Jiminez, M. Molera, M. Rovira, L. Duro and J. Bruno, *Geochim. Cosmochim. Acta*, 1999, **63**, 3097.
17 H. Kleykamp, *J. Nucl. Mater.*, 1985, **131**, 221.
18 H. Kleykamp, *Nucl. Technol.*, 1988, **80**, 412.
19 L. H. Johnson and D. W. Shoesmith, in *Spent Fuel in Radioactive Waste Forms for the Future*, ed. W. B. Lutze and R. C. Ewing, Elsevier, Amsterdam, 1988.
20 M. Razdan, M. Trummer, D. Zagidulin, M. Jonsson and D. W. Shoesmith, *Electrochim. Acta*, 2014, **130**, 29.
21 M. Razdan and D. W. Shoesmith, *J. Electrochem. Soc.*, 2014, **161**(3), H105.
22 D. W. Shoesmith, Nuclear Waste Management Organization Report TR-2007-03, 2007.
23 M. Razdan, D. Hall and D. W. Shoesmith, in *Material Research Society Symposium Proceedings*, ed. R. M. Carranza, G. S. Duffó and R. B. Rebak, Cambridge University Press, 2012, vol. 1475, p. 299.
24 M. E. Broczkowski, D. Zagidulin and D. W. Shoesmith, Nuclear Energy and the Environment, *American Chemical Society Symposium Proceedings*, 2010, vol. 1046, p. 349.
25 M. Jonsson, A. T. S. Jonsson, F. Nielsen, O. Roth, E. Ekeroth, S. Nilsson and M. M. Hossain, *Environ. Sci. Technol.*, 2007, **41**, 7087.
26 M. E. Broczkowski, J. J. Noël and D. W. Shoesmith, *J. Nucl. Mater.*, 2005, **346**, 16.
27 M. E. Broczkowski, P. G. Keech, J. J. Noël and D. W. Shoesmith, *J. Electrochem. Soc.*, 2010, **157**, C275.
28 L. Wu, Y. Beauregard, Z. Qin, S. Rohani and D. W. Shoesmith, *Corros. Sci.*, 2012, **61**, 83.
29 L. Wu, Z. Qin and D. W. Shoesmith, *Corros. Sci.*, 2014, **84**, 85.
30 L. Wu, N. Liu, Z. Qin and D. W. Shoesmith, *J. Electrochem. Soc.*, 2014, **161**, E3259.
31 P. G. Lucuta, R. A. Verrall, H. Matzke and B. J. Palmer, *J. Nucl. Mater.*, 1991, **178**, 48.
32 B. G. Santos, H. W. Nesbitt, J. J. Noel and D. W. Shoesmith, *Electrochim. Acta*, 2004, **49**, 1863.
33 M. Razdan, D. S. Hall, P. G. Keech and D. W. Shoesmith, *Electrochim. Acta*, 2012, **83**, 410.
34 M. Schindler, F. C. Hawthorne, M. S. Freund and P. C. Burns, *Geochim. Cosmochim. Acta*, 2009, **73**, 2471.
35 E. S. Ilton, J. F. Boily and P. S. Bagus, *Surf. Sci.*, 2007, **601**(4), 908.
36 M. Razdan and D. W. Shoemsith, *J. Electrochem. Soc.*, 2014, **161**, H225.
37 M. E. Broczkowski, J. J. Noel and D. W. Shoesmith, *J. Electroanal. Chem.*, 2007, **602**, 8.
38 S. Sunder, N. H. Miller and D. W. Shoesmith, *Corros. Sci.*, 2004, **46**, 1095.
39 J. C. Wren, D. W. Shoesmith and S. Sunder, *J. Electrochem. Soc.*, 2005, **152**, B470.

40 L. Wu and D. W. Shoesmith, *Electrochim. Acta*, 2014, **137**, 83.
41 L. Wu, J. S. Goldik and D. W. Shoesmith, *J. Electrochem. Soc.*, 2014, **161**, C363.41.
42 S. Nilsson and M. Jonsson, *J. Nucl. Mater.*, 2008, **374**, 290.
43 H. He, Z. Ding and D. W. Shoesmith, *Electrochem. Commun.*, 2009, **11**, 1234.
44 L. Desgranges, G. Baldinozzi, P. Simon, G. Guimbretière and A. Canizares, *J. Raman Spectrosc.*, 2012, **43**, 455.
45 H. He and D. Shoesmith, *Phys. Chem. Chem. Phys.*, 2010, **12**, 8108.

Faraday Discussions

PAPER

Water corrosion of spent nuclear fuel: radiolysis driven dissolution at the UO_2/water interface

Ross Springell,*[a] Sophie Rennie,[a] Leila Costelle,[a] James Darnbrough,[a] Camilla Stitt,[a] Elizabeth Cocklin,[b] Chris Lucas,[b] Robert Burrows,[c] Howard Sims,[d] Didier Wermeille,†[e] Jonathan Rawle,[f] Chris Nicklin,[f] William Nuttall,[g] Thomas Scott[a] and Gerard Lander[h]

Received 9th December 2014, Accepted 28th January 2015
DOI: 10.1039/c4fd00254g

X-ray diffraction has been used to probe the radiolytic corrosion of uranium dioxide. Single crystal thin films of UO_2 were exposed to an intense X-ray beam at a synchrotron source in the presence of water, in order to simultaneously provide radiation fields required to split the water into highly oxidising radiolytic products, and to probe the crystal structure and composition of the UO_2 layer, and the morphology of the UO_2/water interface. By modeling the electron density, surface roughness and layer thickness, we have been able to reproduce the observed reflectivity and diffraction profiles and detect changes in oxide composition and rate of dissolution at the Ångström level, over a timescale of several minutes. A finite element calculation of the highly oxidising hydrogen peroxide product suggests that a more complex surface interaction than simple reaction with H_2O_2 is responsible for an enhancement in the corrosion rate directly at the interface of water and UO_2, and this may impact on models of long-term storage of spent nuclear fuel.

[a]*Interface Analysis Centre, University of Bristol, Bristol BS2 8BS, UK. E-mail: phrss@bristol.ac.uk; Tel: +44 (0) 117 331 1176*

[b]*Department of Physics, University of Liverpool, Liverpool L69 7ZE, UK*

[c]*National Nuclear Laboratory, 102B Stonehouse Park, Sperry Way, Stonehouse, Gloucester GL10 3UT, UK*

[d]*National Nuclear Laboratory, Harwell Science and Innovation Campus, Oxfordshire OX11 0QT, UK*

[e]*XMaS, European Synchrotron Radiation Facility, BP220, F-38043 Grenoble Cedex 09, France*

[f]*Diamond Light Source, Harwell Science and Innovation Campus, Harwell OX11 0DE, UK*

[g]*Department of Engineering and Innovation, The Open University, Venables Building, Milton Keynes MK7 6AA, UK*

[h]*European Commission, Joint Research Centre, Institute for Transuranium Elements, Postfach 2340, D-76125 Karlsruhe, Germany*

† Alternative address: Department of Physics, University of Liverpool, Liverpool L69 7ZE, UK.

1 Introduction

The future storage of spent nuclear fuel (SNF) poses some of the most challenging scientific and economic questions.[1–3] With the growing consensus that storage of this material in a deep underground repository is the most viable long term solution, and the likely scenario of containment failure and groundwater contact,[4,5] one of the central problems is to understand the reactions at the interface of the fuel with its surroundings.[4,6–8] Of these, of course, water is the most important, as it can transport radioactive material away from the fuel repository and into the ecosystem.[9]

The predominant component of this fuel is a ceramic oxide of uranium, UO_2, whose solubility in water of its stoichiometric U(IV) form, is very low.[10] However, post burn-up, the UO_2 fuel possesses levels of activity from 10^{14} Bq to 10^{16} Bq at its surface, depending on the reactor type, and this decays by approximately four orders of magnitude over the first 10 000 years of proposed storage, at which point containment failure becomes significantly likely.[11] The strong alpha, beta and gamma radiation fields are sufficient to radiolyse (radiolysis is the dissociation of molecules due to nuclear radiation) water in close proximity.[4,12,13] The radiolysis products comprise short-lived, highly oxidising free radical species, such as $^{\bullet}OH$ and the much longer lived hydrogen peroxide molecule, H_2O_2.[14,15] Under these highly oxidising conditions it is possible to form UO_2^{2+} (uranyl) ions *via* the oxidation of U(IV) to U(VI), whose solubility in water is several orders of magnitude greater than UO_2,[6] leading to accelerated dissolution of the fuel matrix and potential release of radionuclides into the environment. Therefore, an understanding of the water/UO_2 interface, and the ability to predict its long-term behaviour are vitally important.

A stored spent fuel pellet possesses a great deal of complexity, including defects, He bubbles, microscopic cracking and fission daughter products for example. The situation becomes even more complicated when one includes groundwater, containing various ionic species, and further still, if one includes the potential influence from the cladding material. To date, the majority of studies of this complex system have focused on the chemical composition of the dissolution and the electrochemistry of the corrosion mechanism.[4,7,9,10,16] Since one of the most important factors driving oxidising conditions in the groundwater is the radiolysis of the water,[6,17] some studies have gone further and have attempted to replicate the radiolytic conditions electrochemically,[18,19] using external and dopant alpha sources[9,20–22] and by the addition of H_2O_2 to the groundwater solution.[6,10,23] This final method is hotly debated amongst research groups, since it is not clear precisely what the H_2O_2 concentration would be at the fuel surface.

Here, we report a new approach; we aim to remove much of the material complexity and study the corrosion of UO_2 in pure water in the presence of strong radiation fields. Using high-quality single-crystal thin films of UO_2 with atomically smooth surfaces and a thin layer surface tension cell of MilliQ, pure water (nominally pH 7), we expose this model fuel/groundwater interface to an intense, monochromated beam of X-rays from a synchrotron source; an approach first employed on the XMaS beamline, BM28 at the ESRF. This *source-probe* method allows us to simultaneously provide strong radiation fields and probe the structure of the interface. Using a combination of X-ray reflectivity (XRR) and high angle diffraction (XRD) in a specular geometry we are able to probe changes in the interface structure, roughness, electron density, crystallinity and eventual dissolution as a function of exposure time.

It is important to stress the surface sensitivity of these techniques (XRR and XRD), and ask why this has not been observed previously with bulk UO_2 samples. For typical energies and angles of incidence, the penetration of such a photon beam into UO_2 is in the order several microns. Hence, there will be only a minuscule change in the Bragg reflected intensity, since >99.5% of the intensity comes from the undisturbed UO_2 bulk sample. Even if a single-crystal is used, no measurable change will be observed. On the other hand, by using an epitaxial film of 50 Å, there will be enough intensity (at a synchrotron) and the changes will be very substantial, often exceeding 50%. There are also additional features in the diffraction profiles, resulting from finite thickness effects that give even more detailed information about the morphology of the interface. Such films therefore give unprecedented sensitivity to structural changes at the UO_2 interface.

2 Experimental

Single-crystal thin films of UO_2 were grown in a dedicated DC magnetron sputtering facility at the University of Bristol under UHV conditions. Samples were deposited in the three high symmetry directions, [001], [110] and [111], although the majority of the work presented here describes data collected from an [001]-oriented UO_2 sample, deposited onto a single-crystal [001]-YSZ (yttria-stabilised zirconia) substrate of dimensions 1 cm × 1 cm × 0.5 mm.[24] Reactive sputtering was used to deposit uranium in an argon pressure of $\rho_{Ar} = 7.2 \times 10^{-3}$ mbar and an oxygen partial pressure of $\rho_{O_2} = 2 \times 10^{-5}$ mbar, to give a sputtering rate of 1.2 $Å_{UO_2}$ s^{-1} in order to produce a sample of nominal thickness, $t_{UO_2} = 40$ Å. Substrate heating was used to elevate the growth temperature to ~550 °C, providing thermal energy to improve the crystalline quality, monitored using *in situ* reflection high-energy electron-diffraction (RHEED).

Several samples were grown in order to verify the reproducibility of the experiment, and these were characterised using X-ray reflectivity (XRR) and high angle X-ray diffraction (XRD) on a Philips X'Pert Pro MRD, with a Cu Kα source ($\lambda = 1.54$ Å). This same geometry was used in the synchrotron measurements, carried out at the XMaS beamline, BM28, ESRF[25] and I07, Diamond Light Source, as shown in Fig. 1.

X-ray reflectivity is a non-destructive technique particularly well-suited to probing the fine details of surfaces or buried interfaces,[26] where the X-rays probe the electron density perpendicular to the surface normal. In this instance, the diffraction is called specular (or longitudinal) elastic scattering, *i.e.* that the incident and exit wavevectors, $\mathbf{k_i}$ and $\mathbf{k_f}$, respectively, have the same magnitude, and that the angles of incidence and exit, θ_i and θ_f, respectively, are also equivalent. The X-ray intensity is measured as a function of incidence angle, close to the critical angle, θ_C, for total external reflection, which is typically a combination of Fresnel reflectivity (with a $|1/\mathbf{q_z}|^4$ dependence) and a fringe pattern (commonly Keissig fringes), due to constructive interference from scattering at the layer interfaces. Here, $\mathbf{q_z}$ is the wavevector momentum transfer along the surface normal.

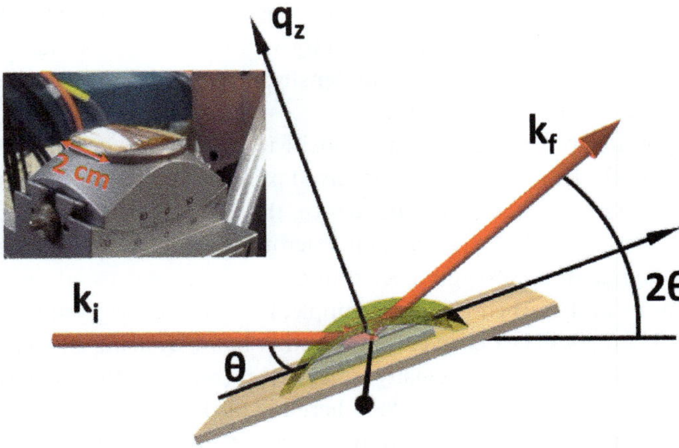

Fig. 1 Schematic of the set-up used in the laboratory characterisation measurements and synchrotron experiments at the ESRF and Diamond Light Sources. Here, the incident and exit wavevectors ($\mathbf{k_i}$ and $\mathbf{k_f}$), angle of incidence (θ), angle with respect to the detector (2θ) and the wavevector momentum transfer ($\mathbf{q_z}$) are labelled. The thin layer surface tension cell can be seen over the sample, which is used to hold a fixed volume of water that covers the entire sample during X-ray irradiation. The insert shows a photograph of the cell, the horizontal dimension is noted along the edge of the cell.

$$\mathbf{q_z} = \mathbf{k_f}\sin\theta_f - \mathbf{k_i}\sin\theta_i = \frac{4\pi}{\lambda}\sin\theta \quad (1)$$

In this case, we have used the Parratt recursion method[27] of calculating transmitted and reflected wave fields, using the GENX computer program, developed by Björck and Andersson,[28] which fits calculated reflectivity profiles to experimental data, using a differential evolution algorithm that can be optimised to avoid local minima; a common problem encountered when modeling X-ray reflectivity. The variables that are used to construct the electron density profile are the densities of materials ($\rho_{\text{substrate}}$, ρ_{UO_2}), the layer thickness (t_{UO_2}) and the roughness of each interface ($\sigma_{\text{substrate}}$, σ_{UO_2}), measured as the root mean squared of the fluctuations in the height of the layer. Also included in the model is a top layer of uranium oxide that is lower in electron density than stoichiometric UO_2, which we label UO_X. This layer is modelled by a series of slices of varying electron density with thicknesses equivalent to a UO_2 monolayer, and this attempts to interpret both the topology and hyperstoichiometric progression of uranium oxide. Later, in Table 1 an average t_{UO_X} is given, that represents the mean overall thickness of this top-layer, assuming half the electron density of UO_2 (*i.e.* the average between the bulk value and air).

High resolution X-ray diffraction provides a host of complementary information to describe the structural composition of thin films. In the case of a single-crystal [001]-UO_2 film on YSZ, a longitudinal measurement across the (002) Bragg peak gives the average *d*-spacing for the thin film lattice parameter along the surface normal, and the finite thickness broadening of the Bragg peak can be used to calculate the number of scattering planes contributing to the intensity, and therefore the thickness of crystalline UO_2. For smooth interfaces, fringes are

Table 1 Table of parameters used in the fitted calculations to model the experimental reflectivity and high angle diffraction data. All values are in Å, where t_{UO_2} is the thickness of UO_2 and σ_{UO_2} is the root mean squared roughness, both have errors of approximately 1 Å, since both high angle XRD and XRR are used to determine these values. t_{UO_x} and σ_{UO_x} are the thicknesses and roughnesses of the top layer of complex oxide, respectively, and have larger errors of ~2 Å

Exposure time (s)	t_{UO_2}	σ_{UO_2}	t_{UO_x}	σ_{UO_x}
0	34	3	10	6
30	27.5	2.5	16	9
90	21	7.6	20	11.5
120	13.5	11	22	13

also present, similar to those observed in X-ray reflectivity, which give information about the electron density profile. A rocking curve at the Bragg peak position produces an observable intensity from planes that are not perfectly parallel. This gives an indication of the crystal mosaicity and the density of dislocation defects. The rocking curve measurement is made by rotating the sample (varying θ), while keeping the detector (2θ) fixed, see Fig. 1.

3 Results

Using a combination of X-ray reflectivity and high resolution diffraction, it is then possible to build up a full structural picture of the thin film and the substrate/film and film/air interface. Fig. 2 shows the X-ray reflectivity (panel (a)) and high resolution diffraction (panel (b)) from a single-crystal [001]-UO_2 film on YSZ, with a nominal thickness of 40 Å. The data are shown as open black circles and fitted

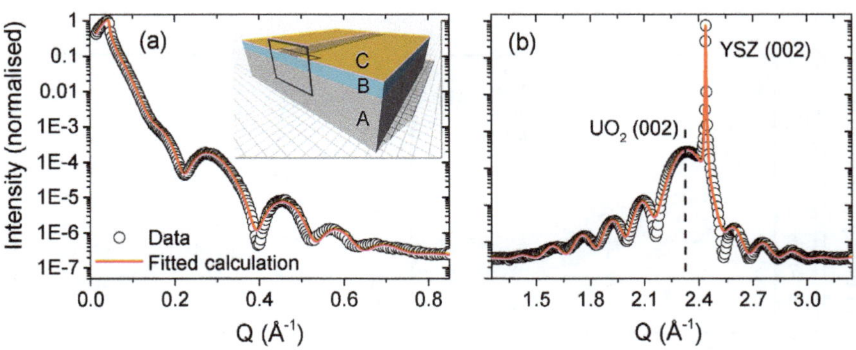

Fig. 2 Panel (a) shows the X-ray reflectivity spectrum for a nominal 40 Å [001]-UO_2 thin film, grown on YSZ. Panel (b) contains a high resolution diffraction spectrum across the UO_2 (002) Bragg peak. The UO_2 (002) thin film and YSZ (002) substrate Bragg peaks have been labelled for clarity. The experimental data are the open black circles and the fitted calculations are the solid red lines. The insert in panel (a) shows the model of a pristine UO_2 film, labelled B, with a complex surface oxide layer, C, grown on a YSZ substrate, A. The copper coloured strip across the centre of the sample indicates the footprint of the X-ray beam at a low angle and the frame indicates the 2D profile used to indicate the corrosion front in Fig. 4.

calculations of the reflectivity and the high angle diffraction are represented by solid red lines.

These data were taken at the I07 beamline of the Diamond Light Source. The photon flux at a synchrotron source is between 4 and 5 orders of magnitude greater than a laboratory source and so allows one to probe the reflectivity far further in $\mathbf{q_z}$ and better resolve the fringes at higher angles. It is also ideal to study buried interfaces, which in our case will be the UO_2/water interaction, since we are interested in the potential dissolution of UO_2 in groundwater.

Table 1 includes the model parameters used to reproduce the calculations. It is clear that even for a pristine sample, not exposed to heavily oxidising conditions, it is necessary to model the thin film with a $\sim$30 Å crystalline UO_2 layer and then a top layer of low electron density oxide. The insert of Fig. 2(a) shows a model of the pristine film, where the substrate layer is labelled A, the UO_2, B, and the top oxide layer, C. Since we are concerned with the corrosion of UO_2 in water, the next step is to repeat these measurements, using synchrotron radiation to simultaneously probe the film structure and to provide the necessary radiolytic products for oxidative dissolution.

In order to probe the change in the uranium oxide film in detail as a function of exposure time, first we set the detector position to the centre of the UO_2 (002) Bragg peak and measured the intensity as a function of time. The following data were taken on the I07 beamline at the Diamond Light Source, UK. The beam energy was monochromated to 17.116 keV (50 eV below the U L_3 absorption edge), which was then focused in the vertical and horizontal directions to give an approximate beam size at the sample position of 200 μm × 200 μm. The incident beam slits were set to 100 μm × 100 μm for all measurements and the scattered photons were detected, using a Pilatus 100K detector.

In Fig. 3(a) and (b) we present data at three points in time, 30 s, 90 s and 120 s. The thin layer surface tension cell was filled with Milli-Q ultra pure type 1 water and the incident slits were set to a 200 micron square. The surface was exposed at

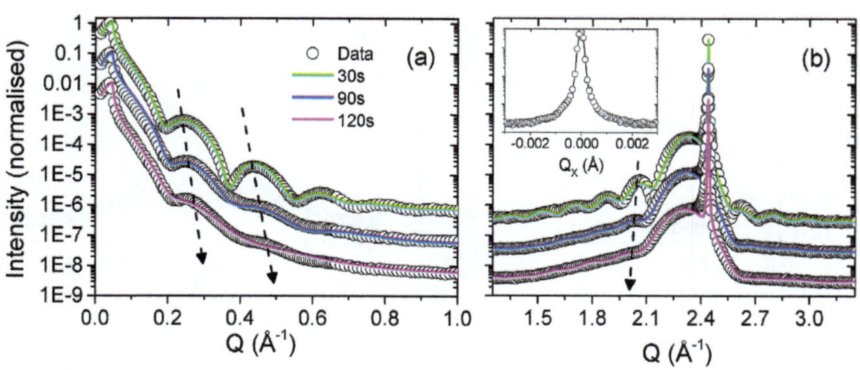

Fig. 3 Panel (a) shows X-ray reflectivity data and panel (b) shows high angle diffraction data, measured at exposure times of 30 s, 90 s and 120 s; the experimental data are represented by the open black circles and the fitted calculations by the solid green, blue and magenta lines, respectively. The insert of panel (b) shows the rocking curve of the (002) Bragg peak for the 30 s exposure. The dashed black arrows indicate an increase in fringe separation as a function of exposure time, which suggests a concomitant loss of material.

an incident angle of ~0.5°, such that the footprint was approx. 2 cm, more than covering the whole length of the sample. The water was then removed, the slits were closed down to 100 μm × 100 μm and the diffraction spectra were recorded. The water removal was carried out using a pipette, and an N_2 gas flow jet was positioned close to the sample surface during measurements. Two simple tests were carried out to confirm the necessary conditions of a surface/water interface and radiation strong enough to drive radiolysis: the first involved exposing the UO_2 surface to water for two hours and then measuring the X-ray reflectivity and diffraction profiles; no changes were detected. The second test involved exposing a sample to an intense X-ray beam for an hour with no water present; again, no changes to the scattered intensity were observed.

A calibrated Si photodiode was placed in the beam at the sample position in order to accurately calculate the number of photons per second per unit area incident for a number of slit settings. Together with the beam energy, these values could then be used to calculate the likely number of oxidising species present in the X-ray beam path in the water, specifically, to calculate the number of long lived, oxidising H_2O_2 molecules. Fig. 3 shows fitted calculations to the experimental data, based on a structural model of the UO_2 film that consists of a layer of crystalline UO_2 with the standard bulk density and a surface layer of reduced electron density, labeled UO_X. Roughnesses for each of the substrate/UO_2 (~2 Å in each case), UO_2/UO_X and UO_X/water interfaces were also computed. Table 1 summarises all of the parameters and Fig. 4 shows a pictorial representation of the corrosion region at each of the 30 s, 90 s and 120 s exposures.

Initially, the intensity of the UO_2 (002) Bragg peak was measured as a function of the exposure time. This is not precisely a measure of the dissolution, since a decrease in intensity will also result from a surface roughening and oxidation, but it does provide a good estimate for the rate of change of the interface structure. This measurement was carried out for thin film samples of [001] (polar surface), [110] and [111] orientations (non-polar surfaces), which due to their different surface energies and water adsorption energies are expected to have significant impact on the rate of dissolution. However, contrary to this assumption, within the errors of this experiment, we saw no evidence to suggest that this may play a significant role. This confirmed earlier measurements carried out at an incident energy of 15 keV at the XMaS beamline, where also the experiment was repeated under alkaline (pH ~ 11) and acidic (pH ~ 2) conditions. As expected, in the case of increased acidity, the rate of corrosion dramatically increased, whereas under

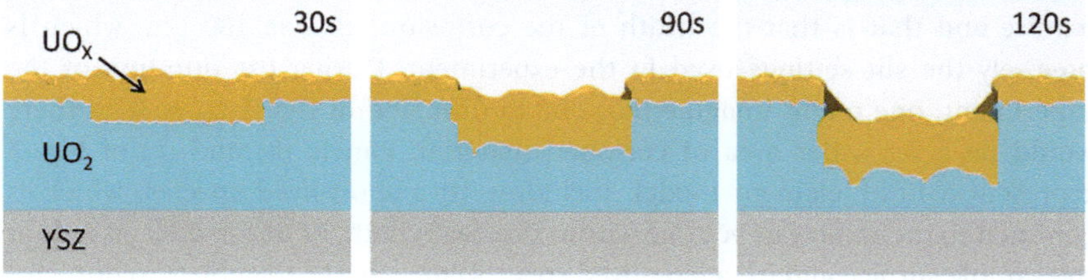

Fig. 4 Pictorial representation of the increase in roughness and UO_X thickness, and the amount of dissolution as the surface undergoes corrosion.

heavily alkaline conditions the rate of change in the UO_2 Bragg peak intensity all but stopped, *i.e.* dissolution was halted.

Furthermore, since we were also interested in any potential surface enhancement of the corrosion by the photocatalytic process, we measured the rate of change in the Bragg peak intensity before (17.116 keV), at (17.166 keV) and after (17.216 keV) the uranium L_3 absorption edge. There is a huge resonant enhancement in the number of electrons excited to the continuum at such an absorption edge, particularly for the U L_3, so this ought to have a pronounced effect on any possible photocatalytic process, however, we did not observe any difference in the rate of the decrease in the Bragg peak intensity, within experimental errors.

4 Discussion

So far we have seen experimentally the effect of radiolytically driven oxidation and dissolution of the UO_2 surface, but it is also possible to calculate the quantity of radiolysis products from first principles, based on the energy and flux of the X-ray source incident on a known water thickness. In the first instance, we are making the same assumption as the vast majority of the literature, that the longest lived and dominant oxidising product is hydrogen peroxide.[6,23] Here, we have $\sim 1 \times 10^{12}$ photons per s of 17.116 keV X-rays incident on a 0.5 mm thick water layer on the UO_2 film surface. The *G*-value, the number of molecules of reactant consumed or product formed (in our case) per unit of incident energy absorbed, is 0.6 molecules of H_2O_2/100 eV.

For this thickness of water at this photon energy, the transmission of photons to the surface is ~59%, which results in $\sim 7 \times 10^{-11}$ moles H_2O_2 produced. Since the volume of the water exposed to the beam is approx. 200 μm wide, 0.5 mm high and 1 cm long, *i.e.* 0.001 cm^3, the H_2O_2 concentration increases by $\sim 7 \times 10^{-5}$ M per second. The equilibrium concentration is in the region of 1×10^{-4} M, which means that it is reached almost instantly, relative to the timescales of this experiment. The question then arises – is H_2O_2 alone enough to drive the changes that we are observing in our experiment? This is the general assumption purported by the literature.[6,23]

Fig. 5 may provide a clue. Panel (a) is a scanning electron microscopy image, obtained using a Zeiss Sigma FEG-SEM; utilising secondary electron detection with electron gun settings of 10 kV and 30 μA. The image shows the area of the sample that has been exposed to the beam. This area is heavily corroded and so is not as conductive as the surrounding UO_2 film. What is observed here is the resultant charging of the corroded region. There is one particularly remarkable feature and that is that the width of the corrosion track is 100 μm, which is precisely the slit settings used in the experiment. During the duration of the experiment, one might imagine that due to diffusion of the H_2O_2 species there would be a far wider area of corroded material. Panels (b) and (c) of Fig. 5 represent a finite element model, including (i) a short-lived species, which is confined to the radiolysis volume within the beam path, or one produced only at the sample surface and (ii) a long-lived species (such as H_2O_2) subject to diffusive transport through the water layer in order to determine the likely footprint of corrosion.

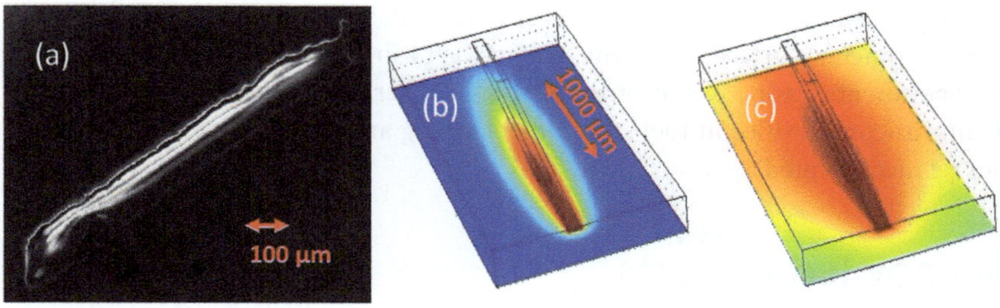

Fig. 5 Panel (a) shows an SEM image of a corroded UO_2 single-crystal thin film, measured at the UO_2 Bragg peak position for 500 s (an incident angle of about 7°) with 100 μm vertical and horizontal slit settings. Panels (b) and (c) are images representing radiolysis product concentrations close to the beam footprint after 500 s, as calculated using finite element modeling, based on diffusion from the beam volume into the bulk liquid for short- and long-lived species, respectively, with the latter being the representative of H_2O_2.

It is clear that a long-lived species, subject to a bulk diffusion, cannot be solely responsible, so this suggests that the corrosion, which is restricted to the beam footprint, is driven by interactions at the surface. There are several candidate propositions that can be explored, for example, it might be due to a photocatalytic effect driven by the high photon flux. As UO_2 has a band gap in the region of 2 eV, the high flux of X-rays may result in the production of electron–hole pairs, that can further enhance the formation of reactive oxygen species. To test this theory we measured the rate of corrosion at a range of X-ray energies spanning the uranium L_3 absorption edge. Crossing through this edge enhances the number of electrons in the valence band, and thus would increase the oxidant species produced *via* photocatalysis. However, on passing through the U L_3 edge, no statistically significant increase in the corrosion rate was observed, indicating that in this case a photocatalytic process is not responsible. While it is unclear why we observe this surface enhanced corrosion, other possibilities may include a significant concentration of short-lived oxidising species, such as OH radicals. Due to the extremely short-lived nature of such species, diffusion outside of the beam footprint would prove unlikely.

5 Conclusions

In these experiments we have demonstrated that we can induce significant oxidation and further, dissolution of a UO_2 surface, by using an intense beam of X-rays, mimicking the radiation fields found at the surface of spent nuclear fuel. Both the X-rays and the water interface are essential ingredients for these changes. We have been able to measure variations in the electron density, surface roughness and rate of dissolution of a radiolysis driven corrosion front in a nuclear fuel material at the Ångström length-scale.

There still remain some open questions regarding the precise mechanism for the observed corrosion, which seems unlikely to be due to hydrogen peroxide alone and may include a more complex surface effect. This could have significant consequences for previous research that has predominantly relied on this assumption to simulate the conditions driven by radiation fields in real SNF.[6,23]

This technique, using thin (<100 Å) epitaxial films of UO_2 gives unprecedented (sub-nanometer) surface sensitivity, and in the future, paves the way for a new set of experiments, using synchrotron X-rays in a series of *source-probe* measurements, as complexities in fuel structure, cladding and groundwater composition can be incorporated.

Acknowledgements

The authors wish to thank Paul Thompson and Laura Glaubes for their assistance in the first experiments on this project at the XMaS beamline BM28, ESRF. Ross Springell would like to thank the Royal Commission for the Exhibition of 1851 for the receipt of a research fellowship. The authors would also like to thank the EPSRC for funding the DISTINCTIVE consortium grant on nuclear waste.

References

1 A. Corner, D. Venables, A. Spence, W. Poortinga, C. Demski and N. F. Pidgeon, *Energy Policy*, 2011, **39**, 4823–4833.
2 C. Pescatore and A. Vari, *J. Risk Res.*, 2006, **9**, 13–40.
3 U. Strandberg and M. Andren, *J. Risk Res.*, 2009, **12**, 879–895.
4 D. W. Shoesmith, *Nuclear Waste Management Organisation*, 2007, Report no. NWMO TR-2007-03.
5 T. E. Eriksen, D. W. Shoesmith and M. Jonsson, *J. Nucl. Mater.*, 2012, **420**, 409–423.
6 D. W. Shoesmith, *J. Nucl. Mater.*, 2000, **282**, 1–31.
7 H. He, M. Broczkowski, K. O'Neil, D. Ofori, O. Semenikhin and D. Shoesmith, *Nuclear Waste Management Organisation*, 2012, NWMO TR-2012-09.
8 F. Garisto, D. Barber, E. Chen, A. Inglot and C. Morrison, *Nuclear Waste Management Organisation*, 2009, NWMO TR-2009-27.
9 S. Sunder, G. D. Boyer and N. H. Miller, *J. Nucl. Mater.*, 2003, **322**, 163–169.
10 D. W. Shoesmith and S. Sunder, *SKB Technical Report*, 1991, SKB-TR–91–63.
11 M. I. Ojovan and W. E. Lee, *An Introduction to Nuclear Waste Immobilisation*, Elsevier, 2005, vol. 1.
12 E. Ekeroth, O. Roth and M. Jonsson, *J. Nucl. Mater.*, 2006, **355**, 38–46.
13 C. Corbel, G. Sattonnay, S. Guilbert, F. Garrido, M.-F. Barthe and C. Jegou, *J. Nucl. Mater.*, 2006, **348**, 117.
14 J. Spinks and R. Woods, *An Introduction to Radiation Chemistry*, John Wiley and Sons, Inc., New York, 3rd edn, 1990.
15 S. LeCaër, *Water*, 2011, **3**, 235–253.
16 N. Rauff-Nisthar, C. Boxall, I. Farnan, Z. Hiezl, W. Lee, C. Perkins and R. Wilbraham, Corrosion in Nuclear Energy Systems: From Cradle to Grave, *ECS Trans.*, 2013, **53**, 95–104.
17 V. Čuba, V. Múčka and M. Pospíšil, in *Radiation Induced Corrosion of Nuclear Fuel*, ed. D. S. T. Revankar, 2012, pp. 27–52.
18 F. Miserque, T. Gouder, D. Wegen and P. Bottomley, *J. Nucl. Mater.*, 2001, **298**, 280–290.
19 A. Seibert, D. Wegen, T. Gouder, J. Ramer, T. Wiss and J. P. Glatz, *J. Nucl. Mater.*, 2011, **419**, 112–121.

20 B. Muzeau, C. Jagou, F. Delaunay, V. Broudic, A. Brevet, H. Catalette, E. Simoni and C. Corbel, *J. Alloys Compd.*, 2009, **467**, 578–589.
21 M. G. Bailey, L. H. Johnson and D. W. Shoesmith, *Corros. Sci.*, 1985, **25**, 233–238.
22 S. Sunder, G. D. Boyer and N. H. Miller, *J. Nucl. Mater.*, 1997, **244**, 66–74.
23 S. Sunder, N. H. Miller and D. W. Shoesmith, *Corros. Sci.*, 2004, **46**, 1095–1111.
24 M. M. Strehle, B. J. Heuser, M. S. Elbakhshwan, X. Han, D. J. Gennardo, H. K. Pappas and H. Ju, *Thin Solid Films*, 2012, **520**, 5616–5626.
25 S. D. Brown, L. Bouchenoire, D. Bowyer, J. Kervin, D. Laundy, M. J. Longfield, D. Mannix, D. F. Paul, A. Stunault, P. Thompson, M. J. Cooper, C. A. Lucas and W. G. Stirling, *J. Synchrotron Radiat.*, 2001, **8**, 1172–1181.
26 P. F. Fewster, *Rep. Prog. Phys.*, 1996, **59**, 1339.
27 L. G. Parratt, *Phys. Rev.*, 1954, **95**, 359.
28 M. Björck and G. Andersson, *J. Appl. Crystallogr.*, 2007, **40**, 1174.

Faraday Discussions

PAPER

Controlling factors in localised corrosion morphologies observed for magnesium immersed in chloride containing electrolyte

Geraint Williams,*[a] Nick Birbilis[b] and H. Neil McMurray[a]

Received 16th December 2014, Accepted 2nd February 2015
DOI: 10.1039/c4fd00268g

The early stages of localised corrosion affecting magnesium (Mg) surfaces when immersed in aqueous sodium chloride (NaCl) solutions involves the propagation of dark regions, within which both anodic metal dissolution and cathodic hydrogen evolution occur. For nominally "pure" Mg, these dark areas can either take the form of discs which expand radially with time, or filiform-like tracks which lengthen with time. For Mg surfaces which display disc-form corrosion features in concentrated NaCl electrolyte, a transition to filiform corrosion (FFC) is observed as the concentration is decreased, indicating ohmic constraints on radial propagation. A similar effect is observed when Mg specimens of different iron impurity are immersed in a fixed, high concentration NaCl solution, where disc-form corrosion is observed on samples having ≥280 ppm Fe, but FFC predominates at ≤80 ppm Fe. An *in situ* scanning vibrating electrode technique (SVET) is used to determine current density distributions within the propagating corrosion features. Cathodic current density values of between −100 and −150 A m^{-2} measured in central areas of disc-like features are sufficient to sustain the radial growth of a local anode at the perimeter of the discs. However, for high purity Mg specimens (≤80 ppm Fe), cathodic current densities of −10 A m^{-2} or less are measured over FFC affected regions, indicating that linear propagation arises when there is insufficient cathodic current produced on the corroded surface to sustain radial growth. The results are consistent with surface control of localised corrosion propagation in concentrated electrolyte, but ohmic control in dilute, lower conductivity NaCl solution.

Introduction

Localised corrosion affecting magnesium (Mg) and its alloys is particularly visually striking since it involves the evolution of black areas which grow with time and eventually consume the entirety of the original lustrous surface.[1–5] Although

[a]Materials Research Centre, College of Engineering, Swansea University, Singleton Park, Swansea, SA2 8PP, UK. E-mail: Geraint.Williams@swansea.ac.uk
[b]Department of Materials Engineering, Monash University, Clayton, Victoria VIC 3800, Australia

numerous publications are devoted to the subject of Mg corrosion, comprehensively reviewed elsewhere,[6–13] studies of surface morphology changes during corrosion have received relatively little attention compared with electrochemical and microscopic investigations. Nevertheless, recently published work has highlighted the significance of the propagating dark areas and the importance of the role that this transformation plays in determining the corrosion rate.[2–4] It is currently accepted that the dark corroded regions, which develop following breakdown of a partially-protective hydr(oxide) film formed upon initial immersion, show considerably enhanced electrocatalytic activity towards cathodic hydrogen evolution when compared with the uncorroded surface.[2–5,14] As a consequence, instantaneous Mg corrosion rates increase with immersion time as a progressively greater area of the original intact exposed surface becomes converted to the darkened, corroded form. Recent studies have shown that this effect is particularly evident under circumstances where the Mg surface is anodically polarised.[4,5,14–17] Progressively higher rates of hydrogen evolution, observed with increasingly positive potentials or higher applied anodic currents, are explained on the basis of a more rapid conversion of the uncorroded Mg to a dark cathodically enhanced surface.[4] In turn, higher growth rates of cathode areas and an associated increase total cathodic current preclude the need to explain the electrochemical behaviour of dissolving Mg surfaces in terms of a contentious mechanism involving an Mg^+ intermediate.[18–23]

In publications where observations on the evolution of Mg or Mg alloy corrosion morphologies are discussed, there appear to be considerable variations in the type of dark features which characterise localised corrosion propagation. Observations of dark, thread-like tracks, which evolve hydrogen and lengthen with time, are described for pure Mg samples immersed in dilute chloride-containing aqueous solutions.[24,25] For a selection of magnesium alloys, including Mg–Al–Zn,[26,27] Mg–Zn–Y,[28] Mg–Li[29] and AZ31B,[30] filiform corrosion (FFC) is reported under conditions of significantly higher chloride ion concentration. However, other investigations involving both pure Mg[2,5,20] and Mg–Nd alloys[31] describe the evolution of dark disc-like areas which expand with time and evolve hydrogen from within their interiors. In this work we seek to elucidate the underlying reasons which dictate localised corrosion behaviour for unpolarised Mg specimens using two different types of approach. In the first a commercially available Mg sample type, which displays disc-form evolution of localised corrosion in concentrated aqueous sodium chloride solution, is subjected to a combination of *in situ* scanning vibrating electrode and close-up time-lapse photography to investigate the influence of electrolyte concentration on localised corrosion morphology. Changes in time-dependent current density distributions associated with the transition from disc-like to filiform corrosion, observed with decreasing sodium chloride concentration, are used to explain the observed transformation in morphology. In the second approach, the same combination of techniques is used in a fixed, high concentration NaCl(aq) electrolyte to explain the change from disc-form to FFC morphology observed for Mg specimens containing varying iron impurity composition. The magnitude of local cathodic current density measured over dark-corroded regions is used to develop theories which demonstrate ohmically limited propagation in dilute NaCl electrolyte and cathodically controlled propagation in concentrated solution.

Experimental details

Materials

Commercial purity (CP) magnesium foil (99.9+%, 25 mm square coupons of 1 mm thickness) was obtained from Goodfellow Metals Ltd, with the main impurities being iron (280 ppm) and manganese (170 ppm). Alternative higher purity Mg specimens were sourced elsewhere. High purity Mg in the form of a section of 10 mm diameter rod, containing a 40 ppm iron impurity level, was supplied by AMAC (Melbourne, Aus), while a Mg specimen comprising 80 ppm Fe was provided as a gift by Prof A. Atrens (University of Queensland). A Mg–Fe alloy of roughly 1% w/w composition was prepared by melting commercially pure Mg with fine Fe powder in a crucible at 700 °C held in a resistance furnace using AM-Cover® (0.2 wt% tetrafluoroethane and 99.8 wt% nitrogen) as a cover gas.[32] The mixture was held in the molten state for several days with intermittent vigorous stirring. Finally, the mixture was homogenised just below the melting point for 3 days. An exact composition of 1.34% w/w for the Mg–Fe alloy was determined independently *via* ICP-AES (by Spectrometer Services, Coburg, Aus).

Methods

Scanning vibrating electrode measurements were carried out using a probe comprising a 125 μm diameter platinum wire sealed in a glass sheath so that the active portion of the probe tip consisted of a 125 μm diameter Pt microdisc electrode and the total tip diameter was ~250 μm. The probe vibration frequency was 140 Hz and the peak-to-peak vibration amplitude was 30 ± 5 μm. Full details of the SVET instrument design, mode of operation and calibration are given elsewhere.[2,33] In this work, the SVET peak-to-peak voltage signal was converted to values of current flux density along the axis of probe vibration (j_z), after galvanostatically checking the calibration using a two-compartment cell containing the relevant NaCl(aq) electrolyte, again as described previously.[2]

For SVET analysis, the various Mg specimens under investigation were cut into conveniently sized coupons, subjected to grinding using silicon carbide paper and polished using an aqueous slurry of 5 μm polishing alumina. Finally the polished surfaces were washed with aqueous detergent and rinsed with distilled water followed by ethanol. Test samples were then prepared by covering the surface with 90 μm thick extruded PTFE 5490 tape (3M Ltd), such that a typical 6 × 6 mm square area in the centre of one face of the coupon was exposed to electrolyte. The Mg samples were completely immersed, exposed area uppermost, in an electrolyte bath containing aqueous sodium chloride electrolyte of varying concentration in the 0.01 to 1 mol dm^{-3} range at pH 6.5. Electrolyte pH was adjusted using appropriate additions of aqueous hydrochloric acid (HCl) or sodium hydroxide (NaOH). Time lapse photography was carried out using a computer-controlled Nikon D70S digital SLR camera fitted with an AF-micro close-up lens. Time-dependent free corrosion potential measurements were made using a Solartron 1280B Electrochemical workstation used in conjunction with a saturated calomel reference electrode.

Results and discussion

Influence of NaCl solution concentration

A series of experiments carried out using CP-Mg (280 ppm Fe impurity) specimens immersed in neutral, aqueous NaCl(aq) electrolyte at concentrations in the range 10^{-2} to 1 mol dm^{-3}, showed significant changes in localised corrosion morphology with decreasing [NaCl]. Localised corrosion at $\geq$0.1 mol dm^{-3} was characterised by the appearance of black spots on the exposed Mg surface within minutes of immersion, which rapidly developed into continuously expanding dark disc-like features, evolving significant streams of rising hydrogen bubbles. Eventually the dark regions impinged on each other and consumed the entirety of the original intact (uncorroded) surface. A detailed description of the localised behaviour of this form of Mg in concentrated NaCl(aq) aqueous solution is given in a previous paper[2] and will not be discussed further in this work. Fig. 1 shows typical corrosion morphologies observed at different [NaCl], where the influence of a 2 order of magnitude concentration change can be seen. At 0.1 mol dm^{-3} the propagation of a hemispherical dark feature is observed, closely resembling the appearance of the corroded surface at higher [NaCl] (see Fig. 1a and b). However, a further 10-fold concentration decrease produced a transformation in the nature of localised corrosion observed, as shown (at a slightly higher magnification) in Fig. 1c. In this case, although several black spots marked the initiation of localised corrosion within 10 min of immersion, propagation was characterised by linear growth of dark areas, where continuously lengthening narrow *ca.* filament-like features, typically between 0.01 and 0.03 mm wide, traversed the exposed surface. Visible evidence of hydrogen evolution was observed over a distance of up to 0.4 mm behind advancing filament front edges.

From close up time-lapse photographic sequences recorded for the three immersion experiments summarised in Fig. 1, propagation rates for both disc-form and filament-like corrosion could be estimated. In the case of the radial expansion of dark corroded regions observed in 1 and 0.1 mol dm^{-3} NaCl(aq), the rates were 1.3×10^{-3} and 1.0×10^{-3} mm s^{-1} respectively. An average rate of advance calculated for a selection of 10 individual corrosion filaments propagating in 10^{-2} mol dm^{-3} electrolyte was also 1.0×10^{-3} mm s^{-1}. The kinetics of corroded area growth in 1 mol dm^{-3} NaCl(aq), has been discussed at length previously,[2] demonstrating that the proportionality of corroded area (A) with the square of time, such that a plot of $(A)^{1/2}$ *vs.* time was a straight line with a slope of 1.9×10^{-3} mm s^{-1}. For linear FFC propagation observed in 0.01 mol dm^{-3} NaCl(aq), the area growth is proportional to time, typically producing a propagation rate of *ca.* 3×10^{-3} mm^2 s^{-1}. As a result of the difference in corroded area growth rate laws, it is therefore difficult to make a direct comparison between disc-form corrosion and FFC in terms of the time-dependence of corroded area evolution. A full description of filiform corrosion (FFC) affecting CP-Mg under immersion in dilute NaCl electrolyte is given elsewhere[25] and will not be elaborated upon further here. The purpose of this work is therefore to explain how both types of corrosion morphology can be exhibited on the same type of Mg surface. To this end, the same set of immersion experiments were performed using *in situ*, repetitive SVET scanning to determine time-dependent changes in local current density patterns under free corrosion conditions. Typical SVET-derived surface

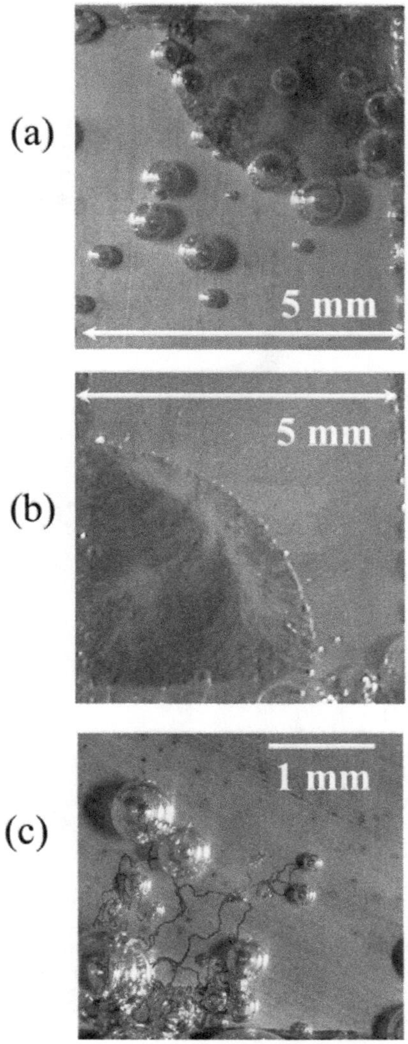

Fig. 1 Photographic images taken *in situ* of corroding CP-Mg surfaces held in (a) 1, (b) 0.1 and (c) 10^{-2} mol dm^{-3} NaCl(aq) at pH 6.5 after 40 min immersion.

maps of current density along the axis of probe vibration (j_z) recorded at 40 min following immersion in 1.0, 0.1 and 10^{-2} mol dm^{-3} NaCl(aq) at pH 6.5 are shown in Fig. 2a–c respectively.

The SVET-derived map in Fig. 2a for a CP-Mg specimen immersed in 1 mol dm^{-3} NaCl(aq) shows that the expanding disc-like dark areas comprise intense local anodes at their perimeter, with central regions becoming highly cathodically enhanced compared with the electrochemically inert, uncorroded surface. Previous work has demonstrated that disc-form corrosion propagation in concentrated NaCl(aq) is cathodically controlled, where anodic j_z values measured at corroded/intact region boundaries increase in proportion to the radius of local cathode regions within the expanding discs.[2] Although the experiment carried out in 0.1 mol dm^{-3} NaCl(aq) again shows similar localisation of anodic and cathodic

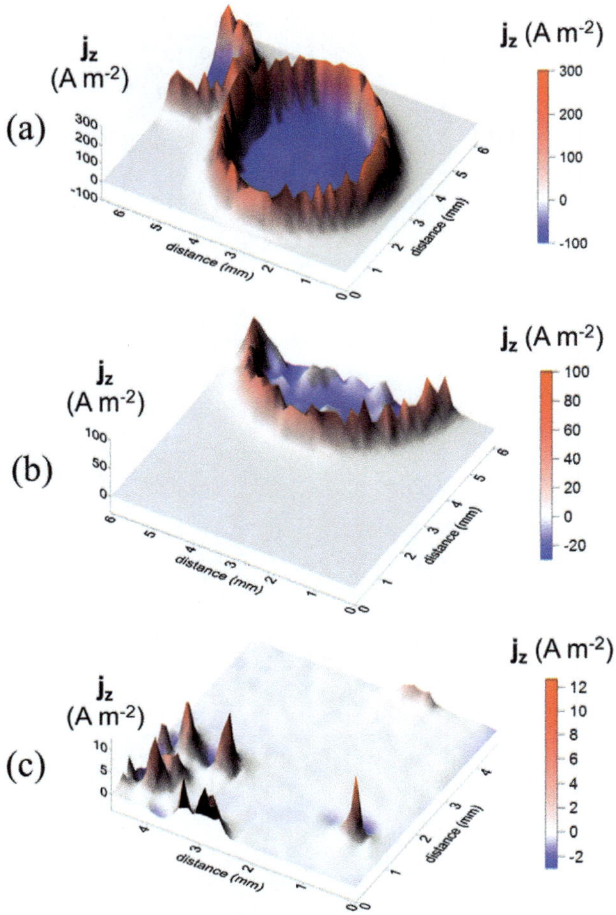

Fig. 2 Surface plots showing the distribution of normal current density j_z above CP-Mg specimens freely corroding in (a) 1.0, (b) 0.1 and 10^{-2} mol dm^{-3} NaCl(aq) after 40 min immersion.

activity, there is a distinct change in behaviour observed by *in situ* SVET analysis. Initial scans, carried out within 30 min of immersion, again showed that local anodic activity was constrained to the perimeter of an expanding corrosion feature with intense cathodic activity located within its centre. However, at longer immersion times it became apparent that some local anodic activity had developed in a region of exposed blackened surface furthest from the anodic front. This is demonstrated in Fig. 2b, where the corroded area on the upper right side of the j_z surface map shows net cathodic activity immediately behind the advancing anodic front, along with several anodic regions located a further distance to the rear. In addition, SVET mapping also revealed that the anodic intensity of the corrosion front did not vary with time as a progressively greater area of dark corroded surface was revealed. This is clearly demonstrated in the series of j_z *versus* distance profiles, taken along the radius of a propagating corrosion feature at different times following immersion in 0.1 mol dm^{-3} NaCl(aq), given in Fig. 3. The magnitude of peak anodic j_z remains unchanged at

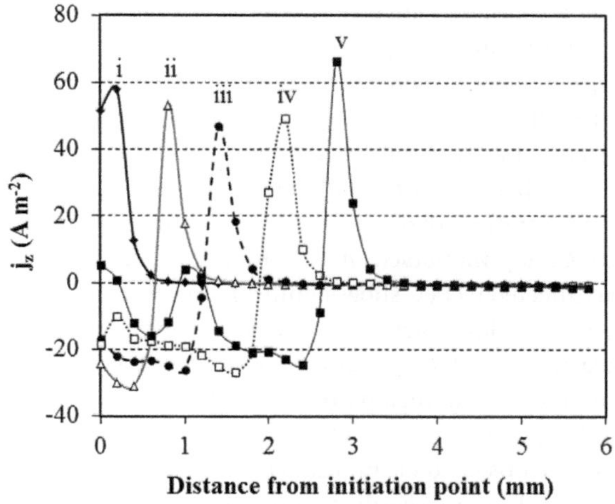

Fig. 3 Normal current density j_z profiles taken along the radius of an expanding semi-circular corrosion feature on unpolarised CP-Mg immersed in 0.1 mol dm^{-3} NaCl(aq) at pH 6.5. Time key: (i) 6 min, (ii) 18 min and at 12 min intervals thereafter.

ca. +60 A m^{-2}, while plots (ii)–(iv) show clearly that the area of maximum cathodic current density of *ca.* −30 A m^{-2} is located immediately behind the 0.5 mm wide anodic front. Fig. 3 also shows that a region immediately in front of the initiation point, which is shown as a strong local cathode over the first 40 min of immersion, becomes transformed into a net local anode as the anodic front moves further forward (see Fig. 3, curve (v)). This suggests that localised corrosion within this zone occurs independently of the apparent ongoing galvanic coupling of the anode at the disc perimeter with a *ca.* 2 mm wide cathodic zone immediately behind.

This observation of the onset of local anodic activity at protracted immersion times within a previously cathodic central region implies that decreased electrolyte conductivity is limiting the distances over which galvanic coupling can occur. As a consequence, the site of primary cathodic activity is thought to become constrained within an inner annular region concentric with the local anode maintained at the circumference of the expanding corrosion feature. This in turn differs from the case of the same Mg surface corroding in 1 mol dm^{-3} NaCl(aq), where the entirety of the interior of the expanding discs remain strong local cathodes over a similar period.[2]

These local anode areas visualised at protracted immersion times appear to be stationary and do not propagate in the same manner to the unified front of anodic activity observed initially. A closer investigation of the original site of localised corrosion initiation for CP-Mg immersed in 0.1 M NaCl(aq), shows that the cathodic current density value at this location declines with time at a rate of 0.8 A m^{-2} min^{-1}, from a starting value of *ca.* −30 A m^{-2}, reaching zero after 40 min immersion. After this point a local anode develops at this location, which grows in intensity with time, reaching a relatively constant value of +20 A m^{-2} at longer immersion times. The rate of expansion of the principal semi-circular corrosion feature remains constant, despite the propagation of local anodes close to the original point of initiation. This implies that these features have no bearing on the anode–cathode coupling which produces the expanding anodic area at the intact-corroded area boundary. Again this is consistent with limited electrolyte throwing power, where ohmic constraints mean that regions which are separated by several mm from the main anodic region corrode independently of each other, where galvanic coupling of local anodes and cathodes can only occur over distances which are dictated by the bulk electrolyte conductivity. For this intermediate NaCl concentration, these distances appear to be in the order of 1–2 mm.

In the case of CP-Mg immersed in 10^{-2} mol dm^{-3} NaCl(aq), Fig. 2c demonstrates that areas affected by FFC show significant localisation of corrosion, where the leading edges of the dark tracks are revealed as regions of intense focal anodic activity with measured j_z values of between +5 and +10 A m^{-2}. As the corrosion tracks propagate and anodic regions move away from their point of initiation, they leave behind regions of net cathodic activity with measured j_z values of up to −3 A m^{-2}. Again, the observed electrocatalytic activation of the corroded surface towards hydrogen evolution, according to reaction (1), is consistent with behaviour observed in higher concentration NaCl electrolyte.

$$2H_2O + 2e^- \rightarrow H_2 + 2OH^- \quad (1)$$

However, here the similarity ends, and the propagation of localised corrosion features involves linear growth of narrow dark tracks, rather than circular expansion of the black corroded surface. Although Fig. 1c shows a "random-walk" type of propagation of filaments over the exposed Mg surface, where an individual track will change direction many times during the course of its advance, there are indications that the directionality is not stochastic. The anodic front edge of a filament cannot re-tread the same path back along its length because of a combination of elevated pH and impurity element enriched surface. There is also evidence in the literature that FFC propagation on Mg in 0.01 mol dm^{-3} NaCl(aq), albeit in the presence of a dilute chromate addition to the corrosive electrolyte, preferentially follows certain crystallographic features of the Mg grains and as a consequence the observed filament directionality is strongly influenced by changes in orientation at grain and twin boundaries.[24]

The principal question to be asked is why the morphology should change so markedly at this lower concentration of corrosive electrolyte. Since it has been demonstrated that the corroded surface is capable of sustaining cathodic current densities which can be as high as −150 A m^{-2} in more concentrated aqueous NaCl solution,[2] then the explanation must lie in the decreased conductivity of the electrolyte which limits the distance over which galvanic coupling can occur. Fig. 4 shows a schematic representation of the proposed development of a circular corrosion feature after the initiation of localised corrosion.

Previous work has shown that black spots, which mark the points of localised corrosion initiation, are intense, highly focal net anodes.[2,4,25] However, within a short space of time, the local anode moves away from its initiation point, leaving behind a cathodically activated region (see Fig. 4a). Forward motion of the local anode is driven by electrocatalytic activation of the previously anodically-attacked surface with respect to reaction (1), while the local corrosion cell is maintained by ionic current flowing in the electrolyte *via* a short path indicated by arrow S. In order to propagate radially, the anode must be capable of moving both forward and laterally as indicated in Fig. 4b. In highly conductive electrolyte, this is made

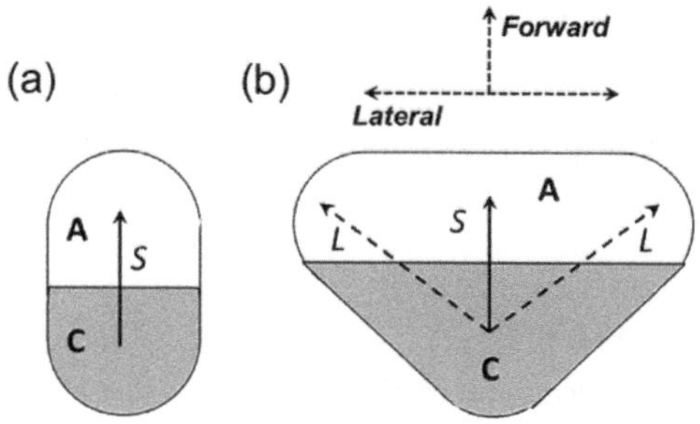

Fig. 4 Schematic representation of the possible stages in the post-initiation evolution of localised corrosion on CP-Mg to give disc-form propagation. Local anode and cathode regions are labelled A and C, respectively.

possible by the ability of ionic current to flow along significantly longer paths, indicated by the arrows marked L. In turn, this allows the local anode to surround the cathodic site as shown in Fig. 5a while simultaneously advancing away from the initiation site, leading to the formation of a circular corrosion feature comprising a cathode interior coupling with an anodic perimeter. However, in the case of dilute NaCl(aq) electrolyte shown in Fig. 5b, where higher resistivity precludes the passage of ionic current over longer paths lengths L, only forward motion of the anode is possible, where the shortest possible ionic current path S completes the local corrosion cell coupling region C to region A. Additionally, in such a low throwing power electrolyte, only a relatively short length of the

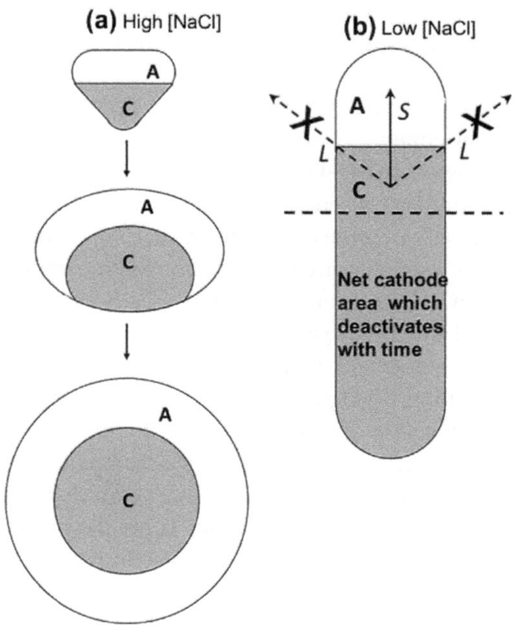

Fig. 5 Schematic representation of the development of localised corrosion features on CP-Mg in (a) high and (b) low concentration NaCl electrolyte.

filament immediately behind the local cathode will be actively coupling with the local anode at its front edge. Therefore as a corrosion filament advances, cathodic j_z values measured along its length some distance behind the leading edge would be expected to progressively decrease with time. This was confirmed experimentally in previously published work.[25]

The effect of iron impurity level

In a separate series of experiments carried out using a test electrolyte comprising concentrated NaCl(aq) at a fixed concentration of 1 mol dm^{-3}, the influence of iron impurity level on both the nature of the observed localised corrosion features and the corresponding current density distributions were investigated. In such electrolyte there should be no ohmic constraints on localised corrosion propagation and galvanic coupling of local anodes and cathodes should be possible over significant distances. Four different types of iron containing Mg specimens were used for this study, spanning a composition range of 40 ppm up to 1.3×10^4 ppm. From preliminary immersion experiments, it became evident that Mg specimens containing ≥280 ppm Fe showed disc-form propagation of the dark corroded surface, while samples containing an iron impurity level below the commonly quoted high corrosion rate threshold of 170 ppm,[34] (in this case two types of high purity Mg containing either 40 or 80 pm Fe) exhibited only filiform-like corrosion. Rapid radial expansion of dark corrosion discs for specimens containing either 1.3% w/w or 280 ppm Fe impurity led to the complete conversion of the original 36 mm^2 exposed area of uncorroded surface to a black, highly corroded surface within 90 min of immersion. The rate is marginally slower than that observed for CP-Mg (280 ppm Fe) specimens, where the entirety of the original surface was typically consumed within 60 min. In contrast, specimens comprising both 40 and 80 ppm Fe showed significantly slower rates of corrosion propagation as estimated by the area percentage of the total area affected by dark, corrosion tracks. Even after a maximum immersion period of 24 h, only *ca.* 75% of the original intact working area had been affected by FFC activity for both types of low Fe impurity Mg specimen.

In situ SVET analysis of the Mg–Fe (1.3%) specimen immersed in 1 mol dm^{-3} NaCl(aq), confirmed similar characteristics to the localised corrosion propagation of CP-Mg (280 ppm Fe) described previously and in detail elsewhere.[2] As demonstrated in Fig. 6, a roughly circular corrosion feature, radially expanding at rate of 0.8×10^{-3} mm s^{-1} and characterised by a cathodic interior and anodic perimeter, is observed on the exposed surface. Peak j_z values of −150 A m^{-2} measured within the cathodically enhanced zone are roughly similar to those established on a corroding CP-Mg surface, despite the 50 fold decrease in iron impurity weight percentage. As observed previously for disc-like propagation, the expansion of the dark corroded region is again accompanied by a progressive increase in anodic j_z values measured at the intact-corroded surface boundary. However, despite the similarities in behaviour, some distinct differences were also observed, most notably the presence of a persistent permanent cathode seen in the lower quarter of Fig. 6a and b, which was evident from the point of initial immersion. An unusual characteristic of local cathode, which evolved a permanent stream of hydrogen bubbles, was that the enhanced cathodic activity was not associated with a region of dark corroded surface and that this particular local

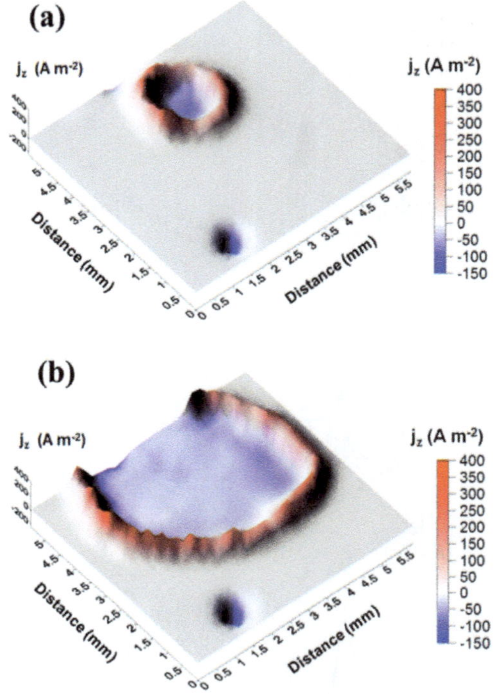

Fig. 6 SVET-derived j_z surface maps showing the evolution of localised corrosion on a Mg–Fe (1.3% w/w) alloy after immersion in 1 mol dm^{-3} NaCl(aq) at pH 6.5 for (a) 15 and (b) 50 min.

cathode remained uncorroded at all times ≤60 min. Although post-corrosion chemical analysis was not carried out, it is proposed that this persistent cathode may have been associated with a region of surface heterogeneity comprising a large number of iron rich particles.

For the highest purity Mg specimen (40 ppm), which was supplied as a section of 10 mm diameter rod set in epoxy resin, the initiation and subsequent propagation of localised corrosion in 1 mol dm^{-3} NaCl(aq), visualised by *in situ* SVET analysis, is summarised in Fig. 7. The principal features are similar to those previously published,[35] but under immersion in more dilute (0.1 mol dm^{-3}) electrolyte, where the point of breakdown after 30 min (Fig. 7a) is revealed as an intense local anode, with the remainder of the uncorroded surface weakly cathodic (j_z up to −1 A m^{-2}). This local anode moves away from its initial location and leaves behind a cathodically activated blackened track, showing significantly higher cathodic j_z values of up to −10 A m^{-2} compared with the intact surface. With time, more local anodes initiate and propagate in this vicinity of the exposed surface, eventually producing a network of dark filaments where the majority of the corroded surface is cathodically enhanced and anodic activity is limited to a few regions at the leading edge of these tracks (see Fig. 7b and c).

The presence of a slightly higher Fe impurity level (80 ppm) did not greatly influence localised corrosion characteristics, as confirmed by Fig. 8 which shows a j_z distribution map obtained after 12 h immersion in 1 mol dm^{-3} NaCl(aq). A comparison with Fig. 7c shows that both types of Mg specimen show a similar extent of corroded area over the same period of immersion. In addition, the

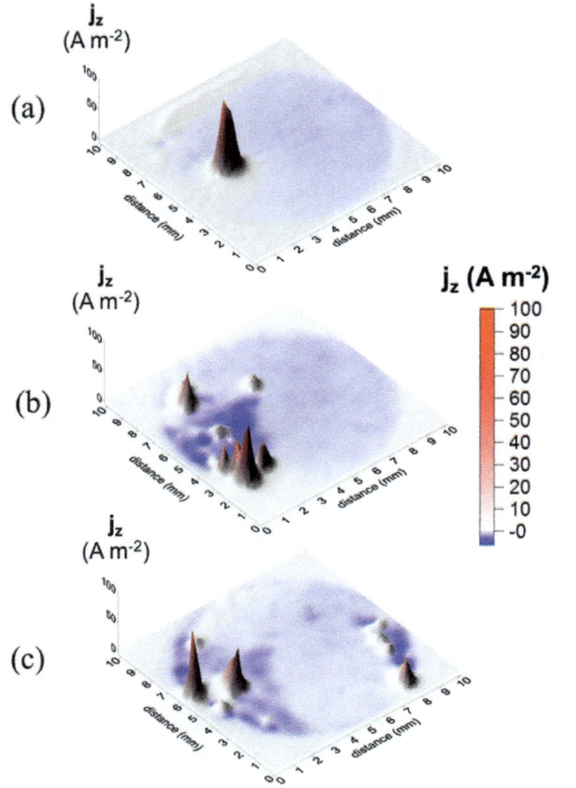

Fig. 7 SVET-derived j_z surface maps showing the evolution of localised corrosion on a high purity (HP) Mg sample (40 ppm Fe) after immersion in 1 mol dm^{-3} NaCl(aq) at pH 6.5 for (a) 30 min, (b) 6 h and (c) 12 h.

magnitude of current density values measured for both anode and cathode areas on both type of surface were also similar, with j_z values of up to +60 and -10 A m^{-2} established for local anode and cathode regions, respectively.

From these results it is evident that the magnitude of the cathodic current density which can be sustained on the dark, corroded surface is the main controlling factor in the propagation of localised corrosion on Mg surfaces in concentrated corrosive electrolyte. High localised corrosion rates are observed on specimens having an iron impurity ≥280 ppm, which exhibit disc-form propagation, where current density values of up to -150 A m^{-2} can be maintained on the corroded surface. In contrast, localised corrosion propagation on high purity Mg, comprising Fe impurity levels well below the commonly-quoted 170 ppm threshold, take the form of linear filaments which advance at a significantly lower rate and can only sustain maximum cathodic current density values of -10 A m^{-2}. The large number of SVET-derived j_z surface maps of the types shown in Fig. 6–8 can be effectively condensed by the estimation of time-dependent, area-averaged anodic current density (Ja_t) as described in previous publications.[2,4,36,37] This is achieved through the numerical area integration of j_z distributions using an approach outlined elsewhere[2,4,36,37] giving a description of the time-dependent evolution of localized corrosion currents in a compact manner. Fig. 9 shows plots

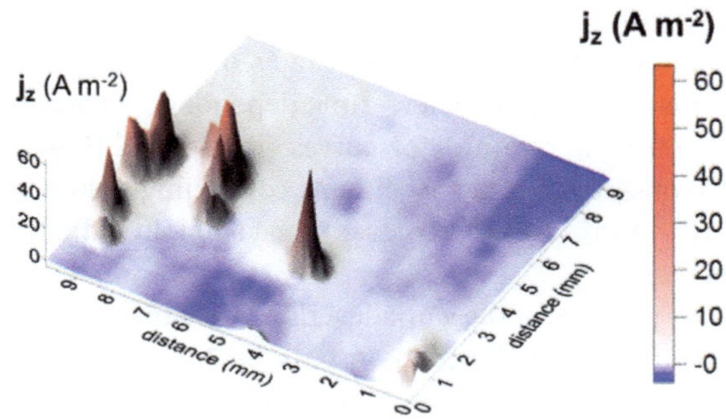

Fig. 8 SVET-derived j_z surface map recorded after 12 h immersion in 1 mol dm^{-3} NaCl(aq), showing localised corrosion patterns on an Mg sample containing 80 ppm Fe impurity.

of Ja_t *versus* time, comparing the behaviour in 1 mol dm^{-3} NaCl of the various Mg specimen types comprising Fe impurity levels in the range 40 ppm up to 1.3% w/w. Ja_t is seen to increase rapidly with immersion time following breakdown for Mg specimens comprising ≥280 ppm (Fig. 9, plots (i) and (ii)), rising from a pre-breakdown value of 1 to *ca.* 40 A m^{-2} after 60 min immersion. The subsequent drop to lower values approaching 10 A m^{-2} occurs because of a transition from localised to more general corrosion as the entirety of the working area is converted to a dark corroding surface. The limitations of SVET in detecting corrosion events under situations where anode–cathode spacing becomes indistinct is discussed at length elsewhere.[37]

However, in the case of both high purity Mg samples (≤80 ppm Fe), initial post-breakdown Ja_t values are an order of magnitude lower and rise significantly more slowly with time as FFC propagates over the exposed Mg surfaces, reaching

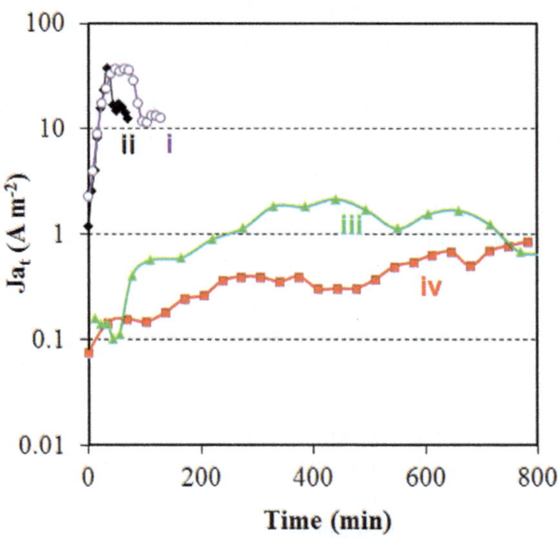

Fig. 9 Area averaged integrated anodic current (Ja_t) plotted as a function of immersion time in 1 mol dm^{-3} NaCl(aq) for Mg specimens containing (i) 1.3 × 10^4, (ii) 280, (iii) 80 and (vi) 40 ppm Fe impurity.

ca., 1 A m^{-2} after periods of immersion of greater than 6 h. Fig. 9 also confirms qualitative observations of greatly increased rates of corrosion at iron compositions of greater than 170 ppm, where time-dependent area-averaged current values rise at *ca.* 1.0 A m^{-2} min^{-1} for curves (i) and (ii). However, the corresponding increase observed for Mg specimens containing ≤80 ppm Fe impurity was roughly 5×10^{-3} A m^{-2} min^{-1}, representing a 500 fold reduction in localised corrosion rate increase.

It is not the main purpose of the work described here to establish why cathodic activation of the dark corroded regions occurs on Mg surfaces, or to offer an explanation about the controlling influence of iron impurity levels on the magnitude of cathodic current density values which can be sustained on these regions. Rather it is to point out that the magnitude of the latter dictates the morphology of localised corrosion propagation, which in turn controls the rate of corrosion. In a situation where ionic current can flow freely between local anode and cathode sites, cathodic current density values of −100 A m^{-2} or higher measured for low purity Mg supports circular propagation of the corroded surface, which in turn produces high rates of localised corrosion. In contrast, the low cathodic current density values of <−10 A m^{-2}, measured on the dark corroded surface produced on high purity Mg surfaces, are insufficient to sustain radial growth even in high conductivity electrolyte, and consequently only linear propagation of narrow dark corrosion tracks is observed, which in turn produces relatively low rates of localised corrosion. These observations on the influence of the magnitude of the iron impurity level tend to support theories which invoke iron-rich particle enrichment as the source of the observed cathodic activation of previously anodically-attacked Mg surface.[2,5,14,16,38]

Influence of solution pH

Finally, it was demonstrated using CP-Mg (280 ppm Fe), that a transition from disc-form to linear corrosion propagation could also be produced in 1 mol dm^{-3} NaCl(aq) by increasing the bulk pH. Experiments performed at bulk pH values of 10 and 12, showed very little change from neutral electrolyte, where disc-form propagation of dark regions was observed, with relatively little decrease in the rate of radial expansion of the localised circular corrosion features. *In situ* SVET analysis again confirmed typical behaviour characterised by the presence of intense local anodes surrounding the discs, coupling with cathodically activated central interiors. At pH 12, typical j_z values measured for local cathodic regions were in the −50 to −80 A m^{-2} range, which is noticeably lower than values determined at pH ≤ 10. However, despite being lower at pH 12, the measured current density for cathodic hydrogen evolution remained sufficient to sustain disc-like corrosion growth, up to a point where the entire of the original "silvery" exposed surface became blackened.

A marked change in behaviour was observed when the bulk pH of the 1 mol dm^{-3} NaCl(aq) electrolyte was further raised to 13. Before immersion experiments were carried out, the pH 13 electrolyte was purged with synthetic air to remove any influence of dissolved carbon dioxide under conditions of high alkalinity. Despite the known increase in passivity of Mg produced at high pH,[39] breakdown of the CP-Mg surface was observed within 6 min of immersion, characterised by the appearance of a small black spot on the exposed surface from which a stream of fine H_2 bubbles was evolved.

In situ SVET analysis of the corroding surface showed that several regions of localised corrosion initiated over the first hour of immersion, but remained active only temporarily. However, after this period a dark region on the left edge of the exposed surface, as shown in Fig. 10 developed into a propagating corrosion filament of roughly 0.5 mm width. As the SVET-derived j_z maps given in Fig. 10a and b demonstrate, the propagating track comprises an intense local anode at its front edge, with the reminder of the track revealed as cathodically activated with respect to the intact (uncorroded) surface. As the track lengthens, there is no evident change in the anodic current density measured at its leading edge. This is confirmed in Fig. 11, which shows profiles of current density *versus* distance taken along the length of the propagating corrosion filament at different immersion times. Peak anodic j_z is reasonably constant at *ca.* +180 A m^{-2}, while cathodic values of up to -20 A m^{-2} are typically highest in a region immediately behind the advancing anode (see Fig. 11, curve (iii)). A velocity of 10^{-3} mm s^{-1} was estimated for the propagating track shown in Fig. 10, which is entirely similar to values determined for localised corrosion features observed on CP-Mg immersed at pH 6.5 in NaCl(aq) electrolyte at concentrations of ≤0.1 mol dm^{-3}. A final question in

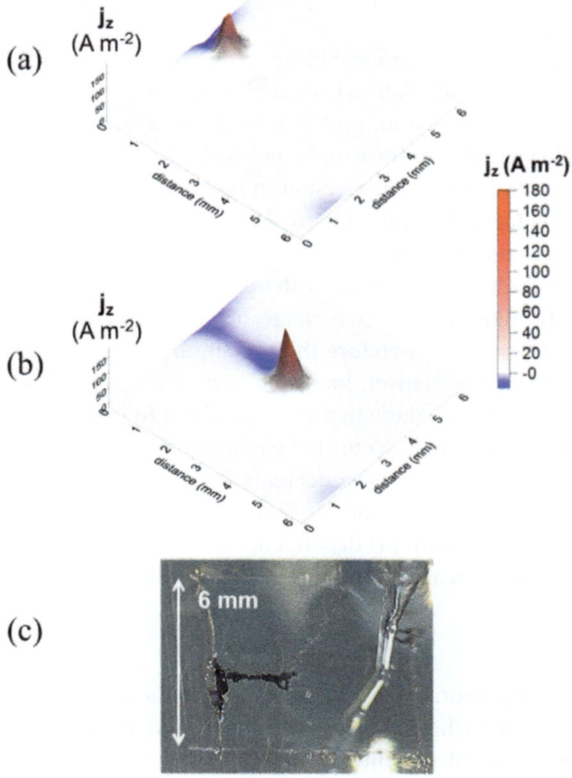

Fig. 10 Surface plots showing the distribution of normal current density j_z above a CP-Mg specimen freely corroding in 1 mol dm^{-3} NaCl(aq) at pH 13, recorded after (a) 60 and (b) 90 min immersion. (c) Shows a photographic image of the corroding surface taken *in situ* after 2 h.

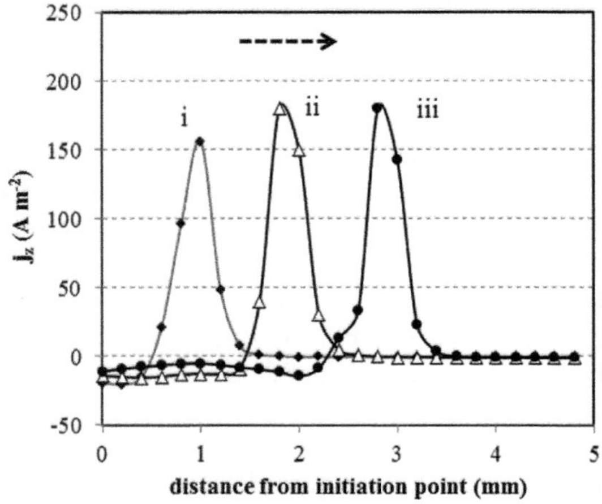

Fig. 11 Normal current density j_z profiles taken along the radius of a propagating corrosion track on unpolarised CP-Mg immersed in 1 mol dm^{-3} NaCl(aq) at pH 13. Time key: (i) 60 min, (ii) 72 min and (iii) 84 min. The arrow indicates the direction of propagation.

this case arises about why radial propagation occurs for CP-Mg in 1 mol dm^{-3} NaCl solution at near neutral values up to pH 12, but transforms to filiform-like propagation at pH 13.

If the observed change in morphology is due to the greater passivity of the uncorroded Mg surface at high pH, then this would be expected to affect the initiation of localised corrosion, probably by significantly lengthening the time delay to breakdown under unpolarised conditions. However, the fact breakdown is observed within minutes of immersion in the absence of external polarisation suggests that this facet of localised corrosion is relatively unaffected by elevated pH and rather it is the propagation phase which is predominately influenced. In this respect, this finding is consistent with observations of the localised corrosion behaviour of Mg alloy AZ31 over a range of bulk pH values in concentrated aqueous NaCl electrolyte.[3] Therefore the transition to linear propagation again seems to derive from the relatively low (up to −20 A m^{-2}) cathodic current that is generated on the corroded, darkened surface, which in turn is below the empirically derived threshold which seems to be required to sustain radial growth over significant distances. A progressive decrease in mean cathodic current density values determined at pH values of 10, 12 and 13 of −140, −80 and −15 A m^{-2} respectively is consistent with pH dependence of the electrochemical potential of the hydrogen evolution reaction (1) with pH given below.[39]

$$E_{H_2} = E^{\circ} - 0.0591\text{pH} \tag{2}$$

The open circuit potential of CP-Mg was shown to be unaffected by an increase in electrolyte pH to 13, where E_{corr} was *ca.* 40 mV more positive than the value of −1.4 V *vs.* SHE measured in neutral 1 mol dm^{-3} NaCl(aq). If it is assumed that local cathode sites are polarised to −1.36 V *vs.* SHE then a progressive increase in pH will lead to a decrease in the magnitude of cathodic polarisation for H_2 evolution from eqn (2). Consequently, it can be calculated that the overpotential for cathodic HER at pH 13 is reduced by 0.42 V compared with neutral conditions.

This appears to be sufficient to markedly de-activate cathodic hydrogen evolution on iron-impurity enriched, corroded regions.

Conclusions

The localised corrosion of unpolarised Mg specimens, comprising varying iron impurity levels, shows propagating features which appear to be governed by the magnitude of the cathodic current density which is maintained on the dark, anodically-attacked surface. *In situ* SVET was used to demonstrate that circular expansion of dark corroded regions in concentrated NaCl(aq) solution occurred under circumstances where cathodic current densities of -100 A m^{-2} or higher were measured within a central portion of a disc-like feature comprising a strong local circumferential anode. Such values were established on Mg specimens which contain iron impurity levels of $\geq$280 ppm and led to high rates of localised corrosion as quantified by numerical area integration of SVET-derived current density maps. In contrast, cathodic current density values of -10 A m^{-2} or lower were measured on the dark corroded surface produced on high purity Mg specimens immersed in the same corrosive electrolyte. These relatively low currents were shown to be insufficient to generate radial expansion of the corroded surface, and consequently only linear propagation to give filament-like features was observed. Unsurprisingly, rates of localised corrosion quantified by the magnitude of time-dependent, SVET-derived integrated anodic currents were over 2 orders of magnitude lower on Mg specimens containing $\leq$80 ppm Fe impurity levels.

In separate experiments using commercial purity Mg containing 280 ppm Fe impurity, the influence of NaCl electrolyte concentration on localised corrosion propagation was investigated. At concentrations $\geq$0.1 mol dm^{-3}, disc-form corrosion is observed, where the expansion of the local anode at the perimeter is sustained by cathode current density values of greater than -30 A m^{-2} within the interior of the disc. However, when the concentration is further decreased by an order of magnitude, only filiform corrosion (FFC) is observed, indicating ohmic constraints which produce linear rather than radial localised corrosion propagation. A similar effect is produced in 1 mol dm^{-3} NaCl(aq) by increasing the bulk pH to 13. In this case, where the concentrated electrolyte should not constrain ionic current paths, the transition from radial to linear propagation resulted from the relatively low cathodic current density of *ca.* -15 A m^{-2} which was measured over the corroded surface.

Acknowledgements

M. A. Gibson (CSIRO/CAST-CRC) is gratefully acknowledged for assistance in preparing the Mg–Fe alloy. Prof A. Atrens (University of Queensland) is thanked for the gift of high purity Mg containing an 80 ppm iron impurity level.

Notes and references

1 M. E. Straumanis and B. K. Bhatia, *J. Electrochem. Soc.*, 1963, **110**, 357.
2 G. Williams and H. N. McMurray, *J. Electrochem. Soc.*, 2008, **155**, C340.
3 G. Williams, H. ap Llwyd Dafydd and R. Grace, *Electrochim. Acta*, 2013, **109**, 489.

4 G. Williams, N. Birbilis and H. N. McMurray, *Electrochem. Commun.*, 2013, **36**, 1.
5 M. Curioni, *Electrochim. Acta*, 2014, **120**, 284.
6 G. L. Makar and J. Kruger, *Int. Mater. Rev.*, 1993, **38**, 138.
7 G. Song and A. Atrens, *Adv. Eng. Mater.*, 1999, **1**, 11.
8 G. Song and A. Atrens, *Adv. Eng. Mater.*, 2003, **5**, 837.
9 E. Ghali, W. Dietzel and K. U. Kainer, *J. Mater. Eng. Perform.*, 2004, **13**, 7.
10 G. Song, *Adv. Eng. Mater.*, 2005, **7**, 563.
11 R. Zeng, J. Zhang, W. Huang, W. Dietzel, K. U. Kainer, C. Blawert and K. E. Wei, *Trans. Nonferrous Met. Soc. China*, 2006, **16**, s763.
12 E. Ghali, *Corrosion resistance of Aluminium and Magnesium alloys*, John Wiley and Sons, New Jersey USA, 2010.
13 N. T. Kirkland, N. Birbilis and M. P. Staiger, *Acta Biomater.*, 2012, **8**, 925.
14 J. Światowska, P. Volovitch and K. Ogle, *Corros. Sci.*, 2010, **52**, 2372.
15 N. Kirkland, G. Williams and N. Birbilis, *Corros. Sci.*, 2012, **65**, 5.
16 N. Birbilis, A. D. King, S. Thomas, G. S. Frankel and J. R. Scully, *Electrochim. Acta*, 2014, **132**, 277.
17 G. S. Frankel, A. Samaniego and N. Birbilis, *Corros. Sci.*, 2013, **70**, 104.
18 J. W. Turrentine, *J. Phys. Chem.*, 1908, **12**, 448.
19 R. L. Petty, A. W. Davidson and J. Kleinberg, *J. Am. Chem. Soc.*, 1954, **76**, 363.
20 G. Song, A. Atrens, D. St John, J. Nairn and Y. Li, *Corros. Sci.*, 1997, **39**, 855.
21 A. Atrens and W. Dietzel, *Adv. Eng. Mater.*, 2007, **9**, 292.
22 T. R. Thomaz, C. R. Weber, T. Pelegrini, L. F. P. Dick and G. Knornschild, *Corros. Sci.*, 2010, **52**, 2235.
23 Z. M. Shi, J. X. Jia and A. Atrens, *Corros. Sci.*, 2012, **60**, 296.
24 P. Schmutz, V. Guillaumin, R. S. Lillard, J. A. Lillard and G. S. Frankel, *J. Electrochem. Soc.*, 2003, **150**, B99.
25 G. Williams and R. Grace, *Electrochim. Acta*, 2011, **56**, 1894.
26 O. Lunder, J. E. Lein, S. M. Hesjevik, T. K. Aune and K. Nisancioglu, *Mater. Corros.*, 1994, **45**, 331.
27 C. Liu, Y. Xin, G. Tang and P. K. Chu, *Mater. Sci. Eng., A*, 2007, **456**, 350.
28 S. Izumi, M. Yamasaki and Y. Kawamura, *Corros. Sci.*, 2009, **51**, 395.
29 Y. W. Song, D. Y. Shan, R. S. Chen and E. H. Han, *Corros. Sci.*, 2009, **51**, 1087.
30 J. R. Kish, G. Williams, J. R. McDermid, J. M. Thuss and C. F. Glover, *J. Electrochem. Soc.*, 2014, **161**, C405.
31 G. Williams, K. Gusieva and N. Birbilis, *Corrosion*, 2012, **68**, 489.
32 S. Simanjuntak, M. K. Cavanaugh, D. S. Gandel, M. A. Easton, M. A. Gibson and N. Birbilis, *Corrosion*, 2015, **71**, 199.
33 G. Williams and H. N. McMurray, *Corrosion*, 2006, **62**, 231.
34 R. E. McNulty and J. D. Hanawalt, *Trans. Electrochem. Soc.*, 1942, **81**, 423.
35 K. D. Ralston, G. Williams and N. Birbilis, *Corrosion*, 2012, **68**, 507.
36 G. Williams, H. N. McMurray and R. Grace, *Electrochim. Acta*, 2010, **55**, 7824.
37 G. Williams, A. J. Coleman and H. N. McMurray, *Electrochim. Acta*, 2010, **55**, 5947.
38 M. Taheri, J. R. Kish, N. Birbilis, M. Danaie, E. A. McNally and J. R. McDermid, *Electrochim. Acta*, 2014, **116**, 396.
39 *Atlas of Electrochemical Equilibria in Aqueous Solutions*, ed. M. Pourbaix, Pergamon Press, London, 1966.

Faraday Discussions

PAPER

Determination of the local corrosion rate of magnesium alloys using a shear force mounted scanning microcapillary method†

Philippe Dauphin-Ducharme,[a] W. Jeffrey Binns,[b] Michael E. Snowden,[a] David W. Shoesmith[b] and Janine Mauzeroll*[a]

Received 23rd December 2014, Accepted 16th January 2015
DOI: 10.1039/c4fd00276h

The successful development of scanning probe techniques to characterize corrosion *in situ* using multifunctional probes is intrinsically tied to surface topography signal decoupling from the measured electrochemical fluxes. One viable strategy is the shear force controlled scanning microcapillary method. Using this method, pulled quartz micropipettes with an aperture of 500 nm diameter were used to resolve small and large variations in topography in order to quantify the local corrosion rate of microgalvanically and galvanically corroded Mg alloys. To achieve topography monitoring of corroded surfaces, shear force feedback was employed to position the micropipette at a reproducible working height above the substrate. We present proof of concept measurements over a galvanic couple of a magnesium alloy (AE44) and mild steel along with a microgalvanically corroded ZEK100 Mg alloy, which illustrates the ability of shear force to track small (1.4 μm) and large (700 μm) topographic variations from high aspect ratio features. Furthermore, we demonstrate the robustness of the technique by acquiring topographic data for 4 mm along the magnesium–steel galvanic couple sample and a 250 × 30 μm topography map over the ZEK100 Mg alloy. All topography results were benchmarked using standard optical microscopies (profilometry and confocal laser scanning microscopy).

1 Introduction

Scanning probe techniques (SPTs) have found extensive applications in corrosion science as tools for quantitative and qualitative analyses that can tackle key issues including: the identification of the corrosion initiation sites, predicting the propagation of corrosion, the effectiveness of smart inhibition coatings and lifetime predictions.[1]

[a]*Department of Chemistry, McGill University, 801 Sherbrooke St W., Montreal, Quebec, H3A 0B8, Canada. E-mail: janine.mauzeroll@mcgill.ca; Fax: +1 514 398-6999; Tel: +1 514 398-3898*

[b]*Department of Chemistry and Surface Science Western, Western University, London, Ontario, N6A 5B7, Canada*

† Electronic supplementary information (ESI) available. See DOI: 10.1039/c4fd00276h

Specifically, scanning electrochemical microscopy (SECM) is a highly relevant electroanalytical method to quantify local chemical fluxes *in situ* and to map the heterogeneous electrochemical activity of corroding surfaces.[1–3] A normal SECM setup uses a microelectrode (ME) sensor, included in a four electrode configuration controlled by a bipotentiostat, attached to a 3-axis translator positioning system. The four electrodes required are two working electrodes: ME and substrate/sample, a counter electrode: CE, and a reference electrode: REF. In corrosion studies, SECM offers an arsenal of modes for the direct detection of ionic/molecular fluxes produced at the corroding interface,[3] to probe corrosion mechanisms,[4] coatings[5,6] and surface modifications.[7] Recently, the potentiometric,[8,9] feedback,[10,11] and generation–collection modes[10,12–14] of SECM have been applied to study the corrosion of Mg alloys which, given their light-weight properties and low cost, are of interest for automotive applications.[15]

An important challenge when performing such SECM measurements is the ability to decouple and resolve variations in topography in order to quantify local corrosion rates. In comparison to the detection of electrochemical fluxes, topography sensing is less sensitive but can still provide local corrosion rates that must be in agreement with the chemical fluxes measured at the ME. Simultaneous topography tracking therefore offers an internal data validation. Several strategies have been developed to decouple topography from surface reactivity.[16] Lee *et.* al.[17] attached a tuning fork to a gold ring disk ME, and monitored the mechanical oscillations of this assembly with a pulsed laser. Tuning fork control loops have been successfully applied to complex biological samples[18,19] with topographical variations in the order of 15 μm. Combined atomic force microscopy (AFM)-SECM probes have also been fabricated to assess surface topography along with electrochemical reactivity.[20–22] A maximal depth of 2 μm was measured in these studies. Another strategy involves alternating current scanning electrochemical microscopy (AC-SECM) that relies on impedance measurements to monitor the tip-to-substrate separation.[4] Using AC-SECM, hard substrates, such as corroding copper,[23] Al alloys,[24] stainless steel[4] and substrates with easily deformed features, *e.g.* cells,[25] of less than 20 μm have been imaged. Finally, shear force (SF) control loops are also available and were originally implemented as a method to maintain a constant tip-to-substrate separation during a SECM scan[26] using optical detection.[27] Nowadays, two piezoelectric elements are attached to the shaft of a SECM probe.[28,29] An alternating potential is applied to the piezo furthest from the ME tip (stimulation piezo) which induces a mechanical oscillation of the ME. The piezo element closest to the ME tip (receiver piezo) monitors the displacement of the ME as an alternating voltage having a characteristic amplitude and phase.[30,31] When the tip-to-substrate distance is less than a few hundred nanometers, the oscillation of the ME is damped by shear forces between the ME tip, solution and substrate surface. The resulting amplitude damping and phase shifts of the receiver piezo are used in a feedback loop in order to maintain a constant tip-to-substrate separation. The overall geometry of the ME significantly affects the SF response,[32] thus requiring the use of a reproducible ME fabrication procedure to ensure comparable SF signals.[29,33] Optimization of the SF conditions has recently allowed height differences of 200–250 μm to be measured whilst simultaneously recording electrochemical data.[34]

To track the extensive depth profiles and accurately quantify bulk and local corrosion rates observed at corroded Mg materials, topography tracking methodologies with an extended range must be developed for SECM applications. In the specific case of pure Mg and its alloys that have standard electrochemical potentials well below −1.5 V *vs.* SCE,[35] severe corrosion rates are expected (in the order of ~10–1000 mm per year[36] when galvanically coupled (depending on the other material) and <10 mm per year[12]as a standalone material), which impedes SECM studies. Weight loss[37,38] and H_2 collection[39] have been employed to measure bulk corrosion rates. They provide absolute amounts of corroded material but lack the ability to produce spatially resolved height profiles of the corroded substrate required in SECM. Microscopy techniques, such as confocal laser scanning microscopy[40] (CLSM) and laser-based profilometry,[41] have also been used to track local corrosion damage but would be difficult to interface with SECM, suffer from diffraction artifacts when a significant amount of corrosion product is present, and have a limited temporal resolution given their stop-go style experiments. Other alternatives include the scanning vibrating electrode technique (SVET), which measures local current densities *in situ* and approximate local corrosion rates using Faraday's law[41] and the scanning droplet cell method (SDC)[42] where the local corrosion rate measurement uses Tafel extrapolation of potentiodynamic polarization curves (PDPs).

We have prioritized the development of shear force (SF) probes because: (i) they have demonstrated excellent topographic performance in several media[43] and over a range of different samples;[19,44,45] (ii) the method is easily adaptable to a range of SECM sensors and (iii) it is capable of tracking the full range of topographic variations expected for corroding Mg materials. Herein, we present evidence that the SF methodology can accurately track topographical features relevant to the study of Mg material corrosion. In our current configuration, we can readily accommodate vertical features as small as 1 nm up to 40 mm. The SF methodology, illustrated here by *ex situ* measurements, can readily be applied during SECM measurements to produce both *in situ* reactivity and topographical data. We demonstrate a large working height range of 700 μm using SF on a severely corroded magnesium alloy-mild steel galvanic couple along with the topography of a microgalvanically corroded ZEK100 Mg alloy presenting microscopic scale features of less than 1.4 μm. The SF topography is benchmarked against conventional imaging techniques (profilometry and CLSM).

2 Experimental

2.1 Fabrication of micropipettes for constant distance imaging

Quartz capillaries of 0.3 mm internal diameter (ID) and 1 mm outer diameter (Sutter Instrument, Novato, CA) were washed in 0.1 M HNO_3 (Caledon Laboratories Ltd., Georgetown, Ont) for 2 hours, and dried overnight in an oven (95 °C). A P-2000 CO_2 laser puller (Sutter Instrument) was used to pull patch pipettes with an aperture of 500 nm diameter and taper length of 2.0 mm (measured along the pipette shaft from the tip to where ID = 0.15 mm).[8]

2.2 Mg alloys selected

Ingots of mild steel (MS), AE44 (composition: 91.6% = Mg, 4% = Al, 4% = rare earths elements, 0.4% other impurities) and ZEK100 (composition: 98% = Mg, 1% = Zn, <0.5% = rare earths elements, <0.5% = Zr) were received from General

Motors India Private Limited or General Motors Canada and cut down to 1.0 cm × 2.0 cm × 1.0 cm (length × width × thickness). AE44 and ZEK100 were selected since they are commercially available and used or envisioned for manufacturing automotive components.[15,46,47] The samples were then mounted in cold mounting epoxy (Epofix) (Struers, Ontario, Canada) and polished following an established procedure.[41] The AE44 and MS samples were placed one next to the other in galvanic contact, exhibiting no measurable resistance (<0.1 Ω), and mounted in epoxy and polished as for the single alloy.

2.3 Scanning droplet cell surface imaging

To perform the scanning droplet cell experiment over the polished ZEK100 sample, a micropipette was filled with an aqueous 1.6 wt% (0.6 M) NaCl solution. A Ag|AgCl reference wire (0.125 mm diameter Ag wire – GoodFellow Metals, Huntingdon, England), fabricated following an established procedure,[48] was inserted at the back of the micropipette. The polished ZEK100 sample was used as the WE in a two-electrode configuration. Upon surface wetting by the micropipette, the corrosion potential was recorded for 10 s. The micropipette was then lifted (10 μm) to break the electrical contact and repositioned to a new *x*,*y* coordinate and approached to the Mg alloy surface to perform a new corrosion potential measurement. This was repeated over a total width of 150 μm.

2.4 Corrosion of the galvanic couple and Mg alloy

Corroded surfaces were obtained by suspending the galvanic couple or ZEK100 face down in an aqueous (18.2 MΩ, Millipore Milli-Q Reference system) 1.6 wt% (0.6 M) NaCl (A.C.S. grade, ACP Chemicals, Montreal, Qc) solution, for 1 day. The accumulated corrosion product was removed through a 20 s immersion in gently agitated 1 M CrO_3 in water. The samples were then sonicated in methanol for 10 s to remove any surface excess of chromium solution and dried using an Ar stream. The alloys were stored in a desiccator to ensure complete dehydration.

2.5 Surface imaging using profilometry and CLSM

Depth profiles of the corroded AE44–MS galvanic couple were acquired with a non-contact profilometer using axial chromatism (CHR-150 – Nanovea 3D, 3.5 mm laser working distance, 30 Hz scanning frequency and 10 μm sample size).

Depth profiles of the ZEK100 sample following corrosion were measured through the detection of the reflected light intensity using an inverted objective Zeiss 510 confocal microscope equipped with a HeNe 633 nm laser. Depth profiles were extracted from the light intensities by considering how many steps of known distance (slices) through the focal plane, in the *z*-direction, were required to reach the deepest region. Maximum depths to achieve a focused image are reached at 30 μm. All samples had reference markers that enabled the same area of the sample to be measured by profilometry, CLSM and SF.

2.6 SF surface imaging parameters

SF tracking of corroded surfaces was achieved using a micropipette equipped with stimulation and receiver piezos (stimulation amplitude: 1 V; stimulation frequency: 274 kHz) and an ElProScan 1 SECM (HEKA, Germany; bipotentiostat model PG340). The optimal stimulation frequency was identified by recording a frequency spectra with a tip-to-substrate separation greater than 500 nm above the substrate and with the tip in contact with the substrate (*vide infra*). To topographically image the corroded surfaces, the z approach was stopped when the SF amplitude decreased to 45% of the SF amplitude measured above the substrate at an approach speed of 10 nm s^{-1}. The height of both the stepper and z piezo were recorded and the tip was retracted 4 μm, before being repositioned to a new x,y coordinate.

3 Results and discussion

3.1 Monitoring microgalvanic corrosion of ZEK100 using a scanning droplet cell

The scanning droplet cell is a useful method to monitor the local corrosion rates of heterogeneous surfaces (Fig. 1a).[49] Mg alloys have a heterogeneous microstructure as secondary phases form during casting due to solubility limitations of

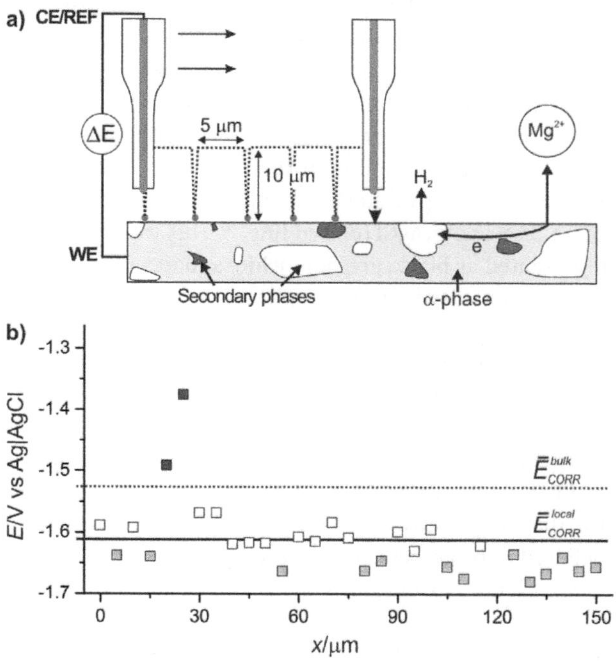

Fig. 1 (a) Schematic representation of the scanning droplet cell experiment performed to characterize the microgalvanic corrosion of Mg alloys. The polished Mg alloy was used as a WE and an Ag|AgCl CE/REF wire was inserted at the back of the micropipette. The red dots represent the electrical contacts between the electrolyte meniscus and the alloy surface. A hopping mode methodology was used with a retraction of 10 μm and motion in x of 5 μm; (b) corrosion potential measurements were performed over a ZEK100 Mg alloy using a filled micropipette with a 1.6 wt% NaCl solution over a line scan of 150 μm with 5 μm step size. The black, white and grey squares represent what is believed to be Zr particles, secondary Mg–Zn–Nd phase and α-phase, respectively. The average local E_{CORR} (full line) is −1.61 V, whereas the average bulk E_{CORR} (dotted line) is −1.53 V *vs.* Ag|AgCl.

the alloying elements in the Mg solid solution. The overall corrosion reaction of Mg materials can be simply described by the reactions,

$$Mg_{(s)} \rightarrow Mg^{2+}{}_{(aq)} + 2e^- \quad (1)$$

$$2H_2O_{(aq)} + 2e^- \leftrightarrows 2OH^-{}_{(aq)} + H_{2(g)} \quad (2)$$

$$Mg^{2+}{}_{(aq)} + 2OH^-{}_{(aq)} \leftrightarrows Mg(OH)_{2(s)} \quad (3)$$

In the case of ZEK100 Mg alloy, the microstructure comprises an α-Mg rich phase, a secondary Mg–Zn–Nd phase and Zr particles randomly distributed across its surface.[47] Some Zr particles have previously shown Fe enrichment, making them preferred cathodes and the location for water reduction (eqn (2)),[12] therefore driving the microgalvanic corrosion of ZEK100. As for the α-Mg matrix, due to its high Mg content, this phase is believed to dissolve producing Mg^{2+} ions (eqn (1)).

To highlight the microstructural heterogeneity of the polished ZEK100 Mg alloy, a scanning droplet cell experiment was performed to measure the local corrosion potential (E_{CORR}) using a micropipette filled with an aqueous 1.6 wt% NaCl solution, Fig. 1a. The measuring cycle entailed bringing the micro droplet into contact with the polished Mg alloy surface and measuring E_{CORR} *vs.* the Ag|AgCl CE/REF for 10 s to obtain an average value, before the contact was broken by retracting the pipette of 10 μm. The micropipette was subsequently displaced laterally by 5 μm to the next location and the measuring cycle was repeated, until a total displacement of 150 μm was reached. The results obtained are presented in Fig. 1b and showed a large variability in the measured E_{CORR}, with an average value of −1.61 V *vs.* Ag|AgCl (full line), which is lower than the reported macroscopic value of −1.53 V *vs.* Ag|AgCl (dotted line).[47] This variation of the measured local E_{CORR} (represented as black, grey and white squares in Fig. 1b) is consistent with an heterogeneous distribution of phases across the ZEK100 Mg alloy surface.[47] Interestingly, the E_{CORR} value of −1.38 V *vs.* Ag|AgCl, that was measured at $x = 25$ μm, is believed to be associated with a Zr particle. The size of the micropipette (500 nm aperture) used to perform this measurement is smaller then the size of these cathodic particles (<5 μm), thus isolating them for a local E_{CORR} measurement. These results are in agreement with previous results obtained by SKPFM, where Zr particles were shown to exhibit a Volta potential difference with the α-Mg-matrix of ~180–250 mV.[50,51] The lower E_{CORR}, white squares, are believed to be associated with the secondary Mg–Zn–Nd phase while the grey squares, with the most negative E_{CORR}, are believed to be associated with the α-Mg matrix.

Given the restricted solvent window of the corroding solution, it is evident that the SDC methodology is useful in tracking the local E_{CORR} of a heterogeneous Mg alloy but lacks the capability to quantify local electrochemical fluxes generated by the ZEK100 microstructure. To track the Mg^{2+} release or molecular H_2 fluxes, micropotentiometric and microamperometric sensors, routinely used in SECM, are required. Currently, quantitative SECM studies of Mg materials are often restricted to corrosion initiation events where variations in topography are in the order of a few microns. This restriction is due to an inability to correlate the material flux measured by the electrochemical probes with the observed local corrosion rates due to the lack of associated topographical information.

This situation is clearly unsatisfactory, so to quantify local corrosion rates occurring during microgalvanic corrosion and for galvanically coupled materials, a SF controlled SECM methodology was developed. This methodology can be easily adapted to a number of SECM uni- or multi-functional probes including amperometric MEs or potentiometric sensors.

3.2 SF controlled scanning micropipette method: positioning and calibration

SF sensitive frequencies are tip-to-substrate distance sensitive, making them ideal to probe surface topography and regulate the tip-to-substrate separation. The identification of these SF sensitive frequencies can be divided in three steps.[16,29]

First, a stimulation piezo and a receiver piezo, positioned 1–2 cm apart, were mounted on a pulled glass micropipette, Fig. 2(a)-(i). Secondly, a sinusoidal potential wave of a given frequency (f) was applied to the stimulation piezo, causing the micropipette to vibrate. With respect to the applied alternating voltage stimulus, of amplitude A_{in}, the response amplitude, A_{out}, and phase shift induced by mechanical oscillations of the micropipette were recorded at the receiver piezo (Fig. 2(a)-(ii)). The entire micropipette setup is brought in gentle contact ($\sim$50 nm) with the surface, where the SF sensitive frequencies are identified. To do so, a spectrum ranging from 250 to 300 kHz was obtained, Fig. 2(a)-(iii) (complete spectra are available in Fig. S1†). Comparison of the amplitude and phase monitored across the frequency spectra with the micropipette in contact with the surface (dashed lines in Fig. 2(a)-(iii)) revealed several frequencies that exhibited a significant decrease in amplitude and change in phase compared to when the micropipette was 500 nm above the surface (full lines in Fig. 2(a)-(iii)). These frequencies were assigned as SF sensitive. For each micropipette, a single SF sensitive frequency was selected and used in subsequent experiments. As schematically presented in Fig. 2b, along with the corresponding measurement, the amplitude and phase response of the selected SF sensitive frequency (in this case 274 kHz) was monitored while the micropipette was approached to the surface. Once in contact with the substrate (this is indicated by the complete attenuation of the amplitude signal), a calibration curve (Fig. 2c and the corresponding schematic representation) of the SF signal with an increasing tip-to-substrate distance was acquired by retracting the tip at 3 nm s^{-1}. From this curve, we selected the height set point used during topography tracking. This provided a quick and reliable approach to accurately monitor the tip-to-substrate separation.

This methodology can readily be applied to monitor topography across hard and soft (Fig. S2†) surfaces, using the 3-axis translator positioning system, and was first used to evaluate the microgalvanic corrosion of the ZEK100 Mg alloy followed by a study of the extent of corrosion damage over an AE44 Mg alloy galvanically coupled to MS, following one day of exposure to 1.6 wt% NaCl solution. For these samples, the set point defining the tip-to-substrate distance was set at 45% of the amplitude measured off the surface, giving a tip-to-substrate separation of less than 10 nm (separation between the attenuated amplitude signal point ($x = 0.039$ μm) and the red dot in Fig. 2c).

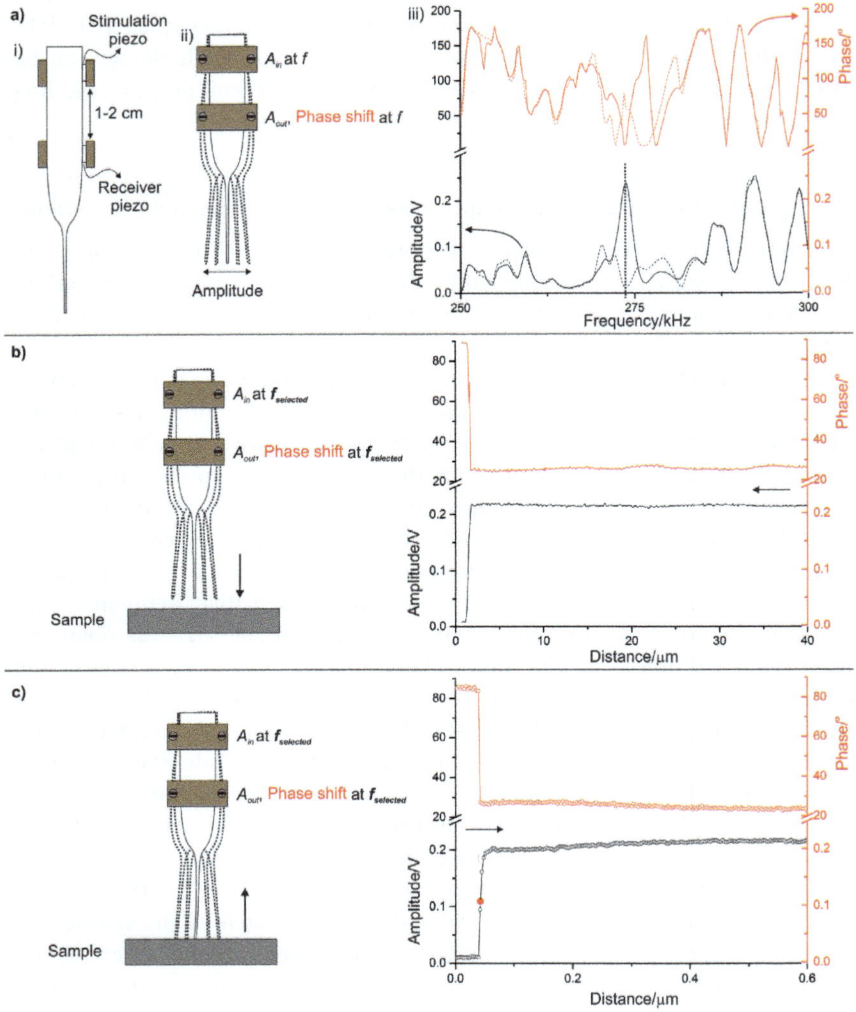

Fig. 2 (a) (i) Schematic representation of piezoelectric elements positioned on the shaft of the micropipette. (ii) Schematic representation of the parameters extracted (A_{in}, A_{out}, phase shift) from a SF spectrum. f represents the stimulation frequency. (iii) SF spectra recorded with a stimulation amplitude of 1 V when the micropipette is placed 500 nm above the surface (full line) and in contact with the surface (dashed line) for the phase (red curves) and amplitude signals (black curves). (b) Schematic representation of a SF approach curve and the corresponding approach curve at 1 μm s^{-1} in amplitude (A_{out}, black curve) and phase (phase shift, red curve) while exciting the stimulation piezo at 274 kHz ($f_{selected}$, dashed line in (a)-(iii)) with a stimulation amplitude of 1 V (A_{in}); (c) schematic representation of a SF calibration curve (pull back) and the corresponding measurement in amplitude (A_{out}, black curve) and phase (phase shift, red curve) recorded at 3 nm s^{-1} while exciting the stimulation piezo at the highlighted frequency and amplitude (A_{in}) in (a)-(iii) (274 kHz, $f_{selected}$); the threshold (45% of the surface amplitude) is highlighted by the dot on the amplitude curve.

3.3 Tracking ZEK100 microgalvanic corrosion depths using the SF constant distance mode

Mg and its alloys not only corrode when galvanically coupled but also as stand-alone materials. Although the severity of corrosion is far less when micro-galvanically corroding, (~10–1000 mm per year[36] when galvanically coupled *vs.* ~<10 mm per year[12]), it is mandatory that the SF methodology, in order to remain a competitive *ex situ* alternative, is able to track subtle microscopic features. To verify if SF has the capability of tracking microgalvanic corrosion features, a ZEK100 Mg alloy was corroded for a day and the corresponding SF results were benchmarked to standard CLSM.

The ZEK100 Mg alloy has previously been shown to exhibit filiform-like corrosion with networks interlinking secondary phases.[47] As observed by CLSM, filiform-like features are observed following a 24 h immersion of the ZEK100 in a 1.6 wt% NaCl aqueous solution, shown in Fig. 3a. From this micrograph, a height profile was extracted and is presented, Fig. 3b (top). Over the corrosion product film (mainly $Mg(OH)_2$, see eqn (3)), diffraction of the laser during CLSM was observed (for example at $x = 40–80$ μm), which limits the achievable resolution by

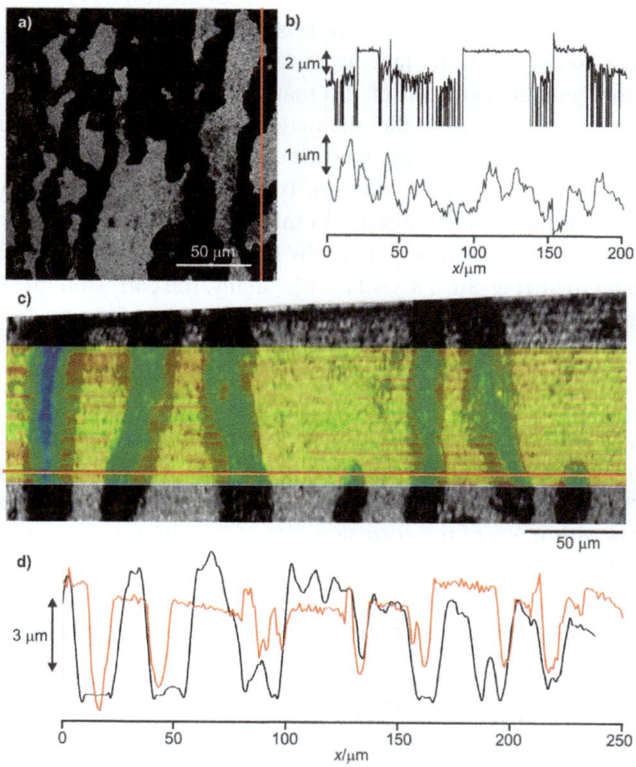

Fig. 3 (a) CLSM micrograph of a corroded ZEK100 Mg alloy corroded for 24 h in a 1.6 wt% NaCl aqueous solution with corrosion products present. (b) CLSM line scan along the red line in (a) (top) and SF line scan recorded across the area imaged by CLSM in (a) (bottom). (c) Topography map of the same area imaged by CLSM (black and white) and SF (color) with corrosion products removed. (d) Comparative SF (red curve) and CLSM (black curve) line scans recorded across the red line in (c).

CLSM. SF topography tracking can avoid those issues. As presented in Fig. 3b (bottom), a SF line scan was performed in the same region as the CLSM micrograph (Fig. 3a). Although it was not possible to relocate exactly where the line scan was taken on the CLSM micrograph, it is still possible to observe, by comparing the overall depth in Fig. 3b, that both line scans show features of similar heights (2–3 μm) and widths (10–20 μm). In addition, the micropipette tip was perfectly capable of tracking the surface covered by porous corrosion products.

To extract a depth profile of the microgalvanic corrosion process of ZEK100, the corrosion products were removed by immersing the sample in a chromium solution for 20 s. SF imaging of the sample was then performed, Fig. 3c (color map). Seven regions of lower heights were located over an area of 250 × 30 μm. These features were ~10–15 μm wide with a relative depth ranging from 3–5 μm, with a smaller feature of less than 1.5 μm deep. The tip of the micropipette was then used to mechanically scratch the sample, which served as a reference marker for subsequent CLSM measurements, Fig. 3c (black and white). Seven features with relative widths of ~10–15 μm were identified by CLSM and show good agreement with the corroded surface map acquired by SF, Fig. 3c. Fig. 3d shows a SF (red curve)/CLSM (black curve) comparative height measurement performed across the red line shown in Fig. 3c. Generally, cavity depths of ~3–5 μm were recorded by CLSM, which is in good agreement with the results obtained by SF. In some instances, depth differences were less than 100 nm (*e.g.* at $x = 140$ μm). However, over other locations (*e.g.* at $x = 55$ μm or $x = 180$ μm), SF underestimated the general corrosion of the material by ~3–5 μm. To improve the resolution obtained by SF, a smaller diameter micropipette and smaller step size should be used. The micropipette aperture has an outer diameter of ~700 nm, which sets a lower limit on the resolution. To properly track the corroded surface with SF, the micropipette is required to completely clear the edge before starting to move into the cavities. This explains why larger valleys are measured in CLSM in comparison to SF (Fig. 3d, $4 < x < 11$ μm). Finally, the corrosion rate (C_R) of the material was determined using:

$$C_R = \Delta z_m / t \quad (4)$$

where Δz_m is the maximum depth of corrosion after a time (t). Topography tracking by SF over the surface revealed that the ZEK100 should experience a corrosion rate in the order of 2.13 mm per year, evaluated from the deepest cavity. This result is within 5% of the corrosion rate of 2.24 mm per year evaluated for this sample by CLSM.

3.4 Tracking AE44–MS corrosion depths using the SF constant distance mode

An AE44–MS couple undergoes galvanic corrosion (Fig. 4a), where the metal with the higher reduction potential is referred to as the cathodic metal, which receives electrons from the metal with the lower reduction potential, the anodic metal. Since Mg has the lowest reduction potential,[52] it will corrode preferentially according to eqn (1)–(3).

The potential gradient is at its maximum at the AE44–MS interface and decays exponentially with increasing distance from that interface. Therefore, the most severe corrosion damage, and largest depth profile (630 μm) was expected at the AE44–MS material interface, with little corrosion damage anticipated on the MS surface.[41]

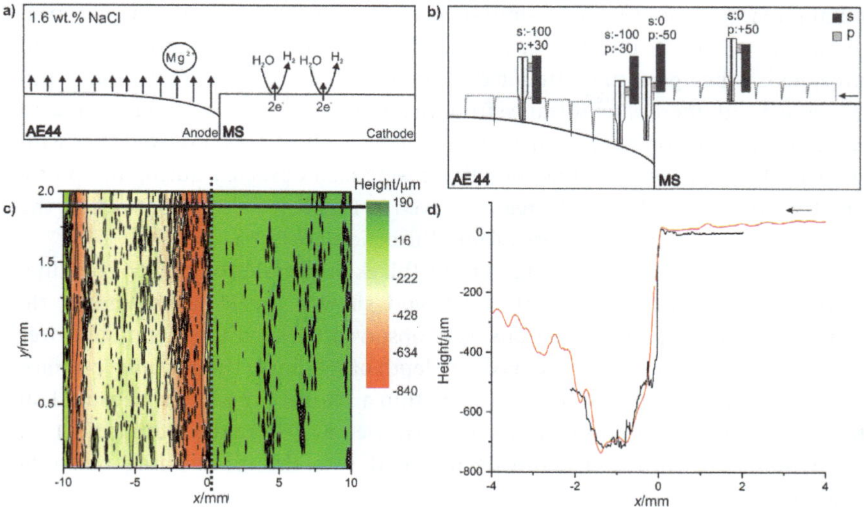

Fig. 4 (a) Schematic representation of an AE44–MS galvanic couple immersed in a 1.6 wt % NaCl solution generating Mg^{2+} at the anode and H_2 at the cathode; (b) stepper (s) and piezo (p) motor position schematic representation (not to scale), as the micropipette is scanned over the mixed material interface; (c) depth profile of a corroded 10–10 mm AE44–MS galvanic couple immersed in a 1.6 wt% NaCl solution for 24 hours, imaged using profilometry; the dashed line represents the interface between AE44 and MS; (d) line scan in SF (black line) and profilometry (red line) selected at $y = 1.95$ mm (highlighted full line shown in (c)). The negative distances represent the AE44 alloy while positive distances represent MS.

First, the topography was mapped by SF by scanning the micropipette from the MS towards the Mg alloy, using a frequency of 274 kHz and a stimulation amplitude of 1 V, as determined by the calibration measurement (Fig. 2c). As schematically represented in Fig. 4b, SF measurements were performed using a hopping mode methodology, where the micropipette was lowered at 10 nm s^{-1} towards the substrate and the approach stopped when the oscillation amplitude measured at the receiver piezo decreased from 0.2 V to 0.11 V. The height was recorded, before the micropipette was retracted 4 μm and moved to the next x,y coordinate. To achieve the large z travel distance suitable for measuring the steep edge due to galvanic coupling, translation of the probe was performed by the combination of the close looped z stepper motor and z piezo motion, providing a maximum z working distance of 40 mm. Here, the piezo motor had a maximum travel distance of 50 μm in the z direction. The approach was primarily performed using the piezo motor, however when it reached its maximal travel distance, the z stepper motor was used to reposition the micropipette at a suitable height for a z piezo approach. In this way changes in topography, which exceed the range of the piezo motor alone, can be accurately tracked (see piezo and stepper motor positions in Fig. 4b).

3.5 Evaluating the corrosion rate of galvanically coupled AE44

The extent of corrosion damage was first measured by profilometry using axial chromatism (Fig. 4c). In the presented case, a 10 μm sampling size was used to obtain a large scan area. A maximum corrosion depth of 840 μm was observed along the AE44–MS interface ($x = 0$), which is in excellent agreement with previous studies performed under the same conditions.[36] The corrosion depth decreased with increasing distance from the interface until $x = -9$ mm where an increase in corrosion rate, consistent with an edge effect, was observed.[41] As with the CLSM experiments, diffraction effects were also observed during the profilometry experiment. The black areas observed in Fig. 4c represent spots where the non-contact profilometer suffered from diffraction artifacts.

Fig. 4d shows line scans acquired by SF (black line) and profilometry (red line) at the same location on the AE44–MS and showed excellent agreement. No significant topographical variations were observed between the techniques over the MS surface ($0 < x < 4$ mm). A maximal depth of 721 μm at the galvanic interface ($0 < x < -1$ mm) is obtained by SF and is within a 5% error of what is measured on the same sample by profilometry. As the tip moves away from the interface ($-1 < x < -4$ mm), the corrosion depth decreased due to the diminishing potential gradient.

In order to achieve a resolution similar to profilometry using SF, it has been reported by Etienne and co-workers that conical electrodes offer better tracking of tiny details present on a rough surface.[33] Following the same strategy, we used a 500 nm internal diameter aperture micropipette, which yielded more details of the scanned surface compared to profilometry, quantitatively providing a 2.7 times improved aspect ratio (3600 for SF *vs.* 1360 for profilometry).

The corrosion rate of AE44 was calculated from the profilometry and SF line scans in Fig. 4d, using eqn (4), for $t = 24$ h.[36] The comparatively large corrosion rate of 263 mm year^{-1} has been attributed to the large corrosion potential difference (ΔE_{CORR}) between AE44 and MS (~1 V), creating a polarized galvanic couple.[36] These data are in agreement with the previously reported corrosion rate of 243 mm year^{-1}.[36]

4 Conclusion

We have demonstrated that the SF methodology can accurately track millimeter and microscale topographical features over corroded samples. A micropipette was positioned at a constant distance above galvanically and microgalvanically corroded Mg alloys. Using a SF control set point to regulate the piezoelectric motors and by combining a long range stepper motor with the fine control of the piezoelectric motor, small (*ca.* 1.4 μm height over 250 μm lateral translation) and large depth profiles (*ca.* 0.72 mm height over 4 mm lateral translation) have been recorded. Importantly, this methodology was successful in tracking high aspect ratio features of a severely corroded AE44–MS galvanic couple. The resulting corrosion rate and depth profiles are consistent within 5% with *ex situ* profilometry and previous literature results. Similar agreement is obtained when comparing the corrosion rate of a ZEK100 Mg alloy, corroded for 24 h in the same solution and obtained by SF topography tracking, with that obtained by CLSM. The SF methodology was shown to be capable of acquiring the overall surface topography of a large area (250 × 30 μm) with excellent spatial resolution, showing the robustness of the technique.

The SF methodology presented can be readily applied during a SECM measurement to produce both *in situ* reactivity and topographical data. Moreover, by using the SF methodology and dual probes, already envisioned for corrosion characterization, it will be possible to measure simultaneously several electrochemical fluxes while concomitantly tracking the surface topography. This concomitant knowledge is a prerequisite to the successful development of future SECM predictive numerical models.

Acknowledgements

This work was supported in part by the Natural Sciences and Engineering Research Council of Canada (NSERC) and General Motors of Canada. We thank Steen B. Schougaard for fruitful discussions. We thank Jon Carter and Justin Gammage from General Motors for providing the samples and helpful discussion. Luis Calzado, Monique Vasseur and Pellumb Jakupi are acknowledged for their technical expertise in profilometry and/or CLSM.

References

1 D. E. Tallman and M. B. Jensen, in *Scanning Electrochemical Microscopy*, CRC Press, 2nd edn, 2012, pp. 451–488.
2 L. Niu, Y. Yin, W. Guo, M. Lu, R. Qin and S. Chen, *J. Mater. Sci.*, 2009, **44**, 4511–4521.
3 M. V. Mirkin, W. Nogala, J. Velmurugan and Y. Wang, *Phys. Chem. Chem. Phys.*, 2011, **13**, 21196–21212.
4 K. Eckhard, M. Etienne, A. Schulte and W. Schuhmann, *Electrochem. Commun.*, 2007, **9**, 1793–1797.
5 M. Pähler, J. J. Santana, W. Schuhmann and R. M. Souto, *Chem.–Eur. J.*, 2011, **17**, 905–911.
6 J. J. Santana, M. Pähler, W. Schuhmann and R. M. Souto, *ChemPlusChem*, 2012, **77**, 707–712.
7 H. Krawiec, V. Vignal, O. Heintz, P. Ponthiaux and F. Wenger, *J. Electrochem. Soc.*, 2008, **155**, C127–C130.
8 P. Dauphin-Ducharme, R. M. Asmussen, D. W. Shoesmith and J. Mauzeroll, *J. Electroanal. Chem.*, 2015, **736**, 61–68.
9 J. Izquierdo, A. Kiss, J. J. Santana, L. Nagy, I. Bitter, H. S. Isaacs, G. Nagy and R. M. Souto, *J. Electrochem. Soc.*, 2013, **160**, C451–C459.
10 S. S. Jamali, S. E. Moulton, D. E. Tallman, M. Forsyth, J. Weber and G. G. Wallace, *Corros. Sci.*, 2014, **86**, 93–100.
11 W. Liu, F. Cao, Y. Xia, L. Chang and J. Zhang, *Electrochim. Acta*, 2014, **132**, 377–388.
12 P. Dauphin-Ducharme, R. Matthew Asmussen, U. M. Tefashe, M. Danaie, W. Jeffrey Binns, P. Jakupi, G. A. Botton, D. W. Shoesmith and J. Mauzeroll, *J. Electrochem. Soc.*, 2014, **161**, C557–C564.
13 U. M. Tefashe, M. E. Snowden, P. Dauphin-Ducharme, M. Danaie, G. A. Botton and J. Mauzeroll, *J. Electroanal. Chem.*, 2014, **720–721**, 121–127.
14 S. S. Jamali, S. E. Moulton, D. E. Tallman, M. Forsyth, J. Weber and G. G. Wallace, *Electrochim. Acta*, 2015, **152**, 294–301.
15 United States Automotive Materials Partnership, *Magnesium Vision 2020, A North American Automotive Strategic Vision for Magnesium*, Department of Energy National Energy Technology Laboratory, USA, 2004.

16 C. Cougnon, K. Bauer-Espindola, D. S. Fabre and J. Mauzeroll, *Anal. Chem.*, 2009, **81**, 3654–3659.
17 Y. Lee, Z. Ding and A. J. Bard, *Anal. Chem.*, 2002, **74**, 3634–3643.
18 H. Yamada, H. Fukumoto, T. Yokoyama and T. Koike, *Anal. Chem.*, 2005, **77**, 1785–1790.
19 Y. Takahashi, H. Shiku, T. Murata, T. Yasukawa and T. Matsue, *Anal. Chem.*, 2009, **81**, 9674–9681.
20 A. Kueng, C. Kranz, A. Lugstein, E. Bertagnolli and B. Mizaikoff, *Angew. Chem., Int. Ed.*, 2003, **42**, 3238–3240.
21 A. Davoodi, J. Pan, C. Leygraf and S. Norgren, *Electrochim. Acta*, 2007, **52**, 7697–7705.
22 A. Davoodi, J. Pan, C. Leygraf and S. Norgren, *J. Electrochem. Soc.*, 2008, **155**, C211–C218.
23 D. Ruhlig and W. Schuhmann, *Electroanalysis*, 2007, **19**, 191–199.
24 J. Izquierdo, S. González and R. M. Souto, *Int. J. Electrochem. Sci.*, 2012, **7**, 11377–11388.
25 P. M. Diakowski and Z. Ding, *Electrochem. Commun.*, 2007, **9**, 2617–2621.
26 K. Borgwarth, D. Ebling and J. Heinze, *Electrochim. Acta*, 1995, **40**, 1455–1460.
27 M. Ludwig, C. Kranz, W. Schuhmann and H. E. Gaub, *Rev. Sci. Instrum.*, 1995, **66**, 2857–2860.
28 A. Hengstenberg, C. Kranz and W. Schuhmann, *Chem.–Eur. J.*, 2000, **6**, 1547–1554.
29 L. Danis, M. E. Snowden, U. M. Tefashe, C. N. Heinemann and J. Mauzeroll, *Electrochim. Acta*, 2014, **136**, 121–129.
30 B. Ballesteros Katemann, A. Schulte and W. Schuhmann, *Chem.–Eur. J.*, 2003, **9**, 2025–2033.
31 B. Ballesteros Katemann, A. Schulte and W. Schuhmann, *Electroanalysis*, 2004, **16**, 60–65.
32 M. Etienne, S. Lhenry, R. Cornut and C. Lefrou, *Electrochim. Acta*, 2013, **88**, 877–884.
33 M. Etienne, J.-P. Moulin and S. Gourhand, *Electrochim. Acta*, 2013, **110**, 16–21.
34 M. Etienne, B. Layoussifi, T. Giornelli and D. Jacquet, *Electrochem. Commun.*, 2012, **15**, 70–73.
35 T. Cain, L. G. Bland, N. Birbilis and J. R. Scully, *Corrosion*, 2014, **70**, 1043–1051.
36 K. B. Deshpande, *Corros. Sci.*, 2010, **52**, 3514–3522.
37 A. D. Sudholz, K. Gusieva, X. B. Chen, B. C. Muddle, M. A. Gibson and N. Birbilis, *Corros. Sci.*, 2011, **53**, 2277–2282.
38 R. Matthew Asmussen, W. Jeffrey Binns, P. Jakupi and D. Shoesmith, *J. Electrochem. Soc.*, 2014, **161**, C501–C508.
39 A. D. King, N. Birbilis and J. R. Scully, *Electrochim. Acta*, 2014, **121**, 394–406.
40 R. M. Asmussen, P. Jakupi, M. Danaie, G. A. Botton and D. W. Shoesmith, *Corros. Sci.*, 2013, **75**, 114–122.
41 D. Trinh, P. Dauphin Ducharme, U. Mengesha Tefashe, J. R. Kish and J. Mauzeroll, *Anal. Chem.*, 2012, **84**, 9899–9906.
42 H. Böhni, T. Suter and A. Schreyer, *Electrochim. Acta*, 1995, **40**, 1361–1368.
43 A. P. Sommer and R. P. Franke, *Micron*, 2002, **33**, 227–231.
44 M. Etienne, E. C. Anderson, S. R. Evans, W. Schuhmann and I. Fritsch, *Anal. Chem.*, 2006, **78**, 7317–7324.

45 M. Etienne, A. Schulte, S. Mann, G. Jordan, I. D. Dietzel and W. Schuhmann, *Anal. Chem.*, 2004, **76**, 3682–3688.

46 J. Min, L. Hector Jr, J. Lin and J. Carter, *J. Mater. Eng. Perform.*, 2013, **22**, 3324–3336.

47 R. M. Asmussen, W. J. Binns, P. Jakupi and D. Shoesmith, *Corrosion*, 2015, **71**, 242–254.

48 T. J. Smith and K. J. Stevenson, in *Handbook of Electrochemistry*, ed. C. G. Zoski, Elsevier, Amsterdam, 2007, pp. 73–110.

49 H. Krawiec, S. Stanek, V. Vignal, J. Lelito and J. S. Suchy, *Corros. Sci.*, 2011, **53**, 3108–3113.

50 M. F. Hurley, C. M. Efaw, P. H. Davis, J. R. Croteau, E. Graugnard and N. Birbilis, *Corrosion*, 2015, **71**, 160–170.

51 A. E. Coy, F. Viejo, P. Skeldon and G. E. Thompson, *Corros. Sci.*, 2010, **52**, 3896–3906.

52 A. S. G82–98, *Annual Book of ASTM Standards, Development and Use of a Galvanic Series for Predicting Galvanic Corrosion Performance*, ASTM International, Philadelphia, 2003.

Faraday Discussions

PAPER

Protein interactions with corroding metal surfaces: comparison of Mg and Fe

Victoria Wagener, Anne-Sophie Faltz, Manuela S. Killian, Patrik Schmuki and Sannakaisa Virtanen*

Received 9th December 2014, Accepted 12th January 2015
DOI: 10.1039/c4fd00253a

The influence of bovine serum albumin (BSA) on the electrochemical behaviour of pure Mg and Fe was studied in simulated body fluid (SBF), in view of the possible application of these materials as biodegradable metals. Results indicate a different trend for the BSA-effect on corrosion for the two metals: for Mg, a strong corrosion-inhibiting effect is observed in the presence of BSA in solution, especially for short-term exposure, whereas for Fe only a slight acceleration of corrosion is caused by the addition of BSA to the solution. For both metals, the protein-effect on the electrochemical behaviour shows a complex time-dependence. Surface analysis indicates that stronger BSA adsorption takes place on Mg than on Fe. Moreover, adsorption experiments with BSA and a second protein (lysozyme) were conducted. The results are discussed in view of electrostatic interactions between differently charged metal oxide/hydroxide surfaces and proteins.

1 Introduction

Low-corrosion resistant metals to be used as biodegradable implant materials have been in the focus of intensive research activities in recent years, see *e.g.* review articles.[1–7] The seemingly simple idea is to exploit the inherent degradation of low-corrosion resistant metals and alloys in biomedical applications, where the device is only required for a certain time of healing and where permanent implants may lead to various long-term complications. Such applications could be for instance cardiovascular stenting or temporary bone healing devices. Due to the required mechanical properties of the materials used for these applications, metals instead of biodegradable polymers are in the focus of interest.

A major part of the investigations on biodegradable metals have been carried out on Mg and its alloys; much less work has been dedicated to Fe based materials. The current knowledge indicates that whereas many Mg alloys seem to show a too fast corrosion rate in biological environments, Fe based materials may

Institute for Surface Science and Corrosion, University of Erlangen-Nuremberg, Martensstrasse 7 (WW4, LKO), 91058 Erlangen, Germany. E-mail: virtanen@ww.uni-erlangen.de; Fax: +499131-8527582; Tel: +499131-8527577

dissolve too slowly.[8] Hence, in both cases, the general challenge is to tailor the materials and surfaces to achieve the desired and controlled corrosion rate for the targeted applications. Therefore, a remarkable amount of work has been carried out – especially on Mg alloys – to explore various approaches of surface modification and coatings for decreased corrosion rates and/or improved biocompatibility.[9–12] A further challenge is to elucidate the influence of the complex biological environment on the corrosion mechanisms of these materials. For instance, it has been frequently reported in the literature that the degradation rate of Mg based alloys is strikingly different in SBF solutions, mimicking the inorganic chemistry of body fluids, than in cell culture medium (*e.g.*, DMEM, with or without addition of serum) that also contains organic components, such as glucose, vitamins and amino acids. In spite of the general agreement on the very significant difference in degradation rates in these media, SBF *vs.* DMEM,[13–15] the role of the different species in the solutions on the degradation mechanisms are far from being understood.

For Mg alloys, specific concerns on biocompatibility arise from possibly too strong hydrogen gas evolution and accompanied alkalization of the electrolyte during Mg dissolution.[5] As these are direct consequences of chemical properties of Mg, it is not possible to completely prevent these effects, as long as Mg is corroding – however, if the temporal gas evolution is not too fast, absorption/transport of hydrogen in tissues may be possible without formation of gas pockets due to local hydrogen accumulation. A further challenge in the application of Mg alloys for biodegradable implants is the corrosion mode: mostly a rather non-uniform corrosion attack takes place, often related to galvanic coupling of the intermetallic phases with the Mg matrix. It therefore seems that the challenges in Mg alloy corrosion, including an in-depth understanding in the underlying mechanisms, are larger than initially expected.

Due to these challenges in Mg alloy corrosion, it may be of interest to look at the behaviour of Fe based alloys in biological environments in more detail. A comparison of the corrosion behaviour of Fe and Mg based alloys in simulated biological environments is not only of practical interest in view of possible biomedical applications of these materials. Also from a mechanistic point of view, this comparison may aid in elucidating mechanisms of interactions between the constituents of the biological environment and the corroding metal surfaces in more detail. It is noteworthy that no direct comparison using identical experimental conditions for biodegradation of Fe and Mg can be found in the literature. Therefore, in the current study we explored the corrosion behaviour of pure Fe and Mg (as model systems for more complex alloys) in simulated body fluid with absence and presence of proteins in the solution. Moreover, surface analytical studies on protein adsorption on Fe and Mg surfaces were carried out.

Proteins in solution can have multiple effects on corrosion reactions: on the one side, protein adsorption layers can block electrochemically active surface sites and therefore slow down corrosion.[16–18] Protein adsorption will be influenced by the surface chemistry of the metals (*e.g.*, type of oxides/hydroxides formed on the surface and the resulting surface wettability and surface charges). It is generally acknowledged that protein adsorption is stronger on hydrophobic than hydrophilic surfaces. However, on hydrophilic surfaces – such as OH^- terminated metal oxides – electrostatic interactions, due to surface charges of the metal oxide and the protein, become more important. On the other side, complexation of metal cations can take place by proteins (and amino acids), thereby reducing the free cation concentration and possibly accelerating corrosion. Also this effect depends on the nature of the metal cations; for Fe and Mg dissolution products a different strength of interactions can be expected.

For Mg, the effect of proteins on the corrosion behaviour has been studied by many authors,[14,19–27] but the results do not show a straightforward trend in that both acceleration and inhibition of corrosion has been observed. Moreover, a complex time-dependent influence on the corrosion rate has been observed:[22] for short-term experiments, modest inhibition of corrosion was observed by albumin in SBF, but after longer exposure times to an albumin-containing solution, an opposite effect of corrosion-acceleration by the presence of albumin in solution was observed. The results suggest that the corrosion-protection effect in the early immersion is due to adsorption of the proteins to the surface blocking the active sites, but this blocking effect is not long-term stable. As the various experiments were carried out on different alloys and using different experimental approaches (for instance, studying the effects of a single protein, such as albumin, or studying the effect of serum addition in simulated body fluids), it is difficult to speculate on the origin of the different findings.

For pure iron, the influence of proteins on corrosion has been less studied than for pure Mg. The effect of proteins on iron released from stainless steels (or generally on corrosion behaviour of stainless steels) has been much more frequently studied than in the case of pure iron,[17,18,28–32] as stainless steels already find applications in biomedicine (and in other applications, where proteins are of relevance). However, as the surface chemistries of stainless steels and iron are far from identical, the relevant protein-induced metal release mechanisms may vary considerably for the two materials. Milosev *et al.* show an increase of corrosion of pure Fe in the presence of a number of bovine serum proteins, whereas the significance of the effect depends on the particular protein.[29] In contrast to this, Hedberg *et al.* have reported on an enhanced release of iron from stainless steel in the presence of albumin or lysozyme, but for the iron metal reduced corrosion rates were observed (it should be noted though that for pure Fe in general a much higher iron release was observed than that for stainless steel, due to the lack of passivity of Fe in the solutions used for the studies).[17]

The available literature hence clearly indicates controversial observations on the influence of proteins on corrosion of metals. A comparison of the findings between the different studies is very difficult, due to different experimental approaches. In the present study, the influence of albumin (BSA) on the electrochemical behaviour of Fe and Mg in simulated body fluids was investigated. In addition, surface analytical studies on adsorption of proteins on Fe and Mg surfaces were carried out. In order to get some insight into the role of surface charges or electrostatic interactions between metal surfaces and proteins, we used two proteins with different isoelectric points and therefore different resulting surface charges, namely albumin (IEP: 4.7–5.2)[33] and lysozyme (IEP: 10.5–11.8).[34–37] For iron oxides/hydroxides IEPs between 6.6 and 8.6, and for Mg oxides/hydroxides IEPs between 10.8 and 14 have been reported.[38] In a neutral simulated body fluid, different electrostatic interactions may therefore be expected between the negatively charged BSA and the positively charged lysozyme, with the rather neutral surface of Fe or with the more strongly positively charged Mg surface. In addition, binding of Fe cations by albumin has been reported to be weak,[39] in contrast to strong binding of Mg cations to albumin.[40] The aim of these experiments is therefore to achieve a better understanding on possible interaction modes of proteins and the corroding Mg and Fe surfaces by carrying out electrochemical and surface analytical studies for both metals exposed to identical conditions.

2 Experimental

The materials used for the experiments were commercially pure Mg (99.9%, Chempur) and commercially pure Fe (99.8%, Chempur). Samples were ground with SiC abrasive paper up to a grid size of 1200 and afterwards sonicated in ethanol and dried with air. In case of Mg, a water free lubricant was used in order to prevent corrosion during sample preparation.

As model proteins, bovine serum albumin (BSA, Sigma) and hen egg white lysozyme (LYS, Sigma) were used, with IEPs of 4.7–5.3 and 11.35 according to the manufacturer.

The SBF solution used for the experiments was "SBF5", according to Müller *et al.*[41]

For the electrochemical measurements, 100 ml of SBF was used as a reference electrolyte. In order to investigate the influence of proteins on the corrosion behaviour of Mg and Fe, 10 g l^{-1} BSA was added to the SBF solution. Mg and Fe were analysed in both different electrolytes.

The electrochemical characterization took place by open circuit potential measurements (OCP), electrochemical impedance spectroscopy (EIS) and polarisation curves (IE). All measurements were conducted by using an electrochemical workstation "IM6eX", a potentiostat, "XPot", and the corresponding "Thales"-Software (Zahner-Elektrik GmbH & Co. KG, Kronach, Germany). The experiments were carried out in an electrochemical cell with a three electrode set-up. The material being analysed acted as working electrode. As counter electrode, a platinum sheet and, as reference, an Ag/AgCl electrode with 3 M KCl were used. The OCP was recorded for 15 min. Afterwards, the impedance was measured at the particular OCP of the samples in the range of 100 kHz to 10 mHz with an excitation amplitude of ±10 mV. Afterwards, polarization curves were recorded, starting −300 mV relative to the particular OCP of each sample. The potential was increased up to 0 V with a slope rate of 3 mV s^{-1}. As abort criterion, a current of 10 mA was used.

For the protein coating experiments, the Mg and Fe samples were passivated by soaking in boiling 1 M NaOH solution for 20 min.

The coating took place in 0.1 mM BSA/SBF solution and in 0.1 mM LYS/SBF solution. Each sample was immersed in 50 ml stirred protein solution for 1 h. As reference, Mg and Fe were stored in pure SBF for 1 h. The pH-values of all solutions were measured before and after immersion.

The surfaces after immersion in the protein solution and in pure SBF, were characterised *via* XPS and ToF-SIMS.

XPS measurements were carried out on a high-resolution X-ray photoelectron spectrometer (PHI 5600 USA) using aluminum Kα radiation (1486.6 eV, 300 W) for excitation. The binding energy of the target elements was determined at a pass energy of 23.5 eV with a resolution of <0.4 eV (values were measured every 0.2 eV for the high resolution spectra and 0.8 for surveying) and a takeoff angle of 45°

with respect to the surface normal. The binding energy of the C1s signal was used to correct the spectra for charging. The background was subtracted using the Shirley method in all spectra. To obtain the molar fractions of each species, the peak areas of the measured XPS spectra were corrected with the photoionization cross sections of Scofield,[42] σ, and the asymmetry parameter, β (orbital geometry),[43] which are contained in the sensitivity factors of the acquisition software (MultiPakV6.1A, 99 June 16, copyright Physical Electronics Inc.,1994–1999).

Due to its surface sensitivity the ToF-SIMS analysis is a predestined method for the characterization of protein adsorption.[44,45] A TOF.SIMS 5 from ION-TOF (Münster) was used for surface analysis. The Mg samples were irradiated with a pulsed Bi_3^+ liquid metal ion beam (E = 25 keV) that was electro dynamically bunched down to <1 ns in order to increase the mass resolution. Thus, positive and negative spectra were recorded in high resolution mode ($\Delta m/m > 8000$ at ^{29}Si). The rastered area was 100 × 100 μm² with 5 measurement spots per sample and polarity. To ensure static conditions, the primary ion dose density (PIDD) was kept at 5 × 10^{11} ions per cm². Signals were identified using the accurate mass as well as the isotopic pattern and calibrated to C^+, CH^+, CH_2^+, CH_3^+ and $C_7H_7^+$ for the positive and CH_2^-, C_2^-, CN^- and CNO^- for the negative spectra. For the integration of the signals, Poisson correction was used.[27]

3 Results and discussion

Fig. 1–5 show a comparison of the electrochemical behaviour of Fe and Mg in SBF, in the absence and presence of albumin. Typical open-circuit potential *vs.* time curves (Fig. 1) show an increase of OCP with time for Mg (Fig. 1a), as frequently found for magnesium and related to the formation of corrosion product layers on the surface. For Fe (Fig. 1b), the decrease of OCP with time indicates the breakdown of the air-formed native passive film on the iron surface in SBF. The OCP behaviour indicates a significant influence of the addition of BSA in SBF in the case of Mg, whereas for Fe, the effect of the BSA addition is very minor.

Potentiodynamic polarization curves indicate that for Mg active dissolution takes place in SBF, both in the absence and presence of BSA (Fig. 2). In the absence of BSA in solution (Fig. 2a), the anodic current densities decrease with longer immersion times before the measurement; this is accompanied with a shift of the corrosion potential to more positive values, which is in agreement

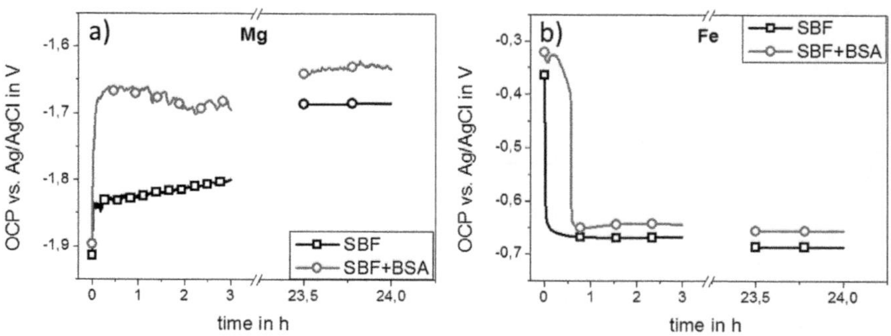

Fig. 1 OCP for Mg (a) and Fe (b) measured for 24 h in SBF and in SBF + BSA.

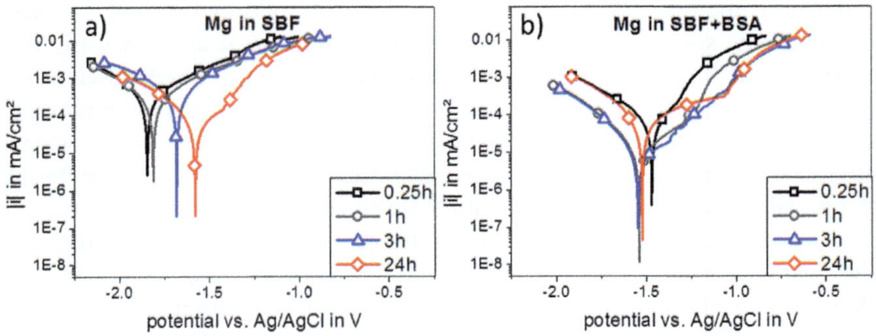

Fig. 2 Polarization curves of Mg after 0.25 h, 1 h, 3 h and 24 h immersion in SBF (a) and SBF + BSA (b).

with the OCP measurements. The corresponding curves measured in SBF + BSA indicate lower i_{corr} values and a more positive E_{corr} from the beginning of exposure, as compared to the SBF solution without BSA addition. Fig. 4a clearly indicates lower corrosion rates for 1 h immersed samples in the SBF + BSA solution. For Fe, an active/passive-transition with a short passive range is observed in SBF, both in the absence and presence of BSA (Fig. 3). Spontaneous passivation is not possible for Fe under the here used experimental conditions. The polarization curves are almost independent of the immersion time as well as of the presence of BSA in solution. This can as well be seen in Fig. 4b, where 1 h immersion in SBF and SBF + BSA is compared.

Impedance spectra measured after various immersion times at the particular OCP show the typical features of two capacitive loops and an inductive loop in the Nyquist plots for Mg (Fig. 5). The shape of the impedance spectra remains similar for all immersion times and independent of the presence of BSA in solution; however, the charge-transfer resistance values and their time-dependence strongly change upon addition of BSA to the solution. The time-dependence of R_{ct} will be discussed in more detail below. For Fe, only one time-constant is seen in the Nyquist plots (Fig. 6). The impedance spectra indicate a significantly lower corrosion rate for Fe as compared to Mg, especially in SBF without the addition of BSA.

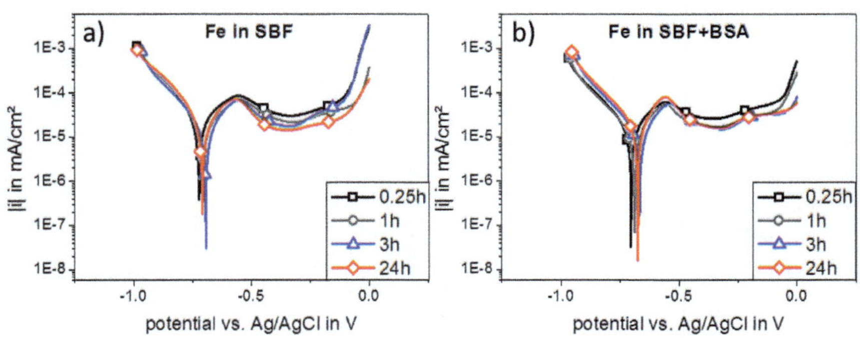

Fig. 3 Polarization curves of Fe after 0.25 h, 1 h, 3 h and 24 h immersion in SBF (a) and SBF + BSA (b).

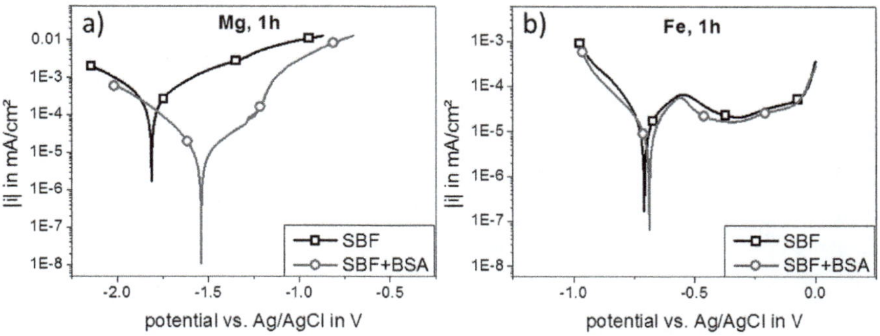

Fig. 4 Comparison of the polarization curves after an immersion time of 1 h in SBF with and without BSA for Mg (a) and Fe (b).

The time-dependence of the R_{ct}-values determined from the impedance spectra is summarized in Fig. 7. For Mg, the addition of BSA in solution leads to a significant increase in the R_{ct}-values, especially in the early stages of immersion (Fig. 7a). Similar to previous findings on a Mg alloy, the corrosion inhibiting effect of BSA in SBF decreases with time.[21] In the case of Fe (Fig. 7b), the effect of BSA addition is much less significant than for Mg. Interestingly, for short-term exposure, the presence of BSA in solution slightly accelerates corrosion of Fe, in contrast to the BSA-inhibiting effect observed for Mg.

The results of the electrochemical characterization therefore clearly indicate that the effect of BSA addition in SBF is significantly stronger for Mg than for Fe. Moreover, the trend for the protein effect is different for both metals: for Mg, a significant corrosion-inhibiting effect is observed by protein addition, especially in the first 3 hours of immersion, whereas for Fe a short-term immersion (up to 3 h) leads to a very modest corrosion-accelerating effect. In order to elucidate the possible reasons for this behaviour, XPS measurements were carried out for samples immersed for 1 and 24 h in SBF as well as in SBF + BSA solutions; the results are summarized in Table 1 for atomic concentrations determined from the N1s, P2p and Ca2p peaks.

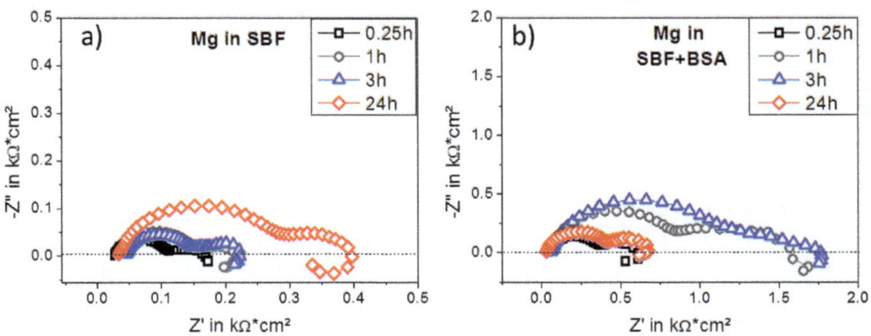

Fig. 5 Nyquist plots for Mg after 0.25 h, 1 h, 3 h and 24 h immersion in SBF (a) and SBF + BSA (b).

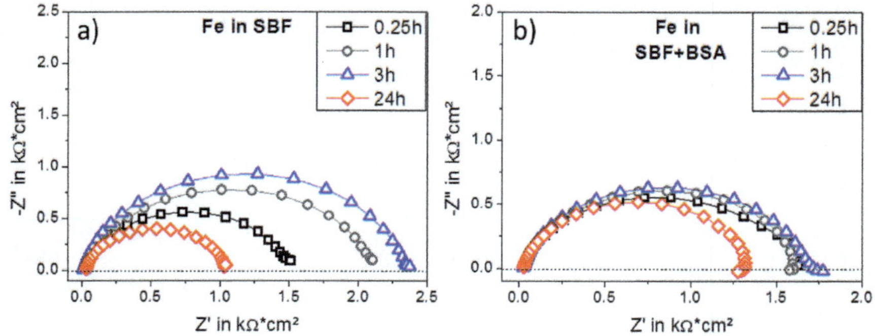

Fig. 6 Nyquist plots for Fe after 0.25 h, 1 h, 3 h and 24 h immersion in SBF (a) and SBF + BSA (b).

In the presence of BSA in solution, a significantly higher amount of N – indicative of adsorbed BSA – is detected on Mg than on Fe. Moreover, the amount of N decreases more strongly with time for Fe than for Mg. The decrease of the N-concentration with immersion time may be related to the continuous corrosion of the samples in SBF and/or to the formation of corrosion product layers on the sample surfaces, thus masking the N signal. However, as Fe corrodes slower than Mg, one would expect this effect to be less significant for Fe. Therefore, the strong decrease in the N-concentration with immersion time observed for Fe may also be an indication of comparably more weakly bound proteins. The lower N-concentration on the surface of Fe immersed in the BSA-containing solution in any case suggests that albumin adsorption is less favoured on Fe than on Mg; this could be a reason for the weak effect of albumin on the electrochemical behaviour of Fe. The P- and Ca-peaks for Mg and Fe after immersion in SBF without or with BSA addition reveal further interesting differences in the behaviour of these two metals. In the absence of BSA, Mg surface immersion in SBF leads to a significant amount of Ca and P detected on the surface, the Ca- and P-concentrations increasing with time. For Fe, on the other hand, the concentrations of Ca and P

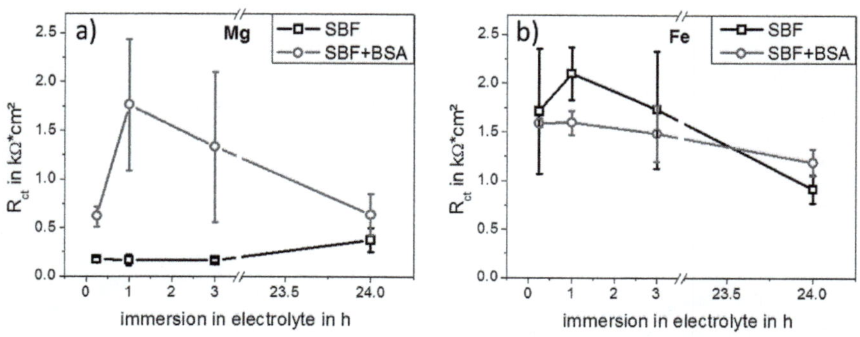

Fig. 7 Comparison of the average charge transfer resistances calculated from the corresponding Nyquist plots; (a) R_{ct} values with the corresponding standard deviation for Mg in SBF with and without BSA; (b) R_{ct} values with the corresponding standard deviation for Fe in SBF with and without BSA.

Table 1 Atomic concentration of N1s, Ca2p and P2p for XPS measurements after 1 h and 24 h immersion of Mg and Fe in SBF with and without BSA

Atomic concentration in at%						
	N (1 h)	N (24 h)	Ca (1 h)	Ca (24 h)	P (1 h)	P (24 h)
Mg SBF	1.03	0.15	8.6	12.06	3.22	5.41
Mg SBF/BSA	16.95	13.33	1.15	1.65	0.35	0.79
Fe SBF	4.19	0	1.52	1.94	1.8	0
Fe SBF/BSA	10.87	3.29	0	0.14	0	0.44

are very low and do not significantly increase upon longer immersion times. The Mg surface hence seems to be more beneficial for precipitation of Ca-phosphate layers from SBF; this may be due to the surface alkalization by Mg dissolution, as the solubility of Ca-phosphates is strongly pH-dependent. It is further noteworthy that only minor amounts of Ca and P are found on the Mg (and Fe) surface in the presence of BSA in solution. This indicates another indirect role of proteins in influencing the corrosion behaviour of Fe and Mg; namely by influencing the formation of Ca-phosphate containing surface layers. This effect will be further investigated in future studies.

A possible reason for the different protein effects on the corrosion of Mg and Fe may be different electrostatic interactions between the negatively charged BSA and the differently charged metal oxide/hydroxide surfaces (see Table 2), which in turn could lead to different amounts of protein adsorption on the surface. In order to study this effect further, additional surface analytical experiments for protein adsorption were carried out with two different proteins, BSA and lysozyme (LYS), which show different surface charges in neutral solutions (Table 2). As the surface analytical data presented above is complicated by simultaneously occurring albumin adsorption and metal dissolution reactions, together with the precipitation of corrosion product layers, these additional surface analyses were carried out on Mg and Fe samples pre-passivated in NaOH. The pre-passivation treatment on the one side leads to a similar surface chemistry on both materials (hydroxide terminated metal oxides) and therefore similar wettability (hydrophilic surfaces). This is advantageous for the elucidation of the role of electrostatic interactions on protein adsorption, as surface wettability is another important factor influencing protein adsorption. On the other side, pre-passivation reduces the reactivity of the Mg and Fe surfaces during immersion in the SBF + protein solutions.

The XPS data for BSA and lysozyme adsorption on passivated Mg and Fe are shown in Fig. 8. Additional results after passivation of the metals in NaOH are shown to elucidate the effect of SBF immersion without proteins. For both metals, an increase of Ca and P can be observed accompanied by a decrease of O, Mg and

Table 2 IEPs for the oxides/hydroxides of Mg and Fe and the proteins, BSA and LYS

	$MgO/Mg(OH)_2$	$FeO/Fe(OH)_2$	BSA	LYS
IEP	10.8–14	6.6–8.6	4.8–5.2	10.5–11.8

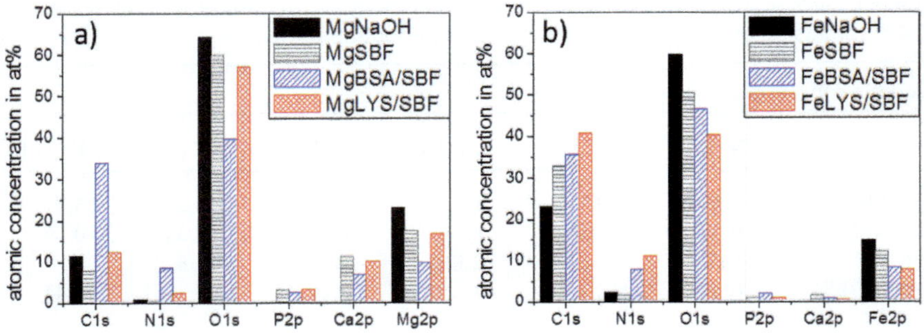

Fig. 8 Results of the XPS measurements of the adsorption experiments. Atomic concentration of several signals after passivation, after immersion in pure SBF and after immersion in SBF with BSA and LYS for Mg (a) and Fe (b).

Fe after SBF treatment, indicating a partial coverage with calcium phosphate compounds of the metal oxide surfaces. Ca and P signals increase much stronger for magnesium than for iron, suggesting a favoured formation of calcium phosphate compounds on magnesium, as was already mentioned above.

As the C1s signal contains contributions from surface contamination as well as from possible carbonate and protein layers on the surface, the amount of protein on the surface can be better determined from the N1s signal. Comparing the atomic concentration of N for Mg immersed in the two protein-containing solutions shows a significantly higher amount of BSA than of lysozyme adsorbed on the surface. For Fe, the trend is reversed in that a higher N-concentration is measured in the lysozyme-containing solution as compared to the BSA-containing solution. Looking at the data in Table 2, the amount of proteins adsorbed indeed seems to follow the expectations based on considering electrostatic interactions: for the strongly positively charged Mg surface, the negatively charged BSA adsorbs in larger amounts than the positively charged lysozyme, whereas for Fe oxides/

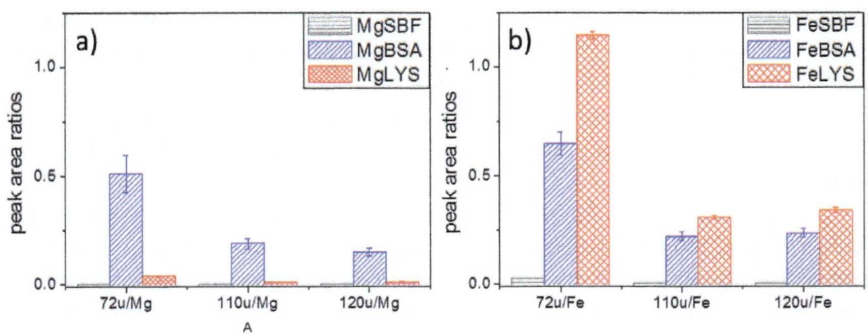

Fig. 9 Peak area ratios calculated from three characteristic protein signals and the substrate signals (Mg and Fe). As signals characterize proteins on the surface, peaks at $m/z = 72$, $m/z = 110$ and $m/z = 120$ corresponding to $C_3H_6NO^+$ (Asn) and $C_4H_{10}N^+$ (Val), $C_5H_8N^+$ (His) and $C_8H_{10}N^+$ (Phe), respectively, were used; (a) peak area ratios for Mg after immersion in pure SBF and after immersion in SBF with addition of BSA and LYS; (b) peak area ratios for Fe after immersion in pure SBF and after immersion in SBF with addition of BSA and LYS.

hydroxides, a stronger electrostatic attraction is expected for lysozyme and indeed a higher amount of protein adsorption is indicated by the XPS data.

The presence of protein adsorption layers on the Mg and Fe surfaces was verified by ToF-SIMS measurements. Fig. 9 shows the peak area ratios of 3 characteristic protein signals and the signals of the particular substrate (Mg respectively Fe). Signals at $m/z = 72$, corresponding to $C_3H_6NO^+$ and $C_4H_{10}N^+$, at $m/z \sim 110$, corresponding to $C_5H_8N^+$ (Asn and Val), at $m/z \sim 110$, corresponding to $C_5H_8N^+$ (His), and at $m/z \sim 120$, corresponding to $C_8H_{10}N^+$ (Phe), were chosen to verify the protein adsorption on the Mg and Fe surfaces. The described peaks are shown in Fig. 10 and 11. For Mg, it can be seen that all peak area ratios increase significantly after immersion in the BSA containing solution compared to the immersion in pure SBF (Fig. 9a). For the immersion in SBF + LYS, only a slight increase of the peak area ratios compared to Mg in pure SBF is detectable, indicating less protein adsorption for LYS than for BSA on Mg. Fig. 10 confirms these results, showing no protein peaks for Mg immersed in pure SBF, clear protein peaks for the addition of BSA to SBF and only small protein peaks for the addition of LYS to SBF.

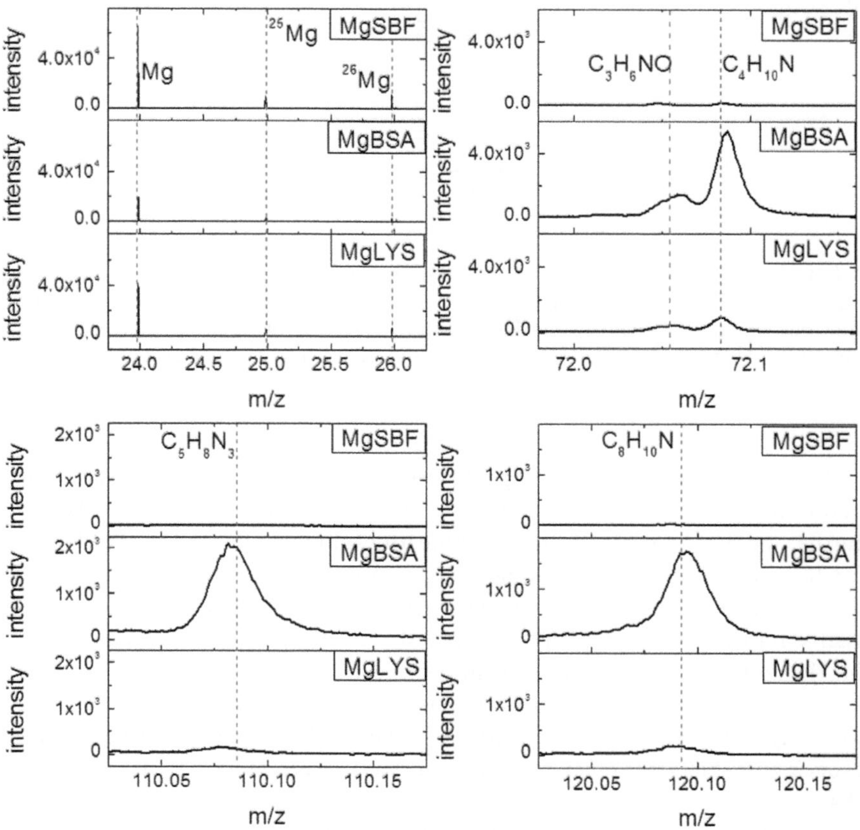

Fig. 10 Signals used for the calculation of the peak area ratios for Mg after immersion in pure SBF and in SBF with BSA and LYS. $m/z = 24$, $m/z = 25$ and $m/z = 26$ corresponding to Mg^+, $^{25}Mg^+$ and $^{26}Mg^+$, respectively, were used as the substrate signals. The amino acid residues at $m/z = 72.05$ (Asn), $m/z = 72.09$ (Val), $m/z = 110.07$ (His) and $m/z = 120.08$ (Phe) were chosen as the characteristic protein signals.

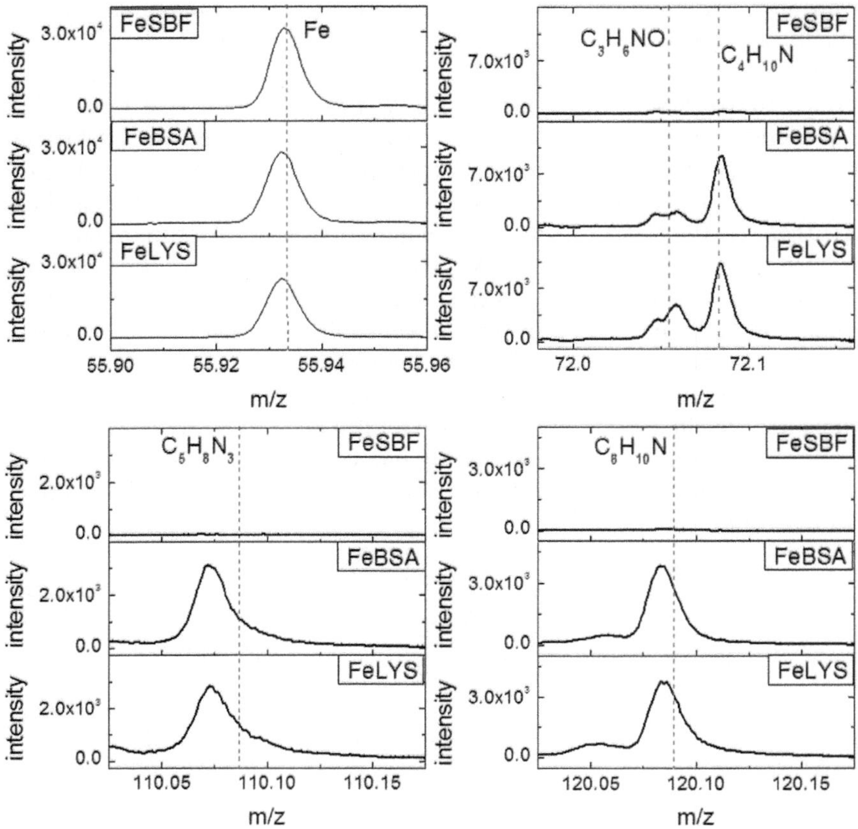

Fig. 11 Signals used for the calculation of the peak area ratios for Fe after immersion in pure SBF and in SBF with BSA and LYS. $m/z = 55.93$ corresponding to Fe^+ was used as the substrate signal. The amino acid residues at $m/z = 72.05$ (Asn), $m/z = 72.09$ (Val), $m/z = 110.07$ (His) and $m/z = 120.08$ (Phe) were chosen as the characteristic protein signals.

For Fe, the peak area ratios (Fig. 9b) also show a significant increase after the addition of BSA to SBF. Unlike for Mg, the addition of LYS to SBF leads to a further increase in peak area ratios, indicating a higher adsorption of LYS than for BSA on the Fe surface. The peaks used for the verification of the protein adsorption on Fe are shown in Fig. 11. There are no characteristic protein peaks visible for the immersion of Fe in pure SBF. For the addition of BSA as well as for the addition of LYS, clear protein peaks can be detected, showing more or less the same intensity for both proteins. A difference for BSA and LYS is the intensity of the Fe signal, indicating a higher surface coverage of Fe by LYS, especially if the fact is taken into account that LYS is a much smaller molecule than BSA.

For Mg as well as for Fe, the ToF-SIMS results verify the presence of protein adsorption on the surface. Even though a direct comparison of Mg and Fe with ToF-SIMS is not possible due to the fact that ToF-SIMS is a semi-quantitative method, the results for Mg and Fe correlate well with the findings of the XPS measurements.

4 Conclusions

Proteins (here: bovine serum albumin, BSA) can inhibit or accelerate metal corrosion. In the present study, the effect of BSA on Mg and Fe corrosion was investigated. The influence of BSA on the electrochemical behaviour of Fe was found to be very minor as compared with Mg. Thus, while BSA was found to show a strong inhibition of Mg corrosion, BSA caused a slight acceleration of Fe corrosion in simulated body fluid for the first hours. Surface analysis with XPS indicates that this may be related to a favoured adsorption of BSA on Mg compared to Fe.

A comparison of the adsorption behaviour of two different proteins, BSA and LYS, suggests that the adsorption behaviour can be explained by considering electrostatic interactions between the charged proteins and metal oxide/hydroxide layers of different surface charges. The extent of these effects depends on the metal cation–protein binding strengths. In the future, metal release studies may further elucidate the influence on proteins on corrosion. These findings are not only of interest in elucidating the effect of proteins on metal corrosion, but may be of general relevance for understanding the role of metal oxide surface charges in corrosion processes, such as localized breakdown of passivity.

Acknowledgements

The authors thank H. Hildebrand for XPS measurements and DFG for funding.

References

1 M. P. Staiger, A. M. Pietak, J. Huadmai and G. Dias, *Biomaterials*, 2006, **27**, 1728–1734.
2 F. Witte, N. Hort, C. Vogt, S. Cohen, K. U. Kainer, R. Willumeit and F. Feyerabend, *Curr. Opin. Solid State Mater. Sci.*, 2008, **12**, 63–72.
3 M. Moravej and D. Mantovani, *Int. J. Mol. Sci.*, 2011, **12**, 4250–4270.
4 Y. Xin, T. Hu and P. K. Chu, *Acta Biomater.*, 2011, **7**, 1452–1459.
5 S. Virtanen, *Mater. Sci. Eng., B*, 2011, **176**, 1600–1608.
6 Y. F. Zheng, X. N. Gu and F. Witte, *Mater. Sci. Eng., R*, 2014, **77**, 1–34.
7 J. Walker, S. Shadanbaz, T. B. F. Woodfield, M. P. Staiger and G. J. Dias, *J. Biomed. Mater. Res., Part B*, 2014, **102**, 1316–1331.
8 M. Peuster, C. Hesse, T. Schloo, C. Fink, P. Beerbaum and C. von Schnakenburg, *Biomaterials*, 2006, **27**, 4955–4962.
9 G. S. Wu, J. M. Ibrahim and P. K. Chu, *Surf. Coat. Technol.*, 2013, **233**, 2–12.
10 H. Hornberger, S. Virtanen and A. R. Boccaccini, *Acta Biomater.*, 2012, **8**, 2442–2455.
11 S. Shadanbaz and G. J. Dias, *Acta Biomater.*, 2012, **8**, 20–30.
12 V. Wagener, M. S. Killian, C. M. Turhan and S. Virtanen, *Colloids Surf., B*, 2013, **103**, 586–594.
13 Y. C. Xin, T. Hu and P. K. Chu, *J. Electrochem. Soc.*, 2010, **157**, C238–C243.
14 X. N. Gu, Y. F. Zheng and L. J. Chen, *Biomed. Mater.*, 2009, **4**, 065011.
15 S. Keim, J. G. Brunner, B. Fabry and S. Virtanen, *J. Biomed. Mater. Res., Part B*, 2011, **96**, 84–90.

16 P. Silva-Bermudez and S. E. Rodil, *Surf. Coat. Technol.*, 2013, **233**, 147–158.
17 Y. Hedberg, X. Wang, J. Hedberg, M. Lundin, E. Blomberg and I. O. Wallinder, *J. Mater. Sci.: Mater. Med.*, 2013, **24**, 1015–1033.
18 Y. Hedberg, M. E. Karlsson, E. Blomberg, I. O. Wallinder and J. Hedberg, *Colloids Surf., B*, 2014, **122**, 216–222.
19 W.-D. Mueller, M. Fernández Lorenzo de Mele, M. L. Nascimento and M. Zeddies, *J. Biomed. Mater. Res., Part A*, 2009, **90**, 487–495.
20 C. L. Liu, Y. C. Xin, X. B. Tian and P. K. Chu, *J. Mater. Res.*, 2007, **22**, 1806–1814.
21 C. L. Liu, Y. J. Wang, R. C. Zeng, X. M. Zhang, W. J. Huang and P. K. Chu, *Corros. Sci.*, 2010, **52**, 3341–3347.
22 R. Rettig and S. Virtanen, *J. Biomed. Mater. Res., Part A*, 2008, **85**, 167–175.
23 A. Yamamoto and S. Hiromoto, *Mater. Sci. Eng., C*, 2009, **29**, 1559–1568.
24 F. Feyerabend, H. Drucker, D. Laipple, C. Vogt, M. Stekker, N. Hort and R. Willumeit, *J. Mater. Sci.: Mater.Med.*, 2012, **23**, 9–24.
25 L. Yang, N. Hort, R. Willumeit and F. Feyerabend, *Corros. Eng., Sci. Technol.*, 2012, **47**, 335–339.
26 P. Wan, X. Lin, L. Tan, L. Li, W. Li and K. Yang, *Appl. Surf. Sci.*, 2013, **282**, 186–194.
27 J. Zhang, N. Kong, Y. Shi, J. Niu, L. Mao, H. Li, M. Xiong and G. Yuan, *Corros. Sci.*, 2014, **85**, 477–481.
28 S. Omanovic and S. G. Roscoe, *Langmuir*, 1999, **15**, 8315–8321.
29 A. Kocijan, I. Milosev and B. Pihlar, *J. Mater. Sci.: Mater. Med.*, 2003, **14**, 69–77.
30 M. L. C. A. Afonso, R. F. V. Villamil Jaimes, E. P. G. Arêas, M. R. Capri, E. Oliveira and S. M. L. Agostinho, *Colloids Surf., A*, 2008, **317**, 760–763.
31 M. Lundin, Y. Hedberg, T. Jiang, G. Herting, X. Wang, E. Thormann, E. Blomberg and I. O. Wallinder, *J. Colloid Interface Sci.*, 2012, **366**, 155–164.
32 I. Frateur, L. Lartundo-Rojas, C. Méthivier, A. Galtayries and P. Marcus, *Electrochim. Acta*, 2006, **51**, 1550–1557.
33 T. Peters Jr, in *All About Albumin*, ed. T. Peters, Academic Press, San Diego, 1995, p. 9-II.
34 T. Imoto, L. N. Johnson, A. C. T. North, D. C. Phillips and J. A. Rupley, in *The Enzymes*, ed. P. Boyer, Academic Press, New York, 1972, vol. 7, pp. 665–868.
35 O. Zschörnig, G. Paasche, C. Thieme, N. Korb and K. Arnold, *Colloids Surf., B*, 2005, **42**, 69–78.
36 W. G. Burton, K. D. Nugent, T. K. Slattery, B. R. Summers and L. R. Snyder, *J. Chromatogr. A*, 1988, **443**, 363–379.
37 N. J. Laible and G. R. Germaine, *Infect. Immun.*, 1982, **36**, 148–159.
38 M. Kosmulski, *Adv. Colloid Interface Sci.*, 2009, **152**, 14–25.
39 R. W. Evans, R. Rafique, A. Zarea, C. Rapisarda, R. Cammack, P. J. Evans, J. B. Porter and R. C. Hider, *J. Biol. Inorg. Chem.*, 2008, **13**, 57–74.
40 Y. C. Guillaume, E. Peyrin and A. Berthelot, *J Chromatogr B*, 1999, **728**, 167–174.
41 L. Müller and F. A. Müller, *Acta Biomater.*, 2006, **2**, 181–189.
42 J. H. Scofield, *J. Chromatogr. B: Biomed. Sci. Appl.*, 1976, **8**, 129–137.
43 R. F. Reilman, A. Msezane and S. T. Manson, *J. Electron Spectrosc. Relat. Phenom.*, 1976, **8**, 389–394.
44 D. S. Mantus, B. D. Ratner, B. A. Carlson and J. F. Moulder, *Anal. Chem.*, 1993, **65**, 1431–1438.
45 M. E. Kargacin and B. R. Kowalski, *Anal. Chem.*, 1986, **58**, 2300–2306.

Faraday Discussions

PAPER

In situ monitoring of corrosion mechanisms and phosphate inhibitor surface deposition during corrosion of zinc–magnesium–aluminium (ZMA) alloys using novel time-lapse microscopy†

James Sullivan,*[a] Nathan Cooze,[a] Callum Gallagher,[a] Tom Lewis,[a] Tomas Prosek[b] and Dominique Thierry[b]

Received 8th December 2014, Accepted 21st January 2015
DOI: 10.1039/c4fd00251b

In situ time-lapse optical microscopy was used to examine the microstructural corrosion mechanisms in three zinc–magnesium–aluminium (ZMA) alloy coated steels immersed in 1% NaCl pH 7. Preferential corrosion of $MgZn_2$ lamellae within the eutectic phases was observed in all the ZMA alloys followed by subsequent dissolution of Zn rich phases. The total extent and rate of corrosion, measured using time-lapse image analysis and scanning vibrating electrode technique (SVET) estimated mass loss, decreased as Mg and Al alloying additions were increased up to a level of 3 wt% Mg and 3.7 wt% Al. This was probably due to the increased presence of MgO and Al_2O_3 at the alloy surface retarding the kinetics of cathodic oxygen reduction. The addition of 1×10^{-2} mol dm^{-3} Na_3PO_4 to 1% NaCl pH 7 had a dramatic influence on the corrosion mechanism for a ZMA with passivation of anodic sites through phosphate precipitation observed using time-lapse image analysis. Intriguing rapid precipitation of filamentous phosphate was also observed and it is postulated that these filaments nucleate and grow due to super saturation effects. Polarisation experiments showed that the addition of 1×10^{-2} mol dm^{-3} Na_3PO_4 to the 1% NaCl electrolyte promoted an anodic shift of 50 mV in open circuit potential for the ZMA alloy with a reduction in anodic current of 2.5 orders of magnitude suggesting that it was acting primarily as an anodic inhibitor supporting the inferences from the time-lapse investigations. These phosphate additions resulted in a 98% reduction in estimated mass loss as measured by SVET demonstrating the effectiveness of phosphate inhibitors for this alloy system.

[a]*College of Engineering, Swansea University, Singleton Park, Swansea SA2 8PP, UK. E-mail: j.h.sullivan@swansea.ac.uk*

[b]*Institut de la Corrosion, 220 Rue Pierre Rivoalon, 29200 Brest, France*

† Electronic supplementary information (ESI) available. See DOI: 10.1039/c4fd00251b

1. Introduction

There is significant interest in steel companies world wide in the production of ternary zinc–magnesium–aluminium (ZMA) galvanised alloy coatings for the cathodic protection of steel substrates. This new class of coating offers potential improvements in corrosion resistance that are of interest to a wide variety of markets from construction to automotive. This increase in corrosion resistance offers improved coating lifetimes for lower coating weights providing a significant cost benefit to the manufacturer. The development of ZMAs follows enhancements to the corrosion resistance of hot dipped galvanised steel coatings over the last 20 years through binary alloy additions such as Al. Al additions of 5 wt% (Galfan) and 55 wt% (Galvalume) have demonstrated significant advances in corrosion resistance over traditional Hot Dipped Galvanised (HDG) coatings.[1,2] The morphology of the microstructure and production conditions of these binary alloys have dramatic effects on the corrosion resistance of Zn/Al alloy coated steels with the corrosion rate related to dendrite size, nucleation rate and volume fraction of phases.[3–5] Attempts have been made to further improve Zn/Al alloy coatings on steel through small ternary additions of Mg of 0.05 wt% but this resulted in negative effects on the cut-edge corrosion resistance due to microstructural changes.[6] Current research in this field has focussed on higher additions of ternary Mg of up to 3 wt% and these alloys have demonstrated increased corrosion protection over standard Zn-0.2 wt% Al galvanised coatings in accelerated corrosion tests.[7–10] The mechanism of the increased corrosion resistance of these alloys, especially in chloride rich environments, has been investigated by a number of authors with regard to stable corrosion product film formation[7,11–15] and the effect of carbon dioxide in the atmosphere.[16] The increased resistance of such coatings to organic coating delamination has also been investigated by focussing on the electrochemical behaviour of $MgZn_2$ and Zn phases that are prevalent in ZMA alloys.[17] The introduction of Mg into these alloys can produce a variety of phases within the microstructure[18,19] and thus exploration of the effect of these phases on the corrosion performance is critical in choosing the best alloy compositions for production. *In situ* aqueous time-lapse microscopy has been previously demonstrated by the authors to show that $MgZn_2$ phases were preferentially attacked in the aqueous corrosion of a ZMA containing 1.5–2 wt% Mg 1.5–2 wt% Al additions.[20] The initial stages of corrosion and preferential phase attack of such alloys has also been assessed at a nanoscopic level in other research.[21]

The development of new ZMA alloy coatings is concurrent with the drive to find replacements for chromate corrosion inhibitors that are being phased out of use through REACH legislation by 2017. Thus, an understanding of the behaviour of these new alloys with chromate replacement technologies is of significant interest within a number of industrial sectors. Phosphate ions have been shown to work effectively as a corrosion inhibitor on galvanised steels in a number of investigations[22–25] but there is currently little research on the evaluation of such ions on ZMA alloy coatings.

This current paper demonstrates a novel time-lapse microscopy technique that enables *in situ* monitoring of the microstructural aqueous corrosion mechanisms of three ZMA alloys with increasing Mg and Al alloy additions. The technique provides valuable mechanistic and rate information regarding preferential phase attack within the microstructure of the alloys allowing the assessment of these alloy additions on the corrosion of the alloys to be determined. The SVET was used to evaluate metal loss over a 24 h period to provide an additional assessment of the performance of the three ZMAs. In addition, a further investigation involving the novel time lapse technique, SVET and polarisation studies was carried out using one ZMA to assess the mechanism of corrosion inhibition by sodium phosphate.

2. Experimental procedure

2.1 ZMA alloys

The ZMA alloys used in this study had increasing levels of Mg and Al levels. The compositions and coating thicknesses of the three alloys are given in Table 1. All alloys were produced under production line conditions from commercial producers.

2.2 *In situ* time-lapse optical microscopy for assessment of microstructural corrosion mechanisms of ZMA alloys

A novel system has been developed that allows imaging of corroding surfaces under immersion conditions to be imaged at a microstructural level. The technique has been iterated based on previous research[20] and encompasses a waterproof shroud made of polyethylene that is placed over the lens of a Meiji MT8000 microscope. The base of the shroud encases a glass window that permits imaging of the sample surface whilst the lens is immersed in the electrolyte. A schematic of the microscope set-up is given in Fig. 1.

Samples of ZMA alloys were prepared by mounting a 20 mm × 20 mm piece of alloy coated steel in phenolic resin and polishing the surface down 2 μm to reveal a flat surface and to remove manufacturing oils and oxides from the surface. The surface was etched using 2% Nital solution to reveal the microstructure, taped with PTFE leaving only a circular feature of diameter 1 mm (area of 0.785 mm^2) exposed. The sample was then securely fixed to a stage in a glass tank using PTFE tape and 250 ml of 1% NaCl pH 7 electrolyte added to the tank so that the sample was immersed. The lens of the microscope was then manoeuvred into place so that imaging of the microstructure under the electrolyte could be achieved. Images were captured every two minutes using the time-lapse function of an infinity 2 digital camera attached to the microscope. The duration of experiments were 17 h for time-lapse analysis of the three ZMA alloys. Time-lapse videos were compiled from the still images after each experiment using Microsoft Movie Maker.

Table 1 Compositions in weight% (wt%) of the three Zn–Mg–Al (ZMA) alloys under investigation

Sample	Zn (wt%)	Mg (wt%)	Al (wt%)	Coating thickness (μm)
ZMA 1	98	1	1	10
ZMA 2	96	2	2	10
ZMA 3	93.3	3	3.7	10

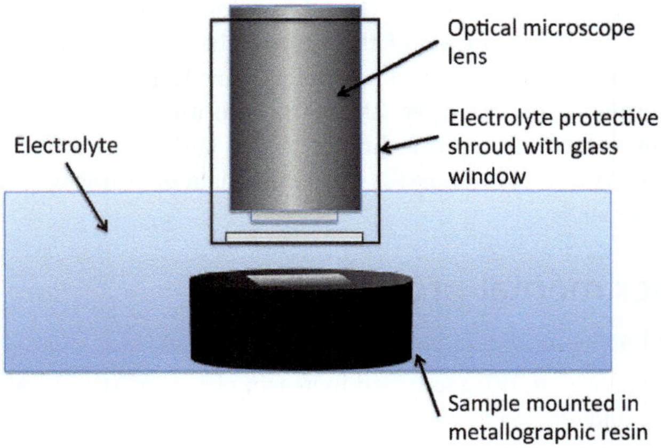

Fig. 1 Schematic diagram of *in situ* optical microscopy configuration that allows imaging of corroding metals under immersion conditions. The microscope lens is shrouded from the electrolyte by a protective cover containing a glass window to allow imaging of corrosion mechanism within the microstructure of the sample.

2.3 *In situ* time-lapse optical microscopy for assessment of the mechanism of phosphate corrosion inhibition of ZMA alloys

The same experimental set-up was used as in Section 2.2. For these tests ZMA 2 was used as the base substrate and experiments were initiated in 1% NaCl to establish initial corrosion activity. After 3 h, additions of Na_3PO_4 were made to the electrolyte to establish a concentration of 1×10^{-2} mol dm^{-3} Na_3PO_4 and time-lapse images taken over a total of 36 h to assess the effect of phosphate on the corrosion mechanisms.

2.4 SVET for the assessment of corrosion performance of ZMA alloys and the effectiveness of phosphate inhibition on ZMA systems

SVET scans were performed on the three ZMA alloys to provide an additional insight into their relative corrosion performance. ZMA samples were prepared as in the same manner for time-lapse experiments as described in Section 2.2 but an area of 100 mm^2 was left exposed on the sample surface. The SVET probe made 50 measurements along the length of the sample and 50 measurements across the sample width generating a matrix of 2500 data points for each scan. The SVET tests were carried out in 1% NaCl pH 7 solution, one scan was taken every hour for 24 h and three repeat tests for each material were carried out. The dissolved oxygen concentration in bulk solution was assumed to be constant at 2.8×10^{-4} mol dm^{-3}, the equilibrium concentration for air saturated water, and all tests were carried out at 25 °C. For the assessment of phosphate inhibition samples of a single alloy coated steel, ZMA 2, were prepared in the same manner and then tested in 1% NaCl containing (a) no inhibitor (b) 1×10^{-4} mol dm^{-3} Na_3PO_4 (c) 1×10^{-2} mol dm^{-3} Na_3PO_4 for 24 h.

2.4.1 SVET apparatus and calibration. The SVET has proved to be a valuable tool in the investigation of cut-edge corrosion in zinc coated and organically coated steels.[3–6,23,26–29] The SVET apparatus is fully described elsewhere.[30–32] The SVET consists of a glass encased, 125 μm diameter platinum wire micro-tip that is vibrated at a constant frequency, amplitude and height above a corroding surface submerged in electrolyte. SVET detects an alternating potential at the vibration frequency which is proportional to the potential gradient in solution in the direction of vibration. Movement of the micro tip is stepwise and controlled by a micro manipulator stage (time and precision) in three dimensions with PC control and a lock-in amplifier (EG & G model 7625) serves to control the frequency of vibration and also measure the signal detected by the probe. Vibration is provided from the lock-in amplifier *via* an amplifier at a frequency of 140 Hz with the drive voltage producing an amplitude of probe vibration of 25 μm.

SVET was used to show the location and intensity of anodes and cathodes on corroding surfaces. The SVET probe was scanned over the surface of a sample in the x, y and z directions at a fixed height of 100 μm. At a site of corrosion potential fields are set-up in solution by ionic current flux passing through the electrolyte emanating from the anodic site. The SVET probe cuts through these lines and registers an alternating potential that is proportional to the potential gradient in the direction of vibration. The SVET signal is thus directly proportional to the component of ionic current density in the direction of vibration at the scan height.

Calibration of the SVET was achieved using a two compartment cell. Each compartment contains a Pt electrode and the electrolyte under investigation and the compartments are linked by a vertically oriented glass tube of 4.8 mm internal diameter. Calibration was achieved by placing the SVET tip inside the vertical bore of the glass tube. Different currents were passed through this glass tube using a nano-galvanostat and the current density calculated for the area of the glass tube (radius $= 2.4 \times 10^{-3}$ m). The tube allows the lines of current flux to align vertically along its length and thus the SVET tip is vibrated in an area of uniform current density. For each current density the SVET voltage was measured and was found to vary linearly with applied current density. The gradient of a plot of current density *versus* SVET voltage provides a calibration factor for the instrument and a typical calibration factor in 1% NaCl was 90 000 A m^{-2} V^{-1}. The use of this tube cell for calibration is rapid and independent of probe height and provides a rapid means of converting voltage data detected from the SVET into A m^{-2}.

2.4.2 Manipulation of SVET data. Data from the SVET can be used to semi-quantitatively estimate metal loss over time by integration of positive current data and the application of Faraday's law and thus can be used to compare the performance of similar materials. The mass loss data is semi-quantitative due to a number of assumptions made in the calculations and thus the metal loss data should not be used as a direct measure of the sample weight loss. It is very useful, however, in comparing the performance of similar materials such as galvanised steel materials with different coatings or processing conditions. By assuming the corrosion activity occurring on the sample remains constant in the interval between scans and that the corrosion generally involves the release of a particular metal ion, *e.g.* Zn^{2+}, it is possible to deliver a value of total metal loss for the 24-h experiment. This is achieved by integrating the area under the current density as measured by the SVET to obtain a total current for each scan. These currents are converted to total charge per scan using the time interval between scans with Faraday's law then applied to calculate a mass loss of the ion of interest *e.g.* Zn. A summation of these values for all 24 scans per sample then provides a total mass loss. The mathematical procedure for these calculations is fully described elsewhere.[4]

2.4.3 Limitations of SVET technique. There are a number of limitations associated with the assessment of metal loss from the SVET and thus it should be considered as a semi-quantitative technique. The SVET detects only normal components of current flux that emanate from an anode and also any current loops that terminate under the plane of scan of 100 μm will not be detected. The SVET resolution is approximately equal to 1.5 times scan height of 100 μm so it will not detect anodes and cathodes separated by less than this distance. The SVET measures a snapshot of the corrosion for each hour and assumes a constant corrosion rate over that hour whereas in reality this rate could vary. Despite these limitations the SVET has proved to be a reliable predictor of corrosion behaviour for Zn coated steel[3,5] and also compares favourably with external weathering Zn run-off tests.[33]

2.5 Polarisation testing of a ZMA alloy in 1% NaCl with additions of Na_3PO_4

Polarisation tests were carried out on samples of alloy ZMA 2 in 1% NaCl containing (a) no inhibitor (b) 1×10^{-4} mol dm^{-3} Na_3PO_4 (c) 1×10^{-2} mol dm^{-3} Na_3PO_4. The potential was swept from −1.15 V to −0.8 V at a rate of 0.1 mV s^{-1} using a Solartron potentiostat. All potentials were measured *versus* a standard calomel electrode (SCE).

3. Results and discussion

3.1 Examination of the microstructures of the ZMA alloys

Optical microscope images of the microstructures of the three ZMA alloys are shown in Fig. 2A–C.

The microstructures were generally composed of primary Zn dendrites and two eutectic phases. The volume fraction of primary Zn rich phase and eutectics varied substantially as alloying additions were increased as shown in Fig. 3.

There was a decrease in primary Zn phase volume from 82% in the 1 wt% Mg 1 wt% Al to 9% in the 3 wt% Mg 3.7 wt% Al with a subsequent increase in eutectic phase from 18 to 91% respectively. Furthermore, the morphology of the microstructure varied from sample to sample with coarse primary Zn rich dendrites observed in the 1% and 2% alloyed samples compared with much finer dendrites in the 3% sample potentially indicative of a faster cooling rate for this sample during production. SEM images of the three alloys are shown in Fig. 2D–F and show the presence of two eutectic phases in the structure, one having coarse lamellae and the other, fine lamellae. EDX analysis of the microstructures is shown in Fig. 2G–I and demonstrates that the coarse eutectic was composed of Zn and Mg and thus a binary eutectic whereas the finer eutectic was composed of Zn, Mg and Al. Previous research describing the solidification sequence of such alloys has shown that the binary eutectic was composed of Zn and $MgZn_2$ lamellae and the ternary eutectic of Zn, $MgZn_2$ and Al nodules for samples produced under production line conditions.[18,20] In addition, the higher magnification SEM images identified small regions of Al rich dendrites that become more prevalent in the higher alloyed samples and this has been previously identified as a face centred cubic Zn/Al dendritic phase.[18]

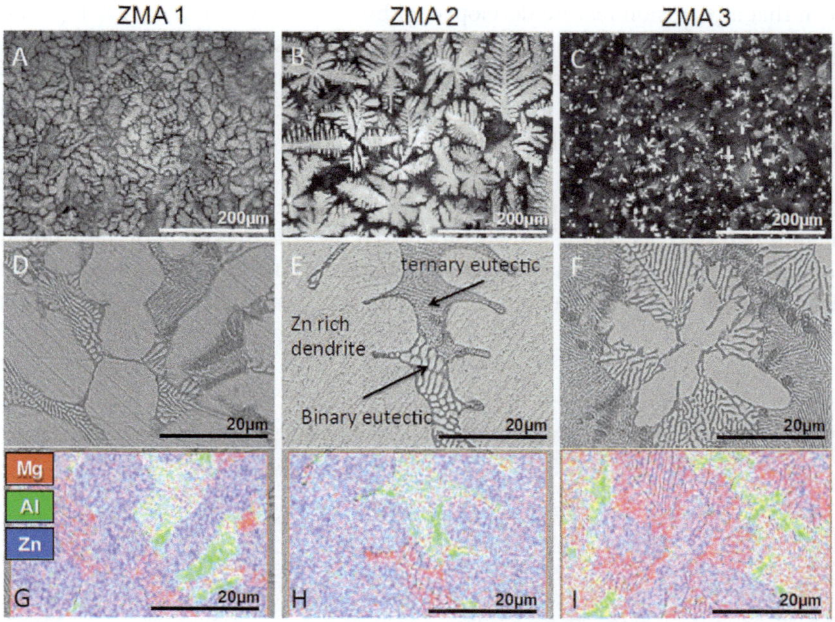

Fig. 2 (A–C) Optical microscope images of the three ZMA samples with increasing alloy contents showing a Zn primary dendritic phase in a matrix of eutectic. An increase in eutectic volume fraction for increasing alloy levels was observed. (D and E) SEM images of the samples showing coarse binary and fine ternary eutectic phases. (G–I) EDX analysis of the distribution of Zn, Mg and Al showing that the fine structure eutectic is of ternary composition.

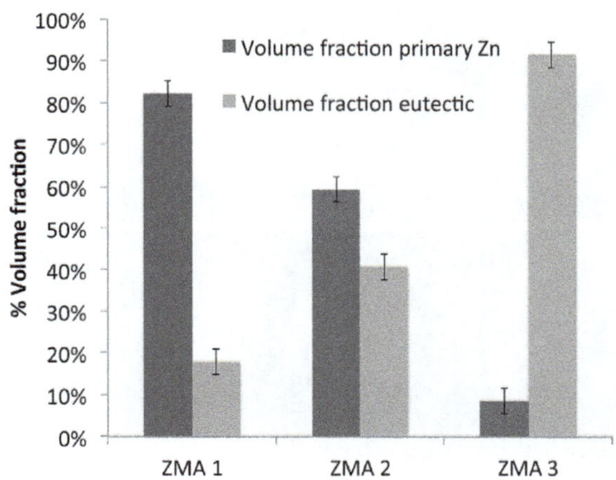

Fig. 3 Volume fraction of primary Zn dendrites and eutectic phases for ZMA alloys with increasing Mg and Al levels showing a large rise in eutectic phase fraction as Mg and Al levels are increased to 3 wt% and 3.7 wt% respectively in ZMA 3.

3.2 Time-lapse corrosion of the ZMA alloys

Fig. 4 images 1–12 show the microstructure of the three ZMA alloys taken at 2 h intervals during *in situ* immersion in 1% NaCl. From Fig. 4 images 1–4 it can be seen that a corrosion feature developed within the microstructure of ZMA 1 and grew in size with time towards the centre of the exposed sample area.

The front of the anodic activity progressed preferentially through the eutectic phases as shown in Video 1.† As anodic growth proceeded a ring of corrosion product was precipitated ahead of the anode indicating the boundary of ionic counter currents of metal and hydroxide ions from the anodic and cathodic activity. The ring of corrosion product was observed to be mobile as the experiment progressed. As anodic activity progressed towards the centre of the exposed sample area, the corrosion product ring dissolved and re-precipitated at a set distance away from the advancing anodic front. This was indicative of pH changes

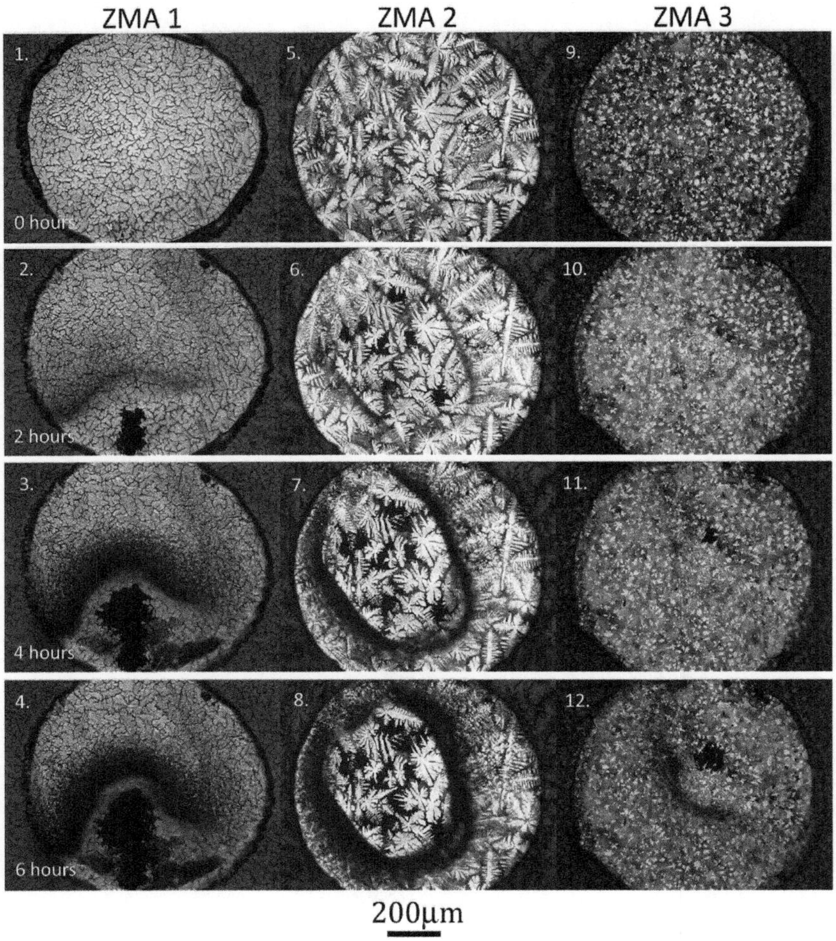

Fig. 4 Optical microscope images of ZMA taken *in situ* under immersion conditions in 1% NaCl. The images shown were taken at 2-h intervals for each alloy. Images 1–4 are of alloy ZMA 1 showing the development and progression of a corroding area and a mobile corrosion product ring. Images 5–8 are of ZMA 2 showing clear preferential attack of the eutectic phases and the formation of a corrosion product ring between the anodic and cathodic areas on the sample surface. Images 9–12 are of ZMA 3 and show no corrosion for the first 2 h followed by preferential attack of the eutectic phases and the formation of a corrosion product ring.

produced as a result of metal dissolution at the anode potentially causing a local reduction in pH that could cause corrosion products to re-solubilise. Fig. 4 images 5–9 show comparative results for ZMA 2. Here, anodic activity clearly initiated and progressed preferentially through the eutectic phases of the microstructure whilst the Zn rich dendrites remained relatively untouched after 6 h. Once again a clear corrosion product ring formed around the anodic sites indicating the transition to areas of cathodic activity on the surface of the alloy. This mechanism is shown in Video 2.† Fig. 4 images 9–12 shows the same experimental conditions but for ZMA 3. Here, there was little corrosion observed in the first 2 h. Between 2 and 4 h (Fig. 4 images 10 and 11) anodic activity initiated and once again progressed through the eutectic phases of the material preferentially leaving islands of primary Zn rich phase. The growth of anodic activity was markedly reduced in comparison to ZMA 1 and 2. This experiment is shown in Video 3.†

Further exploration of the corrosion mechanism was carried out using SEM investigations where a circular area of each sample of 0.785 mm^2 was imaged at 0 h and after 4 h exposure to 1% NaCl. This was performed in order to provide further evidence for the preferential phase attack in the microstructure. The results of the experiment for ZMA 1 are shown in Fig. 5A–D.

Fig. 5A shows the microstructure of ZMA 1 on initial exposure to the electrolyte. Fig. 5B was taken after 4 h exposure to 1% NaCl and areas of anodic activity were seen to have initiated in the upper right quadrant of the circular area with a characteristic corrosion product ring forming a boundary between the anodic and cathodic sites. Higher magnification images of an anodic region are given in Fig. 5C and D and show that anodic activity was focussed on the eutectic phases between the Zn rich dendrites. The highest magnification image, Fig. 5D, demonstrated that regions of binary eutectic had experienced de-alloying of the $MgZn_2$ phase leaving intact regions of Zn lamellar. These images support previous findings that Mg rich phases in ZMA alloys experience preferential attack due to their higher activity and also supports the notion that preferential dissolution in the initial stages of corrosion rapidly develops Mg based corrosion products that provide these alloys with enhanced corrosion resistance compared with standard hot dip galvanised Zn coatings.[15,17,20,21] As corrosion progressed Zn rich dendrites were indeed corroded from the alloy but the progress of anodic fronts was always associated with eutectic attack in the first instance. The same mechanism was observed in ZMA 2 and 3 with eutectic attack in both samples after 4 h in 1% NaCl.

From the time-lapse images it was possible to estimate a corroded area per unit time based on the area of anodic attack. This makes an assumption that the majority of the corrosion progresses laterally across the surface rather than vertically into the specimen. The corrosion was calculated by highlighting the anodic area using Photoshop CS6 on images at hourly intervals from the time-lapse sequences. Photoshop was then used to calculate this area in mm^2 based on a calibration image of a 1 mm standard imaged using identical experimental conditions. Plots of the evolution of corroded area *versus* time and the total corroded area for ZMA 1, 2 and 3 are shown in Fig. 6. From Fig. 6A it can be seen that the growth of corrosion features were relatively linear with time for all the ZMA samples and Fig. 6B shows that total corroded area reduced as alloying levels increased.

Increased alloying additions of Mg and Al changed the phase percentages in the microstructure as shown in Fig. 3. The total corroded area after 17 h was

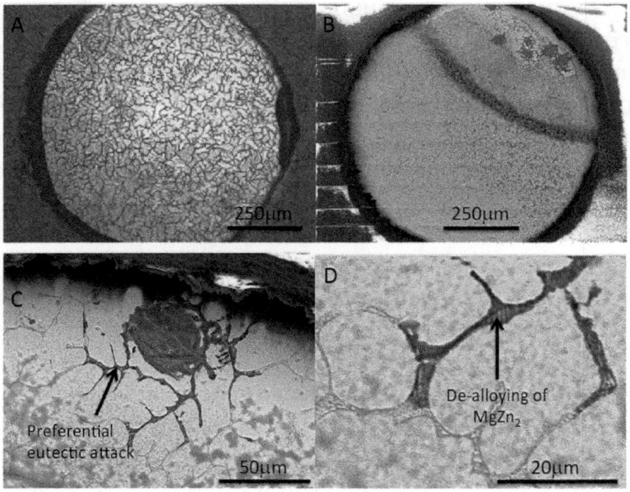

Fig. 5 (A) Microstructure of ZMA 1 immediately on immersion in 1% NaCl. (B) The same sampled as in (A) but imaged after 4 h in 1% NaCl showing anodic features in the upper right quadrant and a corrosion product ring. (C) Magnified anodic region showing preferential corrosion of eutectic phases. (D) Further examination of preferential corrosion sites showing de-alloying of $MgZn_2$ phase from the eutectic structure.

plotted *versus* the volume fraction of eutectic phase in the microstructure and a linear decrease in corroded area was observed as volume fraction of eutectic increased as shown in Fig. 7.

The quantity of eutectic phase dramatically increases from 18% to 91% from ZMA 1 to ZMA 3. It may be counter intuitive that an increased eutectic volume fraction would reduce the corrosion rate given that these phases are preferentially attacked in the microstructure but it would seem that increased levels of Mg and Al in the alloy create more robust oxide layers due to their rapidity of reaction and the passive nature of Al_2O_3 layers. It has been suggested that MgO formation on

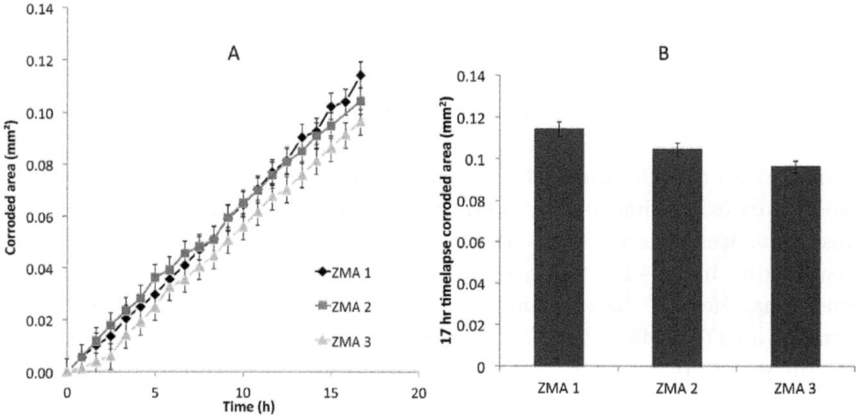

Fig. 6 (A) Size of corroded area *versus* time for the three ZMA samples showing a general linear corrosion for each material in 1% NaCl (B) total corroded area after 17 h in 1% NaCl for each ZMA sample showing reduced corrosion rates with increasing alloy content.

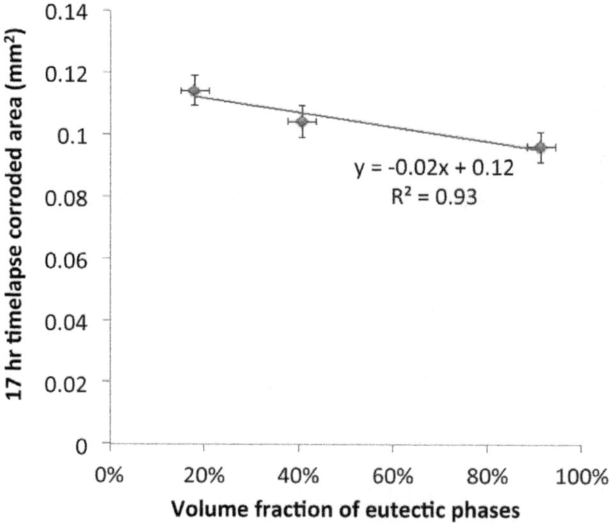

Fig. 7 Graph showing the total corroded area of the three ZMA alloys against the percentage of eutectic phases within their structure showing a linear decrease in overall corrosion as the volume fraction of eutectic phases increased.

the surface of such coatings reduces electron transfer rates due to the wide band gap of this oxide.[34,35] Increasing Al levels within the structure of Zn coatings has been previously shown to reduce the corrosion rate as Al_2O_3 present on the surface of the coatings has a low efficiency for cathodic oxygen reduction.[36] Thus, as alloying levels increase the kinetics of corrosion are reduced but the preferential attack is still thermodynamically more favourable in the $MgZn_2$ regions of the eutectic phases. Visual indicators are provided suggesting this for ZMA 3 in Fig. 4 images 9 and 10 where anodic activity took a number of hours to initiate in the eutectic regions and a perceptible visual brightening of the surface occurred in the initial hours of exposure to 1% NaCl potentially indicating breakdown of passive oxide layers. This was further highlighted by the slow initial rate of anodic growth over the period of 0–3 hours for ZMA 3 in Fig. 6A.

3.3 Scanning vibrating electrode technique (SVET) examination of corrosion behaviour of ZMA alloys

The SVET was used to examine the corrosion of the three ZMA alloys over 24 h in 1% NaCl at pH 7. The SVET was used to estimate the metal loss from the alloys over the 24-h period as described in Section 2.4.2. The metal chosen for use in Faraday's equation was Zn (65.5) as the predominant element in each alloy. It was obviously difficult to predict the contribution to the detected corrosion from each element in the alloy but as the most likely dissolution involved Zn^{2+} and Mg^{2+} ions as demonstrated by the time-lapse investigation, the relative performance relationship between each alloy would be the same if either metal were used in the calculation as each ion has the same charge with a difference arising simply in the ultimate mass loss value. As such the SVET mass estimations have been used in this investigation to simply compare the relative corrosion behaviour of the ZMA

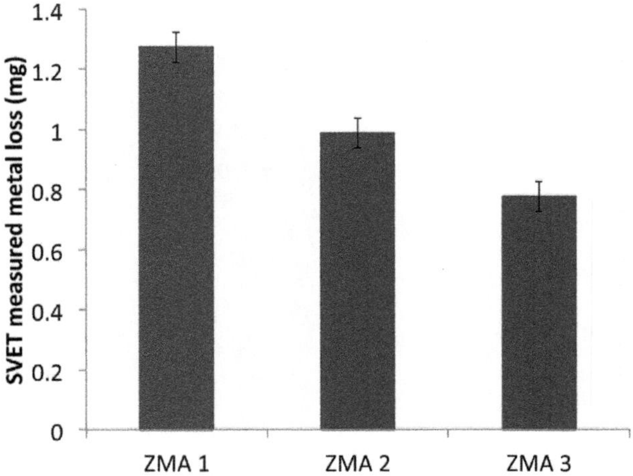

Fig. 8 SVET estimated metal loss for the three ZMA alloys after 24 h exposure in 1% NaCl. A reduction in metal loss was observed for increased levels of Mg and Al alloying additions.

alloys rather than provide a precise value for mass lost from each alloy during experiment.

Mass loss data for the three ZMA alloys as estimated by SVET is presented in Fig. 8.

Fig. 8 once again demonstrated that increasing alloy additions of Mg and Al in the three ZMA alloys had a beneficial effect in terms of corrosion behaviour. For ZMA 2 the total metal loss was 22% lower than ZMA 1 and for ZMA 3 the metal loss was 39% lower than ZMA 1. These represent a significant performance increase. The performance differences in the SVET investigation were more marked than in the time-lapse experiments and this may be due to the limitation of using a 2D area for corrosion measurement in the time-lapse experiments whereas the SVET measurement is dependent on the detection of ionic current flux in the solution above the corroding surface. The SVET measurements were also made over a large area of 100 mm^2 that may also have had an impact in that a number of corrosion features were established over this area over the course of the test thus producing a better average performance indicator for the alloy. Overall though it was encouraging that both experiments provided the same performance ranking for the alloys and in particular provides a degree of confidence in the mass loss predictions from SVET.

3.4 The effect of sodium phosphate additions to the corrosion behaviour of a ZMA alloy

Investigations were carried out to assess the behaviour of phosphate ions on the inhibition of corrosion of one of the ZMA alloys, ZMA 2. The mechanism and extent of phosphate action on these alloys is of much interest as phosphate is rapidly becoming the main choice of inhibitor ion in place of chromate in coated steel products due to imminent legislation changes on the use of chromate.

The mechanism of phosphate action on ZMA 2 was examined using *in situ* time-lapse optical microscopy, SVET and polarisation studies through additions of Na_3PO_4 to 1% NaCl pH 7.

3.4.1 Time-lapse investigation of the mechanism of phosphate action on ZMA alloys. Fig. 9 images 1–12 shows a 36 h time-lapse sequence for ZMA 2 immersed in 1% NaCl pH 7 with 1×10^{-2} mol dm^{-3} Na_3PO_4 added after 3 h of immersion. Image 1 represents the point of inhibitor addition with subsequent images taken at 3 h intervals. The sample was left to corrode for 3 h prior to inhibitor addition in order for some anodic features to become established thus potentially making the mechanism of inhibition easier to observe.

The addition of the phosphate inhibitor dramatically alters the observed corrosion behaviour in comparison to that seen in Fig. 4 for ZMA 2. The time-lapse sequence can be seen evolving in Video 4.† After 3–6 h (Fig. 9 images 2 and 3) a

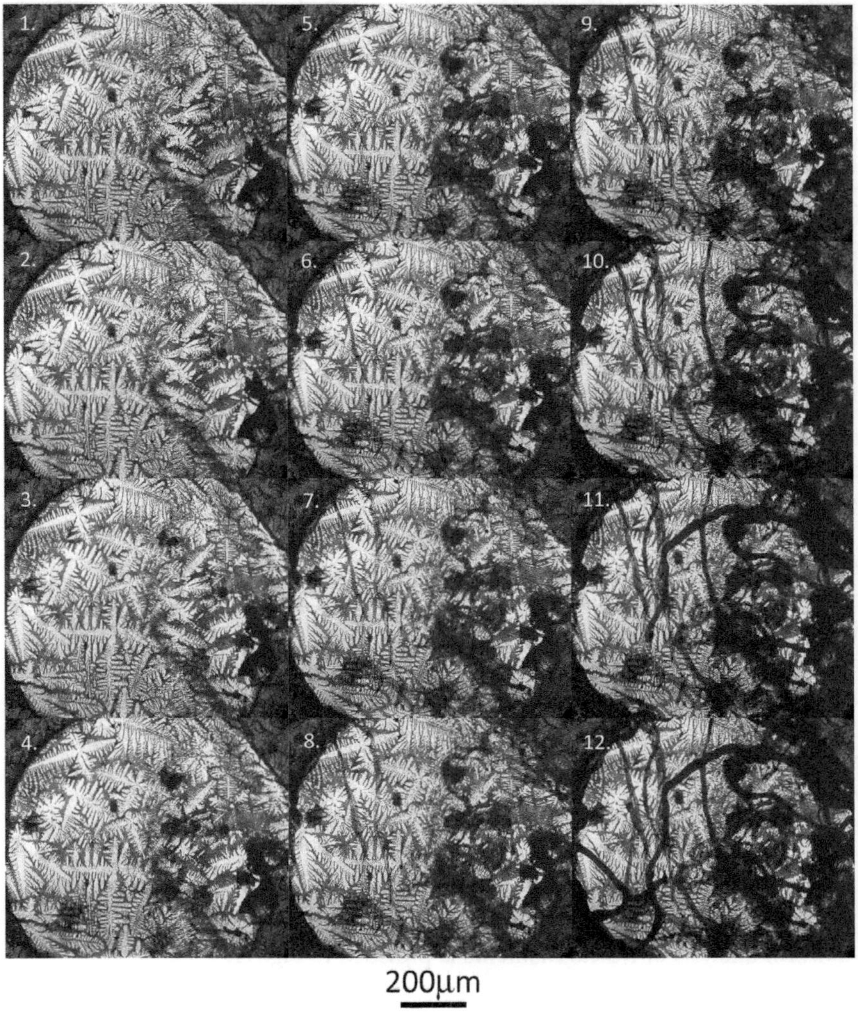

Fig. 9 Images 1–12. Optical microscope images of ZMA 2 taken *in situ* under immersion conditions in 1% NaCl with 1×10^{-2} mol dm^{-3} Na_3PO_4 additions. The images shown were taken at 3 h intervals. Images 1–4 show the initiation and inhibition of anodic features through precipitation local to the anode. Images 5–12 show the development and growth of filamentous precipitates over the surface of the ZMA.

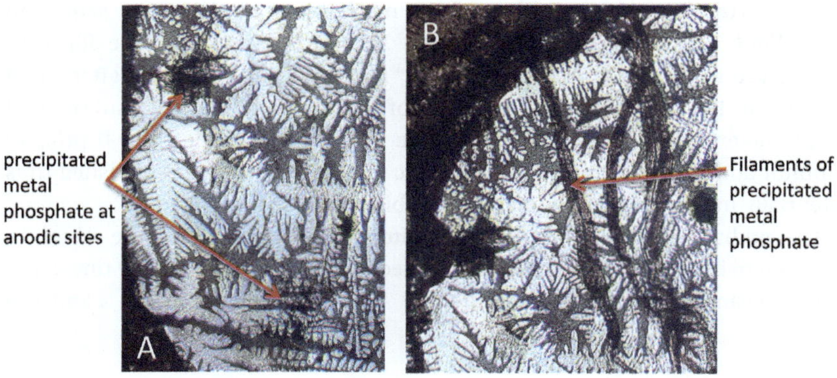

Fig. 10 (A) A region of ZMA 2 microstructure taken *in situ* after 6 h under immersion conditions in 1% NaCl with 1×10^{-2} mol dm^{-3} Na_3PO_4 additions showing precipitation at anodic sites. (B) An image of filamentous precipitates imaged after 30 h exposure to the electrolyte as in (A).

number of anodic features had initiated over the sample surface. Growth of these features was restricted and it was observed that this restricted growth was due to precipitation of a thin solid layer at the anodic site. An image of this deposition at two anodic features is shown in greater detail in Fig. 10A. The profile of these precipitates was triangular tapering away from the anodic site suggesting that a local concentration profile of metallic ions had developed from the anode that subsequently reacted with phosphate ions from solution to form a metal phosphate precipitate. The precipitate was very thin given that the underlying microstructure could still be observed through the deposits. As the experiment progressed (Fig. 9 images 4 and 5) further anodic features initiated but again their growth was somewhat restricted by phosphate deposition at these sites. After 15 h (Fig. 9 image 5) a new morphology of phosphate deposition was observed on the ZMA sample surface. Thin filamentous deposits were seen to rapidly precipitate over large distances on the microstructure. The precipitation initiated from an anodic site but its subsequent growth route was independent of underlying microstructural phases. In fact precipitation was also observed to be independent of the exposed metal on repeated experiments as filaments continued to grow off the metal surface onto the PTFE tape surrounding the exposed area. Initiation and growth of these filamentous deposits continued over the remainder of the 36 h experiment (Fig. 9 images 6–12) with filaments produced towards the end of the experiment being somewhat thicker in depth as the microstructure could no longer be seen through them.

The precipitation and growth of these filaments was unexpected and of some interest. An image of these filaments in greater detail is show in Fig. 10B and SEM/EDX analysis of typical filaments is provided in Fig. 11.

From Fig. 10B it was seen that filaments were often composed of two filaments, one having nucleated on the side of the other and this can be observed happening in the video of the time lapse sequence (Video 4†). It was also seen that the filaments are constrained in their lateral growth whereas forward growth is extremely rapid. Indeed by examination of the individual time-lapse images growth of over 100 μm was observed in the 2 min interval between images. This rapidity of

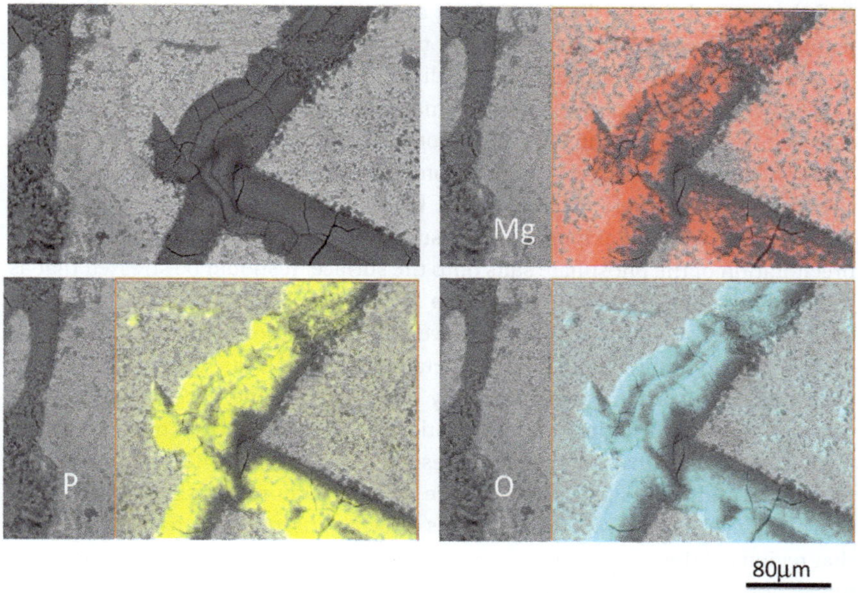

Fig. 11 SEM/EDX analysis of a precipitated filament showing it contained Mg, P and O suggesting the deposition of magnesium phosphate.

precipitation and the independence of precipitation on the surface suggests that nucleation and growth from a locally super saturated solution is occurring. The SEM/EDX data in Fig. 11 showed that the filaments were rich in magnesium, phosphorus and oxygen indicating that they may be precipitates of $Mg_3(PO_4)_2$.

Other EDX data has suggested that some precipitates are composed of zinc phosphate. It is postulated that on exposure to the 1% NaCl solution containing 1×10^{-2} mol dm^{-3} Na_3PO_4 initial anodic attack occurs on the sample surface releasing metallic ions into solution. As the preferential phase attacked was $MgZn_2$ these ions will be Zn^{2+} and Mg^{2+}. The reaction of these metallic species with phosphate anions (PO_4^{3-}) is required for the production of precipitate as $Mg_3(PO_4)_2$ and $Zn_3(PO_4)_2$ both have low solubility in water. Local to the anodic sites, metallic ion concentration will be highest with a decreasing concentration profile of ions with distance away from the site. Thus sufficient metallic ions are present to exceed the solubility limit for Zn or Mg phosphates and precipitation is observed local to the anodic sites with a triangular morphology describing the concentration profile away from the anode. The triangular precipitates are quite intriguing. Diffusion of metal ions from a small anodic feature may be expected to give a radial distribution of precipitate as the concentration profile of ions would be expected to follow a Gaussian distribution with distance away from the site. In this case, such a diffusion profile of metal ions may have been established from the anode with the highest levels of super saturation occurring proximal to the anode thus establishing a preferential site for nucleation and growth of the metal phosphate precipitate. As nucleation and growth of metal phosphate precipitates proceeded local asperities formed at the growing interface. Nucleation of further phosphate precipitates will be more facile on the existing phosphate deposits and diffusion of species to the tip of these features is faster than to the bulk interface and thus growth in the direction of the asperity occurs faster. The width of the feature thus represents a "snapshot" of the local profile of super saturation out into the solution as established by the diffusion profile.

With continued exposure, further small anodic sites were initiated over the surface of the alloy and thus concentrations of metallic ions continued to build in solution. At these sites rapid precipitation of phosphate occurred and new anodic sites initiated elsewhere on the surface. It would seem that these anodic events generate sufficient metal ions to locally super saturate the solution that results in precipitation of the phosphate salts onto the surface. It should be stressed that as the solubility of Mg and Zn phosphate are so low, the amount of metal ions entering solution from these anodic events would be of sufficient quantity to exceed the solubility limit of the metal phosphate salts thus allowing the precipitation to occur but significantly less than for the experiment without inhibitor added due to the rapid passivation of the anodic sites. This seems to be supported by the corrosion test data presented in Sections 3.4.2 and 3.4.3 where phosphate additions significantly reduce the corrosion rate of the alloy.

The nucleation and rapid filamentous growth subsequently observed suggests that regions of the solution have become super saturated with metallic ions. Once nucleation occurs, growth then proceeds rapidly due to the ease of further nucleation on the established solid. This would also explain the subsequent formation of new filaments on previously precipitated filaments forming a double width structure as can be seen in Fig. 10B. The results indicate that super saturation existed at some distance away from anodic features and indeed on the PTFE tape proximal to the exposed metal. It is intriguing as to why lateral growth of the filaments is so restricted. An explanation could be that growth is pH constrained. Previous research has shown that phosphate inhibition on Mg was dependent on pH levels due to solubility differences between insoluble tertiary phosphate species and soluble primary and secondary phosphates.[37] Lower pH values support the formation of soluble primary and secondary phosphate species and thus this restricted lateral growth may be as a result of pH differences in solution associated with the formation of anodic and cathodic sites on the metal surface. At present further time-lapse investigations involving altering the concentration of inhibitor, pH of the electrolyte and stirred electrolyte are ongoing to explore these findings to ascertain whether these postulations are correct.

3.4.2 Polarisation investigations of phosphate on a ZMA alloy. The mechanism of phosphate deposition observed in Section 3.4.1 would potentially suggest that phosphate is acting as an anodic inhibitor due to the precipitation proximal to anodic sites in the first instance preventing their growth. Polarisation experiments were performed in order to test this behaviour for samples of ZMA 2 immersed in 1% NaCl pH 7 with no inhibitor, 1% NaCl pH 7 with 1×10^{-4} mol dm^{-3} Na_3PO_4, and 1% NaCl pH 7 with 1×10^{-2} mol dm^{-3} Na_3PO_4. The results are shown in Fig. 12.

For increasing inhibitor concentrations there was an anodic increase in open circuit potential with a positive shift of 50 mV between the sample with no inhibitor and with 1×10^{-2} mol dm^{-3} Na_3PO_4 additions. The anodic current also dropped by 2.5 orders of magnitude at a potential of −1 V *vs.* SCE for the sample with 1×10^{-2} mol dm^{-3} Na_3PO_4. As inhibitor concentration was increased a region of passivity was produced with this extending for overpotentials of 40 mV

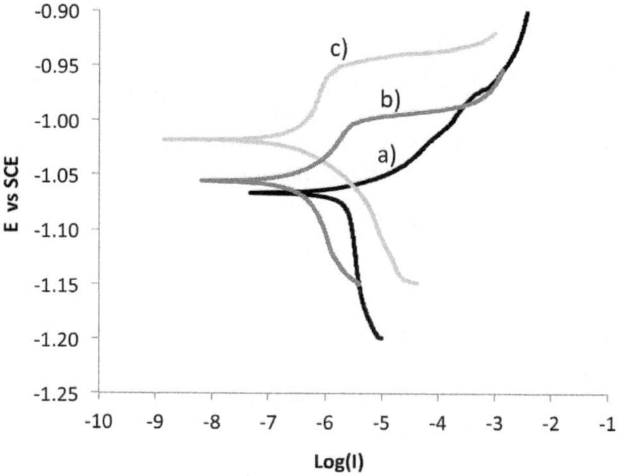

Fig. 12 Polarisation results for ZMA 2 immersed in 1% NaCl pH 7 with (a) no inhibitor (b) 1 × 10^{-4} mol dm^{-3} Na_3PO_4 (c) 1 × 10^{-2} mol dm^{-3} Na_3PO_4.

for a phosphate concentration of 1 × 10^{-2} mol dm^{-3} Na_3PO_4. These data therefore support the findings of the time-lapse images that indeed phosphate was acting predominantly as an anodic inhibitor in this system.

3.4.3 SVET investigation of phosphate inhibition levels on a ZMA alloy. Samples of ZMA 2 were exposed under immersion conditions to 1% NaCl pH 7 with (a) no inhibitor (b) 1 × 10^{-4} mol dm^{-3} Na_3PO_4 (c) 1 × 10^{-2} mol dm^{-3} Na_3PO_4. All samples were exposed to the electrolyte for 24 h with SVET scans of a 100 mm^2 area taken every hour. Once again SVET estimated mass loss was calculated as described in Section 2.4.2. The results are shown in Fig. 13.

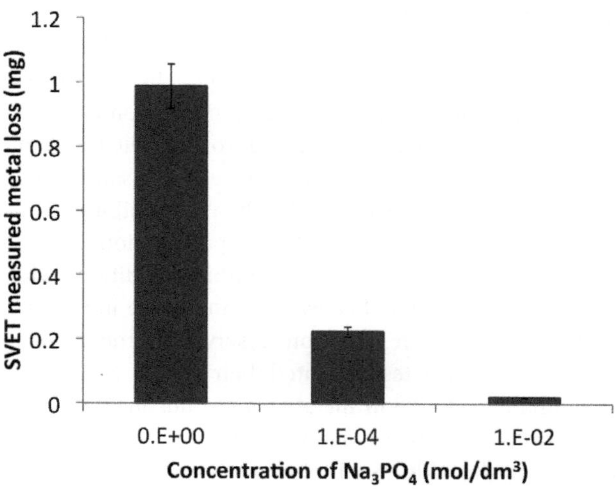

Fig. 13 SVET estimated metal loss for the ZMA 2 after 24 h exposure in 1% NaCl with varying concentrations of Na_3PO_4 inhibitor. A sharp reduction in metal loss was observed for increasing levels of phosphate additions indicating its effectiveness to inhibit corrosion in this alloy.

From Fig. 13 it was seen that the addition of Na_3PO_4 inhibitor to 1% NaCl had a significant impact on the total corrosion mass loss as measured by SVET. For the 1×10^{-4} mol dm^{-3} Na_3PO_4 addition a 77% reduction in metal loss was observed with a 98% reduction in metal loss measured for the 1×10^{-2} mol dm^{-3} Na_3PO_4. These data support the polarisation results in Fig. 12 that showed a progressive decrease in anodic current for increasing inhibitor additions and the images from time-lapse experiments that demonstrated restricted growth of initiating anodic features.

4. Conclusions

The *in situ* optical microscopy time-lapse technique has proved extremely valuable for the assessment of preferential phase attack within the microstructures of three ZMA alloy coatings with increasing Mg and Al levels immersed in 1% NaCl pH 7. It was observed that the Mg rich eutectic phases were preferentially corroded in all three ZMA alloys with initial de-alloying of $MgZn_2$ lamellae from the eutectics before wider scale dissolution of Zn rich phases. The progressing fronts of corrosion were always associated with eutectic attack. Image analysis was used to calculate the area of anodic activity for each image in the time-lapse sequences and this was plotted as a function of time. It was observed that the total corroded area decreased as the alloying additions of Mg and Al increased. This may be due to the increased formation of oxides of Mg and Al that have reduced efficiency for cathodic oxygen reduction. SVET 24 h scans were performed on all three ZMA alloy coatings in 1% NaCl pH 7 and the current density data achieved was integrated and manipulated using Faraday's law to provide an estimated metal loss for each alloy. The same performance trend was observed in the SVET experiments as in time-lapse with increasing alloying additions of Mg and Al decreasing the magnitude of the measured corrosion.

Additions of 1×10^{-2} mol dm^{-3} Na_3PO_4 to 1% NaCl pH 7 in the time-lapse technique had a dramatic effect on the corrosion observed on ZMA 2. Numerous anodic features nucleated on the sample but were passivated by precipitation of a metal phosphate deposit local to the anodic features. After 15 h filament like precipitates of phosphate were rapidly deposited on the metal surface with growth of filaments proceeding over the PTFE masking tape. The rapidity of growth and independence of deposition surface suggested a nucleation and growth mechanism from a super saturated solution produced from anodic loss of metal ions in the electrolyte above the corroding metal surface. Polarisation and SVET tests were carried out on ZMA 2 in 1% NaCl pH 7 with (a) no inhibitor (b) 1×10^{-4} mol dm^{-3} Na_3PO_4 (c) 1×10^{-2} mol dm^{-3} Na_3PO_4. The polarisation tests demonstrated an anodic shift in potential for increasing inhibitor additions with a drop in current consistent with phosphate behaving as an anodic inhibitor. This seemingly supported the mechanism of action observed in the time-lapse images where precipitation at anodic sites prevented their growth. SVET estimated mass loss showed a dramatic decrease in metal loss as phosphate concentration was increased with a drop of 98% for the 1×10^{-2} mol dm^{-3} Na_3PO_4 addition compared with the sample with no inhibitor addition.

References

1 F. Goodwin, *Galfan Galvanising Alloy Technology*, ILZRO, 2nd edn, 1984.

2 A. R. Marder, *Prog. Mater. Sci.*, 2000, **45**, 191.
3 J. Elvins, J. A. Spittle and D. A. Worsley, *Corros. Sci.*, 2005, **47**, 2740.
4 J. Sullivan, C. Weirman, J. Kennedy and D. Penney, *Corros. Sci.*, 2010, **52**, 1853.
5 D. J. Penney, J. H. Sullivan and D. A. Worsley, *Corros. Sci.*, 2007, **49**, 1321.
6 J. Elvins, J. A. Spittle, J. H. Sullivan and D. A. Worsley, *Corros. Sci.*, 2008, **50**, 1650.
7 T. Prosek, A. Nazarov, U. Bexell, D. Thierry and J. Serak, *Corros. Sci.*, 2008, **50**, 2216.
8 T. Prosek, N. Larche, M. Vlot, F. Goodwin and D. Thierry, *Mater. Corros.*, 2010, **61**(5), 412.
9 M. Dutta, A. K. Halder and S. B. Singh, *Surf. Coat. Technol.*, 2010, **205**, 2578.
10 S. Sugimaru, N. Hikita, A. Yoshie, S. Tanaka, H. Ohba and S. Nishida, *Nippon Steel Technical Report*, 2007, vol. 96, p. 34.
11 S. Schuerz, M. Fleischanderl, G. H. Luckeneder, K. Preis, T. Haunschmied, G. Mori and A. C. Kneissl, *Corros. Sci.*, 2009, **51**, 2355.
12 S. Schürz, G. H. Luckeneder, M. Fleischanderl, P. Mack, H. Gsaller, A. C. Kneissl and G. Mori, *Corros. Sci.*, 2010, **52**, 3271.
13 P. Volovitch, C. Allely and K. Ogle, *Corros. Sci.*, 2009, **51**, 1251.
14 N. Hosking, M. Strom, P. Shipway and C. Rudd, *Corros. Sci.*, 2007, **49**, 3669.
15 T. Prosek, D. Persson, J. Stoulil and D. Thierry, *Corros. Sci.*, 2014, **86**, 231.
16 N. LeBozec, D. Thierry, M. Rohwerder, D. Persson, G. Luckeneder and L. Luxem, *Corros. Sci.*, 2013, **74**, 379.
17 R. Hausbrand, M. Stratmann and M. Rohwerder, *Corros. Sci.*, 2009, **51**, 2107.
18 M. Vlot, M. Zuijderwijk, M. Toose, L. Elliott, R. Bleeker and T. Maalman, in *Proceedings Galvatech*, 2007, p. 510.
19 C. Commenda and J. Puhringer, *Mater. Charact.*, 2010, **61**, 943.
20 J. Sullivan, S. Mehraban and J. Elvins, *Corros. Sci.*, 2011, **53**, 2208.
21 J. Duchoslav, M. Arndta, R. Steinbergera, T. Keppertb, G. Luckenederb, K. H. Stellnbergerb, J. Haglerb, C. K. Rienerb, G. Angelib and D. Stifter, *Corros. Sci.*, 2014, **83**, 327.
22 I. M. Zin, S. B. Lyon and V. I. Pokhmurskii, *Corros. Sci.*, 2003, **45**, 777.
23 A. M. Simões, J. Torres, R. Picciochi and J. C. S. Fernandes, *Electrochim. Acta*, 2009, **54**, 3857.
24 K. Aramaki, *Corros. Sci.*, 2001, **43**, 591.
25 S. M. Powell, H. N. McMurray and D. A. Worsley, *Corrosion*, 1999, **55**, 1040.
26 D. A. Worsley, D. Williams and J. S. G. Ling, *Corros. Sci.*, 2001, **43**, 2335.
27 F. Thébault, B. Vuillemin, R. Oltra, K. Ogle and C. Allely, *Electrochim. Acta*, 2008, **53**, 5226.
28 I. M. Zin, S. B. Lyon and A. Hussain, *Prog. Org. Coat.*, 2005, **52**, 126.
29 A. C. Bastos, M. G. S. Ferreira and A. M. Simões, *Prog. Org. Coat.*, 2005, **52**, 339.
30 D. Worsley, H. N. Mcmurray and A. Belghazi, *Chem. Commun.*, 1997, 2369.
31 S. M. Powell and D. A. Worsley, *Br. Corros. J.*, 2001, **36**, 42.
32 S. Böhm, H. N. Mcmurray, S. M. Powell and D. A. Worsley, *Electrochim. Acta*, 2000, **45**, 2165.
33 D. A. Worsley, H. N. McMurray, J. H. Sullivan and I. M. Williams, *Corrosion*, 2004, **60**, 437.
34 R. Hausbrand, M. Stratmann and M. Rohwerder, *Steel Res. Int.*, 2003, **74**, 453.
35 C. Yao, Z. Wanga, S. L. Tay, T. Zhu and W. Gao, *J. Alloys Compd.*, 2014, **602**, 101.
36 H. Dafydd, D. A. Worsley and H. N. McMurray, *Corros. Sci.*, 2005, **47**, 3006.
37 G. Williams, H. N. McMurray and R. Grace, *Electrochim. Acta*, 2010, **55**, 7824.

Faraday Discussions

DISCUSSIONS

Localised corrosion: general discussion†

Gerald Frankel, Geoffrey Thornton, Steven Street, Trevor Rayment, David Williams, Angus Cook, Alison Davenport, Simon Gibbon, Dirk Engelberg, Cem Örnek, Arjan Mol, Philippe Marcus, David Shoesmith, Clara Wren, Kirsi Yliniemi, Geraint Williams, Stuart Lyon, Rob Lindsay, Trevor Hughes, Johannes Lützenkirchen, Su-Ting Cheng, John Scully, Siaw Foon Lee, Roger Newman, Christopher Taylor, Ross Springell, Janine Mauzeroll, Sannakaisa Virtanen, Stephane Heurtault and James Sullivan

DOI: 10.1039/c5fd90046h

Philippe Marcus opened a general discussion of the paper by Roger Newman: Should the passive film not be significantly involved in pit initiation (as you suggest), how would you explain that the time to initiation is very much dependent on the nature of the passive film?

Roger Newman responded: I never said that the passive film is not involved in pit initiation. I understand (of course) that longer passivation gives a longer induction time for pitting. What I say is that the effects of parameters like alloy composition, environment composition, potential, and temperature, are not easily accommodated (at least not predictively) within a passive film breakdown model, but fall out naturally from a modified Galvele type of model that uses pitting potential data (or, if one has the time, lower extremes of pitting potential distributions). Now I don't know whether or not the finest details of film breakdown, detectable at the pA level or lower in electrochemical experiments (not nA to μA – those are already pits), and/or on very pure, flat alloy surfaces, follow the same rules that we find using pitting potentials on industrial or semi-industrial alloys. Those measurements have not been done. Actually I don't think stainless steel is necessarily the best model system for such studies. Under certain conditions, as shown by Bardwell many years ago, iron shows blizzards of pits that are clearly not impurity-particle-related; probably aluminum too. In stainless steel we really don't know whether pit initiation ever occurs without a micro-crevice and/or an impurity particle.

From the viewpoint of practical utility, the Galvele type of approach clearly has the advantage. The Critical Pitting Temperature (CPT) is a propagation-related transition below which metastable pits never become stable at any potential. My group has published extensively on that. One can make a foolproof, if expensive,

† Electronic supplementary information (ESI) available: SVET scans movie. See DOI: 10.1039/c5fd90046h

alloy selection by looking up CPT values for different alloys. Their resistance to passive film breakdown is wholly irrelevant so far as temperature effects are concerned.

John Scully asked: Your paper discusses several important dissolution nuances such as the way Fe and Ni dissolve during corrosion and how that is affected by sulfur. Has our understanding advanced regarding how a nickel or iron atom oxidizes given different first and second nearest neighbor arrangements such as in multi-element solid solutions?

Roger Newman replied: We are dealing here with what Pickering would have called region 3 dealloying – that is, Ni enriches during simultaneous dissolution, without (as far as we know) porosity formation. Some insights have been gained on Pickering's region 1 dealloying (low potential, only one element dissolving – not really accessible for FeNi) through work of Sieradzki and (for an initially ordered alloy) Renner. Also, Erlebacher's MESOSIM simulation can change the "bond energies" and examine the resulting kinetics and morphology.[1–4] But simultaneous dissolution is harder to study. Clearly Ni dissolution limits, in a way, the dissolution of the whole alloy, and if the Ni dissolution is activated by S adsorption, then the alloy will dissolve faster (of course Fe dissolution is also activated). Regarding the effect of Ni content on this surface enrichment and its consequences, I don't believe a systematic study has been done for the FeNi system, and surface analysis of dissolved or dissolving samples is a huge challenge. Obviously one doesn't get exactly the same polarization curve for 5% Ni as for 50% Ni, and when I tried this (on a small scale) I remember having problems with martensite at certain Ni content. Experimentally, it would be nice if there was a binary alloy of more noble metals that could be subjected to either simultaneous or selective dissolution depending on the potential, with no interference from hydrogen evolution, but the usual Au alloys in simple acids are not suitable for that, and many other systems have bad phase diagrams. Possibly one could use a complexing solution like cyanide.

Many years ago we thought we had shown that porosity was formed during simultaneous dissolution of FeNi in HCl solution, and we made a link with Cl-SCC of stainless steels. I still believe that, although the papers were not very convincing.[5]

1 H. W. Pickering, *Corros. Sci.*, 1983, **23**, 1107.
2 I. C. Oppenheim, D. J. Trevor, C. E. Chidsey, P. L. Trevor and K. Sieradzki, *Science*, 1991, **254**, 687.
3 F. U. Renner, A. Stierle, H. Dosch, D. M. Kolb, T.-L. Lee and J. Zegenhagen, *Nature*, 2006, **439**, 707.
4 J. Erlebacher, *J. Electrochem. Soc.*, 2004, **151**, C614.
5 R. C. Newman, R. R. Corderman and K. Sieradzki, Evidence for de-alloying of austenitic stainless steels in simulated stress corrosion crack environments, *Br. Corros. J.*, 1989, **24**, 143–148.

Gerald Frankel said: Others have shown that the semiconducting properties on the passive film on Alloy 800 are a function of the thiosulfate concentration.[1] This behavior correlates with the pitting behavior. In a sense, your measurements are also correlations. How can you rule out the impact of the passive film?

1 J. Luo, personal communication.

Roger Newman responded: I don't know what is meant by correlation in the case of the semiconducting properties. I have seen work of this type where thiosulfate was used, but as I recall (and it may not have been the work of Luo), the authors ignored the framework that we set out in the 1980s. There are many elements of that framework, but the two most important ones are (1) that pitting occurs (with chloride and thiosulfate) at potentials way below where it would ever occur with only chloride, and (2) that pitting occurs in mixtures of sulfate and thiosulfate, but would never occur, at any potential, with only sulfate. I don't know how a "correlation" based on semiconductor properties can accommodate either of these observations, which it would need to do to make any progress at all.

Philippe Marcus asked: Do you have any direct evidence that pitting can be initiated with sulfate/thiosulfate without chlorides (or is scratching necessary to depassivate the surface)? In work we did on FeCr alloys[1] we showed that when both chlorides and thiosulfates are present, passivity breakdown is caused by chlorides and then repassivation is precluded by adsorption of sulphur.

1 C. Duret-Thual, D. Costa, W. P. Yang and P. Marcus, The role of thiosulfates in the pitting corrosion of Fe-17Cr alloys in neutral chloride solution: Electrochemical and XPS study, *Corros. Sci.*, 1997, **39**, 913–933.

Roger Newman replied: It does initiate eventually if you just leave the sample in the susceptible range of potential – that's one reason why we apply an initial passivation in pure sulfate solution (the other reason is that we want a nice low baseline current when we scratch). It does take longer to initiate than when we have thiosulfate and chloride, which can be studied using ordinary voltammetry.[1]To study the sulfate–thiosulfate pitting using voltammetry requires an extremely low scan rate.

I would emphasize (and this is very important) that sulfate–thiosulfate pitting occurs at potentials that are far lower than where ordinary chloride pitting occurs.

Of course we don't know how much of the above is affected by particulate in the alloy, or surface roughness. It may be useful to compare the chloride–thiosulfate and sulfate–thiosulfate systems using very clean alloys, to see if the difference in initiation rate between chloride and sulfate increases. Choice of surface finish(es) would be important in such experiments. It is possible that the sulfate–thiosulfate pitting (or for that matter chloride–thiosulfate pitting) may vanish on surfaces that are extremely smooth and particle-free (see my response to Trevor Rayment for a comment on sputtered thin films and the absence of pitting above a certain Cr content).

1 R. C. Newman, H. S. Isaacs and B. Alman, Effects of sulfur compounds on the pitting behaviour of type 304 stainless steel in near neutral chloride solutions, *Corrosion*, 1982, **38**, 261–265.

Dirk Engelberg asked: I've noticed that all tested microstructures were in the solution annealed condition, which is different to how these alloys are used in plant (*e.g.* thermally treated – TT). Would you expect an effect of microstructure on pit nucleation behaviour and frequency? For example, how would the presence of carbides change the lower/upper bounds of pitting in Table 3/Figure 4?

Roger Newman answered: We use solution annealed alloys because we don't want to be distracted by microstructure effects – our interest is in the solid solution. Regarding the TT treatment, I recall we did a few experiments where we found easier pitting, probably because the Cr depletion is not completely eliminated –perhaps the Cr level next to the grain boundary is still 1 or 2% below the bulk alloy. It might be interesting to pursue that further. For mill-annealed alloys the depletion may be less. But certainly one should pay attention to microstructure in any such study dealing with real plant experience, as we sometimes do – in Canada we still have steam generators with Alloy 600 tubing that is sensitized. One way to preserve that tubing is to reduce secondary-side exposure to oxygenated water during plant outages, as that can contain reduced sulfur species from sludge.

Trevor Rayment commented: This is a question for Roger Newman and Philippe Marcus.

There has been a long standing debate in the corrosion community concerning the mechanism by which molybdenum confers protection against pitting. A simple, if not simplistic, summary of the two schools of thought is a follows. The first school (Philippe Marcus) proposes that the role of Mo is to make the passive film more resistant to pitting, whereas the second school (Roger Newman) proposes that the role of Mo is upon dissolution kinetics after initiation of pitting has taken place.

The debate over the mechanism(s) of pitting protection of molybdenum is not resolved but there has been no discussion of this important topic within this meeting yet. A great deal of experimental data has been published but it does not seem to have convinced all parties to the discussion. A Faraday meeting presents an ideal forum for such a debate.

Therefore I would like to ask Roger and Philippe to propose experimental evidence (new experiments) that others could provide which would persuade them to change their mind on the mechanism for molybdenum protection in corrosion.

Roger Newman answered: I believe many of the critical experiments have been done already. The problem, from where I stand, is that a long list of phenomena in pitting are nicely accounted for (and in several cases quantitatively) by a modified Galvele type of model.[1] Similar success was achieved in studies that addressed the effects of bromide, thiosulfate, temperature, *etc.* The only thing missing from this long list of successes is that (almost always) our definition of pitting was stable pitting. If one's definition is "first onset of detectable current transients due to passive film breakdown," then indeed there is some evidence that thiosulfate takes advantage of spontaneous breakdown events, detectable at the nA level in pitting experiments. But these (perhaps chloride-induced) events are still pits – quite large actually. Burstein and his students looked down to the pA level and found what they considered to be a different type of event, due perhaps to salt forming under the passive film and bursting it open.[2] I don't recall whether the frequency and potential dependence of such events was compared in steels with and without Mo. Such data (although the charges under the transients were still substantial in terms of number of atoms dissolved) may be valuable in determining whether or not Mo affects the nucleation frequency of relevant events. I also don't know whether or not the Burstein type of event occurs, or occurs with different frequency, in steels with lower particulate content than commercial steels. Such delicate experiments have not been done, as far as I know, in support of the "passive film breakdown" hypothesis.

One observation that may address whether the passive film controls pit initiation was that of Ryan *et al.*[3] Pitting vanished above a critical Cr content of the alloy, even under very aggressive conditions. However our interpretation was as follows – these alloys were very smooth, and extreme current densities would be required to initiate any kind of pit. However below the critical Cr concentration, initial dissolution proceeded into the alloy along Fe-rich paths, creating the necessary occlusion for pit nucleation and bringing the observations back into the "Galvele" fold.

The sulfate–thiosulfate pitting is addressed in my response to Dr Marcus. One issue here is that we do not know whether the pitting that we studied (in commercial or home-made alloys) was related to second-phase particles.

1 N. J. Laycock and R. C. Newman, Localized dissolution kinetics, salt films and pitting potentials, *Corros. Sci.*, 1997, **39**, 1771–1790.
2 S. P. Mattin and G. T. Burstein, *Philos. Mag. Lett.*, 1997, **76**, 341.
3 M. P. Ryan, N. J. Laycock, R. C. Newman and H. S. Isaacs, The pitting behavior of Fe–Cr thin film alloys in hydrochloric acid, *J. Electrochem. Soc.*, 1998, **145**, 1566–1571.

Philippe Marcus answered: My response to this question is in three points:

(1) Role of the passive film in protection against pitting.

The role of the passive film in preventing or retarding pitting initiation is extremely important. Indeed it has been demonstrated that on the same alloy and in the same solution enhanced pitting protection is obtained by ageing in the solution, and that this is related to modifications of the chemical composition of the passive film.[1]

An experiment that could contradict the above observation and conclusion would be an experiment in which a well controlled change of composition and structure of the passive film produced on the same alloy and in the same solution would change nothing in the pitting initiation (pitting potential and time to initiation).

(2) Role of molybdenum in both pitting protection by a passive film containing molybdenum, and dissolution kinetics after pitting initiation.

There is no reason to exclude that molybdenum plays both roles, and indeed I think that Mo has two effects: it acts in the passive film resistance against pitting initiation, as well as in the dissolution kinetics after initiation of pitting.

The former effect is likely provided by Mo^{6+} locally enriched on the passivated surface[2] and the latter effect is provided by Mo in the alloy slowing down the dissolution kinetics in the active state (an effect clearly observed in comparing the dissolution kinetics in the active state of 304 and 316 stainless steels, or ferritic stainless steels without and with molybdenum, see *e.g.* Marcus and Olefjord.[3]

(3) Another beneficial effect of molybdenum, originally shown for nickel alloys,[4] is that Mo located on the surface binds to adsorbed sulfur and removes it from the surface (by dissolution), thus improving the corrosion resistance when sulfur species are present.

1 W. P. Yang, D. Costa and P. Marcus, Chemical composition, chemical states, and resistance to localized corrosion of passive films on an Fe–17%Cr alloy, *J. Electrochem. Soc.*, 1994, **141**, 111–116.
2 Vincent Maurice, Hao Peng, Lorena H. Klein, Antoine Seyeux, Sandrine Zanna and Philippe Marcus, *Faraday Discuss.*, 2015, DOI:10.1039/c4fd00231h.
3 P. Marcus and I. Olefjord, A round robin on combined electrochemical and AES/ESCA characterization of the passive films on Fe–Cr and Fe–Cr–Mo alloys, *Corrosion Science*, 1988, **28**, 589–602.
4 P. Marcus and M. Moscatelli, The role of alloyed molybdenum in the dissolution and the passivation of nickel–molybdenum alloys in the presence of adsorbed sulphur, *J. Electrochem. Soc.*, 1989, **136**, 1634–1637.

John Scully enquired: Operating on the assumption that the pitting potential reflects ease of initiation and stabilization with perhaps more weight on stabilization, are there examples where the oxide plays a big role in initiation and thereby affects the pitting potential? In relatively defect free materials, could emphasis shift towards the oxide? One observation has been in the case of amorphous glasses and high purity or relatively defect free stainless steels that there are very few initiation events detected at the instrument limit. These alloys enjoy very high pitting potentials and it seems that the first "detected pit" immediately stabilizes pitting. This stands in contrast to garden variety stainless steels and aluminum alloys with many inclusions and intermetallics where there are many initiation events. You have mentioned that if the defect is on the order of the oxide thickness then stabilization would require extraordinary anodic currents and I agree. But, what was the atomistic event that led to rupture of the oxide in the first place and is oxide engineering then a fruitful avenue for research in relatively defect free materials?

Roger Newman answered: My main point is that the standard of proof is very high if one wants to attribute an improvement in pitting resistance (by alloying) to an increased resistance to passive film breakdown, other parameters such as surface finish and inclusion population, nature and size being kept constant. I don't know of any instance where that standard has been met, whereas for "dirty" alloys, the Galvele type of approach meets it routinely. You are asking whether for clean and/or specially formulated alloys the film might start to dominate and could be engineered for resistance to breakdown. One response to that would be to note that if pitting is not eliminated completely, the pits that do initiate will grow rapidly – this is like "dangerous" inhibition. That is why I would prefer to see passive films engineered to reduce the rate of the cathodic reaction – this is like "safe" inhibition. Assuming that in some applications the oxide could be appropriately engineered on a very clean and smooth substrate (and one would have to consider joining of the material, oxidation, surface damage, surface cleanliness, plastic deformation, *etc.*), I have no doubt that the nucleation frequency of pitting could be kept very low. We saw that ourselves in stainless steel thin films prepared by ion-assisted ion-beam sputtering (see Ryan citation in another answer). Of course that is a kind of coating, so one might more logically see improvement *via* the oxide route as something to pursue using advanced vacuum coating technology applicable to large components. As you know, such considerations apply differently to iron-base and aluminum-base systems, so there is no commonality across the range of our usual materials that suffer from pitting. It has been shown that thin-film aluminum contaminated by oxygen has very high pitting resistance[1] – one might consider this to be some kind of film-reinforcement effect, but I personally consider that this oxygen-alloying effect can be accommodated within a reaction-transport type of picture wherein oxygen dissolution into a proto-pit consumes acid.

1 G. S. Frankel *et al.*, *J. Electrochem. Soc.*, 2014, **161**(4), C195–C200.

Dirk Engelberg opened a general discussion of the paper by Steven Street: For your time-lapse imaging experiment (Fig. 4) do you expect a change in relative humidity over time? Do you take the sample out of the humidity cabinet, and is there then enough time to equilibrate the environment before you take the next images?

Steven Street replied: Samples were in humidity controlled desiccators with clear glass lids through which observations can be made, so there was no exposure to the lab atmosphere during measurements. The time taken to take the image was about 1 min, and the contents of the desiccator helped buffer the temperature so there was little variation here. A Lascar data logger collecting both RH and temperature shows no substantial change in either during optical measurements, maintaining temperature within 3 °C and RH within 3%.

Geraint Williams asked: What role does differential aeration play in the pitting corrosion of the 304 observed under a magnesium chloride containing water droplet? For example, if the surface were to be fully immersed in an aqueous solution of equivalent concentration, would spontaneous pitting still be observed. Is there a lower threshold magnesium chloride concentration observed under a droplet for spontaneous pitting in the absence of external polarisation, than when the surface is fully immersed?

Is there any evidence that local alkalinity at the circumference of the droplet causes precipitation of magnesium hydroxide in the region? Maybe this would explain the ring of apparent metal excavation at the periphery of the droplet as shown in Fig. 4.

Finally, for droplets of different sizes, but fixed concentration, what parameter controls the rate of anodic dissolution? If the pit growth rate is a function of the area of solution in contact with the metal then this would indicate surface/cathodic control rather than there being any ohmic constraints.

Steven Street responded: As noted in another response, the fact that pits do not grow only in the centre of the droplets suggests that the entire surface is vulnerable to pitting initiation in thin droplets. We have started exploratory comparisons with full immersion experiments and have found pitting in concentrations similar to those maintained at the RH used in our atmospheric experiments. This work is in ongoing. Local alkalinity is thought to cause the precipitation of $Mg(OH)_2$ and possibly $MgCO_3$, pinning the droplet perimeter.[1] Figure 4 shows the perimeter of the pit, not the droplet, so we would expect the solution still be to quite acidic here. For fixed CDD and RH (concentration), increasing the droplet diameter tends to give a larger shallow dish diameter, as noted in another answer, so it is likely that this effect is due to the increasing area available for the cathode reaction. Similar increases with droplet size were observed by Mi *et al.*[2] Increasing CDD (*i.e.* droplet height) while maintaining droplet area and concentration also increases the corrosion rate (Fig. 3), which we believe is IR drop related, as discussed in the paper.

1 T. Tsuru, K. Tamiya, and A. Nishikata, Formation and growth of micro-droplets during the initial stage of atmospheric corrosion, *Electrochim. Acta*, 2004, **49**, 2709–2715.
2 N. Mi, S. M. Ghahari, T. Rayment and A. J. Davenport, Use of inkjet printing to deposit magnesium chloride salt patterns for investigation of atmospheric corrosion of 304 stainless steel, *Corros. Sci.* 2011, **53**, 3114–3121.

Gerald Frankel asked: In our work on corrosion of SS under droplets at low RH, the shallow pits smoothly transitioned into attack at smaller areas we called ears, which we attributed to cathodic limitations.[1] Why is that not a good explanation for your observations of pit morphology change?

1 Bastian Maier and G. S. Frankel, Pitting corrosion of bare stainless steel 304 under chloride solution droplets, *J. Electrochem. Soc.*, 2010, **157**, C302–C312.

Steven Street answered: The spiral shaped pits in this work are caused by the "earing" mechanism you described, and this should indeed have been clarified. We agree that the role solution thickness and cathodic current limitation is crucial to the development of these pits, which is supported by our observations of the effects CDD and conductivity (RH). We extended this approach to include the role of mass transport limitation, as we believe this is necessary in explaining the formation of satellite pits. It is interesting that you did not observe satellite pits, possibly due to a combination of slightly higher CDD and finer grinding of the metal surface, which would have removed occluded initiation sites that in our work led to satellites.

Dirk Engelberg asked: Is there interaction between satellite pits and the shallow dish/pit region, since the shallow region seems to have grown into satellite pits (*e.g.* Fig. 3 and Fig. 4)? (how does this affect cathodic *vs.* anodic regions?)

Steven Street answered: Evans' Droplet experiments on steel describe macroscopic general corrosion attack, not localised pitting corrosion. As such, they do not account for microstructure or inclusions common in stainless steel alloys. As Roger Newman pointed out, the model works on the basis of differential aeration which preferentially supplies edges of droplets with oxygen and promotes cathodic reactions there. The observation that usually only one pit is seen in a droplet is expected to be due to the drop in potential in the system associated with pitting. Under full immersion conditions, it is common to observe a sequence of metastable pits that show recovery of the potential after repassivation of previous pits.

David Williams said: I wonder about the temporal sequence of events in your experiments. Could it be that the shallow dishes form as a consequence of the activation of the surface near a pit by corrosion products (specifically sulfides or thiosulfates) that are diffusing out of an active pit? Such effects have been shown before.[1]

1 D. E. Williams, T. F. Mohiuddin, Y. Y. Zhu, Elucidation of a trigger mechanism for pitting corrosion of stainless steels using submicron resolution scanning electrochemical and photoelectrochemical microscopy, *J. Electrochem. Soc.*, 1998, **145**, 2664–2672.

Steven Street responded: We don't doubt the importance of sulphide inclusions in pit initiation in these alloys, and there is a rich literature that supports this. In the discussion section of the paper we suggest that this is not the only factor controlling dish diameter since this varies with position in the droplet. In unpublished work, large droplets of 10 μl have consistently larger shallow dish regions with diameters exceeding 350 μm, and 100 μl droplets with shallow dishes of over 800 μm diameter. If sulphide is the controlling factor we would expect a small variation in dish sizes more similar to each other.

Cem Örnek commented: In 304L austenitic stainless steel the delta ferrite content is about 6–14%, depending on the microstructure processing orientation (TD, RD, ND). The ferrite has significantly different chemical composition in contrast to the austenite matrix which questions the role in localised corrosion. Furthermore, there are MnS and other non-metallic inclusions often present in the microstructure. How are these related to the corrosion morphology during atmospheric corrosion? Are there any preferential initiation sites observed? Can we say from selective corrosion? Can we speak about pitting corrosion if we see odd shaped corrosion morphologies?

Steven Street responded: Our steel contains ~4% delta ferrite, most of which is below the plate surface. We have demonstrated that it influences pit propagation.[1] Our steel also contains sulphide and oxide inclusions which are known to be pit initiation sites. However we have not carried out a systematic study on the influence of different initiation sites.

1 A. J. Davenport, L. Guo, N. Mi, H. Mohammed-Ali, S. M. Ghahari, S. R. Street, N. J. Laycock, T. Rayment, C. Reinhard, C. Padovani and D. Krouse, Mechanistic studies of atmospheric pitting corrosion of stainless steel for ILW containers, *Corros. Eng., Sci. Technol.*, 2014, **49**, 514–520.

Cem Örnek asked: Evans introduced the atmospheric corrosion model under thin-film electrolytes formed by a hemispherical water droplet. Thus the anode is primarily located in the centre of the droplet due to rapid oxygen depletion and impeded oxygen supply and the periphery of the droplet forms the cathode due to the excessive and quick access of oxygen. Preferential corrosion would be expected to occur at or close to the centre due to the development of a large net anodic potential. However, the results shown (and the poster from Angus Cook, University of Birmingham) clearly contradict the model of Evans. Research on the atmospheric corrosion behaviour on the same material and on 2205 duplex stainless steel by my humble self can confirm your observed results. However, there is no clear explanation given as to why the corrosion initiation occurs mostly between the centre and the peripheral area of a hemispherical chloride-containing droplet. Furthermore, the initiation of corrosion is concentrated on a confined region, and, usually, only one or a few corrosion pits are observed which is significantly different to corrosion initiation sites under immersed conditions.

Is there an explanation for this? By the way, corrosion was also observed occurring on the peripheral area and also on secondarily spread chloride regions.

Roger Newman replied: I don't believe the situation is as tidy as the Evans experiment. There are two major differences. First, the stainless steel is spontaneously passive, while the iron is spontaneously active. In the Evans experiment it is the alkalization at the periphery of the drop that passivates the iron and creates the active–passive couple. In the stainless steel experiment there is pit initiation and propagation, which is very different. Second, oxygen reduction on stainless steel is normally activation controlled, except where the stainless steel is in the active state, while in the Evans experiment oxygen reduction is diffusion-limited. As I indicated in another answer, oxygen reduction on stainless steel can also show apparent "limiting current densities" that are too low to be diffusion-limited in the electrolyte.

Roger Newman commented: The reference to Evans' famous drop experiment (on iron) may not be appropriate if oxygen reduction on stainless steel is kinetically rather than diffusion limited. In the drop experiment, oxygen reduction is everywhere under diffusion control.

As a footnote to this comment, it might be interesting to vary the partial pressure of oxygen over the experiment, or in some other way to probe the status of the oxygen reduction reaction.

An old observation is that oxygen reduction on stainless steel sometimes appears to be mass transfer limited, but the limiting current density is too low for boundary-layer diffusion. In that case something within the passive film itself is limiting the reaction. Someone must have explained this.

Steven Street answered: Thank you for your very interesting comments. A further brief discussion on Evans' droplet model has been addressed in answer to another question.

Dirk Engelberg asked: Can you suggest a viable corrosion mitigation strategy to prevent localised corrosion in this system (since you show pitting corrosion occurs with exposure to deliquescence relative humidity, and increasing the relative humidity (*e.g.* 56%) also results in pitting corrosion). Do you accept a chloride deposition density of 10 μm cm^{-2} as a working threshold or have you observed corrosion occurring below this threshold? Can you speculate whether a change of shape during pit growth can be used to affect (and control) pit growth kinetics?

Steven Street responded: The obvious solution is to try to keep chloride concentrations as low as possible and to limit the thickness and continuity of the electrolyte layer. An alternative method would be maintaining a very high RH in storage, to ensure any salts that do deliquesce remain dilute, but these would necessarily pass through an initial highly concentrated stage which may cause corrosion, if only superficially. As initiation sites are seen to not reinitiate pitting once passivated, natural wet–dry fluctuations could be used to exhaust these sites on the surface of the alloy.[1] An improved alloy, such as 316L, would certainly improve performance. 10 μm cm^{-2} CDD is significantly lower than the values used in this paper. A recent study[2] suggests that there is merit to the use of a working threshold.

1 A. J. Davenport, L. Guo, N. Mi, H. Mohammed-Ali, S. M. Ghahari, S. R. Street, N. J. Laycock, T. Rayment, C. Reinhard, C. Padovani and D. Krouse, Mechanistic studies of atmospheric pitting corrosion of stainless steel for ILW containers, *Corros. Eng., Sci. Technol.*, 2014, **49**, 514–520.
2 C. Padovani, R. J. Winsley, N. R. Smart, P. A. H. Fennell, C. Harris and K. Christie, Corrosion control of stainless steels in indoor atmospheres? Practical experience (Part 2), *Corrosion*, 2015, **71**, 646–666.

David Williams asked: I wonder if there are any convection effects associated with surface tension gradients developed as droplets evaporate? Thus, are there any effects that are different at the centre compared with towards the edge of the droplet?

Steven Street replied: Convection effects are something we did not consider and perhaps should have, as the droplet loses up to 95% of its initial volume as it equilibrates with RH within the first hour. We show in Figure 8 that the diameter of the shallow dishes are greater in the centre of the droplet. We attribute this to an increased IR drop and mass transport limitations at the edges. We do not observe any other effect that might be attributed to convection.

Siaw Foon Lee communicated: Under the section of "Experimental", third paragraph – Arrays of 2 µl droplets of 0.27 M $MgCl_2$ were deposited…and Fig. 1(a) showing the relationship between the solution concentration (M) of $MgCl_2$ and relative humidity (%). My question is that could the author shows the calculation steps of how to convert 2 µl droplets of 0.27 M $MgCl_2$ to 5M $MgCl_2$ after it was held in an ECO135 atmospheric chamber for 24 h with the humidity of 33%?

Steven Street communicated in reply: The concentration of the solution is determined by its water activity as it reaches equilibrium with the relative humidity. At equilibrium, the water activity in the air and water phases are equal, regardless of the initial solution concentration. The calculations of equilibrium concentrations are based on Figure 1(a) which was calculated using OLI Systems.

Siaw Foon Lee communicated: Table 2 – the humidity of 33% has a droplet area on deposition of 0.048 cm^2 and has an average droplet CDD of 80 µg cm^{-2}. The humidity of 43% has a droplet area on deposition of 0.048 cm^2 that is same as that of the humidity of 33%, however, it has an average droplet CDD of 790 µg cm^{-2}. I am curious to know how the author works out the calculation since this topic is new to me.

Steven Street communicated in reply: Chloride deposition density is an average measure of mass of chloride over a defined area, in this case the area of the droplet. In our experiments, CDD was altered by varying the deposition solution concentration. As noted in the paper (Table 2) the droplet areas are measured at deposition, as droplet spreading can occur during atmospheric corrosion.

Siaw Foon Lee communicated: Under the section of "Experimental," subsection of "Electrochemical sweeps," second paragraph, starting from the second line. The solutions used were de-aerated by bubbling argon for 20 min…" Could I know about the minimum oxygen concentration of the electrolyte the author achieved using this method? Can the oxygen concentration remain the same during the experiment using this method?

Steven Street communicated in reply: Oxygen concentration was not measured in this experiment. Checks on similar measurements on oxygen reduction rate on Pt wire suggest that this method can maintain concentration.

Dirk Engelberg opened a general discussion of the paper by V. Vivier: Do you keep your micro-capillary very close to the surface (10–15 microns) to initially generate a "crevice-type" environment to induce pitting (crevice)? Do you try to keep the condition (in particular the chloride concentration) in this confined space below the capillary stable during the experiments?

Stephane Heurtault answered: The proximity of the micro-capillary to the stainless steel surface, together with the chosen chloride concentration in the syringe and the release flow rate, helps to create an appropriate environment for pit initiation and propagation. The distance between the micro-capillary and the stainless steel surface was selected to produce a single pit on the steel. 15 microns is the highest distance fulfilling this condition. The chloride concentration injected by the micro-capillary does not necessarily have to be constant during the propagation phase. The supply of chloride has only to be greater or equal to the critical chloride concentration determined in part 3.3.2 of this article.

As you mention crevices, it is well known from the literature that crevices can develop from pits. However after 1 h of propagation, a pit cross section performed in our pit did not show any crevice.

Alison Davenport remarked: Have you carried out a systematic investigation to determine whether the flow rate of the solution influences pit growth? Some of your pits appear to have broken covers: were they damaged during the experiment or afterwards during transfer to the SEM?

Stephane Heurtault replied: The flow rate of the solution is one of the main parameters for the supply of chloride in the environment and it influences the pit growth. The supply of chloride (in mol h^{-1}) can be defined as the product of chloride concentration (in mol L^{-1}) and flow rate (in L h^{-1}). However, we have shown that we can control the amount of chloride independently by changing the chloride concentration or by modifying the flow rate (see for instance the results presented in Fig. 5). With a chloride flow rate of 5.4 μL h^{-1} (as used in most of our experiments presented in this paper) and taking into account the evolution of the pit volume plotted in Fig. 10, only 3 min are required to renew entirely the pit solution after 1 h of propagation, and it will take 30 min after 10 h of propagation. Thus, the convection induced by the micro-capillary may be an important factor that controls the pit propagation. To avoid such a behavior, it is possible to generate and to propagate the pit in two successive steps: first, the pit is generated beneath the micro-capillary and then, it is sustained by controlling the bulk concentration in chloride as was shown possible in the last part of the article.

We believe that the pit cover shown in Fig. 6(a) was not damaged during the experiment but afterwards during transfer to the SEM. Indeed, in another experiment, we succeeded in obtaining a no-broken pit cover using the same experimental conditions. Nevertheless it is possible to break the pit cover during the experiment, for example performing an anodic potential change during the propagation.

Roger Newman asked: Why do you use such a strong sulfate solution? It seems that complications from the mixing of chloride and sulfate could be minimized if you just reduced the sulfate concentration by a factor of 10. Some of your experiments where you vary the chloride might be affected by varying degrees of inhibition by sulfate (since inhibition is related to the chloride/sulfate ratio).

Stephane Heurtault replied: We used such a strong sulfate concentration to be sure that the solution conductivity is high enough, but we know that it affects the pit growth. In our previous work,[1] differences between sulfate and perchlorate solution were investigated. Moreover, some experiments were performed using 6 M SO_4^{2-} and 1.2 M Cl^- instead of 0.5 M SO_4^{2-} and 1.2 M Cl^-. With this increase in sulfate concentration (corresponding to a decrease of the chloride to sulfate ratio), the inhibitor role of sulfate was shown. We agree with this comment that the chloride to sulfate ratio is an important point instead of only indicating the chloride concentration.

1 N. Aouina *et al.*, *Electrochim. Acta*, 2013, **104**, 274.

Gerald Frankel stated: You show that the depth varies with $t^{0.5}$, which would also be found for ohmic control. How can you rule out ohmic effects?

Stephane Heurtault answered: Indeed it should have been mentioned that the variation of pit depth with $t^{0.5}$ also accounts for ohmic control as shown previously.[1] The same dependence with time of pit depth for both ohmic control and diffusion complicates the data interpretation. However for a pit under diffusion control, potential variation has no effect on the pit growth rate, whereas a pit growing under ohmic control depends on potential according Ohm's law. Additional experiments at 20 °C injecting 3 M NaCl + 0.5 M H_2SO_4 with the glass microcapillary were performed by changing the electrode potential during the pit propagation (data not shown in this paper). The single pit was initially formed at 0 V/MSE. After 1 h of propagation, the potential was changed to 0.4 V/MSE. Then the pit depth with this modification of potential was measured and the current density at the pit bottom after changing the potential was calculated using equation (5). It was shown that the modification of potential does not affect the calculated pit bottom current density, indicating a diffusion-control process.

1 G. S. Frankel, *J. Electrochem. Soc.*, 1998, **145**, 2186.

David Williams commented Laycock *et al.* have given a rather complete computational model for evolution of pit shape, including the perforated covers.[1,2] Have you been able to compare your results with such a model? Also, have you been able to see the current fluctuations that should result as the perforations in the pit cover develop? Pit 'death' happens when the cover is detached and the internal processes cannot sustain the critical solution composition against the resultant increased flux out of the pit. This is usually indicated by a sharp increase of current followed by an immediate fall to zero (*e.g.* ref. 3). Have you been able to see such effects in your experiments?

1 N. J. Laycock, S. P. White, J. S. Noh, P. T. Wilson and R. C. Newman, Perforated covers for propagating pits, *J. Electrochem. Soc.*, 1998, **145**, 1101–1108.
2 N. J. Laycock and S. P. White, Computer simulation of single pit propagation in stainless steel under potentiostatic control, *J. Electrochem. Soc.*, 2001, **148**, B264–B275.
3 D. E. Williams, J. Stewart and P. H. Balkwill, The nucleation, growth and stability of micropits in stainless steel, *Corros. Sci.*, 1994, **36**, 1213–1235.

Stephane Heurtault responded: The Laycock model is a great advancement in our understanding of pitting propagation on stainless steel. This model was devised for a disc-shaped pit and accounted for the presence of a salt film at the pit bottom. Due to this salt film, an additional potential drop at the pit bottom exists, which implies a lower current density beneath the salt film than at the pit edges. These results are fully consistent with our experiments (disc-shaped pit and salt film at the bottom of the pit). The Laycock model also accounts for the presence of a pit cover, which is composed of a central hole surrounded by many small perforations forming the lacy-like structure. According to the model, the cover entirely dissolves for long-term pit propagation. In our experiments, using the adequate chloride concentration injected by the micro-capillary, we can get single pits which propagate without any pit cover (this is possible according to the Laycock model) or we can also observe the propagation of a single pit with cover. However, the pit cover obtained in our experiments has only a central hole at the location corresponding to the chloride solution injection by the micro-capillary, and only one or two perforations can be observed in the periphery. This specific behavior may be explained by the presence of the micro-capillary above the pit maintaining the concentration of chloride ions at a high level. We didn't see any current fluctuations resulting from the perforations of the pit cover. This was confirmed by SEM observations. However for both covered and uncovered pits, we noticed small current oscillations which can probably be ascribed to the growth and the breakdown of a thin oxide layer under the salt film.[1,2]

Once the pit propagates, we have never observed repassivation after the collapse of the pit cover. However, it should be mentioned that an increase of the electrode potential during the pit propagation was accompanied by the collapse of the pit cover. In that case, the current increased sharply during the collapse, but the propagation was maintained.

1 R. C. Newman and M. A. A. Ajjawi, *Corros. Sci.*, 1986, **26**, 1057.
2 N. J. Laycock and R. C. Newman, *Mater. Sci. Forum*, 1995, **192**, 649.

Alison Davenport commented: We have recently carried out some radiography work on the growth of 2D pits in stainless steel foils.[1,2] We also found out that growth rate at the bottom of the pit is slower than the sides of the pit. We also attribute this to diffusion-controlled growth of the pit bottom. However, in the presence of a perforated pit cover, the solution concentration at the pit mouth will not be zero as the cover provides an additional diffusion barrier. We found that the growth of the pit width is under ohmic control as the rate increases with solution conductivity.

1 M. Ghahari, *In Situ* Synchrotron X-Ray Characterisation and Modelling of Pitting Corrosion of Stainless Steel, *PhD Thesis*, University of Birmingham, 2012, http://etheses.bham.ac.uk/3269/11/Ghahari12PhD.pdf.
2 M. Ghahari *et al.*, *Corros. Sci.*, submitted.

Alison Davenport opened a general discussion of the paper by David Shoesmith: What are the most effective ways to simulate the effect of radiation on corrosion of UO_2 under realistic conditions?

David Shoesmith answered: A number of experimental approaches can be used.

(a) For γ-radiolysis, direct exposure of the UO_2 to a gamma field with a calibrated dose rate can be used.

(b) For α-radiolysis, UO_2 specimens doped with α-emitters (*e.g.*, Pu(239), U(233)) to provide a known dose rate have been used.[1] This method provides an *in situ* α-radiation field.

(c) An alternative method for α-radiolysis is to use a thin layer electrochemical cell with the electrode (UO_2) brought close to a Au-coated α-source ($\not\leqq$ 25μm) with a known dose rate.[2] This method provides an *ex-situ* source.

(d) Intense beams (*e.g.*, X-rays (Ross Springell *et al.*[3])) can be used, but usually produce extremely high dose rates well beyond those expected inside a failed container.

(e) The influence of γ-radiolysis (and also α-radiolysis) can be simulated by adding the anticipated concentration of hydrogen peroxide chemically. Evidence in support of this approach is well documented.[4]

1 M. Broczkowski *et al., ACS Symposium Proc.*, 2010, **1046**, 34–380.
2 J. C. Wren *et al.*, *J. Electrochem. Soc.*, 2005, **152**, B470–B481.
3 Ross Springell, Sophie Rennie, Leila Costelle, James Darnbrough, Camilla Stitt, Elizabeth Cocklin, Chris Lucas, Robert Burrows, Howard Sims, Didier Wermeille, Jonathan Rawle, Chris Nicklin, William Nuttall, Thomas Scott and Gerard Lander, *Faraday Discuss.*, 2015, DOI:10.1039/c4fd00254g.
4 J. C.Wren, *ACS Symposium Proc.*, 2010, **1046**, 271–295.

Ross Springell replied: Simulating the effects of corrosion of UO_2 under realistic conditions initially requires one to define what these conditions might be, and this can depend on what scenario you choose. The most frequently used method is the addition of small quantities of H_2O_2 to the aqueous environment. This is based on the assumption that any container of spent nuclear fuel will likely fail after the first few hundred years when the gamma fields from the fission products have decayed and the dominant radiation fields are driven by alpha decay; in this case the dominant radiolytic product will be H_2O_2. However, the most likely cause of any containment failure is due to potential defects in the vessel weld. This will be present from day 1 and it is possible that water may ingress way before all of the fission products have decayed. In this instance, strong gamma fields will also be present and this will result in both H_2O_2 and other more oxidising free radicals at the fuel surface. A lot of this discussion really depends on the precise packaging methods and eventual repository conditions. One method to better simulate the radiation fields would be to study changes in the UO_2/water interfaces exposed to an alpha source and a gamma source, but these would have to be *ex-situ* studies with characterisation of the structure with

X-rays at a later time. The advantage of the experimental method outlined in this paper is that we are able to probe the structural changes *in situ*.

John Scully asked: Could you comment on the number of adjustable or disposable parameters that a model like this has, or should be limited to? Of course, it seems that corrosion science might work to define as many "unit processes" as possible so as to eliminate some adjustable parameters when they can be replaced. Is this your view too, or do you see another way forward?

David Shoesmith answered: The primary basis of a model like this one (fuel corrosion model) is a mechanistic understanding of the key reactions required to describe the corrosion process. Our guiding principle is that the model should contain the minimum number of reactions necessary to reasonably represent the overall process. To achieve this the model should contain the minimum number of adjustable or disposable parameters. Ideally, the model should then yield output values which can be compared to trusted experimental data. Unfortunately, required data is commonly not available, unreliable, or determined for inappropriate conditions. Disagreement between calculated (model) and measured values can, however, provide a template for required measurements or theoretical calculations.

Unfortunately, models and calculations can be developed and performed on a time scale which cannot be matched by experimental determinations. This commonly leads to the "over-extension" of models beyond validation, either by comparison to measurements or by theoretical calculations. This is generally the reason for the use of adjustable parameters. The more such parameters are included the more over-extended and unreliable the model becomes. Such over-extensions should be minimized by the iterative process of the model (often sensitivity) calculations, experimental determinations, model redevelopment.

Alison Davenport asked: Could you give further details on the uncertainty in the input parameters to the model, and to the sensitivity of the model to different parameters? Are there any input parameters for which improved measurements would be helpful to reduce the uncertainty in the model?

David Shoesmith responded: There are many uncertainties in the input parameters to the model in its present form. The mechanistic basis is reasonably well established although a detailed understanding of how hydrogen peroxide reacts and interacts with hydrogen on the UO_2 surface has proven elusive. The key uncertainties are in the values of the rate constants required, many of which are uncertain and some of which are unknown. The sources and uncertainties of the input parameters have been discussed elsewhere.[1–3]

The key uncertainty encountered is in the justification that parameter values measured in experimentally accessible concentration ranges can be used in the concentration ranges anticipated inside a failed container which are orders of magnitude lower. Additionally the model cannot be considered fully developed in its present form. While the goal is to develop a transparent and readily comprehensible model with as few complexities as possible, a number of features remain to be elucidated. These include the possibility of corrosion product deposition, especially within fuel fractures. Their presence will lead to uncertain diffusion fields between the two corrosion fronts within the container. The consequences of various failure modes of the fuel cladding is also not included. Model validation is also an on-going issue, and an attempt to compare model calculations to natural analogs is being made.

1 L. Wu, Y. Beauregard, Z. Qin, S. Rohani and D. W. Shoesmith, *Corros. Sci.*, 2012, **61**, 83–91.
2 Linda Wu, Zack Qin and David W. Shoesmith, *Corros. Sci.*, 2014, **84**, 85–95.
3 Linda Wu, Nazhen Liu, Zack Qin and David W. Shoesmith, *J. Electrochem. Soc.*, 2014, **161**, E3259–E3266.

Christopher Taylor asked: Early in the presentation you mention the hypothesis or expectation that the epsilon particles, being noble, could drive anodic processes by acting as a cathodic electron sink by catalyzing HER. However, it appears that instead, your expectation is that they will have an opposite effect, by quenching H_2O_2. Is this the correct interpretation of the role of epsilon particles, or does the competition between these two effects still need to be elucidated?

David Shoesmith responded: We are studying these reactions extensively and many features remain to be elucidated. We have good experimental evidence that these particles act as cathodes to reduce hydrogen peroxide (in the absence of hydrogen), as anodes to oxidize hydrogen (in the absence of hydrogen peroxide), and as a catalysts on which hydrogen oxidation and hydrogen peroxide reduction can couple to yield water. It is this last reaction which can consume hydrogen peroxide and prevent it from driving uranium dioxide corrosion. To date we have not been able to quantitatively determine the kinetic relationships describing these reactions although we have some preliminary data. There are a number of unresolved issues. We do not know the kinetics of hydrogen peroxide decomposition (to oxygen and water) which model sensitivity calculations tell us are important. The influence of alpha radiolysis in "hidden" locations (deep cracks, grain boundary tunnels) is underway but incomplete, and the relative importance of the noble metal particles and the uranium dioxide surface in hydrogen peroxide reactions is uncertain. I could extend the list!

Ross Springell said: You mention in your article that UO_2 has the potential to have a catalytic effect that may influence its corrosion. Could you expand on this point and describe how important the stoichiometry is in this aspect?

David Shoesmith replied: Many oxide catalysts act as electron donor–acceptor catalysts especially p-type oxides.[1] This is attributed to the existence of adjacent metal ions with the surface ion being in a higher oxidation state (*e.g.*, U(V)) than that immediately below (U(IV)). This kind of catalysis is exhibited by Ce (III/IV) oxide. If such a catalysis occurs then an increase in non-stoichiometry (x in $UO_{(2+x)}$) would be expected to enhance catalysis. This effect has been proposed to explain the kinetics of hydrogen peroxide reduction on uranium dioxide.[2,3] However, it is a little more complicated than this and presently not totally resolved.

1 V. Presnov and A. M. Trunov, *Elektrokhimiya*, 1975, **11**, 77.
2 J. S. Goldik *et al.*, *Electrochim. Acta*, 2004, **49** 1699.
3 J. S. Goldik, J. J. Noël and D.W. Shoesmith, *J. Electroanal. Chem.*, **582**, 2005, 241.

Angus Cook commented: In your model system of a failed nuclear waste container (Fig. 2) there is no accounting for any cladding material around the nuclear fuel.

How might cladding change this system?

David Shoesmith answered: Only the electrochemical and chemical reactions were included in the model presented. The Zircaloy cladding is expected to be chemically inert and not involved in the redox processes determining the UO_2 behaviour. Its effect will be as a separating barrier between the site of radiolytic oxidant production at the fuel surface and the site of Fe(II) and H_2 production at the steel container inner surface. It will, therefore, influence the diffusion processes and hence the ability of H_2 to control the redox conditions at the fuel surface by reaction with H_2O_2 and of Fe(II) to consume H_2O_2 by the Fenton reaction. Its efficiency in interfering with these reactions will be determined by the physical nature of the failure in the fuel cladding. We are presently trying to model this effect within our finite element model.

Dirk Engelberg said: With the presence of hydrogen peroxide in the aqueous phase, and/or catalytic reactions expected to occur at the uranium surface/particles (H_2/H^+) wouldn't you expect the formation of uranium hydride?

David Shoesmith answered: The fuel form is uranium dioxide not uranium metal. Hydride formation would only be expected if uranium in the zero oxidation state was present. While H atoms will form on the noble metal particles (an alloy of Ru, Pd and Mo), and some minor absorption into the particle may be possible, this will not lead to reduction of the U(IV) oxide to U metal.

Christopher Taylor remarked: For the water gas shift and reverse water gas shift research efforts, bifunctional catalysts that contain nanoparticles of precious or noble metals dispersed over a redox-active oxide, like ceria, are used to tune the reaction chemistry for $CO_2 + H_2 \rightleftharpoons CO + H_2O$ and potentially form hydrocarbon products. In your case, you have a similar set-up between these epsilon particles, which are noble and catalytic, and the redox active UO_2. By analogy, then we might expect the interfaces between the particles and the urania to be the active regions whereby CO_2 is able to engage in chemistry with the material. Are there other possible insights that could be gained by this analogy?

David Shoesmith responded: The analogy is a good one. In the reaction you describe on ceria you note that the whole oxide matrix is involved not just the dispersed noble metal nanoparticles, but you suspect the coupling within the uranium dioxide will be limited to regions close to the noble metal particles. I agree. The area of the reactive uranium dioxide influenced by the noble metal anode/cathode will be determined by the electrical conductivity of the rare earth doped oxide lattice and the ionic conductivity of the electrolyte it is exposed to; *i.e.*, this is a normal microgalvanic (or internal) coupling process.

I think the analogy can be extended in terms of the behaviour of the uranium and cerium oxides. The hydrogen peroxide we are interested in will react directly on the uranium dioxide surface besides *via* the galvanic coupling process, making the catalytic properties of the oxide also important. Both cerium oxide and uranium dioxide have the same lattice structure and uranium dioxide has the potential to act as a redox catalyst *via* similar mechanisms to those occurring on the much more extensively studied cerium oxide. It is well worth investigating whether this is the case and whether the tuning mechanisms for ceria that you refer to also occur on uranium dioxide.

Clara Wren opened a general discussion of the paper by Ross Springell: The concentration of the radiolytically-produced H_2O_2 that you quoted is very high at ~ 10^{-4} M. What was the energy absorbed by water? Can X-ray induced radiolysis create that high an H_2O_2 concentration? The photoelectrons generated in the UO_2 matrix should lose most of their energy to overcome the binding energy and work function, and, when they are ejected from the UO_2 matrix, will have a relatively low energy, on the order of ~ 100 eV. This is not sufficient to penetrate the adsorbed water layer but is sufficient to ionize water molecules close to the UO_2 surface, affecting the water chemistry close to the surface. If this is the case, the X-ray induced corrosion may be different from that driven by higher energy radiation such as gamma-rays. Can you comment on the potential differences arising in X-ray and γ-ray induced corrosion?

Ross Springell replied: The X-ray energy is 17 keV and for the average 0.5 mm coverage of water that we used in this experiment, the transmission was 59%, which means that 41% of the total photon flux (approx. 10^{12} photons s^{-1}) was absorbed in the beam track through the water. By using the linear energy transfer (LET) of X-rays at this energy it is possible to calculate the number of H_2O_2 molecules produced per 100 eV deposited, per photon. The high photon flux results in the increase of H_2O_2 concentration up to its equilibrium concentration within a fraction of a second. Both the X-rays and gamma rays have a similar LET and produce similar proportions of radiolytic product, but it is certainly the situation that the absorption in the water is much stronger for the X-ray case. In fact, it is this property that we rely on to enable us to measure the corrosion *in situ* during the time frame of a synchrotron experiment; in real spent fuel stores the rates of corrosion are considerably slower.

Geoffrey Thornton stated: We have recorded SXRD data (CTR's) from a water/$TiO_2(110)$ interface using ID32 at the ESRF. The photon flux was also 17 keV, with a photon flux of 10^{12} ph s^{-1}. We see an interface ordered on the atomic scale. In other words, there is no evidence of beam damage. A key difference between your sample and the $TiO_2(110)$ substrate is that the former employs an insulating substrate, whereas the latter was a self-doped n-type semiconductor. Hence, it may be a charging effect that gives rise to the beam damage you observe.

Ross Springell answered: The beam parameters described in the SXRD measurements on TiO_2 at ID32 are remarkably similar to our own, and it's surprising that no beam damage is observed at all. The possible charging of the YSZ insulating substrates as a potential cause of the increased rate of corrosion observed in the UO_2 films is a good suggestion and we will look toward a conducting substrate system in the future to test this theory.

Rob Lindsay commented: As you indicate in the paper, it is somewhat surprising that the (001), (110), and (111) UO_2 surfaces all exhibit similar dissolution rates. How well characterised are these surfaces? For example, have you employed STM to image them in real space? It may be that dissolution is dominated by loss of atoms from step edges or other 'defect sites,' and so rates are more similar than expected for the three surface orientations.

Ross Springell responded: To date our characterisation of the different surfaces has relied solely on X-ray diffraction techniques and in some cases high speed AFM. Diffraction spectra imply large regions with very low roughness (less than 2 monolayers), but in order to be certain of the surface morphology we will look to try a technique with much better real space resolution such as STM, as suggested. In order to further investigate the effect of crystal surface, we are planning an *in situ* high speed AFM study of the corrosion of a number of UO_2 crystal grains on a longer time-scale, using hydrogen peroxide as the oxidant.

David Shoesmith asked: Have you tried to characterize the UO_2 layer before exposure and the oxidized surface layer after exposure using techniques such as Raman spectroscopy and X-ray photoelectron spectroscopy?

Ross Springell responded: We have made some measurements of the oxidised surface layer after exposure, using X-ray photoelectron spectroscopy (XPS), mapping across a corroded region. Our first results indicate a more oxidised hyperstoichiometric surface into the corroded region and then potentially a higher uranium oxide in the very centre of the trough, due to redeposition of material out of the water, on drying. We have beam-time scheduled at I09, Diamond Light Source, to conduct a full investigation of the oxidation of these corroded regions *in situ*, using XPS. Raman spectroscopy is an excellent suggestion; although our films are too thin for this technique to be sensitive in this case, we plan to grow a series of much thicker films and measure the Raman spectra for a range of stoichiometries.

Philippe Marcus asked: Can we really implant in the human body a stent that would produce a lot of hydrogen? What is the conclusion in terms of the effect of iron impurity in magnesium on hydrogen evolution (one of the hypotheses included in the Introductory Lecture presented by Gerald Frankel[1] to explain the "anodic" hydrogen) and what are the implications for the application of Mg alloys for stents (requested purity of Mg)?

1 G. S. Frankel, S. Fajardo and B. M. Lynch, *Faraday Discuss.*, 2015, DOI:10.1039/c5fd00066a.

Sannakaisa Virtanen replied: Certainly the hydrogen gas evolution by Mg based stents is of high concern. There are studies related to the fate of hydrogen gas in different types of *in vivo* situations. The danger of hydrogen gas generation in the biomedical application can be expected to depend on the temporal and spatial rate of its production: *i.e.*, sufficiently slow gas generation can be tolerated (gas can be transported away and local build-up of significant volumes of gas can be avoided). What is typically observed for Mg alloys in simulated *in vivo* conditions, for instance in cell culture medium (but NOT in more simple simulated body solutions, SBFs, which only mimic the inorganic components of blood and do not contain for instance amino acids), is that the initial production of hydrogen gas is quite strong, and in a relatively short time (in the first hours of immersion) the surface is "passivated" by formation of a protective corrosion product layer. Therefore, the key for a safe application from my point of view is to modify the alloy surface with a biodegradable coating, which only offers a weak barrier function, but reduces the initially too high reactivity of the Mg surface. Also, Mg alloy purity certainly is of very high importance in this field.

The corrosion challenge is not only related to hydrogen gas. Also local pH increase can be deleterious for the biological surroundings. But one should consider that most laboratory studies on Mg corrosion are performed under static conditions and this is far from the highly dynamic situation *in vivo*. So there is quite a lot still to do.

Geraint Williams responded: This point is well made. The majority of research devoted to bioresorbable Mg implants largely ignores the fact there may be problems which arise *in vivo* due to high local concentrations of hydrogen caused by corroding magnesium. Most attention has been concentrated on the biocompatibility of the cations which are released upon corrosion. Given the well documented influence of iron impurity level on corrosion rate, closely controlling the trace Fe content of Mg may provide a means of "tuning" the corrosion rate to one which lies in an acceptable time window for optimum *in vivo* performance. However, in reality the situation will be much more complicated since it is likely that nominally pure Mg, with a certain "tuned" Fe trace content, may not possess the requisite mechanical strength (*e.g.* for stent applications), and would need to be alloyed with other elements. There may, in turn, be some interplay of these additional alloying elements with Fe, which could either mitigate (*e.g.* Zr, Al–Mn) or enhance (*e.g.* Si) the influence of Fe on corrosion rate. As a result, there are probably more straightforward methods of limiting corrosion rates, such as the use of an appropriate biocompatible surface treatment or coating.

Stuart Lyon remarked: Should we be much more concerned with the influence of "low" or trace concentrations of elements in alloys and their influence on the surface structure/chemistry and corrosion?

John Scully replied: As mentioned in the closing lecture,[1] I believe that trace and minor alloying element effects are both one of the great triumphs of the corrosion metallurgy field and one of its least well understood aspects.

Consider the effects of Mo and nitrogen on stainless steels as well as Mo and W on Ni–Cr alloys. Consider minor and trace elements effects on corrosion in copper base alloys. Consider the effects of Cu and Ni or others in weathering steels. In the case of amorphous alloys, minor alloying additions are often made to improve quench rate and glass formation but effects on corrosion are not well understood. The need for understanding intensifies further in high entropy alloys.

These effects are not well understood yet provide some of the most productive area of technological accomplishments in corrosion metallurgy. Therefore, I see some of these topics as very worthy for investigation as corrosion science looks towards the future.

1 John R. Scully, *Faraday Discuss.*, 2015, DOI:10.1039/c5fd00075k.

Roger Newman responded: For the arsenic effect on dezincification, its possible analogs, and for the startling detrimental effects of elements like N and P on SCC of austenitic stainless steels (Tsujikawa), yes! By the way, the latter work is still little-known outside Japan. For pitting, and stability of passivity generally, probably not, at those very low levels. An exception is the "anodic segregation" of sulfur studied by Marcus, where the S concentration rises to monolayer levels at the metal-film interface during passivation in acid. He also proposed that similar segregation of N, in stainless steels containing that element, was responsible for resistance to pitting – I would differ on that point, having shown that N markedly reduces the dissolution kinetics in pit-like environments.

Weathering steels are another topic. Those levels are not really "trace" though. In the magnesium world, because everything else "anodically segregates", this is an important topic.

David Shoesmith responded: Yes we should. Their influence is particularly noticeable in the light metal alloys (Mg, Al, Ti) as clearly noted in the introductory address by Frankel[1] and the presentations on Mg alloys. The possible influence of Fe impurities on the anodic formation of hydrogen is not the only example of an "extreme" effect of apparently atomically-distributed impurities. The propagation of crevice corrosion on titanium alloys varies considerably as the minor content of Fe varies.

1 G. S. Frankel, S. Fajardo and B. M. Lynch, *Faraday Discuss.*, 2015, DOI:10.1039/c5fd00066a.

Simon Gibbon communicated: Listening to the discussion today I could believe that there is a real disconnect between the need to do highly defined empirical work and the need to solve less well defined real alloy corrosion problems – there is little connection to modelling. Is this really the state of corrosion science today?

James Sullivan communicated in reply: I would agree that there is probably quite a disconnect between the two and as discussed by Prof. Scully in his Concluding remarks,[1] there needs to be a collective effort by the community to bring these areas closer together in the future. This may enable us to solve and critically predict some of the complex corrosion behaviours of new and current alloys.

1 John R. Scully, *Faraday Discuss.*, 2015, DOI:10.1039/c5fd00075k.

Simon Gibbon communicated: High throughput experimentation and the access to the ability to make a wider range of alloy compositions than could ever be achieved manually, gives an incredible advantage in the search for sweet spots of corrosion resistance, but it is not fundamentally different from Edison's experiments to optimise the light bulb filament, only faster. Surely I must be missing some good examples where modelling has made/is making an explicit quantifiable contribution to overcoming significant corrosion issues?

John Scully communicated in reply: I think that high throughput experiments with higher fidelity data acquisition are certainly an opportunity to advance corrosion technology. However, I agree with the notion of "sweet spots" in many corrosion phenomena which means that a more targeted approach is required where experiments are clustered around potential sweet spots instead of massive testing of a wide range of alloy or coating compositions, for instance. As we all know, the corrosion field is full of examples where strong "thresholds" are seen. The challenge is to find and understand them.

One of the advantages of computational modeling is the ability, if the essence of the corrosion phenomenon is captured in the model, to exercise the model over many conditions to develop a better understanding of what factors really matter. Then these factors can be focused upon in more limited and focused experiments. This is different than the view you express above and in my opinion is not Edisonian.

Regarding ICCME (Computational Corrosion Materials Eng), the success stories are limited and in isolation but include tunable Al-TM-RE glass and nanocrystalline coatings that our group invented,[1–4] ultra high strength stainless steels that do not require heavy metal coatings (Questek), and iron based BMG which, maybe by accident, combine all the tools in the MSE tool box to make very corrosion resistant materials (significant Cr and Mo in solid solution, beneficial minor alloying elements such as C, B, Y, lack of inclusions or large defects *etc.*). Another example might be grain boundary engineering of polycrystalline alloys for SCC resistance. These examples stand in isolation and do not yet signal a broad movement.

1 N. R. Tailleart, B. Gauthier, S. Eidelman and J. R. Scully, Metallurgical and physical factors controlling the multi-functional corrosion properties of pulsed thermal-sprayed Al–Co–Ce coatings, *Corrosion*, 2012, **68**(3), 035006-1–035006-26.
2 J. R. Scully, F. Presuel-Moreno, M. Goldman, R. G. Kelly and N. Tailleart, User-selectable barrier, sacrificial anode, and active corrosion inhibiting properties of Al–Co–Ce alloys for coating applications, *Corrosion*, 2008, **64**(3), 210–229.
3 M. A. Jakab and J. R. Scully, On-demand release of corrosion-inhibiting ions from amorphous Al–Co–Ce Alloys, *Nat. Mater.*, 2005, **4**(9), 667–670.
4 F. Presuel-Moreno, M. A. Jakab, N. Tailleart, M. Goldman and J. R. Scully, Corrosion-resistant metallic coatings, *Mater. Today*, 2008, **11**(10), 14–23.

Janine Mauzeroll communicated in reply: On the atomic scale, *ab initio* and DFT based models have been used to understand already known corrosion inhibition, but also to predict the effect of doping. For example, models were used to study the effect of chloride ions on passive NiO films,[1] the bonding of benzotriazole corrosion inhibitor on copper and copper oxide surfaces[2] and the effect of different doping elements on the thermodynamic corrosion properties of Ni based alloys.[3] The earlier has inspired much more research and interaction between modeling and experiments on passive films and their decomposition.[4] Increasing scale size to microscopic phases in alloys and their coupling, models are still being developed. One often cited work in this field is by Kiran Deshpande[5] on the Microgalvanic coupling of alpha and beta phase in AZ91 magnesium alloys.

Most direct impact has finite element modeling of macro galvanic corrosion. Frequently used in engineering applications to estimate the effect of galvanic coupling are numerical models based on the ionic conductivity of the electrolyte, *i.e.* the Laplace equation of the electric field.[6–8]

1 Asmae Bouzoubaa, Boubakar Diawara, Vincent Maurice, Christian Minot and Philippe Marcus, *Corros. Sci.*, 2009, **51**, 2174–2182.
2 Anton Kokalj, *Faraday Discuss.*, 2015, DOI:10.1039/c4fd00257a.
3 Bai-Lin Lü, Guo-Qing Chen, Wen-Long Zhou, Hui Su and Rui Liu, *J. Nucl. Mater.*, 2011, 286–291.
4 Toni Massoud, Vincent Maurice, Lorena H. Klein and Philippe Marcus, *J. Electrochem. Soc.*, 2012, **160**, C232–238.
5 Kiran B. Deshpande, *Electrochim. Acta*, 2011, **56**, 1737–1745.
6 Nicolas Murera, Roland Oltra, Bruno Vuillemin and Olivier Néel, *Corros. Sci.*, **52**, 130–139.
7 Dao Trinh, Philippe Dauphin Ducharme, Ushula Mengesha Tefashe, Joseph R. Kish and Janine Mauzeroll, *Anal. Chem.*, 2012, **84**, 9899–9906.
8 Daniel Höche, *J. Electrochem. Soc.*, 2015, **162**, C1–C11.

Christopher Taylor opened a general discussion of the paper by Geraint Williams: As I understand your interpretation of the Mg corrosion, at higher concentrations of NaCl, you expect that the solution can support a higher throwing power, and so the radial growth of the corrosion front occurs, whereas in lower concentrations of NaCl, the throwing power is lower, thus only 1-D spreading of corrosion occurs, giving this filiform appearance. Is this process amenable to modeling? It is not clear to me how changing the throwing paper should lead to a shift in geometry, only in rate, so I would find a mathematical model convincing. Can you describe how the geometry of the corrosion front changes in the intermediate case, as one moves from higher concentration (radial) to lower concentration (1D)?

Geraint Williams responded: I would be more than happy to see the phenomenon modelled and willing to supply the relevant experimental data to any volunteer who felt like taking on such an endeavour. For the intermediate case (between 0.1 M and 0.01 M NaCl), the breakdown of the Mg surface causes initial propagation of circular features, but as they grow radially, the anodic region at their perimeters starts to fragment. Consequently, filiform-like tracks emanate from these points of fragmentation, as shown in the photograph (see Fig. 1) of a post-corrosion Mg surface after immersion in 0.02 M NaCl (aq), and disc-like corrosion propagation stops.

Alison Davenport asked: The growth velocity of the corrosion filaments can be converted into a current density. Could you comment on the range of current density values found, and in particular if there is a minimum current density below which filament growth cannot take place?

Geraint Williams answered: Yes it is correct that velocity can be converted to a current density value, although this requires some assumptions to be made regarding the depth of anodic dissolution and the width of individual filaments. In a previous paper,[1] a more rigorous characterisation of filiform corrosion affecting the same type of Mg under similar chloride ion concentrations was described. In this work the minimum filament width observed was 0.1 mm and a depth of attack of *ca.* 1 μm was estimated by AFM mapping. Using an average filament velocity of 1 μm s^{-1} and assuming the front of the "track" comprises a 3 dimensional rectangle of width 0.1 mm, length 10^{-3} mm and depth 10^{-3} mm,

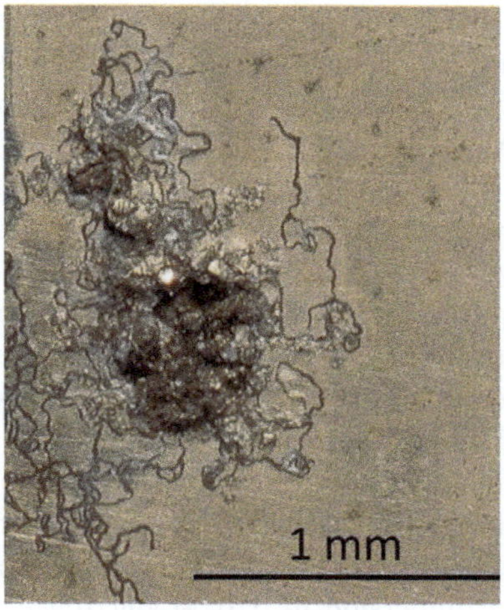

Fig. 1 Photograph of a post-corrosion Mg surface after immersion in 0.02 M NaCl (aq).

then it can be calculated that a mass of Mg equating to 1.8×10^{-10} g is excavated every second (using density (Mg) = 1.8 g cm^{-3}). Using Faraday's law this works out as a current of 1.4 μA and if it is further assumed that the area from which it emerges comprises the rectangle face (width 0.1 mm, length 10^{-3} mm), then the current density becomes 14 A m^{-2}. Given that filament velocity values never fall below 0.7–0.8 μm s^{-1} and that the minimum width is *ca.* 0.1 mm, then it appears that current density values of at least 10 A m^{-2} are required to sustain propagation. This correlates well with SVET-derived anode current density values of between +5 and +15 A m^{-2} reported in the aforementioned publication.

1 G. Williams and R. Grace, *Electrochim. Acta*, 2011, **56** 1894.

Gerald Frankel asked: It seems that you should be able to understand the phenomenon by assessing the cathodic activation after corrosion using polarization curves. This might vary with the chloride concentration similar to the effect of the prior applied anodic current density. In that case, the corrosion morphology might be controlled by the amount of available cathodic current and galvanic coupling rather than by the ohmic resistance.

Geraint Williams responded: Yes, this is certainly a possibility that should be considered. However, there will be some difficulty in reproducibly generating a working electrode area which is fully corroded, especially where filiform attack is prevalent and there will be some "islands" of uncorroded surface remaining amongst the network of dark tracks, even after protracted immersion. Also, there may be some issues in effectively accounting for *iR* drops in carrying out potentiodynamic experiments on fully corroded surfaces over a range of different chloride concentrations, even in the presence of a Luggin capillary. One would also have to think of a physical description of why chloride ion concentration would influence the electrocatalytic efficiency of the corroded surface toward cathodic hydrogen evolution.

David Shoesmith asked: You elegantly demonstrate the relationship between anodes and cathodes and show how corrosion tracks across the surface. Is the substrate microstructure at all involved in determining the direction of propagation?

Geraint Williams responded: There is no real indication, especially at concentrations of 0.1 M NaCl (aq) and above, that propagation is influenced by microstructure for the pure Mg studied here. There are reports in the literature,[1] that indicate filiform corrosion (FFC) on pure Mg can preferentially follow certain crystallographic features of the Mg grains. These authors showed that directionality of individual filaments appeared to be influenced by changes in orientation at grain and twin boundaries. However, the investigation was carried out in chromate inhibited dilute chloride electrolyte, which is not the case for the present work disseminated in this Faraday discussion. The random walk FFC patterns observed upon immersion in 0.01 M electrolyte (see Fig. 1c), for the rather impure form of Mg studied in our work, suggests that microstructural influence on filament directionality is minimal in the absence of dissolved chromate inhibitor.

1 P. Schmutz, V. Guillaumin, R. S. Lillard, J. A. Lillard and G. S. Frankel, *J. Electrochem. Soc.*, 2003, **150**, B99.

Clara Wren opened a general discussion of the paper by Janine Mauzeroll: Your sensors use microcapillaries for local corrosion potential measurements. Mg corrodes very fast and, as far as I can tell, this corrosion is often limited by the solution transport rate. Do you expect the same solution transport behaviour for redox species in the microcapillaries and in bulk water? How do you ensure that what you are measuring is not the transport limited corrosion potential inside the microcapillary?

Janine Mauzeroll answered: The question is referring to the measurement of corrosion potential of the different microstructures using the microcapillary technique presented in Figure 1 of our paper. For this measurement, the meniscus of the filled micropipette is brought in electrical contact with the substrate where the open circuit potential is recorded. This experiment does not consist of any measurement of substrate kinetics and we do not extract any rate information where solution transport is important, as previously mentioned.[1] When performing a substrate polarization curve, we assume that the mass transport within the micropipette will be governed by diffusion only within the confines of the micropipette. The comparison of numerical simulations for mass transport governed by diffusion only within micropipettes (~500 nm) is in excellent agreement with experimental data.[2]

1 S. Thomas and H. Bohni, Analytical Methods in Corrosion Science and Engineering. CRC Press, Boca Raton, FL, USA, 2005, pp. 649–696.
2 Williams *et al., Anal. Chem.*, 2009, **81**, 2486–2495.

David Williams remarked: The shear force-controlled scanning vibrating capillary seems a very powerful general tool to me, analogous to the SNOM. I wonder what other probe methods you have considered that might utilise this platform – for example scanning ion conductance microscopy to map solution conductivity changes or using the capillary as a waveguide to implement optical microscopy or perhaps Raman. I wonder also if the modulation of the signal caused by the vibration could be used to advantage.

Janine Mauzeroll answered: The shear force controlled scanning vibrating capillary can be adapted for many situations in corrosion science simply by changing the nature of the probe.

In its unmodified state, the scanning droplet cell remains versatile towards electrochemical techniques that can be performed locally such as potentiodynamic polarization or voltammetry to extract local kinetics through accurate corrosion modeling. The wide variety of electrochemical techniques also include, local chronoamperometry that can probe ions released during the corrosion process[1,2] or even perform local modification of surfaces.[3,4]

It can be easily adapted to measure local impedances, making it very useful in the study of ageing and failure of protective coatings.[5–8]

It can be adapted to quantify several chemical fluxes generated during the corrosion. For example, corrosion product deposition affects the SECM feedback regeneration of different dissolved redox mediators such as FcMeOH when an alloy is immersed in a corroding solution.[9] It can also be used to directly track hydrogen fluxes from the alloy microstructure.[10] It can also be modified through the use of an ion-selective microelectrode to track Mg^{2+} fluxes.[11] Finally, with minor microelectrode manipulations, it is possible to pull fiber optics and to monitor combined spectroscopy and electrochemistry as reported by A. J. Bard in the early 2000s.[12] I cannot conclusively comment on if the modulation of the signal caused by the vibration could be used to advantage.

1 K. A. Lill, A. W. Hassel, G. Frommeyer and M. Stratmann, *Electrochim. Acta*, 2005, **51**, 978–983
2 M. M. Mennucci, M. Sanchez-Moreno, I. V. Aoki, M. C. Bernard, H. G. Melo, S. Joiret, and V. Vivier, *J. Solid State Electrochem.*, 2012, **16**, 109–116.
3 M. M. Lohrengel, A. Moehring and M. Pilaski, *Fresenius' J. Anal. Chem.*, 2000, **367**, 334–339.
4 M. Pilaski and M. M. Lohrengel, *Electrochim. Acta*, 2003, **48**, 1309–1313.
5 F. Zou and D. Thierry, *Electrochim. Acta*, 1997, **42**, 3293–3301.
6 D. Snihirova, L. Liphardt, G. Grundmeier and F. Montemor, *J. Solid State Electrochem.*, 2013, **17**, 2183–2192.
7 A. M. Mierisch and S. R. Taylor, *J. Electrochem. Soc.*, 2003, **150**, B303–B308.
8 M. W. Wittmann, R. B.Leggat, S. R. Taylor, *J. Electrochem. Soc.*, 1999, **146**, 4071–4075.
9 W. Liu, F. Cao, Y. Xia, L. Chang and J. Zhang, *Electrochim. Acta*, 2014, **132**, 377–388.
10 U. M.Tefashe, M. E. Snowden, P. Dauphin-Ducharme, M. Danaie, G. A. Botton and J. Mauzeroll, *J. Electroanal. Chem.*, 2014, **720–721**, 121–127.
11 P Dauphin Ducharme, R. M. Asmussen, D. W. Shoesmith and J. Mauzeroll, *In situ* Mg^{2+} release monitored during magnesium alloy corrosion, *J. Electroanal. Chem.*, 2015, **736**, 61–68.
12 Y. Lee, S. Amemiya and A. J. Bard, Scanning electrochemical microscopy. 41. Theory and characterization of ring electrodes, *Anal. Chem.*, 2001, **73**, 2261–2267.

Stuart Lyon opened a general discussion of the paper by Sannakaisa Virtanen: Given the low net atomic number of Mg-based implant alloys, how can they be easily imaged by X-rays?

Sannakaisa Virtanen answered: They cannot be easily imaged by X-rays. So I assume one would try to use some X-ray markers. In the case of coated Mg-based implants, the coating could provide also a means for imaging. I am not aware if there is a state-of-the-art in the field, or simply some exploratory approach.

Rob Lindsay said: From reading your paper, I believe you have assumed a homogenous substrate for quantification of the XPS data. Clearly, if this is true, this substrate model does not represent reality particularly well, in my opinion, as there will be discrete layers of material, with the metallic substrates lower most. How do think your approach impacts upon your interpretation of the XPS data (*e.g.* data shown in Fig. 8)?

Sannakaisa Virtanen replied: The XPS data presented simply shows the at% determined, and is not further interpreted for instance in view of a homogeneous or partial coverage or in view of the different layers from the metallic substrate, to passive film, to the protein adsorption layer. We do not believe that we have a well-ordered and complete monolayer of albumin in this case. The XPS data should be considered only in a qualitative manner, *i.e.* in comparison of the different systems (substrate material, time, electrolyte). In order to perform a full analysis, more experimental data would be required, including information on coverage and thickness of the different layers.

Kirsi Yliniemi remarked: In addition to the nature of surface charges on either Mg or Fe surfaces, could you comment how the surface charge density affects the protein adsorption as it may also influence your results? There exist models which suggest that at low surface charge densities polyelectrolytes adsorb on a more flat orientation (2D films) while at higher surface charge densities thicker 3D film starts to grow (see for example the review[1]).

1 A. Dobrynin and M. Rubinstein, *Prog. Polym. Sci.*, 2005, **30** 1049–1118.

Sannakaisa Virtanen replied: Thank you very much for the important comment and helpful literature. Yes, indeed not only the type of surface charge but also the charge density should play a role. Until now, we did not carry out any detailed studies on this, and the surface charge considerations in our paper are simply based on IEP values found in the literature.

But indeed we would like to know more about the type of the protein adsorption layers, so in addition to addressing the question on the amount of protein adsorbed on different surfaces, we would like to study the orientation of the proteins on the surface (with additional experiments and a more detailed data analysis) and also their functionality (for instance, are they in their native conformation or possibly denaturated, depending on the surface they are adsorbing on).

In the case of Mg some of the studies we would like to do are challenged by the high reactivity of the Mg surfaces in aqueous solutions, so we often have the problem of not only adsorption of proteins taking place but at the same time substantial corrosion of the substrate (this is also briefly mentioned in the paper).

Johannes Lützenkirchen communicated: I have a comment concerning the remark in the introduction of the paper, where you mention that from your references 14, and 19–27, "both acceleration and inhibition of corrosion has been observed" with respect to the action of proteins related to Mg-materials. Your interpretation is based on the charging of the adsorbing surface and that of dissolved protein. The opposite sign of charge should favor adsorption and cause inhibition. In that respect you quoted from your reference 38 a compilation of points of zero charge for Mg-related surfaces. Indeed the points of charge of particles of these materials have been reported to be high (*i.e.* above pH 10), but a scan of references 14, and 19–27 of your paper shows that usually flat surfaces have been used in the adsorption and corrosion studies. It has been previously observed that such flat surfaces may have points of zero charge that are much lower than those of the corresponding particles.[1–4] It is one issue we discuss in our paper. Fig. 2 shows unpublished zeta-potential measurements for a flat MgO surface in 5 mM NaCl. The system has been titrated under purified and humified Argon atmosphere, using HCl and NaOH solutions. We used freshly prepared NaOH solution to exclude carbon dioxide. The MgO surface is expected to hydroxylate and expose a $Mg(OH)_2$ like surface to the solution. Even for this "MgO"-surface, we retrieve a drastic difference between particles and flat plates. It has been shown that this feature also occurs on alumina specimens of different orientation,[2,3] so that it is not related to the surface chemistry.

The isoelectric point (IEP) of the sample was initially at pH 4 as shown in Fig. 2. On aging the IEP increases, but remains below pH 5 (data not shown), *i.e.* still much lower than the powder values from ref. 38 of our paper. It is important to note that data on sapphire-c surfaces show that roughness decreases the IEP of this plane to even lower values (Lützenkirchen *et al.*, manuscript in preparation).

Furthermore, the wide variation of the IEPs in Kosmulski's collection (ref. 38) of many samples would make conclusions based on literature data alone arbitrary for many cases. Thus the use of published particle IEPs does not necessarily give an appropriate interpretation. Even for nominally identical solids, the IEP may vary widely. Therefore, it is favorable to measure it for the conditions used. In particular since adsorption of some proteins may occur even under electrostatically unfavorable conditions.[5] Although in the present case, no contradiction occurs between the charge-interpretation and the experimental data, in general the point of zero charge of a given sample should be determined to avoid ambiguity.

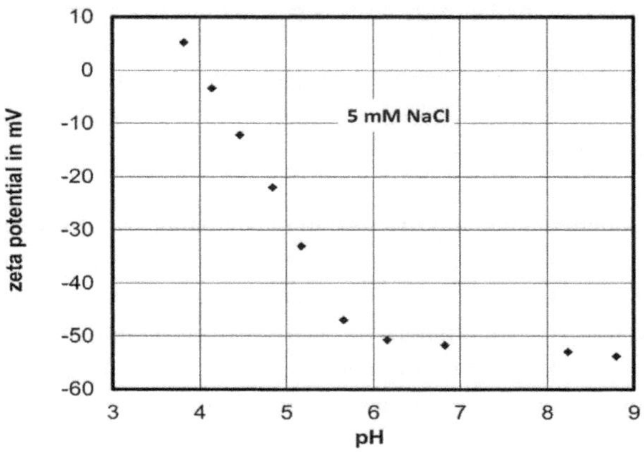

Fig. 2 Zeta-potential of a flat MgO surface in 5 mM NaCl.

A related question concerns the iron samples. Does the entry in Table 2 invoke that ferrous iron is the relevant form of iron? The background of the question is that, while ferric oxides, hydroxides and oxyhydroxides are easily studied and the IEPs can usually be determined quite well, iron(II) oxide and hydroxide minerals are much more difficult to investigate. One major difficulty is that it is crucial to avoid oxidation of iron(II). Very few studies on these iron(II) systems are therefore available and only recently it has been reported[6] that the IEPs of most of the iron(II) samples under investigation in that study exhibited IEPs of about 10 or higher.[6] The substrates also showed transformation phenomena. The data from ref. 38 (in your paper) are all for pure iron(III) systems or magnetite. How would the interpretation be affected if you considered the higher IEPs from the recent study?[6]

1 G. V. Franks and Y. Gan, *J. Am. Ceramic Soc.*, 2007, **90**, 3373–3388.
2 G. V. Franks and L. Meagher, *Colloid Surf. A-Physicochem. Eng. Asp.*, 2003, **214**, 99–110.
3 J. P. Fitts, X. Shang, G. W. Flynn, T. F. Heinz and K. B. Eisenthal, *J. Phys. Chem. B*, 2005, **109**, 7981–7986.
4 E. Mączka, J. Luetzenkirchen and M. Kosmulski, *J. Colloid Interface Sci.*, 2013, **393**, 228–233.
5 T. Arai and W. Norde, *Colloids Surf.*, 1990, **51**, 1–15.
6 E. Maczka, E. Jartych and M. Kosmulski, *Colloid Surf. A-Physicochem. Eng. Asp.*, 2014, **441**, 326–330.

Sannakaisa Virtanen communicated in reply: Thank you for the very interesting comments and questions and also for sharing some unpublished data. I completely agree that our approach is a rough and simple one, as certainly the IEP values should be measured for the sample surfaces under the conditions of the study, to have a better and more realistic view. Also, I agree that surface charge is not the only factor influencing protein adsorption, for instance surface wettability is also of importance. If other driving forces are stronger than electrostatic interactions, then adsorption could take place as you mention also for electrostatically unfavorable conditions. In our case, as both Fe and Mg surface are hydrated and show a similar hydrophilic behaviour, according to the literature electrostatic interactions can be more significant.

As regards the nature of the Fe surface: The passive film formed in NaOH should be an oxide with a mixed valency between Fe_3O_4 and Fe_2O_3 (but not FeO). We refer to the values for FeO/$Fe(OH)_2$ in Table 2, as we assume a hydroxide deposit layer on top of the Fe-oxide layer. Again, a determination of IEP values for the given case should be carried out.

The discussion in the manuscript is based on the fact that our hypothesis on the relevance of the surface charge argument seems to hold for two different proteins. But it still remains a hypothesis.

Clara Wren opened a general discussion of the paper by James Sullivan: Your corrosion rate decreases with the aluminium content in the alloy. Have you tried to normalise the corrosion rate with respect to the same zinc–magnesium or magnesium content? Is the change in corrosion rate simply a function of the relative amount of magnesium available, or is there any synergistic effect of the composition-induced alloy-microstructure on the corrosion rate?

James Sullivan replied: We haven't tried to normalise the data to date but that is an interesting suggestion. I think the ultimate change in corrosion rate of these alloys is dependent on both alloy additions (Mg and Al) especially in atmospheric corrosion conditions where the establishment of Al and Mg corrosion products are critical for the performance of these alloys. I feel the initial corrosion rate differences observed here maybe more dependent on increasing Al levels with a greater coverage of alumina present as the eutectic volume fraction increases that is more resistant to corrosion initiation in the earliest of stages.

Stuart Lyon asked: What is the specific composition of the various components of the eutectic in these complex Zn–Al–Mg alloys? How does the area/volume fraction of eutectic influence the overall corrosion performance?

James Sullivan answered: We haven't performed a detailed experimental analysis of the various phase compositions. MTDATA simulations were performed for ZMA 2 during its development that suggested that the binary eutectic contained $MgZn_2$ and Zn lamellae with around 1% Al in the Zn lamellae and 0.6–0.8% Al in the $MgZn_2$. In the ternary eutectic the same phases were present with addition of an Al fcc phase containing 66% Al, 0.02% Mg and the remainder Zn. The volume fraction of eutectic and its influence on corrosion performance are given in the presented paper in Figures 3 and 7 and show a fairly linear relationship between increased corrosion performance with increasing eutectic volume fraction.

Gerald Frankel said: Why doesn't the attack in the eutectic region go deep into the structure and why does it initiate at only a few places and then spread across the surface?

James Sullivan responded: The attack does also penetrate vertically into the sample and we have seen this from examination of cut samples post corrosion although it would seem that the rate of lateral spreading is greater. The reason for the corrosion spreading through the eutectic laterally at the surface may be due to the surface being more eutectic rich as the solidification sequence starts with nucleation of Zn rich dendrites at the steel substrate. Therefore the volume fraction of primary Zn is greater at the coating/steel interface with a subsequent reduction of primary Zn and increase in eutectic as the coating is traversed to the surface. As the $MgZn_2$ in the eutectic is the preferential phase for initial corrosion attack then lateral attack through this eutectic rich surface may dominate.

Arjan Mol asked: First, I would like to thank James Sullivan for presenting this very interesting work and paper during this Faraday Discussions session, I greatly enjoyed it. My question is the following: can you please show, explain and/or comment on the positions and evolution of (local) cathodic and anodic areas in the SVET measurements with time and their correspondence to those in the time-lapse optical microscopy videos considering a.o. the different lateral resolutions of the techniques?

James Sullivan responded: That is an interesting question. The SVET samples are much larger, 10×10 mm compared with the 1mm diameter circle area of the time lapse samples. The SVET has a practical resolution of around 150 microns. On the SVET maps we see localised anodes that are well spaced apart (>1 mm) and have a radius of approximately 500 microns. It must be remembered that this reflects the current flux from the anodic features that the SVET detects in its plane of scan 100 microns above the sample surface. As the SVET scans progress the anodic features actually spread out as small rings from the initiation point. I have attached a movie to show this.† It is somewhat difficult to correlate this with the time-lapse movies due to the differences in sample area and resolution of the techniques but the spreading of the anodes as rings may reflect the lateral growth of the anodic features shown in the time-lapse videos. Obviously, the growth of features in the time-lapse experiments maybe more constrained due to the smaller sample area exposed.

Steven Street asked: The three different ZMA alloys had quite different dendrite sizes, indicating different thermal gradients when solidifying. As corrosion seemed to propagate in the interdendritic regions, is there any value in trying to refine the size of these dendrites? Would this improve the corrosion behaviour of these alloys?

James Sullivan responded: These experiments at a range of cooling rates were carried out using a 1.6 wt% Mg 1.6 wt% Al ZMA alloy during the development phase of this alloy and little difference was observed during industrial corrosion tests. It would be interesting to see if any effects of morphology could be identified using this technique. Within the group we have certainly seen that morphological differences within the microstructure induced by cooling rate changes seem to influence the corrosion behaviour of Zn - 4.8 wt% Al alloys.[1]

1 J. Elvins, J. A. Spittle and D. A. Worsley, *Corros. Eng., Sci. Technol.*, 2003, 197–204.

Stuart Lyon asked: Corrosion resistant metallic coatings on steel are much thinner than they used to be. How does this influence the corrosion process/resistance at the cut edge?

James Sullivan answered: These ZMA coatings are certainly thinner than more traditional HDG type coatings but in industrial salt spray tests they certainly outperform the HDG materials during surface testing. We also see this improvement in corrosion performance when examining surface corrosion. At the cut edge we have comparable performance to that of HDG for the thinner ZMA coating in SVET and external run-off testing that is encouraging considering the change in anode/cathode ratio. However, the performance at the cut edge is not nearly as improved as the surface performance in comparison to HDG, that may reflect this change in anode/cathode ratio.

Clara Wren remarked: What is the geometry of the viewing cell, the diameter and the depth of the solution, with respect to that of the rest of the corrosion cell? I noticed in your video images that the corrosion front always starts close to the boundary between the viewing area and the rest of the corrosion cell. Could your observation indicate that the Mg corrosion is solution transport limited?

James Sullivan answered: The distance between the viewing window and the sample is around 1 mm so there is a 1 mm body of electrolyte between the sample and the window. The diameter of the protective shroud on the microscope is 20 mm. The total volume of electrolyte in the cell is 250 ml and electrolyte is accessible to the sample from 360° from the bulk volume into the 1 mm gap as shown in Figure 1 in the paper. The corrosion often starts near the periphery of the exposed area as the circular area is masked using Teflon tape. The slight geometry change near the tape is often sufficient to induce anodic activity here and we also see this fairly regularly in SVET experiments where again a sample area is masked off using tape. We plan to carry out some tests involving stirring to assess the effect on the corrosion mechanism and to assess if diffusion of species is restricted into the gap above the sample but I feel with a 1 mm gap there maybe sufficient space for electrolyte diffusion especially as it can access the sample from the full circumference around the shroud.

David Williams commented: I'm interested in the patterns and timing of appearance of the dark shadows that correspond to precipitation of the corrosion products – for example the sudden appearance of a ring that fades and then reappears. There seems to be some rich information here about the development of the anodic and cathodic sites and convection–diffusion–reaction in the immediate vicinity of the interface. Are you able to comment on this?

James Sullivan replied: I too feel that this is potentially showing some interesting information about the nature of anodic and cathodic sites, the ionic diffusion between the two electrodes and pH gradients established between them. The pH effects have been demonstrated in some previous work[1] using a phenolphthalein indicator and is visible due to the dissolution of Mg corrosion products as the front of anodic activity moves, presumably as the pH drops local to the anodic site. I think it may also be interesting to see the effect of electrolyte concentration on the formation of the corrosion product rings with regards to the distance from the anode to assess the effect of the electrolyte's throwing power on anode-cathode spacing.

1 James Sullivan, Shahin Mehraban and Jon Elvins, *Corros. Sci.*, 2011, 53(6), 2208–2215.

Su-Ting Cheng said: In Fig. 2, the red area increases largely from ZMA1 to ZMA3. Since the bulk concentrations of Zn (green) and Mg (red) are kept the same, a different colour distribution could indicate a surface segregation of Mg. Have you ever measured the concentration of elements throughout the whole depth of the specimens? We have studied the Mg–Zn system with DFT, and the results tell us the surface of the $MgZn_2$ intermetallic compound is more stable with a higher surface Mg concentration, either in vacuum or in contact with oxygen. It would be nice to see if this could be also observed in experiments.

James Sullivan responded: I may have misunderstood the question but the quantity of Mg increases from 1–3 wt% from ZMA 1–3 and the Zn level drops as the Mg and Al levels increase. This I believe accounts for the increase in red as the volume fraction of eutectic increases in the microstructure from ZMA 1–3. At present we have not measured the depth profile of the elements.

Rob Lindsay stated: I was wondering if you believe that the filamentous precipitates you observe (as shown in Fig. 10 (B)) are important for controlling corrosion or are just an interesting phenomenon?

James Sullivan replied: The filaments are an interesting phenomenon but I think they can offer an important role in controlling corrosion as the deposition of the filaments occludes the surface from further attack potentially. I think the initial stages of precipitation local to the anodic features also demonstrates quite nicely the formation of inhibitor films at the anode. Also, these conditions may reflect those occurring in drying electrolyte films where concentration of salts may become saturated.

Simon Gibbon commented: Yesterday Mira Todorova[1] told us about the application of point defect models to give new insights into the defects formed on zinc surface which could lead to corrosion. In your paper you have considered the corrosion behaviour of commercially important zinc alloys. I would be interested to know if you have considered applying such models to help in the understanding of your systems?

1 Mira Todorova and Jörg Neugebauer, *Faraday Discuss.*, 2015, DOI:10.1039c/4fd00238e.

James Sullivan answered: This is not something we have considered to date but it would be of interest to assess how the model could be applied to these alloys to provide a prediction of corrosion behaviour with regards to preferential attack.

Trevor Hughes communicated: A very interesting paper on the use of *in situ* time-lapse optical microscopy to compare the corrosion behaviour of Zn–Mg–Al alloys in 1% NaCl at pH 7. In the paper it is stated "it may be counterintuitive that an increased eutectic volume would reduce the corrosion rate given that these phases are preferentially attacked in the microstructure but it would seem that increased levels of Mg and Al in the alloy create more robust oxide layers due to their rapidity of reaction and the passive nature of Al_2O_3." Question 1: Have the authors looked at brines with different pH, presumably the passive oxide layers would not be stable in alkaline (pH 10–11) brines? What are the pH limits of such passive oxide layers. In addition, for the neutral (pH 7) system, what is the effect of elevated temperatures?

The paper also shows that the corrosion rate of the ZMA2 alloy (1% NaCl, pH 7) decreases in the presence of sodium phosphate (0.0001, 0.01 mol L^{-1} Figure 13). Question 2: How does this response to phosphate compare for the 3 alloys studied and for pure Zn?

James Sullivan communicated in reply: We are currently undertaking a matrix of tests that include varying the pH with pH levels of 3, 7 and 10 coming under examination. To date I cannot give a detailed report on the results as they are fairly preliminary but the initial indications do indeed show visual differences to the progression of corrosion and formation of corrosion products. We have not carried out any tests as of yet at an elevated temperature and to date we have only carried out the Phosphate tests on ZMA2. However, a pure Zn test is in the matrix of tests planned.

Faraday Discussions

PAPER

Ab initio modeling of the bonding of benzotriazole corrosion inhibitor to reduced and oxidized copper surfaces

Anton Kokalj*

Received 11th December 2014, Accepted 2nd February 2015
DOI: 10.1039/c4fd00257a

The bonding of benzotriazole—an outstanding corrosion inhibitor for copper—on reduced and oxidized copper surfaces is discussed on the basis of density functional theory (DFT) calculations. Calculations reveal that benzotriazole is able to bond with oxide-free and oxidized copper surfaces and on both of them it bonds significantly stronger to coordinatively unsaturated Cu sites. This suggests that benzotriazole is able to passivate the reactive under-coordinated surface sites that are plausible microscopic sites for corrosion attack. Benzotriazole can adsorb in a variety of different forms, yet it forms a strong molecule–surface bond only in deprotonated form. The bonding is even stronger when the deprotonated form is incorporated into organometallic adcomplexes. This is consistent with existing experimental evidence that benzotriazole inhibits corrosion by forming protective organometallic complexes. It is further shown that adsorption of benzotriazole considerably reduces the metal work function, which is a consequence of a large permanent molecular dipole and a properly oriented adsorption structure. It is argued that such a pronounced effect on the work function might be relevant for corrosion inhibition, because it should *diminish* the anodic corrosion reaction, which is consistent with existing experimental evidence that benzotriazole, although a mixed type inhibitor, predominantly affects the anodic reaction.

1 Introduction

There are various ways of corrosion protection and one of them is by means of corrosion inhibitors, which are substances that have the ability to considerably slow down the corrosion of metals and alloys by decreasing the rate of corrosion processes. Among efficient corrosion inhibitors are also organic molecules. It is commonly believed that strong interaction between an organic inhibitor molecule and a substrate is very important for achieving the inhibitory effect.[1–3] In particular, Bockris stated that *organic molecules must be adsorbed to become inhibitors*.[3]

Department of Physical and Organic Chemistry, Jožef Stefan Institute, Jamova 39, SI-1000 Ljubljana, Slovenia. E-mail: tone.kokalj@ijs.si; Web: http://www.ijs.si/ijsw/K3-en/Kokalj; Fax: +386-1-251-93-85; Tel: +386-1-477-3523

Despite the soundness of the *strong adsorption* premise, DFT (density functional theory) calculations have revealed—contrary to what might have been intuitively expected—that some outstanding inhibitor molecules, such as benzotriazole and several other azoles, interact only weakly in their intact form with substrates.[4–6] For example, an intact benzotriazole molecule bonds only by about 0.5 eV to a Cu(111) surface,[5,6] which is similar in strength to solvation interaction between benzotriazole and water.[7,8] This implies that the net adsorption energy at the water/solid interface should be rather small, which may seem surprising, because if the bonding is that weak then aggressive corrosive species would easily replace benzotriazole from the surface (if it would adsorb at all). How then, can benzotriazole act against corrosion?

Benzotriazole has been used for a long time as one of the most efficient copper corrosion inhibitors[1,9–11] and there are numerous studies of its inhibition action on copper. Nevertheless, Finšgar and Milošev stated in a recent review:[11] "... *the exact mechanism of benzotriazole action on copper materials still remains to be elucidated. It is important to clarify why benzotriazole, and not other organic molecules with similar electronic structure, imparts corrosion protection. The answer to this question will provide the basis for predicting new and more effective corrosion inhibitors*."

The aim of this paper is to discuss the adsorption behavior of benzotriazole—as elucidated by DFT calculations—on reduced and oxidized copper surfaces (for the latter only the Cu(I) oxidation state is currently considered). The issue seems relevant, because controversial suggestions exist in the literature,[11] *i.e.*, according to some researchers the presence of oxides is important,[12–14] while others claimed that their presence is not needed for the adsorption of benzotriazole.[15,16] It has to be emphasized that adsorption behavior is not synonymous with the mechanism of corrosion inhibition, yet it may nevertheless provide some useful information and insight. The literature is replete with numerous suggestions about possible adsorption mechanisms of benzotriazole on copper.[11] There are at least two reasons of why this is so: the first is due to complexity of the investigated system. The interpretation of experimental data therefore involves various degree of assumptions and different assumptions may lead to different conclusions. The second reason is due to—as is argued in this paper—nontrivial adsorption behavior of benzotriazole, because it can adsorb in a variety of different forms. It all depends on the details and perhaps this is one of the strengths of benzotriazole and other inhibitor molecules; depending on different conditions they will adopt one of several possible forms and thus sustain various situations.

There are several important points to keep in mind when considering the results presented in this paper. The most important is that due to obvious modeling reasons the presented results refer to the solid/vacuum interface (solvent effects are estimated *a posteriori* when explicitly stated so), whereas in the context of corrosion the solid/water is the relevant interface. Apart from the fact that solvent considerably affects the energetics of adsorption, there is another distinction between the adsorption at solid/vacuum and solid/water interfaces. The adsorption at the latter is competitive (or substitutional), because the surface is always covered with solvent molecules and other possible species. So a given molecule will adsorb only if its adsorption is competitive enough to substitute other species from the surface. In contrast, at the solid/vacuum interface the surface is clean and the molecule adsorbs readily (say that surface temperature is low enough) unless its interaction with the surface is repulsive.

The paper is organized as follows: Section 2 contains a brief description of computational method along with some definitions, while Section 3 deals with bare metallic copper surfaces and reviews the pertinent computational DFT studies. These data are put into a new context and the relevance of the work function reduction upon adsorption of inhibitor is emphasized. Oxidized copper surfaces are considered in Section 4, where significant new results are presented. Section 5 finally contains the conclusions.

2 Technical details

2.1 Definitions

Term "copper surface" shall implicitly indicate the bare (reduced) metallic copper surface (oxidation state of 0); the oxidized surface shall be explicitly referred to as "oxidized". The labels BTAH and BTA indicate the intact and dissociated (without H1 atom/proton) benzotriazole molecule, respectively (atom numbering and skeleton structures of neutral BTAH molecule and deprotonated BTA^- anion are shown in Scheme 1). $BTAH_{(g)}$, $BTAH_{(solv)}$, $BTAH_{(ads)}$ stand for gas-phase, solvated, and adsorbed molecule, respectively (and analogously for other species). Notation BTA–Cu refers to organometallic complex, while the stand-alone chemisorbed form is designated as $BTA_{(ads)}$. Other labels are defined when first used.

2.2 Computational

DFT calculations were performed with the plane-wave pseudo-potential method as implemented in the *PWscf* code from the *Quantum ESPRESSO* distribution,[17] using the ultrasoft pseudopotentials[18] and plane wave basis set with a kinetic energy cut-off = 30 Ry (240 Ry for the charge-density), while molecular graphics were produced by the *XCRYSDEN*[19] graphical package. Calculations were performed with PBE[20] and reparametrized PBE-D′ energy functionals; the former was used for chemisorption and the latter for parallel physisorption modes. PBE-D′ includes a semi-empirical dispersion correction of Grimme,[21,22] which was reparametrized[23] to match the experimental adsorption energy of benzene on Cu(111). Cu surfaces were modeled by periodic slabs with the in-plane lattice spacing fixed at the calculated equilibrium Cu bulk lattice parameter (3.67 Å for PBE[24] and 3.64 Å for PBE-D′[23]). For Cu_2O surfaces, the PBE calculated equilibrium bulk lattice parameter of 4.35 Å was used. Spin polarization was used when needed. For further computational details see our previous publications, in particular ref. 8 and also ref. 7.

Molecular adsorption energies (at solid/vacuum interface) were calculated as:

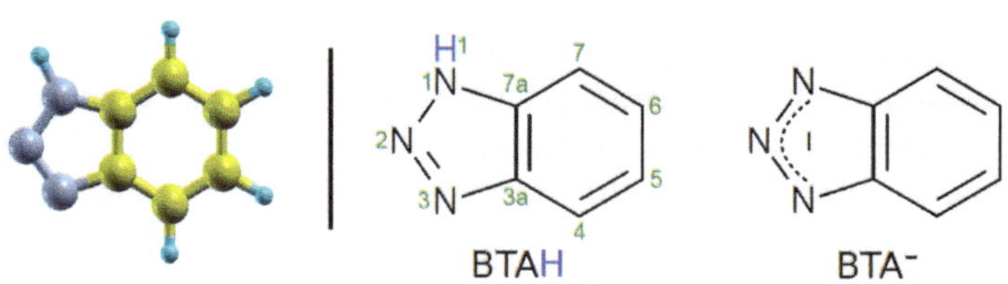

Scheme 1 Skeleton structures of neutral BTAH (middle) and its deprotonated BTA^- form (right). Numbering of atoms is also indicated. Ball-and-stick model of BTAH is shown on the left to indicate the respective atom coloring.

$$E_{\mathrm{ads}} = E_{\mathrm{A/surf}} - (E_{\mathrm{surf}} + E_{\mathrm{A}}). \tag{1}$$

where the subscript A stands for adsorbate; E_A, E_{surf}, and $E_{A/surf}$ are the total energies of isolated adsorbate (molecule or atom), Cu or Cu_2O slab, and adsorbate/slab system, respectively. The use of charged species was avoided in calculations employing periodic boundary conditions, where the long range Coulomb interactions dictate the use of charge neutrality. The adsorption (binding) energies of deprotonated BTA^-, Cl^-, and H^+ are therefore calculated with respect to isolated radicals ($BTA^{\odot}$, Cl, and H) and are designated as $E_b^{\odot}$. Symbol $\odot$ is used to indicate the radical nature of the initial state. Subscript "b" is used in favor of "ads" to remind that binding energy of $BTA_{(ads)}$ is measured with respect to isolated $BTA^{\odot}$ (and not BTAH) and that of $Cl_{(ads)}$ and $H_{(ads)}$ with respect to isolated Cl and H (and not Cl_2 and H_2 molecules). The $E_b^{\odot}$ therefore measures the binding energy of these radicals to the surface. On metal surfaces, the binding energy can be recalculated with respect to anion in the initial state as: $E_b^{(-)} = E_b^{\odot} + EA^{\odot} - \Phi$, where $EA^{\odot}$ is the electron affinity of corresponding radical and Φ is a work function (for cations the equation is: $E_b^{(+)} = E_b^{\odot} - IP^{\odot} + \Phi$, where $IP^{\odot}$ is the ionization potential or respective radical). Analogously to $E_b^{\odot}$, also the E_{ads} of $BTAH_{(ads)}$ is occasionally labeled as E_b, in particular, when the molecule–surface binding is of primary concern.

3 Adsorption of benzotriazole on reduced copper surfaces: bare Cu(*hkl*) and defects thereon

Adsorption structures of benzotriazole on bare copper surfaces have been studied in detail by means of DFT calculations.[5–8,23,25–28] The main findings of these studies are briefly explained below (Section 3.1 and, in part, also 3.2) to put them in context for the current discussion paper.

3.1 Description of various DFT identified adsorption modes

Benzotriazole can adsorb either molecularly as $BTAH_{(ads)}$ or dissociatively as $BTA_{(ads)}$ + $H_{(ads)}$ (cleavage of N1–H bond). While the $BTAH_{(ads)}$ bonds weakly, the $BTA_{(ads)}$ is strongly chemisorbed on copper surfaces. The calculated activation energy of dissociation for weakly chemisorbed BTAH is about 1.1 eV on Cu(111), but on more open surfaces and step-edge defects the activation energy decreases below 1 eV.[27] Dissociative adsorption of benzotriazole is experimentally well established, because it has often been observed that the H1 atom of benzotriazole is removed upon chemisorption on copper surfaces even under ultra high vacuum conditions.[15,28–31]

Structures of different identified adsorption modes of $BTAH_{(ads)}$ and $BTA_{(ads)}$ are shown in Fig. 1 and binding energies as a function of coordination number of surface Cu atoms involved in the adsorption site are shown in Fig. 2.

3.1.1 Adsorption forms of $BTAH_{(ads)}$. BTAH can either weakly chemisorb in an upright geometry ($BTAH_{\perp}$) *via* triazole nitrogen atoms with the N–Cu distances of about 2.1 Å (Fig. 1a) or physisorb† nearly parallel ($BTAH_{\parallel}$) to copper surfaces

† Terms chemisorption and physisorption refer to the type of molecule–surface interaction and not to the scale of the interaction energy. Molecule–surface bond length is a good descriptor to distinguish between the two. For chemisorption the benzotriazole–surface distances are about 2 Å, whereas for physisorption the distances are about 3 Å as inferred from respective covalent and van der Waals radii.[23]

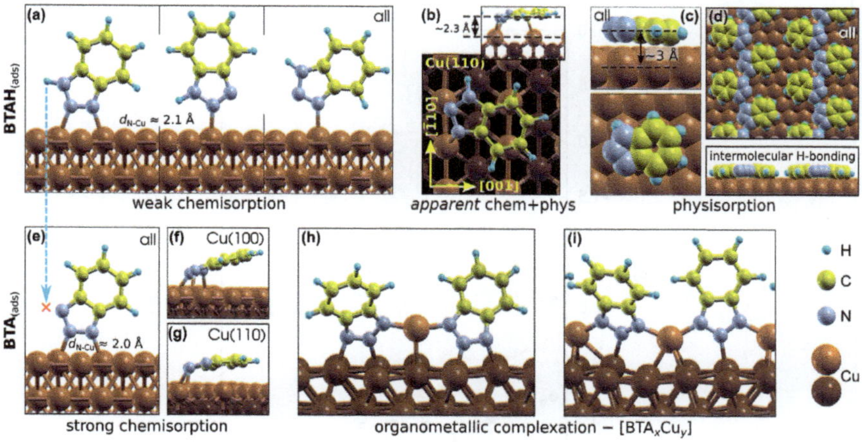

Fig. 1 Various adsorption forms of benzotriazole on copper surfaces identified by means of DFT calculations[5–8,23,25] (label "all" indicates that the form exists on all (several) Cu surfaces). Intact molecular forms ($BTAH_{(ads)}$, top row) and dissociated forms ($BTA_{(ads)}$, bottom row). (a) Weakly chemisorbed modes of BTAH bonded perpendicularly to the surface *via* triazole nitrogen atoms (d_{N-Cu} stands for N–Cu bond length).[5,6] (b) More strongly bonded parallel adsorption mode of BTAH on Cu(110) that is roughly oriented along the [001] direction; this mode was ascribed as apparent chemisorption + physisorption mode (labeled as "apparent chem + phys").[6] (c) Physisorption mode with BTAH molecular plane nearly parallel to the surface[6,7] and (d) intermolecular H-bonded networks of physisorbed BTAH;[7] this lateral H-bonding stabilizes the physisorbed BTAH by about 0.5 eV. (e) Strongly chemisorbed $BTA_{(ads)}$ bonded perpendicularly to the surface;[7,8] on Cu(100) and Cu(110) also highly tilted (f) and almost parallel (g) variants were identified.[8] (h and i) Benzotriazole–copper organometallic complexes: (h) [BTA–Cu–BTA] dimer[25] and (i) $[BTA–Cu]_n$ polymer.[7]

(Fig. 1c) with the molecule–surface height ranging from about 2.6 Å on Cu(110) to 3 Å on Cu(111).[6] While the adsorption bonding of $BTAH_{\perp}$ becomes stronger as passing from densely packed Cu(111) to more open surfaces and low coordinated defects, the physisorption energy is similar on all three low Miller index surfaces (Fig. 2);[6] a notable exception for the latter is the so-called *apparent* chemisorption + physisorption mode on Cu(110) (Fig. 1b), which adsorbs significantly stronger (Fig. 2).[6]

3.1.2 Adsorption forms of $BTA_{(ads)}$. In contrast to intact $BTAH_{(ads)}$, dissociated $BTA_{(ads)}$ strongly bonds to copper surfaces and the magnitude of chemisorption energy increases as the coordination number of surface Cu atoms involved in the adsorption site decreases (Fig. 2).[8] Dissociated benzotriazole can also form organometallic complexes on surfaces of copper (Fig. 1h and i); these complexes are specifically addressed in Section 3.4.

3.2 Adsorption characteristics

Adsorption forms of benzotriazole can be classified into four types: (i) weakly chemisorbed $BTAH_{(ads)}$, (ii) physisorbed $BTAH_{(ads)}$ (including the H-bonding networks of Fig. 1d), (iii) stand-alone strongly chemisorbed $BTA_{(ads)}$, and (iv) BTA–Cu adcomplexes. These four types display rather different adsorption features and

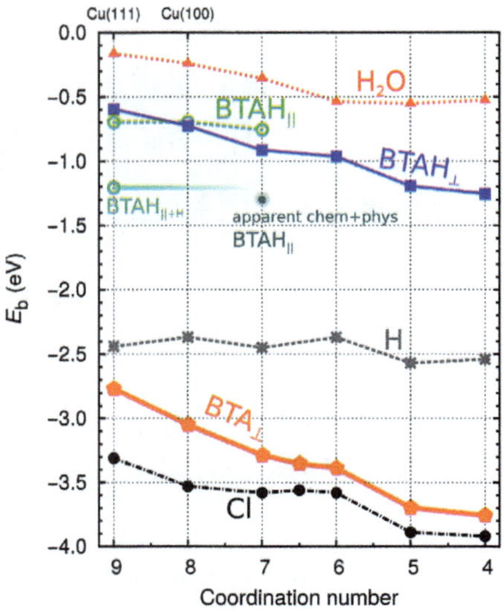

Fig. 2 Dependence of low coverage adsorption (binding) energies of BTAH,[6] BTA, Cl, H,[8] and H_2O-molecule on coordination number of surface Cu atoms at adsorption site (parallel BTAH modes are calculated with the PBE-D′ functional, while for all the others the PBE is used). Only the stablest identified adsorption energies at a given coordination number are reported [for BTAH, BTA, Cl, and H see ref. 6 and 8, whereas for H_2O the points correspond to adsorption on (111), (100), (110), and adtetramer, adtrimer, and addimer defects on Cu(111)].

for the first three types several adsorption related properties are listed in Scheme 2 (the BTA–Cu adcomplexes are treated separately later).

3.2.1 Molecule–surface bond strength and role of surface geometry. Although the intact $BTAH_{(ads)}$ bonds weakly and dissociated $BTA_{(ads)}$ bonds strongly to copper surfaces, the $BTAH_{(ads)} \rightarrow BTA_{(ads)} + H_{(ads)}$ dissociation is

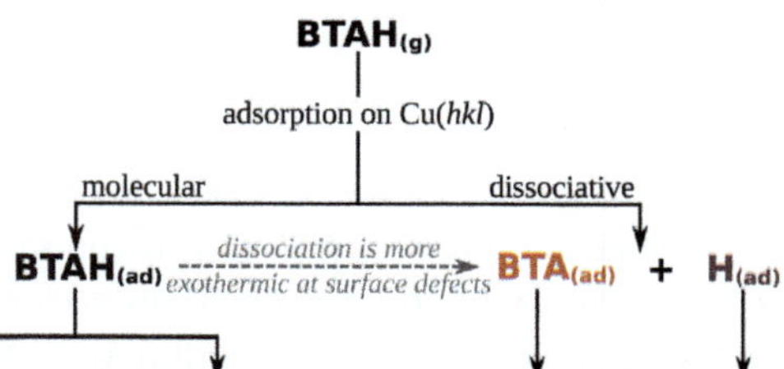

	physisorption	weak chemisorption	strong chemisorption	
molecule-surface bond strength	weak	weak	strong	strong
sensitiveness to surface geometry	weak	strong	strong	weak
sensitiveness to surface coverage	weak	strong	weak	weak
intermolecular dipole-dipole interactions	weakly attractive	strongly repulsive	weakly repulsive	weak
adsorption induced work function change	weak reduction	strong reduction	weak reduction	insignificant

Scheme 2 Characteristics of molecular and dissociative adsorption of benzotriazole on copper surfaces.

slightly endothermic on Cu(111) in the limit of zero coverage. The reason is the cost for the breakage of the N1–H bond; namely, the dissociation energy can be written as:

$$\Delta E_{\text{diss}} = E_{\text{b}}^{\text{BTA}^{\ominus}} + E_{\text{b}}^{\text{H}} - E_{\text{b}}^{\text{BTAH}} + D_{\text{N1-H}}, \tag{2}$$

where $E_{\text{b}}^{\text{BTA}^{\ominus}}$, E_{b}^{H}, and $E_{\text{b}}^{\text{BTAH}}$ are binding energies of respective species to copper surface; $D_{\text{N1-H}}$ is the N1–H bond strength (4.7 eV). For Cu(111) the respective numbers in the limit of zero-coverage are: $\Delta E_{\text{diss}} = -2.8 + (-2.4) - (-0.6) + 4.7$ eV $= 0.1$ eV.

In order for dissociation to be exothermic, the following criterion has to be met (*cf.* eqn (2)):

$$E_{\text{b}}^{\text{BTAH}} - E_{\text{b}}^{\text{BTA}^{\ominus}} > E_{\text{b}}^{\text{H}} + D_{\text{N1-H}}. \tag{3}$$

The bonding of H is rather insensitive to surface details and on Cu(*hkl*) the $E_{\text{b}}^{\text{H}} \in [-2.4, -2.5]$ eV (see Fig. 2). Hence the $BTA_{(ads)}$ has to bond by about 2.3 eV stronger to the surface than $BTAH_{(ads)}$ for dissociation to be exothermic. This criterion is met on open surfaces and low coordinated defects, because the enhancement of the molecule–surface bond strength is larger for $BTA_{(ads)}$ than for $BTAH_{(ads)}$ as the coordination number of surface Cu atoms involved in the adsorption site decreases (see Fig. 2); consequently dissociation energy becomes more exothermic in the same direction (Fig. 3).

In addition to $BTAH_{(ads)}$, $BTA_{(ads)}$, and $H_{(ads)}$, Fig. 2 also plots the dependence of adsorption (binding) energy of $Cl_{(ads)}$ and $H_2O_{(ads)}$ on the coordination number of surface Cu atoms involved in the adsorption site; the former may be seen as a prototype corrosive species and the latter is relevant due to the importance of water/metal interface for corrosion (though there is a large difference between a film of liquid water and a single water molecule).

3.2.2 Role of molecular dipole on lateral intermolecular interactions. Among the species considered in Fig. 2, the neutral BTAH displays by far the strongest

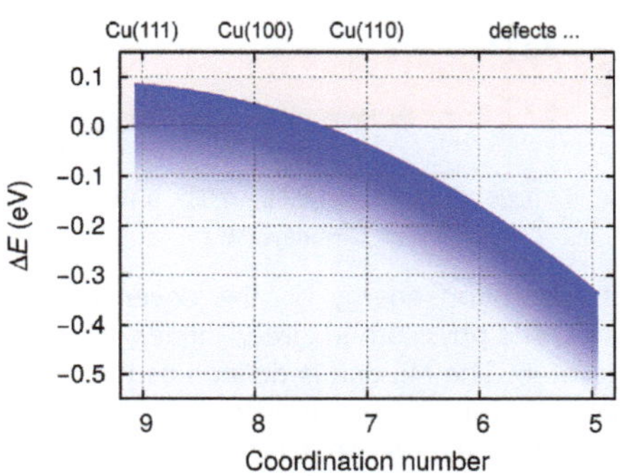

Fig. 3 Schematically shown dependence of benzotriazole dissociation energies for stand-alone $BTAH_{\perp}$ ($BTAH_{(ads)} \rightarrow BTA_{(ads)} + H_{ads}$, cleavage of N1–H bond) on coordination number of surface Cu atoms involved in the adsorption site; based on data of ref. 27. Dissociation energy (ΔE) is very sensitive to surface coverage due to a pronounced intermolecular long range repulsion between adsorbed $BTAH_{\perp}$ (*cf.* Fig. 4). The dependence of ΔE on coverage is roughly indicated by the thickness of the blue band; the larger the coverage the more exothermic the ΔE.

dependence of adsorption energy on surface coverage. This dependence is due to a large permanent dipole moment of BTAH (4.1 D), which results in repulsive and long range dipole–dipole interactions between perpendicularly adsorbed BTAH molecules, while for parallel physisorbed BTAH the lateral dipole–dipole interactions are far less important and slightly attractive at higher coverages (see Fig. 4).[6,32] On the other hand, lateral dipole–dipole interactions are not that important for $BTA_{(ads)}$ due to considerable adsorption induced charge transfer, which greatly reduces the dipole.[7]

Due to the strong lateral intermolecular repulsion between weakly chemisorbed $BTAH_{(ads)}$ and much weaker lateral interactions for $BTA_{(ads)}$ the dissociation of benzotriazole becomes more favorable as the surface coverage increases (this effect is indicated schematically by the width of the blue band in Fig. 3).

3.2.3 Role of molecular dipole on adsorption induced work function change. In addition to long-range lateral intermolecular interactions, a large permanent molecular dipole can also lead to a significant adsorption induced dipole moment (μ), which alters the metal work function by $\Delta\Phi = -4\pi\theta\mu$ (in atomic units), where θ is the surface coverage (in molecule/bohr^2 unit); a positive value of μ stands for an outward-pointing dipole with the negative end at the surface and the positive end pointing outward (*i.e.*, $\mu = \boldsymbol{\mu}\cdot\boldsymbol{n}$, where $\boldsymbol{n}$ is the surface normal and $\boldsymbol{\mu}$ is the dipole of adsorbed molecule). The adsorption induced dipole moment due to a weakly chemisorbed $BTAH_{\perp}$ is remarkably large (Table 1). Fig. 5 illustrates the dependence of $\Delta\Phi$ of Cu(111) on the molecular coverage for various forms of adsorbed benzotriazole. Even at rather small molecular coverage the reduction in the work function is appreciable for $BTAH_{\perp}$. Also the $[BTA\text{–}Cu]_n$ and [BTA–Cu–BTA] organometallic complexes significantly affect the work function, whereas stand-

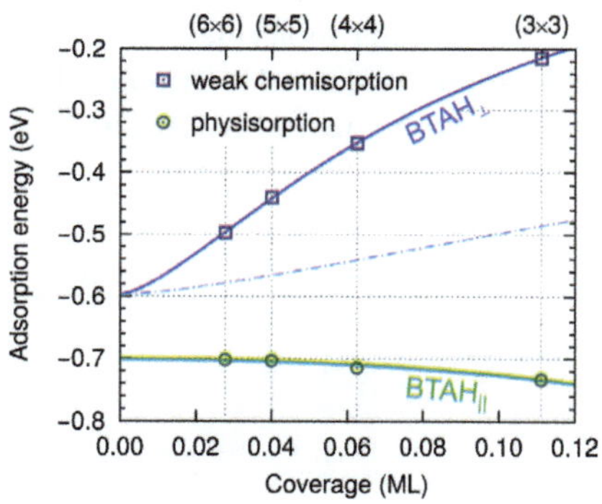

Fig. 4 Dependence of adsorption energy on the coverage for stand-alone weakly chemisorbed (blue squares) and physisorbed (green circles) BTAH on Cu(111); adsorption energies are taken from ref. 6. The ML unit is defined as the number of molecules per surface Cu atom. Curves and zero-coverage adsorption energies are obtained with a polarizable point-dipole model extrapolation of ref. 32. The dependence is plotted up to a coverage of 0.12 ML, where the physisorbed $BTAH_{\parallel}$ reaches optimum;[23] at larger coverage the molecules start to feel intermolecular Pauli repulsion. At the water/solid interface the long range lateral intermolecular interactions would be screened by the solvent molecules; the effect is roughly estimated by a thin dash-dotted blue line, which approximates the lateral dependence for the permittivity of $\varepsilon = 6$, which is a typical value for water in the double-layer[33] (note that continuum dielectric approximation is not expected to be valid at short intermolecular distances and also the magnitude of adsorption energy would be significantly altered by solvent effects).

Table 1 Adsorption induced dipole moments (μ) and work function changes ($\Delta\Phi$) on Cu(111) due to adsorption of benzotriazole. For comparison, values for $Cl_{(ads)}$, $H_{(ads)}$, and adsorbed H_2O-molecule are also listed

Species	Adsorption mode	Shown in	Coverage (ML)	μ (D)	$\Delta\Phi$ (eV)
$BTAH_{\perp}$	N2 + N3[a]	Fig. 1a	1/9	3.15	−2.25
			1/16	4.72	−1.90
$BTAH_{\parallel}$	stand-alone	Fig. 1c	1/9	0.84	−0.62
	H-network	Fig. 1d	1/10	0.96	−0.64
$BTA_{(ads)}$	N2 + N3[a]	Fig. 1e	1/16	0.51	−0.21
$[BTA–Cu]_n$	N1 + N2 + N3[a]	Fig. 1i	2/23	1.91	−1.07
			1/5	1.18	−1.52
$Cl_{(ads)}$	fcc	—	1/9	−0.34	+0.24
$H_{(ads)}$	fcc	—	1/4	−0.03	+0.04
$H_2O_{(ads)}$	top	—	1/9	+0.68	−0.49

[a] Notation indicates which N atoms bond to Cu. The N1 and N3 atoms are symmetry equivalent for BTA.

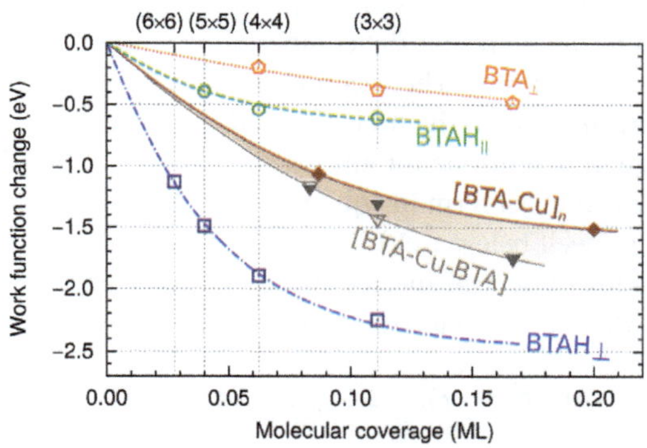

Fig. 5 Dependence of $\Delta\Phi$ of Cu(111) on the molecular coverage for various forms of adsorbed benzotriazole. The curves are drawn by fitting the data points with the polarizable point-dipole model of ref. 32. For organometallic complexes the dependence is emphasized by the brownish band.

alone-$BTA_{(ads)}$, $Cl_{(ads)}$, and $H_{(ads)}$ do not (Table 1). Altered work functions affect the adsorption of ions and this issue is discussed in Section 3.3.1.

In this respect it should be noted that a large permanent molecular dipole of the inhibitor molecule does not automatically lead to a reduction in the work function. It depends on the adsorption structure of the inhibitor and for a weakly chemisorbed BTAH on copper, the adsorption structure is such that this is true.

3.3 Adsorption at water/metal interfaces

In the earlier discussion the adsorption of benzotriazole was considered at vacuum/copper interfaces, but in the context of corrosion inhibition the water/metal interface is more relevant. The presence of aqueous solvent considerably affects the energetics of adsorption due to an interplay of several competitive effects, *i.e.*, molecule–metal, molecule–water, and metal–water interactions. In order to understand these effects, the presence of water has to be either implicitly or explicitly considered in calculations. The explicit modeling is, however, rather tedious. For this reason, the involved effects have been only roughly estimated[7,8] by following an approach that utilizes a thermodynamic cycle,[34] which is shown in Scheme 3. The adsorption energy of deprotonated BTA^- or any other anion from the solution ($A^-_{(solv)} \rightarrow A_{(ads)} + e^-_{metal}$) can be written as:‡

$$\Delta G_{ads}^{(aq)} \approx \left(E_b^{\odot} - \Delta G_{solv}^{A^-}\right) + \left(EA^{\odot} - \Phi^*\right) + \Delta\Delta G_{solv}^{A\|metal}, \tag{4}$$

where:

$$\Delta\Delta G_{solv}^{A\|metal} = \Delta G_{solv}^{A|metal} - \Delta G_{solv}^{metal}, \tag{5}$$

and Φ^* is the work function of the metal/water/vacuum system, which can be written as:

$$\Phi^* = \Phi + \Delta\Phi, \tag{6}$$

where Φ is the metal work function and $\Delta\Phi$ is its variation due to a thick layer of liquid water.[35] For the meaning of other terms, refer to Scheme 3. The terms were deliberately grouped in parentheses in eqn (4) to illustrate the competitive effects. The first grouped terms represent the competition for the molecule between the metal surface and the solvent. The second grouped terms represent the competition for the electron between the molecule and the metal surface. Each of these four terms is large in magnitude for anions such as BTA^- and Cl^- (*e.g.*, a few to several eV), whereas the last term in eqn (4) is expected to be small in magnitude for copper surfaces (well below 1 eV).[7,8,36]

For the adsorption of neutral species ($A_{(solv)} \rightarrow A_{(ads)}$) the corresponding eqn is simpler, because there are no electron terms. Hence:

$$\Delta G_{ads}^{(aq)} \approx (E_b - \Delta G_{solv}^{A}) + \Delta\Delta G_{solv}^{A\|metal} \tag{7}$$

In this case the E_b and ΔG_{solv}^{A} of BTAH are much smaller in magnitude than the corresponding terms for deprotonated BTA^- (or any other anion).

A large mutual cancellation between the competitive terms in eqn (4) and (7) results in moderate adsorption energies. As a consequence the huge difference between the gas-phase adsorption energies of $BTAH_{(ads)}$ and $BTA_{(ads)}$ is greatly reduced in aqueous-phase. Solvent effects also relatively favor the adsorption of BTA^- compared to Cl^-, because Cl^- is much smaller than BTA^- and solvates by about 0.6 eV stronger; Cl also displays by $\approx$0.15 eV larger electron affinity than $BTA^{\odot}$. While in gas-phase Cl adsorbs stronger than $BTA_{(ads)}$ (see Fig. 2), the two just mentioned effects seem to reverse the adsorption energy trend and $BTA_{(ads)}$ *wins* over $Cl_{(ads)}$ in aqueous-phase.[8]

Solvent effects also diminish the adsorption energy dependence on the coordination of surface metal atoms because the adsorbate has to displace water molecule(s) from the surface during specific adsorption and water displays a

‡ Eqn (4) is a straight blend of energy contribution ($E_b^{\odot}$) and solvation free energy terms; neither the zero-point energy correction nor the concentration (translational) and roto-vibrational entropy contributions to free energy are taken into account. The equation is therefore very approximate, but it should suffice to illustrate the discussed points.

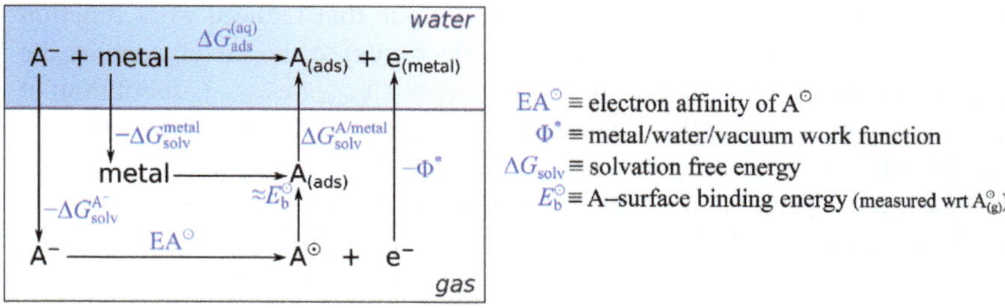

Scheme 3 Thermodynamic cycle used to obtain a rough estimate for the adsorption energy of deprotonated molecule (or anion) in the aqueous phase (mnemonic: A stands for adsorbate or anion).

similar bonding enhancement trend as benzotriazole and $Cl_{(ads)}$ (Fig. 2). This effect is captured by the $\Delta\Delta G_{solv}^{A\|metal}$ term in eqn (4) and (7). For anionic species (BTA^- and Cl^-) there is an additional effect that is even more pronounced and is due to dependence of Φ^* on the surface geometry. Namely, the reduction of Φ^* on going from (111) to (110) is so large§ that it reverses the adsorption trend, *i.e.*, estimated magnitudes of aqueous-phase adsorption energies of BTA^- and Cl^- follow the (111) $\gtrsim$ (100) > (110) sequence.[8] This finding is in agreement with the analysis of electrochemical experiments of Trasatti *et al.*,[38,39] who reported the same crystal-face specificity trend for the adsorption of halide anions on Ag.

3.3.1 Role of inhibitor induced work function change. Weakly chemisorbed BTAH considerably reduces the work function (Table 1). Such a pronounced effect on the work function should, according to eqn (4), drastically affect the

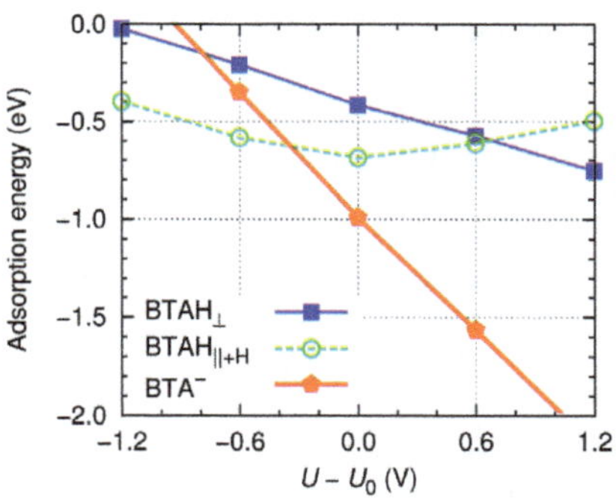

Fig. 6 Roughly estimated dependence of adsorption energy of benzotriazole on copper surfaces (*i.e.*, values are averaged as $\approx\frac{1}{3}$[Cu(111) + Cu(100) + Cu(110)]) on the electrode potential relative to *static* potential of zero charge, U_0.¶ The following species are considered: weakly chemisorbed BTAH (labeled as $BTAH_{\perp}$, left structure shown in Fig. 1a), physisorbed BTAH intermolecularly associated into a H-bonded network ($BTAH_{\|+H}$, Fig. 1d), and deprotonated BTA^- (Fig. 1e).

§ Both Φ and $\Delta\Phi$ [*cf.* eqn (6)] reduce with reducing the coordination of surface Cu atoms; the PBE calculated Φ values are 4.84, 4.60, and 4.39 eV for Cu(111), Cu(100), and Cu(110), respectively,[37] while the experimentally estimated $\Delta\Phi$ at potentials of zero charge are −0.54, −0.57, and −0.65 eV, respectively,[38] which result in Φ^* values of 4.29, 4.03, and 3.74 eV for (111), (100), and (110) faces, respectively.[8]

adsorption of anions. On this basis one can argue that reduced work function should *diminish* any reaction that donates electron(s) to the metal, such as the adsorption of corrosive chloride anions ($Cl^-_{(solv)} \rightarrow Cl_{(ads)} + e^-_{(metal)}$), dissolution of metal (*e.g.*, $Cu_{(solid)} \rightarrow Cu^{n+}_{(solv)} + ne^-_{(metal)}$), or oxidation of metal (*e.g.*, $2Cu + H_2O \rightarrow Cu_2O + 2H^+ + 2e^-$).

By comparing Fig. 4 and 5, it can be seen that the effect of the molecular dipole on the work function is larger than on the adsorption energy. The effect of the molecular dipole on the inhibitor–surface bonding has often been discussed in studies that correlate inhibitor electronic parameters with their inhibition efficiency, but the effect on the work function has not been appreciated.

3.3.2 Electric field effects in the double-layer. The electric field ($\boldsymbol{E}$) in the double-layer influences the adsorption due to several effects, which according to the current model can be decomposed into: (i) variation of the molecule–surface bonding, (ii) variation of contribution due to electron transfer from anion to electrode (or to cation from electrode), and (iii) variation of the $\Delta\Delta G_{\rm solv}^{\rm A\|metal}$ term. These effects were roughly estimated in ref. 7¶ and here this method is utilized to discuss the effect of electrode potential on various adsorbed forms of benzotriazole. In particular, Fig. 6 plots the estimated dependence of the aqueous-phase adsorption energy on the electrode potential for the weakly chemisorbed BTAH (labeled as $BTAH_{\perp}$), physisorbed BTAH intermolecularly associated into H-bonding network ($BTAH_{\|+H}$), and deprotonated BTA^-.

This figure should be understood only qualitatively, because the presented adsorption energies are subjected to significant uncertainty due to the nature of involved approximations. Nevertheless, the qualitative trends (*i.e.*, the slopes/shapes of the curves) should be correctly captured. There are a few things to notice from the figure: (i) $BTAH_{\|+H}$ displays the well known *parabolic* dependence of simple organic molecules, with the strongest adsorption at the potential of zero charge.[33] (ii) The adsorption energy of $BTAH_{\perp}$ becomes more exothermic as the

¶ The effects (i) and (ii) were estimated by putting the slab model (either adsorbate/copper or water-film/copper) in homogeneous electric field $\boldsymbol{E}$ and performing relaxation calculations at several different values of electric field. The relation between the electric field and electrode potential (U) was treated within the Helmholtz–Perrin parallel plate capacitor model of double layer, *i.e.*:

$$U = U_0 + d\boldsymbol{n}\cdot\boldsymbol{E}, \tag{8}$$

where U_0 is the electrode potential corresponding to $\boldsymbol{E} = 0$, which can be seen as a *static* potential of zero charge (spzc), but notice that spzc can be different from the measurable potential of zero charge.[40] The proportionality constant d is the thickness of a double layer, which was approximated by the often used value of $d = 3$ Å.[41–43] The adsorption energies were then calculated as:

$$\Delta G_{\rm ads}^{\rm (aq)}(\boldsymbol{E}) = \Delta G_{\rm ads}^{\rm (aq)}(0) + \delta E_{\rm b}^{\odot}(\boldsymbol{E}) + \delta\Delta\Delta G_{\rm solv}^{\rm A\|metal}(\boldsymbol{E}) + ze(U - U_0), \tag{9}$$

and

$$\delta E_{\rm b}^{\odot}(\boldsymbol{E}) = E_{\rm b}^{\odot}(\boldsymbol{E}) - E_{\rm b}^{\odot}(0),$$

$$\delta\Delta\Delta G_{\rm solv}^{\rm A\|metal}(\boldsymbol{E}) = \Delta\Delta G_{\rm solv}^{\rm A\|metal}(\boldsymbol{E}) - \Delta\Delta G_{\rm solv}^{\rm A\|metal}(0),$$

where e is the unit charge and z is the charge of species in solution (-1 for BTA^- and 0 for BTAH). For more details see the ESI of ref. 7.

electrode potential becomes more positive (almost linearly). This distinctive behavior can be attributed to the orientation of molecular dipole of the two forms because the variation of molecule–surface bond with electric field is roughly proportional to $-\boldsymbol{\mu}\cdot\boldsymbol{E}$; for physisorbed BTAH the dipole points largely parallel to the surface, hence $-\boldsymbol{\mu}\cdot\boldsymbol{E} \approx 0$ and the parabolic dependence stems from the enhancement of water–metal interaction with increasing $\boldsymbol{E}$. On the other hand, the dipole of $BTAH_{\perp}$ points outward and in this case $\partial E_b/\partial U \approx \mu/d = 0.3e$ (if $d = 3$ Å and μ is taken from Table 1).|| (iii) Finally, the slope of the BTA^- line is largely dominated by the $-e(U - U_0)$ term, eqn (9), and is thus close to one.

According to Fig. 6, the $BTAH_{\parallel+H}$ form is the stablest among the three forms at low values of potential ($U - U_0 \lesssim -0.3$ V), but at larger values the deprotonated form becomes the stablest. The curves in Fig. 6 correspond to the following reactions:

$$BTAH_{(solv)} \rightarrow BTAH_{(ads)} \quad \text{and} \tag{10}$$

$$BTA^-_{(solv)} \rightarrow BTA_{(ads)} + e^-_{(metal)}, \tag{11}$$

which implies that $BTAH_{(ads)}$ and $BTA_{(ads)}$ are treated as independent. This may not always be the case due to a dissociation reaction at the surface:

$$BTAH_{(ads)} \rightleftharpoons BTA_{(ads)} + H_{(ads)}. \tag{12}$$

Alternatively the H1 proton can be abstracted by a neighboring water molecule (which may then diffuse away from the surface):

$$BTAH_{(ads)} \rightleftharpoons BTA_{(ads)} + H^+ + e^-. \tag{13}$$

The argument applies to both cases. Consider, for example, the desorption of $BTA_{(ads)}$ under low pH conditions.** If the barrier for the reverse of reaction (12) or (13) is not too large, it may be more convenient for $BTA_{(ads)}$ to desorb as $BTAH_{(ads)}$; namely, the larger is the coverage of $H_{(ads)}$ (or concentration of H^+) the more to the left the reaction (12) [or (13)] is pushed. This argument, together with Fig. 6, suggests that $BTA_{(ads)}$ is more relevant at higher potentials and higher pH, while $BTAH_{(ads)}$ may dominate at low potentials and low pH (relatively speaking); this is in fact self-evident from reaction (13). With respect to corrosion inhibition, it would be desirable to make reaction (12) *irreversible* or at least to push the dissociation reaction to the right as much as possible, because $BTA_{(ads)}$ bonds much stronger to the surface than $BTAH_{(ads)}$ and strong bonds imply larger desorption barriers than weak bonds. This may be the case on low coordinated surface defects, because Fig. 3 reveals that dissociation is more exothermic there. A step further in this direction is achieved by the BTA–Cu organometallic complexation, which is an issue that we turn to now.

|| This implies that E_b changes by ≈ 0.3 eV if potential is altered by 1 V. The actual calculations reveal that the variation is even larger due to molecular polarization, but this extra effect is largely cancelled by the enhancement of the water–metal interaction with increasing $\boldsymbol{E}$ (note that water has a considerably smaller dipole moment than BTAH); hence the slope of the $BTAH_{\perp}$ line in Fig. 6 is close to $0.3e$.

** Although the argument is similar for both reactions, (12) and (13), the atomistic mechanism is different. In the reverse direction of reaction (12) $BTA_{(ads)}$ reacts with $H_{(ads)}$ (Langmuir–Hinshelwood mechanism), while in the reverse of reaction (13) the H^+ comes from solution and reacts with $BTA_{(ads)}$ directly (Eley–Rideal mechanism).

3.4 Role of BTA–Cu organometallic complexes

The ability of benzotriazole to inhibit corrosion is often attributed to formation of protective BTA–Cu surface complexes.†† The arguments presented earlier may provide a sound rationale of why this is so. According to DFT calculations,[7,8] the BTA–Cu organometallic complexes are thermodynamically more stable than stand-alone chemisorbed $BTA_{(ads)}$. For example, the two organometallic complexes—shown in Fig. 1h and i—are by about 0.1 to 0.2 eV per BTA more stable than the stand-alone form. This difference is not large, but it includes the cost for the Cu adatoms formation [0.76 eV per atom on Cu(111)] that are incorporated into the complex (for more details, see ref. 8). This implies that the gross bonding of BTA within the organometallic complex is much stronger than the BTA–surface bond of the stand-alone form. Moreover, within the complex all the BTA's N atoms are bonded to Cu atoms and are thus saturated (*cf.* Fig. 1h and i). Hence, the removal of BTA or the fusion of H into the BTA are more unlikely for the complex than for the stand-alone form. The organometallic complexes are therefore more resistant than the stand-alone $BTA_{(ads)}$ forms in the reverse of reaction (12) or (13), which eventually lead to desorption of benzotriazole. But they are still susceptible to potential and pH, according to the reaction of Youda:[47] $n\mathrm{BTAH}_{(ads)} + n\mathrm{Cu} \rightleftharpoons [\mathrm{Cu{-}BTA}]_n + n\mathrm{H}^+ + ne^-$.

According to some researchers the formation of protective BTA–Cu complex is faster on bare metallic surfaces,[16] while others emphasized the importance of surface oxides in their formation.[14,46] The latter is one amongst other reasons that the adsorption of benzotriazole on oxidized copper surfaces is considered below.

4 Adsorption of benzotriazole on oxidized copper surfaces

Up to this point the bonding of benzotriazole with reduced (oxide-free) copper surfaces was considered, which are more relevant at acidic pH, but in other conditions the copper surfaces are oxidized. It should be noted that bare metallic surfaces are structurally and electronically simpler than oxidized surfaces, which is the reason that the bonding of benzotriazole to oxidized copper surfaces has been considered by DFT calculations to a sufficiently smaller extent than to bare copper surfaces. Jiang and Adams[48] studied the bonding of BTAH on stoichiometric $Cu_2O(111)$, while Peljhan and Kokalj[49,50] extended the calculations and considered also the bonding on thermodynamically stablest so-called $Cu_2O(111)$-w/o-Cu^{CUS} and $Cu_2O(110):CuO$. Currently, the adsorption of benzotriazole is considered in more detail on a broader set of potential surface sites on oxidized copper surfaces.

Because oxidized surfaces may display a plethora of different structures and stoichiometries, some of which might be rather complicated, a reductionistic approach is followed. The surfaces of Cu_2O are well characterized in gas-phase and these are taken as a starting model of oxidized copper surface (Section 4.1).

†† Linear polymeric structure consisting of alternating Cu^+ and BTA^- ions in 1 : 1 ratio that form a -Cu–BTA–Cu–BTA- bidentate structure is usually proposed;[9,10,44–46] a corresponding model of such $[BTA{-}Cu]_n$ adstructure on bare copper surface, as obtained by DFT calculations, is shown in (Fig. 1i).[7] Recently, Chen and Häkkinen proposed, on the basis of DFT calculations, a [BTA–Cu–BTA] dimer adstructure (Fig. 1h).[25] This adcomplex was recently observed experimentally by STM (scanning tunneling microscopy) under ultra-high-vacuum conditions on Cu(111).[28,31]

Subsequently also simple models of hydroxylated Cu_2O surface are considered (Section 4.2), because several studies indicate that surface oxide layer is likely hydroxylated in aqueous solution.[51–54] Due to obvious modeling reasons, the reported results again refer to the solid/vacuum interface and all the models are based on cuprous oxide [the Cu(II) oxidation state is not considered]. In this respect it should be noted that the Cu_2O is more relevant than CuO for formation of the protective film of benzotriazole.[13]

4.1 Cu_2O as a model of oxidized copper surfaces

4.1.1 Stability of various Cu_2O surfaces. The stability of various Cu_2O surfaces in an ambient oxygen atmosphere was characterized in detail by Soon *et al.*, by means of DFT calculations.[55,56] Two surfaces display notably lower surface free energies than the others; one of them is the $Cu_2O(111)$-w/o-Cu^{CUS} (Fig. 7c). It differs from stoichiometric $Cu_2O(111)$ (Fig. 7a and b) by the absence of coordinately unsaturated (CUS) copper atoms; the label "$Cu_2O(111)$-w/o-Cu^{CUS}" thus stands for $Cu_2O(111)$ without Cu CUS atoms. The STM study of Önsten *et al.*,[57] seems to have identified the $Cu_2O(111)$-w/o-Cu^{CUS} structure (their proposed model B). According to their nomenclature this surface was labeled as

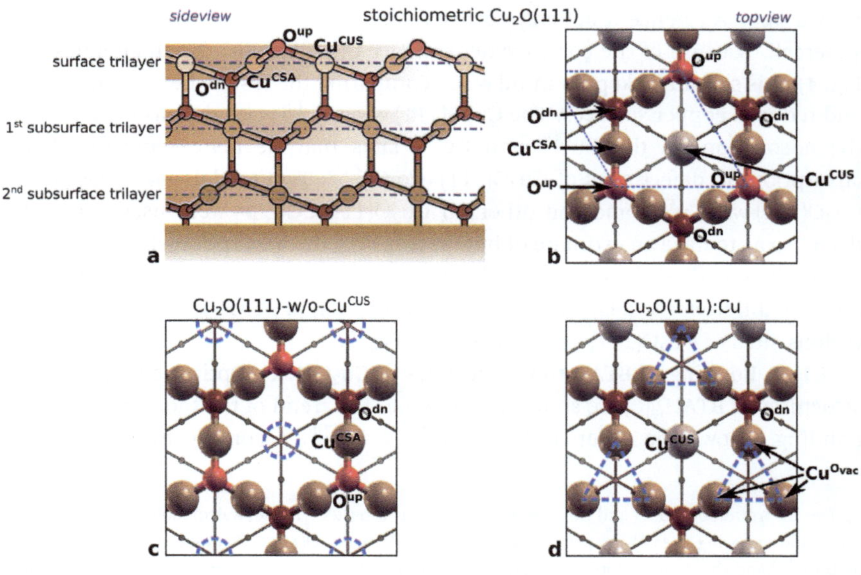

Fig. 7 Structures of various considered Cu_2O surfaces (copper and oxygen atoms are colored brownish and reddish, respectively). For clarity of presentation non-relaxed (non-reconstructed) structures are plotted and surface trilayers are emphasized in the top-view plots. (a and b) Side- and top-view of stoichiometric $Cu_2O(111)$; note the stacking of O–Cu–O trilayers shown in the side view. There are six atoms in the surface trilayer per (1 × 1)–unit cell [indicated by blue dashed parallelogram in (b)]: 3 × Cu^{CSA}, 1 × Cu^{CUS}, 1 × O^{up}, and 1 × O^{dn} (note that O^{dn} is CSA and O^{up} is CUS). (c) Non-stoichiometric $Cu_2O(111)$ without Cu^{CUS} atoms (labeled as $Cu_2O(111)$-w/o-Cu^{CUS}); Cu^{CUS} vacancies are indicated by blue dashed circles. (d) Cu terminated $Cu_2O(111)$ surface (labeled as $Cu_2O(111)$: Cu), which can be seen as stoichiometric $Cu_2O(111)$ without O^{up} atoms; O^{up} vacancies are indicated by blue dashed triangles and the surrounding Cu atoms are labeled as $Cu^{O_{vac}}$ (note that $Cu^{O_{vac}}$ is CUS).

(1 × 1). In addition, they also observed the $(\sqrt{3} \times \sqrt{3})R30°$ reconstructed surface, where one third of surface oxygens (O^{up} as defined in Fig. 7) are missing. A characteristic of the $(\sqrt{3} \times \sqrt{3})R30°$ is the presence of the $Cu^{O_{vac}}$ sites that are described below. Under electrochemical conditions the $Cu_2O(111)$ surface can be Cu terminated (labeled as $Cu_2O(111):Cu$) and stabilized by adsorbed hydroxyls.[51,52,58] Adsorption of benzotriazole on hydroxylated $Cu_2O(111)$ is considered in Section 4.2.

4.1.2 Description of considered adsorption sites. Fig. 7 shows the structure of stoichiometric $Cu_2O(111)$, $Cu_2O(111)$-w/o-Cu^{CUS}, and Cu terminated $Cu_2O(111):Cu$. It also defines the naming of various atoms (ions) in the near surface region. Some atoms (ions) in the surface region are coordinatively saturated (CSA) and others are coordinatively unsaturated (CUS); caption of Fig. 7 explicitly lists all the CSA and CUS atoms. Note that the naming of atoms does not follow the coordinative CSA/CUS description (only the surface Cu atoms of *stoichiometric* $Cu_2O(111)$ are named Cu^{CUS} and Cu^{CSA}), but a rather more structural-like description, *e.g.*, O^{up} and O^{dn} indicate that they are above and below the surface Cu layer, respectively. The Cu atoms, which surrounds the O^{up} vacancy (Fig. 7d) are CUS, but they are named $Cu^{O_{vac}}$ as to differentiate them from Cu^{CUS} of $Cu_2O(111)$.

Calculations on $Cu_2O(111)$ surfaces were performed with slabs that consist of four trilayers. A (2 × 2) supercell was used. Its size would correspond to Cu(111)–(4 × 4), but the lattice parameter of Cu_2O bulk (4.35 Å, PBE value) is larger than for Cu (3.67 Å), hence it actually corresponds to the size of Cu(111)–(4.7 × 4.7) supercell. Because of the coverage dependence of BTAH adsorption energies (*cf.* Fig. 4), this should be kept in mind when comparing the results between oxidized and reduced copper surfaces. The $Cu_2O(111)$-w/o-Cu^{CUS} was taken as a reference; the adsorption on the Cu^{CUS} and $Cu^{O_{vac}}$ sites may be hence considered as adsorption on defect sites of $Cu_2O(111)$-w/o-Cu^{CUS}. Two models—one based on $Cu_2O(111)$-w/o-Cu^{CUS} and the other on $Cu_2O(111):Cu$‡‡—were used for each defect so as to have an estimate of how susceptible the adsorption characteristics are on the specifics of the given model (the results were however not too different between different models). The results for adsorption of $BTAH_{(ads)}$ and $BTA_{(ads)}$ on various sites of $Cu_2O(111)$ are presented in Fig. 8.

4.1.3 Bonding of BTAH to various sites of $Cu_2O(111)$. Optimized adsorption structures of $BTAH_{(ads)}$ are shown in Fig. 8a–c. The trend of the molecule–surface bonding follows the order: $Cu^{CSA} < Cu^{O_{vac}} < Cu^{CUS}$. According to Fig. 8a, the

‡‡ The adsorption on Cu^{CUS} and $Cu^{O_{vac}}$ sites of $Cu_2O(111)$-w/o-Cu^{CUS} model was modeled by adding one Cu^{CUS} and removing one O^{up} ion per (2 × 2) supercell, whereas $Cu_2O(111):Cu$ model already consists of both Cu^{CUS} and $Cu^{O_{vac}}$ sites. Note, however, that $Cu_2O(111):Cu$ is a polar terminated surface of Tasker type-3,[59] hence it undergoes a reconstruction.[52] In particular, the O^{dn} ions upshift to the plane of surface Cu ions, the three $Cu^{O_{vac}}$ ions (marked by blue dashed triangles in Fig. 7d) move closer to one another and the Cu^{CUS} relaxes laterally as to bond with $Cu^{O_{vac}}$. There are multiple ways of how this can occur and different reconstruction patterns display significantly different stability. For this reason the adsorption of benzotriazole on $Cu_2O(111):Cu$ is modeled using the following procedure: (i) the structure of $Cu_2O(111):Cu$ is optimized, (ii) benzotriazole is added to the structure-(i) and relaxation is performed, and (iii) benzotriazole is removed and the surface is reoptimized. The adsorption energy is then calculated by using the total energies of (ii) and (iii). The reason for step (iii) is that step (ii) lowers the symmetry of the system (*i.e.*, the molecule perturbs the system) and the surface may find a new relaxation pattern with lower energy; step (iii) therefore assures that structures (ii) and (iii) have a compatible relaxation pattern. A more sophisticated (but computationally much more expensive) approach would be to use a simulated annealing technique as was described by Islam *et al.*[52]

bonding of BTAH to Cu^{CSA} is thus similar to that of Cu(111) (keeping in mind that compatible coverages have to be compared), whereas on coordinatively unsaturated $Cu^{O_{vac}}$ and Cu^{CUS} the bonding is considerably enhanced and it is even stronger than on low coordinated defects on bare Cu surfaces (*cf.* Fig. 2). Indeed, the bonding to Cu^{CUS} is so large that it overcompensates the thermodynamic deficiency of Cu^{CUS} and, consequently, the BTAH@Cu^{CUS} is thermodynamically more stable than BTAH@Cu^{CSA}.§§

4.1.4 Bonding of BTA to various sites of $Cu_2O(111)$. Adsorption structures of $BTA_{(ads)}$ are shown in Fig. 8d–f. It is evident that on Cu_2O surfaces the $BTA_{(ads)}$ bonds considerably stronger than $BTAH_{(ads)}$, but the difference is smaller than on bare Cu surfaces. The bonding is the weakest on Cu^{CSA} sites, −2.0 eV, while on $Cu^{O_{vac}}$ and Cu^{CUS} sites it is enhanced to −3.0 and −2.8 eV, respectively.

4.1.5 Dissociation of BTAH on various sites of Cu_2O. According to eqn (3), the $BTA_{(ads)}$ has to bond by about 2.3 eV stronger to bare Cu surfaces than $BTAH_{(ads)}$ in order for the dissociation to be exothermic. On Cu_2O surfaces this difference is smaller (1.7 eV), because H binds stronger to Cu_2O (−3.0 eV) than to Cu surfaces.¶¶ Nevertheless, the $E_b^{BTAH} - E_b^{BTA^\ominus} > 1.7$ eV condition is met only on $Cu^{O_{vac}}$ sites, which implies that dissociation of $BTA_{(ads)}$ is thermodynamically favored only thereon (the issue is confirmed by separate $BTA_{(ads)} + H_{(ads)}$ coadsorption calculations). It should be noted, however, that under non-acidic conditions, the BTA^- can also be supplied from solution; the BTAH's pK_a constant at 25 °C is 8.4 for $BTAH \rightleftharpoons BTA^- + H^+$,[60] hence at pH = 7 the $BTA^-_{(solv)} : BTAH_{(solv)}$ ratio is 4 : 96.

4.1.6 BTA–Cu organometallic complexes on Cu_2O. The rationale behind the formation of BTA–Cu organometallic complexes on bare Cu surfaces is the stronger bonding of $BTA_{(ads)}$ to low coordinated Cu sites.[8] It was shown above that benzotriazole also binds stronger to coordinatively unsaturated sites on Cu_2O surfaces. For this reason the issue of BTA–Cu complexation on oxidized copper surfaces is considered from a DFT perspective. The aim is not to identify structures that actually exist—this seems insurmountable without any experimental input—but rather to investigate whether organometallic complexes of benzotriazole are favored over stand-alone $BTA_{(ads)}$ on oxidized copper surfaces, which is affirmative according to experiments.[13,14,46,61] Several complexes were identified that are more stable than stand-alone adsorbed $BTA_{(ads)}$ forms, and Fig. 9 shows exemplar monomer, dimer, and polymer adcomplex structures. Gross binding of BTA within the monomer, dimer, and polymer complexes is calculated to be −4.3, −3.8, and −3.7 eV per BTA, respectively. If the cost for the formation of Cu^{CUS} and

§§ The BTAH bonding enhancement on passing from Cu^{CSA} to Cu^{CUS} is 1.1 eV, while the cost for the Cu^{CUS} formation is between 0.43 and 1.06 eV, taking into account the range of Cu chemical potential between that of Cu and Cu_2O bulk, *i.e.*, $\mu_{Cu}^{max} = E_{Cu}^{bulk}$ and $\mu_{Cu}^{min} = \frac{1}{2}\left[E_{Cu_2O}^{bulk} - \frac{1}{2}E_{O_2}\right]$, where E_{Cu}^{bulk}, $E_{Cu_2O}^{bulk}$ and E_{O_2} are DFT total energies of Cu atom in the Cu-bulk, formula unit of Cu_2O bulk, and oxygen molecule, respectively.

¶¶ Note that on Cu_2O surfaces, H binds stronger to O than to Cu ion, hence the dissociation reaction should be correspondingly written as:

$$BTAH_{(ads)} + O_{(Cu_2O)} \rightarrow BTA_{(ads)} + OH_{(ads)}, \tag{14}$$

where $O_{(Cu_2O)}$ is the O of the Cu_2O surface.

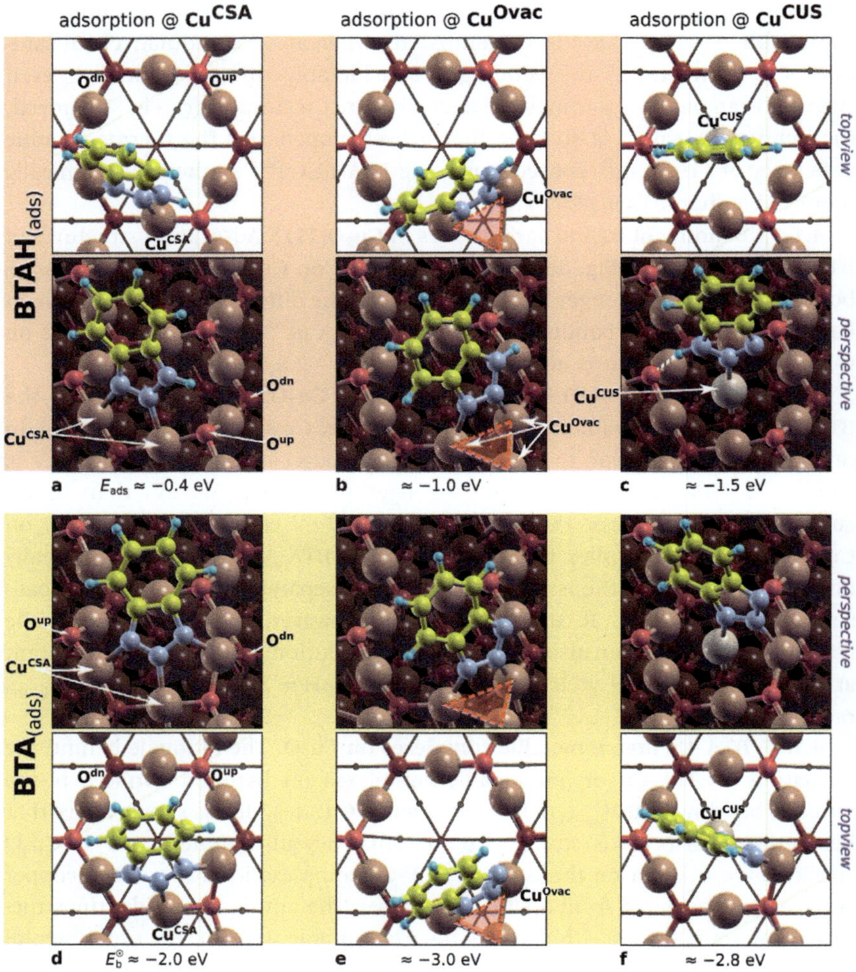

Fig. 8 Top-view and perspective plots of representative structures of $BTAH_{(ads)}$ (top) and $BTA_{(ads)}$ (bottom) on various considered sites of $Cu_2O(111)$; partially based on ref. 49 and 50. Adsorption (binding) energies are also reported. From left to right: molecular bonding to Cu^{CSA}, $Cu^{O_{vac}}$, and Cu^{CUS} sites. The O^{up} vacancy with the surrounding $Cu^{O_{vac}}$ atoms is indicated by reddish triangle. Note that in addition to direct N–Cu bonds, the BTAH adsorbed at Cu^{CUS} also forms the N1–H···O^{up} hydrogen bond.

Cu^{ad} atoms is taken into account (Cu^{ad} stands for Cu adatom), then the formation energies are in the range of [−2.7, −1.4], [−2.7, −1.8], and [−2.6, −1.7] eV per BTA for monomer, dimer, and polymer complex, respectively, where the lower number corresponds to μ_{Cu}^{max} and the upper number to μ_{Cu}^{min} (chemical potentials μ_{Cu}^{max} and μ_{Cu}^{min} are defined in the footnote of Section 4.1.3). For comparison, the formation energy of $BTA_{(ads)}$ bonded to Cu^{CUS} is in the range of [−2.4, −1.7,] eV.

4.2 Hydroxylated $Cu_2O(111)$ surfaces

According to current calculations the OH bonds by about 1.9 eV to Cu^{CSA}, 3 eV to Cu^{CUS}, and 3.6 eV to $Cu^{O_{vac}}$ (with respect to OH radical in the gas-phase). Bonding

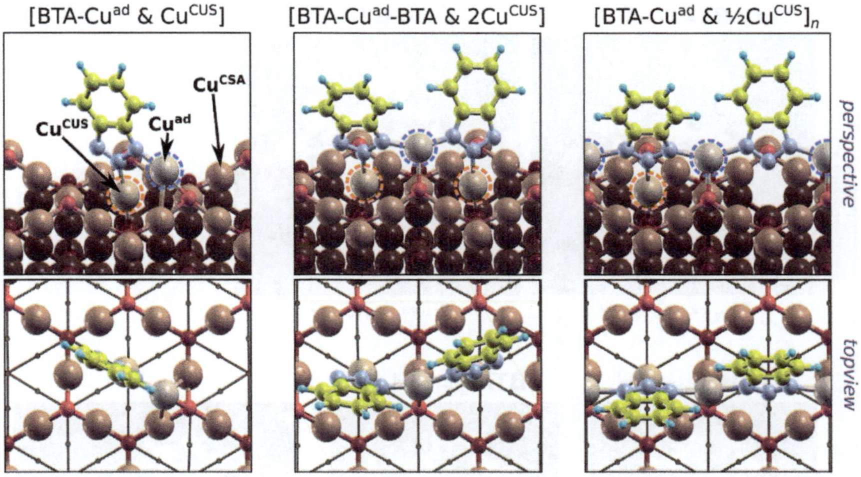

Fig. 9 Various BTA–Cu organometallic complexes on $Cu_2O(111)$-w/o-Cu^{CUS} (perspective and top-view). From left to right: [BTA–Cu^{ad} & Cu^{CUS}] monomer, [BTA–Cu^{ad}–BTA & $2Cu^{CUS}$] dimer, and $\left[\mathrm{BTA}-\mathrm{Cu}^{ad}\ \&\ \frac{1}{2}\mathrm{Cu}^{CUS}\right]_n$ polymer. Cu^{ad} stands for Cu adatom that is located above the surface and is incorporated into the complex. Note that, in addition to Cu^{ad}, complexes also involve the Cu^{CUS} atom(s). The naming of complexes reveals the stoichiometry between BTA, Cu^{ad}, and Cu^{CUS}. The Cu^{ad} and Cu^{CUS} are highlighted by dashed blue and orange circles, respectively.

enhancement is thus considerable—a trend similar to $BTA_{(ads)}$ (*cf.* Section 4.1.4). This enhancement helps us understand why, in contrast to gas-phase ambient, at water/solid interface the Cu_2O surface can contain appreciable amounts of coordinatively unsaturated Cu ions, which are stabilized by adsorbates such as hydroxyls[52] or inhibitor molecules (if present).

Two simple models of the hydroxylated surface of Cu_2O were used, one derived from $Cu_2O(111)$-w/o-Cu^{CUS} and the other from $Cu_2O(111):Cu$.|||| The first can be seen as a model of hydroxylated surface without (or an insignificant amount of) unsaturated Cu ions and the second as a model with considerable amounts of unsaturated Cu ions. Both models give rather similar benzotriazole–surface binding energies, which are about −1.0 eV for $BTAH_{(ads)}$ (a value similar to the one obtained on $Cu^{O_{vac}}$ sites of plain $Cu_2O(111)$ surface) and about −3.8 eV for $BTA_{(ads)}$ (Fig. 10a and b). $BTA_{(ads)}$ thus binds considerably stronger to the hydroxylated surface than to plain Cu_2O. Consequently, the dissociation of benzotriazole is favored on hydroxylated surfaces. Calculations further reveal that benzotriazole forms hydrogen bonds with the $OH_{(ads)}$ more efficiently than with the O^{up} ions of plain Cu_2O. The reason is that hydroxyls are much more flexible than O ions and can easily adjust so as to form hydrogen bonds with benzotriazole's N atoms (see Fig. 10). The flexibility of OH groups

|||| The $Cu_2O(111)$-w/o-Cu^{CUS} derived model was made hydroxylated by substituting all O^{up} and all O ions below the Cu^{CUS} vacancies with OH groups; the model contains 8 hydroxyls per (2 × 2) supercell. For the $Cu_2O(111):Cu$ the model as used by Asthagiri *et al.*[58] was used, where the $Cu_2O(111):Cu$-(2 × 2) supercell contains 4 hydroxyls, which bridge the Cu^{CUS} and $Cu^{O_{vac}}$ ions.

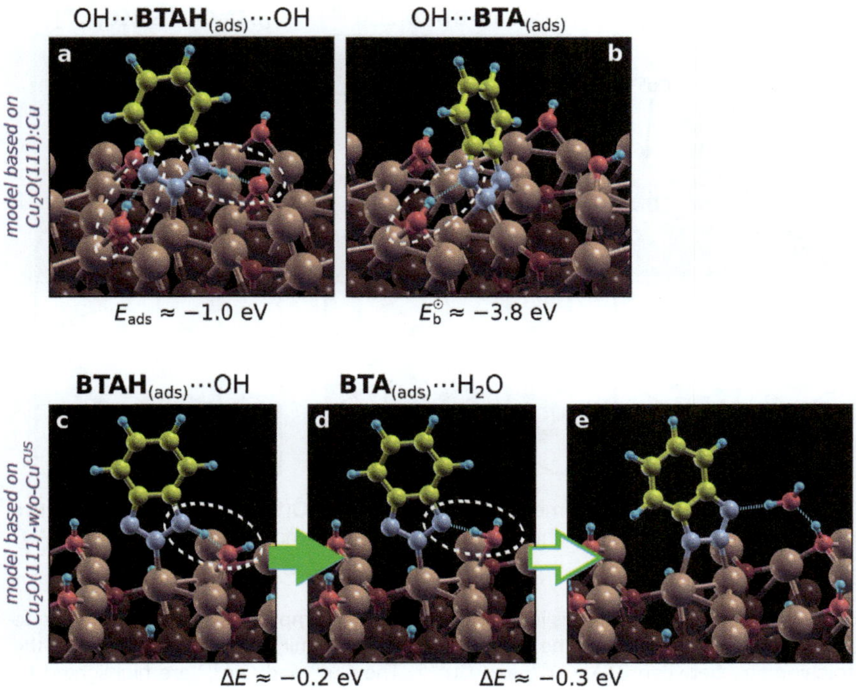

Fig. 10 Adsorption structures of $BTAH_{(ads)}$ and $BTA_{(ads)}$ on two different models of hydroxylated $Cu_2O(111)$ surface (refer to the text). The adsorption (binding) energies are also reported in (a and b). Note that in addition to direct N–Cu bonding, benzotriazole also forms hydrogen-bonds with adsorbed hydroxyls (*i.e.*, OH···N and/or NH···OH bonds), which are indicated by white dashed ellipses. (c and d) Dissociation of adsorbed BTAH by the proton transfer from its N1 atom to $OH_{(ads)}$; the corresponding process is exothermic by 0.2 eV. (e) The $BTA_{(ads)}\cdots H_2O_{(ads)}$ can further stabilize by 0.3 eV by forming another N–Cu bond and $H_2O\cdots OH$ hydrogen bond (but the $H_2O\cdots Cu$ bond is lost).

also leads to a new dissociation mechanism of benzotriazole. Namely, in addition to dissociation reaction (14), benzotriazole can also dissociate by interacting with the hydroxyl and forming a water molecule in the course of dissociation:

$$BTAH_{(ads)} + OH_{(ads)} \rightarrow BTA_{(ads)} + H_2O. \tag{15}$$

A snapshot of this reaction is shown in Fig. 10c and d. The corresponding elementary dissociative step is exothermic by 0.2 eV, but notice that the resulting $BTA_{(ads)}$ structure (Fig. 10d) is not the stablest possible; the $BTA_{(ads)}$ then further stabilizes by 0.3 eV by forming another N–Cu bond. Water formation has been proposed to form during BTAH dissociation in the literature, although due to interaction with oxygen.[16,62]

5 Conclusions

Controversial suggestions exist in the literature, concerning the role of copper oxides for the adsorption of benzotriazole.[11] To shed some new light onto this question, the adsorption behavior of benzotriazole on reduced and oxidized copper surfaces was discussed on the basis of DFT results. DFT calculations reveal that benzotriazole is able to bond with oxide-free and oxidized copper surfaces. Its bonding is considerably enhanced on passing from Cu(111) to low coordinated defects on oxide-free surfaces. The bonding of aggressive chloride anions enhances in the same direction, but to a lesser extent. On oxide surfaces benzotriazole bonds considerably stronger to coordinatively unsaturated Cu sites. This suggests that benzotriazole is able to passivate the reactive under-coordinated surface sites that are plausible microscopic sites for corrosion attack.

The adsorption behavior of benzotriazole is non-trivial, because it can adsorb (at least in principle) in a variety of different forms. Perhaps this is one of the strengths of benzotriazole; depending on different conditions it adopts one of several possible forms and thus sustains various situations. However, benzotriazole chemisorbs strongly only in the deprotonated form. Gross bonding is even stronger when BTA is incorporated into BTA–Cu organometallic adcomplexes. DFT results are therefore consistent with experimental evidence that the ability of benzotriazole to inhibit corrosion is due to the formation of organometallic complexes. This readily explains why benzotriazole is more effective at higher pH values. Namely, at low pH (and low potentials) the BTA transforms to $BTAH_{(ads)}$,[47] which is bonded considerably weaker to the surface and may eventually desorb; according to the calculations the most strongly bonded adsorption form under such conditions is the BTAH physisorbed parallel to the surface and intermolecularly connected into a H-bonding network with other BTAH molecules.***

Strong inhibitor–surface bonding, as important as it may be, is by no means the only relevant factor for corrosion inhibition. $BTAH_{\perp}$, [BTA–Cu–BTA] and $[BTA–Cu]_n$ considerably reduce the metal work function, which is a consequence of a large permanent molecular dipole and a properly oriented adsorption structure. It has been argued that such a pronounced effect on the work function should make any reaction that donates electron(s) to the metal less favorable (*e.g.*, $Cl^-_{(solv)} \rightarrow Cl_{(ads)} + e^-_{(metal)}$ or $Cu_{(solid)} \rightarrow Cu^{n+}_{(solv)} + ne^-_{(metal)}$). This reasoning is consistent with the fact that benzotriazole predominantly affects the anodic reaction, although it is a mixed type inhibitor.[11,63] A large permanent molecular dipole is a general characteristic of azoles [*e.g.*, imidazole (3.8 D), triazole (4.4 D), tetrazole (5.4 D)].[4] A mechanism based on the reduction of work function may be therefore operative also for other azole inhibitors. There are, however, significant differences between imidazole and triazole/tetrazole based inhibitors. Our recent DFT calculations suggest that imidazoles are active against corrosion in neutral molecular form, while tetrazoles—similar to triazoles—inhibit the corrosion in their deprotonated form.[36] The reason for the different behaviour of imidazoles is two fold; the first is due to their more basic nature and the second is related to

*** Only deprotonated BTA^- and neutral BTAH were considered in this paper, because protonated $BTAH_2^+$ seems not to be competitive. According to DFT calculations, $BTAH_2^+$ bonds only marginally stronger than the neutral BTAH to Cu surfaces,[26] but it solvates considerably stronger (for further DFT-based analysis of the adsorption of deprotonated, neutral, and protonated azoles, see our recent study[36]). This implies that $BTAH_2^+$ prefers being solvated rather than adsorbed. It was reported that it is the dominant adsorbed form only at pH $\lesssim$ 0.[62]

their molecular spatial structure, because of which they can form only one strong N–Cu bond, while triazoles and tetrazoles can form at least two.

An ability to form soluble complexes with dissolved metal ions as well as the solubility of stand-alone inhibitor are also relevant in the context of corrosion. A formation of soluble complexes between inhibitors and metal ions would help increase the dissolution of metal, thus having the effect of promoting the corrosion. This effect has been characterized for several imidazole based inhibitors in our parallel study.[64] As for the solubility, it is known that the more soluble the molecule is, the smaller is its (relative) affinity for adsorption.[3] Ideas of a similar kind were very recently used by Mondal and Taylor,[65] who described a clever approach towards a rational design of corrosion inhibitors.

Last but not least, the ability of a molecule to inhibit corrosion of a metal is also given by the chemical nature of the metal itself, because molecules bond differently to various metals. For example, while Cu and Al are not reactive enough to interact with the azoles' π system,††† Fe can readily do so.[26] This can have a pronounced effect on the structure of the adsorbed molecule.

Acknowledgements

This work has been supported by the Slovenian Research Agency (Grant no. P2-0148). The author is indebted to several researchers, first and foremost to Prof. Ingrid Milošev for bringing the subject of corrosion inhibitors to his attention. Sebastijan Peljhan and Nataša Kovačević are acknowledged for their many calculations and useful discussions on the subject azoles on metal surfaces. Prof. Phillipe Marcus is acknowledged for enlightening discussion about the growth of oxides on metal surfaces and surface hydroxylation. Dunja Peca and Matic Poberžnik are acknowledged for their careful reading of the manuscript.

References

1 M. M. Antonijević and M. B. Petrović, *Int. J. Electrochem. Sci.*, 2008, **3**, 1–28.
2 G. Schmitt, *Br. Corros. J.*, 1984, **19**, 165–176.
3 J. O. Bockris and A. K. N. Reddy, *Modern Electrochemistry*, Kluwer Academic/Plenum Publishers, New York, Boston, Dordrecht, London, Moscow, 2nd edn, 2000, vol. 2B.
4 N. Kovačević and A. Kokalj, *J. Phys. Chem. C*, 2011, **115**, 24189–24197.
5 Y. Jiang and J. B. Adams, *Surf. Sci.*, 2003, **529**, 428–442.
6 S. Peljhan and A. Kokalj, *Phys. Chem. Chem. Phys.*, 2011, **13**, 20408–20417.
7 A. Kokalj, S. Peljhan, M. Finšgar and I. Milošev, *J. Am. Chem. Soc.*, 2010, **132**, 16657–16668.
8 S. Peljhan, J. Koller and A. Kokalj, *J. Phys. Chem. C*, 2014, **118**, 933–943.
9 J. B. Cotton and I. R. Scholes, *Br. Corros. J.*, 1967, **2**, 1–5.
10 G. W. Poling, *Corros. Sci.*, 1970, **10**, 359–437.
11 M. Finšgar and I. Milošev, *Corros. Sci.*, 2010, **52**, 2737–2749.
12 D. Chadwick and T. Hashemi, *Corros. Sci.*, 1978, **18**, 39–51.
13 R. F. Roberts, *J. Electron Spectrosc. Relat. Phenom.*, 1974, **4**, 273–291.

††† Cu can do it only exceptionally: an example is the apparent chem + phys form of Fig. 1b and even then the interaction is driven by the dispersion van der Waals forces.[6]

14 V. Brusic, M. A. Frisch, B. N. Eldridge, F. P. Novak, F. B. Kaufman, B. M. Rush and G. S. Frankel, *J. Electrochem. Soc.*, 1991, **138**, 2253–2259.
15 B.-S. Fang, C. G. Olson and D. W. Lynch, *Surf. Sci.*, 1986, **176**, 476–490.
16 G. Xue, J. Ding, P. Lu and J. Dong, *J. Phys. Chem.*, 1991, **95**, 7380–7384.
17 P. Giannozzi, S. Baroni, N. Bonini, M. Calandra, R. Car, C. Cavazzoni, D. Ceresoli, G. L. Chiarotti, M. Cococcioni, I. Dabo, A. D. Corso, S. de Gironcoli, S. Fabris, G. Fratesi, R. Gebauer, U. Gerstmann, C. Gougoussis, A. Kokalj, M. Lazzeri, L. Martin-Samos, N. Marzari, F. Mauri, R. Mazzarello, S. Paolini, A. Pasquarello, L. Paulatto, C. Sbraccia, S. Scandolo, G. Sclauzero, A. P. Seitsonen, A. Smogunov, P. Umari and R. M. Wentzcovitch, *J. Phys.: Condens. Matter*, 2009, **21**, 395502.
18 D. Vanderbilt, *Phys. Rev. B: Condens. Matter Mater. Phys.*, 1990, **41**, 7892–7895.
19 A. Kokalj, *J. Mol. Graphics Modell.*, 1999, **17**, 176–179.
20 J. P. Perdew, K. Burke and M. Ernzerhof, *Phys. Rev. Lett.*, 1996, **77**, 3865–3868.
21 S. Grimme, *J. Comput. Chem.*, 2006, **27**, 1787–1799.
22 V. Barone, M. Casarin, D. Forrer, M. Pavone, M. Sambi and A. Vittadini, *J. Comput. Chem.*, 2009, **30**, 934–939.
23 A. Kokalj and S. Peljhan, *Langmuir*, 2010, **26**, 14582–14593.
24 S. Peljhan and A. Kokalj, *J. Phys. Chem. C*, 2009, **113**, 14363–14376.
25 X. Chen and H. Häkkinen, *J. Phys. Chem. C*, 2012, **116**, 22346–22349.
26 N. Kovačević and A. Kokalj, *Mater. Chem. Phys.*, 2012, **137**, 331–339.
27 A. Kokalj, S. Peljhan and J. Koller, *J. Phys. Chem. C*, 2014, **118**, 944–954.
28 F. Grillo, D. W. Tee, S. M. Francis, H. A. Früchtl and N. V. Richardson, *J. Phys. Chem. C*, 2014, **118**, 8667–8675.
29 J. F. Walsh, H. S. Dhariwal, A. Gutirrez-Sosa, P. Finetti, C. A. Muryn, N. B. Brookes, R. J. Oldman and G. Thornton, *Surf. Sci.*, 1998, **415**, 423–432.
30 J. O. Nilsson, C. Tornkvist and B. Liedberg, *Appl. Surf. Sci.*, 1989, **37**, 306–326.
31 F. Grillo, D. W. Tee, S. M. Francis, H. Fruchtl and N. V. Richardson, *Nanoscale*, 2013, **5**, 5269–5273.
32 A. Kokalj, *Phys. Rev. B: Condens. Matter Mater. Phys.*, 2011, **84**, 045418.
33 J. O. Bockris, A. K. N. Reddy and M. Gamboa-Aldeco, *Modern Electrochemistry*, Kluwer Academic/Plenum Publishers, New York, Boston, Dordrecht, London, Moscow, 2nd edn, 2000, vol. 2A.
34 M. T. M. Koper and R. A. van Santen, *Surf. Sci.*, 1999, **422**, 118–131.
35 S. Trasatti, *Electrochim. Acta*, 1991, **36**, 1659–1667.
36 N. Kovačević and A. Kokalj, *Corros. Sci.*, 2013, **73**, 7–17.
37 A. Kokalj, *Chem. Phys.*, 2012, **393**, 1–12.
38 S. Trasatti and L. M. Doubova, *J. Chem. Soc., Faraday Trans.*, 1995, **91**, 3311–3325.
39 L. Doubova and S. Trasatti, *Electrochim. Acta*, 1997, **42**, 785–791.
40 C. D. Taylor, S. A. Wasileski, J.-S. Filhol and M. Neurock, *Phys. Rev. B: Condens. Matter Mater. Phys.*, 2006, **73**, 165402.
41 M. J. Weaver, *Appl. Surf. Sci.*, 1993, **67**, 147–159.
42 M. T. M. Koper, R. A. van Santen, S. A. Wasileski and M. J. Weaver, *J. Chem. Phys.*, 2000, **113**, 4392–4407.
43 S. A. Wasileski, M. T. M. Koper and M. J. Weaver, *J. Phys. Chem. B*, 2001, **105**, 3518–3530.
44 J. B. Cotton, *Proceedings of the 2nd International Congress on Metallic Corrosion*, New York, USA, 1963, pp. 590–596.

45 J. C. Rubim, I. G. R. Gutz, O. Sala and W. J. Orville-Thomas, *J. Mol. Struct.: THEOCHEM*, 1983, **100**, 571–583.
46 D. Chadwick and T. Hashemi, *J. Electron Spectrosc. Relat. Phenom.*, 1977, **10**, 79–83.
47 R. Youda, H. Nishihara and K. Aramaki, *Electrochim. Acta*, 1990, **35**, 1011–1017.
48 Y. Jiang, J. B. Adams and D. Sun, *J. Phys. Chem. B*, 2004, **108**, 12851–12857.
49 S. Peljhan, Ph.D. thesis, University of Ljubljana, 2012.
50 A. Kokalj and S. Peljhan, *J. Phys. Chem. C*, 2015, DOI: 10.1021/acs.jpcc.5b01677.
51 V. Maurice and P. Marcus, *Electrochim. Acta*, 2012, **84**, 129–138.
52 M. M. Islam, B. Diawara, V. Maurice and P. Marcus, *Surf. Sci.*, 2009, **603**, 2087–2095.
53 J. Kunze, V. Maurice, L. H. Klein, H.-H. Strehblow and P. Marcus, *J. Phys. Chem. B*, 2001, **105**, 4263–4269.
54 V. Maurice, H.-H. Strehblow and P. Marcus, *Surf. Sci.*, 2000, **458**, 185–194.
55 A. Soon, M. Todorova, B. Delley and C. Stampfl, *Phys. Rev. B: Condens. Matter Mater. Phys.*, 2007, **75**, 125420.
56 A. Soon, M. Todorova, B. Delley and C. Stampfl, *Phys. Rev. B: Condens. Matter Mater. Phys.*, 2007, **76**, 129902.
57 A. Önsten, M. Göthelid and U. O. Karlsson, *Surf. Sci.*, 2009, **603**, 257–264.
58 X. Nie, G. L. Griffin, M. J. Janik and A. Asthagiri, *Catal. Commun.*, 2014, **52**, 88–91.
59 P. W. Tasker, *J. Phys. C: Solid State Phys.*, 1979, **12**, 4977–4984.
60 L. D. Hansen, B. D. West, E. J. Baca and C. L. Blank, *J. Am. Chem. Soc.*, 1968, **90**, 6588–6592.
61 M. Finšgar, J. Kovač and I. Milošev, *J. Electrochem. Soc.*, 2010, **157**, C52–C60.
62 H. Y. H. Chan and M. J. Weaver, *Langmuir*, 1999, **15**, 3348–3355.
63 M. Finšgar, A. Lesar, A. Kokalj and I. Milošev, *Electrochim. Acta*, 2008, **53**, 8287–8297.
64 N. Kovačević, I. Milošev and A. Kokalj, *Corros. Sci.*, 2015, submitted.
65 S. K. Mondal and S. R. Taylor, *J. Electrochem. Soc.*, 2014, **161**, C476–C485.

Faraday Discussions

PAPER

Understanding corrosion inhibition with van der Waals DFT methods: the case of benzotriazole

Chiara Gattinoni and Angelos Michaelides*

Received 20th December 2014, Accepted 8th January 2015
DOI: 10.1039/c4fd00273c

The corrosion of materials is an undesirable and costly process affecting many areas of technology and everyday life. As such, considerable effort has gone into understanding and preventing it. Organic molecule based coatings can in certain circumstances act as effective corrosion inhibitors. Although they have been used to great effect for more than sixty years, how they function at the atomic-level is still a matter of debate. In this work, computer simulation approaches based on density functional theory are used to investigate benzotriazole (BTAH), one of the most widely used and studied corrosion inhibitors for copper. In particular, the structures formed by protonated and deprotonated BTAH molecules on Cu(111) have been determined and linked to their inhibiting properties. It is found that hydrogen bonding, van der Waals interactions and steric repulsions all contribute in shaping how BTAH molecules adsorb, with flat-lying structures preferred at low coverage and upright configurations preferred at high coverage. The interaction of the dehydrogenated benzotriazole molecule (BTA) with the copper surface is instead dominated by strong chemisorption *via* the azole moiety with the aid of copper adatoms. Structures of dimers or chains are found to be the most stable structures at all coverages, in good agreement with scanning tunnelling microscopy results. Benzotriazole thus shows a complex phase behaviour in which van der Waals forces play an important role and which depends on coverage and on its protonation state and all of these factors feasibly contribute to its effectiveness as a corrosion inhibitor.

1 Introduction

The oxidation and corrosion of metals affects many areas of industry, technology, and everyday life. Indeed, it plays a critical role in the failure of metal parts in engineering, electrochemical and catalytic devices, as well as of metal construction parts such as piping and roofing, to name but a few. The atomistic details of the processes of metal corrosion are still unclear, however an effective solution to

Thomas Young Centre, London Centre for Nanotechnology and Department of Chemistry, University College London, London, WC1H 0AH, UK. E-mail: angelos.michaelides@ucl.ac.uk

the problem of corrosion has been found in the use of organic molecules as corrosion inhibitors, which work by interacting with metal/oxide surfaces and forming a protective film.[1,2] Widely-used organic corrosion inhibition systems, whose effectiveness has been empirically verified, are amines or zinc dithiophosphates on steel and benzotriazole (BTAH[3]) on copper. However, the chemical details of the inhibiting action and the structure of the protective layer these molecules form against metal surfaces are still unresolved. In this work, state-of-the-art computational methods are applied to the BTAH/Cu copper system, which is the most studied both experimentally[4] and computationally.[5–10] The aim is to identify the structures BTAH forms on the copper surface and obtain insights into how they affect its function as a corrosion inhibitor.

Most experimental studies on organic corrosion inhibitors are macroscopic studies involving the immersion of a metal specimen in a solution containing corrosive agents and organic inhibitors. Information about inhibition can be gathered during sample immersion, monitoring the evolution of the material using electrochemical or spectroscopic techniques (see *e.g.* ref. 11–13) or after immersion, by measuring the mass change of the specimen or the amount of metal solvated in the solution (see *e.g.* ref. 14). These studies reveal insight into the effectiveness of different chemicals. An alternative approach is to examine well-defined copper single-crystal surfaces under ultra-high vacuum (UHV) using X-ray absorption spectroscopy (XAS) and scanning tunnelling microscopy (STM). These techniques, used mainly by the surface science community, reveal structural details of the adsorbed system. The inhibitor molecules can be deposited onto the surface either from gas phase or *via* an aqueous solution (which is subsequently evaporated). Discrepancies are seen for different experimental conditions, with zigzag structures of flatly-adsorbed molecules having been postulated[11–13] (for molecules deposited from aqueous solutions), as well as a structure of vertically-adsorbed BTAs[15] (for evaporated molecules). In this work we only compare our results with experiments performed on evaporated BTAH molecules on copper substrates in UHV, since the computational work presented here consists of molecules adsorbed on a Cu(111) surface in vacuum conditions. Previous studies have shown that BTAH deprotonates when adsorbed on copper surfaces, revealing that it can deprotonate not only by interacting with high-pH environments (its pK_a is 8.4 at 25 °C) but also through its interaction with the surface. BTA was seen to form nearly-upright organometallic surface complexes involving bonds between the azole nitrogen atoms and copper adatoms.[15–18] In particular STM studies showed the formation of strings of stacked dimers at low coverage and of more complex structures, also composed of BTA–Cu_{ad}–BTA dimers, at high coverage.

A number of computational studies, using density functional theory (DFT), have looked at the stability of BTAH and BTA structures on copper. Benzotriazole is a challenging molecule to study with DFT as it combines a strongly electronegative azole moiety, which preferably interacts with the surface through chemisorption, with a benzene-like ring which can interact with the surface *via* van der Waals (vdW) forces. Traditionally, vdW dispersion forces have been a challenge for DFT methods and are not accounted for in the most widely used exchange–correlation functionals, such as Perdew–Burke–Ernzerhof (PBE),[19] that have been used in most previous studies of BTAH. Thus the role of vdW forces in this type of system is still largely unexplored. Isolated BTAH was found, in calculations using PBE, to chemisorb weakly to a Cu(111) surface in an upright geometry, forming two N–Cu bonds *via* the triazole moiety.[5,6] Physisorption, with the molecule lying flat onto the surface, was found to yield a very small binding energy (~0.1 eV) with PBE.[6] The addition of an empirical van der Waals correction led instead to a much stronger bond (~0.72 eV).[5] At higher coverages,[9] BTAH was found to form hydrogen-bonded (HB) chains with the molecules lying parallel to the surface. The dehydrogenated BTA molecule was found to bind strongly to the surface in an upright configuration[5] when isolated, and to form Cu-bonded organometallic complexes of vertical or tilted molecules at higher coverages. However, the structure of the organometallic BTA–Cu_{ad} complexes is still debated. Chen and Hakkinen[10] found that deprotonated BTA–Cu_{ad}–BTA dimers are more stable than $[BTA–Cu_{ad}]_n$ chains thus agreeing with the experimental results of Grillo *et al.*,[18] and in disagreement with the DFT results of Kokalj *et al.*[8]

Here we report the most extensive DFT study performed to date and we explore in depth the importance of using vdW approaches in the study of adsorbate–surface interactions. Intramolecular interactions and coverage effects are also examined in detail; they are found to be particularly relevant for the systems formed by fully protonated BTAHs, with hydrogen bonding dominating in the low-coverage structures and vdW and electrostatic forces at high coverage. Moreover, the energetics of the dissociation process and of the formation of complex structures with Cu adatoms were investigated and linked to experimental conditions. The obtained structures and adsorption energy trends are comparable to the experimental results in UHV and a link with the effectiveness of the molecule as a corrosion inhibitor is discussed.

The remainder of the paper is organised as follows. The computational methodology and set-up is presented in the next section (Section 2), followed by the results for the protonated (Section 3) and deprotonated (Section 4) molecules adsorbed on Cu(111). Finally, a discussion and conclusions are presented in Section 5.

2 Methodology and computational set-up

Calculations of inhibitor molecules adsorbed on copper surfaces were performed by means of DFT using the VASP code.[20–23] In the present work most results have been obtained with the optB86b-vdW functional, a modified version of the non-local vdW density functional[24] which explicitly accounts for dispersion based on the electron density. Indeed, optB86b-vdW has been shown to perform best in comparison to experiment for several adsorption problems,[25,26] including the (relevant for this work) adsorption of benzene on Cu(111).[27] A number of other functionals were also considered for testing and comparison with previous results: PBE and a number of vdW-inclusive functionals (vdW-DF,[24] optB86b-vdW,[28] PBE-D2[29] and PBE-TS[30]). For all functionals the calculated value for the lattice constant a and of the bulk modulus B are within 10% of the experimental value (see Table 1), and in good agreement with previous theoretical results.[28]

Slabs of 3–6 layers were tested with optB86b-vdW, and the surface energy was seen to converge at a 4-layer thickness to $\gamma_{Cu(111)}^{optB86b\text{-}vdW} = 0.102$ eV Å^{-2}, in good agreement with the experimental value[31] ($\gamma_{Cu(111)}^{exp} = 0.112$ eV Å^{-2}). In adsorption calculations the atoms in the bottom layer of the slab were kept fixed at the bulk positions and a vacuum of ~20 Å in the non-periodic direction was added to

Table 1 Lattice constants and bulk moduli for all the exchange–correlation functionals used in this work. All the values are in good agreement with the experimental values corrected for the zero-point anharmonic expansion[31] and with previous computational results[28]

	a [Å]	B [GPa]
PBE	3.64	137
vdW-DF	3.71	113
optB86b-vdW	3.60	148
PBE-D2	3.57	148
PBE-TS	3.55	168
Expt.	3.49	137

prevent interactions between molecules in neighbouring cells. All calculations were performed using the PAW method[32,33] and a kinetic energy cut-off on the plane wave basis set of 400 eV. The Brillouin-zone integration was performed using Monkhorst-Pack grids of (12 × 12 × 12) k-points for the conventional unit cell of copper bulk and of (12 × 12 × 1) k-points for a 1 × 1 Cu(111) surface with the vacuum in the z direction. A dipole correction as implemented in the VASP code[34,35] was added in all cases, and calculations of the dissociated molecule (into BTA and H) in the gas-phase were spin-polarized. The inhibitor (BTAH and BTA) molecules were adsorbed onto the surface at a variety of coverages, between 1/16 to 1/4 of a monolayer (ML), where 1 ML corresponds to 1 adsorbed molecule per surface copper atom. Experimental studies looking at the sub-ML regime were performed at 1/16 ML[18] and at 1/3 ML,[16] close to the coverages studied in this work. A variety of hexagonal and orthorhombic cells were used and a large number of initial configurations have been chosen and optimised, in order to explore phase space as comprehensively as possible.

The adsorption energies, E_{ads}, were calculated as follows:

$$E_{\text{ads}} = (E_{\text{sys}} - E_{\text{slab}} - N_{\text{mol}} \times E_{\text{mol}})/N_{\text{mol}}, \quad (1)$$

where E_{sys} is the total energy of the adsorbed complex, E_{slab} is the total energy of the substrate (a Cu(111) slab with or without adatoms), E_{mol} is the total energy of the isolated molecule and N_{mol} is the number of molecules in each cell. A negative value of E_{ads} indicates a stable adsorbed structure.

The formation of complex systems, such as deprotonated BTAH on a copper surface with defects, includes endothermic (bond breaking and defect formation) and exothermic (adsorption of the molecules on the surface) processes. In order to account for all these energy contributions, an energy of formation, E_{form}, is defined here and used as an indicator of the likelihood of formation of the system. The following two stoichiometric equations are relevant to the formation of a system of adsorbed BTA molecules. In the first one, n BTAH molecules dissociate and adsorb on a Cu(111) surface where m copper adatoms Cu_{ad} are already present, to form the nBTA/Cu(111)/mCu$_{\text{ad}}$ organometallic complex:

$$n\text{BTAH} + \text{Cu(111)}/m\text{Cu}_{\text{ad}} \rightarrow n\text{BTA/Cu(111)}/m\text{Cu}_{\text{ad}} + \text{H}_{\text{sys}} \quad (2)$$

The second process is similar, however the copper surface is assumed to initially be atomically flat Cu(111) and m copper adatoms Cu_{ad} are formed from the bulk:

$$n\text{BTAH} + \text{Cu(111)} + m\text{Cu}_{\text{bulk}} \rightarrow n\text{BTA/Cu(111)}/m\text{Cu}_{\text{ad}} + \text{H}_{\text{sys}}. \quad (3)$$

In eqn (2) and (3), the term H_{sys} indicates either of two possible scenarios for the dissociated hydrogen atoms: they can either adsorb onto the copper surface or combine to form gas-phase hydrogen molecules. The formation energy, E_{form}, is defined as the difference in total energy between the final (right-hand side of eqn (2) and (3)) and the initial system (left-hand-side of eqn (2) and (3)). Note that the adsorption energy E_{ads} of BTAH on clean Cu(111) can be considered as a formation energy obtained from the stoichiometric equation:

$$n\text{BTAH} + \text{Cu(111)} \rightarrow n\text{BTAH/Cu(111)}. \quad (4)$$

For certain overlayer structures, energy decompositions have been performed in order to estimate the strength of hydrogen bonds, of non-HB intermolecular interactions and of adsorbate–substrate bonding. These are performed by taking the optimised system and performing single point calculations of the individual system components. The strength of a hydrogen bond, E_{HB}, is obtained by performing single point calculations for isolated (gas phase) HB chains or dimers and for each gas phase BTAH molecule:

$$E_{\text{HB}} = \left(E_{N\times\text{BTAH}}^{\text{gas}} - \sum_{N} E_{\text{BTAH}} \right) \Big/ N, \quad (5)$$

where $E_{N\times\text{BTAH}}^{\text{gas}}$ is the total energy in gas phase of the HB system of $N \times$ BTAH molecules, and E_{BTAH} is the total energy of each BTAH molecule in the chain. Similarly, non-HB interactions are estimated using:

$$E_{\text{non-HB}} = \left(E_{N\times\text{BTAH}}^{\text{unit}} - \sum_{N} E_{\text{BTAH}} - E_{\text{HB}} \right) \Big/ N, \quad (6)$$

where $E_{N\times\text{BTAH}}^{\text{unit}}$ is the total energy obtained by removing the copper surface from the unit cell and performing a single point calculation, E_{BTAH} is the total energy of each BTAH molecule in the system and E_{HB} is the strength of the HB interactions (if any) calculated using eqn (5). The interaction energy of the molecules with the substrate is obtained using:

$$E_{\text{mol/Cu(111)}} = (E_{\text{sys}} - E_{N\times\text{BTAH}}^{\text{unit}} - E_{\text{Cu(111)}})/N, \quad (7)$$

where E_{sys} is the total energy of the adsorbed system, $E_{N\times\text{BTAH}}^{\text{unit}}$ and $E_{\text{Cu(111)}}$ the total energies obtained by performing a single point calculation after removing, respectively, the surface and the molecules from the unit cell. Energy decomposition schemes entail a certain level of arbitrariness,[36] however they are also useful in providing a semi-quantitative understanding of the relative importance of hydrogen bonding *versus* adsorbate–substrate bonding.

The nature of the bonding between the adsorbates and substrate was investigated by looking at electron density difference plots. They were constructed by subtracting from the charge density of an optimised adsorbed system the charge density of the adsorbates and substrate. As for the energy decompositions, single point calculations are performed on the components *i.e.* they are kept to the structure they assume in the adsorption system.

3 BTAH on Cu(111)

The adsorption of benzotriazole on Cu(111) was studied first for a variety of coverages. The Cu(111) surface was chosen since it has the lowest surface energy among the copper surfaces, and therefore it would be the most prevalent surface even on a copper polycrystalline sample. Moreover, it is the most widely studied surface experimentally and computationally.

Although experiments in vacuum have shown that BTAH in contact with a copper surface deprotonates,[16–18] it is still useful to look at the adsorption of the fully protonated molecules for at least two reasons. First, the drive to deprotonation can depend on the environment and in acidic environments the molecule can be expected to be protonated. Indeed, a strong dependence on pH has been found on the effectiveness of benzotriazole as a corrosion inhibitor[37] (with worse inhibition in acidic media), suggesting that the molecular structure changes with pH and, thus, with the BTAH/BTA ratio present in the system. Second, the BTAH/Cu(111) system is well-studied and it provides the possibility of validating the results obtained here with previous work.

A large number of DFT calculations (summarized in Fig. 1–5) was performed for BTAH on Cu(111) considering molecules at different adsorption sites, different orientations and different coverages. In this section we are going to look first at the behaviour of isolated molecules (Section 3.1), and subsequently at supramolecular complexes (Section 3.2). In particular, we are going to highlight the role of hydrogen bonding, van der Waals dispersion and chemical bonding in the coverage dependence of BTAH on Cu(111).

3.1 Isolated BTAH on Cu(111)

The most stable adsorption configurations for a single BTAH on Cu(111) at low coverage (1/16 ML) were established using both PBE and the vdW functionals mentioned in Section 2. Four stable structures were obtained and they are shown in Fig. 1. Two upright configurations were found: the 'Up' structure, which binds to the surface *via* the N2 triazole atom and the 'Side' structure which binds *via* the N2 and N2 triazole atoms. A physisorbed flat configuration ('Flat' in Fig. 1) was also seen. Finally, a not-previously seen tilted configuration ('Tilted' in Fig. 1) which combines chemisorption through two N–Cu bonds (as shown in Fig. 2), with physisorption through π-bonding *via* the benzene-like ring, was identified.

The comparison of their adsorption energies obtained with different functionals highlights the importance of choosing the most suitable methodology for the system under study. Indeed, the graph in Fig. 1 shows that the relative adsorption energies of the four structures changes when they are optimised with different functionals. For example, the least stable structure with the PBE functional is the 'Flat' configuration, since PBE is unable to describe physisorption. This is however one of the most stable structures for all the other functionals. The adsorption energy of the flat structure with PBE is $E_{\text{ads}} = -0.10$ eV, consistent with ref. 38. It is also in line with PBE calculations of benzene on Cu(111)

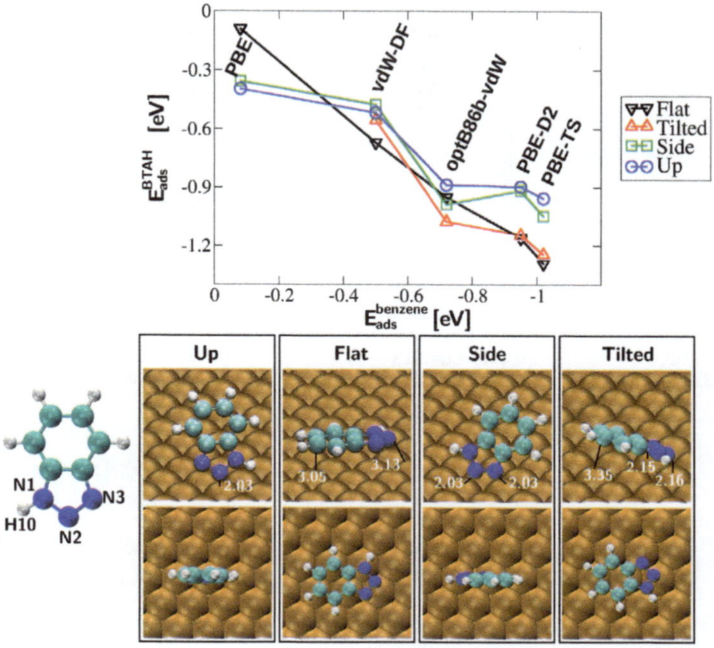

Fig. 1 Top: adsorption energies for the Flat, Tilted, Side and Up structures calculated with five different functionals. They are plotted as a function of the adsorption energy of benzene on Cu(111), calculated with the corresponding functional. The adsorption energies of benzene on Cu(111) are taken from ref. 9 and 28. Bottom: structures of the four lowest energy configurations for a single BTAH adsorbed on Cu(111). Selected bond distances are given in Å. The molecule and labelling scheme are shown on the left.

($E_{ads}^{benzene} = -0.08$ eV)[28] which also showed almost no interaction of the molecule with the surface. When dispersion interactions are accounted for, the stability of the system increases. In addition, we find that the adsorption energy of the 'Flat' structure is linearly related to that of the benzene ring (Fig. 1), and it was found to be between $E_{ads}^{vdW\text{-}DF} = -0.66$ eV to $E_{ads}^{PBE\text{-}TS} = -1.29$ eV. The nature of the interaction of the molecule with the surface, *i.e.* physisorption, was confirmed by looking at electron density difference plots, shown, for selected functionals, in Fig. 2. No interaction between the surface and the molecule is observed *via* the triazole ring, and only a small amount of charge rearrangement is observed for functionals which yield a stronger binding of benzene to Cu(111) (PBE-TS and optB86b-vdW). The molecule-surface distance, when optimizing with functionals which underbind the benzene ring (PBE and vdW-DF), is constant across the molecules (3.55–3.65 Å). When functionals which strongly bind the benzene ring are used, a small amount of tilting is observed, in good agreement with ref. 39.

The most stable configurations with PBE are the upright chemisorbed configurations ('Up' and 'Side' in Fig. 1). Their PBE adsorption energies ($E_{ads} = -0.40$ eV for 'Up' and $E_{ads} = -0.36$ eV for 'Side') were found to be in excellent agreement with ref. 9. An inversion of the energy trend between these two structures, *i.e.* $E_{ads}^{Side} < E_{ads}^{Up}$, can be seen for optB86b-vdW, PBE-D2 and PBE-TS, but not for vdW-DF, thus showing that the use of different van der Waals interactions can also affect results involving chemisorption.

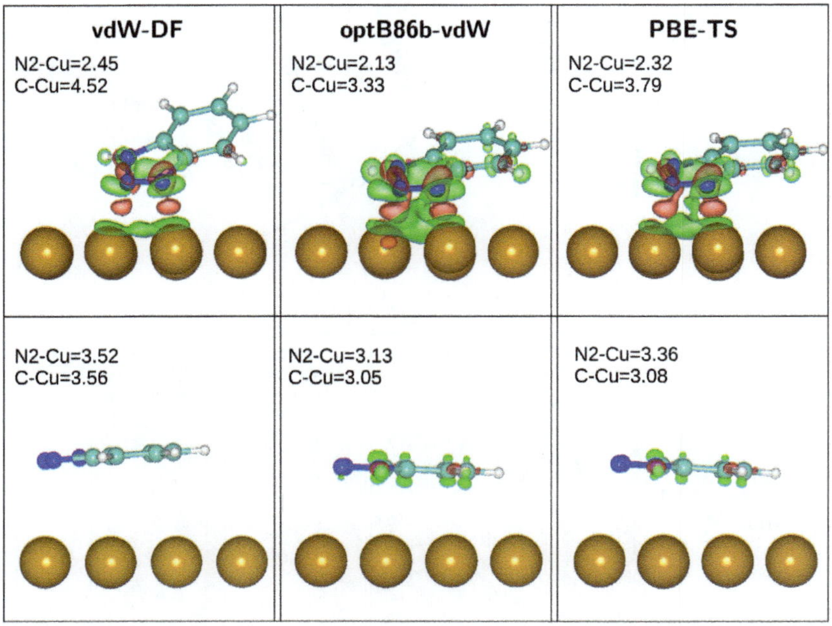

Fig. 2 Electron density difference plots for the 'Tilted' (top row) and 'Flat' (bottom row) BTAH/Cu structures using three representative vdW-inclusive functionals. Green represents the region of charge density depletion and red regions of accumulation. The isosurface level is $0.002e/a_0{}^3$ for all structures. The distance between the azole N2 and the surface and between the centre of the benzene-like ring and the surface are given in Å.

The inclusion of dispersion interactions in the optimisation of all the structures treated in this work allowed for the identification of the tilted structure ('Tilted' in Fig 1). This was found to be the most stable configuration using optB86b-vdW ($E_{\text{ads}} = -1.07$ eV) and degenerate to the 'Flat' structure with PBE-D2 ($E_{\text{ads}} = -1.15$ eV) and PBE-TS ($E_{\text{ads}} = -1.29$ eV). It is however unstable when optimised with PBE (the 'Flat' configuration is recovered) and unfavourable with vdW-DF, the adsorption energy being ~0.1 eV higher than for the 'Flat' configuration. This result is in good agreement with previous work on functionalized benzene rings with PBE and vdW-inclusive functionals: it was seen that the inclusion of dispersion interactions is the driving element behind the chemisorption of molecules which would otherwise weakly physisorb using PBE.[39,40] From Fig. 2 it can be clearly seen that there is substantial rearrangement of the charge density at the azole end of the molecule and the Cu surface atoms to which it is bonded. This is indicative of the formation of a weak chemical bond between adsorbate and substrate. Therefore, from the results obtained for the 'Flat' and 'Tilted' configurations with PBE and the vdW-inclusive functionals it is clear that dispersion stabilises these two configurations by bringing the molecule closer to the surface. This reinforces the chemical interaction of the molecule to the surface *via* the azole group, in the case of the 'Tilted' structure and *via* π bonding of the benzene ring, in the case of the 'Flat' structure.

It can be noticed from the graph in Fig. 1 that the difference between the absolute adsorption energies obtained with the functional yielding the weakest

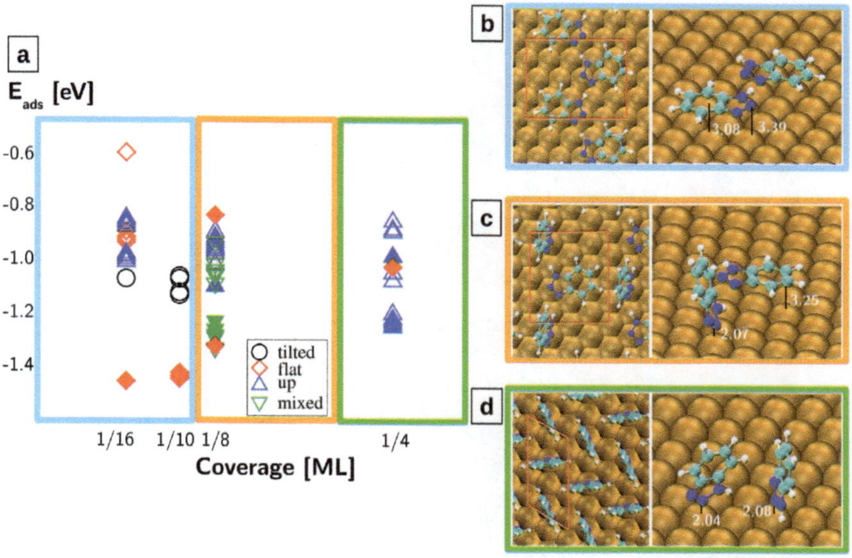

Fig. 3 (a) Adsorption energy of BTAH on Cu(111) as a function of coverage. Each point corresponds to a fully optimised adsorption structure. The filled symbols indicate a system where hydrogen bonding is present. The three coloured boxes define the three observed regimes: light blue for the physisorbed regime at low coverage, orange for the mixed adsorption regime at intermediate coverage and green for the chemisorption regime at the highest coverage. (b–d) Top and side view of the most stable BTAH/Cu(111) structure for each regime. In the top view the red box indicates the unit cell and periodic images are shown. In the side view the periodic images are not shown for clarity. Selected distances are given in Å. (b) Physisorbed regime: BTAH forms HB chains of flat molecules on Cu(111). (c) 1/8 ML coverage: mixed arrangement of HB flat and upright structures. (d) 1/4 ML coverage: representative low-energy structure with vertically adsorbed molecules at a 60° angle with respect to one another.

adsorption energies (PBE) and the strongest (PBE-TS) is ~0.9 eV. In general, absolute adsorption energies will inevitably vary between exchange–correlation functionals.[26] However, studies of benzene on Cu(111) have shown that the optB86b-vdW functional yields a binding energy in very good agreement with experiment, with the PBE-D2 and PBE-TS functionals overbinding and vdW-DF underbinding.[27] This makes optB86b-vdW the best candidate to study the relatively similar BTAH, and it has thus been chosen to obtain the results presented in the next sections.

3.2 Coverage dependence of BTAH on Cu(111)

Having established the most stable configurations for the isolated BTAH molecule on Cu(111), the stability of an extensive set of structures was evaluated with the optB86b-vdW functional as a function of coverage. The results, for 1/16 ML, 1/10 ML, 1/8 ML and 1/4 ML coverage, are shown in Fig. 3. Each point of Fig. 3a corresponds to a fully optimized adsorption structure for BTAH on Cu(111). Three regimes of adsorption can be observed: at low coverage (up to 1/10 ML), hydrogen-bonded physisorbed structures were found to be dominant, at intermediate coverage (1/8 ML) an adsorption mode consisting of a mixture of physisorption

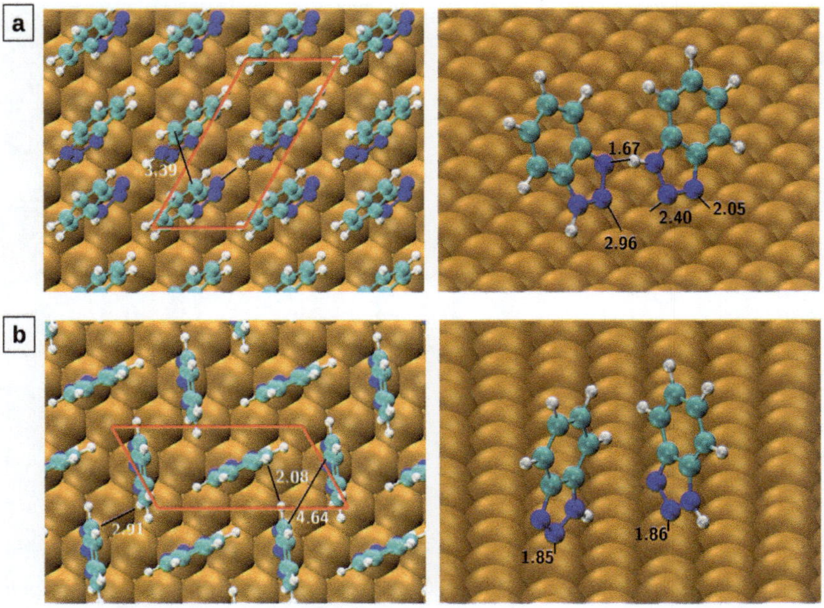

Fig. 4 Top (left-hand-side) and side (right-hand-side) views of the most stable structures for BTAH at 1/4 ML coverage (a) HB dimer structure. (b) Non-HB upright structure. The red box shows the unit cell. Selected distances are shown in Å. In the top views the unit cell and periodic replicas are shown. In the side view only the molecules in the unit cell are shown for clarity.

and chemisorption is found, and at high coverage (1/4 ML) pure chemisorption is favoured. Hydrogen bonding is present in all low-energy configurations. It is dominant at low-coverage and it competes with other forms of intermolecular interaction at high coverage.

In the range of pure physisorption, BTAH molecules form regular patterns in the form of hydrogen bonded chains with the molecules parallel to the surface with the N2 atom on the top copper site (Fig. 3b). The 'Tilted' structure in Fig. 1 has an adsorption energy ~0.40 eV higher (less stable) than this HB configuration ($E_{ads} = -1.46$ eV). This shows that the study of the interaction of isolated molecules on the surface gives only a limited description of the behaviour of the system and that the study of supramolecular aggregates is necessary to establish the full picture, even at low coverages. The adsorption energy of the HB chain at 1/16 ML can be decomposed into a contribution of ~−0.92 eV from the physisorption of BTAH onto the copper surface, and a ~−0.59 eV contribution from hydrogen bonding (Fig. 5). It is interesting to note that even with PBE, where the adsorption energy of a flat isolated molecule is very low, the flat hydrogen bonded structure is the most stable adsorption structure (with adsorption energy $E_{ads} = -0.61$ eV which is ~0.20 eV lower than for the 'Up' structure). Fig. 5 shows that this is due to the strength of the hydrogen bond ($E_{HB} \sim -0.51$ eV) which contributes to more than 90% of the adsorption energy.

Hydrogen bonding dominates adsorption also at 1/8 ML, however the closer packing of the molecules makes steric repulsive interactions between the hydrogen atoms of the benzene-like rings relevant, and forces the system in a

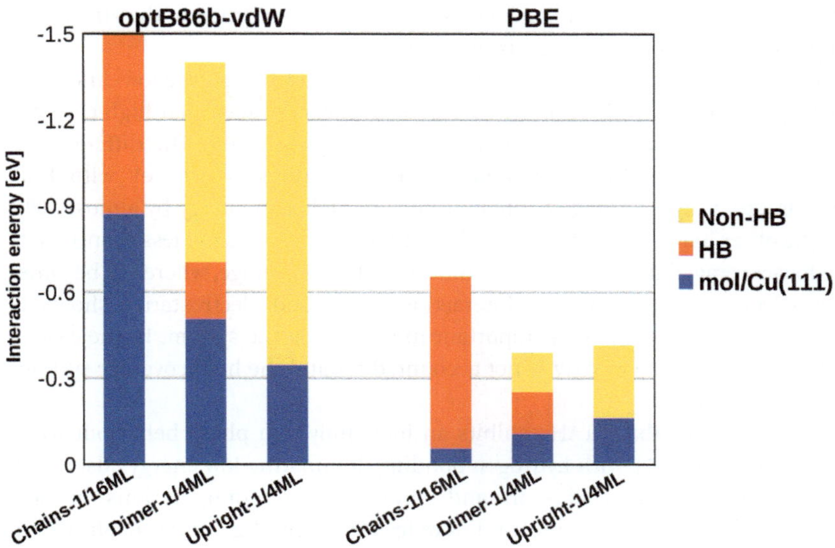

Fig. 5 Interaction energies for low coverage chains, high coverage dimers and upright molecules. The contributions to the interaction energy are represented by different colours: in blue are the contributions from the interaction of the molecules with the surface, in red interactions from hydrogen bonds and in yellow intramolecular interactions not from HB (van der Waals and electrostatic interactions). The components of the decompositions were calculated using eqn (5) and (6).

mixed flat/upright configuration (Fig. 3c). In these non-flat configurations, the highly-directional HBs become weaker ($E_{HB} \sim 0.3$–0.5 eV according to the torsion) and other forms of interaction, such as N–Cu or π–π bonding, become relevant. The flat HB structure seen for 1/16 ML and 1/10 ML, is still among the low energy structures (Fig. 3a), with the chains sliding on top of one another because of the space restrictions. The π-bonds between the vertically stacked benzene-like rings add ∼−0.10 eV to the total energy of the structure, however the interaction of the molecules with the substrate diminishes and the BTAH–Cu contribution reduces to ∼−0.67 eV on average. This results in an overall destabilisation of the HB chains, from $E_{ads} = -1.46$ eV at 1/16 ML to $E_{ads} = -1.33$ eV.

The same packing issues are relevant at 1/4 ML where the molecules adsorb in an upright configuration (a representative structure is shown in Fig. 3d). At this coverage, HB and non-HB configurations (in the form of HB dimers and upright molecules, respectively) coexist in the low-energy regime. In the HB structure (Fig. 4a), two upright BTAH molecules form a strong HB to the detriment of their interaction with the surface, with most N–Cu bonds much larger than the optimal length (∼2.00 Å). In the non-HB case (Fig. 4b), the BTAHs are preferentially packed at a 60° angle with respect to one another, in a trade-off between the most stable anchoring of the upright molecules to the surface (with N2 and H10 on a copper top site) and avoiding lateral steric interactions between the H atoms in the benzene-like ring. It can be seen in Fig. 3a that, for 1/4 ML, there are two HB structures with $E_{ads} \sim -1.00$ eV and they both correspond to HB chains (with one HB per molecule). The flat lying chain has already been discussed, while the chain of upright molecules owes its reduced stability to the distortion of the HBs whose strength for this configurations is only ~0.1 eV.

Fig. 5 highlights the differences obtained when optimising the low- and high-coverage structures with optB86b-vdW or PBE. It can be seen that high-coverage packing is less favourable with PBE than with optB86b-vdW. The difference in energy between the low- and high-coverage structures is ~0.3 eV with PBE, meaning that high-coverage configurations are higher in energy by almost 50%. On the other hand, the difference with optB86b-vdW is ~0.2 eV, less than 15% of the low-coverage interaction energy. Indeed, at high-coverage, where HB becomes less important, intramolecular interactions (vdW and electrostatic), shown in yellow in Fig. 5, become very important in stabilising the system. In the case of PBE the contribution of vdW is not accounted for and the high coverage systems are thus less stable.

Overall we see that BTAH exhibits an incredibly rich phase behaviour when adsorbed on Cu(111), with hydrogen bonding dominating low energy adsorption before packing effects kick in and non-H-bonded configurations become competitive. Moreover, significant differences between PBE and optB86b-vdW in the description of the adsorption behaviour of the molecules were seen for isolated molecules at low coverage and for high-coverage configurations. Van der Waals forces increase the interaction between the molecules and the substrate and between the molecules in the overlayer, meaning that more strongly bound and higher coverage overlayers can be obtained. This suggests that the use of vdW-inclusive functionals is necessary when studying the coverage dependence of molecular overlayers where intermolecular π bonding is possible.

4 BTA on Cu(111)

As already mentioned, the study of protonated BTAH is likely to be relevant for acidic environments, whereas experimental evidence suggests that BTAH is deprotonated in alkaline solutions and in UHV.[4,16,18] It has been suggested that BTA forms organometallic structures on copper surfaces, *i.e.* ordered networks of BTA molecules connected to one another *via* copper adatoms, *in lieu* of the lost hydrogen atom which allows the formation of hydrogen bonded networks (see *e.g.* ref. 11 and 18 and other references in ref. 4). Thus, understanding the structure of deprotonated benzotriazole on the Cu(111) surface is important from the point of view of applications (in alkaline environments) and as a direct link and validation with experimental work performed in vacuum conditions.

The most stable structures formed by BTA on Cu(111) are presented in Section 4.1 and energetic considerations regarding the formation of the system, with respect to systems of fully protonated BTAH molecules are detailed in Section 4.2. All calculations were performed with the optB86b-vdW functional.

4.1 Structures and adsorption energies

The adsorption of single deprotonated molecules on Cu(111) and the formation of BTA–Cu_{ad} organometallic complexes on the surface were considered. The most stable low-coverage (1/20–1/16 ML) structures are shown in Fig. 6, for a single BTA (panels a and b), for a BTA–Cu_{ad} complex (panel c), for a BTA–Cu_{ad}–BTA dimer (panel d) and for a [BTA–Cu_{ad}]$_n$ chain (panel e). Their adsorption energies (calculated using eqn (1) with BTA as a reference) are shown in the top part of Table 2. It can be seen that BTA interacts significantly more strongly (>2.0 eV) with the surface than BTAH. No appreciable charge transfer was seen between the H atoms and the BTA molecule when modelled in the same unit cell, suggesting the molecule undergoes a dehydrogenation process (rather than deprotonation).

For the isolated BTA, two degenerate low energy structures ($E_{ads} = -3.36$ eV) were found: a heavily tilted structure (Fig. 6a), and an upright structure (Fig. 6b). In the tilted structure the bonding to the surface is obtained *via* all three nitrogen atoms (resulting in a 0.05 Å stretch in the N2–N3 bond) and physisorption through the benzene-like ring. Only the N2–Cu bond has the optimal length of 1.96 Å; N2 and N3 form weaker bonds with a length ~2.20 Å. The upright structure is similar to the 'Side' structure for BTAH, and the interaction with the surface occurs *via* two N–Cu bonds.

In the BTA–Cu$_{ad}$ complex (Fig. 6c, $E_{ads} = -3.94$ eV) the BTA molecule is tilted ~70° with respect to the surface and it forms three N–Cu bonds, two with the surface and one with the adatom (in the bridge site). Upright BTA–Cu$_{ad}$ structures were found to have higher energy by ~0.2 eV, although they were previously shown to be the most stable structure with PBE.[8]

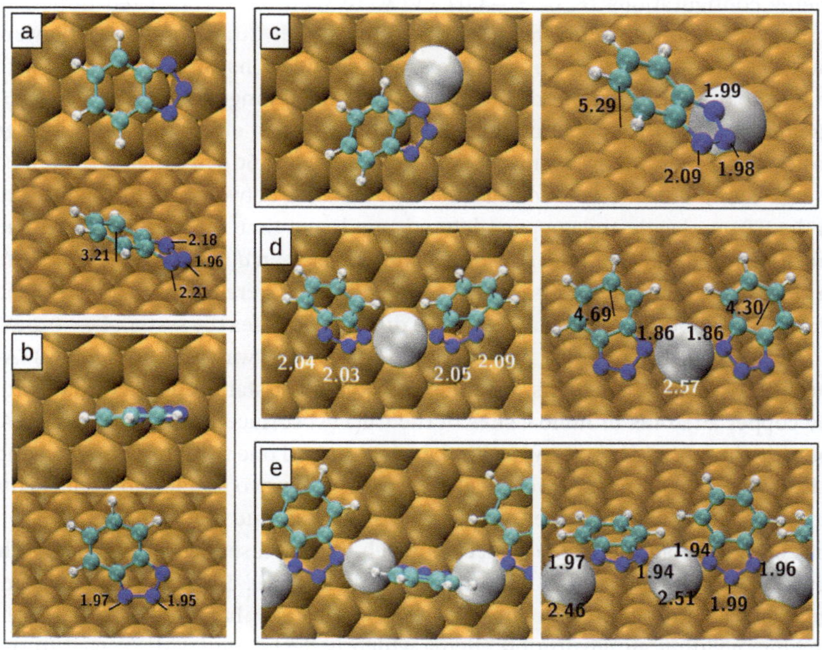

Fig. 6 Top and side view of the most stable deprotonated BTA structures on Cu(111) at low coverage. The Cu adatom is shown in grey. (a) Tilted configuration of BTA on Cu(111). (b) Upright configuration of BTA on Cu(111). (c) BTA–Cu$_{ad}$ complex. The most stable position for the copper adatom was found to be the bridge site. (d) BTA–Cu$_{ad}$–BTA dimer. This is a representative structure of a low-energy configuration dimer with the copper adatom on the copper top site and a heavily tilted BTA. Other low-energy configurations include the copper adatom on the bridge site and upright BTA molecules. (e) [BTA–Cu$_{ad}$]$_n$ chain. The copper adatoms are found on the hollow sites. Selected distances are given in Å and, for clarity, periodic replicas are not shown.

Table 2 E_{ads}^{BTA} for the BTA/Cu(111) structures shown in Fig. 6. All values were calculated using eqn (1) and are in eV. Also reported are adsorption energies of two typical corrosive atoms, Cl and S. These adsorption energies (also in eV) are with respect to gas phase Cl and S atoms

Inhibitor	E_{ads}^{BTA}
BTA (Fig. 6a and b)	−3.36
BTA–Cu_{ad} (Fig. 6c)	−3.94
BTA–Cu_{ad}–BTA (Fig. 6d)	−4.02
BTA–Cu_{ad}–BTA stacked	−4.24
$[BTA–Cu_{ad}]_n$ (Fig. 6e)	−4.55
Corrosive atoms	E_{ads}
S	−4.79
Cl	−3.54

Thus, as with BTAH, we find that vdW forces alter the adsorption structures formed. Isolated and stacked (with a distance of ~4.4 Å between successive copper adatoms) dimers were investigated. A number of almost degenerate low energy configurations (E_{ads} = −4.02 eV) were seen for the isolated dimer, in contrast with PBE results where upright dimers with Cu_{ad} at the top side were found to be the most stable structures. The inclinations of the molecule with respect to the surface in these structures was found to range from upright to ~45° inclination, and the copper adatom was found to sit in either the top or bridge position (a representative configuration is shown in Fig. 6d). In general, the lowest energy configurations were found when the distance between the nitrogen atom in the BTA molecule and copper adatom is not larger than $d(N–Cu_{ad}) \sim 1.9$ Å and the distance of Cu_{ad} with the substrate is not larger than $d(Cu_{ad}–Cu(111)) \sim 2.5$ Å. In the case of stacked dimers, upright configurations were found to be preferred and the π–π bonding between successive dimers stabilises the system by ~0.2 eV.

The strongest adsorption energy, $E_{ads}^{BTA} = -4.55$ eV, was found for the $[BTA–Cu_{ad}]_n$ organometallic chains. Fig. 6e shows the most stable configuration, with the copper adatoms in hollow sites of the Cu(111) surface and one vertical BTA, forming three N–Cu bonds, two with the adatoms and one with the substrate, and a flat lying BTA, forming two N–Cu bonds with the two copper adatoms. This structure is in good agreement with the structures found with PBE,[8,10] however, with a higher rotation of the molecule which, in our case, lies almost flat while with PBE it is tilted by ~60° with respect to the surface. Tests of stacked organometallic chains show that they do not significantly benefit from increased packing, yielding an adsorption energy which is degenerate with respect to the isolated chain. Indeed, the flat-lying molecule in Fig. 6e has to rotate at a ~45° angle with respect to the surface in order to allow for stacking, and the π bonding with the surface is replaced with π bonding with the benzene-like ring of the BTA in the parallel chains.

When the adsorption energies for the BTA systems are compared to the adsorption energy of atoms like sulfur or chlorine (bottom of Table 2), well-known active agents in the corrosion process of copper, it can be seen that their adsorption energies are very close, especially for chlorine. Although this is a fairly crude comparison (corrosive agents in a realistic system are likely to be sulfur- or chlorine-containing molecules rather than isolated atoms), these values provide a hint that competitive adsorption could be an important factor in the corrosion inhibition process of BTA on copper.

4.2 Energies of formation

A direct comparison between the stability of the systems discussed in this and in Section 3 is given by their formation energies, calculated using eqn (2) and (3) for BTA and $E_{form} = E_{ads}$ for BTAH (see Section 2). Four scenarios for the formation of the BTA systems are shown in Fig. 7 (each point corresponds to a fully optimised BTA–Cu(111) structure). In Fig. 7a formation energies according to eqn (2) are shown, *i.e.* it is assumed that the surface contains defects and that copper adatoms are already present on the surface (in the hollow site) before the adsorption of the BTAH molecules. The organometallic chains are the most stable system in this case. In Fig. 7b formation energies according to eqn (3) are shown, *i.e.* it is assumed that the surface is initially atomically flat and that the adatoms are extracted from the bulk. The number of adatoms used in eqn (3) is $m = 0$ for the BTA/Cu system, $m = 1$ for the BTA–Cu_{ad} complex and for the BTA–Cu_{ad}–BTA dimers and $m = 2$ for the chains. In this case the stacked BTA–Cu_{ad}–BTA structure have the lowest energy, ~0.2 eV more stable than the organometallic necklaces. The two y axes report energies for the two systems formed by the dissociated hydrogen atom: adsorbed on the copper surface (left-hand-side) and forming H_2 molecules (right-hand-side).

In Table 3 the formation energies for the most stable structures in all four cases are reported. These are compared to the formation energy of the most stable BTAH/Cu(111) structure, the HB chains, and the bold text indicates the deprotonated structures which are more stable than the BTAH/Cu(111) system ($E_{form} < E_{form}^{BTAH} = -1.46$ eV). When the formation energy of the copper adatom is not taken into account (top half of Table 3) most deprotonated structures, except for the isolated BTA molecule, are found to be more stable than the protonated HB chains. If the formation energy of the copper adatom is considered (bottom half of Table 3), only the organometallic chains and the stacked dimers are found to be more stable that the BTAH HB chains, and only if the hydrogen atom is assumed to adsorb on the surface. If the H atoms are assumed to associate into H_2 molecules the only competitive structure to the BTAH HB chains are the stacked dimers ($E_{form} = -1.38$ eV). This shows that benzotriazole has a higher drive to deprotonate on defective surfaces with Cu adatoms than on atomically flat surfaces.

The results obtained from these four extreme cases can be used to give an approximate description of the behaviour of a real experimental system. In the work of Grillo *et al.*[18] the copper surface is reconstructed and therefore the mobility of the surface copper atoms is low thus making this system close to the case in Fig. 7b, *i.e.* the copper adatoms need to be extracted from the bulk. In this case (Table 3) the stacked BTA–Cu_{ad}–BTA dimers, with the dissociated H atoms adsorbed on the surface, are the lowest energy configuration with $E_{form} = -1.59$. This is in excellent agreement with the STM experimental results, where stacked dimers are observed at low coverage and where the reconstruction of the Cu(111)

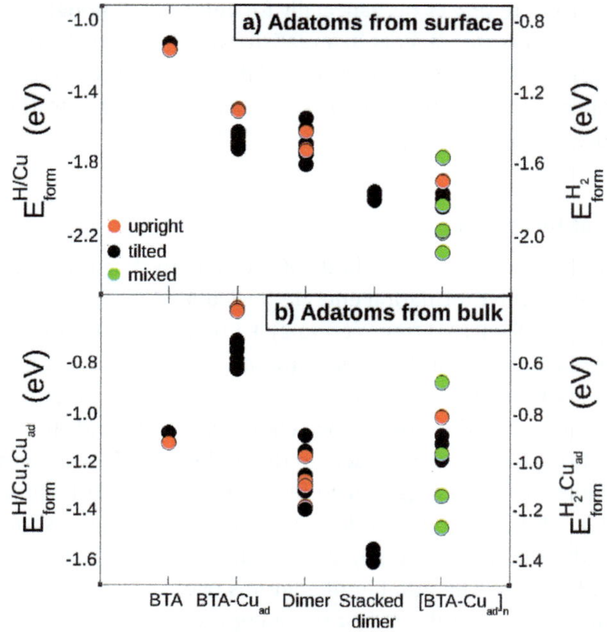

Fig. 7 Formation energies for an isolated BTA, for a BTA–Cu_{ad} complex, for an isolated BTA–Cu_{ad}–BTA dimer, for stacked dimers and for a [BTA–Cu_{ad}]$_n$ necklace. The formation energies are calculated assuming that the H atom adsorbs on the surface (left-hand-side *y*-axis) or forms a gas-phase H_2 molecule (right-hand-side *y*-axis). (a) Formation energies for a BTAH dissociating and adsorbing on a surface where adatoms are already present, following eqn (2). (b) Formation energies for a BTAH dissociating and adsorbing on a surface where adatoms are extracted from the bulk, following eqn (3) with $m = 0$ for BTA, $m = 1$ for BTA–Cu_{ad}, dimer and stacked dimer, $m = 2$ for [BTA–Cu_{ad}]$_n$.

surface, which is generally known not to reconstruct,[41] indicates the presence of impurities[42,43] (such as H atoms) on the surface.

5 Discussion and conclusion

Results for the structures and energies of BTAH and BTA on Cu(111) are presented here. The adsorption of fully protonated BTAH is relevant to applications where the environment is acidic, whereas the adsorption of BTA on copper is relevant to alkaline environments and for comparison with experiments of BTAH deposition on copper in vacuum conditions, where the molecule is found to deprotonate. Benzotriazole is a complex molecule which requires treatment with a theoretical methodology capable of simultaneously describing chemisorption and physisorption. In the present work DFT with a vdW-inclusive functional has been employed to optimise a large number of protonated and deprotonated structures on Cu(111).

We find that dispersion forces significantly alter the relative stabilities of adsorbed BTAH structures. In addition to this, in the lowest energy tilted structure an interesting interplay with chemical bonding is found wherein dispersion forces bring the molecule close to the surface thus enhancing the chemical bonding of the molecule with the surface *via* the triazole group. Whilst this is

Table 3 E_{form} for the BTA/Cu(111) structures shown in Fig. 6. The top part corresponds to the data shown in Fig. 7a. The values are calculated using eqn (2) *i.e.* the formation energy of the adatoms is not taken into account. The bottom part corresponds to the data shown in Fig. 7b *i.e.* it is assumed that the copper adatoms are extracted from the bulk [eqn (3)]. The left column shows the formation energies when assuming that the dissociated H atoms adsorb on Cu(111), the right column when assuming they form gas phase H_2 molecules. All values are in eV. The values highlighted in bold are those where $E_{form} < -1.46$ eV, the formation energy of the BTAH flat HB structure

Structure	$E_{form}^{H/Cu(111)}$	$E_{form}^{H_2}$
BTA (Fig. 6a and b)	−1.12	−0.93
BTA–Cu_{ad} (Fig. 6c)	**−1.72**	**−1.51**
BTA–Cu_{ad}–BTA (Fig. 6d)	**−1.82**	**−1.61**
BTA–Cu_{ad}–BTA stacked	**−2.02**	**−1.81**
[BTA–Cu_{ad}]$_n$ (Fig. 6e)	**−2.31**	**−2.10**
Structure	$E_{form}^{H/Cu(111),Cu_{ad}}$	$E_{form}^{H_2,Cu_{ad}}$
BTA (Fig. 6a and b)	−1.12	−0.93
BTA–Cu_{ad} (Fig. 6c)	−0.86	−0.65
BTA–Cu_{ad}–BTA (Fig. 6d)	−1.38	−1.18
BTA–Cu_{ad}–BTA stacked	**−1.59**	−1.38
[BTA–Cu_{ad}]$_n$ (Fig. 6e)	**−1.45**	−1.24

interesting and a potentially general effect, we have also shown that a single absorbed molecule provides limited insight into the behaviour of the overall system. Indeed, while isolated BTAH preferentially adsorbs on copper *via* the azole nitrogen atoms and π-bonding of the carbon ring with the surface, the lowest energy structure at low coverage are HB chains of flat-lying molecules.

Overall we have found an incredibly rich coverage-dependent phase diagram for BTAH on Cu(111). Three regimes were identified for the adsorption of BTAH on Cu(111), as a function of coverage: a low coverage hydrogen-bonded regime, where the molecules preferentially adsorb flat on the surface, an intermediate regime, where mixed flat and upright structures are observed, and a high coverage regime, where the molecules adsorb upright. Steric interaction drives the change in configuration from the flat-lying physisorbed to the upright chemisorbed configuration of the BTAH.

The lowest energy configurations seen for BTA are either stacked BTA–Cu_{ad}–BTA dimers or organometallic chains, according to whether the formation energy of the copper adatom is taken into account when calculating the stability of the complexes. Good agreement is seen with experimental results in UHV, where the molecule was found to adsorb *via* the azole moiety in a vertical or near-vertical manner.[15,16,18] In particular, the stacked dimer configuration was observed using STM to form on a reconstructed Cu(111) surface, where the mobility of the copper atoms is likely to be low and therefore copper adatoms require a large amount of energy to form. In this case, the conditions of Fig. 7b apply, and indeed the expected configuration from calculations are stacked BTA–Cu_{ad}–BTA dimers.

The use of a suitable exchange–correlation functional was found to be important for all the systems considered here. Indeed, for an isolated BTAH on

Cu(111) the 'Flat' configuration was found to be favourable only when vdW dispersion forces were accounted for, and not in the case of PBE. Moreover, the 'Tilted' low coverage structure, which is the most stable with the optB86b-vdW functional and is favourable with all vdW functionals tested, is instead unstable when optimized with PBE (the 'Flat' configuration is retrieved instead). At high coverage, the lack of any description of π-π bonding in PBE leads to a larger equilibrium distance between two BTAH molecules and thus to (comparatively) weaker adsorption energies for the high-coverage structures. In the BTA–Cu systems it has been seen that PBE favours upright adsorption, because of the lack of dispersion interactions between the benzene-like ring and the surface. When vdW interactions are accounted for a more complex behaviour is uncovered, especially for the BTA–Cu_{ad}–BTA dimers where many degenerate low-energy structures are seen.

Both BTAH and BTA offer the possibility of forming fairly close packed layers on Cu(111) with little cost to the adsorption energy, and, for the case of BTA, with a gain in energy when the dimers are stacked, thus in principle offering a physical barrier to incoming corrosive molecules or atoms. There is however a large difference (~2 eV) in the adsorption energy of the molecule with the surface between BTAH and BTA. Since benzotriazole performs better as an inhibitor in alkaline conditions, where the likelihood of deprotonation is higher, there might be a link between the strongest interaction of the molecule with the surface and inhibition. Indeed, the adsorption energy of BTA (~4.00 eV) is fairly close to the adsorption energy of *e.g.* two well known corrosive agents for copper. It is likely that the actual corrosive agents in a corrosive solution are sulfur- or chlorine-containing molecules, rather than isolated atoms. However, our calculations on chlorine and sulfur atoms give a ball-park estimate which suggests that competitive adsorption could be the key here for the success of benzotriazole as a corrosion inhibitor.

Acknowledgements

A.M.'s work is partly supported by the European Research Council under the European Unions Seventh Framework Programme (FP/2007–2013)/ERC Grant Agreement no. 616121 (HeteroIce project) and the Royal Society through a Wolfson Research Merit Award. The authors are grateful for computational resources to the London Centre for Nanotechnology and to the U.K. Car-Parrinello Consortium UKCP (EP/F036884/1), for access to HECToR. C.G. would like to thank Dr Federico Grillo for useful discussions.

References

1 J. N. Breston, *Ind. Eng. Chem.*, 1952, **44**, 1755–1761.
2 M. M. Antonijevic and M. B. Petrovic, *Int. J. Electrochem. Sci.*, 2008, **3**, 1.
3 Procter & Gamble, Ltd., British Patent 652339, 1947.
4 M. Finsgar and I. Milosev, *Corros. Sci.*, 2010, **52**, 2737.
5 A. Kokalj and S. Peljhan, *Langmuir*, 2010, **26**, 14582.
6 Y. Jiang and J. B. Adams, *Surf. Sci.*, 2003, **529**, 428.
7 S. Peljhan and A. Kokalj, *Phys. Chem. Chem. Phys.*, 2011, **13**, 20408–20417.
8 S. Peljhan, J. Koller and A. Kokalj, *J. Phys. Chem. C*, 2014, **118**, 933–943.

9 A. Kokalj, S. Peljhan, M. Finsgar and I. Milosev, *J. Am. Chem. Soc.*, 2010, **132**, 16657.

10 X. Chen and H. Hakkinen, *J. Phys. Chem. C*, 2012, **116**, 22346.

11 I. Dugdale and J. B. Cotton, *Corros. Sci.*, 1963, **3**, 69.

12 R. F. Roberts, *J. Electron Spectrosc. Relat. Phenom.*, 1974, **4**, 273.

13 J. C. Rubim, I. G. R. Gutz, O. Sala and W. J. Orvillethomas, *J. Mol. Struct.*, 1983, **100**, 571.

14 S. Rathgeber, R. Bauer, A. Otto, E. Peter and J. Wilde, *Microelectron. Reliab.*, 2012, **52**, 2452–2456.

15 B. S. Fang, C. G. Olson and D. W. Lynch, *Surf. Sci.*, 1986, **176**, 476.

16 J. F. Walsh, H. S. Dhariwal, A. Gutierrez-Sosa, P. Finetti, C. A. Muryn, N. B. Brookes, R. J. Oldman and G. Thornton, *Surf. Sci.*, 1998, **415**, 423.

17 F. Grillo, D. W. Tee, S. M. Francis, H. A. Fruechtl and N. V. Richardson, *Nanoscale*, 2013, **5**, 5269.

18 F. Grillo, D. W. Tee, S. M. Francis, H. A. Fruechtl and N. V. Richardson, *J. Phys. Chem. C*, 2014, **118**, 8667.

19 J. P. Perdew, K. Burke and M. Ernzerhof, *Phys. Rev. Lett.*, 1996, **77**, 3865.

20 G. Kresse and J. Hafner, *Phys. Rev. B: Condens. Matter*, 1993, **47**, 558.

21 G. Kresse and J. Hafner, *Phys. Rev. B: Condens. Matter*, 1994, **49**, 14251.

22 G. Kresse and J. Furthmueller, *Comput. Mater. Sci.*, 1996, **6**, 15.

23 G. Kresse and J. Furthmueller, *Phys. Rev. B: Condens. Matter*, 1996, **54**, 11169.

24 M. Dion, H. Rydberg, E. Schroder, D. C. Langreth and B. I. Lundqvist, *Phys. Rev. Lett.*, 2004, **92**, 246401.

25 J. Carrasco, B. Santra, J. Klimeš and A. Michaelides, *Phys. Rev. Lett.*, 2011, **106**, 026101.

26 J. Klimeš and A. Michaelides, *J. Chem. Phys.*, 2012, **137**, 120901.

27 J. Carrasco, W. Liu, A. Michaelides and A. Tkatchenko, *J. Chem. Phys.*, 2014, **140**, 084704.

28 J. Klimeš, D. R. Bowler and A. Michaelides, *Phys. Rev. B: Condens. Matter Mater. Phys.*, 2011, **83**, 195131.

29 S. Grimme, *J. Comput. Chem.*, 2006, **27**, 1787.

30 A. Tkatchenko and M. Scheffler, *Phys. Rev. Lett.*, 2009, **102**, 073005.

31 P. Haas, F. Tran and P. Blaha, *Phys. Rev. B: Condens. Matter Mater. Phys.*, 2009, **79**, 085104.

32 P. E. Blochl, *Phys. Rev. B: Condens. Matter*, 1994, **50**, 17953.

33 G. Kresse and D. Joubert, *Phys. Rev. B: Condens. Matter Mater. Phys.*, 1999, **59**, 1758.

34 J. Neugebauer and M. Scheffler, *Phys. Rev. B: Condens. Matter*, 1992, **46**, 16067.

35 G. Makov and M. C. Payne, *Phys. Rev. B: Condens. Matter*, 1995, **51**, 4014.

36 A. Michaelides, A. Alavi and D. A. King, *Phys. Rev. B: Condens. Matter Mater. Phys.*, 2004, **69**, 113404.

37 M. M. Musiani, G. Mengoli, M. Fleischmann and R. B. Lowry, *J. Electroanal. Chem.*, 1987, **217**, 187–202.

38 Y. Jiang and J. B. Adams, *Surf. Sci.*, 2003, **529**, 428.

39 N. Atodiresei, V. Caciuc, P. Lazic and S. Bluegel, *Phys. Rev. Lett.*, 2009, **102**, 136809.

40 W. Liu, S. N. Filimonov, J. Carrasco and A. Tkatchenko, *Nat. Commun.*, 2013, **4**, 2569.

41 Z. Crljen, P. Lazic, D. Sokcevic and R. Brako, *Phys. Rev. B: Condens. Matter*, 2003, **68**, 195411.
42 M. A. Muoz-Mrquez, G. S. Parkinson, P. D. Quinn, M. J. Gladys, R. E. Tanner, D. P. Woodruff, T. C. Q. Noakes and P. Bailey, *Surf. Sci.*, 2005, **582**, 97.
43 K. Moritani, M. Okada, Y. Teraoka, A. Yoshigoe and T. Kasai, *J. Phys. Chem. C*, 2008, **112**, 8662.

Faraday Discussions

PAPER

Multiphysics modelling, quantum chemistry and risk analysis for corrosion inhibitor design and lifetime prediction

C. D. Taylor,*[ab] A. Chandra,[a] J. Vera[a] and N. Sridhar[a]

Received 11th November 2014, Accepted 7th January 2015
DOI: 10.1039/c4fd00220b

Organic corrosion inhibitors can provide an effective means to extend the life of equipment in aggressive environments, decrease the environmental, economic, health and safety risks associated with corrosion failures and enable the use of low cost steels in place of corrosion resistant alloys. To guide the construction of advanced models for the design and optimization of the chemical composition of organic inhibitors, and to develop predictive tools for inhibitor performance as a function of alloy and environment, a multiphysics model has been constructed following Staehle's principles of "domains and microprocesses". The multiphysics framework provides a way for science-based modelling of the various phenomena that impact inhibitor efficiency, including chemical thermodynamics and speciation, oil/water partitioning, effect of the inhibitor on multiphase flow, surface adsorption and self-assembled monolayer formation, and the effect of the inhibitor on cathodic and anodic reaction pathways. The fundamental tools required to solve the resulting modelling from a first-principles perspective are also described. Quantification of uncertainty is significant to the development of lifetime prediction models, due to their application for risk management. We therefore also discuss how uncertainty analysis can be coupled with the first-principles approach laid out in this paper.

Materials sustainability through the molecular control of corrosion

Corrosion control plays a significant role in materials sustainability as it provides one means for the optimization of the stability and longevity of structural materials exposed to either the common elements of nature or the more extreme chemical and/or high-pressure/high-temperature environments encountered in industrial chemical, metallurgical and energy production.[1] Effective corrosion management, including the appropriate selection and application of organic corrosion inhibitors, allows the use of abundant and inexpensive mild steel for the majority of our energy infrastructure (*i.e.* pipelines, refineries), reduces the

[a]*DNV GL, Dublin, OH 43017, USA. E-mail: Christopher.taylor@dnvgl.com*
[b]*Fontana Corrosion Center, Materials Science & Engineering, Ohio State University, Columbus, OH 43210, USA*

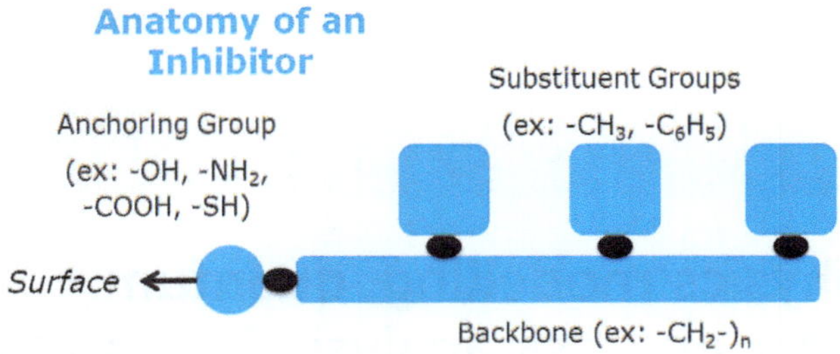

Fig. 1 Inhibitor molecule structure provided along with some example chemical fragments for each component.

incidence of costly environmental and safety impacts that would result from corrosion failures, and prevents the need for sourcing more costly and scarce metals that would be required to produce intrinsically corrosion resistant alloys.[2]

Molecular systems deployed for corrosion inhibition are usually selected through intensive experimental testing and then combined with chemical surfactants and biocides to produce a commercial package that is then injected into the pipeline.[2] Successfully designed inhibitor molecules interfere with mechanisms of corrosion in such a way as to reduce the rate of corrosion by at least two orders of magnitude. The inhibitor efficiency (I.E.) quantifies the extent to which corrosion rate without inhibition (C_r^0) is reduced to the corrosion rate with inhibition (C_r^i), expressed as the percentage decrease:

$$\text{I.E.} = 100\% \times (C_r^0 - C_r^i)/C_r^0, \tag{1}$$

Over time the efficiency of the inhibitor is reduced due to equilibrium desorption, mechanical shearing or chemical reactions. Inhibitors are therefore injected either in continuous or batch mode to maintain a protective environment. Inhibitors can also be applied using slow-release agents – polymers that release low-levels of the inhibitor molecules over time – or embedded into coatings and primers.[3,4]

An inhibitor molecule is typically constituted of three parts: a functional group anchor, a hydrocarbon backbone, and functionalized substituents (Fig. 1).[2] From the perspective of a chemist, the design and selection of effective inhibitors should be viewed as the activity of constructing the optimal (anchor, backbone, and substituent) triplet for a given material/environment combination.[5] Whereas some effort has been devoted to this problem, primarily using the quantitative-structure–activity-relations approach (QSAR),[6,7] our intent here is to present an alternative framework based on combining the principles of corrosion science and engineering, quantum chemistry, computational materials science, as well as electrode kinetics, statistical mechanics and thermodynamics. First, however, we provide an overview of the current state-of-the-art in corrosion inhibitor theory and modeling at the molecular scale.

Theory and modeling of corrosion inhibitors

The molecular structure of an inhibitor will influence its mechanism of action, suggesting that there may exist an optimal molecular design for a particular application. Characterization by atomic force microscopy, quartz crystal microbalance and electrochemical impedance measurements has shown that corrosion inhibitor molecules adsorb onto metal surfaces, and it is generally believed that the formation of this adsorption layer provides the corrosion protection.[8–11] The extent to which the metal surface is covered by the inhibitor is called the surface coverage, θ, which usually ranges from 0 to 1, although it is entirely possible for multilayers to form too. θ is related to the inhibitor concentration in solution and the Gibbs free energy of adsorption, ΔG_{ads}, by the appropriate adsorption isotherm (*i.e.* constant temperature relation). The most common isotherm for describing adsorption is the Langmuir isotherm, but alternatives exist, such as the Temkin isotherm and the Freundlich isotherm.[12] The distinction between the various models is related to the nature of the interactions between the adsorbing molecules. Assuming non-interaction, the Langmuir isotherm is frequently applied to fit observed inhibitor behavior,[13] with general form:

$$K = \frac{\theta}{(1-\theta)C_{\text{A}}} = \exp[-\Delta G_{\text{ads}}/RT], \tag{2}$$

This isotherm model relates the coverage, θ, to the concentration of inhibitor molecule A, C_{A}, and the adsorption equilibrium constant, K. RT is the product of the ideal gas constant, R, with temperature, T.

Given a particular material/environment combination, the inhibitor efficiency can be considered an intrinsic property of the inhibitor molecule itself (for a given concentration) and so it should be derivable from quantum theory. According to quantum mechanics, all the information one needs to know about a system can be obtained from the wavefunction, which, in turn, is obtained as the solution to the Schrodinger equation: $\mathscr{H}\Psi = E\Psi$, in which the quantum mechanical operator, $\mathscr{H}$, contains the physical terms that contribute to the system energy and E is an eigenvalue.[14] Ψ is the quantum mechanical wavefunction that is related to the probability density of the electrons in the system (this density, ρ, is given by $|\Psi|^2$). Any observable quantity, $\langle A\rangle$, can be determined by taking the expectation value of the appropriate quantum mechanical operator, $\mathscr{A}$, applied to the wavefunction *via* the equation:

$$\langle A\rangle = \int_{-\infty}^{+\infty} \Psi^{*}\mathscr{A}\Psi \mathrm{d}x = \langle\Psi|\mathscr{A}|\Psi\rangle. \tag{3}$$

The angular brackets on the right introduce Dirac notation and the * indicates the complex conjugate.[14]

Over the latter half of the 20th century, powerful techniques were derived alongside the development of microcomputers that allowed the wavefunctions of molecular systems to be obtained with great accuracy.[15–18] Unfortunately, deducing inhibitor efficiency directly from the wavefunction is non-trivial: the operator representing the inhibitor efficiency must somehow capture all the relevant information regarding the target material and environment! Instead, Vosta and Eliasek (and then, later, many others)[6,19–24] introduced the approach utilized in drug design of attempting to correlate readily computable molecular descriptors (such as the dipole moment or energies of the frontier molecular orbitals) with the inhibitor efficiencies obtained from an experiment using the method of linear (and, later, non-linear) regression (Fig. 2). Following this approach inevitably leads to a "black-box" approach. The systems of equations derived that relate inhibitor efficiency to molecular properties are generally opaque to the underlying physics involved in the inhibition. Consequently, they are of little utility for inhibitor prediction, have very poor transferability between different classes of inhibitors, cannot be applied to different material/environment combinations, and the uncertainties are not quantified.

Around the year 2000, the quantum mechanical approaches that were so successful in the molecular realm were becoming capable of simulating surfaces with significant complexity, including metal surfaces with adsorbed molecular species.[26–30] The standard technique for solving the quantum mechanical equations for solid-state systems is known as density functional theory (DFT),[31,32] and requires the electron density of a material (or material/molecular system), ρ, to be self-consistently solved according to the Hohenberg–Kohn theorem, which computes the energy E from ρ:[33–35]

$$E = \int \rho(r)v(r)\mathrm{d}r + T[\rho] + \frac{1}{2}\iint \frac{\rho(r_1)\rho(r_2)}{r_{12}}\mathrm{d}r_1\mathrm{d}r_2 + E_{\mathrm{xc}}[\rho] \tag{4}$$

$v(r)$ is the electrostatic field of atomic nuclei, $T[\rho]$ is the kinetic energy functional, the double-integral provides the electron–electron interaction (although a self-interaction term is incurred when calculated this way), and $E_{\mathrm{xc}}[\rho]$ is a correction for the errors in the kinetic energy functional and the electron–electron double-integral.

The approaches for quantum chemistry (*i.e.* molecules) and DFT (*i.e.* solids and surfaces) have reached a high degree of convergence, such that it is common to find molecular systems treated with DFT and solids treated with quantum chemistry (such as the linear combinations of atomic orbitals).[36] Yet, DFT has some important shortcomings that cannot be systematically improved using our current understanding of many-body physics. These shortcomings follow from uncertainty regarding the correct form for the corrections captured by the term

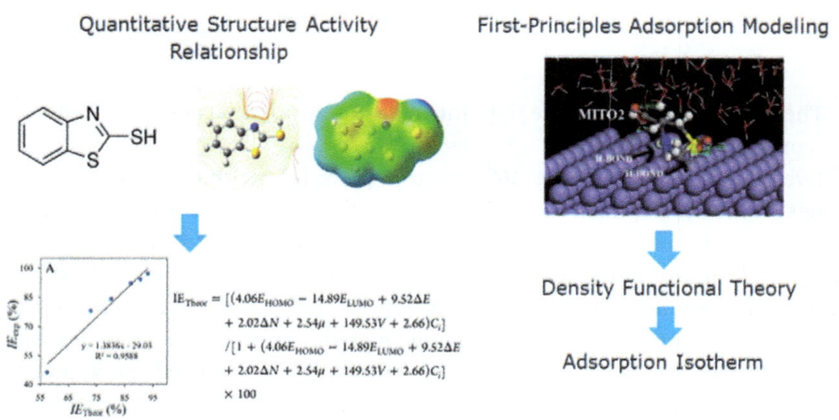

Fig. 2 The QSAR approach compared to the first-principles adsorption modeling approach *via* DFT/atomistic simulation and the construction of an isotherm.[19,25]

$E_{xc}[\rho]$.[37] Because of this, all DFT predictions contain an as yet unquantifiable error, and, although much effort has been expended to identify approximate electron correlation functionals that perform well under certain sets of conditions,[28,38] there is no standard means for estimating the accuracy of a given quantity computed from the self-consistent densities obtained *via* DFT, even when experimental data are used for "calibration".[39]

Despite this significant issue, which is an active area of research, DFT is commonly used and considered to have close to "chemical accuracy". With regards to the particular context of corrosion and interfacial science, density functional theory can be used for predicting the adsorption behavior of materials/molecular systems.[26] For example, DFT has also begun to be applied to explore the mechanisms of inhibition: Kokalj *et al.* considered the adsorption of a series of molecules on the surface of copper and computed the binding energies directly from DFT, using them to estimate an adsorption isotherm for the inhibitor and making parallels to the mechanism of inhibition (Fig. 2).[40]

One mechanistic simplification that has been applied to relate both experimental and theoretical quantities to the inhibitor efficiency is to assume that the inhibitor efficiency is equivalent to the surface coverage, θ. This identity implies that a surface completely covered with an inhibitor (*i.e.* one monolayer, in adsorption parlance) must have a 100% inhibitor efficiency, that is, a corrosion rate of zero:[13]

$$\text{I.E.} = 100\% \times \theta \tag{5}$$

As observed in Fig. 3, a metal surface that is entirely covered with an inhibitor may not have an identically zero corrosion rate.[7,8,41,42] The extent to which corrosion is inhibited will depend on the efficiency with which the surface is packed: molecules having a bulky topology may leave "gaps" through which metal atoms can dissolve.[7] This "packing factor" will also be a function of the microstructure of the exposed material surface. Furthermore, the adsorbed inhibitor molecules are in *dynamic* equilibrium with the solution, leaving transient opportunities for the underlying metal atoms to dissolve, possibly even to dissolve

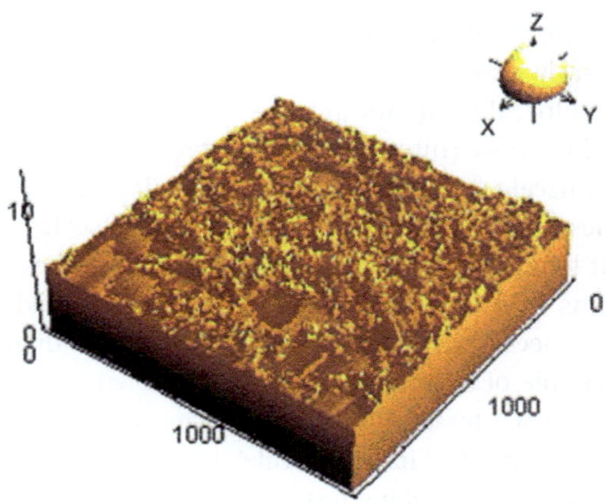

Fig. 3 3D Atomic Force Microscope (AFM) image of mild steel with an inhibitor, showing loose packing of inhibitor molecules.[10] The scale is in nm.

while tethered to an anchoring group. It has also been proposed that some inhibitors act not by blocking the surface dissolution reaction, but instead by blocking the corresponding cathodic reduction reaction, which supplies the counterbalancing electronic and ionic currents that allow dissolution.[2] In the case of pipeline corrosion control, inhibitors have been shown to modify multiphase flow, decreasing the tendency for water to contact the surface.[43] Consequently, there remain many chemical and physical factors that need to be included in a predictive, mechanistic model for corrosion inhibition. In this paper, we argue that such a predictive model can be achieved through an integrated multiphysics modeling approach that includes quantum chemical evaluation of fundamental molecular properties, but also goes beyond that to include aspects of surface adsorption, material microstructures, and environmental factors such as pH, co-dissolved species and the global properties of the system, such as % water cut and flow rate. Such an approach will be inherently multiphysics in nature, and so we now proceed to briefly introduce the advent of multiphysics modeling and simulation in the materials and chemical sciences.

Trends in multiphysics modeling of corrosion

Material failures, including corrosion, frequently occur due to the compounding effects of multiple phenomena rather than from one single physical or chemical process acting alone. For this reason, models based on an entirely reductionist philosophy will always be insufficient for the purpose of modelling risk, and, in this case, predicting the extent to which risk may be reduced by the application of corrosion inhibitors. Examination of the recent modelling literature in corrosion and related failure modes (such as stress corrosion cracking) reveals the following main modelling approaches in use today:

- Thermodynamic models.
- Fluid dynamics.
- Electrochemistry and mass transport.
- Atomistic and quantum chemistry.
- Microstructural evolution.
- Damage mechanics.
- Empirical modelling.

While these methods are most frequently used in isolation (since scientists tend to have narrowly defined fields of expertise), it is widely recognized that interdisciplinary multiscale and multiphysics approaches are required to provide effective engineering tools (Integrated Computational Materials Engineering, ICME).[44,45] The multiscale modelling paradigm is built upon the recognition that the types of physics occurring at small time and length scales are distinct from those occurring at longer time and length scales. Multiphysics models combine phenomena such as heat and mass transport, or electrode kinetics with fluid mechanics (*i.e.* not necessarily at different scales and so distinct from multi-scaling). As an example of a multiphysics problem-solving approach, molecular dynamics simulations of interactions between the molecules that constitute an oil reservoir were used to predict its megascale thermodynamic properties.[46] The problem of stress-corrosion cracking has seen a lot of attention using multiscale methods that connect long-range stress fields treated *via* finite element continuum mechanics with atomistic simulations.[47]

Multiphysics and empirical approaches form the subspace of the much larger set of integration tasks that are required to build effective models for materials performance.[48,49] The circos diagram in Fig. 4 embeds these technical scientific and empirical approaches into the larger space of modelling tasks that includes uncertainty quantification and informatics, and draws some connections between them to demonstrate possible integration pathways. Uncertainty quantification includes activities such as the application of Bayesian inference relations,[50] experimental testing and validation, field data collection, damage accumulation models,[51] Monte Carlo simulation,[52] first-order and second-order reliability methods (FORM/SORM),[53] and the identification of knowledge gaps.[54] The informatics tasks include multiphysics simulation,[55] open source code sharing and distribution, machine learning,[52] expert team construction, and data-handling protocols.[56]

Given this trend towards integrated computational materials engineering, how does one go about constructing a comprehensive integrated model? Due to the interdependence between chemical environments, flowing oil–water mixtures, material microstructures, mechanical effects and stochastic behavior, the prediction of the materials/environment response is a non-facile problem that requires a multiphysics approach. Furthermore, the environmental and microstructural conditions in the field will not directly coincide with those in the laboratory, and hence modeling is required to extrapolate laboratory knowledge into applications for the field with the provision of appropriate bounds on the uncertainty. One approach for beginning to deal with multiphysics phenomena was proposed by Staehle in the microprocess-centric approach, which divides the

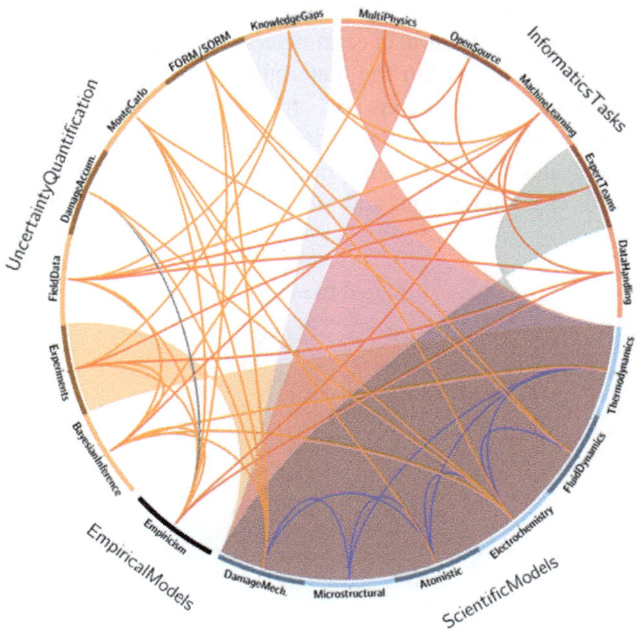

Fig. 4 The construction of useful models for materials performance, including models for the design of effective corrosion inhibitor systems, requires the integration of scientific, empirical, uncertainty quantification and informatics tasks.

physical system into a set of *domains* and *microprocesses*.[57] By categorizing the phenomena that lead to the accumulation of corrosive behavior and changes in the material integrity across the domains of the global system – the bulk environment, near-surface environment, the passive film, the near-surface metal and the bulk metal (Fig. 5) – a systematic framework was established for predicting the rates of corrosion and the incidence of failures, including those which may not have yet been experienced in the field (unknown unknowns).

By necessity such a framework requires the integration of high-level characterization tools, systems monitoring, experts across the fields of chemistry and materials science, experimental testing and high-fidelity modeling from the atomic scale through to the systems level. Examples were provided in Staehle's original paper for how this approach could be used to predict hitherto unforeseen failure modes from sulfur and lead-assisted stress corrosion cracking in a light water reactor environment.[57]

Despite the apparent appeal of attempting to rigorously model a complex system like a corrosion pipeline entirely from first-principles (like quantum chemistry or atomistic simulation, for example), even given the most optimistic estimates of future available computing power, the task itself would be entirely intractable.[58,59] Considering that 1 km of pipeline could include approximately 10^{15} micron-sized grains and precipitates, with a similar order of magnitude of grain boundaries, and each grain comprising $\sim 10^{15}$ atoms, even a simulation that operates at the mesoscopic level of grains would be too complex to be computationally tractable. Instead, it makes more sense to decompose the corrosion phenomena into distinct length scales according to their heterogeneity (Fig. 6). For instance, megascale conditions, such as local climates might vary on the length scale of 10s of kilometers (10^4 m). Soil quality, on the other hand, might vary in the order of meters (10^0 m). Galvanic cells and other electrochemical conditions would vary on the level of centimeters to millimeters (10^{-3} to 10^{-2} m). Microstructural variations within the material occur in the length of the grain size, sub-microns to tens of microns (10^{-7} to 10^{-5} m). Atomic scale heterogeneities, including point-, line- and planar-defects take place over nanometer and sub-nanometer scales (10^{-10} to 10^{-8} m). Thus, it would suffice to explore, using advanced characterization and *ab initio* models, the contributors of

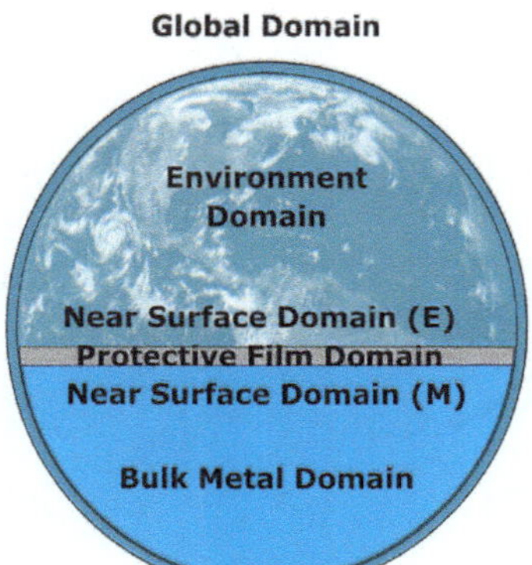

Fig. 5 Schematic illustration of the decomposition of a global corrosion *via* the domains–microprocesses sequence approach.[57]

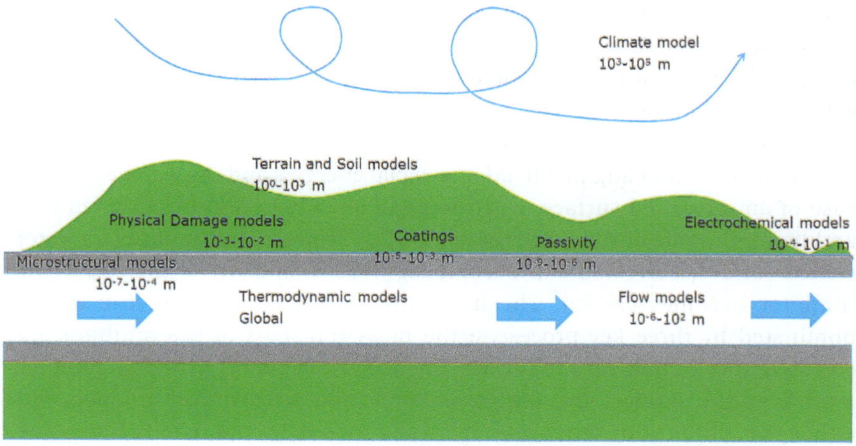

Fig. 6 Illustration of the integrated modeling approach to pipeline corrosion, showing a number of contributing factors that are amenable to modeling at various length scales.

microstructural effects under the possible set of variations that may be introduced at the higher length scales. Since the material microstructure is likely to remain similar at points along the multi-kilometer stretches of material, appropriate parameters from the lower length scale could be used in the higher length scale models. Thus, the hierarchical approach to multiscale modeling provides an efficient pathway to obtaining an integrated approach. A similar breakdown and integrated multiscale approach has been proposed by Cole and Marney, using the four sub-categories of electrochemical activity, oxide scales, local macro-environmental parameters (*e.g.* soil pH), and general external environmental factors (*e.g.* seasonal variations), drawing from the extensive models developed by soil scientists and hydrologists.[60]

An integrated multiphysics corrosion inhibition model

In this section, we now apply the domains and microprocesses approach to develop a multiphysics framework for the prediction of inhibitor efficiencies and eventual inhibitor design. In this way we will construct the phenomenological links between the inhibitor performance as a corrosion suppressor and the molecular properties of the inhibitor in a way that also incorporates the role of the materials and environment conditions.

We first identify the relevant domains, with reference to the schematic provided in Fig. 5. The environment domain is the content of the pipeline, in this context a hydrocarbon oil of complex composition with some fractional water cut, which may in turn contain carbonates, sulphides, chlorides and other inorganic and organic aqueous neutral and ionic species, including the applied inhibitor. Processes that are relevant to the inhibitor performance in this domain will include partitioning of the inhibitor molecule to either the aqueous or hydrocarbon phases, acid–base equilibria in the aqueous phase, the association of ions to form salts or dissolved ion-pairs, and the aggregation of species to form micelles. The fluid is also in a state of mass transport by convection and so flow processes are also relevant, including the entrainment of water by the hydrocarbon under certain conditions of temperature, flow velocity and water cut.[61–65]

The near surface domain will have two possible natures depending on the relative metal–water and metal–oil interfacial energies and the water cut. In the case of an oil-wetted surface, corrosion will already be significantly reduced, and hence the role of the inhibitor will be less relevant. In the case of a water-wetted surface, corrosion can become significant, and hence the role of the inhibitor becomes critical. The near surface domain in this case will be dominated by three key processes: the mass transport of the inhibitor and corrosive species towards the surface, the mass transport of corrosion products away from the surface, and the possible reaction of corrosion products with either the inhibitor or other molecules and ions in solution, which may reprecipitate as scale-formers.[66,67]

The surface domain is the domain in which the inhibitors may adsorb or desorb and form a corrosion inhibiting film. The film will retard chemical reaction by occupying metal sites through the formation of chemisorbed or physisorbed layers. The film properties, such as the nature of dynamic equilibrium and the permeability of the film to cation and anion transport, will determine the extent to which corrosion is reduced. Inhibitors may adsorb as neutral molecules, or as dipolar films, consisting of a charged inhibitor (anionic or cationic) with a counteranion pair.[63,68–70]

Globally, the system will be subject to certain temperatures and pressures, morphological features, such as joins, welds and stress-states, and Galvanic effects.

On the metallic side, processes in the near-surface domain will dictate what types of surfaces are exposed to the environment. The material microstructure will determine the relative distribution of cathodic and anionic corrosion sites, for example. Heterogeneities in the surface can drive certain localized corrosion modes, such as pitting.[67,71,72]

These domains and processes relevant to the overall mechanism of the inhibitor action are presented schematically in Fig. 7, and tabulated in Table 1.

We now proceed to outline some of these processes in more detail and provide ways in which the suite of modelling tools presented in Fig. 4 can be applied to develop a comprehensive molecules-to-materials approach for corrosion inhibitor modelling and design.

Speciation of the inhibitor molecule in the environment

Inhibitors can be ionized by reaction with water according to their acid–base properties, which is quantified by the acid dissociation constant, pKa. Cheminformatics techniques, based on functional group recognition, can predict acid dissociation constants for a given molecular structure with a high level of accuracy. Quantum chemical methods based on calculating the free energy of acid dissociation are also available.[73]

Ionized inhibitors can also form ion pairs that affect the oil–water partition coefficient in the presence of anions like chloride. The extent of ion pair formation is also predictable through quantum chemical calculations of the free energy of ion association.[74]

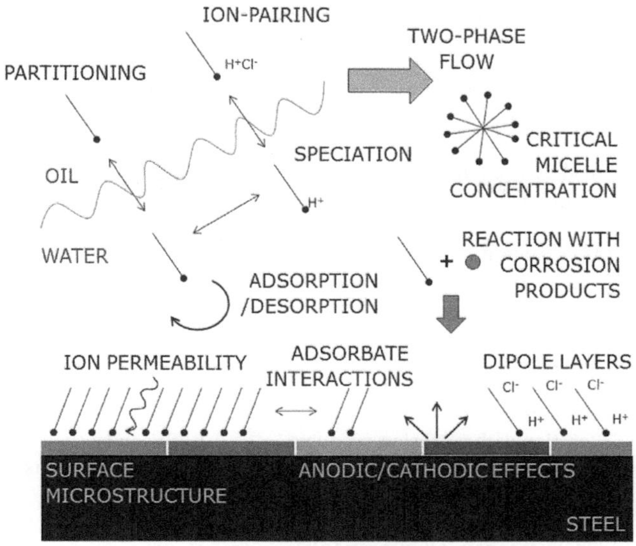

Fig. 7 Processes that affect the overall efficiency of a corrosion inhibitor across the environment, near-surface environment, surface and near-surface metal domains.

Inhibitors in neutral or ion-pair states can also form micelles that will limit their ability to form self-assembled monolayers on exposed metal surfaces.[61,75] This latter effect is measured by the critical micelle concentration and can be predicted to some extent from classical molecular dynamics or quantum chemical calculations, although such tools are not widely available.[76–82]

Partitioning of the inhibitor species between the oil and water phases

Inhibitor molecules partition unequally between water and oil, depending on their molecular hydrophobicity/hydrophilicity. In order to inhibit aqueous corrosion, there must be a sufficient inhibitor molecular concentration to maintain an equilibrium surface coverage over the metal. This tendency is quantified by the oil–water partition coefficient, log P. A higher value of log P will mean fewer inhibitor molecules are available to act against corrosion in the water phase. Numerous methods are available to compute log P, but this is primarily in the context of *n*-octanol as a solvent; more polar than most hydrocarbon media. Quantum chemical techniques have been developed for a much larger group of solvents, including numerous hydrocarbons of relevance to the oil and gas industry.[83,84] Further work needs to be performed to determine whether or not correlations from single component hydrocarbons can be made to the crude oil environment as a whole, or what descriptors can be taken from a crude assay, for example, that would enable effective partition coefficients to be estimated from first-principles, or some key experimental values in one, two or a few simpler solvent mixtures.

Ultimately the extent of partitioning will be a composite quantity of the water cut, the log P quantity, the pKa, the extent to which micelles form and ion-pairs migrate across the oil–water interface. Chemical thermodynamics models can be

Table 1 Domains, processes and primary components

Domain	Micro-process	Constituents
Environment	Partitioning	Hydrocarbon oil of complex composition
	Acid–base equilibria	Water
	Ion-pairing	Carbonates, sulphides, chlorides and other inorganics
	Micellization	Organic aqueous neutral and ionic species
	Convection	
	Solubility of gases	
	Entrainment	
Near surface environment	Oil *versus* water wetting	Oil
	Mass transport	Water
	Reaction with corrosion products	Diffusion layers
		Corrosion products
	Precipitation	Scales
Surface	Adsorption (chemisorption, physisorption)	Inhibitor
	Desorption	Metal ions
	Diffusion	Oxygen
	Corrosion	Hydrogen (H^+, H_2)
	Cathodic, anodic processes	Water
Near-surface metal	Localized corrosion	Intermetallics and precipitates
	Defect mobility	Passive films
	Galvanic coupling	Scales
	Dealloying	Defects (grain-boundaries, dislocations)
		Depletion layers
	Corrosion products	Residual stress
Global	Macroscopic variations in temperature, pressure and inhibitor concentration	Joins
	Cathodic protection	Welds
		Pipe stress
	Galvanic coupling/cladding	Cladding

used to integrate these variables and obtain the ultimate concentration of the inhibitor present in the aqueous *versus* crude oil phases.[85]

Impact of the inhibitor upon multiphase flow

The separation of oil and water into a two-phase flow depends on the surface tensions between the oil–water, water–metal, and oil–metal interfaces.[65] Each of these surface tensions can be affected by small concentrations of inhibitor species, since their bifunctional nature tends to lead to concentration at the interfaces. These surface tensions can be predicted from molecular dynamics simulations or the classical density functional theory applied to liquids.[86–88] Once the surface and interfacial tensions are determined from first-principles or experiments then the values can be inserted into fluid mechanics expressions to predict whether or not water entrainment or multiphase flow occurs, and whether or not the surface becomes oil wet or wetted with water.

Migration of the inhibitor molecule to the metal surface across the hydrodynamic boundary layer

Under flowing conditions, the transport of the inhibitor from the bulk solution phase to the metal–water interface will depend upon the diffusion of the molecule across the hydrodynamic boundary layer. Thus, the flow rate and the molecular structure of the inhibitor, as well as the surface quality, will influence the effective concentration of the inhibitor near the surface.[89] Molecular dynamics simulations can be used to generate diffusivity data that can then be used in the effective near-surface models. Similar factors apply to the diffusivity of corrosion reactants, products, and potential products of the reaction of the inhibitor molecules with the corrosion products. Diffusion constants can then be applied in mesoscale models that combine reaction kinetics with mass transport, such as reaction-kinetic models or Lattice–Boltzmann simulations.[90–92]

Identification of the solid surface phases available for inhibitor adsorption

Mild steel in an oil and gas environment will have surface phases that could consist of oxides, oxy-hydroxides, hydroxides, sulphides, carbides, carbonates and, under severe acidic corrosion, bare iron surfaces.[29,67,71,73] Other solid surfaces that may be present include sand and formation fines, and organic deposits, such as asphaltenes and paraffins.[61,93] The inhibitor–surface interaction will be different for each phase, and so will the overall inhibitor efficiency. Thermodynamic methods can be used to predict stable surface phases, but these may need to be coupled with kinetic stability diagrams that take metastability into account, since it is known that the most thermodynamically stable phase is not always the phase that is observed to form in the field.[94] Characterization, thermodynamic prediction of microstructures, and microstructural models, like phase field theory, could be used as predictive tools to generate the surface microstructures most likely to prevail under distinct materials/environment conditions.[95–97]

Formation of self-assembled monolayers on the solid surface phases

The physics of formation of chemisorbed layers on solid surfaces is described by the theory of adsorption isotherms, most commonly the Langmuir isotherm, although other variants also exist.[12] The key parameter controlling the formation of these layers is called the Gibbs' free energy of adsorption. This value can be inferred from experimental analysis (given certain assumptions) or predicted directly from first principles calculations using density functional theory. The latter method has been reviewed earlier in this paper and is a field of current research activities, due to the recent appearance of robust codes that can compute the structure and energy associated with chemisorbed phases at solid–fluid interfaces with considerable accuracy. Another approach that has been applied is to approximate chemisorption energies by replacing the quantum chemical calculation of a surface with the quantum chemical calculation of a representative metal-oxy or oxy-hydroxide coordination complex interacting with the inhibitor molecule.[98] In some cases these may lead to faster computations, although with a reduction in the information that may be needed to fully capture the surface–inhibitor interaction. The same authors also used the log P (*n*-octanol/water partition coefficient) to estimate the change in entropy associated with the inhibitor adsorption.

Prediction of the corrosion rates of the metallic surface phases with the inhibitor self-assembled layers

In adsorption-centric studies of the mechanism of the chemical inhibition of corrosion, it is commonly assumed that the extent of the surface coverage by the inhibitor molecule corresponds exactly to the extent by which corrosion is reduced. However, we propose herein that this may not exactly be the case, due to porosity of the inhibitor film as a result of imperfect packing, and the dynamic nature of adsorption–desorption of the molecules composing the surface film.[7,41] We propose herein that instead there is some reduction in the corrosion rate for the covered metal surface, which is not necessarily 100% effective. The reduced rates of corrosion in the presence of the inhibitor films can be investigated by contrasting different techniques for measuring inhibitor adsorption (quartz crystal microbalance for example) with changes in capacitance and/or the corrosion current. First principles and atomistic-based models of inhibitor packing efficiency on the surface can also be used to estimate the extent to which a maximally covered surface provides total coverage of the metal atom sites. Newly emerging simulation techniques, like reactive force field, may also be able to simulate the kinetic processes associated with the dynamic dissolution and reformation of the protective inhibitor surface layer, and thus provide direct estimation of the corrosion rates in the presence of an inhibitor *versus* no inhibitor.[99] Inhibitor molecules may also reduce corrosion through other mechanisms, such as the reaction with corrosion products to form scales that act as diffusion barriers, thus limiting corrosion.[66]

Model evaluation and uncertainty quantification

One rarely discussed feature of quantum chemical and atomistic modelling is the uncertainty in the evaluated molecular descriptors, such as the energy, geometry and other derived quantities. This challenge has been recognized recently, and some new density functional theory approaches devised that directly incorporate uncertainty *via* Bayesian error estimation.[100] Furthermore, statistical uncertainty can be derived in a natural way from molecular dynamics and Monte Carlo simulations, since they are inherently ensemble methods.[101] For the purpose of lifetime prediction and comparing potential new inhibitor candidate molecules, it is necessary to consider the meaningful differences between the predictions that would be made by slight variations in the choice of theoretical methods, and what are the inherent uncertainties based upon the model approach used and the assumptions contained therein.

The foremost means of quantifying the accuracy of a model is to compare the predictions of a model with a reliable validation dataset in order to evaluate the error in the model prediction. This method should be augmented with more sophisticated treatments, such as exploring the nature of error propagation through the model. This should be based on the model's set of assumptions, uncertainties in the data provided as model inputs, uncertainties in the data used for validation, and fundamental uncertainties that arise due to limitations in our understanding of the physics of the problem (known as *epistemic* or knowledge-based uncertainty).[51,100,101] These latter categories of error can be more challenging to quantify, and are rarely investigated due to the academic emphasis on

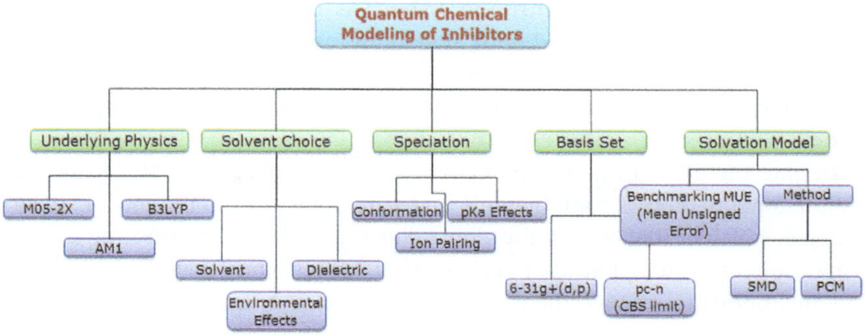

Fig. 8 Sources of uncertainty that arise from different model design decisions that must be made when simulating molecular properties in a solvated environment from first-principles.

making models deeper and more theoretically rigorous rather than more relevant to engineering use.

Uncertainty quantification has been incorporated in the integrated multiscale modelling pathway shown in Fig. 4. Some methods used in recent literature to evaluate model uncertainties include Bayesian inference,[102] comparing model predictions with experiments and available field data, integration into damage accumulation models,[51] and examination through the use of Monte Carlo and/or first- and second-order reliability methods (FORM/SORM).[53]

With respect to quantum chemical-based prediction of the molecular contributions to corrosion inhibition, the methods used to compute the properties of the inhibitors are subject to various critical decisions that need to be made in setting up the problem. For example, the basis function sets used to represent electron density can be made more or less complete, depending upon how much computer time or memory is available to perform the computation.[103] In addition, there are competing semi-empirical approaches that have been developed as a way to emulate the effects of a surrounding organic or aqueous medium.[104] A list of the possible decisions that could be made when constructing such a model is shown in Fig. 8. These include the underlying physics model, the choice of solvent, the molecular conformation, the electronic basis set, and the representation of solvent. In the ideal case, transferability between different decisions should lead to similar outcomes but this is not guaranteed. When these sources of variation are taken into account, it is possible to gain an appreciation for the uncertainty in the predictions made by these otherwise highly rigorous theoretical methods.

Summary and conclusions

To conclude, this paper presents a new approach to the molecular-based modeling and design of corrosion inhibitors that is based upon a phenomenological breakdown of the critical factors that affect the efficiency of a corrosion inhibitor. The objective was to focus the molecular search by providing constraints upon the molecular suitability based upon the mechanisms of corrosion inhibition, spanning the various domains and processes *via* which the inhibitor interacts with the corrosion system. As most of the factors outlined above are able to be modeled with state-of-the-art modeling and simulation techniques, we propose that a team-wise effort across these categories of mechanistic response be applied to construct an integrated multiphysics inhibitor model.

Despite the effort to be comprehensive, we acknowledge that there may be several domain–process combinations that were omitted from the model. As an example, consider that, within the global domain, some factors affecting a molecule's suitability for application as a chemical inhibitor may include the toxicity and environmental impact, the cost of synthesis, and its interaction with other chemicals that may be part of the chemical package (scale inhibitors, demulsifiers, hydrate inhibitors and biocides). Along those lines, the intrinsic molecular stability regarding thermal decomposition or reaction with water should also be given consideration and be used to constrain the design search. Such examples demonstrate the importance of creativity and team-building in reviewing the list of possible processes to incorporate within the integrated multiscale model.

Given that the modern computational solution of molecular level problems has become a tractable and relatively inexpensive affair, the integrated multiscale modeling approach should be highly promising in future models of inhibitor efficiency and an attractive, science-based alternative to the proliferation of regression based theory–experiment correlations abundant in the literature today.

References

1 H. H. Uhlig, *Corrosion and Corrosion Control*, John Wiley & Sons, Inc., New York, 2nd edn, 1971.

2 S. Papavinasam, in *Corrosion Control in the Oil and Gas Industry*, Elsevier, London, 2014, pp. 374–399.

3 S. Ramachandran, K. Miner, M. Greaves, J. Thomas and V. Jovancicevic, *presented in part at the Corrosion 2010*, San Antonio, TX, 2010.

4 P. A. Sorensen, S. Kiil, K. Dam-Johansen and C. E. Wienell, *J. Coat. Technol. Res.*, 2009, **6**, 135–176.

5 G. Schmitt, *Br. Corros. J.*, 1984, **19**, 165–176.

6 G. Gece, *Corros. Sci.*, 2008, **50**, 2981–2992.

7 A. Edwards, C. Osborne, S. Webster, D. Klenerman, M. Joseph, P. Ostovar and M. Doyle, *Corros. Sci.*, 1994, **36**, 315–325.

8 Y. Xiong, B. Brown, B. Kinsella, S. Nesic and A. Pailleret, *Corrosion*, 2014, **70**, 247–260.

9 J. Zajac, J. L. Trompette and S. Partyka, *Langmuir*, 1996, **12**, 1357–1367.

10 E. E. Oguzie, Y. Li and F. H. Wang, *J. Colloid Interface Sci.*, 2007, **310**, 90–98.

11 D.-S. Kong, S.-L. Yuan, Y.-X. Sun and Z.-Y. Yu, *Surf. Sci.*, 2004, **573**, 272–283.

12 K. Y. Foo and B. H. Hameed, *Chem. Eng. J.*, 2010, **156**, 2–10.

13 R. Fuchs-Godec, *Colloids Surf., A*, 2006, **280**, 130–139.

14 I. N. Levine, *Quantum Chemistry*, Prentice HAll, Upper Saddle River, NY, 5th edn, 2000.

15 J. Bernholc, *Phys. Today*, 1999, 30–39.

16 G. Kresse and J. Furthmuller, *Phys. Rev. B: Condens. Matter Mater. Phys.*, 1996, **54**, 11169–11186.

17 G. Kresse and J. Hafner, *Phys. Rev. B: Condens. Matter Mater. Phys.*, 1993, **47**, 558–561.

18 G. Kresse and D. Joubert, *Phys. Rev. B: Condens. Matter Mater. Phys.*, 1999, **59**, 1758–1775.

19 M. Gholami, I. Danaee, M. H. Maddahy and M. RashvandAvei, *Ind. Eng. Chem. Res.*, 2013, **52**, 14875–14889.

20 W. Durnie, R. De Marco, A. Jefferson and B. Kinsella, *J. Electrochem. Soc.*, 1999, **146**, 1751–1756.

21 W. Durnie, R. De Marco, B. Kinsella, A. Jefferson and B. Pejcic, *J. Electrochem. Soc.*, 2005, **152**, B1–B11.

22 L. C. Murulana, A. K. Singh, S. K. Shukla, M. M. Kabanda and E. E. Ebenso, *Ind. Eng. Chem. Res.*, 2012, **51**, 13282–13299.

23 E. E. Ebenso, M. M. Kabanda, L. C. Murulana, A. K. Singh and S. K. Shukla, *Ind. Eng. Chem. Res.*, 2012, **51**, 12940–12958.

24 J. Vosta and J. Eliasek, *Corros. Sci.*, 1971, **11**, 223–229.

25 K. F. Khaled, *J. Solid State Electrochem.*, 2009, **13**, 1743–1756.

26 J. K. Norskov, F. Abild-Pederson, F. Studt and T. Bligaard, *Proc. Natl. Acad. Sci. U. S. A.*, 2010, **108**, 937–943.

27 R. A. van Santen and M. Neurock, *Molecular Heterogeneous Catalysis: A Conceptual and Computational Approach*, Wiley-VCH, Weinheim, 2006.

28 B. Hammer, L. B. Hansen and J. K. Nørskov, *Phys. Rev. B: Condens. Matter Mater. Phys.*, 1999, **59**, 7413–7421.

29 R. Armiento and A. E. Mattsson, *Phys. Rev. B: Condens. Matter Mater. Phys.*, 2005, **72**, 85108–85112.

30 A. E. Mattson and D. R. Jennison, *Surf. Sci.*, 2002, **520**, L611–L618.

31 J. Hafner, *Acta Mater.*, 2000, **48**, 71–92.

32 J. Hafner, C. Wolverton and G. Ceder, *MRS Bull.*, 2006, **31**, 659–668.

33 P. Hohenberg and W. Kohn, *Phys. Rev.*, 1964, **136**, B864–B871.

34 W. Kohn and L. Sham, *Phys. Rev.*, 1965, **140**, A1133–A1138.

35 M. C. Payne, M. P. Teter, D. C. Allan, T. A. Arias and J. D. Joannopoulos, *Rev. Mod. Phys.*, 1992, **64**, 1046–1097.

36 J. A. Torres, E. Artacho, J. M. Cela, J. Gale, A. Garcia, J. Junquera, R. M. Martin, P. Ordejon, D. Sanchez-Portal and J. M. Soler, *SIESTA*, http://www.icmb.es/siesta.

37 J. P. Perdew, J. A. Chevary, S. H. Vosko, K. A. Jackson, M. R. Pederson, D. J. Singh and C. Fiolhais, *Phys. Rev. B: Condens. Matter Mater. Phys.*, 1992, **46**, 6671–6687.

38 I. D. Prodan, G. E. Scuseria and R. L. Martin, *Phys. Rev. B: Condens. Matter Mater. Phys.*, 2007, **76**, 33101.

39 A. E. Mattsson, R. Armiento, J. Paier, G. Kresse, J. M. Wills and T. R. Mattsson, *J. Chem. Phys.*, 2008, **128**, 84714–84724.

40 A. Kokalj, S. Peljhan, M. Finsgar and I. Milosev, *J. Am. Chem. Soc.*, 2010, **132**, 16657–16668.

41 S. Ramachandran, B. L. Tsai, M. Blanco, H. J. Chen, Y. Tang and W. A. Goddard III, in *New Techniques for Characterizing Corrosion and Stress Corrosion*, ed. R. H. Jones and D. R. Bear, The Minerals, Metals and Materials Society, 1996, pp. 117–127.

42 E. E. Oguzie, Y. Li, S. G. Wang and F. H. Wang, *RSC Adv.*, 2011, **1**, 866–873.

43 X. Tang, S. Richter and S. Nesic, *presented in part at the Corrosion 2013*, Orlando, FL, 2013.
44 J. Allison, D. Backman and L. Christodoulou, *JOM*, 2006, **58**, 25–27.
45 J. Allison, *JOM*, 2011, **63**, 15–18.
46 A. Lucia, B. M. Bonk, R. R. Waterman and A. Roy, *Comput. Chem. Eng.*, 2012, **36**, 79–98.
47 W. A. Curtin and R. E. Miller, *Modell. Simul. Mater. Sci. Eng.*, 2003, **11**, R33–R68.
48 G. B. Olson, *Acta Mater.*, 2013, **61**, 771–781.
49 J. P. Holdren, *Materials Genome Initiative for Global Competitiveness*, National Science and Technology Council, Washington, D.C., 2011.
50 F. Ayello, N. Sridhar, G. Koch, V. Khare, A. W. Al-Mathen and S. Safri, *presented in part at the Corrosion (2014)*, San Antonio, TX, 2014.
51 G. Engelhardt and D. D. Macdonald, *Corros. Sci.*, 2004, **46**, 2755–2780.
52 T. M. Mitchell, *Machine Learning*, McGraw-Hill, Boston, 1997.
53 B. Saassouh and Z. Lounis, *Cem. Concr. Compos.*, 2012, **34**, 1082–1093.
54 N. N. Taleb, *The Black Swan: The Impact of the Highly Improbable*, Random House, New York, NY2007.
55 R. de Borst, *Comput. Mater. Sci.*, 2008, **43**, 1–15.
56 P. Murray-Rust and D. Murray-Rust, in *Implementing Reproducible Research*, eds. V. Stodden, F. Leisch and R. D. Peng, CRC Press, Boca Raton, FL, 2014.
57 R. W. Staehle, *The Banff Center*, Calgary, Alberta, Canada, 2004.
58 E. P. Benedictis, *The most generous estimates of the fundamental limitations on computing imply that a full atomistic simulation via a technique likely molecular dynamics of even a picogram of material is an unobtainable goal*, Assoc. Comput. Machinery, 2005, p. 391.
59 G. Egan, *Permutation City Millenium*, Orion Books Ltd., Great Britain, 1994.
60 I. S. Cole and D. Marney, *Corros. Sci.*, 2012, **56**, 5–16.
61 J. A. Dougherty, *presented in part at the Corrosion*, 1998.
62 J. A. Dougherty and Y. S. Ahn, *presented in part at the Corrosion*, 1999.
63 R. H. Hausler, *Corrosion*, 1986, **42**, 729–739.
64 R. H. Hausler, D. W. Stegmann and R. F. Stevens, *Corrosion*, 1989, **45**, 857–870.
65 C. Li, S. Richter and S. Nesic, *presented in part at the Corrosion 2013*, Orlando, FL, 2013.
66 V. Jovancicevic, Y. S. Ahn, J. A. Dougherty and B. Alink, *presented in part at the Corrosion*, 2000.
67 R. DeMarco, W. Durnie, A. Jefferson and B. Kinsella, *Corrosion*, 2001, **57**, 9–18.
68 S. Sun, D. Zhang, C. Li, Y. Wang and Q. Yang, *Chem. Eng. J.*, 2014, **258**, 128–135.
69 J. O. Mendes, E. C. da Silva and A. B. Rocha, *Corros. Sci.*, 2012, **57**, 254–259.
70 N. Hackerman, *Corrosion*, 1962, **18**, 332–337.
71 C. Sorriano, R. Oltra, A. Zimmer, B. Vuillemin and C. Borkowski, *Surf. Interface Anal.*, 2013, **45**, 1649–1653.
72 A. Anderko and P. J. Shuler, *Comput. Geosci.*, 1997, **23**, 647–658.
73 S. Sastre, R. Casasnovas, F. Munoz and J. Frau, *Theor. Chem. Acc.*, 2013, **132**, 1310–1317.
74 J. Sangster, *Octanol–water Partition Coefficients: Fundamentals and Physical Chemistry*, John Wiley & Sons, Chichester, 1997.

75 P. Xue, Z. Fu, X. Chu, Y. Zhang and Z. Yang, *Appl. Surf. Sci.*, 2014, **317**, 752–759.
76 Z. Yang, D. Wang, L.-C. Xu, X. Liu, X. Li and B. Xu, *Int. J. Hydrogen Energy*, 2014, **39**, 19621–19629.
77 R. Gao, D.-B. Cao, Y. Yang, Y.-W. Li, J. Wang and H. Jiao, *Appl. Catal., A*, 2014, **475**, 186–194.
78 K. Honkala, *Surf. Sci. Rep.*, 2014, **69**, 366–388.
79 J. Zhang, Z. Wang and Z. Zhu, *J. Power Sources*, 2014, **255**, 65–69.
80 S. S. Rath, N. Sinha, H. Sahoo, B. Das and B. K. Mishra, *Appl. Surf. Sci.*, 2014, **295**, 115–122.
81 J. Radilla, G. E. Negrón-Silva, M. Palomar-Pardavé, M. Romero-Romo and M. Galván, *Electrochim. Acta*, 2013, **112**, 577–586.
82 N. Y. Acelas, S. M. Mejia, F. Mondragón and E. Flórez, *Comput. Theor. Chem.*, 2013, **1005**, 16–24.
83 A. V. Marenich, C. J. Cramer and D. G. Truhlar, *J. Phys. Chem. B*, 2009, **113**, 6378–6396.
84 G. Scalmani and M. J. Frisch, *J. Chem. Phys.*, 2010, **132**, 114110–114123.
85 U. Domanska, M. K. Kozlowska and M. Rogalski, *J. Chem. Eng. Data*, 2002, **47**, 456–466.
86 M. Talreja, I. Kusaka and D. L. Tomasko, *Fluid Phase Equilib.*, 2012, **319**, 67–76.
87 A. von Muller and K. Leonhard, *Fluid Phase Equilib.*, 2013, **356**, 96–101.
88 B. D. Marshall, K. R. Cox and W. G. Chapman, *J. Phys. Chem. C*, 2012, **116**, 17641–17649.
89 J. A. Dougherty, *presented in part at the Corrosion*, 1997.
90 J. Ozbolt, G. Balabanic and M. Kuster, *Corros. Sci.*, 2011, **53**, 4166–4177.
91 J. Ozbolt, G. Balbanic, G. Periskic and M. Kuster, *Construction and Building Materials*, 2010, **24**, 1638–1648.
92 H.-Y. Chang, Y.-S. Park and W.-S. Hwang, *J. Mater. Process. Technol.*, 2000, **103**, 206–217.
93 J. Rajasankar and N. R. Iyer, *Eng. Fract. Mech.*, 2006, **73**, 553–570.
94 E. Deltombe, N. de Zoubov and M. Pourbaix, 1974.
95 I. Simonovski and L. Cizelj, *J. Nucl. Mater.*, 2011, **414**, 243–250.
96 M. A. Arafin and J. A. Szpunar, *Comput. Mater. Sci.*, 2010, **47**, 890–900.
97 Y. Wang and J. Li, *Acta Mater.*, 2010, **58**, 1212–1235.
98 S. Mondal and S. R. Taylor, *presented in part at the Corrosion (2014)*, San Antonio, TX, 2014.
99 B. Jeon, S. K. R. S. Sankaranarayanan, A. C. T. van Duin and S. Ramanathan, *J. Chem. Phys.*, 2011, **134**, 234706.
100 A. Strachan, S. Mahadevan, V. Hombal and L. Sun, *Modell. Simul. Mater. Sci. Eng.*, 2013, **21**, 65009–65028.
101 A. Grossfield and D. M. Zuckerman, *Annu. Rep. Comput. Chem.*, 2009, **5**, 23–48.
102 G. Koch, F. Ayello, V. Khare, N. Sridhar and A. Moosavi, *presented in part at the International Pipeline Conference*, Calgary, Alberta, Canada, 2014.
103 F. Jensen, *J. Chem. Phys.*, 1999, **110**, 6601–6605.
104 A. V. Marenich, C. J. Cramers and D. G. Truhlar, *J. Phys. Chem. B*, 2209, **113**, 6378–6396.

Faraday Discussions

ROYAL SOCIETY OF CHEMISTRY

PAPER

Corrosion protection of iron using porous anodic oxide/conducting polymer composite coatings

Yoshiki Konno,[a] Etsushi Tsuji,[ab] Yoshitaka Aoki,[ab] Toshiaki Ohtsuka[b] and Hiroki Habazaki*[ab]

Received 24th November 2014, Accepted 22nd December 2014
DOI: 10.1039/c4fd00232f

Conducting polymers (CPs), including polypyrrole, have attracted attention for their potential in the protection of metals against corrosion; however, CP coatings have the limitation of poor adhesion to metal substrates. In this study, a composite coating, comprising a self-organized porous anodic oxide layer and a polypyrrole layer, has been developed on iron. Because of electropolymerization in the pores of the anodic oxide layer, the composite coating showed improved adhesion to the substrate along with prolonged corrosion protection in a NaCl aqueous corrosive environment. The anodic oxide layers are formed in a fluoride-containing organic electrolyte and contain a large amount of fluoride species. The removal of these fluoride species from the oxide layer and the metal/oxide interface region is crucial for improving the corrosion protection.

1 Introduction

The use of conducting polymers (CPs) for the corrosion protection of metals has attracted considerable attention over the past decades. Oxidative properties of CPs maintain the potential of the substrate in the passive state.[1,2] Polypyrrole (PPy), one of the CPs, is often studied for corrosion protection of metals and can be readily deposited on steel by electropolymerization in aqueous electrolytes. During the electropolymerization of pyrrole, electrolyte anions are incorporated into the PPy film, and the nature of the incorporated anions significantly influences the properties of the resultant PPy films.[3–5] The incorporated anions are released during the reduction of PPy. Inhibitor anions are preferable as incorporated species because they protect metals from corrosion. In fact, PPy films with incorporated molybdate[6–10] and molybdophosphate[11–13] have shown good corrosion protection of metals. Release of such inhibitor anions into corrosive solutions can be suppressed by the introduction of an outer PPy layer with

[a]Graduate School of Chemical Sciences and Engineering, Hokkaido University, Sapporo, Hokkaido 060-8628, Japan. E-mail: habazaki@eng.hokudai.ac.jp; Fax: +81-11-706-6575; Tel: +81-11-706-6575
[b]Faculty of Engineering, Hokkaido University, Sapporo, Hokkaido 060-8628, Japan

incorporated large organic anions, which inhibit the decomposition and release of molybdophosphate ions in the inner layer.[12] Such bi-layer coating further improves the corrosion protection properties.

Although the CPs are promising for corrosion protection and extensive studies have been conducted regarding their corrosion protection properties, one of the important issues, which must be addressed to fulfil the requirements of high-performance corrosion-protection coatings under widely varying practical conditions, is the poor adhesion of the coatings to the metal substrate. In the present study, self-organized porous anodic films are introduced between the iron substrate and the PPy layer to further improve the adhesion and corrosion protection properties.

Anodizing of aluminium in acid electrolytes develops self-organized porous anodic films, which have been widely used in industry for corrosion protection, wear resistance and colouring of aluminium and its alloys. The formation of such self-organized nanoporous and nanotubular anodic films has been recently extended to other metals such as titanium, zirconium, niobium, tantalum and tungsten.[14]

Anodizing of iron in aqueous electrolytes at high potentials results in the formation of soluble ferrate ions,[15,16] such that no porous anodic films can be formed on iron. However, recent reports disclosed that nanoporous and nanotubular anodic films were formed on iron in an ethylene glycol or glycerol electrolytes containing fluoride and small amounts of H_2O.[17–22] The anodic films thus formed are mainly amorphous and highly contaminated with fluoride species; however, they are readily converted to nanoporous or nanotubular α-Fe_2O_3, which is of potential interest as a photoanode for water splitting,[17,20,23] as a photoelectrocatalyst[24] and as electrodes in electrochemical capacitors.[25] In the pores of the anodic oxide on iron, it is expected that PPy is deposited by the electropolymerization of pyrrole, similar to electropolymerization in porous anodic alumina templates.[26] In this case, the PPy layer prepared on the anodized iron substrate should exhibit improved adhesion due to an anchor effect.

In the present study, PPy/anodic oxide composite coatings have been prepared. The anodic films on iron have been prepared by anodizing iron in an ethylene glycol electrolyte containing NH_4F and H_2O. The resultant films contain large amounts of fluoride and dissolve readily in aqueous solutions. Thus, thermal treatment is required to remove fluoride species and to form crystalline iron oxide phases. The conditions of thermal treatment for the removal of fluoride species have also been examined. Electropolymerization to produce a PPy layer has been performed in a phosphoric acid aqueous electrolyte containing pyrrole monomer and molybdophosphate, which acts as an inhibitor as well as a dopant.

2 Experimental

2.1 Formation of anodic films

A high purity (99.99%) iron plate was used as a substrate, which was mechanically polished with 1500 grit SiC polishing paper. After degreasing in acetone ultrasonically, the iron specimens were anodized at a constant voltage of 35 V in an ethylene glycol electrolyte, containing 0.1 mol L^{-1} NH_4F and 0.5 mol L^{-1} H_2O, at

313 K for 300 s. After anodizing, the resultant anodic film was dissolved under cathodic polarization conditions of 200 A m^{-2} in 0.1 mol L^{-1} HCl solution. Then, the iron specimen was re-anodized under the same conditions. This two-step anodizing was performed to obtain a smoother substrate surface than that obtained after the first anodizing step. After anodizing, the specimen was ultrasonically washed in ethylene glycol followed by acetone.

The thermal treatment of the anodized specimens was conducted either in air or argon at temperatures between 523 K and 673 K for 1.8 ks. The heating rate at the selected temperature was 2.5 K min^{-1}. A relatively slow heating rate was selected to avoid cracking of the anodic oxide layer. After the thermal treatment, the specimen was gradually cooled to room temperature in the furnace.

2.2 Formation of the PPy layer

The PPy layer was formed on the anodized specimens and subsequently on thermally-treated iron and passivated iron specimens. The passivation treatment of iron was performed at 0.8 V *vs.* Ag/AgCl in 0.06 mol L^{-1} Na_2MoO_4 aqueous solution for 15 min. For the electropolymerization of PPy, a three-electrode cell with an Ag/AgCl reference electrode and a platinum foil counter electrode was used. The electrolyte was 0.2 mol L^{-1} H_3PO_4 solution containing 0.1 mol L^{-1} pyrrole monomer and 5 mmol L^{-1} $H_3PMo_{12}O_{40}$. The solution was prepared in deionized milli-Q water and then deoxygenated by nitrogen bubbling for 30 min prior to use. The electropolymerization was performed at a constant current density of 10 A m^{-2} for 30 min at room temperature to deposit an approximately 5 μm-thick PPy layer.

2.3 Characterization of the coatings

The surfaces and cross-sections of the coated specimens were observed by a JEOL JSM-6500F field emission-scanning electron microscope (FE-SEM) operated at 10 kV. Some cross-sections of the anodized and thermally treated iron specimens were also observed by a JEOL JEM-2010 transmission electron microscope (TEM) operated at 200 kV and a Hitachi HD-2000 scanning transmission electron microscope (SEM) operated at 200 kV. Electron-transparent sections were prepared by a Hitachi FB-2100 FIB system employing a Ga^+ ion beam. Phases in the porous anodic layers were identified using a Rigaku RINT2000 X-ray diffractometer (XRD) using Cu Kα irradiation. Both θ–2θ and α–2θ modes with $\alpha = 1°$ were used to obtain information regarding the depth distribution of the individual phases.

Elemental depth profile analysis was performed using a Jobin-Yvon 5000 RF glow discharge optical emission spectroscopy (GDOES) instrument in a neon atmosphere of 1100 Pa by applying an RF of 13.56 MHz and a power of 50 W. Neon gas was used to excite emissions from fluorine because no excitation is observed with the conventional argon gas.[27] Light emissions of characteristic wavelengths were monitored throughout the analysis with a sampling time of 0.1 s to obtain depth profiles. The wavelengths of the spectral lines used were 121.567 nm for hydrogen, 165.701 nm for carbon, 130.217 nm for oxygen, 685.602 nm for fluorine, 178.287 nm for phosphorus, 385.991 nm for iron and 317.035 nm for molybdenum. The signals were detected from a circular area of approximately 4 mm diameter.

The adhesion of the coatings to the iron substrate was evaluated by a cross-cut test (ISO-2409). The corrosion tests of the coated specimens were performed by immersing them in 3.5 wt% NaCl solution at room temperature and continuously recording the corrosion potential.

3 Results and discussion

3.1 Formation of a nanoporous anodic film

Nanoporous anodic films on iron can be prepared in ethylene glycol electrolytes containing NH_4F and H_2O.[21,28] In this study, we developed the anodic film by anodizing at 35 V in an ethylene glycol electrolyte containing 0.1 mol L^{-1} NH_4F and 0.5 mol L^{-1} H_2O at 313 K for 300 s. The current transient during anodizing (Fig. 1a) reveals an initial current decay to a steady-state current density of approximately 90 A m^{-2} within 60 s. The SEM image of the surface of the anodized specimen (Fig. 1b) discloses the formation of a nanoporous anodic film with a pore size of ~30 nm. The cross-sectional observation (Fig. 1c) reveals the development of cylindrical nanopores in the anodic film. These nanopores are grown in the direction normal to the metal/film interface, as in the case of the well-known porous anodic alumina films formed in acidic aqueous electrolytes, such as sulphuric acid, oxalic acid and phosphoric acid.[29] The thickness of the anodic film is 1.2 μm, and a thin barrier layer (~60 nm) is also observed at the bottom of the anodic film, corresponding to the ratio of the thickness of the barrier layer to the formation voltage of 1.7 nm V^{-1}. This value is slightly smaller than that observed at 293 K (1.9 nm V^{-1}).[21] The ratio usually increases with an increase in electrolyte temperature because of the reduction of field strength at higher temperatures. This reduction may be associated with a change in film composition, such as the amounts of incorporated electrolyte species in the anodic film, although a more detailed study is required for clarification. In the fractured cross-section (Fig. 1c), the anodic film is separated from the iron substrate, which probably occurs during the preparation of the cross-section. The anodic film is highly contaminated with fluoride species, as discussed later, and fluoride-rich anodic films often have poor adhesion to metal substrates.[30,31] The formation of a fluoride-containing anodic film could be one of the reasons for the detachment of the anodic film from the substrate.

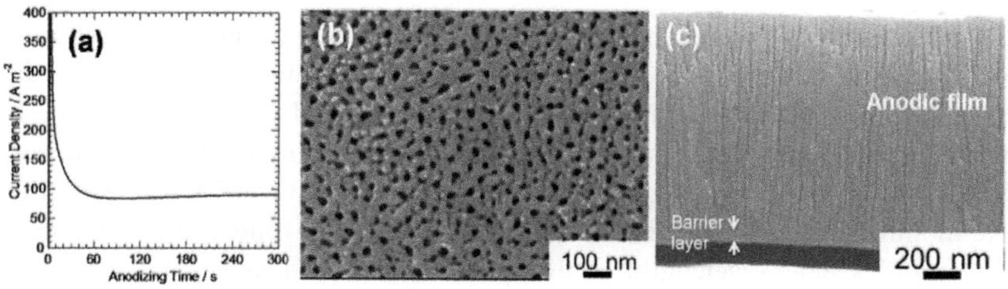

Fig. 1 (a) Current transient of iron during anodizing at 35 V in an ethylene glycol electrolyte containing 0.1 mol L^{-1} NH_4F and 0.5 mol L^{-1} H_2O at 313 K for 300 s; scanning electron micrographs of (b) the surface and (c) the fracture cross-section of the resultant anodic film.

3.2 Thermal treatments to remove fluoride species

The anodic film formed on iron in the present electrolyte contains a large amount of fluoride species, and the film is soluble even in water. To remove the fluoride species and form more chemically stable oxide films, thermal treatments of the anodized specimens were performed. SEM images of the specimens that were thermally treated up to 673 K, both in air and argon, revealed a nanoporous surface morphology similar to Fig. 1b. No obvious change in surface pore morphology occurred during the thermal treatments.

Fig. 2 shows the SEM images of the cross-sections of the anodized specimens after thermal treatments at different temperatures in air and argon. Vertically aligned nanopores are still present even after thermal treatment, although nanogranular materials are developed at higher temperatures. At the bottom of the anodic film, the growth of thermal oxide appears to proceed during thermal treatment. When the thermal treatment is performed at 673 K, the thermal oxide layer consists of two layers, an inner compact layer and an outer porous layer. The pore walls in the outer thermal oxide layer are significantly thicker than those in the anodic porous layer, which is located above the thermal oxide layer.

The change in the film thickness with the temperature of the thermal treatment depends on the atmosphere (Fig. 3). The total thickness of the oxide layer is almost constant even after thermal treatment in air at several temperatures, although thermal oxide layers tend to thicken with the temperature of thermal treatment. This is possibly due to the outward diffusion of the iron species in the thermal oxide layer during thermal treatment; the outward diffusing iron species form a new oxide material at the pore base, implying that the oxide formation at the oxide/substrate interface is negligible. In contrast, the total film thickness increases from 1.2 μm to 1.5 μm after thermal treatment in argon at 523 K, without a significant change in the thickness of the porous layer. A thermal oxide layer (~300 nm thick) appears to be formed without filling pores, *i.e.* at the oxide/metal interface due to the inward diffusion of anions. A trace amount of impurity

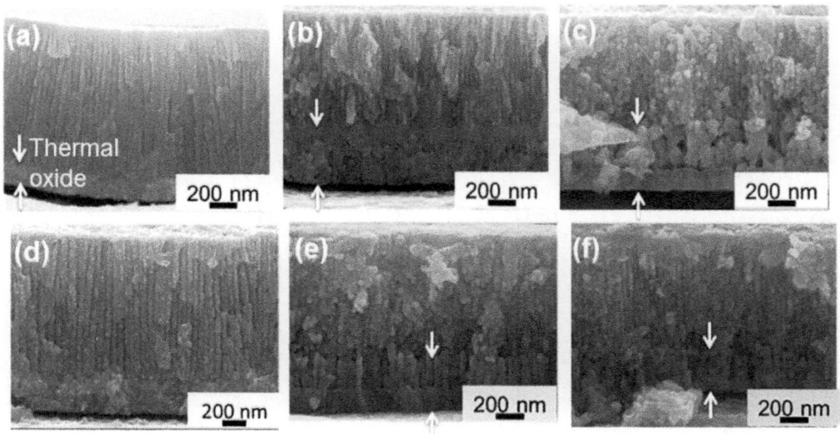

Fig. 2 Scanning electron micrographs of the cross-sections of the anodized iron specimens after thermal treatment in air at (a) 523 K, (b) 623 K and (c) 673 K and in argon at (d) 523 K, (e) 623 K and (f) 673 K.

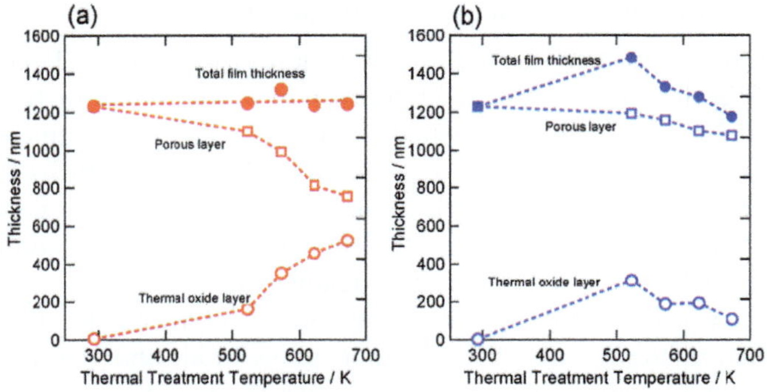

Fig. 3 Change in the thickness of the oxide layers on the anodized iron specimens with the temperature of the thermal treatment in (a) air and (b) argon atmospheres.

oxygen in the argon atmosphere may contribute to the formation of thermal oxide.

The total film thickness decreases gradually from 1.5 μm to ~1.2 μm with an increase in the temperature of the thermal treatment, mainly due to the thinning of the thermal oxide layer. The reduction of the thickness of the porous layer is limited compared to the thermal treatment in air.

The removal of the fluoride species from the anodic film by thermal treatment was examined by GDOES depth profile analysis. It is clear from Fig. 4a that the as-anodized specimen contains a relatively large amount of fluoride species throughout the anodic film. The film also contains carbon and nitrogen species which are probably derived from ethylene glycol and ammonium ions, respectively. Incorporation of such species increases the solubility of the anodic film in water.

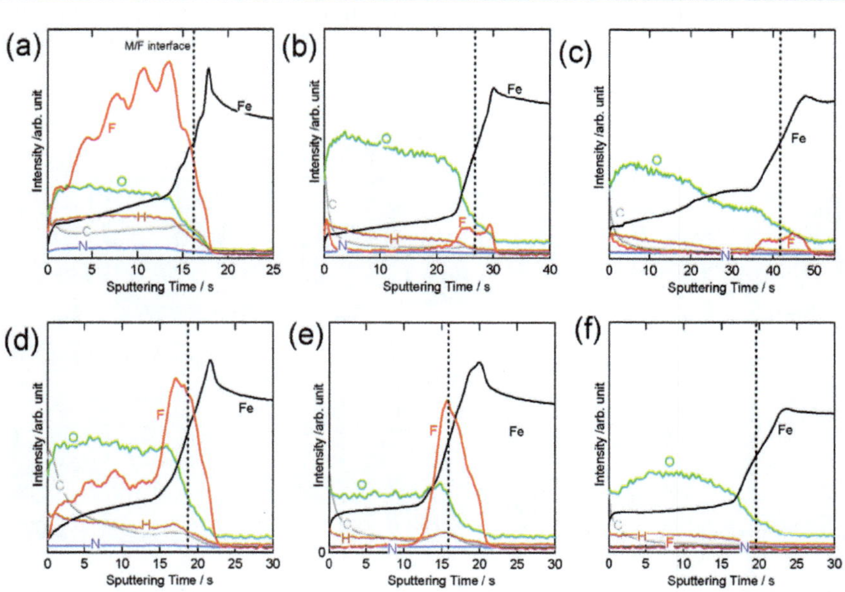

Fig. 4 GDOES depth profiles of the anodized specimens (a) before and (b–f) after thermal treatments in air at (b) 523 K and (c) 673 K and in argon at (d) 523 K, (e) 623 K and (f) 673 K.

After thermal treatment in air at 523 K (Fig. 4b), the fluoride species in the anodic film become negligible and the intensities of carbon and nitrogen are remarkably reduced; the anodic film is converted to an oxide-based layer. However, the fluoride species are still present at a depth close to the metal/oxide interface. The fluoride species located near the metal/oxide interface could not be fully removed even after thermal treatment at 673 K in air (Fig. 4c).

As discussed above, iron species diffuse outward in the thermal oxide layer during its growth in air. A new iron oxide is developed at the pore base. The formation of such oxides above the fluoride-containing layer may impede the removal of fluoride during thermal treatment.

Thus, we also carried out thermal treatment in an inert atmosphere. When thermal treatment is performed in argon at 523 K, a considerable amount of the fluoride species still remain in the anodic film (Fig. 4d). Because the amount of the fluoride species is negligible in the anodic film, except near the metal/oxide interface, after thermal treatment in air at the same temperature (Fig. 4b), it can be concluded that the fluoride removal from the anodic film is accelerated in an air atmosphere compared to that in an argon atmosphere. The removal of the carbon species is also not sufficient at 523 K in argon. The fluoride species are removed from the anodic film after thermal treatment in argon at 623 K, but not from the region near the metal/oxide interface. However, after thermal treatment in argon at 673 K, the presence of the fluoride species near the interface becomes negligible, even though thermal treatment in air cannot remove the fluoride species near the interface. The suppression of the outward diffusion of iron species to form an oxide layer above the fluoride-containing layer may be a necessary condition for the removal of fluoride from the metal/oxide interface region.

Phase transformation of the anodic film by thermal treatment was examined by XRD measurements. Fig. 5 shows the θ–2θ mode XRD patterns of the anodized iron specimens before and after thermal treatment under various conditions. The as-anodized specimen reveals reflections only from the iron substrate, indicating that the anodic film is amorphous or poorly crystalline. Reflections of crystalline oxides appear when the thermal treatment in air is performed at 573 K. The reflections become more intense with an increase in temperature. Reflections of both Fe_3O_4 and α-Fe_2O_3 are observed after thermal treatment at or above 623 K. From a comparison with the respective α–2θ mode XRD pattern, in which the relative intensity of the α-Fe_2O_3 reflections is enhanced, the porous layer formed by anodizing appears to be converted mainly to α-Fe_2O_3, whereas Fe_3O_4 is mainly formed by the thermal oxidation of the iron substrate.

When thermal treatment was performed in an argon atmosphere, both Fe_3O_4 and α-Fe_2O_3 are formed even at 523 K, although fluoride species are remained as shown in Fig. 4d. Since no oxide phases are observed at this temperature in air (Fig. 4a), it can be said that an amorphous-to-crystalline transition is accelerated in an argon atmosphere at a low oxygen partial pressure, compared to in air. At a higher temperature of 673 K, at which fluoride species are removed, only Fe_3O_4 reflections are observed. However, in the α–2θ mode XRD pattern ($\alpha = 1°$), weak α-Fe_2O_3 reflections were also detected, suggesting that the outer α-Fe_2O_3 layer may become thinner in argon.

The anodized iron specimens thermally treated at 673 K in air and argon were further examined by TEM. Fig. 6a shows the bright field transmission electron

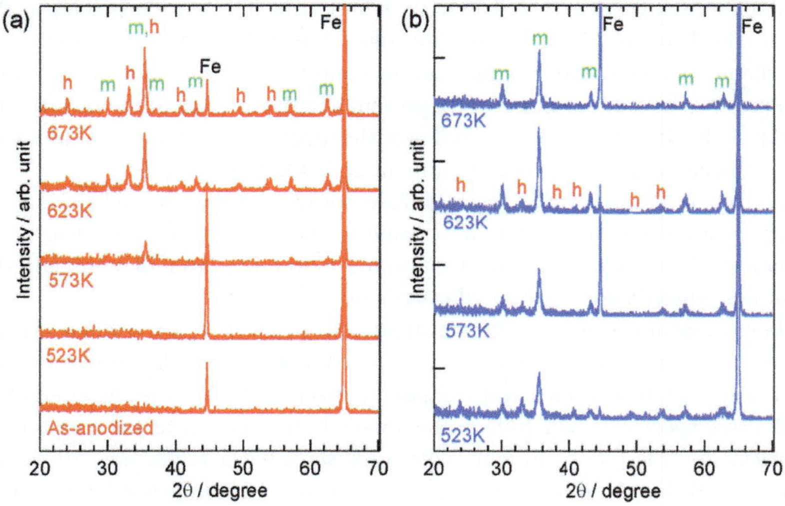

Fig. 5 XRD patterns of the anodized iron specimens before and after thermal treatment in (a) air and (b) argon at several temperatures; m: Fe_3O_4, h: α-Fe_2O_3.

micrograph of an FIB cross-section of the anodized specimen after thermal treatment in air at 673 K. Obviously, vertically aligned nanopores are developed in agreement with the SEM observations (Fig. 2c). Apparently, the oxide layer is divided into two regions from contrast. From the HAADF image (Fig. 6b), it is clear to see that each oxide layer has a different composition. A selected area electron diffraction (SAED) pattern of the outer layer with a dark appearance in the HAADF image discloses the presence of α-Fe_2O_3 in this layer. The remaining inner dark layer in Fig. 6a, corresponding to the thermal oxide layer shown in Fig. 2c, mainly consists of Fe_3O_4, derived from the SAED pattern. Because of its higher oxygen content, the outer layer shows a relatively dark appearance in the HAADF image. It can be confirmed that the anodic porous layer is converted to α-Fe_2O_3, and the Fe_3O_4 phase is mainly formed in association with the thermal oxidation of the iron substrate.

In contrast, the thermal treatment in argon results only in the formation of α-Fe_2O_3 in the outer ~300 nm thickness of the porous layer, as shown in Fig. 6d–f. The main part of the oxide layer consists of Fe_3O_4, which is in agreement with the XRD patterns. The grain size of α-Fe_2O_3 is relatively small, *i.e.* less than 50 nm. This size is limited by the thickness of the pore walls. Larger grains of Fe_3O_4 are developed in agreement with Fig. 2. The grain size of each phase is almost independent of the atmosphere of the thermal treatment.

3.3 Electropolymerization of PPy

A PPy layer was formed on the anodized iron specimens after thermal treatment. Fig. 7a shows the potential transient during electropolymerization at a constant current density of 10 A m^{-2}. The electropolymerization proceeded at a steady-state potential of 0.6 V *vs.* Ag/AgCl on the passivated iron substrate without an anodic film. The iron specimen anodized and thermally treated in argon at 673 K shows a similar electropolymerization potential. In contrast, the steady-state

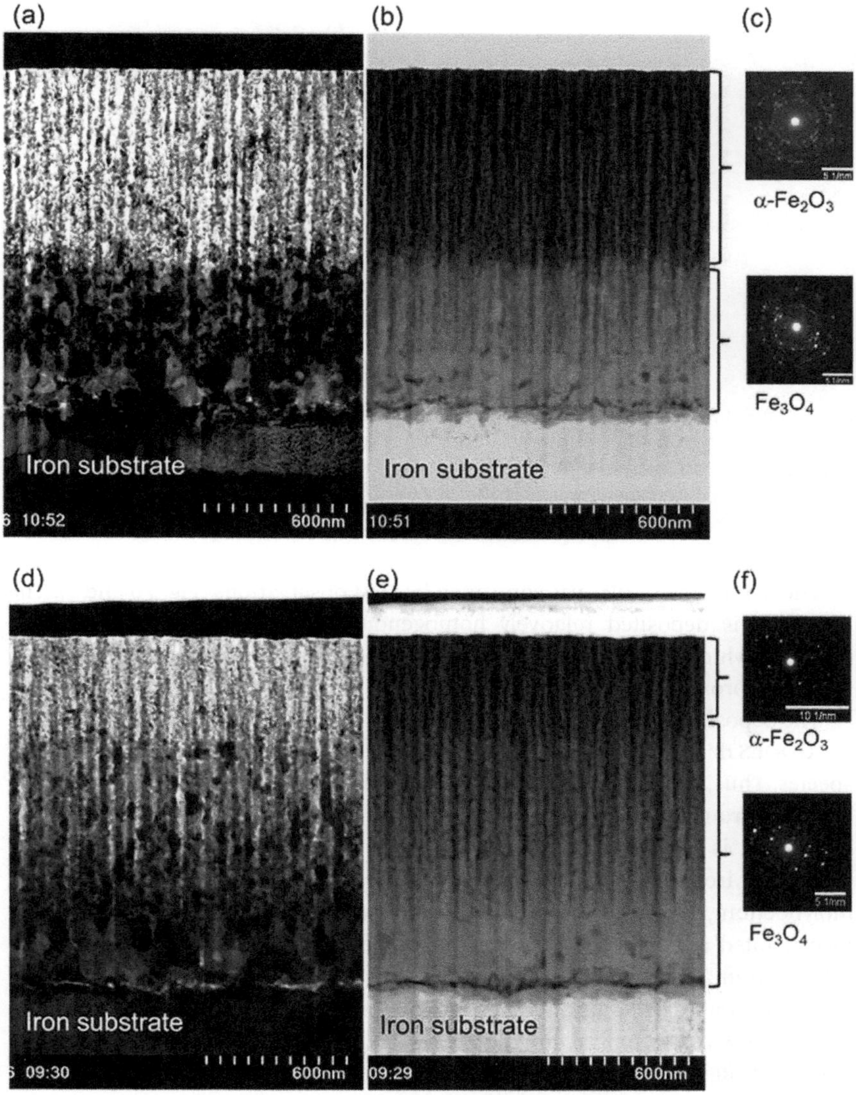

Fig. 6 Scanning transmission electron micrographs of FIB cross-sections of the anodized iron specimens after thermal treatment in (a–c) air and (d–f) argon at 673 K; (a and d) bright field images, (b and e) HAADF images and (c and f) selected area electron diffraction patterns.

potential increases to more than 1.4 V for iron specimens that were anodized and thermally treated in air. The potential increases gradually with the temperature of the thermal treatment. Such high electropolymerization potentials may be associated with the high electric resistivity of the oxide layer formed in air at elevated temperatures. When the anodized and thermally treated, in argon at or below 623 K, iron specimens were used, electropolymerization was unsuccessful to form a PPy layer. In the electropolymerization electrolyte, the oxide layer was detached because of the presence of relatively high amounts of fluoride species near the metal/oxide interface.

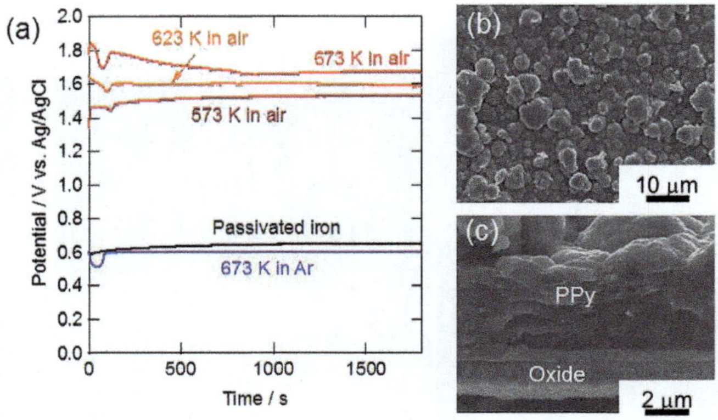

Fig. 7 (a) Potential transients during electrosynthesis of PPy on iron specimens with and without porous oxide layers thermally treated at several temperatures in air and argon; scanning electron micrographs of (b) the surface and (c) the cross-section of the PPy/porous oxide-coated specimen after thermal treatment in argon at 673 K.

The deposition of PPy was confirmed by SEM observations (Fig. 7b and c). The PPy layer is deposited relatively homogenously with a globular morphology (Fig. 7b), which is typical of electrosynthesized PPy.[12] The thickness of the PPy layer is approximately 5 μm (Fig. 7c).

The deposition of PPy in the nanopores of the iron oxide layer was evident from the GDOES depth profiles (Fig. 8). The deposited PPy contains incorporated anion species. Thus, the penetration of PPy into the nanopores of the oxide layer can be evaluated from the presence of molybdenum and phosphorus species in the oxide layer. This is evident from the depth profile of the PPy layer formed on the passivated iron (Fig. 8a). The PPy layer contains carbon, nitrogen, hydrogen, molybdenum, phosphorus and oxygen with the latter three derived from the incorporated anion species. The depth profile of phosphorus is different from molybdenum, suggesting that phosphate derived from phosphoric acid may also be incorporated in addition to molybdophosphate anions. For the anodized specimens with thermal treatments in argon at 673 K (Fig. 8b), it is obvious that PPy penetrates into the pores of the oxide layer from molybdenum, carbon and phosphorus profiles. The absence of these species in the inner oxide layer is because of the formation of the thermal oxide layer. A comparison of the relative intensities of nitrogen, molybdenum and phosphorus in Fig. 8b is interesting. With respect to the intensity of nitrogen, derived from PPy, the intensity of molybdenum, derived from incorporated anions, becomes lower in the porous anodic oxide layer compared to the outer PPy layer; the intensity of molybdenum in the outer PPy layer is higher than that of nitrogen, whereas the intensity of the former becomes approximately half of the latter in the porous anodic oxide layer. Similar result is obtained for the phosphorus intensity with respect to the intensity of nitrogen. The depth profile of the outer PPy layer in Fig. 8b is similar to that of the PPy layer in Fig. 8a, which suggests that anion incorporation is reduced in the nanopores of the anodic oxide layer.

The electropolymerization of pyrrole proceeds *via* the reactions shown in Scheme 1. The pyrrole monomer is oxidized on the anode to form a neutral

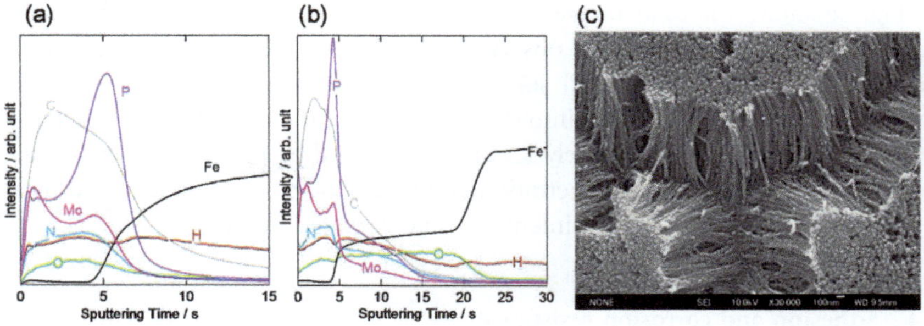

Fig. 8 GDOES depth profiles of the iron specimens with (a) PPy coating and (b) PPy/anodic oxide composite coating with thermal treatment in argon at 673 K and (c) a scanning electron micrograph of the PPy nanofibers formed in the porous anodic layer thermally treated in argon at 673 K. The nanofibers were observed after chemical dissolution of iron and the anodic oxide layer in HCl solution.

polymer. This polymer is further oxidized to produce positive charges which are compensated by doped anions. Because the neutral polymer does not have a high conductivity, the doping process is indispensable for increasing the conductivity of the polymer. A lower concentration of doping anions in the nanopores of the porous anodic oxide layer, in comparison to the outer PPy layer, suggests that either the available anions are limited in the nanopores electrolyte or only limited doping occurs during electropolymerization in the nanopores.

Different concentrations of anions in the nanopores and bulk solution were recently demonstrated in porous silicon.[32] Depending on the concentration and size of the anions, the deposition of platinum in the nanopores of silicon was markedly influenced. Therefore, it is possible that the concentration of the anions in the nanopores of the porous anodic layer in the present study is also different from that in the bulk solution. For a more detailed understanding, the influence of the type of anions, pore size and pore wall hydrophobicity is under investigation.

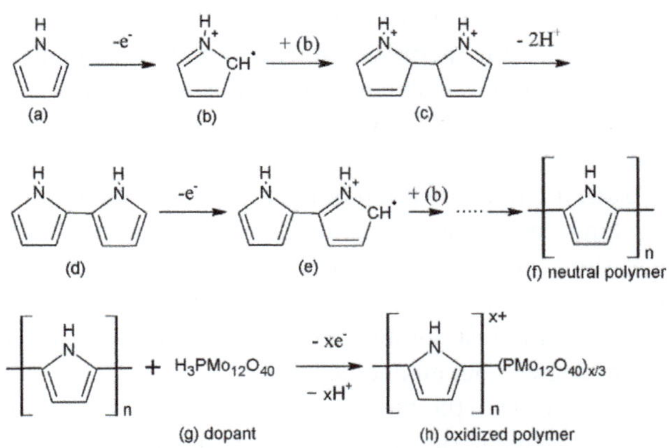

Scheme 1 Reactions during anodic oxidation of the pyrrole monomer in the presence of $H_3PMo_{12}O_{44}$.

Fig. 8c shows the SEM image of the PPy nanofibers deposited in the oxide nanopores. The nanofibers were observed after chemical dissolution of the oxide layer and iron substrate in HCl solution. The presence of the PPy nanofibers indicates the deposition of PPy into the nanopores of the oxide layer. The length of the nanofibers is approximately 1.2 μm, corresponding to the thickness of the porous oxide layer. A similar penetration of PPy into the porous anodic oxide layer was also confirmed for the specimens thermally treated in air.

3.4 Adhesion and corrosion resistance

The adhesion of the coatings to the iron substrate was examined by a cross-cut test. The PPy layer deposited on the passivated iron revealed detachment in more than 60% of the area, indicating insufficient adhesion of the PPy layer to the substrate. For the PPy/anodic oxide composite coatings thermally treated in air, adhesion is improved but detachment always occurred at the substrate/oxide layer interface mainly due to the presence of the fluoride species near the interface. In contrast, detachment at the substrate/oxide layer interface never occurred for the coatings thermally treated at 673 K. As discussed above, the presence of fluoride in this specimen is negligible, suggesting that the removal of the fluoride species is of crucial importance for improving the adhesion of coatings.

The corrosion tests of the PPy and PPy/anodic oxide coated iron specimens were examined in 3.5% NaCl solution. The open circuit potential (OCP) was measured during the immersion of the coated and non-coated specimens (Fig. 9). The non-coated iron specimen shows an initial potential of −0.22 V *vs.* Ag/AgCl, which is within the active dissolution region. This potential further decreases to a steady-state potential of −0.65 V *vs.* Ag/AgCl. In contrast, the PPy-coated specimen exhibits a considerably higher potential of 0.2 V *vs.* Ag/AgCl at the commencement of immersion, which is in the passive region. This relatively high potential is because of the oxidative properties of the PPy layer. The high OCP of >0.1 V *vs.* Ag/AgCl is maintained for a certain period of time, during which no obvious corrosion occurs. Corrosion occurs after an abrupt decrease in the OCP to −0.2 V *vs.* Ag/AgCl. Thus, the durability of the PPy coatings can be evaluated from the duration for which a high OCP is maintained.

The PPy-coated, passivated iron specimen maintains the passive and protective state for ~300 ks. The protective duration is shorter for the PPy/anodic oxide composite coatings when the thermal treatment is performed in air at or below 623 K. The highest durability is obtained for the composite coating thermally treated in argon at 673 K, with negligible fluoride species. Because all the composite coatings with thermal treatment in air contain the fluoride species near the metal/oxide interface, their removal may be one of the important factors influencing the durability of the coating. In the presence of the fluoride species, the adhesion of the coatings is not directly correlated with the durability of corrosion protection because the composite coating with thermal treatment in air at 623 K, showing the shortest durability, exhibited better adhesion in comparison to the PPy coating without an anodic layer.

The composite coating with thermal treatment in argon at 673 K exhibits a twice as long durability compared to the PPy coating without an anodic oxide layer. The improved adhesion of the coating on the substrate may impede the penetration of the corrosive solution into the coating/metal substrate interface. In addition, the

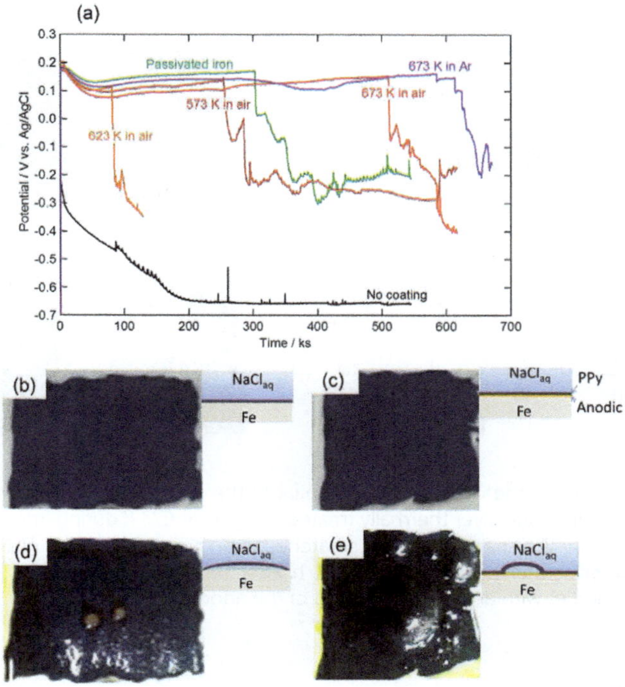

Fig. 9 (a) Change in the open circuit potential of the non-coated and coated iron specimens during immersion in 3.5 wt% NaCl solution; PPy coating was performed on anodized iron substrates with thermal treatment in air and argon at several temperatures and a passivated iron substrate. Specimen photos and schematic illustrations (b and d) before and (c and e) after immersion of (b and c) the PPy-coated specimen and (d and e) the PPy/anodic oxide composite coated specimen with thermal treatment in argon at 673 K.

presence of the thermal oxide barrier layer and the porous oxide layer together with the PPy layer may also improve the barrier nature of the coating. As shown in Fig. 9b and c, the PPy-coated specimen without an anodic oxide layer blisters over the entire surface, whereas blisters form in a limited area for the composite coating thermally treated in argon at 673 K. The presence of the adherent anodic oxide layer appears to suppress the expansion of corrosion over the entire surface.

During immersion in the NaCl solution, the PPy-coated specimen and the PPy/anodic oxide composite-coated specimen with thermal treatment in argon at 673 K were scratched (Fig. 10). After scratching, the OCP decreases steeply and corrosion commenced at the scratched region for both the specimens. However, a significant difference appears on the coatings within 1 h after scratching. For the PPy-coated specimen without anodic oxide, the coating is detached from the substrate on the entire surface exposed to the corrosive solution (Fig. 10b). Thus, the corrosion expanded on the entire surface within 1 h after scratching. Because of insufficient adhesion of PPy, the corrosive solution may penetrate into the coating/metal interface, accelerating the detachment and corrosion. In contrast, the composite coating is still adherent to the substrate even after scratching and the corrosion is apparently limited to the scratched region (Fig. 10c). The extension of corrosion on the entire surface is effectively suppressed by the composite coating with better adhesion to the substrate.

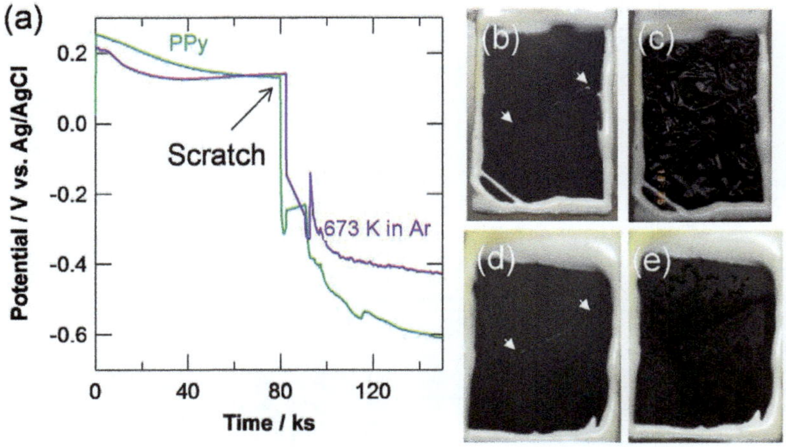

Fig. 10 (a) Change in the open circuit potential of the PPy-coated specimens with and without the anodic oxide layer thermally treated in argon at 673 K during immersion in 3.5 wt% NaCl solution. The specimens were scratched during immersion. The photos show (b and d) just after scratching and (c and e) 1 h after the scratching for the PPy-coated specimens (d and e) with and without (b and c) the anodic oxide layer.

4 Conclusions

In summary, the present study demonstrates the beneficial role of the porous anodic oxide layer in corrosion protection of iron by PPy coating. The PPy/anodic oxide composite coating on iron remarkably improves the durability of the protective coating in NaCl solution compared to the PPy single coating. This improved adhesion to the substrate may be one of the reasons for the improved durability. The nanoporous anodic films were formed in ethylene glycol electrolyte containing NH_4F and H_2O. The resultant anodic film is amorphous and contains a relatively high concentration of fluoride, such that the as-formed film is chemically unstable. Thermal treatment converts the amorphous material to a mixture of α-Fe_2O_3 and Fe_3O_4, with the former forming the outer layer. However, thermal treatment in air cannot remove the fluoride near the metal/oxide interface. The formation of thermal oxide by outward diffusion of iron species inhibits the removal of the fluoride present beneath the developed oxide layer. Almost complete removal of fluoride is achieved by thermal treatment in argon at 673 K. The composite coating with this thermal treatment exhibits the longest durability for corrosion protection.

Acknowledgements

A part of this work was supported by the "Nanotechnology Platform" Program of the Ministry of Education, Culture, Sports, Science and Technology (MEXT), Japan.

References

1 B. Wessling, *Adv. Mater.*, 1994, **6**, 226–228.
2 N. Ahmad and A. G. MacDiarmid, *Synth. Met.*, 1996, **78**, 103–110.

3 J. C. Wu and J. Pawliszyn, *Anal. Chim. Acta*, 2004, **520**, 257–264.
4 A. Bhattacharya, A. De and S. Das, *Polymer*, 1996, **37**, 4375–4382.
5 T. Ohtsuka, T. Wakabayashi and H. Einaga, *J. Electroanal. Chem.*, 1994, **377**, 107–114.
6 F. Li, G.-x. Li, J. Zeng and G.-h. Gao, *J. Appl. Polym. Sci.*, 2014, **131**, 40602.
7 M. B. Gonzalez and S. B. Saidman, *Corros. Sci.*, 2011, **53**, 276–282.
8 I. L. Lehr and S. B. Saidman, *Electrochim. Acta*, 2006, **51**, 3249–3255.
9 U. Rammelt, L. M. Duc and W. Plieth, *J. Appl. Electrochem.*, 2005, **35**, 1225–1230.
10 G. Ilangovan and K. C. Pillai, *J. Solid State Electrochem.*, 1999, **3**, 474–477.
11 D. Kowalski, M. Ueda and T. Ohtsuka, *J. Mater. Chem.*, 2010, **20**, 7630–7633.
12 D. Kowalski, M. Ueda and T. Ohtsuka, *Corros. Sci.*, 2007, **49**, 1635–1644.
13 T. Ohtsuka, M. Iida and M. Ueda, *J. Solid State Electrochem.*, 2006, **10**, 714–720.
14 A. Ghicov and P. Schmuki, *Chem. Commun.*, 2009, 2791–2808.
15 F. Beck, R. Kaus and M. Oberst, *Electrochim. Acta*, 1985, **30**, 173–183.
16 M. Pourbaix, *Atlas of Electrochemical Equilibria in Aqueous Solutions*, NACE, Houston, 1974.
17 H. E. Prakasam, O. K. Varghese, M. Paulose, G. K. Mor and C. A. Grimes, *Nanotechnology*, 2006, **17**, 4285–4291.
18 S. P. Albu, A. Ghicov and P. Schmuki, *Phys. Status Solidi RRL*, 2009, **3**, 64–66.
19 T. J. LaTempa, X. J. Feng, M. Paulose and C. A. Grimes, *J. Phys. Chem. C*, 2009, **113**, 16293–16298.
20 R. R. Rangaraju, A. Panday, K. S. Raja and M. Misra, *J. Phys. D: Appl. Phys.*, 2009, **42**, 135303.
21 H. Habazaki, Y. Konno, Y. Aoki, P. Skeldon and G. E. Thompson, *J. Phys. Chem. C*, 2010, **114**, 18853–18859.
22 R. R. Rangaraju, K. S. Raja, A. Panday and M. Misra, *Electrochim. Acta*, 2010, **55**, 785–793.
23 S. K. Mohapatra, S. E. John, S. Banerjee and M. Misra, *Chem. Mater.*, 2009, **21**, 3048–3055.
24 Z. H. Zhang, M. F. Hossain and T. Takahashi, *Appl. Catal., B*, 2010, **95**, 423–429.
25 K. Y. Xie, J. Li, Y. Q. Lai, W. Lu, Z. A. Zhang, Y. X. Liu, L. M. Zhou and H. T. Huang, *Electrochem. Commun.*, 2011, **13**, 657–660.
26 Q. Xu, G. Meng, F. Han, X. Zhao, M. Kong and X. Zhu, *Mater. Lett.*, 2009, **63**, 1431–1434.
27 K. Wagatsuma, K. Hirokawa and N. Yamashita, *Anal. Chim. Acta*, 1996, **324**, 147–154.
28 Y. Konno, E. Tsuji, P. Skeldon, G. E. Thompson and H. Habazaki, *J. Solid State Electrochem.*, 2012, 1–10.
29 G. E. Thompson, *Thin Solid Films*, 1997, **297**, 192–201.
30 H. Habazaki, K. Fushimi, K. Shimizu, P. Skeldon and G. E. Thompson, *Electrochem. Commun.*, 2007, **9**, 1222–1227.
31 S. Yoriya, M. Paulose, O. K. Varghese, G. K. Mor and C. A. Grimes, *J. Phys. Chem. C*, 2007, **111**, 13770–13776.
32 K. Fukami, R. Koda, T. Sakka, Y. Ogata and M. Kinoshita, *J. Chem. Phys.*, 2013, **138**, 094702.

Faraday Discussions

PAPER

Graphene as an anti-corrosion coating layer

Line Kyhl,[a] Sune Fuglsang Nielsen,[b] Antonija Grubišić Čabo,[a] Andrew Cassidy,[b] Jill A. Miwa[b] and Liv Hornekær*[ab]

Received 12th December 2014, Accepted 19th December 2014
DOI: 10.1039/c4fd00259h

Graphene, a single layer of carbon atoms arranged in an aromatic hexagonal lattice, has recently drawn attention as a potential coating material due to its impermeability, thermodynamic stability, transparency and flexibility. Here, the effectiveness of a model system, a graphene covered Pt(100) surface, for studying the anti-corrosion properties of graphene, has been evaluated. Chemical vapour deposition techniques were used to cover the single crystal surface with a complete layer of high-quality graphene and the surface was characterised after exposure to corrosive environments with scanning tunnelling microscopy (STM) and Raman spectroscopy. Graphene covered Pt samples were exposed to: (i) ambient atmosphere for 6 months at room temperature and 60 °C for 75 min, (ii) Milli-Q water for 14 hours at room temperature and 60 °C for 75 min, and (iii) saltwater (0.513 M NaCl) for 75 min at room temperature and 60 °C. STM provides atomic resolution images, which show that the graphene layer and the underlying surface reconstruction on the Pt(100) surface remain intact over the majority of the surface under all conditions, except exposure to saltwater when the sample is kept at 60 °C. Raman spectroscopy shows a broadening of all graphene related peaks due to hybridisation between the surface Pt d-orbitals and the graphene π-bands. This hybridisation also survives exposure to all environments except saltwater on the hot surface, with the latter leading to peaks more representative of a quasi free-standing graphene layer. A mechanism explaining the corrosive effect of hot saltwater is suggested. Based on these experiments, graphene is proposed to offer protection against corrosion in all tested environments, except saltwater on a hot surface, and Raman spectroscopy is proposed as a useful method for indirectly assessing the chemical state of the Pt surface.

1 Introduction

Since the first demonstration of the remarkable electronic properties of graphene,[1] research efforts have expanded rapidly to include graphene in a range of

[a]*Interdisciplinary Nanoscience Center, Aarhus University, Gustav Wieds Vej 14, 8000 Aarhus C, Denmark. E-mail: line1@inano.au.dk; Tel: +45 871 55695*

[b]*Department of Physics and Astronomy, Aarhus University, Ny Munkegade 120, 8000 Aarhus C, Denmark. E-mail: liv@phys.au.dk; Tel: +45 871 56336*

applications, for example, ultrafast electronics,[1] sensors with single molecule sensitivity[2] or highly efficient solar cells.[3,4] More recently, research has expanded to include graphene-based anti-corrosion coatings.[5]

An ideal material for anti-corrosion coatings for metals should possess a range of properties including being lightweight, atomically thin, impermeable, inert, wear resistant and mechanically strong. Graphene, a one-atom thin layer of carbon atoms arranged in a honeycomb lattice, satisfies all of these requirements.

Graphene transmits 97% of irradiated light across a broad range of the electromagnetic spectrum. This can be considered an advantage in anti-corrosion coating applications where maintaining the surface optical appearance is desirable. Graphene exhibits high flexibility,[6] which can follow the curvature or roughness of the surface of an underlying substrate to a high extent. Flexibility is useful not only in coating applications but also for flexible electronics.[7,8] Graphene is also thermodynamically stable, with stability arising from long range delocalisation of the π-electron cloud, giving rise to an extended aromatic network of C=C bonds across the entire basal plane. This chemical inertness is one of the strongest drivers justifying the use of graphene as an anti-corrosion coating. For example, graphene exposed to high pressures of superheated water, a highly corrosive cocktail, has proven more chemically stable than diamond exposed to the same conditions.[9] Graphene is able to resist reaction, breakage and penetration by superheated water when this is encapsulated in highly pressurized nano-bubbles formed at a graphene–diamond interface. The basal plane is not only chemically inert but is also impermeable. Not even He atoms can penetrate a single layer of defect-free graphene.[10]

The combination of properties described above has ignited research efforts aimed at testing the limits of graphene coatings. A range of metallic substrates has been considered and, similarly, a range of methods has been used for producing the graphene coating. Examples of studies where the graphene is shown to have a definite positive effect on the protection of metals from corrosive environments, such as air, seawater, O_2 or CO, include Cu,[11–14] Pt,[15–17] Ru,[18] Ni,[12,19,20] Fe[21] and stainless steel 304.[22] These investigations have included both graphene grown directly on the substrate by CVD and graphene transferred after growth. Consequently, graphene layers of differing quality are often compared in the literature, producing somewhat contradictory results regarding the use of "graphene" as an anti-corrosion coating. For example, reduced graphene oxide, often considered to be low-quality graphene, has been shown to protect Cu from corrosion upon exposure to a saline environment.[23] Yet, in a separate study, it has been suggested that a defected graphene coating on Cu can lead to enhanced corrosion of a Cu surface after long term exposure to air.[24,25] The authors of the latter study suggested that oxides on an uncovered Cu surface block electrical contact between unreacted areas and inhibit wet corrosion. The graphene coating, however, acted as a continuous source of electrons and provided an electrical contact, enabling corrosion of the entire surface. Furthermore, the Cu oxide layer below the coating was stressed, since it developed with a non-uniform thickness and this led to cracks, enabling pitting and crevice corrosion. In this context it is interesting to know that graphene oxide, an insulator, has been shown to protect both Cu[26] and mild carbon steel[27] from oxidation. This may be because graphene oxide, although chemically reactive, opaque and stiff, does not conduct the electrons responsible for some of the galvanic corrosion, which has been observed to occur, in some cases, for conducting graphene coatings.

The majority of the graphene produced for coating applications is produced following chemical vapour deposition protocols on metals, since this gives large area graphene often of high quality and with good substrate adhesion. Such graphene sheets can also be transferred to yield a graphene coating layer on other desired surfaces.[5,28,29] Often, however, the transferring method deteriorates the graphene quality by introducing numerous defects. Graphene coatings can be improved by growing or transferring more layers on top of each other to seal the defected areas. Thus, the possibility of using multilayer graphene as a coating is an ongoing area of study.[5,30,31] Such multilayer systems, however, are not suitable for all substrates and it can be difficult to accurately reproduce the same quality of graphene coating repeatedly.

The synthesis of a high quality graphene coating, using conventional chemical vapour deposition (CVD) methods, requires substrate temperatures >800 °C. For more industrially relevant scenarios, recent studies have shown the synthesis of graphene coatings directly on steel and Fe surfaces to be possible.[22,32,33] The high growth temperatures are, however, in many cases problematic when dealing with alloys and steels, due to phase changes and resulting reductions in mechanical properties. Research has therefore focused on finding an optimal low temperature growth procedure in order to improve the applicability for commercial applications in the coating industry. In the pursuit of a low temperature growth mechanism the use of large carbon based molecular building blocks, such as coronene or naphthalene, has proven to be advantageous.[20,34–37] Nonetheless, controlled direct graphene growth on Fe-based substrates is difficult compared to more catalytically active transition metals, and often compromises in reproducibility and graphene quality have to be made.

DFT studies have shown that a pristine, defect-free graphene layer offers an impenetrable barrier to oxygen molecules and atoms.[38] The authors calculated a lower energy barrier for reversibly binding oxygen atoms to graphene than for the same atoms to diffuse through the graphene basal plane and reach the metal substrate below. However, the diffusion barrier through the graphene could be significantly lowered by the presence of defects, such as holes or atomic vacancies, in the graphene lattice. In line with this, experimental studies demonstrate that oxygen atoms can intercalate beneath a high quality graphene layer grown on an Ir(111)[39–42] substrate. The intercalation requires slightly elevated temperatures that are thought to provide the energy required to overcome the diffusion barrier at defect sites and graphene edges allowing the oxygen atoms to diffuse beneath the graphene sheet at these areas.

Several studies highlight that the protective properties of a single graphene sheet are compromised by the presence of wrinkles, grain boundaries and atomic defects.[5,43,44] These areas constitute reactive sites where elements or molecules have an increased possibility of intercalating, thus reaching the substrate surface. For example, it has been shown that oxygen molecules can dissociate and bind to carbon atoms at defect sites, if those carbon atoms constitute an incomplete hexagon in the graphene lattice.[38] Once oxygen atoms are bound, the energy barrier preventing the oxygen to reach and bind to the surface below the graphene is low, and thereafter intercalation proceeds rapidly. Similarly, wrinkles in a graphene sheet stretch the carbon–carbon bonds in these regions of high curvature, thus increasing the chemical reactivity of the graphene π-bands in these areas. Breaking a C=C bond under strain is energetically more feasible than reactions on the flat basal plane. The thermal expansion coefficient of metallic substrates is larger than that of the graphene and wrinkles frequently arise during cooling of a system after the high temperature growth, typical of chemical vapour deposition techniques. The challenge, therefore, is to prevent the formation of sites facilitating the intercalation of oxidizing species, since this may lead to a reaction with the substrate below and thereby increase the rate of corrosion.

While reducing the defect density and/or passivating vacancy sites that act as intercalation pathways is key to increasing corrosion protection with graphene coatings,[45] it is not trivial to predict the quality of graphene required. The mechanism of protection will differ with the substrate–graphene interaction and with the environment to which the system is to be exposed.[17]

The challenge, then, is to combine high resolution characterisation methods, such as scanning tunnelling microscopy with large scale, higher throughput characterisation methods, such as Raman spectroscopy. STM can provide atomic resolution of the graphene lattice and at the same time provides a topographic overview of the surface, which can be used for assessing physical changes as a function of surface reactivity. Raman spectroscopy gives data averaged over a larger area, limited in this case by the spot size of the laser on the sample. Although Raman spectroscopy on metal surfaces is usually not favourable because of the screening effect of the metal, graphene samples on metal substrates are well studied and several graphene–metal systems are characterised in the literature. When combined, the three graphene peaks, the D, G and 2D peaks, provide a chemical fingerprint of the graphene lattice, and their position and intensities are strongly correlated to the chemical environment of the graphene sheet. In some cases, when the graphene sheet strongly interacts with the underlying metal surface, such as graphene on Ir, the Raman signal is completely annihilated because the Raman active graphene modes hybridise too strongly with the transition metal surface d-orbitals.[46] In cases where this hybridisation is not as strong, such as graphene on Cu, Raman spectroscopy can be used to determine the quality of a graphene layer. The graphene–Pt(100) system exhibits fairly strong hybridisation and Raman peaks from graphene modes that, although visible, appear strongly attenuated and blue-shifted. In this article we propose to use this effect to establish the chemical state of the Pt(100) surface beneath a graphene layer. Since lifting of the Pt(100) reconstruction is associated with the formation of surface oxides, this lifting will simultaneously reduce hybridisation between the Pt surface d-orbitals and the graphene π-band. Hence, graphene becomes more free-standing and thus the Raman peaks will be red-shifted.

Previous work from our group has investigated the interaction between the Pt(100) surface and graphene with scanning tunnelling microscopy, and shown the system to be useful for investigating the anti-corrosion properties of graphene on an atomic level.[16,17] This is facilitated by the distinct striped surface reconstruction, which serves as a fingerprint of the most stable form of a clean unreacted Pt(100) surface.[47] Upon oxidation of the surface, the reconstruction is lifted and the excess Pt atoms form clusters lying on top of a truncated bulk (100) arrangement of the remaining surface atoms. The surface reconstruction and its lifting can be followed by scanning tunnelling microscopy (STM), even when covered by a graphene layer.[16,17,48] Previous STM studies on this system have demonstrated that graphene acts as an effective anti-corrosion coating against O_2 and H_2S, while CO was shown to intercalate at the graphene–Pt(100) interface for pressures above 10^{-6} mbar.[16,17] Here, we test the stability of the graphene–Pt(100) system towards air, water and saltwater while keeping the sample at either room temperature or at 60 °C. For the first time, we also relate the lifting of the Pt(100) surface reconstruction, following the oxidation of the Pt surface, to changes in peak positions and intensities in Raman spectroscopy. The graphene coating was observed to be effective in protecting the Pt surface from reactions with air and water, while it was observed to fail under saltwater at 60 °C, where a lifting of the Pt(100) surface reconstruction was observed by STM and *via* a shift in Raman peak intensity and position.

2 Experimental

Graphene samples were prepared in an ultra-high vacuum chamber with a base pressure of 5×10^{-10} mbar. Pt(100) single crystal samples were cleaned by several cycles of argon sputtering followed by flash annealing to 900 °C. For graphene growth, the sample was maintained at 700 °C and exposed to ethylene gas at a partial pressure of 5×10^{-7} mbar for 40 min. The sample was subsequently annealed to 800 °C in UHV for 5 min. Low-energy electron diffraction (LEED) (Omicron Nanotechnology GmbH) was performed to check the quality of the prepared graphene. Scanning tunnelling microscopy was performed in air using an Aarhus-type STM[49] with a cut Pt/Ir tip. The distances measured by STM are accurate to within 10% of the quoted values. These uncertainties are due to piezo creep and/or thermal drift.

Raman spectra were obtained using a Renishaw Invia Reflex Raman Spectroscopy System. All spectra were obtained using a 457 nm Ar-ion laser with a 10 s acquisition time and 50 accumulations per spectrum through a 100× objective. Spectra were taken at several points for each sample and the spectra shown here are representative thereof.

To test the stability of graphene as an anti-corrosion coating, the samples were exposed to: (i) ambient conditions (22 °C in air), (ii) ultra-pure water (type 1, Milli-Q grade), and (iii) a seawater analogue prepared by dissolving pure NaCl in Milli-Q water (0.513 M NaCl). Where samples were heated to 60 °C, a hot plate was used. A k-type thermocouple mounted on the back of a dummy sample of the same thickness, also placed on the hot plate, was used to measure the temperature.

For short term exposures to solutions, *i.e.* <75 min, the solutions were continuously added to maintain a full surface coverage. For longer term exposures to Milli-Q water, *i.e.* 14 hours, a humidity chamber was built and used to reduce the rate of water evaporation during the experiment.

3 Results

A graphene layer was grown on a clean Pt(100) surface following the recipe described in the Experimental section. This pristine graphene–Pt system was characterised and the results are shown in Fig. 1. Fig. 1a shows a low-energy electron diffraction (LEED) pattern collected immediately after growth under ultra-high vacuum conditions. Fig. 1b shows a representative scanning tunnelling microscopy image and Fig. 1c a Raman spectrum, both collected under ambient conditions. In the LEED pattern in Fig. 1a both the striped pattern, resulting from Hex-R0.7° reconstruction of the Pt(100) surface, and a ring, arising from the graphene layer, are visible. The weak interaction between graphene and the reconstructed Pt(100) surface does not drive a preferred angular orientation between different graphene domains, resulting in the circular diffraction pattern visible in the LEED experiments.[50] The STM image in Fig. 1b also depicts the Pt surface reconstruction and the graphene layer, the latter being visible with atomic resolution in the inset image. The Pt surface reconstruction appears as rows of atoms running from top to bottom in the larger scale image. This is also visible to some extent in the inset. The visibility of the reconstruction under ambient conditions demonstrates that the graphene layer protects the underlying Pt surface from immediate reactions with the ambient environment. Raman spectroscopy, taken of the same sample under the same conditions, reveals four distinct peaks, see Fig. 1c. Three of these peaks are related to the graphene layer:

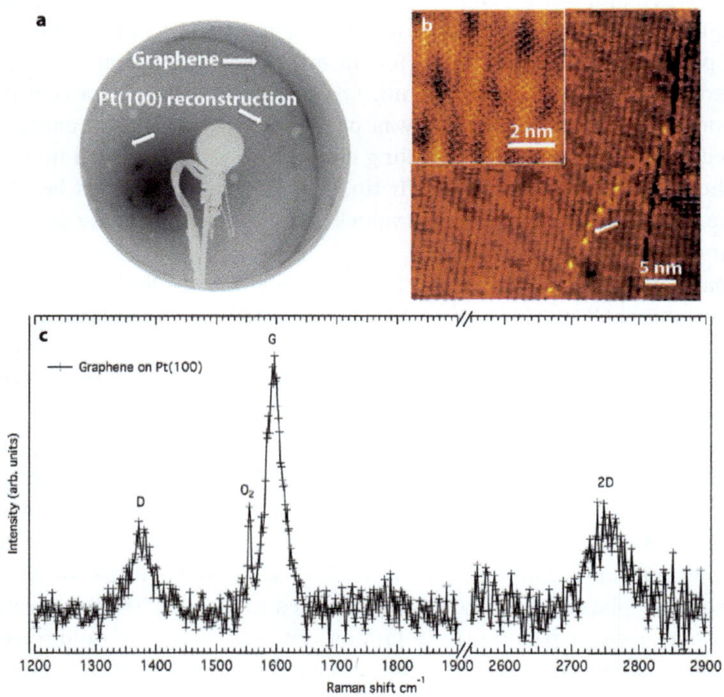

Fig. 1 Characterization of a pristine graphene–Pt(100) sample. (a) LEED pattern ($E_0 = 75$ eV) for a graphene covered Pt(100) surface recorded under UHV. Points representing the striped Hex-R0.7° reconstruction are marked with arrows. The dark circle representing graphene domains is also marked with an arrow. (b) An STM image recorded under ambient conditions showing the Pt(100) surface reconstruction depicted as lines running almost vertically. A graphene grain boundary is marked by an arrow; $I_t = 0.640$ nA, $V_t =$ 275.3 mV. The inset shows an atomic resolution image of the graphene covering the reconstructed Pt surface; $I_t = 1.080$ nA, $V_t = 59.2$ mV. (c) A Raman spectrum of the pristine graphene–Pt(100) sample. The spectrum shows three peaks relevant for the graphene: D, G and 2D, and one peak originating from O_2 present in the air. The peak positions are 1375 cm^{-1}, 1596 cm^{-1}, 2753 cm^{-1} and 1556 cm^{-1}, respectively.

(i) a low intensity D peak at 1376 cm^{-1}, (ii) a G peak at 1596 cm^{-1}, and (iii) a 2D peak at 2753 cm^{-1}. The last peak at 1556 cm^{-1} can be attributed to O_2 from the air. These peaks have been assigned on the basis of calculations performed on graphene on a Pt substrate.[46] By fitting the peaks with Lorentzian functions, a D/G intensity ratio of 0.32 was measured. Using the empirical relation $L_D^2(nm^2) = (1.8 \pm 0.5) \times 10^{-9}\lambda_L^4(I_D/I_G)^{-1}$, where L_D is the distance between defects, λ_L is the laser wavelength in nm, and I_D and I_G are the peak intensities of the D and G peaks, respectively, yields $L_D = 16$ nm.[51] The equation is valid assuming a defect distance larger than 10 nm, which is supported by the analysis of the STM images.

Following the preparation and characterisation of the pristine graphene–Pt(100) system, stability tests were performed in order to test the limits of graphene as an anti-corrosion protective barrier. Experiments may be divided into three categories testing against: (i) Air: a 6 month study at room temperature and a 75 min test at 60 °C; (ii) water: a 14 hour test at room temperature and a 75 min test at 60 °C; (iii) saltwater: a 75 min test at room temperature and a 75 min test at 60 °C. Scanning tunnelling microscopy was performed for all experiments (Fig. 2–4) and Raman spectra were obtained from samples exposed to 75 min tests (Fig. 5). The coating was deemed successful if the STM images revealed the survival of the Pt(100) surface reconstruction over most of the surface and the Raman shift for graphene peaks did not move to lower wave numbers, relative to the shifts measured for the pristine graphene–Pt(100) system. As will be shown below, the graphene coating was successful in all cases aside from the 75 min 60 °C saltwater test.

Fig. 2 shows the STM results from the stability test against air. A pristine graphene–Pt(100) system was left in ambient air for a period of 6 months and was periodically characterised. Fig. 2 includes representative images after (a) 8 days, (b) 34 days, and (c) and (d) 6 months. In Fig. 2a–c it is evident that the reconstruction of the Pt(100) surface, represented in the images as vertical lines running from top to bottom, remains predominantly intact. In Fig. 2b and c the bright and dark areas represent two different terraces, separated by a step edge running from top to bottom, on the Pt surface. In Fig. 2c there are some areas were the Pt reconstruction has begun to lift, represented by the bright protrusions in the image. An atomically resolved image of an area with lifted reconstruction is shown in Fig. 2d. From this image it is evident that the graphene carbon lattice remains intact despite the lifting of the reconstruction. The larger light and dark areas of contrast in this image reflect the topography of the underlying Pt surface. The Pt(100) surface increases in roughness in response to the lifting of the reconstruction.[16] In a separate experiment, a graphene-Pt(100) sample was heated under ambient air conditions to 60 °C. The STM results are shown in Fig. 2e and f. In Fig. 2e multiple flat terraces on the graphene covered Pt(100) surface are shown. The reconstruction is observed to be intact, visible as lines running vertically on the terraces. In Fig. 2f the atomic resolution of the graphene covering the intact reconstruction is shown.

The stability of the graphene coating against water was also tested and the STM results are presented in Fig. 3. Fig. 3a shows a representative STM image following exposure of a pristine graphene–Pt(100) sample to Milli-Q water for 14 hours at room temperature. The image has been filtered to remove background noise. A spatial fast fourier transform (FFT) of the (unfiltered) image is included in the inset with white arrows pointing out the spots arising from the intact

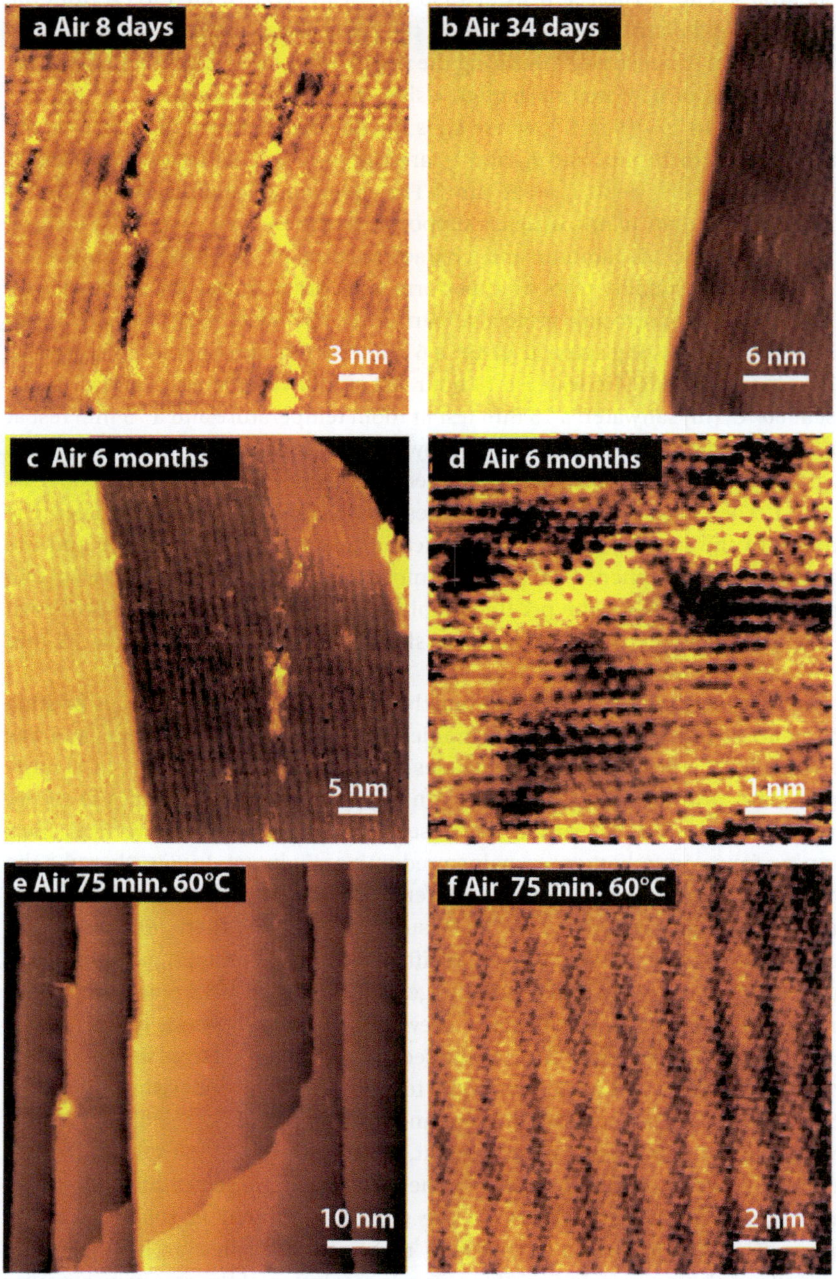

Fig. 2 STM images of graphene–Pt(100) following air exposure at room temperature (a–d) and at 60 °C (e and f). (a) After 8 days: $I_t = 0.960$ nA, $V_t = 4.0$ mV; (b) after 34 days: $I_t = 0.350$ nA, $V_t = 55.8$ mV; (c and d) after 6 months: (c) $I_t = -0.570$ nA, $V_t = -630.0$ mV and (d) shows an atomic resolution image in an area of lifted reconstruction; $I_t = -1.080$ nA, $V_t = -374.1$ mV. (e) After 75 min air exposure at 60 °C: showing intact reconstruction running as vertical lines on the Pt terraces; $I_t = 0.280$ nA, $V_t = 131.8$ mV. (f) Atomic resolution of the graphene coating covering the Pt(100) reconstruction; $I_t = 0.750$ nA, $V_t = 24.4$ mV.

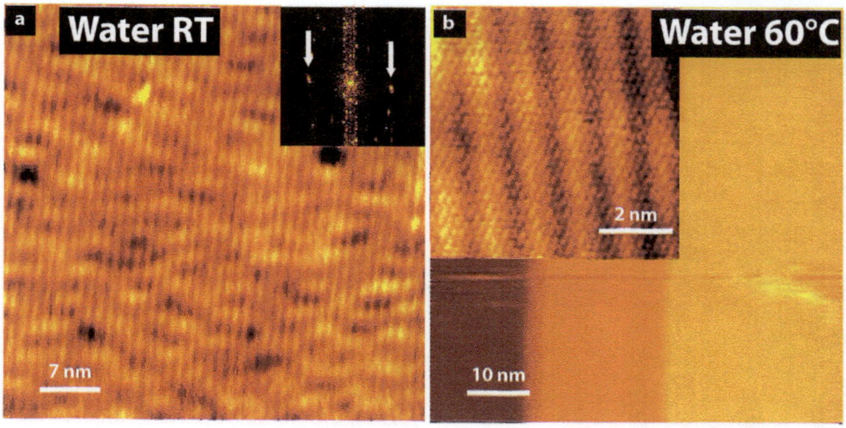

Fig. 3 STM image of graphene–Pt(100) exposed to water for (a) 14 hours at room temperature; $I_t = 0.260$ nA, $V_t = 18.6$ mV. The noise has been filtered from this image to reveal the Pt(100) reconstruction. A fast fourier transform of the image is depicted in the inset (upper right corner) with arrows marking the points corresponding to the reconstruction. (b) STM image of graphene–Pt(100) exposed to water for 75 min at 60 °C showing the intact reconstruction running as vertical lines on the Pt terraces in the image; $I_t = -0.260$ nA, $V_t = -1277.1$ mV. The inset shows, with atomic resolution, the graphene coating covering the intact Pt reconstruction; $I_t = 0.680$ nA, $V_t = 220.9$ mV.

reconstruction. The intact reconstruction confirms that the graphene protects the Pt(100) surface from reaction with water at room temperature for at least 14 hours. In a separate experiment, a pristine graphene–Pt(100) sample was exposed to Milli-Q water at 60 °C, and a representative STM image of the resulting surface is included in Fig. 3b. The reconstruction remains visible on the surface, indicating efficient coating performance under these conditions.

Finally the stability of the graphene coating was tested against saltwater. A saltwater solution (0.513 M) was prepared as an analogue to the North Sea (0.56 M). Two 75 min experiments were performed, shown in Fig. 4a and b with the sample at room temperature and in Fig. 4c with the sample at 60 °C. In Fig. 4a and b the reconstruction, running as vertical lines in the images, remains intact. Fig. 4b shows the atomic resolution of the graphene layer with the intact reconstruction visible beneath. The inset in Fig. 4b shows a close-up of the graphene honeycomb lattice. The images show that the surface remains unperturbed and the graphene coating protects the Pt surface.

Fig. 4c shows a representative STM image following the second experiment, where the sample was heated to 60 °C while exposed to the saltwater solution. The surface roughness has increased, this is especially visible in the left side of the image. This roughness represents the lifting of the Pt(100) surface reconstruction below the graphene layer, suggesting that the graphene coating has failed to protect the Pt surface from reaction with hot saltwater.

Fig. 5 shows Raman spectra collected from samples exposed to (a) (green squares) 75 min in air at 60 °C, (b) (yellow circles) 75 min in water at 60 °C, (c) (red triangles) 75 min in saltwater at room temperature, and (d) (blue crosses) 75 min in saltwater at 60 °C. There are four peaks visible in each Raman spectrum and they are labelled as they were for the pristine graphene–Pt(100) system in Fig. 1c;

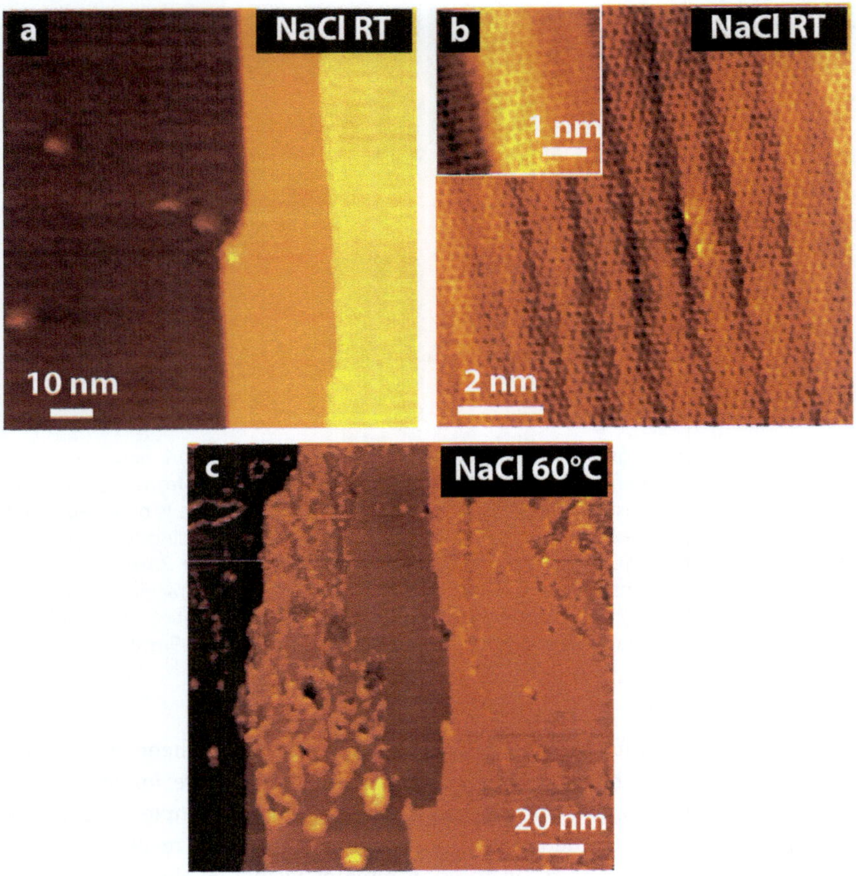

Fig. 4 (a–c) STM results, following the 75 min exposure of graphene–Pt(100) to saltwater (0.513 M), (a and b) at room temperature; (a) $I_t = -0.260$ nA, $V_t = -611.3$ mV. (b) Atomic resolution of the graphene coating covering the intact Pt reconstruction; $I_t = -0.970$ nA, $V_t = -2.7$ mV. The inset shows a zoom on the graphene coating depicting the hexagonal pattern of the graphene; $I_t = -0.950$ nA, $V_t = -2.7$ mV. (c) STM result at 60 °C; $I_t = 0.470$ nA, $V_t = 1250.0$ mV. The image shows a roughening of the surface consistent with what is expected from a lifted Pt reconstruction.

three peaks corresponding to the graphene layer (D, G and 2D peaks) and an O_2 peak from the oxygen molecules present in the air. The peak positions and intensities were determined by fitting with Lorentzian functions. The Raman spectra can be divided into two distinct groups, the first group shows peak positions similar to the pristine graphene–Pt(100) system shown in Fig. 1c. These data relate to the 75 min ambient saltwater test, the 75 min 60 °C air test, and the 75 min 60 °C water test. The I_D/I_G ratios for this group yield an average value of 0.20 ± 0.016, giving a defect distance of $L_D = 20$ nm, indicating a lower defect density than that for pristine graphene–Pt(100), presented in Fig. 1c (where $L_D = 16$ nm). This indicates that the defect density may vary between different growth cycles and/or may not be homogeneous across the graphene layer.

The Raman spectrum, relating to the 75 min, 60 °C, saltwater test, is obviously different from the spectra in group one; peak positions are red-shifted and the

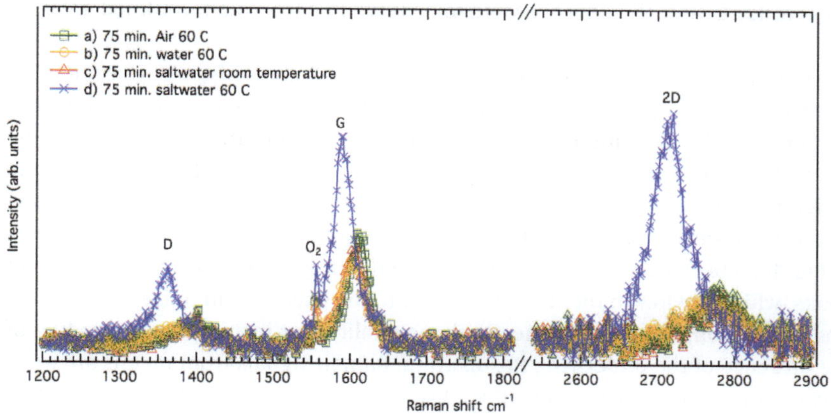

Fig. 5 Raman spectra corresponding to samples exposed to: (a) 75 min, air, 60 °C, (b) 75 min, water, 60 °C, (c) 75 min, saltwater, room temperature, and (d) 75 min, saltwater, 60 °C, which are plotted with a (a) green square, (b) yellow circle, (c) red triangle and (d) blue cross. Spectra a–c have similar peak shapes and positions while in spectrum d the peaks have shifted and increased in intensity.

relative as well as the absolute intensities of the individual peaks have changed. The D peak has shifted to 1361 cm^{-1}, the G peak to 1590 cm^{-1}, and the 2D peak to 2715 cm^{-1}. The I_D/I_G ratio has increased to 1.1, corresponding to an estimated defect distance of $L_D = 8$ nm. It was not possible to obtain STM data with sufficient resolution to confirm this L_D, which is below the threshold that would allow use of the empirical relation between I_D/I_G and L_D.[51]

4 Discussion

In this work it has been demonstrated that graphene protects the surface reconstruction on the Pt(100) surface from reactions with oxidising species under ambient conditions for up to 6 months, from reactions with water for up to 14 hours, and from reactions with air and water at 60 °C for at least 75 min. However, during the long term stability test in air (Fig. 2) the Pt surface reconstruction was observed to be lifted in some small, confined areas on the surface after 6 months (Fig. 2d). It is expected that these areas are related to small scale defects such as grain boundaries, which were already present on the pristine graphene–Pt(100) system following the initial graphene growth. As discussed in the introduction, it is anticipated that it is only at these sites that oxidising species can penetrate the graphene coating layer. After this initial penetration, the lifting of the surface reconstruction is diffusion limited as impurities intercalate between the graphene sheet and the Pt surface. This was illustrated by the atomic scale lattice presented over the area of lifted reconstruction, shown in Fig. 2d. The graphene lattice remains intact despite the lifting of the underlying reconstruction. Since the perfect graphene layer should be impenetrable for the oxidising species and since no etching of the graphene at this site is observed, the observations indicate that the oxidising species arrived at this site by diffusing beneath the graphene layer after intercalating at some defect site outside of the imaged area. Interestingly, the diffusion process does not seem to be very efficient in this case, since the reconstruction was only observed to be lifted in a few small, confined areas on the surface.

While STM only gives local information on small areas of the surface, Raman spectroscopy is a global technique frequently used to assess the quality of a graphene layer on a multitude of substrates. Typically the ratio between the intensities of the 2D/G peaks is used to assess the number of layers of graphene which have been produced and the quality of the graphene layer can be assessed by the ratio between the intensities of the D/G peaks. The Raman spectrum shown in Fig. 1c, representing a pristine graphene–Pt(100) system, shows relatively broad peaks, which are low in intensity compared to the background noise level. This is not a reflection, however, on the graphene quality, but instead is a by-product of the presence of the underlying Pt substrate. Raman signals from metals are weak because of the metal-screening effect. Even so, graphene samples on very weakly interacting metallic substrates, like Cu, produce well-resolved Raman spectra. In the graphene–Pt(100) system, however, hybridisation of the metallic Pt d-states with the graphene π-orbitals leads to a large reduction in Raman intensity and to a blue-shift in graphene-related peaks.[46] Calculations can be used to compare the change in the 2D peak wave number and the level of hybridisation.[46] Thus, for quasi free-standing graphene, where the overlap between the Pt d-states and the graphene π-orbitals is reduced, the 2D peak moves to lower wave numbers and becomes more narrow. In the Raman spectrum shown in Fig. 1c, which represents pristine graphene, and the Raman spectra shown in Fig. 5a–c, corresponding to unreacted graphene–Pt(100) systems, the 2D peak occurs at a higher wave number, is broad and is reduced in intensity due to hybridisation effects. The D and G peaks in these Raman spectra are similarly blue-shifted and reduced in intensity. Once the Pt substrate has been oxidised following the exposure to saltwater at 60 °C, the D, 2D and G peaks are red-shifted, and gain in absolute intensity. The intensity gain is probably caused both by an increased separation between the metal surface and the graphene sheet, reducing the metal-screening effect, and by a reduction in hybridisation between the Pt surface d-orbitals and the graphene π-band.[46] The observed shift in peak position is also expected from this change in hybridisation. However, a change in graphene strain, caused by the formation of a rough Pt oxide layer on the surface, has also been suggested to lead to a shift in Raman peak positions.[52] Hence, the Raman data can be used to indirectly assess the chemical state of the Pt substrate. The Raman spectrum recorded from high temperature, salt-water exposed graphene–Pt(100) showed that the sample more closely resembles free-standing graphene. This concurs with the conclusion from the STM data, namely that the reconstruction has lifted. This red-shift in the Raman-active graphene modes was accompanied by both an absolute and relative change in peak intensities, see Fig. 5d. An attempt was made to extract the defect density from the new I_D/I_G ratios, but the result was inconclusive. Since the intensities of the Raman peaks are influenced by the hybridisation with the surface, it is unclear if the change in I_D/I_G ratio resulted from an increase in defect density or relates to the change in hybridisation of the graphene π-band.

The combination of heat and saltwater was observed to lead to large scale coating breakdown and reaction with the Pt surface. Previous DFT calculations have examined the edge state of graphene islands on a Pt surface and concluded that in the presence of OH^- groups an open-edge configuration is favoured.[17] This open-edge configuration sees the OH– groups react with and separate edge-C atoms on the graphene sheet from the underlying Pt surface atoms. The resulting open-edge configuration allows atoms and molecules to intercalate beneath the graphene layer and facilitates reaction with the Pt surface. A temperature of 60 °C is not sufficient to break the bond between the graphene C-edge atoms and the underlying Pt atoms, without involving edge-reactions with external species. The Na^+ and Cl^- ions in the saltwater may extend the lifetime of transient OH^- and H^+ groups in solution when compared to the pure water solution. These charged species may increase the possibility for reactions involving edge-C atoms at defects and grain boundaries in the graphene sheet. Since no coating breakdown is observed for room temperature saltwater, we speculate that the activation energy barrier for these reactions makes them inefficient at room temperature.

5 Conclusion

A continuous single layer of graphene was prepared on a single crystal Pt(100) surface. Raman spectroscopy in combination with STM was used to detect the conditions under which this coating was compromised. The graphene layer successfully protected the Pt surface, as determined by the survival of the Pt(100) surface reconstruction, from oxidising in air for the 6 months experiment, water for the 14 hour test period and NaCl for the 75 min test period at room temperature. When the sample was exposed to saltwater at 60 °C, the Pt(100) surface reconstruction was observed to lift. STM imaging provided high resolution data, and this was combined with Raman spectroscopy to study the changes to the surface structure and chemistry. STM images revealed that the graphene carbon lattice can remain intact, even if the underlying Pt surface has been oxidised. This suggests that oxidising species do not etch intact areas on the graphene basal plane, but instead diffuse beneath the graphene layer having first intercalated at defects in the graphene lattice. Such defects probably originate during the original growth procedure, predominantly at grain boundaries. It is noteworthy, however, that the Pt surface was not oxidised in air for an extended period of time. Even after 6 months of exposure, the areas where the reconstruction was lifted were few and small in size. An uncovered Pt(100) surface oxidises in seconds following exposure to air, meaning that the graphene layer offers significant protection.

Atomic resolution in STM experiments was not achieved on samples exposed to a combination of heat and saltwater. This means that it was not possible to discern whether or not the graphene basal plane was compromised under these conditions. Certainly, the underlying Pt surface changed both structurally, as determined from STM images, and chemically, as determined from Raman spectroscopy. It is proposed that these changes represent a lifting of the Pt(100) surface reconstruction, heralding the oxidation of the Pt metal surface.

References

1 K. S. Novoselov, A. K. Geim, S. V. Morozov, D. Jiang, Y. Zhang, S. V. Dubonos, I. V. Grigorieva and A. A. Firsov, *Science*, 2004, **306**, 666–669.

2 T. Kuila, S. Bose, P. Khanra, A. K. Mishra, N. H. Kim and J. H. Lee, *Biosens. Bioelectron.*, 2011, **26**, 4637–4648.

3 Y. Z. Zhang, T. Liu, B. Meng, X. H. Li, G. Z. Liang, X. N. Hu and Q. J. Wang, *Nat. Commun.*, 2013, **4**, 1811.
4 Y. L. Chen, X. B. Feng and D. D. Hou, *Acta Phys. Sin.*, 2013, **62**, 187301.
5 D. Prasai, J. C. Tuberquia, R. R. Harl, G. K. Jennings and K. I. Bolotin, *ACS Nano*, 2012, **6**, 1102–1108.
6 C. Gomez-Navarro, M. Burghard and K. Kern, *Nano Lett.*, 2008, **8**, 2045–2049.
7 S. K. Lee, H. Y. Jang, S. Jang, E. Choi, B. H. Hong, J. Lee, S. Park and J. H. Ahn, *Nano Lett.*, 2012, **12**, 3472–3476.
8 C. Sire, F. Ardiaca, S. Lepilliet, J. W. T. Seo, M. C. Hersam, G. Darnbrine, H. Happy and V. Derycke, *Nano Lett.*, 2012, **12**, 1184–1188.
9 C. Lim, A. Sorkin, Q. L. Bao, A. Li, K. Zhang, M. Nesladek and K. P. Loh, *Nat. Commun.*, 2013, **4**, 1556.
10 R. R. Nair, H. A. Wu, P. N. Jayaram, I. V. Grigorieva and A. K. Geim, *Science*, 2012, **335**, 442–444.
11 S. S. Chen, L. Brown, M. Levendorf, W. W. Cai, S. Y. Ju, J. Edgeworth, X. S. Li, C. W. Magnuson, A. Velamakanni, R. D. Piner, J. Y. Kang, J. Park and R. S. Ruoff, *ACS Nano*, 2011, **5**, 1321–1327.
12 N. T. Kirkland, T. Schiller, N. Medhekar and N. Birbilis, *Corros. Sci.*, 2012, **56**, 1–4.
13 J. H. Huh, S. H. Kim, J. H. Chu, S. Y. Kim, J. H. Kim and S. Y. Kwon, *Nanoscale*, 2014, **6**, 4379–4386.
14 Y. Q. Liang, L. L. Yu, Z. D. Cui, S. L. Zhu, Z. Y. Li and X. J. Yang, *Sci. Adv. Mater.*, 2014, **6**, 545–549.
15 R. T. Mu, Q. Fu, L. Jin, L. Yu, G. Z. Fang, D. L. Tan and X. H. Bao, *Angew. Chem., Int. Ed.*, 2012, **51**, 4856–4859.
16 L. Nilsson, M. Andersen, R. Balog, E. Laegsgaard, P. Hofmann, F. Besenbacher, B. Hammer, I. Stensgaard and L. Hornekaer, *ACS Nano*, 2012, **6**, 10258–10266.
17 L. Nilsson, M. Andersen, B. Hammer, I. Stensgaard and L. Hornekaer, *J. Phys. Chem. Lett.*, 2013, **4**, 3770–3774.
18 B. Borca, F. Calleja, J. J. Hinarejos, A. L. V. de Parga and R. Miranda, *J. Phys.: Condens. Matter*, 2009, **21**, 134002.
19 Y. S. Dedkov, M. Fonin and C. Laubschat, *Appl. Phys. Lett.*, 2008, **92**, 52506.
20 U. Mogera, N. Kurra, D. Radhakrishnan, C. Narayana and G. U. Kulkarni, *Carbon*, 2014, **78**, 384–391.
21 Y. S. Dedkov, M. Fonin, U. Rudiger and C. Laubschat, *Appl. Phys. Lett.*, 2008, **93**, 22509.
22 R. John, A. Ashokreddy, C. Vijayan and T. Pradeep, *Nanotechnology*, 2011, **22**, 165701.
23 S. C. Sahu, A. K. Samantara, M. Seth, S. Parwaiz, B. P. Singh, P. C. Rath and B. K. Jena, *Electrochem. Commun.*, 2013, **32**, 22–26.
24 M. Schriver, W. Regan, W. J. Gannett, A. M. Zaniewski, M. F. Crommie and A. Zettl, *ACS Nano*, 2013, **7**, 5763–5768.
25 F. Zhou, Z. T. Li, G. J. Shenoy, L. Li and H. T. Liu, *ACS Nano*, 2013, **7**, 6939–6947.
26 K. Krishnamoorthy, A. Ramadoss and S. J. Kim, *Sci. Adv. Mater.*, 2013, **5**, 406–410.
27 J. H. Park and J. M. Park, *Surf. Coat. Technol.*, 2014, **254**, 167–174.
28 G. Libo, N. Guang-Xin, L. Yanpeng, L. Bo, A. H. Castro Neto and L. Kian Ping, *Nature*, 2014, **505**, 190–194.

29 J. Lee, Y. Kim, H. J. Shin, C. Lee, D. Lee, S. Lee, C. Y. Moon, S. C. Lee, S. J. Kim, J. H. Ji, H. S. Yoon and S. C. Jun, *ACS Appl. Mater. Interfaces*, 2014, **6**, 12588–12593.
30 V. Migkovic-Stankovic, I. Jevremovic, I. Jung and K. Rhee, *Carbon*, 2014, **75**, 335–344.
31 Y. Su, V. G. Kravets, S. L. Wong, J. Waters, A. K. Geim and R. R. Nair, *Nat. Commun.*, 2014, **5**, 4843.
32 A. Hyosub, L. Won-Jun and J. Jongwan, *Curr. Appl. Phys.*, 2011, **11**, S81–S85.
33 R. S. Edwards and K. S. Coleman, *Acc. Chem. Res.*, 2013, **46**, 23–30.
34 W. Tianru, D. Gugiao, S. Honglie, W. Haomin, S. Lei, Z. Yun, J. Da and X. Xiaoming, *Nanoscale*, 2013, **5**, 5456–5461.
35 Z. C. Li, P. Wu, C. X. Wang, X. D. Fan, W. H. Zhang, X. F. Zhai, C. G. Zeng, Z. Y. Li, J. L. Yang and J. G. Hou, *ACS Nano*, 2011, **5**, 3385–3390.
36 A. Zangwill and D. D. Vvedensky, *Nano Lett.*, 2011, **11**, 2092–2095.
37 B. Zhang, W. H. Lee, R. Piner, I. Kholmanov, Y. P. Wu, H. F. Li, H. X. Ji and R. S. Ruoff, *ACS Nano*, 2012, **6**, 2471–2476.
38 M. Topsakal, H. Sahin and S. Ciraci, *Phys. Rev. B: Condens. Matter Mater.Phys.*, 2012, **85**, 155445.
39 R. Larciprete, S. Ulstrup, P. Lacovig, M. Dalmiglio, M. Bianchi, F. Mazzola, L. Hornekaer, F. Orlando, A. Baraldi, P. Hofmann and S. Lizzit, *ACS Nano*, 2012, **6**, 9551–9558.
40 S. Ulstrup, M. Andersen, M. Bianchi, L. Barreto, B. Hammer, L. Hornekaer and P. Hofmann, *2D Materials*, 2014, **1**, 025002.
41 J. C. Johannsen, S. Ulstrup, M. Bianchi, R. Hatch, D. Guan, F. Mazzola, L. Hornekaer, F. Fromm, C. Raidel, T. Seyller and P. Hofmann, *J. Phys.: Condens. Matter*, 2013, **25**, 94001.
42 E. Granas, J. Knudsen, U. A. Schroder, T. Gerber, C. Busse, M. A. Arman, K. Schulte, J. N. Andersen and T. Michely, *ACS Nano*, 2012, **6**, 9951–9963.
43 I. Wlasny, P. Dabrowski, M. Rogala, P. J. Kowalczyk, I. Pasternak, W. Strupinski, J. M. Baranowski and Z. Klusek, *Appl. Phys. Lett.*, 2013, **102**, 111601.
44 Y. H. Zhang, H. R. Zhang, B. Wang, Z. Y. Chen, Y. Q. Zhang, B. Wang, Y. P. Sui, B. Zhu, C. M. Tang, X. L. Li, X. M. Xie, G. H. Yu, Z. Jin and X. Y. Liu, *Appl. Phys. Lett.*, 2014, **104**, 143110.
45 Y. P. Hsieh, M. Hofmann, K. W. Chang, J. G. Jhu, Y. Y. Li, K. Y. Chen, C. C. Yang, W. S. Chang and L. C. Chen, *ACS Nano*, 2014, **8**, 443–448.
46 Q. Zhou, S. Coh, M. L. Cohen, S. G. Louie and A. Zettl, *Phys. Rev. B: Condens. Matter Mater. Phys.*, 2013, **88**, 235431.
47 H. B. Lyon and G. A. Somorjai, *J. Chem. Phys.*, 1967, **46**, 2539–2550.
48 L. Nilsson, M. Andersen, J. Bjerre, R. Balog, B. Hammer, L. Hornekaer and I. Stensgaard, *Surf. Sci.*, 2012, **606**, 464–469.
49 E. Laegsgaard, F. Besenbacher, K. Mortensen and I. Stensgaard, *J. Microsc.*, 1988, **152**, 663–669.
50 S. Ulstrup, L. Nilsson, J. A. Miwa, R. Balog, M. Bianchi, L. Hornekaer and P. Hofmann, *Phys. Rev. B: Condens. Matter Mater. Phys.*, 2013, **88**, 125425.
51 L. G. Cancado, A. Jorio, E. H. M. Ferreira, F. Stavale, C. A. Achete, R. B. Capaz, M. V. O. Moutinho, A. Lombardo, T. S. Kulmala and A. C. Ferrari, *Nano Lett.*, 2011, **11**, 3190–3196.
52 X. Yin, Y. Li, F. Ke, C. Lin, H. Zhao, L. Gan, Z. Luo, R. Zhao, T. Heinz and Z. Hu, *Nano Res.*, 2014, **7**, 1613–1622.

Faraday Discussions

PAPER

The corrosion protection of AA2024-T3 aluminium alloy by leaching of lithium-containing salts from organic coatings

Peter Visser,*[ac] Yanwen Liu,[b] Xiaorong Zhou,[b] Teruo Hashimoto,[b] George E. Thompson,[b] Stuart B. Lyon,[b] Leendert G. J. van der Ven,[c] Arjan J. M. C. Mol[a] and Herman. A. Terryn[ad]

Received 28th November 2014, Accepted 12th January 2015
DOI: 10.1039/c4fd00237g

Lithium carbonate and lithium oxalate were incorporated as leachable corrosion inhibitors in model organic coatings for the protection of AA2024-T3. The coated samples were artificially damaged with a scribe. It was found that the lithium-salts are able to leach from the organic coating and form a protective layer in the scribe on AA2024-T3 under neutral salt spray conditions. The present paper shows the first observation and analysis of these corrosion protective layers, generated from lithium-salt loaded organic coatings. The scribed areas were examined by scanning and transmission electron microscopy before and after neutral salt spray exposure (ASTM-B117). The protective layers typically consist of three different layered regions, including a relatively dense layer near the alloy substrate, a porous middle layer and a flake-shaped outer layer, with lithium uniformly distributed throughout all three layers. Scanning electron microscopy and white light interferometry surface roughness measurements demonstrate that the formation of the layer occurs rapidly and, therefore provides an effective inhibition mechanism. Based on the observation of this work, a mechanism is proposed for the formation of these protective layers.

1 Introduction

The development of alternative surface pretreatments and pigmented coatings to eliminate the health hazards and toxicity of soluble hexavalent chromium has been pursued for many years. Especially in the aeronautical industry, the replacement or elimination of hexavalent chromium has been and still is a major challenge.[1] High strength aluminium alloys such as AA2024-T3 and AA7075-T6

[a]Delft University of Technology, Department of Materials Science and Engineering, Delft, The Netherlands. E-mail: p.visser-1@tudelft.nl
[b]School of materials, The University of Manchester, Manchester, UK
[c]AkzoNobel, Automotive and Aerospace Coatings, Sassenheim, The Netherlands
[d]Vrije Universiteit Brussel, Group of electrochemistry and surface engineering, Bussels, Belgium

are most commonly used in aircraft design because of their favourable strength to weight ratio. However, these alloys are very susceptible to localized corrosion such as pitting and intergranular corrosion.[2–4] For decades, chromate-containing organic coatings and chromate-based conversion coatings have been used for the protection of these high strength aluminium alloys, with superior performance. The corrosion protective scheme employs chromate in many steps, including pre-cleaning,[5] conversion coating/anodizing,[6–8] and the organic coatings.[9] These chemical conversion coatings and organic coatings provide active corrosion inhibition toward the substrate and have the ability to release hexavalent chromium in the case of a local defect to provide the "self-healing" protection.[10,11] The unique properties of hexavalent chromium make the replacement of chromate inhibitor pigments in organic coatings a difficult task.[9,12] Numerous investigations have been performed to find promising candidates to replace chromates in organic coatings. Ideally, these compounds should have a similar performance as chromates, *i.e.* providing corrosion protection to the substrate and serving as a reservoir of corrosion inhibitor, which has the ability to be released from the coating when damages occur and heal the defect by the formation of a protective or passivation layer.[13,14] This leaching of soluble chemical species is the main corrosion inhibiting strategy used in corrosion inhibiting coating technology for the protection of high strength aluminium alloys.

Over the last few years, several inhibitors have been identified as promising replacement for chromates. These include oxo-anionic, cationic, metal, metal oxide, and organic inhibitors. The oxo-anionic inhibitor salts include molybdates,[15] metavanadates,[16,17] permanganate,[18] and phosphates,[19–21] The cationic inhibitors include rare-earth compounds[22] such as cerium[23] and praseodymium.[24] Magnesium[25] and zinc[26] metals are also found to be effective when used as pigments in organic coatings. Furthermore, several organic inhibitors have shown promising results.[27–29] Whilst many approaches have been attempted to achieve acceptable alternatives, relatively few systems have demonstrated similar effectiveness as chromate containing systems.[30] Therefore, the search for suitable alternatives for chromates is ongoing.

Recently, the use of lithium salts has been proposed as a potential replacement for chromate in organic coatings.[31] In the 1990s, alkaline lithium solutions were studied as possible replacements of the traditional chromated chemical conversion process. Lithium salts, including lithium carbonate,[32–36] lithium sulfate,[32] lithium hydroxide[37] and lithium chloride[32] in aqueous alkaline environment have been shown to form a continuous polycrystalline layer on super pure aluminium,[33,34] AA1100,[35,36] AA6061,[32,35] AA2024,[35,37] and AA7075[35] aluminium alloys. The lithium-based conversion coating greatly increased the corrosion resistance of the various selected substrates by increasing the pitting potential and reducing the corrosion current density. The conversion coated super pure aluminium and AA6061 aluminium alloy passed the standard salt spray tests.[35] The protective properties provided by these lithium conversion layers are suggested to originate from the formation of a lithium-aluminium-hydroxide-carbonate-hydrate ($Li_2[Al_2(OH)_6]_2 \cdot CO_3 \cdot nH_2O$) layer[35] with lithium aluminate as a pore filler.[34] Lithium is the only monovalent cation that is known to intercalate in aluminium hydroxide to form these layered double hydroxides.[38]

Although lithium based chemical conversion layers have been studied using electrochemical approaches and corrosion tests, limited studies have been carried out with a focus on obtaining detailed morphological information, especially cross-sectional observation of the protective layer. Further, the mechanisms of layer formation remain unclear. Additionally, no detailed investigation has focused on the use of lithium-containing species as leachable corrosion inhibitor incorporated into organic coatings.

In the present study, lithium-containing salts, *i.e.* lithium carbonate or lithium oxalate, were used as corrosion inhibitors in an organic model coating for the protection of AA2024-T3. The coatings were damaged and the scribe was examined by electron microscopic techniques, before and after exposure to the neutral salt spray corrosion test (ASTM B-117). This is the first investigation of such a protective layer that was generated on AA2024-T3, from an organic coating containing lithium-salt as leachable corrosion inhibitor. Specific and careful sample preparation combined with low-voltage scanning electron microscopy (SEM) cross-sectional work and high-resolution transmission electron microscopy (TEM) and electron energy loss spectroscopy (EELS) were used to reveal morphological and compositional information of the protective layer. In addition, the formation of the layer has been studied as a function of time by white light interferometry and cross-sectional SEM. A mechanism is proposed for the formation and the protective nature of this layer, based on the observations of this study.

2 Experimental

2.1 Materials

Model formulations based on polyurethane chemistry were designed using a polyester resin (Desmophen® MPA) and an aliphatic polyisocyanate (Tolonate® HDB 75 MX) with a NCO/OH ratio of 0.75 and a pigment volume concentration (PVC) of 30% as described in Table 1.

Tartaric–sulfuric acid (TSA) anodized AA2024-T3 unclad sheets were used as the substrate in this study (dimensions of 70 × 70 × 0.8 mm). The AA2024-T3 substrate (2024-T3 QQ-A250/5) was purchased from Alcoa and the TSA

Table 1 Composition of model organic coatings

	Supplier	Non-inhibiting	Lithium carbonate	Lithium oxalate
Component A				
N-Butylacetate	Sigma Aldrich	75.0 g	75.0 g	75.0 g
Desmophen 650MPA	Bayer materials science	47.7 g	47.7 g	47.7 g
Lithium carbonate	Sigma Aldrich		23.6 g	
Lithium oxalate	Sigma Aldrich			32.0 g
Magnesium oxide	Sigma Aldrich		16.4 g	16.4 g
Tioxide TR 92	Huntsman		5.9 g	5.9 g
Blanc Fixe N ($Ba(SO_4)$)	Sachtleben		15.4 g	15.4 g
Component B				
Tolonate HDB 75 MX	Vencorex	28.5 g	28.5 g	28.5 g
Dynasilan Glymo	Evonik	5.2 g	5.2 g	5.2 g

pretreatment was performed according to aerospace requirements (AIPI 02-01-003) at Premium AEROTEC, Bremen Germany.

2.2 Preparation and application of the model coating

The pigmented coatings were prepared according to the following procedure:

All components of Component A were added under stirring in a 370 mL glass jar. After addition of the final component, 400 g of Zirconox® pearls (1.7–2.4 mm) were added to the mixture. The pigments were dispersed to a particle size smaller than 25 μm by 20 min shaking on a Skandex® paint shaker. After this procedure, the mixtures were filtered to remove the pearls. The ingredients of component B were mixed separately.

Component B was added to Component A and the mixture was stirred to a homogeneous mixture. The paints were applied with a HVLP spray gun on the AA2024-T3 unclad TSA panels, 30 min after mixing. The coatings were cured at 80 °C for 16 h. The measured dry film thickness was 20–25 μm.

A mechanical milling device was used to artificially damage the coating with a U-shaped scribe, before exposing the samples to the neutral salt spray test (ASTM-B117). The panels were scribed from corner to corner, leaving a 1 mm wide scribe that penetrated 100–150 μm deep into the metal.

2.3 Experimental techniques

The samples were exposed to the neutral salt spray (ASTM-B117) test for varying periods from 30 min up to 168 h. After exposure, the process was quenched by a 2-min rinse with flowing demineralised water to remove any residual chlorides and the panels were air dried.

2.3.1 Scanning electron microscopy (SEM). The cross-sectional observation of the oxide layers, in the scribe, was carried out by scanning electron microscopy (SEM) using a ZEISS Ultra 55 instrument with an acceleration voltage of 0.5 kV. The composition of the oxide films was analysed by energy dispersive X-ray analysis on FEI Magellan HR FEG-SEM.

2.3.2 Transmission electron microscopy (TEM-EELS). The layers were also examined by transmission electron microscopy (TEM) using a JEOL FX 2000 II instrument operated at 120 kV, with a nominal sample thickness of 15 nm, prepared by a Leica EM UC6 ultra-microtome with a diamond knife. Electron energy loss spectroscopy (EELS) was conducted on a TECNAI TF30 TEM fitted with a Gif2001 detector, operated at 300 kV Mapping images were acquired with 1024 × 1024 pixels, this results pixel size of 0.56 × 0.56 nm for the selected magnification.

2.3.3 White light interferometry (WLI). The surface roughness of the scribe was measured with WLI, using a Veeco NT1100 profileometer. The high-resolution measurements were carried out using a 20× objective and multiple measurements were stitched together to sample a 2 mm length of the scribe. The data were de-trended by removal of any planar tilt. In order to remove the low frequency spatial waves associated with scribing a high pass Fourier filter was used with a frequency cut-off of 5 mm^{-1}. Visibly corroded samples were subjected to a cleaning step (1 h, 1 M nitric acid) to dissolve the corrosion products and reveal the attacked morphology of the substrate.

3 Results and discussion

3.1 General morphological observations in the scribed area before and after salt spray testing

The scribed regions were examined with an optical microscope. Fig. 1a shows the scribe before exposure to the accelerated corrosion test. The pristine scribe is 1 mm wide and 100 μm deep into the alloy. The panels were exposed to 168 h neutral salt spray (ASTM B-117). Fig. 1b shows the detrimental effect of the corrosive conditions when a coating does not contain an effective corrosion inhibitor. The coating remains intact but the scribe is heavily affected and contains many areas with white corrosion products. In contrast to this, the lithium-salt loaded coatings remained free of corrosion products (Fig. 1c and d). Both, lithium carbonate and oxalate show a similar protective behaviour. These lithium salts were selected for this study because of their moderate solubility compared to most other lithium salts, respectively 1.3 g/100 mL for lithium carbonate and 8 g/100 mL for lithium oxalate, and ease of paint formulation.

These observations and visual comparison of the scribed areas of coated AA2024-T3 before and after salt spray testing indicates the important role of the presence of lithium-containing salts as corrosion inhibitor in organic coatings, leading to a detailed investigation of the protection mechanism of the AA2024-T3 by lithium-containing salts, as discussed below.

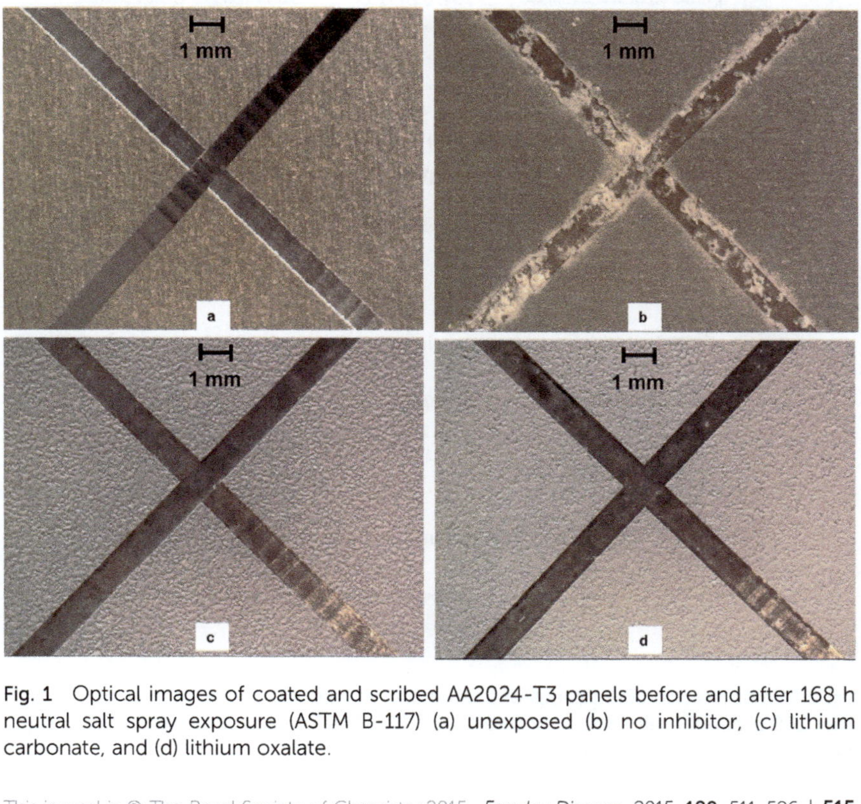

Fig. 1 Optical images of coated and scribed AA2024-T3 panels before and after 168 h neutral salt spray exposure (ASTM B-117) (a) unexposed (b) no inhibitor, (c) lithium carbonate, and (d) lithium oxalate.

3.2 Detailed morphological observations in the scribed area before and after salt spray testing

The scribed region was examined in more detail by studying cross-sections using scanning electron microscopy. The cross-sections of the scribed area of the alloy substrate showed a "U" shape in the middle of the images, with the organic coated alloy substrate evident on both sides (Fig. 2a). After 168 h salt spray exposure, the scribed alloy with the non-inhibiting coating showed the presence of a large amount of corrosion products with a typical thickness of about 25 μm (Fig. 2b). However, when lithium carbonate or lithium oxalate was incorporated as a corrosion inhibitor in the organic coating, no apparent damage is observed in the scribed area, as shown in Fig. 2c and d.

3.2.1 Non-inhibited coating. The scribed area of the non-inhibited coating has been examined after 168 h neutral salt spray exposure. The SEM cross-section clearly showed the hydrated aluminium oxide formed during salt spray exposure; Fig. 3a displays the full scribe with the corrosion products. Fig. 3b shows the cross-section of the corrosion products at higher magnification, with pores providing access to the metal surface for aggressive ions. Although the metal/oxide or hydroxide interface is relatively flat, the corrosion products are homogeneously distributed in the scribe. Fig. 3a indicates a relatively uniform corrosion process. However, localised corrosion, such as pitting and intergranular, was

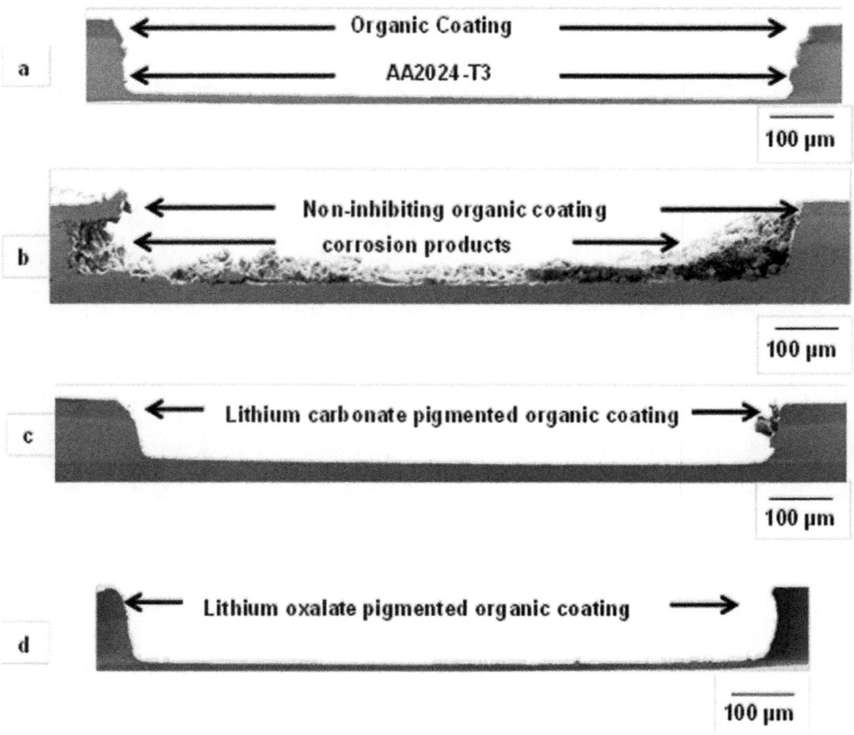

Fig. 2 Cross-section SEM images of coated and scribed AA2024-T3 panels before and after 168 h neutral salt spray exposure (ASTM B-117) (a) unexposed, (b) no inhibitor, (c) lithium carbonate, and (d) lithium oxalate.

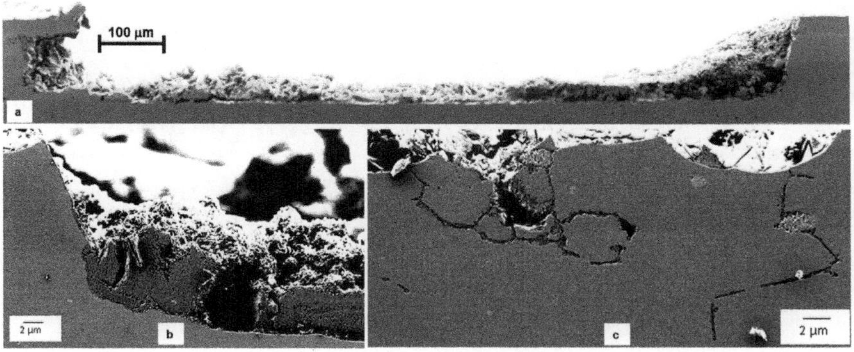

Fig. 3 SEM crosscut images of non-inhibited coating on AA2024-T3. (a) full scribe with corrosion products, (b) detailed corrosion products, (c) intergranular corrosion.

observed as well (Fig. 3c). It is evident that the grain boundary of the alloy is vulnerable to corrosion attack, especially when second phase particles are present at the grain boundaries. EDX analyses revealed that the corrosion products contained mainly aluminium and oxygen, with traces of copper and magnesium present.

According to Foley,[39] corrosion of aluminium commences according to a multistep process which leads to oxide thinning and finally dissolution of the aluminium. This dissolution is followed by a fast hydrolysis of the Al^{3+} ion to form $Al(OH)_3$ that transforms into $Al_2O_3 \cdot H_2O$. Further, EDX analyses of the corrosion products indicated that the corrosion products are mainly hydrated aluminium oxide ($Al_2O_3 \cdot H_2O$) or aluminium hydroxide ($Al(OH)_3$), with oxygen to aluminium ratios in the ranging from of 2 : 1 up to 4 : 1. This correlates with various hydrated aluminium oxides, which are common corrosion products of aluminium occurring in chloride containing solutions.[40]

3.2.2 Lithium-salt containing coatings. In contrast to the scribed area of the non-inhibited coating, the lithium-salt loaded organic coatings did not show this type of corrosion and the scribed areas were without any significant amount of corrosion products after 168 h neutral salt spray exposure. For both samples, SEM investigations reveal that the substrate of the scribe is covered with a new layer.

Fig. 4 shows the SEM cross-sectional examination of the protective layer generated from a lithium carbonate loaded coating. A thin deposited layer, of about 0.5 to 1.5 µm thickness, developed throughout the scribed surface area (Fig. 4b and c). Higher magnification images reveal a typical detailed morphology of the layer formed (Fig. 4d). The layer comprises three distinct regions: a dense barrier region at the metal/deposited layer interface, a porous region in the middle, and a columnar outer region. This outer region can be correlated with the surface as observed with a planar view SEM image (Fig. 5c).

Fine textures were observed on the overall surface at increased magnification (Fig. 5c and d). Both the lithium carbonate and lithium oxalate pigmented primers show this particular texture. In some regions, the texture is more significant, while in other regions the surface is still relatively flat. The textures comprised fine flakes, with some oriented almost vertical to the surface (Fig. 5d) while others are randomly orientated. It is unclear whether this different top-view

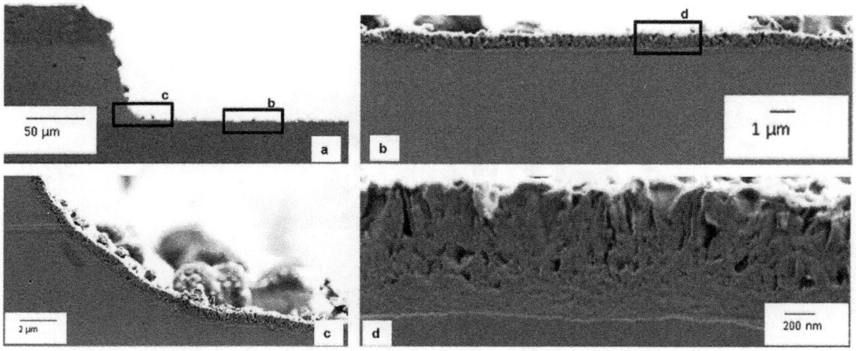

Fig. 4 SEM cross-cut images of protective layers generated from a lithium carbonate loaded coating after 168 h ASTM B-117 exposure (a) scribed area, (b) middle section of the scribe, (c) curved area of bottom of the scribe, (d) detailed morphology of layer (middle section of the scribe).

morphology or texture is a result of the anion or different lithium ion concentrations due to the higher solubility of lithium oxalate. This study did not focus on the anion effect. However, these observations justify further studies on the effect of the anion on the formation of these protective layers.

The lithium oxalate loaded model coating demonstrates similar deposition behaviour in the scribed region after salt spray testing (Fig. 6). SEM cross-sectional examination revealed the same three distinct morphologies of a dense barrier region, porous region, and columnar outer region, similar to the deposited layer of the lithium carbonate sample. Although the flakes formed are finer and denser compared with the morphologies of the film formed from the coated alloy pigmented with lithium carbonate.

The observed inhibitive effect must be a result of this layer, which was generated *in situ* in the damaged area. The presence of the columnar layer at the

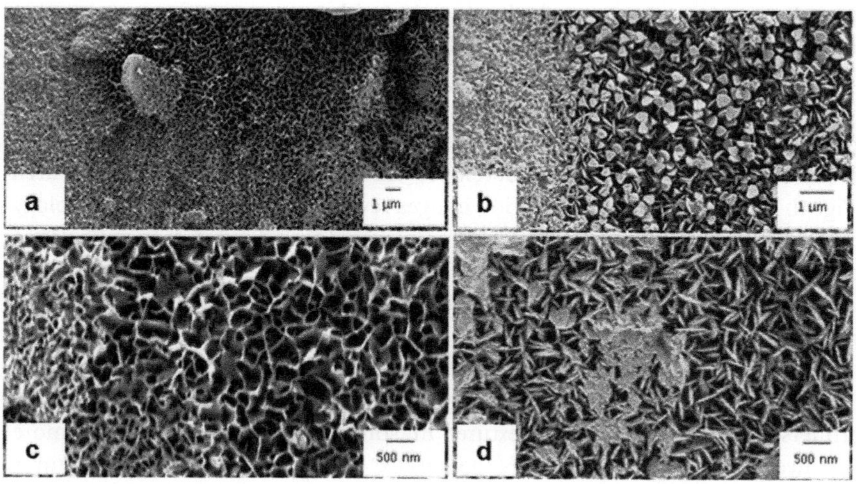

Fig. 5 Top view appearance of protective layer with increasing magnification after 168 h ASTM B-117 exposure. (a) and (c) lithium carbonate and (b) and (d) lithium oxalate.

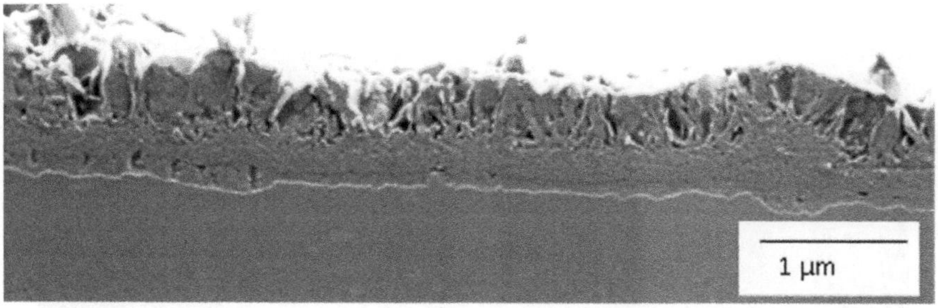

Fig. 6 SEM cross-sectional image of the typical morphology of the protective layer generated from a lithium oxalate loaded coating.

outer region of the cross-section and the microstructure of the surface could indicate the possible formation of a lithium-aluminium-carbonate-hydroxide hydrate as reported by Drewien *et al.*[36] They generated these protective poly-crystalline layers on aluminium from saturated aluminate ($Al(OH)_4^-$) alkaline lithium (pH 11.5–13) solutions under well-controlled conditions. Besides the poly-crystalline layer, a thin very fine or poorly crystalline layer was found underneath the poly-crystalline layer. However, the performance of these layers was not sufficient for the protection of AA2024-T3.

Layers with a similar morphology as found in this study, have been observed when exposing aluminium to (weak) alkaline solutions.[41,42] The ratio of the thickness of the inner barrier layer *versus* the non-barrier outer layer was found to depend on the pH of the solution. It was suggested that the layer was initially formed by initially an aluminium hydroxide gel, which aged to boehmite and pseudoboehmite and later to bayerite. However, these layers did not provide long-term corrosion protection.

The results from this study suggest that lithium-containing species in the organic coating provide fast and effective corrosion protection to the aluminium alloy. The layered morphology is significantly different from the layers formed in the conversion process. The inhibitive properties of this layer might be derived from the presence of the dense non-porous layers near the aluminium substrate.

3.3 Composition of the protective layer

The composition of the deposited layer has been analysed with transmission electron microscopy equipped with electron energy loss spectroscopy (TEM-EELS) in order to detect the lithium species in the layer. The TEM image and corresponding EELS maps of the layer generated from the lithium carbonate based coating are shown in Fig. 7a–d. These maps demonstrated the presence of aluminium, oxygen, and lithium throughout the deposited layer. The bright regions in Fig. 7d indicate the uniform distribution of lithium throughout the entire layer. Similar results were also obtained from the scribed area of the AA2024-T3 aluminium alloy with lithium oxalate pigmented coating. The TEM images showed the presence of the barrier film on the metal surface followed by fine porous and flake network regions.

The presence of lithium throughout the protective layer including the dense layer shows that lithium-ions leached from the organic coating matrix into the

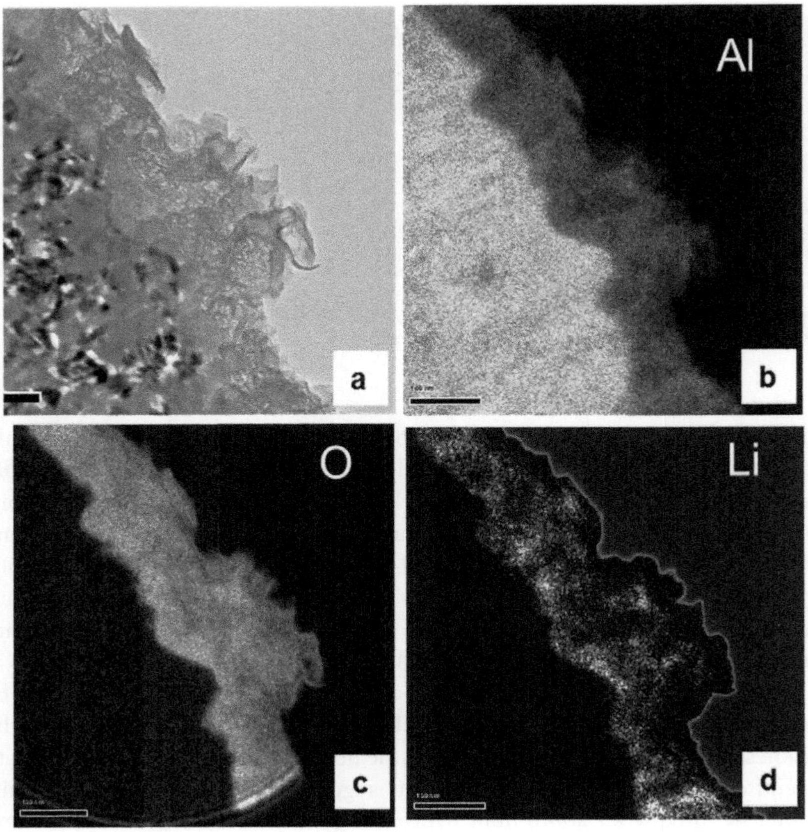

Fig. 7 TEM image and TEM-EELS elemental mapping of the layer generated from the lithium carbonate loaded coating (a) TEM image, (b) aluminium, (c) oxygen, and (d) lithium distribution.

scribe and formed a protective layer in the scribe that consists of lithium, aluminium and oxygen. Whereas the hydrated aluminium oxides of the non-inhibited sample are not able to prevent the aluminium matrix from further dissolution and pitting corrosion, these results indicate that the morphology and composition of these lithium-based layers provide long-term corrosion protection.

3.4 Rate of the protective layer formation

3.4.1 Surface roughness development in time. White light interferometry (WLI) has been used to quantify the degree of corrosion by measuring the surface roughness (s_a) of the scribed areas after different periods of salt spray exposure. Fig. 8 shows the development of the surface roughness of the samples over time when exposed to the corrosive salt spray conditions. An unexposed scribe has a typical surface roughness of about 170 nm (Fig. 9a).

The detrimental effect of the corrosive conditions on the non-inhibiting sample is evident. The surface of the scribe is severely damaged due to pitting corrosion and the dissolution of aluminium. The surface roughness of the non-

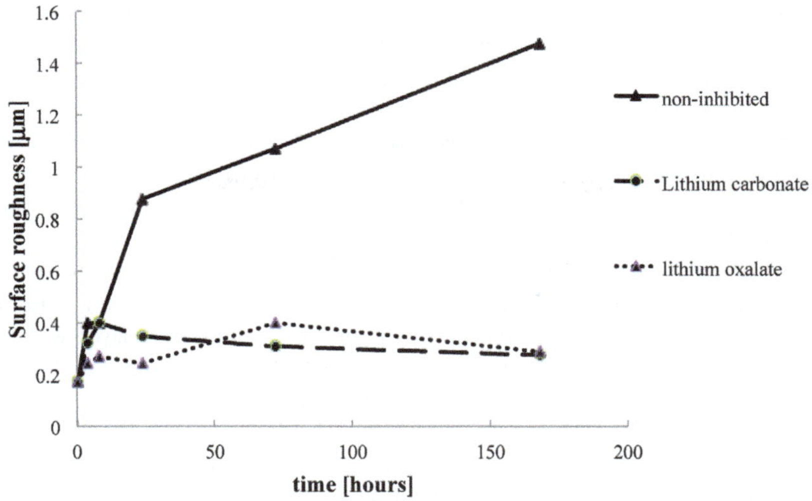

Fig. 8 Development of the surface roughness in the scribe measured by white light interferometry after different periods of neutral salt spray exposure.

inhibiting sample increased rapidly after the aluminium oxide layer had been destabilised and localized corrosion processes were initiated. After 24 h exposure the roughness increased up to 0.9 µm and this increased further to 1.5 µm after 168 h. In contrast to this, the surface roughness of the lithium-salt containing coatings increased to 0.3–0.4 µm and remained relatively constant from then onwards. This demonstrates that slow or ineffective inhibition mechanisms result in a rapid increase of the surface roughness in the scribe due to pitting corrosion and aluminium dissolution processes,[43] as illustrated for the non-inhibited sample in Fig. 9b. Compared to the unexposed sample, the WLI scan of non-inhibited sample visualizes the degree of damage caused by corrosion in the first

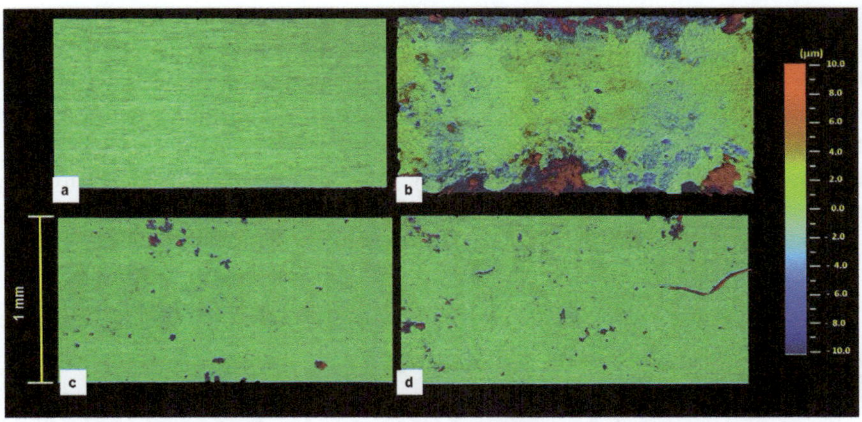

Fig. 9 High-resolution WLI images of the scribed area before and after 168 h exposure to neutral salt spray conditions, (a) unexposed scribe, (b) non-inhibited coating, (c) lithium carbonate coating and, (d) lithium oxalate coating.

168 h. On the other hand, the lithium-based samples show only limited pitting corrosion and a superficial attack on the surface which is illustrated by Fig. 9c and d. Therefore it can be concluded that the lithium-based coatings provide fast and effective inhibition on AA2024-T3.

3.4.2 Generation of the protective layer. The WLI results indicate that the inhibition mechanism must be fast and complete within 48 h. SEM crosscuts were made after 4 and 8 h with the aim of observing the generation of the protective layer at an early stage.

Fig. 10 and 11 show the formation of the protective layer after 4, 8 and 168 h neutral salt spray exposure. After 4 h, both the lithium carbonate and lithium oxalate loaded coatings (Fig. 10a and 11a) provide a protective layer with a thickness between 200 and 400 nm in the scribe. These layers have a similar morphology and thickness as the dense layers as found underneath the porous columnar layer after 168 h exposure. After 4 and 8 h, there is some columnar material present on top of the porous layer. According to these observations, it seems that the columnar layer is formed at a later stage in the process and increases the thickness of the protective layer to 1–1.5 μm.

3.5 Proposed mechanism of passive film formation

Lithium species have leached from the organic coating and a three-layered protective layer has been formed on AA2024-T3. However, the mechanism of the formation is not yet fully understood. SEM cross-sectional and TEM-EELS microscopic techniques have provided information about the protective layer in more detail. Based on this information, a mechanism is proposed for the formation of such a layer.

The SEM-cross-sectional results reveal the significant different morphology for these layers compared to the poly-crystalline layers generated at high pH and saturated aluminate solutions in the chemical conversion process of Drewien *et al.*[36] This can be explained by possible concentration and pH gradients during the leaching process. The formation of the layer appears to occur *via* a competitive growth and dissolution process of aluminium hydroxide at the aluminium interface as suggested by Tabrizi *et al.*[41] Oxide layers with a dual morphology were generated on aluminium, consisting of an inner barrier layer and an outer non-barrier crystalline layer, in long-term exposure to alkaline media. Similar layers were observed when aluminium was subjected to weak alkaline conditions by Hurlen and Haug.[42] The thickness of the barrier layer was decreasing with increasing pH. Considering the relatively thick dense inner layer, it can be assumed that the formation of the protective layer from these lithium-based

Fig. 10 SEM cross-cut images of the protective layer generated from the lithium carbonate loaded coating after (a) 4 h, (b) 8 h and, (c) 168 h neutral salt spray conditions.

Fig. 11 SEM cross-cut images of the protective layer generated from the lithium oxalate loaded coating after (a) 4 h, (b) 8 h and, (c) 168 h neutral salt spray conditions.

organic coatings occurs at weak alkaline pH. In addition to this process, the role of lithium needs to be considered.

The proposed mechanism of layer formation consists of several processes (Fig. 12). First, oxide thinning will occur due to chloride-ion adsorption in the oxide layer[39] when the aluminium is exposed to the sodium chloride solution of the salt spray test. At the same time, the leaching process starts and lithium-salts will be released from the organic coating matrix. Both lithium carbonate and lithium oxalate have an alkaline nature in solution. In addition to this, the magnesium oxide in the coating can act as an alkaline buffer and will provide an alkaline pH in the bulk solution. Since there are no significant amounts of magnesium found in the elemental analysis, it can be assumed that magnesium is not active in the inhibition process.

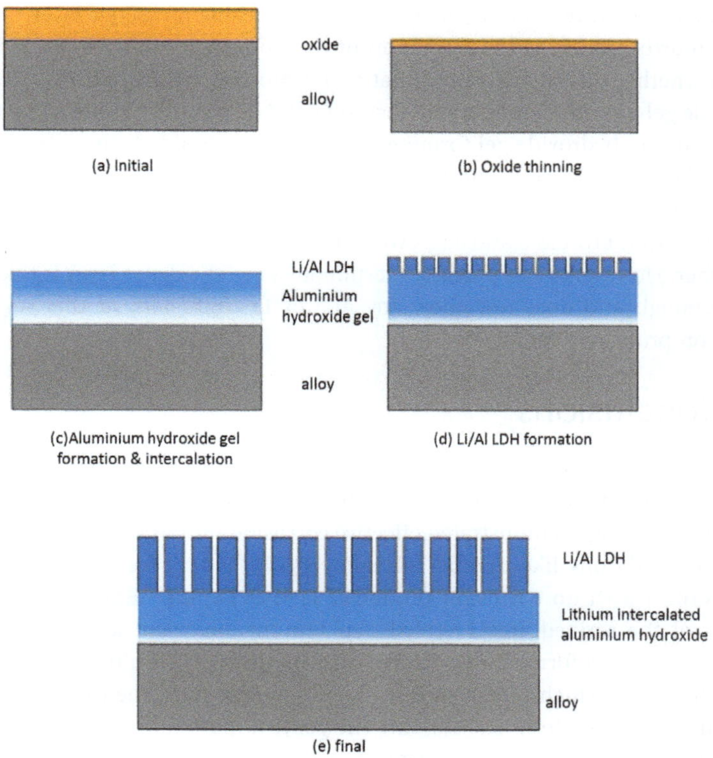

Fig. 12 Schematic illustration of the steps in the formation.

The second process is the destabilisation of the hydrated aluminium oxide in the alkaline chloride-containing environment. This process transfers the initial hydrated aluminium oxide into an aluminium hydroxide gel at the surface. Once the oxide layer is sufficiently thinned, the corrosion process commences *via* the direct attack and dissolution of the aluminium. In this alkaline chloride-containing environment, the Al^{3+} ion hydrolyses and forms amorphous $Al(OH)_3$ at the aluminium substrate interface (growth process).[40]

The third process is the dissolution reaction of the aluminium hydroxide gel at the oxide electrolyte interface. The aluminium hydroxide gel that was formed in the second process will dissolve into ionic $Al(OH)_4^-$ at the aluminium hydroxide gel/electrolyte interface (dissolution process). $Al(OH)_4^-$ is the main species present in alkaline environment.[44,45] The equilibrium or dissolution rate depends on the pH.[46] Therefore, at high pH, the solubility of $Al(OH)_4^-$ is high and it is expected that the aluminium hydroxide gel will be thin.

The SEM results show the formation of a dense layer in the early stages of salt spray exposure. In addition, EELS data reveal the presence of lithium in the dense barrier layer. Rangel and Travassos[34] suggest formation of lithium aluminate ($LiAlO_2$) *via* the AlO_2^- ion. However, this seems unlikely since this dehydrated form of $Al(OH)_4^-$ does not exist in significant concentrations according to Sipos.[45] The lithium ion is known for its facile intercalation into aluminium hydroxides, such as Bayerite and Gibbsite, forming Layered double hydroxides (LDH).[47] Moreover, amorphous aluminium hydroxides are known as effective sorbents of lithium salts.[48] Therefore, it can be assumed that lithium intercalates with the aluminium hydroxide gel to generate very small grain size particles or even an amorphous structure resulting in a dense impermeable protective layer.

The fourth process is the formation of the non-barrier outer layer. It is still unclear whether this step is a precipitation reaction or an aging process from the hydroxide gel process. It is assumed that $Al(OH)_4^-$ dissolves from the outer side of the aluminium hydroxide gel forming a Li/Al layered double hydroxide from the reaction between $Al(OH)_4^-$ and Li^+. The SEM cross-sectional results suggest that the precipitation or aging reaction occurs at a later stage in the formation of the protective layer. Process 3 and 4 (growth and dissolution) are in competition with each other. However, these processes terminate when the dense layer is thick and dense enough and does not allow any Al^{3+} or Li^+ diffusion. At this stage the inhibition process is complete.

4 Conclusions

Lithium-containing salts were used as leachable corrosion inhibitors in model organic coatings for the protection of AA2024-T3 aluminium alloys. These lithium salt loaded coatings demonstrated effective corrosion inhibition when exposed to corrosive conditions like neutral salt spray (ASTM B-117) for 168 h, providing clean scribes with no corrosion products. TEM-EELS and SEM cross-sectional analysis of the scribed area revealed that lithium ions were leached from the coating matrix and formed a protective layer on the 2024-T3 aluminium alloy of the scribed area, which protected the alloy from corrosion. The thickness of the layer varied, ranging from 0.5–1.5 μm. The general morphology of the protective layer salts showed three distinct regions, comprising a dense barrier layer near the metal/passive layer interface, a porous layer in the middle, and an outer columnar layer. The layer comprised aluminium, oxygen and lithium, as identified by EELS analysis and lithium was relatively uniform distributed in the layer. The poly-crystalline outer layer could be an Al/Li layered double hydroxide. The dense barrier layer is composed of lithium and aluminium hydroxide. However, it remains unclear in which form. The formation of this layer is a multistep process controlled by competitive film growth and dissolution processes.

Acknowledgements

The authors would like to thank Dr Simon Gibbon and Gerard Smyth for their support and contributions. Dr Derek Graham for the white light interferometry, and Dr Yaiza Gonzalez Garcia. This work was performed under the collaboration agreements of AkzoNobel with The University of Manchester and Delft University of Technology.

References

1 R. L. Twite and G. P. Bierwagen, *Prog. Org. Coat.*, 1998, **33**, 91–100.
2 N. Birbilis and R. G. Buchheit, *J. Electrochem. Soc.*, 2005, **152**, B140–B151.
3 R. G. Buchheit, R. P. Grant, P. F. Hlava, B. McKenzie and G. L. Zender, *J. Electrochem. Soc.*, 1997, **144**, 2621–2628.
4 A. Boag, A. E. Hughes, A. M. Glenn, T. H. Muster and D. McCulloch, *Corros. Sci.*, 2011, **53**, 17–26.
5 A. E. Hughes, R. J. Taylor, K. J. H. Nelson, B. R. W. Hinton and L. Wilson, *Mater. Sci. Technol.*, 1996, **12**, 928–936.
6 M. Kendig, S. Jeanjaquet, R. Addison and J. Waldrop, *Surf. Coat. Technol.*, 2001, **140**, 58–66.
7 D. Chidambaram, C. R. Clayton, M. W. Kendig and G. P. Halada, *J. Electrochem. Soc.*, 2004, **151**, B613–B620.
8 S. J. Garcia-Vergara, P. Skeldon, G. E. Thompson and H. Habazaki, *Surf. Interface Anal.*, 2007, **39**, 860–864.
9 J. Sinko, *Prog. Org. Coat.*, 2001, **42**, 267–282.
10 L. Xia, E. Akiyama, G. Frankel and R. McCreery, *J. Electrochem. Soc.*, 2000, **147**, 2556–2562.
11 F. H. Scholes, S. A. Furman, A. E. Hughes, T. Nikpour, N. Wright, P. R. Curtis, C. M. Macrae, S. Intem and A. J. Hill, *Prog. Org. Coat.*, 2006, **56**, 23–32.
12 M. W. Kendig and R. G. Buchheit, *Corrosion*, 2003, **59**, 379–400.
13 H. Wang, F. Presuel and R. G. Kelly, *Electrochim. Acta*, 2004, **49**, 239–255.
14 S. A. Furman, F. H. Scholes, A. E. Hughes and D. Lau, *Prog. Org. Coat.*, 2006, **56**, 33–38.
15 O. Lopez-Garrity and G. S. Frankel, *J. Electrochem. Soc.*, 2014, **161**, C95–C106.
16 M. Iannuzzi, T. Young and G. S. Frankel, *J. Electrochem. Soc.*, 2006, **153**, B533–B541.
17 R. L. Cook Jr, *Corrosion*, 2000, **56**, 321–333.
18 A. E. Hughes, J. D. Gorman, T. G. Harvey, A. Galassi and G. McAdam, *Corrosion*, 2006, **62**, 773–780.
19 J. A. Hill, T. Markley, M. Forsyth, P. C. Howlett and B. R. W. Hinton, *J. Alloys Compd.*, 2011, **509**, 1683–1690.

20 D. Ho, N. Brack, J. Scully, T. Markley, M. Forsyth and B. Hinton, *J. Electrochem. Soc.*, 2006, **153**, B392–B401.
21 T. Stimpfling, F. Leroux and H. Hintze-Bruening, *Eur. J. Inorg. Chem.*, 2012, 5396–5404.
22 M. Forsyth, M. Seter, M. Y. Tan and B. Hinton, *Corros. Eng., Sci. Technol.*, 2014, **49**, 130–135.
23 A. J. Aldykewicz Jr, H. S. Isaacs and A. J. Davenport, *J. Electrochem. Soc.*, 1995, **142**, 3342–3350.
24 M. O'Keefe, B. Fahrenholtz, E. Morris and R. Albers, *Corrosion resistant polymer coatings containing rare earth compounds*, National Association for Surface Finishing Annual Conference and Trade Show, SUR/FIN, Chicago, IL, 2013, 501–513.
25 M. E. Nanna and G. P. Bierwagen, *JCT Res.*, 2004, **1**, 69–80.
26 P. Plagemann, J. Weise and A. Zockoll, *Prog. Org. Coat.*, 2013, **76**, 616–625.
27 S. V. Lamaka, M. L. Zheludkevich, K. A. Yasakau, M. F. Montemor and M. G. S. Ferreira, *Electrochim. Acta*, 2007, **52**, 7231–7247.
28 M. L. Zheludkevich, K. A. Yasakau, S. K. Poznyak and M. G. S. Ferreira, *Corros. Sci.*, 2005, **47**, 3368–3383.
29 T. G. Harvey, S. G. Hardin, A. E. Hughes, T. H. Muster, P. A. White, T. A. Markley, P. A. Corrigan, J. Mardel, S. J. Garcia, J. M. C. Mol and A. M. Glenn, *Corros. Sci.*, 2011, **53**, 2184–2190.
30 G. Bierwagen, R. Brown, D. Battocchi and S. Hayes, *Prog. Org. Coat.*, 2010, **67**, 195–208.
31 P. Visser and S. A. Hayes, WO2010112605-A1, 2010.
32 J. Gui and T. M. Devine, *Scr. Metall.*, 1987, **21**, 853–857.
33 C. M. Rangel and M. A. Travassos, *Corros. Sci.*, 1992, **33**, 327–343.
34 C. M. Rangel and M. A. Travassos, *Surf. Coat. Technol.*, 2006, **200**, 5823–5828.
35 R. G. Buchheit, M. D. Bode and G. E. Stoner, *Corrosion*, 1994, **50**, 205–214.
36 C. A. Drewien, M. O. Eatough, D. R. Tallant, C. R. Hills and R. G. Buchheit, *J. Mater. Res.*, 1996, **11**, 1507–1513.
37 M. R. S. Castro, J. C. Nogueira, G. P. Thim and M. A. S. Oliveira, *Thin Solid Films*, 2004, **457**, 307–312.
38 M. Frenkel, A. Glasner and S. Sarig, *J. Phys. Chem.*, 1980, **84**, 507–510.
39 R. T. Foley, *Corrosion*, 1986, **42**, 277–288.
40 R. T. Foley and T. H. Nguyen, *J. Electrochem. Soc.*, 1982, **129**, 464–467.
41 M. R. Tabrizi, S. B. Lyon, G. E. Thompson and J. M. Ferguson, *Corros. Sci.*, 1991, **32**, 733–742.
42 T. Hurlen and A. T. Haug, *Electrochim. Acta*, 1984, **29**, 1133–1138.
43 P. Visser, *Chrome free corrosion inhibiting coating technology for the protection of high strength aluminum alloys*, National Association for Surface Finishing Annual Conference and Trade Show, SUR/FIN, Chicago, IL, 2013, 514–526.
44 L. A. Carreiea, V. A. Maroni, J. W. Swaine Jr and R. C. Plumb, *J. Chem. Phys.*, 1966, **45**, 2216–2220.
45 P. Sipos, *J. Mol. Liq.*, 2009, **146**, 1–14.
46 C. F. Baes and R. E. Mesmer, *The hydrolysis of cations*, Wiley, 1976.
47 G. R. Williams and D. O'Hare, *J. Phys. Chem. B*, 2006, **110**, 10619–10629.
48 V. P. Isupov, *J. Struct. Chem.*, 1999, **40**, 672–685.

Faraday Discussions

PAPER

Mapping water uptake in organic coatings using AFM-IR

S. Morsch,*[a] S. Lyon,[a] P. Greensmith,[b] S. D. Smith[c] and S. R. Gibbon[d]

Received 21st November 2014, Accepted 28th January 2015
DOI: 10.1039/c4fd00229f

The long-term failure of seemingly intact corrosion resistant organic coatings is thought to occur *via* the development of ionic transport channels, which spontaneously evolve from hydrophilic regions on immersion, *i.e.*, as a result of localized water uptake. To this end, we investigate water uptake characteristics for industrial epoxy–phenolic can coatings after immersion in deionized water and drying. Moisture sorption and the changing nature of polymer–water interactions are assessed using FTIR for dry and pre-soaked films. More water is found to be absorbed by the pre-soaked coatings on exposure to a humid environment, with a greater degree of hydrogen-bonding between the polymer and water. Furthermore, morphological changes are then correlated to localized water uptake using the AFM-IR technique. Nanoscale softened regions develop on soaking, and these are found to absorb a greater proportion of water from a humid environment.

Introduction

Paints are widely used to isolate metals from corrosive environments; therefore the long-term breakdown in protection afforded by organic coatings is a fundamentally important process in corrosion science. For organic coatings on steel, whilst corrosion initiation mechanisms are well-established in the presence of a macroscopic defect, the deterioration of seemingly intact coatings is not fully understood.[1–3] Currently, the prevailing model describing failure of intact organic coatings entails the spontaneous formation of conducting pathways through the coating, followed by corrosion initiation (*i.e.*, the base of the ionic pathway behaves as a macroscopic defect), Scheme 1. Nguyen proposed that these ionic transport pathways are formed by the action of water during immersion; *via* hydrolysis

[a]*Corrosion and Protection Centre, School of Materials, The University of Manchester, The Mill, Sackville St, Manchester, M13 9PL, UK. E-mail: suzanne.morsch@manchester.ac.uk; Tel: +44 (0)161 306 2914*

[b]*Material Science Centre, School of Materials, The University of Manchester, Oxford Road, Manchester, M13 9PL, UK*

[c]*AkzoNobel Chemicals bv, Supply Chain, Research & Development, P.O. Box 10, 7400 AA Deventer, The Netherlands*

[d]*AkzoNobel, Supply Chain, Research & Development, Stoneygate Lane, Felling, Gateshead, Tyne & Wear, NE10 0JY, UK*

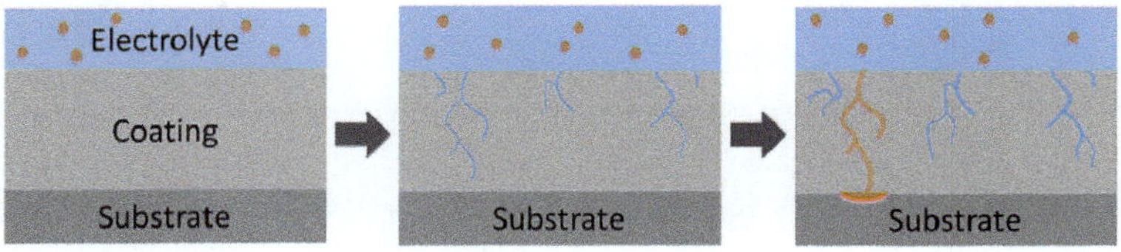

Scheme 1 The proposed mechanism for breakdown of an intact organic coating in contact with electrolyte: formation of hydrophilic pathways and the subsequent generation of an ionic transport channel leading to a localised corrosion spot.

reactions, leaching of unreacted material or else the gradual formation of hydrophilic channels traversing the film.[4] In the case of epoxies, (which are widely used as binders in corrosion resistant organic coatings) little evidence exists for hydrolytic breakdown under ambient conditions. Furthermore, for food-contact epoxy–phenolic coatings (such as those studied here) resistance to hydrolysis reactions is well documented and polymer leaching can be considered negligible.[5–8] Thus, ionic transport pathways are considered to evolve from widening water transport channels within the coating, *i.e.*, as a result of localised water uptake.

Water transport in epoxy resins has been extensively investigated, since epoxies are employed in a host of applications (*e.g.*, aerospace engineering composites,[9–11] marine coatings,[12] microelectronics,[13,14] and packaging)[15] where the deleterious effects of absorbed moisture are problematic.[16,17] Water uptake has been evaluated by numerous techniques including NMR,[18,19] FTIR,[20–24] fluorescence,[25] dielectric spectroscopy,[26,27] simulation[28] and gravimetric analysis.[29,30] In general, epoxy materials have been found to be hygroscopic, and water uptake has been shown to follow Fickian diffusion kinetics, indicating that diffusion of the penetrant (water) is driven by a concentration gradient across the polymer film. The apparent equilibrium water content has been correlated firstly to the polarity of the polymer network,[31–35] and secondly, in the absence of polarity effects, to the available free volume.[36–40] Soles *et al.* suggested a model encompassing these results whereby water uptake is dominated by transport through the available free volume, which only allows access to neighbouring polymeric hydrogen-bonding sites.[39] However, in addition to water uptake into the available free-volume, numerous studies have demonstrated that after a pseudo-equilibrium state is reached, a considerably slower, non-Fickian process drives further sorption.[41–44] This is thought to be brought about by polymeric relaxation induced by water plasticization; the slow segmental motions of polymer chains allow water to interact with additional hydrogen-bonding sites and thereby penetrate the glassy resin, swelling the network.[45–47] In reality, both processes (polymer relaxation and Fickian water uptake into the free volume) are likely to be active during immersion, but the kinetics of water uptake are often interpreted as purely Fickian behaviour.[48] This is partly because uptake into the free volume dominates at short times for thin films. In addition, for thick samples de-convolution of the two superimposed processes can be problematic, since diffusion into the free volume may occur on a comparable timescale to polymeric relaxation (the time required for concentration driven diffusion into the film is thickness dependent, whereas polymer relaxation is not).[41]

In a recent publication we reported time-resolved FTIR analysis of water uptake into an industrial epoxy–phenolic can coating, and found that water residing in the films after short-term exposure to a humid environment was primarily associated with the free volume, in keeping with Soles' proposed mechanism.[49] In the present contribution we use FTIR to assess water uptake into thin films of the same epoxy–phenolic can-coating before and after prolonged immersion in deionised water. Moreover, nanoscale morphological changes induced by immersion are evaluated using AFM topographical imaging, and localised water sorption is monitored using AFM-IR in order to establish whether these correspond to the formation of hydrophilic pathways, in support of Nguyen's mechanism for the breakdown of intact organic coatings.

Experimental

Sample preparation

Steel substrates were degreased by sonic cleaning in ethanol (Fisher Scientific, >99%). Thin films of epoxy–phenolic Vitalure 345 Lacquer (AkzoNobel Packaging Coatings, Birmingham, UK) or a 1 : 2 mixture of Vitalure 345 Lacquer and thinner (50 : 50 Solvesso100 : methoxy propanol, AkzoNobel Packaging Coatings, Birmingham, UK) were then cast onto electrolytic chrome-coated steel pieces (4 cm^2) or polished carbon steel pieces (1 cm^2) respectively at 2000 rpm for 30 s (Headway Research Inc., 1–10 000 rpm). Samples were cured by placing in an oven maintained at 200 °C for 10 min.

Film characterisation

Infrared spectra are the result of 16 accumulations collected in the reflectance mode using an FTIR spectrometer (Spectrum 2000, Perkin Elmer), operating at 4 cm^{-1} resolution across the 700–4000 cm^{-1} range. In order to assess water uptake, saturated NaCl solution was warmed to 50 °C and placed in recrystallization dishes at the bottom of the sample chamber, which was fitted with a portable temperature and humidity data logger (Lascar Electronics). Humidity was allowed to equilibrate for 20 min prior to collection of the background spectrum (electrolytic chrome-coated steel substrate). Within 5 min of salt solution insertion, humidity within the sample chamber was measured to be 80 ± 1% RH, and remained stable throughout the experiment. FTIR spectra were gathered continuously for 60 min following insertion of the coated samples (4 cm^2 electrolytic chrome coated steel covered with 7.4 ± 0.6 μm epoxy–phenolic coatings).

Sessile contact angle analysis was performed at 20 °C using video capture apparatus (FTA188 Tensiometer, FTA Europe) and 5 μL of high purity water.

For modulated DSC, epoxy–phenolic coatings were mechanically removed from the substrate and 1–3 mg placed in closed aluminium pans. DSC thermograms were obtained over a temperature range of −90 °C to 300 °C under nitrogen, using a heating/cooling rate of 3 °C min^{-1} with a modulation period of ±1 °C min^{-1} (Q100 modulated DSC, TA Instruments).

Atomic force microscope images were obtained using a Multimode 8 (Bruker, Santa Barbara) operating in Peakforce Nanomechanical mode using a Pt–Ir coated SCMPIT probe (Bruker, nominal spring constant 2 N m^{-1}, nominal resonance frequency of 80 kHz). Measurements were made within a custom made humidity chamber equipped with a portable temperature and humidity data logger (Lascar Electronics). Humidity was maintained at 70% RH by controlling the rate at which

Scheme 2 AFM-IR with top-down illumination.

air was bubbled through tap water into the chamber, which envelops the microscope scanner and head.

AFM-IR

AFM-IR was performed on a NanoIR2 system (Anasys Instruments). For AFM-IR, the sample was illuminated by a tunable infrared source, which was pulsed rapidly, Scheme 2. On absorbance of IR radiation, abrupt thermal expansion of the sample excites an AFM probe in contact with the surface to oscillate at its resonant frequency. It has been shown that the amplitude of this oscillation is proportional to IR absorbance, so that plotting amplitude as a function of IR wavelength yields spectra closely matched to those obtained by macroscopic transmission-mode FTIR.[50] Furthermore, the IR pulse (10 ns duration), thermal expansion, and damping down of the induced oscillation occurs on a faster timescale than the feedback electronics of the AFM, enabling simultaneous contact-mode topographical measurement and localised IR absorbance mapping at a given wavelength.[51–54] AFM-IR images were collected in contact mode at a scan rate of 0.04 Hz using a gold-coated silicon nitride probe (Anasys Instruments, 0.07–0.4 N m^{-1} spring constant, 13 $\pm$ 4 kHz resonant frequency). An optical parametric oscillator was used as the source of IR radiation incident on the sample, which was subjected to pulses of 10 ns duration at a repetition rate of 1 kHz. The amplitude of induced cantilever oscillations were then mapped using 32 co-averages per 1024 points per 300 scan lines. The specimen and AFM head were contained within a sample chamber equipped with a portable temperature and humidity logger. In order to raise humidity, saturated NaCl solution was placed in a recrystallizing dish within the sample chamber and allowed to equilibrate for 24 h. After mapping under raised humidity, the salt solution was removed from the chamber and replaced with silica desiccant. Samples were then allowed to dry for 48 h prior to imaging. Ratio images were generated from overlapping regions of IR amplitude maps, cross-correlated using the corresponding height images.

Results

FTIR water uptake

In order to investigate the effect of prolonged exposure to water, epoxy–phenolic coatings were immersed in deionised water for prescribed time intervals (1 h to 7 days), and then dried in a desiccator for 3 days. Water uptake by the dried coatings was then assessed by continuously monitoring FTIR spectra under raised humidity, Fig. 1. Following immersion and drying, infrared absorbance in the fingerprint region (700–2000 cm^{-1}) was found to be unchanged, and remained unaffected by exposure to a high humidity (80% RH) environment, confirming that no disruption of macromolecular structure had occurred due to hydrolysis reactions. During exposure to raised humidity, an incremental increase of the broad absorbance centred around 3416 cm^{-1} was observed in all cases, Fig. 1. This band is attributed to the OH stretch of hydroxyl groups within the epoxy–phenolic coating, with the OH stretch of absorbed water superimposed upon it.[55] Moisture uptake is therefore directly proportional to the total absorbance increase in this region, and was quantified by integration of the 3416 cm^{-1} band at each time point. For direct comparison of water sorption between samples, this value was normalised to the unchanged area of the CH stretch band (centred on 2892 cm^{-1}) and the initial reading (corresponding to 0.7 min) was set to zero, Fig. 2. The time required to acquire each data point is 0.7 min since each spectrum corresponds to 16 accumulations. It can be seen that the pre-soaked samples absorbed more water from the humid environment.

Furthermore, plotting the absorbance increase with respect to $t^{1/2}$ revealed initially linear correlations for both dry and pre-soaked samples in keeping with Fickian diffusion kinetics, Fig. 3.[56] In the case of coatings dried after immersion for less than 3 days, it can be clearly seen that water uptake deviates from ideal Fickian behaviour at later stages (*i.e.*, it does not level off to an equilibrium uptake value). This is consistent with the proposed mechanism of polymer relaxation in the presence of adsorbed water, leading to continued slow water sorption after saturation of the readily available free volume. In contrast, no clear transition between Fickian and non-Fickian kinetics was observed for coatings immersed for longer periods (3 days or 7 days), which instead show a linear dependence throughout. Pre-soaked samples display increasingly fast water sorption kinetics with soaking time (the gradient of the linear portion increases, note the difference in *z*-scales, Fig. 3). This increase in kinetics and overall absorption can be understood by considering that the network rearrangement which takes place during immersion is frozen into position during drying (due to the removal of

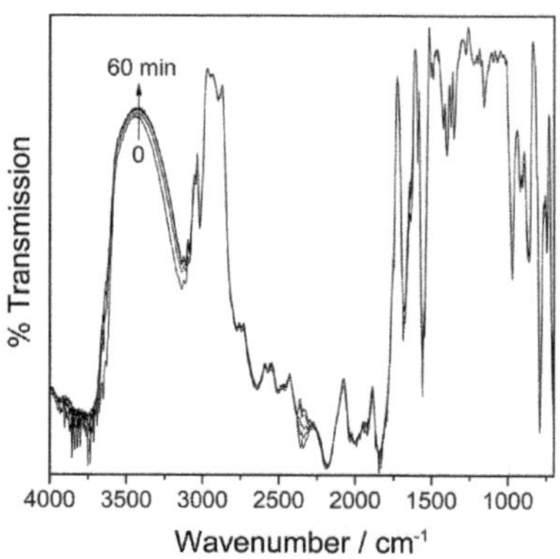

Fig. 1 Mid-IR spectra of spin-coated epoxy–phenolic coating after exposure to an 80% RH environment for 0, 10, 20 and 60 min.

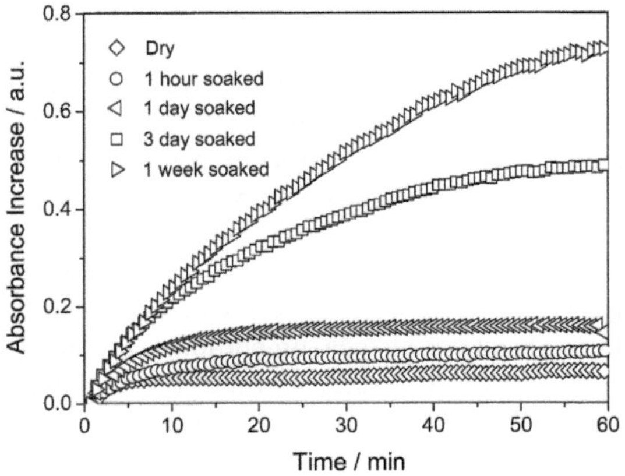

Fig. 2 Absorbance increase of the OH stretch band (normalised to the CH stretch band) as a function of exposure time to 80% RH for epoxy–phenolic coatings as deposited and following 1 h, 1 day, 3 days and 7 days immersion in deionised water and drying.

plasticizing water), so that a more open structure is presented for faster re-adsorption of water from humid air.

Further evidence for rearrangement of the coating was obtained from difference spectra corresponding to the absorbed water. These were generated by subtraction of the initial FTIR spectrum (at 0.7 min exposure time) from those collected after 60 min exposure to humid conditions. Inspection of the OH stretch absorbance band revealed a complex envelope with a fine structure, indicating that absorbed water exists in a variety of environments within the epoxy–phenolic films, as has previously been reported,[49,57] Fig. 4. Notable features of the OH stretch band for sorbed water include a sharp absorbance in the 3600–3700 cm^{-1} region, characteristic of dangling OH bonds uninvolved in hydrogen-bonding,[24]

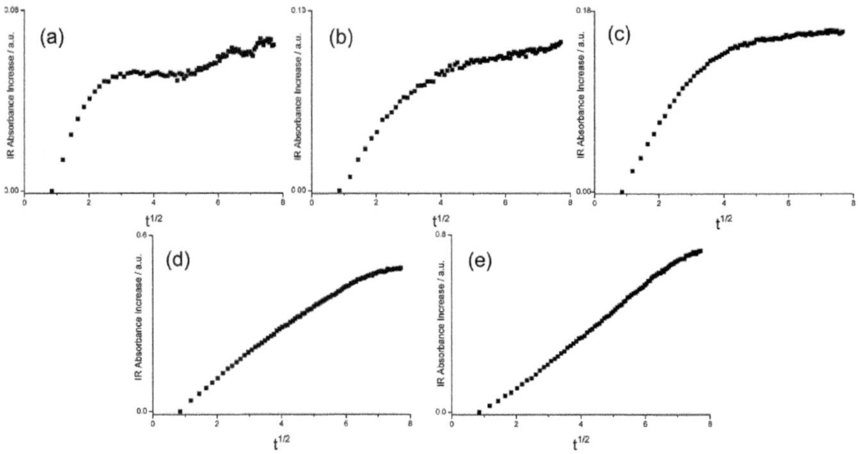

Fig. 3 Absorbance increase of the OH stretch band (normalised to the CH stretch band) as a function of the square root of time exposed to 80% RH for epoxy–phenolic coatings (a) as deposited; and following (b) 1 h; (c) 1 day; (d) 3 days and (e) 7 days immersion in deionised water and drying.

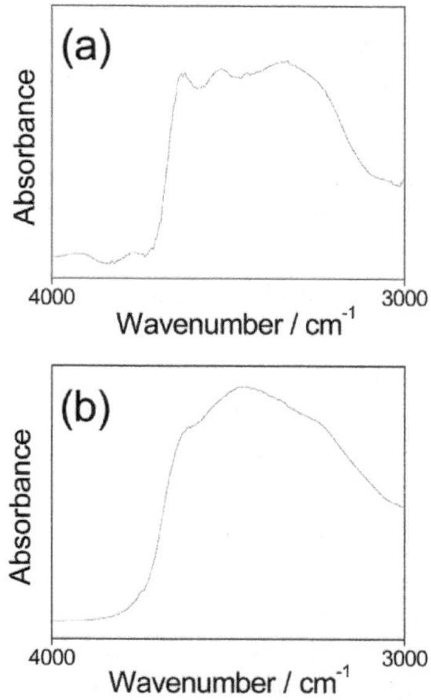

Fig. 4 Water OH stretch band obtained from difference spectra for epoxy–phenolic coating after 60 min exposure to 80% RH for (a) as deposited, dry coating and (b) coating after 7 days soak in deionised water and drying.

whereas absorbance in the lower wavenumber region around 3100–3250 cm^{-1} is indicative of strongly hydrogen-bonded water; typically highly-orientated, multiple hydrogen-bonded species, corresponding to clustered water.[58] The intermediate region, centred on ~3400 cm^{-1}, corresponds to water weakly hydrogen-bonded to polymeric sites. Difference spectra generated after 60 min exposure to raised humidity clearly show enhanced absorbance in this central region for the 7 day pre-soaked sample when compared to the spectrum generated for the as-deposited coating, Fig. 4. This demonstrates that a greater proportion of the water taken up by pre-soaked samples exists in a form weakly hydrogen-bonded to the polymer network.

In addition, sessile contact angle analysis indicated that samples became more wettable after prolonged immersion and drying, thereby providing evidence that polymer relaxation during soaking results in more favourable conformations for interaction with water, Table 1. This is unsurprising in light of a reported SFG vibrational spectroscopy study, which demonstrated rapid reorientation of

Table 1 Static contact angle data for spin coated epoxy–phenolic coatings. Errors correspond to 1 s.d.

Sample	T_g (°C)	Contact angle (°)
Dry (as deposited)	89.3	90 ± 1
1 hour immersed in H_2O	90.5	89 ± 2
1 day immersed in H_2O	88.1	87 ± 2
3 days immersed in H_2O	87.7	83 ± 1
7 days immersed in H_2O	84.7	83 ± 1

phenolic functional groups at the surface of phenolic resin after short-term exposure to a humid environment,[59] however the timescales involved in the present case are much longer. Overall, the results from FTIR difference spectra and contact angle analysis point towards significant network rearrangement during prolonged soaking, after which more hydrogen-bonding sites become available to water re-absorbed into the coating and also to water droplets residing at the sample surface. Since enhanced wettability coincides with changes in water sorption kinetics (*i.e.*, wettability and kinetics change most dramatically in the case of 3 day and 7 day soaked samples, Table 1 and Fig. 3), it can be inferred that such changes in the hydrogen-bonding capability occur *via* the same mechanism, *i.e.*, water-plasticized relaxation of polymer segments.

The plasticizing effect of absorbed water was confirmed using modulated DSC analysis, Table 1. After immersion, the coating was mechanically removed from steel substrates, and stored under humid conditions until analysis. Measured T_g values were found to decrease as a function of immersion time. After heating to 100 °C for 1 h (*i.e.*, above the glass transition), the measured T_g value for a sample immersed for 7 days not only recovered, but increased to 95.1 °C (*i.e.*, higher than the T_g measured for a pristine sample, indicating that drying and further curing was induced by warming).

AFM-IR water uptake

In order to establish whether water uptake was heterogeneous, the samples were analysed using AFM. To ensure good resolution for AFM-IR measurements, thin films (500–600 nm thick) of epoxy–phenolic coating were prepared on polished mild steel substrates. For long-term immersion experiments these samples can be considered representative of the thicker coatings used for bulk FTIR analysis

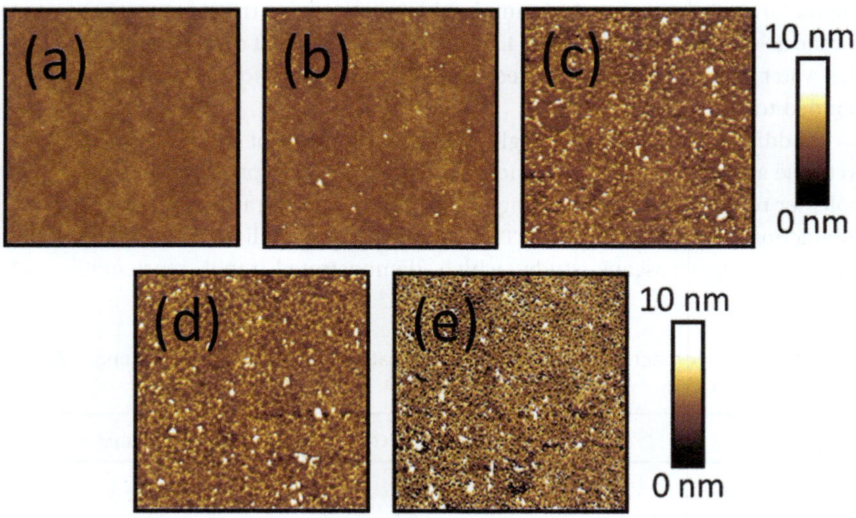

Fig. 5 4 µm × 4 µm peak force tapping mode AFM height images of spin-coated epoxy–phenolic thin films (a) as deposited; (b) after 1 h immersion in deionised water; (c) after 1 day immersion in deionised water; (d) after 3 days immersion in deionised water and (e) after 7 days immersion in deionised water.

since the kinetics of polymer relaxation should be independent of film thickness (whereas thickness-dependent Fickian diffusion into the free volume was shown to be complete in under 10 min by FTIR, Fig. 3). Morphological changes were initially measured using the peak force tapping mode to avoid distortion of the topography. Samples were removed from water, air dried and immediately transferred to the microscope, where the topography was measured under humid conditions (70% RH), Fig. 5. It can be seen that surface roughness increased after immersion in water, with raised bumps and a 'dimpled' morphology emerging after several days.

To determine whether these topographic changes were due to chemical heterogeneity, a 7 day pre-soaked specimen was transferred to the AFM-IR, and infrared spectra were acquired in numerous positions, Fig. 6. No change in the position or relative intensity of the peaks was observed, indicating that the film is chemically homogeneous, and ruling out the possibility that raised regions are due to sample contamination.

AFM-IR mapping was then performed at 2964 cm^{-1}, 3296 cm^{-1} and 3420 cm^{-1} under humid conditions (60% RH), Fig. 7. Following this, desiccant was placed in the sample chamber and this process was repeated over the same region after 48 h at 30% RH. At both humidity levels, the raised bumps were found to correspond

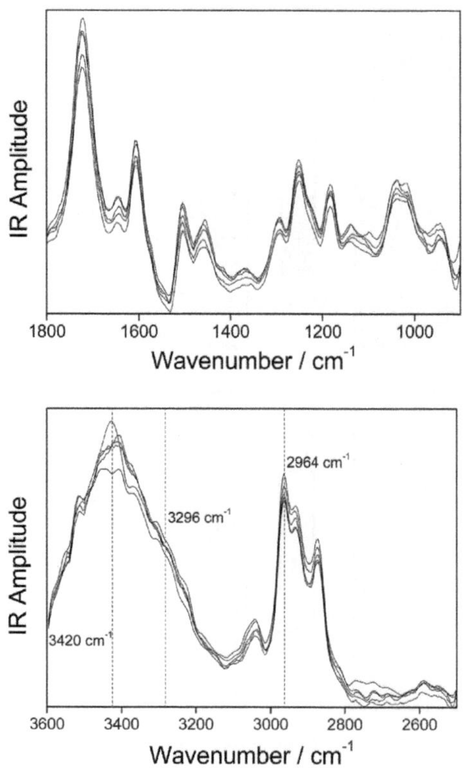

Fig. 6 Induced amplitude of the AFM probe as a function of excitation wavenumber for the mid-IR fingerprint region and higher wavenumber CH and OH stretch band region, taken at various locations on the 7 day pre-soaked epoxy–phenolic coating.

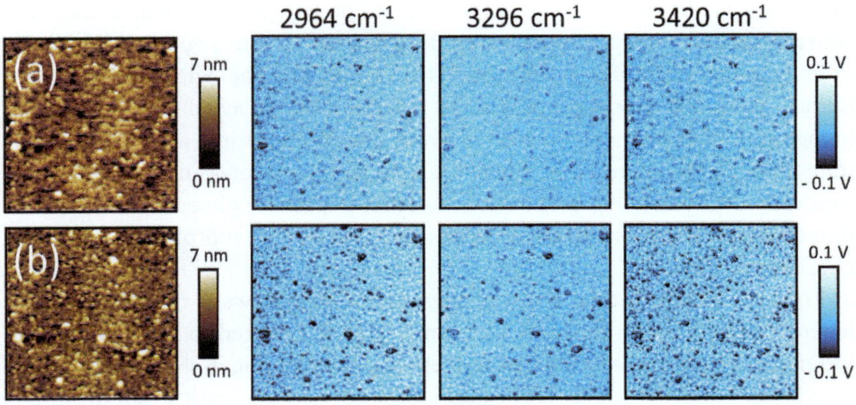

Fig. 7 2 μm × 2 μm AFM-IR height and corresponding IR amplitude images of epoxy–phenolic coating after 7 days immersion under (a) dry conditions (30% RH) and (b) raised humidity (60% RH). AFM-IR amplitude maps have been flattened around zero for clarity, where the *z*-scale (Volts) corresponds to the AFM photodetector response used to detect IR induced amplitude.

to lower IR-induced amplitude (and hence IR absorbance) at each wavelength, Fig. 7. However, when ratio images were generated from maps taken at 60% RH and 30% RH (thus corresponding to the location of adsorbed water, comparable

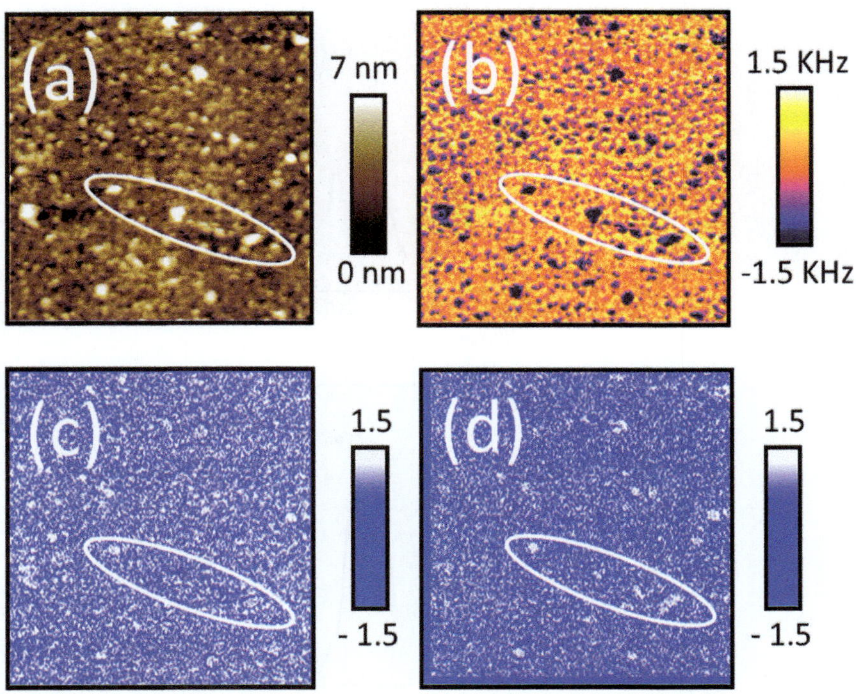

Fig. 8 2 μm × 2 μm AFM-IR maps of the epoxy–phenolic coating after 7 days immersion (a) height and (b) corresponding AFM-IR frequency map gathered under raised humidity at 2964 cm^{-1} illumination; (c) IR ratio image generated from IR amplitude maps at 3420 cm^{-1} under 60% RH/30% RH and (d) IR ratio image generated from IR amplitude maps at 3296 cm^{-1} under RH 60%/30% RH. Circled regions are presented as a guide to the eye.

to difference spectra generated using bulk FTIR), increases in absorbance were found to be localised on raised regions, indicating enhanced water uptake, Fig. 8c and d. This is particularly notable for absorbance at 3296 cm^{-1}, Fig. 8d. Furthermore, AFM-IR mapping simultaneously generates an image corresponding to the frequency of the induced oscillation. Since the tip is continually in contact with the surface, this frequency is a measure of local hardness. From frequency images corresponding to the soaked sample surface, it can be seen that the raised regions which take up more water are also softer than the surrounding coating, Fig. 8b.

Further evidence for enhanced absorbance of water into raised polymer regions was found in the IR ratio image generated from the overlapping IR amplitude maps at 3420 cm^{-1}/2964 cm^{-1} (weakly hydrogen-bonded OH stretch normalised to the CH stretch) at 60% RH. Areas of bright contrast should in this case correspond to the location of weakly hydrogen-bonded water, and from this image it is apparent that a greater degree of water is present on all raised regions of the polymer, Fig. 9b. The ratio image generated for 3296 cm^{-1}/2964 cm^{-1} (strongly hydrogen-bonded OH/CH stretch) corresponds to the location of multiple hydrogen-bonded species, and displays comparatively enhanced contrast, indicating that clustered water is more heavily localised to some of the larger raised patches, Fig. 9c. This is further confirmed by the ratio image

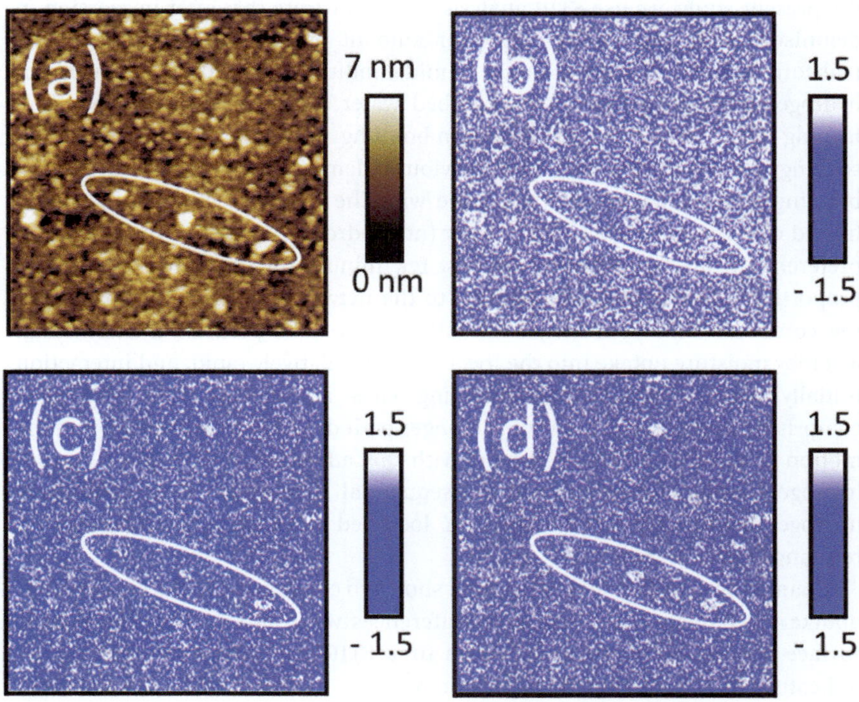

Fig. 9 2 μm × 2 μm AFM-IR (a) height and (b) corresponding IR ratio image generated from IR amplitude maps obtained under 60% RH at 3420 cm^{-1}/2964 cm^{-1}; (c) 3296 cm^{-1}/2964 cm^{-1} and (d) 3296 cm^{-1}/3420 cm^{-1}. Circled regions are presented as a guide to the eye.

generated from maps at 3296 cm^{-1}/3420 cm^{-1} (strongly hydrogen-bonded OH/ weakly hydrogen-bonded OH), Fig. 9d.

Discussion

It is has long been known that water diffusion into glassy polymer networks follows 'two-stage' kinetics, corresponding to two distinct absorption mechanisms. The first stage can be approximated by Fickian diffusion kinetics, indicating that it corresponds to the movement of water molecules driven by a concentration gradient.[60,61] During this initial stage transport occurs through the frozen free volume, and water binds only to neighbouring hydrogen-bonding sites (because further hydrogen-bonding sites within the network are inaccessible). For epoxies, this initial stage of relatively rapid water uptake (diffusion coefficients at 35 °C have previously been calculated based on purely Fickian kinetics to be in the range of 2×10^{-9} cm^2 s^{-1} to 4×10^{-9} cm^2 s^{-1} for a wide range of epoxies[38]) has been demonstrated to occur alongside a slower non-Fickian process, thought to correspond to network relaxation processes plasticized by water. This mechanism has previously been proposed based on gravimetric methods used to monitor kinetics, and thermal analysis on bulk samples.[41–44,46,47,61,62] For example, it has repeatedly been shown that water desorption and resorption occurs on a faster timescale than the initial absorption, due to irreversible (below the T_g) rearrangement during water uptake (because on desorption of water, the network is no longer plasticized).[41,46,47] In the present study, we use FTIR analysis to demonstrate that after immersion in deionised water and drying, a greater amount of moisture is more rapidly reabsorbed into the coatings under humid conditions. Moreover, the degree of hydrogen-bonding between the absorbed water and network increased, confirming that relatively more hydrogen-bonding sites became available after soaking. For dry samples, we have previously demonstrated that the hydrogen-bonding sites available for water scale with the available free volume under humid conditions, and that 'free' water (not hydrogen-bonded to the resin) was preferentially absorbed during the first few minutes of exposure (whereas the proportion of water hydrogen-bonded to the network gradually increased over the course of an experiment).[49] Overall, these results point to a mechanism whereby moisture uptake into the free volume is relatively rapid, and interaction initially occurs with hydrogen-bonding sites adjacent to the microvoids comprising the free volume. Then, at longer periods, it can be envisaged that the motion of polymer chains in contact with this adsorbed water exposes further hydrogen-bonding sites and the consequential disruption of the polymeric hydrogen-bonding network results in localised swelling, plasticization and rearrangement, Scheme 3.

Changes in sample topography were shown to correspond to the slower water uptake mechanism, since minimal differences were observed on the sample surface after 1 h immersion, at which time FTIR water sorption experiments indicate that a transition from Fickian kinetics to slower water sorption has occurred. Absorption driven by polymer relaxation is associated with network swelling and plasticization, because long-range macromolecular motions allow a degree of disruption to the polymer–polymer hydrogen-bonding network. Furthermore, this process is thought to allow access to isolated pockets of free

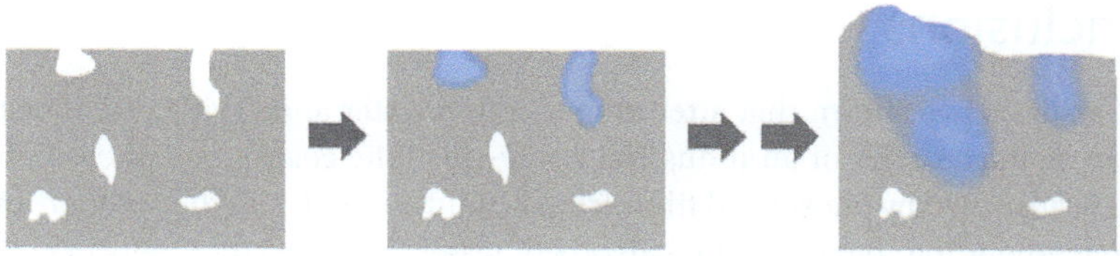

Scheme 3 Diffusion of water into the available free volume accompanied with hydrogen-bonding to sites on walls of the microvoids comprising the free volume, followed by polymer relaxation, eventually allowing access to isolated pockets of free volume and hydrogen-bonding sites in the bulk, leading to localised swelling.

volume trapped within the network, resulting in a more open, interconnected structure.[61,63] Thus, the observation of raised, softened regions able to absorb more water supports heterogeneous long-term water uptake. Attenuated IR amplitudes were consistently measured for these raised regions at every wavelength mapped. Given that the normalised spectra gathered display no differences in relative peak intensity, it can be inferred that a correspondingly lowered IR absorbance would be detected at every wavelength corresponding to polymeric functional groups. This is consistent with a locally lowered density of material in contact with the AFM probe tip (*i.e.*, a more open structure, also consistent with detected local softening). Furthermore, the IR amplitude signal measured for strongly hydrogen-bonded water is particularly concentrated in these regions. This can be interpreted as water–water hydrogen-bonding, *i.e.*, the formation of clustered water within regions of more open, connected material, which itself corresponds to highly swollen regions generated during immersion. The location of such regions may correspond to local chemical/physical heterogeneities within the polymer network (*e.g.*, a higher concentration of hydrophilic groups, or enhanced polymer mobility) which induces localised swelling during immersion. Since epoxy networks are highly cross-linked glassy systems, polymer mobility is limited and therefore swelling is expected to produce minimal topographic changes. It is notable that since the size of softened regions is much smaller than the overall coating thickness (lateral dimensions < 100 nm, height < 10 nm, 500–600 nm coating thickness) thus it seems unlikely that these regions penetrate the coating after immersion for one week.

These findings are significant in the field of corrosion protection since in the absence of a defect, the breakdown of corrosion-resistant coatings is thought to occur *via* the formation of hydrophilic pathways, which eventually allow the passage of ions between the electrolyte and underlying substrate.[4] Since the second, slower stage of water uptake generally proceeds on a timescale more commensurate with corrosion initiation, it is likely that any formation of such transport pathways would take place *via* this polymer relaxation mediated mechanism. Furthermore, for ion transport to occur, the ion and its hydration shell must be able to traverse the coating, so channels are required which hold significant amounts of free water (not hydrogen-bonded to the polymeric network). In light of this, the formation of highly localised low density, softened regions with the ability to absorb a greater volume of free water corroborates the coating breakdown model.

Conclusions

FTIR analysis has shown that after immersion in water and drying, the moisture uptake characteristics of an industrial epoxy–phenolic coating are altered; more water is absorbed by pre-soaked films from humid air, with a higher proportion of hydrogen-bonding between the adsorbed water and polymer network. This coincides with increasing wettability of the coatings, a drop in the measured T_g and the appearance of nanoscale raised regions on the surface. This is interpreted in terms of a water plasticized macromolecular relaxation mechanism active during immersion, which results in the formation of nanoscale hydrophilic regions.

Acknowledgements

S. Morsch would like to thank AkzoNobel for materials and financial support.

References

1 M. Stratmann, R. Feser and A. Leng, *Electrochim. Acta*, 1994, **39**, 1207–1214.
2 H. Leidheiser, W. Wang and L. Igetoft, *Prog. Org. Coat.*, 1983, **11**, 19–40.
3 B. Reddy, M. J. Doherty and J. M. Sykes, *Electrochim. Acta*, 2004, **49**, 2965–2972.
4 T. Nguyen, J. B. Hubbard and J. M. Pommersheim, *J. Coat. Technol.*, 1996, **68**, 45–56.
5 Z. W. Wicks, F. N. Jones, S. P. Pappas and D. A. Wicks, *Organic Coatings Science and Technology*, Wiley-Interscience, Hoboken, NJ, 3rd edn, 2007.
6 T. Yoshida, M. Horie, Y. Hoshino and H. Nakazawa, *Food Addit. Contam.*, 2001, **18**, 69–75.
7 C.-M. Chang, C.-C. Chou and M.-R. Lee, *Anal. Chim. Acta*, 2005, **539**, 41–47.
8 J.-H. Kang and F. Kondo, *Res. Vet. Sci.*, 2002, **73**, 177–182.
9 A. Toldy, B. Szolnoki and Gy. Marosi, *Polym. Degrad. Stab.*, 2011, **96**, 371–376.
10 G. Williams, R. Trask and I. Bond, *Composites, Part A*, 2007, **38**, 1525–1532.
11 J. Njuguna and K. Pielichowski, *Adv. Eng. Mater.*, 2003, **5**, 769–778.
12 B. Chen, M. Guizar-Sicairos, G. Xiong, L. Shemilt, A. Diaz, J. Nutter, N. Burdet, S. Huo, J. Mancuso, A. Monteith, F. Vergeer, A. Burgess and I. Robinson, *Sci. Rep.*, 2013, **3**, 1177.
13 P. Rafiee, G. Khatibi, N. Nelhiebel and R. Pelzer, *Microelectron. Reliab.*, 2013, **53**, 1563–1567.
14 C. Zhang, J. Hankett and Z. Chen, *ACS Appl. Mater. Interfaces*, 2012, **4**, 3730–3737.
15 J. E. White, H. C. Silvis, M. S. Winkler, T. W. Glass and D. E. Kirkpatrick, *Adv. Mater.*, 2000, **12**, 1791–1800.
16 M. K. Antoon and J. L. Koenig, *J. Polym. Sci., Polym. Phys. Ed.*, 1981, **19**, 197–212.
17 R. F. Fedors, *Polymer*, 1980, **21**, 713–715.
18 S. Luo, J. Leisen and C. P. Wong, *J. Appl. Polym. Sci.*, 2002, **85**, 1–8.
19 J. Zhou and J. P. Lucas, *Polymer*, 1999, **40**, 5505–5512.
20 Y. Ngono and Y. Maréchal, *J. Polym. Sci., Part B: Polym. Phys.*, 2001, **39**, 1129–1136.

21 P. Musto, G. Ragosta, G. Scarinzi and L. Mascia, *J. Polym. Sci., Part B: Polym. Phys.*, 2002, **40**, 922–938.
22 P. Musto, L. Mascia, G. Ragosta, G. Scarinzi and P. Villano, *Polymer*, 2000, **41**, 565–574.
23 J. Mijović and H. Zhang, *Macromolecules*, 2003, **36**, 1279–1288.
24 S. Cotugno, G. Mensitieri, P. Musto and L. Sanguigno, *Macromolecules*, 2005, **38**, 801–811.
25 J. González-Benito, J. Bravo, F. Mikes and J. Baselga, *Polymer*, 2003, **44**, 653–659.
26 I. D. Maxwell and R. A. Pethrick, *J. Appl. Polym. Sci.*, 1983, **28**, 2363–2379.
27 C. Grave, I. McEwan and R. A. Pethrick, *J. Appl. Polym. Sci.*, 1998, **69**, 2369–2376.
28 J. Mijović and H. Zhang, *J. Phys. Chem. B*, 2004, **108**, 2557–2563.
29 M. Liu, P. Wu, Y. Ding and S. Li, *Phys. Chem. Chem. Phys.*, 2003, **5**, 1848–1852.
30 D. Hayward, E. Hollins, P. Johncock, I. McEwan, R. A. Pethrick and E. A. Pollock, *Polymer*, 1997, **38**, 1151–1168.
31 Y. Ding, M. Liu, S. Li, S. Zhang, W.-F. Zhou and B. Wang, *Macromol. Chem. Phys.*, 2001, **202**, 2681–2685.
32 J. Wang, J. Gong, Z. Gong, X. Yan, B. Wang, Q. Wu and S. Li, *Nucl. Instrum. Methods Phys. Res., Sect. B*, 2010, **268**, 2355–2361.
33 L. Li, M. Liu and S. Li, *Polymer*, 2004, **45**, 2837–2842.
34 L. Li, S. Zhang, Y. Chen, M. Liu, Y. Ding, X. Luo, Z. Pu, W. Zhou and S. Li, *Chem. Mater.*, 2005, **17**, 839–845.
35 L. Li, Y. Yu, Q. Wu, G. Zhan and S. Li, *Corros. Sci.*, 2009, **51**, 3000–3006.
36 M. Jackson, M. Kaushik, S. Nazarenko, S. Ward, R. Maskell and J. Wiggins, *Polymer*, 2011, **52**, 4528–4535.
37 C. L. Soles, F. T. Chang, B. A. Bolan, H. A. Hristov, D. W. Gidley and A. F. Yee, *J. Polym. Sci., Part B: Polym. Phys.*, 1998, **36**, 3035–3048.
38 C. L. Soles, F. T. Chang, D. W. Gidley and A. F. Yee, *J. Polym. Sci., Part B: Polym. Phys.*, 2000, **38**, 776–791.
39 C. L. Soles and A. F. Yee, *J. Polym. Sci., Part B: Polym. Phys.*, 2000, **38**, 792–802.
40 M. T. Aronhime, X. Peng, J. K. Gillham and R. D. Small, *J. Appl. Polym. Sci.*, 1986, **32**, 3589–3626.
41 Y. C. Lin and X. Chen, *Polymer*, 2005, **46**, 11994–12003.
42 S. Roy, W. X. Xu, S. J. Park and K. M. Liechti, *J. Appl. Mech.*, 2000, **67**, 391–396.
43 M. R. Vanlandingham, R. F. Eduljee and J. W. Gillespie, *J. Appl. Polym. Sci.*, 1999, **71**, 787–798.
44 A. Mubashar, I. A. Ashcroft, G. W. Critchlow and A. D. Crocombe, *J. Adhes.*, 2009, **85**, 711–735.
45 C. Wu and W. Xu, *Polymer*, 2007, **48**, 5440–5448.
46 T. C. Wong and L. J. Broutman, *Polym. Eng. Sci.*, 1985, **25**, 521–528.
47 L.-R. Bao, A. F. Yee and C. Y.-C. Lee, *Polymer*, 2001, **42**, 7327–7333.
48 A. R. Berens and H. B. Hopfenberg, *Polymer*, 1978, **19**, 489–496.
49 S. Morsch, S. Lyon, P. Greensmith, S. D. Smith and S. R. Gibbon, *Prog. Org. Coat.*, 2015, **78**, 293–299.
50 B. Lahiri, G. Holland and A. Centrone, *Small*, 2013, **9**, 439–445.
51 A. Dazzi, F. Glotin and R. Carminati, *J. Appl. Phys.*, 2010, **107**, 124519.
52 K. Kjoller, J. R. Felts, D. Cook, C. B. Prater and W. P. King, *Nanotechnology*, 2010, **21**, 185705.

53 B. Van Eerdenbrugh, M. Lo, K. Kjoller, C. Marcott and L. S. Taylor, *Mol. Pharmaceutics*, 2012, **9**, 1459–1469.
54 C. Mayet, A. Dazzi, R. Prazeres, J.-M. Ortega and D. Jaillard, *Analyst*, 2010, **135**, 2540–2545.
55 D. Lin-Vien, N. B. Colthup, W. G. Fateley and J. G. Grasselli, *The Handbook of Infrared and Raman Characteristic Frequencies of Organic Molecules*, Academic Press Ltd., London, 1991.
56 S. J. Barton and G. Pritchard, *Polym. Adv. Technol.*, 1994, **5**, 245–252.
57 S. Cotugno, D. Larobina, G. Mensitieri, P. Musto and G. Ragosta, *Polymer*, 2001, **42**, 6431–6438.
58 P. Musto, G. Ragosta and L. Mascia, *Chem. Mater.*, 2000, **12**, 1331–1341.
59 X. Lu, J. Han, N. Shephard, S. Rhodes, A. D. Martin, D. Li, G. Xue and Z. Chen, *J. Phys. Chem. B*, 2009, **113**, 12944–12951.
60 E. Bagley and F. A. Long, *J. Am. Chem. Soc.*, 1955, **77**, 2172–2178.
61 T. C. Wong and L. J. Broutman, *Polym. Eng. Sci.*, 1985, **25**, 529–534.
62 P. Johncock and G. F. Tudgey, *Br. Polym. J.*, 1983, **15**, 14–18.
63 A. Apicella, L. Nicolais, G. Astarita and E. Drioli, *Polymer*, 1979, **20**, 1143–1148.

Faraday Discussions

DISCUSSIONS

Corrosion control: general discussion

Angus Cook, Gerald Frankel, Alison Davenport, Trevor Hughes, Simon Gibbon, David Williams, Hendrik Bluhm, Vincent Maurice, Stephen Lyth, Philippe Marcus, David Shoesmith, Clara Wren, Julian Wharton, Gregory Hunt, Stuart Lyon, Tom Majchrowski, Rob Lindsay, Geraint Williams, Beatriz Rico Oller, Mira Todorova, Sonja Nixon, Su-Ting Cheng, John Scully, Ann Wilson, Frank Renner, Ying-Hsuan Chen, Christopher Taylor, Hiroki Habazaki, Angelos Michaelides, Suzanne Morsch, Peter Visser, Line Kyhl and Anton Kokalj

DOI: 10.1039/c5fd90047f

Gregory Hunt opened a general discussion of the papers by Anton Kokalj, Angelos Michaelides, and Christopher Taylor: Can you comment on how you apply DFT to non-aqueous systems (hydrocarbons)? How does the non-aqueous system influence the surface corrosion mechanism?

Angelos Michaelides answered: The application of a solvent to DFT can be done by co-adsorbing a very thin film of *e.g.* hydrocarbons on top of the slab and running molecular dynamics simulations. This is the so-called *ab initio* molecular dynamics approach.[1] We have carried out many *ab initio* molecular dynamics studies of solid-liquid interfaces, *e.g.* ref. 2 and ref. 3. Although these two papers cited deal with water/solid interfaces, the same approach can in principle be applied to liquid hydrocarbon/solid interfaces. The main and very important drawback with these approaches is their enormous computational cost and if we were to consider the solvent we wouldn't have been able to perform such an extensive set of simulations of different adsorption structures as we have shown in our paper. For complex systems like the one considered here it is probably more appropriate to study a hydrocarbon solvent with a classical method, since good parameterised force fields exist for hydrocarbon molecules.

1 R. Car and M. Parrinello, *Phys. Rev. Lett.*, 1985, **55**, 2471.
2 G. Tocci, L. Joly and A. Michaelides, *Nano Lett.*, 2014, **14**, 6872.
3 G. Tocci and A. Michaelides, *J. Phys. Chem. Lett.*, 2014, **5**, 474.

Anton Kokalj responded: Technically the treatment of aqueous or non-aqueous solvents is similar. This implies that non-aqueous systems can be and have been studied using DFT methods. However, I have no experience with non-aqueous systems, hence I cannot comment on your second question.

Christopher Taylor replied: There are two approaches that make sense. One is to use implicit solvation models where the DFT calculation is embedded in a "medium" that mimics the effect of a solvent-hydrocarbon or otherwise, and so the charges on atoms, the energies for bond-breaking, bond-making, *etc.* are modified to take into account the solvent. The second method uses "explicit solvation" and means that the DFT calculations use atomic and molecular representations of the solvent molecules and include them as either direct or indirect participants in the reaction mechanism. The surface corrosion mechanisms in hydrocarbon media will be different from those in aqueous corrosion, particularly in that they may be predominantly chemical and not electrochemical.

Rob Lindsay commented: Our mechanistic understanding of corrosion inhibition is still very limited. With some notable exceptions, experimental work to date has largely failed to reliably address key questions, *e.g.* coverage of inhibitor molecules required for effective inhibition. In my opinion, the three papers presented in by Anton Kokalj, Angelos Michaelides and Christopher Taylor demonstrate that cutting edge theoretical modelling is ahead of experiments in this area. Experimentalists need to catch up and provide essential data to direct further modelling work, *e.g.* identify the chemical state of the inhibited interface, so that theoreticians can undertake modelling on the most likely substrate termination.

Alison Davenport addressed Anton Kokalj, Angelos Michaelides and Christopher Taylor: There has been quite a bit of experimental work done on the Cu/BTA system. For example, it has been proposed by Gewirth and coworkers (*e.g.* ref. 1) that under certain conditions, inhibition is provided by a Cu(I)-BTA polymer film, which may be templated by the presence of sulphate ions adsorbed on the copper surface. It is therefore important for modellers and experimentalists to work together to ensure that the models match the experimental evidence for the nature of the system.

1 N. R. Honesty and A. A. Gewirth, *J. Raman Spectrosc.*, 2012, **43**, 46.

Angelos Michaelides responded: Corrosion protection provided by BTA(H) at the interface between aqueous systems and oxides is a very complex process involving a large number of factors. As computational chemists we are working to gradually increase the complexity of the system under study, in order to be able to reproduce and explain the past and current experimental work on these systems. However, we need to start from the metal/vacuum interface in order to ensure that we understand the most basic behaviour of the absorbed molecule. So far, we find good agreement with experiments performed in these experimental conditions. We agree on the importance of experimentalists and modellers working together, and experiments performed on these less complex systems would be of great help for the early stages of the modelling work.

Anton Kokalj answered: I completely agree with this comment. The experimentally provided information is crucial. In particular, any atomic scale details, provided by experiments, are invaluable. Otherwise, modellers can only address some general issues on idealistic model systems, such as, low Miller index metal surfaces.

Philippe Marcus addressed Anton Kokalj and Angelos Michaelides: I would like to raise three points on DFT calculations on inhibitors:

- Surface reconstruction: metal surfaces can be reconstructed (*e.g.* Cu (111) is reconstructured under the effect of OH adsorption at high pH).
- DFT calculations on metals covered by oxide films *versus* oxides without metal underneath.
- The binding energy as an indicator of inhibitor efficiency. This is risky because a large binding energy can be associated to an adsorbed species that would promote corrosion instead of inhibition (this is the case for example for S adsorbed on different metals).

Angelos Michaelides responded: In our present work we were looking to study the simplest possible adsorption system for benzotriazole in order to understand its general behaviour. This means that we were working at the metal/vacuum interface where no reconstruction is expected. Further work aiming to understand this molecule should surely add a liquid interface and the related surface reconstruction of the metal. Indeed we have already started work exploring adsorption on copper oxide surfaces and are exploring the possibility of using force fields to account for a liquid film above the surface.

Regarding your second question, DFT is a costly methodology and oxides have a complex electronic structure whose study requires a larger amount of computer time than the bare metal. The first approximation to study a metal film is to model a few layers of the metal oxide only with the bottom layer of the slab fixed at the bulk positions. This should be a good enough approximation for thick enough oxide films, and indeed the literature shows that the thickness of a native film of copper oxide formed in air is a few-nm thick. For the case of a very thin film this model would probably not work and one would have to take into account the lattice mismatch with the metal and migration of electrons at the interface.

Regarding your third question, we completely agree with this statement. Surely, a "strong enough" binding energy is necessary in order to ensure that the inhibitors are competitive with corrosive molecules when adsorbing on the surface. However, this is only one of many factors which determine the effectiveness of a corrosion inhibitor and more work needs to be performed in order to identify all of them.

Anton Kokalj replied: Let me merge the first two questions into a single answer. The adsorption of inhibitor on non-reconstructed Cu(111)—which is the most stable among the low Miller index surfaces—and on surfaces of bulk Cu_2O can be seen as the two end-member models. But actual systems relevant for corrosion, such as the reconstructed surfaces and thin oxide films supported on metals that you mentioned, may be seen to lie in between these two limits. Very thin supported oxide films indeed display very different chemical reactivity from surfaces of bulk oxides, as you showed in your recent study of Al-oxide film supported on Al.[1] As for the reconstructed surfaces, this is one of the reasons that, in addition to Cu(111), the adsorption was studied also on Cu(100), Cu(110), and low coordinated defects thereon. While this is not reconstruction, it may at least give an idea of the possible span of values that one may expect on more "realistic" copper surfaces. These two questions demonstrate why it is very important that experimentalists and modellers work together—as already emphasized by Alison

Davenport. Any atomic scale details that experiments can provide are invaluable for the relevant modelling.

As for the third question, I completely agree. One of the points that I make in the Conclusions of our paper is that strong inhibitor–surface bonding, as important as it may be, is by no means the only relevant factor for corrosion inhibition. Both corrosive species and inhibitors may interact strongly with the surface. So why is it that the former activate and the latter inhibit the corrosion. One answer is that corrosive species form stable soluble complexes and hence accelerate metal dissolution. In contrast, corrosion inhibitors had to form insoluble complexes (or none at all, but instead adsorb stand-alone) and thus build a protective film over metal surface. Perhaps there is a range of "good" binding energies, *i.e.*, not too weak (or the inhibitor may desorb too easily or not even adsorb) and not too strong; but I expect this to be system specific. Another factor that may be relevant, and is discussed in our paper, is the effect of inhibitor on metal work function. For further relevant factors, see also Christopher Taylor's paper (in particular Fig. 7; DOI: 10.1039/C4FD00220B) as well as, *e.g.*, the paper of W. A. Goddard *et al.*[2], who constructed a physically sound model using several different descriptors, binding energy being one among them.

1 D. Costa, T. Ribeiro, F. Mercuri, G. Pacchioni and P. Marcus, *Adv. Mater. Interfaces*, 2014, **1**, 1300072.
2 S. Ramachandran, B.-L. Tsai, M. Blanco, H. Chen, Y. Tang and W. A. Goddard, *Langmuir*, 1996, **12**, 6419–6428.

Simon Gibbon addressed all: Big Data is a current big push in science, seen as a grand challenge and the solution to a number of societal needs. What we have been discussing is also a grand challenge, also aligning to a number of societal needs and can certainly be enhanced by the use of Big Data. Is there an opportunity here in conjunction with the Big Data push to access significant funding to bring together all these currrently disparate methodologies which are needed to address corrosion computationally?

John Scully responded: I agree that there is an opportunity here for corrosion. We should pay attention to developments in other fields such as health. There are challenges for the corrosion field such as a lack of data, empirical data, and issues such as "how portable" is the data? That is, should data from slightly different systems be considered in a new analysis of another condition. How do we decide what data to "include" and what data to "exclude"?

Alison Davenport addressed Christopher Taylor, Anton Kokalj and Angelos Michaelides: One of the most crucial events in localised corrosion is the process of repassivation, in which a surface that is dissolving suddenly starts to form a passive film and the corrosion is switched off. This implies that there is competition between dissolution and passivation on a corroding surface, and there are one or more critical factors that favour one over the other. What are the prospects for tackling this question through modelling? This will certainly be challenging since it involves the effect of an electric field at the interface as well as information on the local chemistry of the electrolyte.

Angelos Michaelides answered: Understanding dissolution and repassivation is a major challenge for modelling since, as already stated in the question, many factors are at play. The modelling community is working to achieve the capability of solving such complex challenges, through a number of different approaches such as multi scale methods which can combine the accuracy of quantum mechanics with the possibility of modelling larger scale systems. Moreover, enhanced sampling techniques are making it possible to understand 'rare' events such as the dissolution of a surface atom. First principles approaches are also available to explore the role of electric fields and many modellers in the electrochemistry community do such simulations already. We have for example looked at how electric fields can alter the dynamics of water molecules at surfaces[1] and we have also been able to explore a dissolution event purely from first principles.[2] With similar approaches and the multi-scale methods mentioned above, it should be possible to make more progress with our atomic level understanding of corrosion. However, in general it is true to say that relatively little is known of the atomistic details of corrosion and passivation, and simplifications are generally made when tackling these systems with modelling.

1 H. Gawronski, J. Carrasco, A. Michaelides and K. Morgenstern, *Phys. Rev. Lett.*, 2008, **101**, 136102.
2 L.-M. Liu, A. Laio and A. Michaelides, *Phys. Chem. Chem. Phys.*, 2011, **13**, 13162.

Anton Kokalj responded: Issues of this kind have been studied by DFT modelling (*e.g.* see the review of Maurice and Marcus[1] and Taylor[2]). Also the effect of the electric field can be and has been modelled either by charging a slab or applying a homogeneous external field perpendicular to the surface within the simulation (*e.g.*, see ref. 3 and the references therein).

1 V. Maurice and P. Marcus, *Electrochim. Acta*, 2012, **84**, 129–138.
2 C. D. Taylor, *Int. J. Corros.*, 2012, **2012**, 204640.
3 C. D. Taylor, S. A. Wasileski, J. S. Filhol and M. Neurock, *Phys. Rev. B: Condens. Matter Mater. Phys.*, 2006, **73**, 165402.

Christopher Taylor replied: I believe that this field is ripe for more modeling. In fact, already atomistic modeling has been applied to begin to uncover the nature of some of the fundamental processes occurring during such events. For instance, the competition between adsorption of chlorides and hydroxides on steel surfaces has been studied by Taylor,[1] and predictive models developed that can apply to different temperatures and chemical concentrations. The atomistic process by which chloride might displace hydroxide from a passivated nickel surface has also been studied by the Marcus group.[2] As more investigations of these kinds are performed, I believe we will have a fuller picture of localized corrosion and passivity, although there is much work to be done – both *in situ* experimentally and *ab initio*.

1 C. D. Taylor, *Corros. Sci.*, 2012, **68**, 591–599.
2 N. Pineau, C. Minot, V. Maurice and P. Marcus, *Electrochem. Solid-State Lett.*, 2003, **6**, B47–B51.

Frank Renner addressed Angelos Michaelides and Anton Kokalj: Benzotriazole or thiols are often described as corrosion inhibitors for Cu and Cu alloys. Mainly single-molecule systems are currently addressed. Do you see potential benefits for forming stable adsorption layers by offering a multi-component system, *i.e.* for examples two molecules? I would also appreciate a response in regard of making a step towards a higher complexity of corrosion scenarios.

Angelos Michaelides commented in reply: It is certainly conceivable that co-adsorption of more than one inhibitor molecule could lead to interesting results. This would make a very interesting project.

We completely agree that the more realistic and complex we can make the modelling the better. In a similar manner the more well defined corrosion experiments that can be carried out, which provide precise atomic level information, the better.

Anton Kokalj responded: Indeed, many researchers, myself included, modelled only single-inhibitor systems. The reason is that atomistic modelling of corrosion inhibitors is at an early stage stage—here only studies that explicitly model pertinent interactions are meant, thus excluding numerous studies that rely on HOMO–LUMO type inferences based on inhibitor molecules alone. Mainly basic issues were studied so far, *e.g.*, how inhibitor molecules bind with metal surfaces and how this binding depends on the type of molecule, type of metal, surface terminations, and oxidation state of metal. Solvent effects on adsorption bonding were also estimated. But we have much to learn yet. As far as I know, there are many examples where multi-component systems (corrosion inhibitor formulations) display superior inhibition performance over single-component systems. Although I am ignorant of multi-component systems and how they work, I anticipate the following. There are various types of corrosion and in multi-component systems some molecules may be more effective for inhibiting one type of corrosion and other molecules more effective for inhibiting another type of corrosion. Moreover, some molecules may better inhibit cathodic and other anodic reactions. A mixture of two or more inhibitor molecules that differ in size and shape may also better tile the surface thus reducing the area of unprotected surface.

I agree that we need to make a step towards a higher complexity of corrosion scenarios.

Simon Gibbon continued a general discussion of the paper by Anton Kokalj: Fig. 7 in your paper shows a range of different species all following the same pattern of showing stronger adsorption on copper sites with decreasing copper co-ordination number, whether it is water, chlorine, benzotriazole or deprotonated benzotriazole. Naively, I could believe that this figure is telling us more about the chemistry of the copper surface than the adsorbing species.

Have any of your simulations shown a molecule which is more strongly bound to high co-ordination number copper sites?

For corrosion inhibition a molecule having the strongest adsorption on low co-ordination number sites (defects) is required. For lubrication the opposite is true, where the molecule needs to preferentially adsorb to high co-ordination number sites (flat surfaces). If such molecular specificity/separation did not occur in practice in systems formulated for both lubrication and inhibition, then multi-layer structures would be required.

Is the DFT approach also able to help me select a lubricant molecule?

Anton Kokalj answered: I would say that Fig. 2 of our paper is telling us about the chemistry of both involved constituents, that is, adsorbate and copper surface. That the molecule–surface bonding becomes stronger on passing from densely packaged Cu(111) to low coordinated surface defects is not surprising for late-transition metal surfaces. This can be rationalized in terms of the Hammer–Nørskov chemisorption model,[1] where the d-band center is a principal descriptor, *i.e.*, the metal d-band center upshifts as the coordination decreases which in turn results in stronger bonding. However, not all presented species follow the same trend: atomic H is rather insensitive to surface details and the physisorption modes are less sensitive to surface details than chemisorption modes. Among the presented species deprotonated benzotriazole (BTA) displays the strongest dependence on the coordination of copper atoms.

As for the inhibitor/lubricant issue, I believe that the requirement that lubricant should bind stronger to flat facets than defects is too strong. Adsorption of inhibitor at low coordinated defects and lubricant at flat facets can be achieved even if the lubricant prefers the defect sites. What is required is that the lubricant displays a weaker preference to defect sites than the inhibitor. Let me make a detour to heterogeneous catalysis as to make a supporting example: recently we studied a co-adsorption of ethylene and CO on stepped Cu(410).[2] The CO shows a much weaker dependence on the copper coordination number than ethylene and only marginally favors the step-edge over the flat facets. As a result, ethylene is adsorbed at step-edges and CO at terraces. This was experimentally verified.

Furthermore, in a recent study of adsorption of several azoles (among them also mercapto-azoles) on copper surfaces,[3] we found that molecular bonding *via* S is much less sensitive to surface details than the molecular bonding *via* N. Indeed, the N heteroatom is particularly sensitive to surface details. This, together with your premise that lubricant should adsorb on flat facets, suggests that N is not a good heteroatom for lubricant, and S would be better.

DFT is a very good tool to understand such type of bonding issues. But DFT has also been used to model other relevant aspects of a lubricant, such as friction.

1 B. Hammer and J. K. Nørskov, *Adv. Catal.*, 2000, **45**, 71–129.
2 T. Makino, M. Okada and A. Kokalj, *J. Phys. Chem. C*, 2014, **118**, 27436–27448.
3 N. Kovačević, I. Milošev and A. Kokalj, *Corros. Sci.*, 2015, doi:10.1016/j.corsci.2015.05.041.

John Scully asked: So what would a bad inhibitor "look like" in terms of the parameters that you modeled? What are the features of a bad inhibitor in contrast with a good inhibitor for corrosion? How does the corrosion field sort out the plethora of inhibitors using these modeling advances? Would one factor be poor adsorption relative to water and/or chloride?

Anton Kokalj replied: To the last question, concerning the "poor" adsorption, I answer affirmatively.

Molecular modelling of corrosion inhibitors has indeed become very fashionable in corrosion inhibition studies and a quick literature search reveals that in the last decade the number of such studies increased rapidly. By far the prevalent approach in the literature is based on considering only inhibitor molecules and making HOMO–LUMO type inferences. I don't believe this approach is very useful, *i.e.*, it is a kind of black-box.[1] I completely agree with the statement made recently by Robert Lindsay *et al.* that "in terms of predictive power, such an approach has at best limited value, and is potentially simply misleading".[2] A step forward is to rigorously model interactions between components of corrosion system and recently several attempts in this direction appeared. In such studies the most utilized descriptor is the molecule–surface binding energy, typically calculated at metal/vacuum interface. While this parameter is likely important, it is nevertheless only one among the relevant factors. Aqueous-phase adsorption energy is a more relevant parameter, but harder to calculate. For example, a given molecule may form a strong adsorption bond and also hydrates strongly: if the latter is stronger than the former the molecule may not even adsorb (that would be a bad inhibitor). Hence the solubility should be neither too large nor so weak as to make an inhibitor insoluble. A given reactive molecule may display (i) strong adsorption bonding, (ii) strong hydration as well as well as (iii) the ability to form soluble molecule–metal complexes. So it is the "differential" bonding that matters, *i.e.*, a good inhibitor molecule should favour (i) over (ii) and disfavour (iii). But such "differential" interplays are much harder to anticipate and ascertain, because they depend on "details". Furthermore, a good inhibitor should have a higher affinity towards under-coordinated defect sites than corrosive species; *e.g.*, metal dissolution likely proceeds *via* kink sites, hence an inhibitor should "protect" them. Another way the inhibitor molecule may act against corrosion is *via* some non-steric inductive or indirect effects, *e.g.*, the adsorption of an inhibitor may affect not only adsorption sites but also neighbouring sites, which become less susceptible to corrosion reaction.[3] A non-local factor of this type that may be relevant, and is discussed in our paper, is the effect of an inhibitor on the metal work function; adsorption of benzotriazole considerably reduces the work function of copper. Although under realistic conditions several factors would counteract the work function change, some effect may still survive. The reduction of the metal work function should make anodic reaction less favourable and *vice versa* for cathodic reaction. Hence this effect is a "double-edged sword", depending on which reaction is rate limiting. These are some hints on parameters that a good inhibitor should display (and *vice versa* for a bad inhibitor), some of which were considered in the paper. But I believe that these alone are insufficient to be predictive. Much work needs to be done in this direction. In this respect the general comment made by Robert Lindsay that "our mechanistic understanding of corrosion inhibition is still very limited" is right on target.

Last but not the least, it should be noted that atomic-scale models represent only a part of the whole picture. I agree with Christopher Taylor, who argued in his presentation, that predictive models will need to address phenomena on different length (and time) scales (DOI: 10.1039/C4FD00220B).

1 A. Kokalj, *Electrochim. Acta*, 2010, **56**, 745–755.
2 P. Morales-Gil, M. Walczak, R. Cottis, J. Romero and R. Lindsay, *Corros. Sci.*, 2014, **85**, 109–114.
3 P. Kutej, J. Vosta, J. Pancir, J. Macak and N. Hackerman, *J. Electrochem. Soc.*, 1995, **142**, 829–834.

Stuart Lyon enquired: The calculations refer to adsorption from vacuum. How could this approach be modified to a technically important aqueous system?

Anton Kokalj replied: The majority of calculations indeed refers to adsorption from vacuum. The reason for this is that vacuum calculations are both conceptually as well as technically simpler. In a reductionistic manner one first starts with a simpler system and elaborates later. For example, adsorption from an aqueous system is due to an interplay of several competitive effects, *i.e.*, molecule–surface, molecule–water, and surface–water interactions. Hence, it is important to understand the individual effects in order to interpret the numbers that result from sophisticated simulations. In Section 3.2 of our paper, the influence of the solvent is roughly estimated using the implicit solvent model and thermodynamic cycle. The use of the latter enables the understanding of individual involved effects. A more sophisticated approach would be to treat the solvent explicitly atom by atom; its liquid character is then modelled by performing molecular dynamics simulations. The third option for modelling aqueous systems is a hybrid approach where several water molecules are treated explicitly and the rest of the solvent implicitly.

Mira Todorova commented: 1) I am not sure whether I completely understand which quantities within the thermodynamic cycle (Scheme 3, p. 11 of your paper) were obtained by density-functional theory calculations and how the calculations were performed. This relates in particular to the $\Delta\Delta G_{solv}$ term appearing in eqn. 4 and 5 of your paper. Were the terms related to this formula actually calculated by performing explicit calculations for a "surface/water" and a "surface/adsorbate/water" supercell (maybe by using an implicit water *via* some solvation model, rather than explicit water) or are these components within the thermodynamic cycle approximated by the calculated binding energy $E_b^{\ominus}$? Are the used EA (electron affinities) taken from thermodynamic tables?

2) Concerning the CuO_2 system: what is your opinion on how important the correct description of the band gap is in the context of the adsorption studies?

Anton Kokalj answered: Almost all quantities in eqn. 4 and 5 were obtained with DFT calculations. The only exception is the work function of the metal/water/vacuum system (Φ^*, eqn. 6), where the experimental values were taken for $\Delta\Phi$;[1] $\Delta\Phi$ is a variation of the metal work function due to a thick layer of liquid water (for more details, see the footnote § in the paper). All solvation terms were calculated with an implicit solvent model (for more details see the ESI of ref. 2 and 3), whereas for Fig. 6 the dependence of the $\Delta\Delta G_{solv}$ terms of metal slabs on the electric field (eqn. 9) were estimated from the behaviour of adsorbed H-up and H-down water bilayers on the applied electric field (see the ESI of ref. 2).

Electron affinities were calculated as the difference between total energies of neutral and anionic forms ($E_X - E_{X^-}$) as obtained by molecular calculations using local Gaussian-type orbital basis set. BTA and Cl have positive electron affinities: PBE calculated values are 3.53 and 3.69 eV, respectively, whereas the experimental value for Cl is 3.62 eV. (Even for molecules with negative electron affinities, pragmatic DFT calculations using small local basis sets often give reasonable results.[4])

The issue of the band gap of Cu_2O is certainly important and indeed plain GGA functionals considerably underestimate it (PBE gives 0.41 eV, whereas the experimental value is 2.14 eV). In your paper (DOI: 10.1039/C4FD00238E) you stated that defect formation energies calculated with plain GGA functionals can be way off (I've encountered similar problems in studies of Li-ion battery materials, where the +U correction is needed to remedy the situation). But I would anticipate that correct description of band gap is less important for adsorption studies than for defect formation energies and other issues where band alignment is crucial. Furthermore, it is reasonable to assume that adsorption trends turn out better than individual adsorption energies (and in the paper the emphasis is on the trends). The adsorption of benzotriazole was studied in considerably less detail on Cu_2O than on Cu surfaces and one among future improvements of the Cu_2O model is to take the issue of the band gap into account either with the +U correction or by using hybrid functionals. This would quantify how important the issue of the band gap is, in the context of adsorption of the inhibitor.

1 S. Trasatti and L. M. Doubova, *J. Chem. Soc., Faraday Trans.*, 1995, **91**, 3311–3325.
2 A. Kokalj, S. Peljhan, M. Finšgar and I. Milošev, *J. Am. Chem. Soc.*, 2010, **132**, 16657–16668.
3 S. Peljhan, J. Koller and A. Kokalj, *J. Phys. Chem. C*, 2014, **118**, 933–943.
4 M. C. Kim, E. Sim and K. Burke, *J. Chem. Phys.*, 2011, **134**, 171103.

Gerald Frankel communicated: It has been shown that BTA is more effective on Cu_2O than on bare metal or CuO. In fact, there seems to be some specificity of BTA for systems that form monovalent ions. For example, it is very effective for Ag, but not for Fe. You did not present results for CuO, but did show a figure comparing Cu and Cu_2O. The binding energy was stronger for Cu than for Cu_2O. Is there anything in your calculations that might explain this behavior of BTA? It might be useful both in validating your results and for providing a basis for the development of new inhibitors.

Anton Kokalj communicated in response: This is a very good point, *i.e.*, the specificity of BTA for systems that form monovalent ions. Unfortunately, we haven't looked at that yet. What I do know is the following: calculations indeed suggest that the bonding of deprotonated BTA is stronger to Cu than Cu_2O surfaces. Also the $BTAH_{(ads)} \rightarrow BTA_{(ads)} + H_{(ads)}$ dissociation is more favourable on the former (exception to this are hydroxylated Cu_2O surfaces). The pH is likely important here: at higher pH, copper surfaces are oxidized and solvated benzotriazole exists in deprotonated BTA^- form (its pK_a constant is about 8), whereas at lower pH the situation is the opposite (reduced copper surfaces and neutral BTAH form). But only deprotonated BTA bonds strongly to the surfaces of copper! This may perhaps explain the relevance of Cu_2O. As for the bonding to Fe systems, it should be noted that benzotriazole bonds differently to Cu and Fe surfaces.[1] While Cu and Ag are not reactive enough to interact with the benzotriazole's π-system (Cu can do it only as an exception and even then the interaction is driven by the dispersion van der Waals forces), Fe can readily do so. Hence benzotriazole binds to Cu (and Ag) surfaces predominantly *via* σ-bonding and to Fe surfaces predominantly *via* its π-system. This implies different adsorption geometries of benzotriazole on Cu and Fe surfaces, which in turn entails different adsorption induced work function changes on the two metals.

1 N. Kovačević and A. Kokalj, *Mater. Chem. Phys.*, 2012, **137**, 331–339.

Vincent Maurice communicated to Anton Kokalj and Angelos Michaelides: One of the aspects discussed to evaluate the corrosion inhibition efficiency was the bond strength. However, some elements known as very detrimental to the corrosion resistance such as Cl and S also bond strongly to metallic surfaces and I agree that the adsorbate bond strength is a good criterion to discuss competitive adsorption. However, one secondary effect of strong adsorbate bonding is weakening of the bond strength of the metallic atoms which are nearest neighbours to the ones directly bonded to the adsorbate. As a result, metal dissolution is promoted in the presence of adlayers of elements such as Cl and S, as shown in previous studies.[1–3] Could you speculate on the implication for the mechanism of inhibition if the inhibitor is strongly bonded?

1 P. Marcus and J. Oudar, *Appl. Surf. Sci.*, 1979, **3**, 48–67.
2 S. Ando, T. Suzuki and K. Itaya, *J. Electroanal. Chem.*, 1996, **412**, 139–146.
3 O. M. Magnussen, L. Zitzler, B. Gleich, M. R. Vogt and R. J. Behm, *Electrochim. Acta*, 2001, **46**, 3725–3733.

Angelos Michaelides communicated in reply: It is not clear yet when a strongly bound molecule results in a good or bad inhibitor. We have verified with the benzotriazole systems, especially those presenting adatoms that chelation is not likely, since it requires a large amount of energy (over 5 eV) to remove a molecule/adatom complex from the surface. More comparative work is needed between strongly bound molecules which are good and bad inhibitors in order to be able to tell if there is an *a priori* way of assessing when a strongly bound molecule will result in the dissolution of the surface.

Anton Kokalj communicated in response: I completely agree with this comment. As already mentioned in the answer to Philippe Marcus' question, the pertinent difference between corrosive species, such as Cl and S, and corrosion inhibitor might be that the former form soluble and the latter insoluble complexes. The first thus promote the metal dissolution, whereas the later forms a protective insoluble film on the metal surface.

I anticipate two factors that may be relevant in this respect. The first is whether the formed complexes are charged or neutral: the former hydrate strongly and would be soluble and the latter hydrate weakly and would be insoluble. The second factor is related to molecular size: *e.g.*, the Cl^- is smaller than BTA^-, which is why the former hydrates stronger. Analogously, the $CuCl_2^-$ complex is much smaller and more soluble than $[BTA_x–Cu_y]^{\delta\pm}$ complexes (for generality let us assume that the complex may be non-stoichiometric and thus charged).

Vincent Maurice further communicated: The weaker efficiency of benzotriazole at acidic pH is discussed in terms of weaker adsorption of the molecule in the hydrogenated form (BTAH). However, copper dissolves at acidic pH with Cu adatoms available for forming BTA–Cu complexes which might be more relevant to discuss.

Anton Kokalj communicated in reply: Although at acidic pH Cu ions are available, the benzotriazole is in neutral BTAH form. But for the formation of BTA–Cu complexes, the BTA form (*i.e.* benzotriazole stripped from its H1 atom/

proton) is needed. Hence there are pros and cons. Nevertheless, this is a useful suggestion that should be considered.

That the argument of the weaker adsorption of BTAH seems relevant is corroborated by the fact that 1-methyl-benzotriazole is not effective in preventing Cu corrosion, because the substitution of the H1 atom by a methyl group prevents the formation of BTA. This strongly suggests that benzotriazole is active against corrosion in the BTA form, which binds strongly.

Tom Majchrowski communicated: In the proposed DFT calculations, how would the presence of alloying additions alter the adsorption of the benzotriazole, would there be a competitive effect observed? Also, what is the effect of the interactions of the adsorbed species?

Anton Kokalj communicated in response: This is a very interesting question. Benzotriazole interacts differently with different metals, *e.g.*, it forms a stronger chemisorption bond to transition metals with an open d-band than to transition metals with a fully occupied d-band. On transition metals with an open d-band, the inhibitor prefers to chemisorb parallel to the surface with a pronounced π-d hybridization, whereas on transition metals with fully occupied d-band and on sp-metals, benzotriazole prefers to chemisorb through σ-molecular orbitals.[1] Hence benzotriazole would bind differently to copper than to alloying additions, but specifics depend on the actual alloy system.

As for the second question, I am not sure I understand what you meant by "the effect of the interactions of the adsorbed species", hence I will assume that you meant the lateral interactions between adsorbates. Benzotriazole displays a rich variety of lateral interactions. First, there are long range lateral dipole–dipole interactions due to a large permanent dipole moment of benzotriazole (4.1 D). These are repulsive for perpendicular BTAH adsorption modes and slightly attractive for parallel BTAH adsorption modes; this behaviour is due to orientation of molecular dipole in the adsorbed state. For deprotonated BTA, these dipolar interactions are far less important due to considerable adsorption induced charge transfer, which greatly reduces the dipole. At higher coverage (say, intermolecular distances shorter than about 5 Å) the lateral van der Waals dispersion interactions become important (see ref. 2 and 10.1039/C4FD00273C); these are always attractive. At short intermolecular distances, properly oriented BTAH molecules also form intermolecular N1–H···N hydrogen bonds. And at even higher coverage Pauli repulsion occurs due to steric hindrance (molecular overlap). The presence of water solvent affects the lateral interactions, in particular water molecules screen the long range dipole–dipole interactions and diminish the "importance" of intermolecular BTAH–BTAH hydrogen bonding, because water molecules can also form hydrogen bonds with BTAH molecules.[3]

1 N. Kovačević and A. Kokalj, *Mater. Chem. Phys.*, 2012, **137**, 331–339.
2 A. Kokalj, *Phys. Rev. B: Condens. Matter Mater. Phys.*, 2011, **84**, 045418.
3 A. Kokalj, S. Peljhan, M. Finšgar and I. Milošev, *J. Am. Chem. Soc.*, 2010, **132**, 16657–16668.

Gregory Hunt communicated: How do you use Fig. 2 in your paper to select corrosion inhibitors? How would a non-aqueous system affect your conclusions?

Anton Kokalj communicated in reply: I do not think that Fig. 2 can be used to select corrosion inhibitors. Only one inhibitor (benzotriazole) along with some other species (Cl, H, H_2O) are considered therein. Nevertheless, Fig. 2 tells something important, *i.e.*, benzotriazole displays the largest bonding enhancement on passing from densely packed Cu(111) to low-coordinated defects among the presented species, which indicates its ability to passivate reactive under-coordinated surface sites. This bonding enhancement is also a thermodynamic driving force for the formation of the BTA–Cu organometallic complexes; according to experimental evidence the ability of benzotriazole to inhibit corrosion is due to the formation of these complexes.

As for the second question, I will instead comment how aqueous solvent affects the results of Fig. 2, because I do not have experience with non-aqueous solvents. The presence of a solvent considerably reduces the magnitude of adsorption energies. This is because in vacuum non-dissociative adsorption can be described solely as bond-making, whereas in a solvent phase it consists of bond-breaking and bond-making. Namely, prior to adsorption the molecule must get rid, at least partially, of its solvation shell and must also empty adsorption site by displacing solvent molecule(s) from the surface; this is bond-breaking and costs energy. Only then the adsorption bond can form (energy gain). Fig. 2 shows that in vacuum Cl binds stronger to Cu surfaces than standalone BTA, but in an aqueous solvent the just mentioned bond-breaking effects reverse the trend, because Cl^- is much smaller than BTA^- and hydrates by about 0.6 eV stronger. The solvent also diminishes adsorption energy dependence on the coordination of surface metal atoms because the adsorbate has to displace solvent molecule(s) from the surface during specific adsorption and a water molecule displays a similar bonding enhancement trend (although less pronounced) as benzotriazole and Cl.

How the non-aqueous solvent affects the results depends on the solvent characteristics. Yet some of the above explained effects may still be relevant, in particular for polar solvents.

Simon Gibbon continued a general discussion of the paper by Angelos Michaelides: How would you select the best corrosion inhibiting molecule?

Angelos Michaelides answered: Unfortunately it is not completely clear yet what makes for an effective corrosion inhibitor. Compared to fields like chemical catalysis (see *e.g.* ref. 1) our understanding of what the molecular level "descriptors" are of a good corrosion inhibitor have yet to be identified. However, a few factors are likely to contribute: molecules which bind strongly to the surface and form uniform overlayers are good candidates. Crucially though further computational and experimental work is needed on well defined model systems in order to identify all the factors which contribute to good inhibition efficiency.

1 A. Michaelides, Z.-P. Liu, C. J. Zhang, A. Alavi, D. A. King and P. Hu, *J. Am. Chem. Soc.*, 2003, **125**, 3704.

Anton Kokalj commented: For the time being, I would heavily rely on a "guided" experimental trial-and-error approach. Only experiments can tell whether a given molecule is a good inhibitor for a given application. I don't believe the corrosion inhibitor modelling has already reached the predictive stage (but we are moving forward toward this goal). Nevertheless, modelling can provide some useful hints and insights. This means that the above mentioned trial-and-error approach would not be completely blind but would be driven, not only by researcher's knowledge and chemical intuition, but also by insights gained by modelling. In order to select a good inhibitor for a given application, I would pursue a "learning" iterative approach of experiments and modelling, where each new iteration bases on the knowledge gained in previous iteration(s).

Christopher Taylor added to the previous replies: I propose in my paper that first a model be developed that describes the key phenomena controlling the inhibitor performance. The next step would be to couple this model with a "design tool", that takes a class of molecules, perhaps broken into fragments (head, tail, substituent, for example) and begins to iteratively compute the anticipated performance using this inhibition model. The design tool would employ a "machine learning" strategy to mix and match different head/tail/substituent combinations in order to arrive at the best inhibitor. Some additional chemistry modules should probably be included to screen out combinations that would be expected to be too difficult/expensive to synthesize or those that may be environmentally toxic, for example. Cheminformatics tools can be coupled in to perform these last couple of functions, and perhaps other aspects of the model that were addressed in the paper.

Mira Todorova said: 1) Four different van-der-Waals functionals are tested in this work. With respect to their construction the D2 and the TS can be considered to be empirical (in the sense, that they present a correction to the calculated energy, but are not contained within the xc-functional), while the vdW-contribution within the other two functionals is contained within the xc-functional term. There are several instances within your paper, where significant differences in the description of the studied system are observed, depending on which functional is used. For example:

- the energy range found for the adsorption of the 'flat' BTAH geometry, which is almost doubled by using vdW-TS (−1.29 eV) compared to the −0.66 eV of vdW-DF (page 7).
- the comment on page 9 which states that vdW-DF usually under-binds while vdW-TS and vdW-D2 usually over-bind.

Can you comment on what features of these four functionals are responsible for such differences in performance? Should one expect in general a better description when using the optB86b-vdW functional?

2) In Section 4.1, where the adsorption behaviour of BTA is examined, it is suggested that BTAH undergoes "dehydrogenation" rather than "deprotonation". Wouldn't you expect this to be different (the opposite) in the presence of water?

Angelos Michaelides answered: This is an interesting and important question. The short but not very deep answer is that the performance of DFT xc-functionals depends on a balance of the exchange and correlation terms. For a fuller answer that goes into the details, including several of the functionals considered in our article, we have recently written a review article: see ref. 1. The optB86b-vdW functional and the related optB88-vdW and optPBE-vdW functionals[2] were developed based on the realisation that the exchange component used in the original Dion *et al.* functional[3] was too repulsive which very often leads to underbinding. This underbinding is largely cured in the modified vdW-DFs and for a broad range of systems we do see excellent performance with these functionals, *e.g.* solids and ice,[4–6] adsorption.[7,8] However all functionals have problems and weaknesses in certain areas and for some adsorption systems we see overbinding with our revised vdW-DFs (*e.g.* water on h-BN)[9]. So the answer to the question is that yes, optB86b-vdW is in this system likely to be performing the best out of those considered, but in general, one must be careful and critical when discussing adsorption energies. Yes, if liquid water was present there is certainly the possibility (depending on the precise conditions) that deprotonation rather than dehydrogenation would be observed.

1 J. Klimeš and A. Michaelides, *J. Chem. Phys.*, 2012, **137**, 120901.
2 J. Klimeš, D. R. Bowler and A. Michaelides, *J. Phys.: Condens. Matter*, 2010, **22**, 022201.
3 M. Dion, H. Rydberg, E. Schröder, D. C. Langreth and B. I. Lundqvist, *Phys. Rev. Lett.*, 2004, **92**, 246401.
4 J. Klimeš, D. R. Bowler and A. Michaelides, *Phys. Rev. B: Condens. Matter Mater. Phys.*, 2011, **83**, 195131.
5 G. Graziano, J. Klimeš, F. Fernandez-Alonso and A. Michaelides, *J. Phys.: Condens. Matter*, 2012, **24**, 424216.
6 B. Santra, J. Klimeš, A. Tkatchenko, D. Alfè, B. Slater, A. Michaelides, R. Car and M. Scheffler, *J. Chem. Phys.*, 2013, 139, 154702.
7 J. Carrasco, B. Santra, J. Klimeš and A. Michaelides, *Phys. Rev. Lett.*, 2011, **106**, 026101.
8 J. Carrasco, W. Liu, A. Michaelides and A. Tkatchenko, *J. Chem. Phys.*, 2014, **140**, 084704.
9 Y. S. Al-Hamdani, M. Ma, D. Alfè, O. A. von Lilienfeld and A. Michaelides, *J. Chem. Phys.*, 2015, **142**, 181101.

Su-Ting Cheng asked a follow up question to MiraTodorova's question for Angelos Michaelides: It is interesting to see that the van der Waals correction can change the adsorption geometry to this extent. The hydrogen bonding is shown to be crucial in this system. Another example for hydrogen bonding being deterministic would be probably water, which could be also the solvent for BTAH. I am curious, could the adsorption geometry be altered again or changed drastically, with the presence of water molecules (the solvent) in the calculation?

Angelos Michaelides replied: This is an interesting question but we haven't looked at water adsorption along with BTAH or BTA on Cu, so we don't know the impact it will have.

Anton Kokalj commented: I find the change of adsorption geometry due to inclusion of van der Waals (vdW) correction both intuitive and expected. I guess the most interesting issue in this respect is the emergence of stable tilted adsorption modes. This can be rationalized as follows: dispersion interactions prefer to maximize molecular contact with the surface, thus favouring the parallel adsorption mode. However, chemisorption bonding *via* σ molecular orbitals favours the upright bonding. Yet we have shown in ref. 1 that tilting of an upright BTAH molecule is extremely soft even without the vdW correction, *i.e.*, there is almost no energy penalty up to 20° tilt, whereas the penalty for 45° tilt is mere 0.08 eV (PBE data). But at such large tilt there is a gain due to dispersion vdW interaction, which further increases the tilt. Hence the tilted structure is a compromise between the upright structure favoured by chemisorption and parallel structure favoured by dispersion interactions.

As for the effect of solvent water molecules on the hydrogen bonding between BTAH molecules, you are certainly right. Water molecules can form hydrogen bonds with BTAH and thus diminish the "importance" of BTAH–BTAH hydrogen bonding. However, the BTAH–BTAH hydrogen bonds are stronger than the BTAH–H_2O bonds. Hence, the net effect of hydrogen bonding stabilization is reduced from 0.5 eV/molecule in the gas-phase to about 0.1 eV/molecule in the aqueous phase, as deduced from approximate calculations.[2]

1 S. Peljhan and A. Kokalj, *Phys. Chem. Chem. Phys.*, 2011, **13**, 20408–20417.
2 A. Kokalj, S. Peljhan, M. Finšgar and I. Milošev, *J. Am. Chem. Soc.*, 2010, **132**, 16657–16668.

Clara Wren enquired: The DFT calculations which investigate the stability of one single molecule on alloy slabs are not very useful. Have you expanded your calculations to benzotriazole–$(H_2O)_n$ cluster systems? Will this change the bond energy or van der Waals energy? The dipole moments of the water molecules can influence the Coulombic interaction. How do you incorporate this effect in your DFT calculations?

Angelos Michaelides replied: In our paper we report results for single molecules and overlayers of molecules on the surface. It is not restricted to single molecules – please consult the paper. We do not show results for any alloy surfaces, our study focuses on copper. Again please consult the paper on this point. Even if our paper was restricted to single molecules, it would still be of value as this is surely the first step in understanding any adsorption system.

We have not looked at BTAH–water complexes and so cannot comment on this. It would be interesting and nice to explore this in the future. We have, however, extensively looked at water adsorption on surfaces, see. *e.g.* ref. 1.

DFT is an electronic structure technique that explicitly captures (from the electron density) dipolar and Coulombic interactions. For an introduction to DFT approaches see some recent reviews in the Journal of Chemical Physics.[2–4]

1 J. Carrasco, A. Hodgson and A. Michaelides, *Nat. Mater.*, 2012, **11**, 667.
2 K. Burke, *J. Chem. Phys.*, 2012, **136**, 150901.
3 A. Michaelides and J. Klimeš, *J. Chem. Phys.*, 2012, **137**, 120901.
4 A. D. Becke, *J. Chem. Phys.*, 2014, **140**, 18A301.

Anton Kokalj replied: The adsorption of benzotriazole was modelled also at the metal/water interface. The corresponding modelling is, however, not trivial and the effect of water was only roughly estimated using the implicit solvent model (see Section 3.2 in our paper). Indeed, the presence of aqueous solvent considerably affects the energetics of adsorption due to an interplay of several competitive effects, *i.e.*, molecule–metal, molecule–water, and metal–water interactions. Consequently, the net adsorption energies are much reduced in the aqueous phase with respect to the gas-phase.

The dipole moment of water indeed influences Coulombic interactions. For example, water molecules screen the long range lateral dipole–dipole interactions between adsorbed molecules (an estimate is shown by a thin dash-dotted line in Fig. 4 in our paper). Coulombic interactions are treated naturally in DFT calculations. This is true even when the solvent is present, provided that the model is sophisticated enough.

Simon Gibbon addressed Angelos Michaelides and Anton Kokalj: If my understanding is correct, your two papers have used different approaches to studying the effect of copper co-ordination on the interaction of benzotriazole with a copper surface, is this correct and has it resulted in differences in your results?

Angelos Michaelides responded: Both papers studied the adsorption behaviour of benzotriazole (BTAH) on copper surfaces using density functional theory (DFT). In DFT an appropriate "functional" needs to be chosen, and the choice of functional can determine the adsorption configurations obtained (see. *e.g.* ref. 1). In the two papers different functionals were chosen: we used a self-consistent van der Waals-inclusive functional,[2] while in the other paper a combination of an empirical van der Waals-inclusive functional and other functionals which do not account for dispersion interactions were used. While we obtained some differences in the results, the overall picture on the behaviour of BTAH is consistent in the two studies.

1 J. Klimeš and A. Michaelides, *J. Chem. Phys.*, 2012, **137**, 120901.
2 J. Klimeš, D. R. Bowler and A. Michaelides, *Phys. Rev. B: Condens. Matter Mater. Phys.*, 2011, **83**, 195131.

Anton Kokalj answered: The approaches in the two papers are rather similar, but the focus is to some extent different. Gattinoni and Michaelides (10.1039/C4FD00273C) studied the adsorption of benzotriazole on Cu(111) and also the BTA–Cu organometallic complexes on Cu(111)—by "complex" I mean that BTA molecules are linked *via* Cu atoms (ions) that are extraneous to the copper lattice—whereas I presented the adsorption on Cu(111), Cu(100), Cu(110) and low coordinated defects thereon as well as on several surfaces of Cu_2O, including the BTA–Cu complexes.

Perhaps, the largest difference between the two studies is the use of different functionals. Gattinoni and Michaelides used the optB86b-vdW, whereas I used the PBE for upright adsorption modes (chemisorption) and PBE-D′ for parallel adsorption modes (physisorption); PBE-D′ includes a Grimme's empirical van der Waals correction, reparametrized to reproduce experimental adsorption energy of benzene on Cu(111). Due to the use of different functionals there are, of course, quantitative differences. But qualitatively the results of the two papers are compatible. For example, both show that (i) benzotriazole displays reach chemistry on Cu surfaces, (ii) dissociated BTA bonds considerably stronger than neutral BTAH, and that (iii) the bonding of BTA is even stronger in the BTA–Cu organometallic complexes. Another difference is that Gattinoni and Michaelides focused on high-coverage BTAH phases, whereas I focused on long range dipole–dipole interactions between BTAH at lower coverage.

Vincent Maurice communicated: In the discussion of the BTA inhibition in alkaline solution, the presence of the copper surface oxide should be taken into account.

Angelos Michaelides communicated in reply: In the study of the interaction of BTA with copper we have started with the simplest possible system in order to obtain a general understanding of the adsorption behaviour of the molecule. Doing so also allows us to compare with well defined surface science experiments performed under the same conditions. A more complex system, such as a BTA/oxide system, certainly needs to be studied and we are working on this at the moment. As modellers, we welcome any input from experimentalists regarding the fundamental aspects of the corrosion process that need to be taken into account.

Clara Wren continued a general discussion of the paper by Christopher Taylor: You mentioned that you have expanded your calculations to investigate clusters using molecular dynamic simulations. In your MD calculations, how large was the size of the clusters in your unit cell, and how many unit cells were considered? How do the size of the cluster and the optimum size of the calculation matrix affect the bonding energy and the corrosion inhibition behaviour of, for example, benzotriazole?

Christopher Taylor replied: Using classical molecular dynamics around 10 000 molecules can be simulated on a single multi-core desktop PC. This enables larger diversity of molecular arrangements and solvation environments to be studied. When the simulations become more complex than, say, the DFT models of single molecules on clean surfaces, it is important to apply statistical sampling to get an average picture of the behavior of inhibitor self-assembled monolayers, their structure, and their ability to protect the surface against corrosion. These studies are in progress, so I do not have conclusive answers to your questions as yet.

David Shoesmith said: We have tried many modelling approaches to model/assess fuel wasteform and nuclear waste container corrosion ranging from deterministic models to Monte Carlo simulations but have not found a method to interlink them sensibly without extreme complexity. You appear to have a format for interlinking these approaches. Can you discuss how close to such a synthesis you are?

Christopher Taylor answered: The formation of a "multiscale model" is often complex as you mention. The model developed for inhibitors has addressed several factors, but is still far from a complete integration. The current model combines thermodynamic quantities – for inhibitor speciation and partitioning, and the formation of self-adsorbed monolayers – with kinetic terms for the corrosion rates. Some of the pieces have yet to be determined, such as the kinetic rate constants in the presence of the inhibitor. In the case you mention, my recommendation would be to consider how the two types of simulations (deterministic and Monte Carlo) can complement one another to form a third, higher-level model, rather than to try to directly interlink them, which would be computationally complex, and perhaps, not the ideal approach.

David Williams addressed Christopher Taylor and Simon Gibbon: There is in corrosion studies a large amount of "fuzzy" data – qualitative or semi-quantitative knowledge, much of it derived from field experience or simple laboratory tests –

about what works as an inhibitor and what does not. Also, as Prof Taylor has discussed in his paper, corrosion overall is a complex process. I wonder to what extent theoretical models such as those described by Kokalj and Michaelides can be incorporated into a prediction framework along with fuzzy data, perhaps using Bayesian inference methods or techniques similar to the "land use regression" methods used in environmental science? More generally, how might the "fuzzy" data be collected and systematised so that it might be used in some such modelling?

Christopher Taylor replied: This is an excellent suggestion and I believe it is one that demands further inquiry. The use of data analytics methods, such as Bayesian inference, could be of value and is highly aligned with materials science initiatives currently in vogue, such as "Materials Informatics".

Simon Gibbon answered: Currently insufficient use is made of modeling across corrosion studies, due to the data not have been collected at all, or not in a systematic manner. Considerable practical progress could be made by a better statistical analysis of properly collected systematic "fuzzy" data on its own. Modeling corrosion is an inherently complex challenge, as highlighted during these discussions. So there is no question that "fuzzy" data has value in modeling corrosion processes if it reduces the complexity of the modeling challenge. Bayesian inference, by allowing the incorporation of prior understanding gained from the "fuzzy" data, would provide some degree of simplification. However, the *meta*-data that goes with the "fuzzy" data also needs to be incorporated into any enhanced modeling approaches. So understanding what the "fuzzy" data relates to and the conditions under which it was acquired, may be as important to incorporate into any modeling as the "fuzzy" data itself. The applicability of modeling studies could be enchanced by using field experience to define the modeling conditions more precisely or using "fuzzy" data as a test set against which to check the trends of results.

A multi-scale modeling approach would be the best framework within which to incorporate "fuzzy" data, where inputs from quantum, molecular, fluid dynamics and other appropriate models could be used to provide a complete picture of the complex corrosion process, much as outlined in Taylor's paper. Current Bayesian modeling approaches such as those of H Wang *et al.*[1] looking at pipeline corrosion have not achieved the integration from the atomic scale up, but have shown excellent correlation to inspections and it is not obvious that atomic integration is necessary.

1 H. Wang, A. Yajima, R. Y. Liang and H. Castaneda, *Computer-Aided Civil and Infrastructure Engineering*, 2015, **30**, 300–316.

Trevor Hughes communicated: Excellent paper in terms of summarising key aspects of corrosion inhibitor design and proposing the use of a "multiphysics" modelling approach to link the various domains and microprocesses (Fig. 7 in your paper). I have two comments.

Comment 1: Perhaps the feasibility of such a modelling approach could be explored for a specific ("pilot") system which has already been extensively studied by experimentalists and modellers. A potential candidate could be the benzotriazole/copper system for which there is an extensive published literature and active research on modelling the inhibitor adsorption (*e.g.* refer to papers 10.1039/C4FD00257A and 10.1039/C4FD00273C). Another candidate might be acid corrosion inhibitors (*e.g.* refer to posters "*Effect of duplex steel elemental composition and microstructure on the adsorption behaviour and efficiency of acid corrosion inhibitors*, by Man Yi Ho, Evgeny Barmatov, Jill Geddes, Lynne Crawford and Trevor Hughes", "*Acid corrosion inhibitor film formation kinetics: experimental methods & effects of inhibitor chemistry*, by Trevor Hughes, Evgeny Barmatov, Jill Geddes and Michaela Nagl", and "*Effect of corrosion products on efficiency of film-forming corrosion inhibitors in hydrochloric acid*, by Evgeny Barmatov and Trevor Hughes") but this area is less specific in that it brings in a broad range of chemical structures, synergistic interactions and a broad range of temperature conditions.

Comment 2: Referring to the summary and conclusions section of your paper, it is stated "we acknowledge that there may be several domain-process combinations that were omitted from the model". The fact that the corrosion inhibitor may be part of a chemical package is very important in the application of production chemicals in the oilfield. It is stated that "the intrinsic molecular stability regarding thermal decomposition or reaction with water should be given consideration" – I agree with this but I would also add that reactive (polymerisable) inhibitors are enabling in terms of the corrosion inhibitor packages used in high temperature matrix acidizing applications.

Christopher Taylor communicated in reply: In reference to Comment 1. I am in agreement that, having now presented a case for integrating across many different domains, some "signature problems" need to be selected collectively by a group of interested parties and the "expert teams" approach advocated in this paper applied as a demonstration of the power of the "integrated multiscale" philosophy. In reference to Comment 2. Depending on the specifics of the inhibitor problem being modeled, a set of "core physics/chemistry" modules will need to be implemented first, and then "additional" modules introduced incrementally. Deciding which physical/chemical effects are "core" and which are "additional" will have to be the prerogative of the expert team assembled to tackle the problem. I agree that, very broadly, all physical and chemical effects deemed to be significant be thrown into the mix during the initial brainstorming, and from there the team will need to downselect those effects expected to have the highest impact on the inhibitor performance. If, for example, a quantifiable expression exists for how chemicals in the package interact and modify the inhibitor effect or speciation, there seems to be no problem to incorporating it into the model.

Gerald Frankel opened a general discussion of the paper by Hiroki Habazaki: Does the conducting polymer still provide protection to the substrate by the mechanisms known for such polymers even though the coating is separated from the substrate by a thick oxide film?

Hiroki Habazaki responded: The oxide film consists of two layers; an outer porous layer with cylindrical pore structure and an inner thin barrier layer. After thermal treatment, the inner layer is transformed from amorphous to a Fe_3O_4 structure, as confirmed by TEM observations. Fe_3O_4 is not an insulator, showing relatively high conductivity. In fact, the electropolymerization potential on the anodized iron after thermal treatment in argon was similar to that on passivated iron without a thick oxide layer. This suggests that the conducting polymer layer and metal substrate are electrically connected such that the conducting polymer can provide anodic protection to the substrate even if the thick oxide layer is present.

Geraint Williams commented: What role does cyclic re-oxidation of the reduced polypyrrole play in maintaining "ennoblement" of the iron substrate? In the presentation which accompanied the paper, an equation was displayed which represented the interaction of the oxidised form of the PPy with iron (giving Fe(II) and reduced PPy). However, this process, in isolation, would not produce the sustained high potentials (ca +0.2 V *vs.* Ag/AgCl) shown in Fig. 9 and 11 of your paper. Without re-oxidation of the reduced PPy, the open circuit potential (OCP) would be expected to decline with time. The fact that OCP remains time-independent upon immersion suggests one of two things. Either (i) the anodic film is stopping direct contact of the iron with the PPy and that it is simply acting as a barrier coating, or (ii) there is sufficient conductivity through the anodic film to maintain electrical contact between the iron and PPy. For the latter, the sustained OCP of +0.2V would, by implication, mean that the activity of dissolved oxygen in the electrolyte is enough to allow re-oxidation of any reduced PPy. A relatively simple experiment would allow you to differentiate between the two possibilities. If the solution is de-oxygenated (*e.g.* by sparging with nitrogen or argon) then the observation of little or no change in OCP would support option (i), while a significant decrease in OCP provides evidence in support of (ii).

Hiroki Habazaki replied: Thank you very much for your important comment on the role of dissolved oxygen in re-oxidation of reduced PPy. We would like to examine the immersion test under de-aerated conditions. In reduction of PPy, release of doped anions and/or incorporation of cations must occur. In fact, we found both the release of doped anions from the outer part of the PPy coating and the incorporation of sodium ions into the inner part of the PPy coating during immersion. Both processes contribute to the reduction of PPy and a decrease in the conductivity. These processes may not be reversible such that it is not clear whether re-oxidation of reduced PPy occurs and the conductivity of PPy is recovered. However, the correlation of electrochemistry and ion migration in PPy will be important for more detailed understanding of corrosion protection by conducting polymers.

Beatriz Rico Oller enquired: In your paper, you describe a two-step anodizing process for iron before thermal treatment in order to get a smoother substrate surface.

How different is the surface after just 1 anodizing step when compared to the one you have used? Is the adhesion of the PPy-layer that different when the iron surface has not been re-anodized? Does a second anodising not reduce the anchor effect?

Hiroki Habazaki responded: The first step anodizing and its film dissolution were used just as a replacement of mechanical polishing using alumina paste.

Then, we were able to get reproducible results. Without any surface pretreatment, the surface appearance of the anodized specimens was not uniform; Pores with non-uniform pore sizes were obtained. After two-step anodizing, pores of ~30 nm diameter were developed uniformly on the entire surface, providing sufficient anchor effect to the PPy layer.

David Williams asked: Could this coating advantageously be used as a primer for painting, so that corrosion inhibitor could be released below the paint coat?

Hiroki Habazaki answered: Although we have not yet studied our composite coating as a primer for painting, there have been several studies demonstrating a beneficial role of a conducting polymer layer as a primer to improve the corrosion protection. Some references are as follows: ref. 1–3.

1 K. Kamaraj, V. Karpakam, S. S. Azim and S. Sathiyanarayanan, *Synth. Met.*, 2012, **162**, 536.
2 T. Y. Pan and Z. Y. Wang, *Microsc. Res. Tech.*, 2013, **76**, 1186.
3 S. U. Rahman, M. A. Abul-Hamayel and B. J. A. Aleem, *Surf. Coat. Technol.*, 2006, **200**, 2948.

Stuart Lyon remarked: Does the corrosion resistance depend on the oxide thickness in any way?

Hiroki Habazaki responded: The thickness of the porous layer did not influence the corrosion protection a lot, with a thickness of up to 2 micrometers, while the thickness of the conducting polymer layer changed the durability of the coating a lot.

Julian Wharton enquired: The anodization electrolyte contains a small amount of water (0.5 mol L^{-1} H_2O), as such there is likely to be some residual water at the metal/anodic film interface. Please comment on the influence of this residual water on the fluoride diffusion and also how it may affect the thermal oxide growth during heat treatment in both argon and air.

Hiroki Habazaki answered: The water concentration in anodizing electrolytes influences significantly the growth of porous anodic films on iron. At relatively high water concentrations, gas generation is enhanced and the efficiency for film growth is reduced. At high water concentrations, crystalline oxide nanocrystals, which may act as sites for gas generation, are more readily formed in the barrier layer of the anodic films on iron.[1] In the barrier layer, a high electric field is applied such that water content is negligible at the metal/anodic film interface. In addition, fluoride migrates inwards faster than oxide ions, resulting in the formation of FeF_3 just above the metal/anodic film interface, regardless of water concentration. For thermal oxide growth, no large influence of water concentration may be expected, although we have not yet conducted the detailed study.

1 H. Habazaki, Y. Konno, Y. Aoki, P. Skeldon and G. E. Thompson, *J. Phys. Chem. C*, 2010, **114**, 18853.

Angus Cook commented: In Fig. 4(a) of your paper, the GDOES depth profile of fluorine (F) seems to have a very regular intensity variation before reaching the M/F interface. The same variation is not seen in any of the other surveyed species.

Could you comment on the reason for this intensity variation within the applied film?

Hiroki Habazaki replied: Wavy profiles in GDOES are obtained sometimes when transparent films on metallic substrates are analyzed. The wavy profile is an artefact, originating from interference of light emitted from fluorine in the anodic film. So, this intensity variation does not reflect the compositional variation.

Ann Wilson communicated: My question follows on from Prof Williams' question. I have attempted some adhesion studies for PPy at mild steel and found that while the deposition of an oxide layer on the metal surface resulting from flaming of the metal electrode improves adhesion of PPy, often there appears to be no contact between the metal surface and the electroactive polymer film, leaving me to think that PPy acts only as a barrier film and does not allow for electro-protection of the metal. What is your experience as far as this is concerned with the PPy films over the anodized layer?

Hiroki Habazaki communicated in reply: We believe the electric contact between the PPy and metal substrate from the following findings.

1. The oxide located between the PPy in pores and iron substrate is Fe_3O_4, which is well-known to show high electric conductivity, not an insulator.
2. The potential of electropolymerization on the anodized iron after thermal treatment in argon is similar to that on the passivated iron, although the electropolymerization occurred at higher potential on the anodized iron after thermal treatment in air.
3. For the anodized iron after thermal treatment in argon, the electropolymerization occurs on the entire pore wall surface, not only at the pore bottom. This means that pore walls, mainly consisting of Fe_3O_4, have sufficiently high electric conductivity for electropolymerization of PPy. From these findings, at least for the anodized iron after thermal treatment in argon, the PPy layer should be electrically connected sufficiently to the metal substrate, and anodic protection must be contributing the corrosion protection, in addition to the barrier nature of PPy.

Simon Gibbon opened a general discussion of the paper by Line Kyhl: Please could you explain how graphene on diamond is able to trap super critical steam on the diamond surface such that the steam destroys the diamond leaving the graphene intact,[1] but in your work graphene was not able to protect platinum from salt water at 60 °C? Would edge functionalisation of the graphene give better resistance to salt water?

1 C. H. Lim, A. Sorkin, Q. Bao, A. Li, K. Zhang, M. Nesladek and K. P. Loh, *Nat. Commun.*, 2013, **4**, 1556.

Line Kyhl replied: In their work on graphene–diamond systems, Lim *et al.* concluded that after heating to 1275 K, carbon atoms at the graphene–diamond interface could form strong covalent bonds. As the graphene sheet lay buckled on the diamond surface, the formation of these covalent bonds led to the evolution of nanobubbles, somewhat like buckyballs, with an sp2 hybridised cage (graphene layer), with the lower part of the ball consisting of weaker sp3 hybridised bonds; the diamond surface. Water molecules were trapped inside these cages. After heating the system above the super critical limit of water, Lim *et al.* reported an etching of the diamond and not of the graphene. This indeed proves that the sp2 carbon allotropes are extremely inert and impermeable. It was the strong interfacial bonds that were key in this experiment. These interfacial bonds effectively sealed the edges of the graphene and inhibited the trapped water from diffusing out of the nanobubbles, creating a pressure cooker, with super-heated water molecules trapped inside. These super-heated water molecules then began to etch the diamond surface. Thus, the graphene-diamond system prepared by Lim *et al.* offered a unique possibility to study the behavior of small confined areas of defect-free, edge-sealed graphene.

Taking our experiments with graphene on Pt(100) into perspective, the graphene edges here form carbon–metal bonds. These bonds will be weaker than the strong carbon–carbon interfacial bonds formed by graphene and diamond. The weaker C–metal bonds may more easily be compromised, allowing corrosion-inducing reactants to intercalate and gain access to the Pt surface.

Regarding the second part of your question: if the mechanism for coating breakdown depends on the ability of intercalants to diffuse beneath a graphene domain edge or defect site, then one would expect that an increased interaction between the graphene edge and the substrate would prevent intercalation. This effect may be provided by functionalization and would thus offer an increased corrosion protection. Functionalizing the graphene basal plane may lead to a stronger interaction with the substrate. This effect is seen, for example, with hydrogen functionalized graphene on Ir(111),[1] and is thus expected to hinder the diffusion of intercalated species.

1 R. Balog, M. Andersen, B. Jørgensen, Z. Sljivancanin, B. Hammer, A. Baraldi, R. Larciprete, P. Hofmann, L. Hornekær and S. Lizzit, *ACS Nano*, 2013, 7, 3823–3832.

Hendrik Bluhm asked: The stability of the graphene layer on Pt(100) against intercalation by gas phase molecules is surprising, given that, *e.g.*, oxygen and CO intercalation has been observed already at millibar or lower pressures for graphene layers grown on other metal single crystal surfaces. What explains the increased stability of the graphene layer on Pt(100) against intercalation?

Line Kyhl answered: CO has been reported to intercalate both grapene on Pt(111) [1] and Pt(100) [2] at partial pressures of 10^{-6} mbar and above. In the case of graphene on Ir(111) where the interaction between the graphene layer and the metal surface is relatively strong, a CO pressure in the mbar regime is necessary to force intercalation.[3] In this latter case, the adsorption pattern which CO forms on Ir(111) below the graphene was found to be identical to the adsorption pattern of CO adsorbed on bare Ir(111), when exposed to the same partial pressure of CO in the mbar regime. This shows that a certain coverage of CO on the substrate is needed to make intercalation favorable. Oxygen has been shown to intercalate graphene on Pt(111), but only through channels of delaminated graphene, which were fabricated by first intercalating CO. On graphene on Ir(111) elevated sample temperatures are needed for oxygen intercalation to happen.[4] There is no evidence from our results that graphene on Pt(100) has increased nor decreased stability towards oxygen intercalation, when compared to the above findings for graphene on Pt(111) or graphene on Ir(111). The stability towards CO intercalation, however, is lower for this system (Pt(100)) than for graphene on Ir(111) in the sense that a lower partial pressure of CO is needed for intercalation to proceed.[2] It is peculiar that CO intercalates graphene on metal surfaces, at room temperature, at a partial pressure which is substrate dependent. The mechanism for oxygen intercalation must be different to that for CO intercalation, since the same dependence on partial pressure is not observed, and instead the substrate temperature appears to be the relevant parameter.

1 R. Mu, Q. Fu, L. Jin, L. Yu, G. Fang, D. Tan and X. Bao, *Angew. Chem.*, 2012, **124**, 4940–4943.
2 L. Nilsson, M. Andersen, R. Balog, E. Lægsgaard, P. Hofmann, F. Besenbacher, B. Hammer, I. Stensgaard and L. Hornekær, *ACS Nano*, 2012, **6**, 10258–10266.
3 E. Grånäs, M. Andersen, M. A. Arman, T. Gerber, B. Hammer, J. Schnadt, J. N. Andersen, T. Michely and J. Knudsen, *J. Phys. Chem. C*, 2013, **117**, 16438–16447.
4 E. Grånäs, J. Knudsen, U. A. Schröder, T. Gerber, C. Busse, M. A. Arman, K. Schulte, J. N. Andersen and T. Michely, *ACS Nano*, 2012, **6**, 9951–9963.

Stephen Lyth commented: Recently it has been shown that graphene and graphene oxide are permeable to protons.[1–3]

Could ion transport through the graphene coating be responsible for corrosion in NaCl solution at elevated temperature? Repeating the corrosion studies especially in acidic conditions may shed light on this possible mechanism.

1 J. L. Achtyl, R. R. Unocic, L. Xu, Y. Cai, M. Raju, W. Zhang, R. L. Sacci, I. V. Vlassiouk, P. F. Fulvio, P. Ganesh, D. J. Wesolowski, S. Dai, A. C. T. van Duin, M. Neurock and F. M. Geiger, *Nat. Commun.*, 2015, **6**, 6539.
2 S. Hu, M. Lozada-Hidalgo, F. C. Wang, A. Mishchenko, F. Schedin, R. R. Nair, E. W. Hill, D. W. Boukhvalov, M. I. Katsnelson, R. A. W. Dryfe, I. V. Grigorieva, H. A. Wu and A. K. Geim, *Nature*, 2014, **516**, 227–230.
3 T. Bayer, S. R. Bishop, M. Nishihara, K. Sasaki and S. M. Lyth, *J. Power Sources*, 2014, **272**, 239–247.

Line Kyhl responded: It has been calculated that an open configuration of the graphene edge is favorable in the presence of OH^- groups.[1] For a basic solution with a surplus of OH^- ions, the above open edge configuration may occur, allowing sodium and chloride ions and water to intercalate. Other mechanisms, such as the direct diffusion of OH^- or H^+ ions through the graphene basal plane may be a possibility. A full study using solutions of varying pH would be interesting, as you suggest. If the scenario involving OH^- groups binding to the graphene edges is correct, one should observe a higher degree of protection against the acidic solutions.

1 L. Nilsson, M. Andersen, B. Hammer, I. Stensgaard and L. Hornekær, *J. Phys. Chem. Lett.*, 2013, **4**, 3770–3774.

Vincent Maurice commented: I find the work quite interesting but wonder if the same conclusions on the protectiveness of the graphene layer would apply to a different substrate. In the present case of the platinum surface, stability of the graphene layer and corrosion protection is concluded as long as surface reconstruction is not lifted, which is a 2D surface transformation. Would the graphene layer be as protective on a substrate, where exposure to oxygen or water leads to a 3D growth of an oxide layer?

Line Kyhl commented in reply: When oxygen or water penetrate a graphene layer to react with the underlying substrate this occurs through defects or the edges of graphene domains. In the case of graphene on Pt(100) the effectiveness of the coating was in our work probed by detecting the lifting of the surface reconstruction below the graphene. Observation of a lifted surface reconstruction was taken as evidence that reactants had penetrated the graphene sheet to react with the underlying Pt surface, a 2D transformation. Our experiment was thus designed to study the effectiveness of graphene as an impenetrable barrier. Hence, on this basis, there will be no difference in the graphene barrier properties between a substrate with the possibility for 3D corrosion processes and the Pt(100) model substrate. With the proviso that 3D corrosion, as it is not necessarily a surface process, may not be as readily detectable as the 2D transformation which leads to the lifting of the Pt surface reconstruction.

On the other hand there is no doubt that a graphene layer will bond differently to different substrates depending on the adhesion between the substrate and the graphene layer. On Pt(100) the graphene-edge C-atoms bind down to nearby Pt-atoms, thereby closing for the immediate penetration of water or oxygen.[1] This favorable, closing-off effect, where graphene edges are sealed with the underlying substrate, may vary between different metal surfaces. Furthermore, any evolution of an oxide layer below the graphene may lead to a delamination of the graphene coating.

1 L. Nilsson, M. Andersen, B. Hammer, I. Stensgaard and L. Hornekær, *J. Phys. Chem. Lett.*, 2013, **4**, 3770–3774.

Rob Lindsay remarked: In your paper, you mention that transferring a graphene layer to another substrate introduces numerous defects. Could you please expand on this comment, indicating the type of defects introduced. This issue also suggests that engineering application of a graphene coating for corrosion control (*e.g.* for protection of an oil pipeline) may be impractical, as the environment will inevitably be much harsher. What is your opinion?

Line Kyhl replied: If domain boundaries are present from the graphene growth as, *e.g.*, is the case of graphene on Cu films, the sheet is susceptible to rupture along these domain boundaries during transfer processes. This gives rise to a transferred film consisting of flakes rather than one continuous sheet. Point defects on the graphene basal plane may be introduced during transfer methods involving electrochemical reactions. This may compromise the corrosion protection as the diffusion barrier for corrosive molecules or atoms may be lower at these defect sites. As an example, the diffusion barrier for oxygen at such a site has been calculated to become significantly lowered relative to oxygen diffusing through an intact hexagonal ring of the graphene basal plane.[1] Often the transfer methods additionally include using a polymer resist, which can be difficult to remove from the transferred graphene. Whether or not these polymer residues have an influence on corrosion protection is unclear.

Generally, graphene grown on substrates with a stronger interaction between the graphene and metal gives more uniformly oriented nucleation sites and better epitaxial growth. This leads to better quality of the graphene. This high degree of interaction gives rise to high adhesion between the graphene sheet and the substrate, which poses challenges for any transfer process. Therefore a high quality graphene usually demands a transfer scheme involving harsher electrochemical conditions in order to detach it from its substrate.

On the other hand, transferring high quality graphene sheets from graphite, where there is poor adhesion between the sheets, can be done gently without introducing defects. However, only small flakes can be produced in this way. Transferring several layers of graphene on top of each other to a metal surface has been shown to decrease the corrosion rate when compared to a single transferred layer.[2] The authors concluded that defects from the transferred layers are unlikely to be in registry and the metal surface therefore essentially becomes entirely covered by intact graphene. The lowest corrosion rate found in the study, however, was that of a metal with a graphene coating directly grown on the surface. Non-destructive transfer methods are continuously being developed. This may enable the future application of using transferred graphene as a protective coating material. However, the direct growth of a graphene coating on the metal surface may show to be more advantageous.

1 M. Topsakal, H. Şahin and S. Ciraci, *Phys. Rev. B: Condens. Matter Mater. Phys.*, 2012, **85**, 155445.
2 D. Prasai, J. C. Tuberquia, R. R. Harl, G. K. Jennings and K. I. Bolotin, *ACS Nano*, 2012, **6**, 1102–1108.

Tom Majchrowski requested: This is an interesting application of Raman spectroscopy for corrosion studies. In the presented Raman spectra there is a significant shift (approx 50 cm^{-1}) in the position of the D, G and 2D signals after exposure to saltwater at 60 °C. In the paper you explain the shift in the position of the signal by considering hybridization or change in the graphene strain. I think there is significant change taking place in the chemical environment around that vibrating bond. Would you be able to present your thoughts on this in more detail?

Line Kyhl answered: A sample temperature of 800 °C was reached during the preparation of the graphene coating. The thermal expansion coefficient of the metal is much larger than that of the graphene. Because of this, the graphene will experience a compressive stress when the sample is cooled. This compressive stress shifts the vibrational energies of the raman active phonons towards higher wavenumbers, when compared to freestanding, relaxed graphene. Furthermore a hybridization of the graphene pz orbitals and Pt d orbitals leads to a distortion of the carbon lattice, again increasing the vibrational energy of the phonons. Moreover, the intensity of the Raman signal is screened by the metal surface leading to a significant decrease in intensity compared to what is observed for graphene on insulating substrates. Upon exposure to hot saltwater the D, G and 2D graphene peaks shift towards lower wavenumbers and increase in intensity. The peaks thus change to resemble those of a more freestanding graphene, less influenced by the metal surface below. Under the saltwater exposure the Pt surface reconstruction becomes lifted, but there is no evidence that the graphene lattice is chemically modified from this exposure. We therefore conclude that the graphene relaxes and becomes quasi-freestanding.

Christopher Taylor asked: As a surface becomes progressively more rough, is there a limit at which point graphene can not be smoothly applied to the surface? As a separate question, what developments have been made in the application of graphene oxide, reduced graphene oxide, and functionalized graphene/graphene-oxide surfaces to corrosion protection of metals?

Line Kyhl responded: It is possible to bend and stretch the graphene lattice ~20% away from its equilibrium, free-standing, state.[1] Studies of graphene growth on rough surfaces are scarce. The usual growth mechanisms may be inhibited by a rough surface, but a transferred graphene film should essentially be unaffected by a surface roughness on the μm-scale. A defect-free single crystalline, single layer graphene sheet as an anti-corrosion coating is unlikely to be obtainable. One possible method to increase the protective properties of a defected coating is to apply several layers on top of each other, to cover defects. A cheap and easy method for obtaining multilayers of graphene-based sheets on a substrate is by applying graphene oxide (GO) or reduced graphene oxide (rGO). GO is impermeable to most atoms and molecules, even helium, but under humid conditions water can readily penetrate.[2] Reducing the GO, however, using hydroiodic acid has been shown to lead to leak-tight coatings.[3] The water can penetrate the GO multilayers through capillaries formed in the interlayer spacing with a width of ~10 Å.[2] After reduction, the interlayer spacing approaches that of graphite and none of the tested gases, liquids, salt or acids could penetrate. The reduction method is key to obtain a high quality rGO anti-corrosion coating. Thermally reduced GO films contain numerous structural defects, which compromise the mechanical strength of the sheet. Furthermore, a high density of non-reduced moieties allows for the penetration of water. To summarize, the work of Su *et al.* indicates a possible route for creating an almost defect free corrosion protective coating, superior in quality to CVD graphene, using GO as a starting point.

1 C.-L. Wong, M. Annamalai, Z.-Q. Wang and M. Palaniapan, *J. Micromech. Microeng.*, 2010, **20**, 115029.
2 R. R. Nair, H. A. Wu, P. N. Jayaram, I. V. Grigorieva and A. K. Geim, *Science*, 2012, **335**, 442–444.
3 Y. Su, V. G. Kravets, S. L. Wong, J. Waters, A. K. Geim and R. R. Nair, *Nature Commun.*, 2014, **5**, 4843.

Sonja Nixon enquired: Can you speculate about future industrial feasible applications of graphene? Are there any limitations for industrial applications to be expected especially when we consider the PVD high temperature deposition process?

Line Kyhl replied: A discussion of the outlook for future applications is not trivial. For each application one must consider the quality or type of graphene needed. This leads to the question of what graphene is, which in the literature has a wide span. For example multilayer sp2 carbon flakes or heavily functionalized sheets are often referred to as graphene in the same contexts as a single layer, single crystalline sp2 carbon sheet with low defect density. Each grade or type of graphene might be useful for different application scenarios and is accompanied by its own set of qualities, limitations and challenges. High growth temperatures in the CVD and PVD processes are usually undesirable for industrial applications. For applications in anti-corrosion coatings, phase changes, during the heating processes, in the material on hand may compromise the desired mechanical properties. Another challenge is applying the carbon precursors to the hot material in a controlled manner. Nonetheless research into alternative low temperature growth methods is a hot topic.[1]

1 T. Wu, G. Ding, H. Shen, H. Wang, L. Sun, Y. Zhu, D. Jiang and X. Xie, *Nanoscale*, 2013, **5**, 5456–5461.

Sonja Nixon opened a general discussion of the paper by Peter Visser: We can control very precisely the pH during chemical conversion coating processes. During the corrosion process the pH depends on many factors. How sensitive is the corrosion protection mechanism by the lithium salts and how can it be controlled? Does the coating or the interface to substrate degrade during the leaching and the corrosion process? In the figure of your paper, the coating beside the scribe looks opaque after exposure to the salt spray.

Peter Visser answered: As demonstrated from the literature cited in the paper, the protective layer will be formed at alkaline pH. In contrast to the chemical conversion process, the pH must be controlled with the leachable species from the coating. In this study the uninhibited coating does not contain pigments and is transparent. The lithium-salt loaded coatings contain a pigmentation package and therefore they are opaque.

Julian Wharton asked: Have the authors quantified the lithium-ion leach rates from the polymeric coatings? In addition, please give the rationale for using a 100 μm deep scribe in the AA2024-T3 substrate, as opposed to just coating removal down to the metal surface. Were the panels in a standard orientation within the neutral salt spray testing and if so were there any morphological differences in inhibition performance between the top and bottom of the panels?

Peter Visser replied: For both model systems the leach rate from the polymeric coatings has been determined. Due to the difference in solubility of lithium oxalate and carbonate a large effect has been observed on the leach rate. The concentrations of the lithium oxalate coatings are about 10 to 20 times higher compared to the lithium carbonate. However, both species provide similar protective layers. The rationale for the 100 μm deep scribe originates from the specification requirements of the aerospace industry. The panels were exposed according to the ASTM-B117 requirements at a 6° angle. No performance difference between the top and bottom of the panels was observed.

Gerald Frankel questioned: Have you studied the throwing power? How far from a cut edge can be protected? Have you calculated or modeled the transport of the Li ions across the surface? The protection must be a function of time and distance? What concentration is needed for protection?

Peter Visser answered: This particular study represents the first observations of the protective film when using lithium leaching salts in an organic coating. The throwing power of this system has not been studied yet. Indeed, the inhibitor efficiency is an important parameter when considering their potential use. The throwing power will be an important factor in this inhibitor efficiency, which is influenced by the amount of inhibitor, the type and thickness of the coating. Another factor would be the presence of a topcoat influencing the water flux and inhibitor diffusion from the coating. No modeling or further calculations were performed so far. In prior art the protective layers were generated from high pH (>11) and saturated aluminate solutions. These results demonstrate the activity of the lithium ion at much lower concentrations and pH values. Hence such throwing power, modeling and detailed calculations will be focus points of a research study in the near future.

Rob Lindsay asked: Have you attempted to study what happens if you re-scratch an inhibited scribed area, *i.e.* is this area also inhibited?

Peter Visser responded: No, this was not done for this system. However, some unpublished work revealed inhibition of a new scribe right next to (*i.e.* 2 mm from) a protected scribe after saltspray exposure was resumed.

John Scully commented: Given the Li treatment described and the coverage of the bare Al alloy scratch shown, do you think that the treatment enables cathodic inhibition, anodic inhibition or both? A series of electrochemical experiments on both the alloys and separately on the constituent particles present in the alloy might help address this question. It seems that the thick film shown could provide a means of cathodic inhibition of transport of oxygen to the Al-Cu-Fe, Al-Cu, or Al-Cu-Fe-Mn co-constituent particles, which otherwise may see fast O_2 transport in thin electrolyte films. Have you considered adding another chemical species to provide a second inhibitor function?

Peter Visser replied: Prior literature on the lithium based conversion coatings and some unpublished electrochemical work indicate that the inhibition due to the lithium based inhibitor leaching does not exhibit clear anodic or cathodic inhibition behavior as some well-known inhibitors (*e.g.* benzotriazole) of aluminium alloys. The thick protective film is essentially formed by the grace of the alkaline corrosion reaction and follows upon the formation of the aluminium hydroxide layer and is exclusively due to the presence of the lithium ion. Indeed, the oxygen reduction reaction has been inhibited due the layer formed on the entire surface of the alloy, including the intermetallic particles. Possible damage to and local defects in the protective layer can potentially lead to further local corrosion.

Finally, yes we are investigating the addition of additional species to the coating formulation to even further enhance the corrosion inhibition.

Geraint Williams remarked: Most lithium salts are highly water soluble, so I'm slightly surprised that lithium carbonate and oxalate salts can be used as sparingly soluble pigments in a primer coating. Is there a danger of the lithium salts deliquescing and causing osmotic blistering of the protective coating during long-term immersion? Has the use of ion-exchange based minerals, so-called "intelligent release" pigments, such as bentonites and zeolites, been considered as a possible alternative method of storing and releasing the Li(I) cations? Finally, although salt-spray testing is a useful benchmark for the comparison of the protective ability of various coatings, a more realistic evaluation of the corrosion resistance of coatings for AA2024 aerospace alloy is typically achieved using a Lockheed test, where coating failure *via* filiform corrosion (anodic disbondment) predominates. Is there any evidence to suggest that in-coating lithium cations would inhibit this type of underfilm corrosion?

Peter Visser answered: Lithium carbonate and lithium oxalate are the two least soluble pigments next to lithium phosphate. Indeed, coating formulation must be carefully balanced to prevent the effect of osmotic blistering. This particular study focused on the primary effects of the use of lithium salts in coatings. Alternative delivery methods of Li cations are being considered for further research. Both saltspray and filiform corrosion tests are widely used and required to assess the corrosion resistance of coatings for AA 2024-T3. Both types of corrosion have different mechanisms. In this study filiform corrosion has not been evaluated, but shall be in the near future.

Trevor Hughes communicated: This is an interesting study leading to a proposed mechanism for the formation of a protective layer (Fig. 12 of your paper). My comment is based on my experience of studying aluminium hydroxide gels in the context of their use as delayed gels for blocking excessive water production in oilfield reservoirs. The successive addition of OH^- ions to an acidic aqueous solution of Al chloride leads to the formation of an aluminium hydroxide gel whose strength mainly depends on the total concentration of Al in the system. The pH at which such gels begin to form is around 5.5. In terms of understanding the composition and properties of the gel, the onset of gelation is more conveniently defined in terms of an OH/Al ratio of the system. Beyond the OH/Al ratio required for the onset of gelation, with further addition of OH^-, the average OH/Al ratio of the system increases further towards a value of 3 (corresponding to the thermodynamically stable phase gibbsite $Al(OH)_3$) and beyond to the non-gel regime dominated by $Al(OH)_4^-(aq)$. The kinetic stability of aluminium hydroxide gels depends on the OH/Al ratio of the system – those with OH/Al ratio < 3 are more stable than those at OH/Al = 3. There is a large body of literature on the characterisation of such gels in terms of constituent poly-nuclear Al species and inorganic polymers formed from the tridecamer, $Al_{13}O_4(OH)_{24}(H_2O)_{12}^{7+}$ (commonly known as the "Al_{13}" cation). In the context of the present paper, it would be interesting to characterise Li-doped aluminium hydroxide gels in terms of their pH limits and their kinetic stability – as mentioned in the paper, the nature and properties of the associated anion (carbonate, oxalate,...) will also have important effects on the properties and stability of the gel.

Peter Visser communicated in reply: Thank you very much for this comment and suggestions. This literature can be of great value. The interaction of lithium with the aluminium hydroxide gel seems to be very important to obtain the inhibitive effect observed. However, the detailed role or function of lithium in the stabilization of the aluminium hydroxide gel is not well understood as yet. The characterization of these Li-doped aluminium hydroxide gels could provide more insight.

Julian Wharton opened the discussion of the paper by Suzanne Morsch: High-purity water can rapidly pick-up contaminants such as carbon dioxide that can affect its pH. Was the pH or resistivity assessed during the pre-exposures in deionised water to establish that the water maintained its high quality over the 7 day immersion?

Suzanne Morsch responded: The pH was not assessed during a 7 day immersion. However, epoxy-phenolic resins are renowned for chemical resistivity, so that slight changes in pH are not expected to degrade the coating directly. Contamination of the water may however be expected to affect water sorption as a consequence of reduced water activity. We recently examined coatings exposed to 5% (w/w) aqueous NaCl solutions and found that morphological changes were attenuated under these conditions.

Sonja Nixon questioned: Does the investigated coating contain any pigments? Can the method for determination of water uptake be applied for pigmented coatings?

Suzanne Morsch replied: The investigated coating did not contain pigments, it is a simplified epoxy-phenolic lacquer (thermoset polymer). The *in situ* FTIR experiment described could be applied to measure water sorption for a pigmented coating, provided that sufficient transmission is possible to acquire a reflection mode spectrum. We have found this to be highly dependent on the coating thickness. In the case of AFM-IR, we have previously analysed pigmented primers and are able to acquire localised spectra and images with relative ease. One possible complication arises for such coatings, where the effective sampling depth varies locally as a result of the underlying pigment particles. This can be circumvented by mapping an IR-induced amplitude at several wavelengths (corresponding to peaks in the IR spectrum), and then generating a ratio image (since any variation in the signal due to the film thickness should cancel out).

Julian Wharton asked: How long did it take to perform a typical AFM-IR area scan and also what were the criteria for assessing AFM cantilever tip contamination?

Suzanne Morsch answered: The AFM-IR images acquired for this work were scanned within approximately 2 h. In this particular case 32 co-averages were taken for the photothermally induced ring-down at each point, in order to enhance sensitivity towards small changes in IR absorbance accompanying water uptake, and this requires an unusually slow scanning speed. Typically, AFM-IR scans may be performed in as little as 20 min. However, this does not include the time required for the system set-up, *i.e.*, to perform laser alignment and IR background scans. Since AFM-IR imaging is necessarily performed in contact mode, tip contamination can be an issue. Generally speaking, tip contamination is apparent from streaking or blurring effects in height scans, particularly over small scan areas, as in conventional AFM imaging. If a contaminant is IR active, contamination may also be apparent from IR spectra obtained using the probe.

David Shoesmith remarked: You demonstrate clearly that the coating develops bubbles due to the accumulation of water in void locations within the coating. Does this introduce a physical stress in the coating, and, if so, what are the consequences for the stability of the coating? Having developed such voids, can you explain how they lead subsequently to ionic pathways through the coating? Is it possible to deposit a coating without any void space?

Suzanne Morsch responded: In polymers, a proportion of the free volume is known to exist as nanovoids associated with chain ends, the size and distribution of which is such that some are capable of housing water molecules. For conventional organic coatings, such as epoxy networks, such voids are known to be present, and appear to be an inherent feature of amorphous polymer networks. Since we propose that nanovoids are enlarged by spontaneous polymeric relaxation in the presence of water, this process is considered to be a consequence of increased overall thermodynamic stability. It is expected that during prolonged immersion swelling stresses generated in the coating would eventually restrict any further growth of such voids, so that an equilibrium morphology/water content is eventually reached. Indeed recent examination of epoxy-phenolic samples immersed in deionised water for 3 weeks, and also after immersion for 10 months, revealed comparable morphologies to those we reported here for samples soaked for 1 week. We have not directly measured stress, although coating delamination is never observed. Long term degradation of the corrosion protection provided by intact organic coatings has previously been proposed to occur *via* hydrophilic channels which eventually widen to form ionic pathways. This work represents some of the first direct evidence cases for the long-term formation of such hydrophilic pathways, since voids contain clustered, bulk-like water. It can be envisioned that since polymers are dynamic systems, established void regions may migrate/deform during prolonged immersion, so that ionic transport through the coating becomes possible, although we have no direct evidence for this at present.

Simon Gibbon commented: People don't normally think about the water that polymers contain. In reality all polymers absorb an equilibrium quantity of water dependent on their specific chemistry and the environment. However, coatings protect against corrosion despite containing water. This work is aimed at understanding what changes cause corrosion to occur after a long period of time.

Ying-Hsuan Chen asked: Have you tried immerse the sample again in DI water after drying the sample? A second cycle of drying?

Suzanne Morsch answered: We have not done this experiment. Several aging studies have however reported that cycled immersion and drying yields faster water sorption kinetics together with a greater volume of sorbed water for epoxy resins. This is presumably because polymeric relaxation in the presence of water is frozen into place on drying. We suspect that after prolonged immersion or numerous cycles an equilibrium conformation would be reached whereupon no further increases in kinetics/water content would be detected, however this is a subject of on-going research.

Stuart Lyon remarked: The challenge for this and related work is to consider and validate (or exclude) hypotheses relating to the performance of coatings – some of these derive from the 1960s and, while intellectually compelling, have never been experimentally verified. Working below the diffraction limit allows, for the first time, spectroscopic identification of functional group chemistry (*i.e.* arising either from intrinsic differences in the polymer or from the segregation of small molecular mass species, such as water) in polymers at a length scale of 50–100 nm.

Christopher Taylor opened the discussion of the Concluding remarks by John Scully: A new cost of corrosion study is underway, in partnership between DNV GL and APQC. The new study will extend beyond previous studies – limited to the US – to consider the global impact of corrosion, and strategies to mitigate these impacts and costs (see the NACE press release at the URL shown in ref. 1).

1 http://www.nace.org/Newsroom/Press-Releases/NACE-International-Institute-Selects-DNV-GL-and-APQC-as-Research-Partners-on-Global-IMPACT-Study/

Alison Davenport responded: Studies of this type are vital to make the case to the government for further investment in research and developing skilled people. It is particularly valuable that this study considers the global impact of corrosion. It would be very helpful if such studies were also conducted in the UK and other countries to highlight the economic and societal impact of corrosion.

Faraday Discussions

PAPER

Corrosion chemistry closing comments: opportunities in corrosion science facilitated by *operando* experimental characterization combined with multi-scale computational modelling

John R. Scully*

Received 12th May 2015, Accepted 1st June 2015
DOI: 10.1039/c5fd00075k

Recent advances in characterization tools, computational capabilities, and theories have created opportunities for advancement in understanding of solid–fluid interfaces at the nanoscale in corroding metallic systems. The Faraday Discussion on Corrosion Chemistry in 2015 highlighted some of the current needs, gaps and opportunities in corrosion science. Themes were organized into several hierarchical categories that provide an organizational framework for corrosion. Opportunities to develop fundamental physical and chemical data which will enable further progress in thermodynamic and kinetic modelling of corrosion were discussed. These will enable new and better understanding of unit processes that govern corrosion at the nanoscale. Additional topics discussed included scales, films and oxides, fluid–surface and molecular–surface interactions, selected topics in corrosion science and engineering as well as corrosion control. Corrosion science and engineering topics included complex alloy dissolution, local corrosion, and modelling of specific corrosion processes that are made up of collections of temporally and spatially varying unit processes such as oxidation, ion transport, and competitive adsorption. Corrosion control and mitigation topics covered some new insights on coatings and inhibitors. Further advances in *operando* or *in situ* experimental characterization strategies at the nanoscale combined with computational modelling will enhance progress in the field, especially if coupling across length and time scales can be achieved incorporating the various phenomena encountered in corrosion. Readers are encouraged to not only to use this *ad hoc* organizational scheme to guide their immersion into the current opportunities in corrosion chemistry, but also to find value in the information presented in their own ways.

Center for Electrochemical Science and Engineering, Department of Materials Science and Engineering, SEAS, University of Virginia, Charlottesville, VA, USA 22904. E-mail: jrs8d@virginia.edu; Fax: +1-434-982-5799; Tel: +1-434-982-5786

1 Introduction

Recently, there has been substantial progress in the development of experimental and theoretical methods for probing metals, scale or passive films, and solid–fluid interfaces at the nanoscale.[1] The field of corrosion science, focusing on degradation of metallic materials exposed to liquids and gases, can capitalize on these recent advances with the ultimate goal of relating nanoscale understanding to engineering behaviour. The technological needs and challenges associated with corrosion are particularly acute as our highly engineered civilization moves towards nanoscale devices dependent on the stability of every atom, the use of engineered materials in increasingly complex and harsh environments, increased reliability over the long term epitomized by very diverse challenges such as health care appliance reliability and engineered nuclear waste packages, rapid material replacement and insertion, and the reinvention of the last century's disposability materials-based society with a more sustainable one.[1] The aim of this Faraday Discussion was to bring together experimentalists and theoreticians concerned with understanding interfacial chemical processes at the nanoscale relevant to corrosion and its control.[1]

Corrosion chemistry conducted from April 13–15, 2015 provided a snap shot of the current status of the corrosion field touching upon many of the issues mentioned above. Given the task of reflecting on the many papers, posters, and discussion comments, it is important to note that no closing lecture or written summary could give justice to the papers, posters and many valuable points made in the discussions. The organizational scheme adapted herein represents "my take" on the overall themes of this diverse and multi-disciplinary field. Other organizational structures may serve as well or better. Each paper and poster contained many facets illuminating corrosion advancements and remaining issues as well as touching upon a number of the themes listed above. To borrow an example from the internet, web browsers learn, adapt and then reposition the presentation of topics based on the visitation records and popularity of a particular site or topic. In a similar fashion, the enduring scientific value of each paper and what aspects are seen as the most beneficial to the field will be voted upon over time by the corrosion community and cannot be determined *a priori* by a closing lecturer. My primary hope is that the closing comments made below will point readers towards papers that might not have otherwise been considered for deep immersion and consideration.

This closing comment is organized by the themes above as depicted in Fig. 1; these include underlying fundamentals, unit processes and phenomena, and corrosion science and engineering. Fig. 1 illustrates the hierarchical nature of corrosion. Fundamentals include thermodynamic and kinetic data of value towards corrosion computations. Unit processes included films and scales, solid–fluid interfaces and molecular interactions with surfaces. Corrosion science and engineering included modelling of defined aspects of corrosion processes, localized corrosion and complex alloy dissolution, as well as corrosion control. Each paper presented contained several facets and reasonably fell into several of the areas of this hierarchical scheme. Moreover, corrosion is an inherently multi-scale process where nanoscale chemical, physical and electrical events govern unit processes which in turn assemble to regulate or control larger-scale corrosion

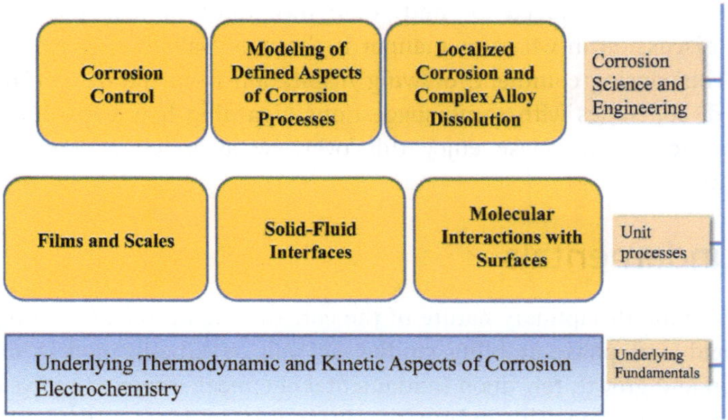

Fig. 1 Corrosion chemistry research areas discussed at the RSC Faraday Discussion on Corrosion Chemistry in April 2015. Each paper presented could be organized into one or more of these areas. The underlying fundamentals inform the various unit processes which together form sets of coupled processes that contribute to the understanding of multi-disciplinary corrosion science and engineering problems addressed by the corrosion field.

processes (Fig. 2). Fundamental laws of thermodynamics of scales and bulk materials, interfaces and electrolytes are critical inputs to crucial unit processes in corrosion such as transport, adsorption, redox reaction, and precipitation. In turn, these unit processes operate in concert to regulate the events taking place during corrosion as well as the engineering consequences. These processes may interact at the nanometre, micrometre or millimetre scales to regulate and control engineering scale corrosion processes. Often corrosion events themselves may occur in sequence or in series. For instance, coating or inhibitor film breakdown may be followed by local corrosion or dealloying which is followed by stress corrosion or corrosion fatigue. Such an assemblage requires the linking of corrosion stages; each containing its own unit processes. Only at this point can corrosion control be considered; ideally guided by nanoscale understanding of all the contributing processes.

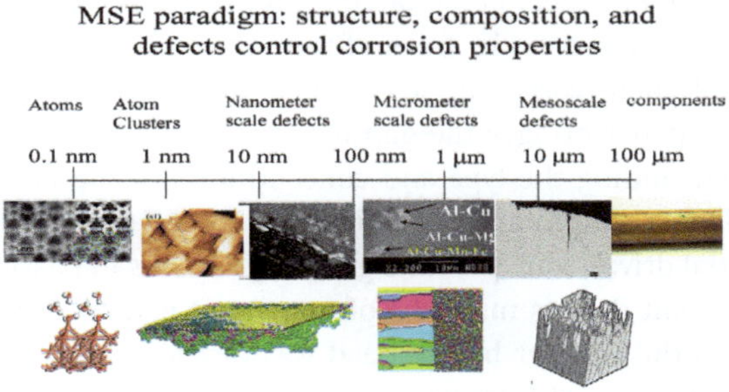

Fig. 2 Hierarchy of the multi-scale corrosion problem starting at the atomic scale and transitioning to the micrometre scale and above. Critical processes at different length scales may factor in to regulate engineering scale corrosion of the copper pipe observed at the far right. It is also noted that the various model length scales from first principles, Monte Carlo, and continuum scale may or may not match the physical or microstructural scale. (Adapted from ref. 60.)

Readers are encouraged to "find" their own value in each paper and the ensuing discussion in whatever manner is the most valuable to their research efforts. This closing comment, drawing inspiration from the insightful papers presented, concludes with some suggestions regarding how corrosion science might be advanced. Please enjoy this overview as much as I enjoyed the conference.

2 Fundamentals

Given the multi-disciplinary nature of the corrosion field, scientific progress in phase field and phase stability studies as well as kinetic models of phase formation and growth rely upon fundamental chemical, structural, physical and electronic information. This information is a necessary input to model the thermodynamics and kinetics of individual unit processes, which when combined attempt to explain various corrosion phenomena. Therefore, fundamental thermodynamic and kinetic parameters must continue to be defined, calculated, reported and perhaps further refined after experimental scrutiny in order to enable accurate modelling. Design of safe windows of operation, computational corrosion materials design, as well as corrosion control strategies will also rely on such information. Phase field construction by computational modelling requires fundamental thermodynamic information as inputs. Papers presented demonstrated that it is possible to develop multiple levels of understanding of phase fields beyond the classical potential-pH regions of oxide stability, and metal dissolution. For instance, it is now possible to compute the type of oxide defects present in a phase field where a specific oxide is stable, as well as equilibrium concentrations of each type of defect considered.[2] The effects of defects on the boundaries of the phase fields and E–pH diagrams, effects of applied stress, surface adsorption and other factors can now be computed using first principles methods as a function of external variables and will likely add further insights into corrosion.[3–8]

Many oxide covered metal interfaces that control corrosion are either monolithic semiconductors or p–n semiconductor junctions in conformal layers. Noteworthy was a paper connecting the field of semi-conductor defect physics and electrochemical concepts.[2] The most thermodynamically stable point defects as a function of electron chemical potential and proton activity or potential and pH were predicted for a given oxide. In the future this could be extended to other compounds involved in corrosion scales such as hydroxides, carbonates, and phosphates. This paper bridges the gap between semiconductor "physics" and corrosion. Understanding the type and concentration of the most stable defect type in oxides is essential to and sheds light on the effects of electric field and chemical potential driven transport and growth processes in passive films. It also suggests which point defects might be of interest to control transport and ultimately oxide growth. Another bridge head was developed by linking corrosion relevant oxide properties with mineral scale expertise at the nanoscale.[9] Further progress can enlighten understanding of how defect sequestration, oxide dopants or other mitigation strategies might or might not be advantageous towards corrosion.[2] As pointed out, there is cross-cutting interest in geochemistry, colloid science, catalysis, nanotechnology, soil and clay chemistry and corrosion regarding the properties of oxidized surfaces where information is sought in multiple dimensions such as surface structure and macroscopic surface charge.[9] In the specific example presented, the atomic and electronic structures of aged and fresh hematite were considered over exposure time and related to the isoelectric point of the oxide surface over a range of pH. A working framework of the structure, composition and local electronic density of states of valence (conduction) band electrons was constructed. This detailed understanding has applicability to all oxide covered metals and would provide improved ability to elucidate how the scaled surface might interact with the ionic electrolyte and adsorbing molecular species as well as polar molecules during corrosion. A third advance was to consider phase stability and oxygen adsorption in the presence of compressive and tensile stress.[4] It was shown that the oxygen adsorption energy changed with tensile and compressive strain. Such data shows that the strain induced by oxygen adsorption can be offset by external stress and that under potential, surface oxidation may be affected by tensile strain.

In total these papers shed light on how a greater basic insight into the fundamental aspects of surfaces, interfaces, and phases enabled by modelling will move the field of corrosion science forward.[2,4,9]

3 Critical unit processes in corrosion

There are many unit processes at interfaces and across films that govern or mediate corrosion. Unit processes across films and interfaces discussed included film and scale formation, and growth, as well as molecular interactions and reactions at surfaces/solid–fluid interfaces. All of these play a role to control or otherwise mediate electrochemical corrosion reactions including anodic and cathodic half-cell reactions. While these individual processes do not themselves describe everything about a corrosion process, they are critical unit processes. For instance, ionic transport across a scale or how a scale controls an electrochemical reaction at the metal/scale interface are of paramount importance in corrosion. Progress must be made in understanding these individual processes in order to move the field forward.

3.1 Films and scales

An enduring theme in corrosion is the need to understand oxides, films, and scales. Corrosion in aqueous solutions, upon atmospheric exposure and at high temperature is often mediated and controlled not only by conformal nanometre scale oxides, but by thick scales and films adherent to metallic substrates. Corrosion scales and passive films can both significantly affect corrosion (*e.g.*, a passive film can reduce corrosion to a negligible rate).[1] Experimental/theoretical studies concerning both engineering and model (*e.g.*, single crystal) substrates were presented. Relationships between the composition–structure–chemistry–corrosion resistance of scales and films were addressed, together with insight into processes governing their initiation and growth, including the impact of environmental conditions.[1]

The identity and structure of these layers is dictated by the underlying materials, the solution, and the molecular and colloidal species in the solution and/or all of these. Therefore, the exact nature of the protective scale is hard to predict *a priori*. Crucial to progress in the corrosion field is the need to obtain *in situ* or *operando* chemical, structural and mechanical information about these films and scales. Particularly exciting was progress in understanding the chemical, oxidation state, and chemical environment of these layers and films accessed only recently by X-ray photoelectron spectroscopy (XPS) through thin water layers. The chemical environment and oxidation behaviour of metal oxides in dry, humid, and wet conditions was investigated.[10] This information was combined with electrochemical treatments to produce different oxides interrogated by *operando* XPS.[10] Continuing on the theme of examination of scales and films *in situ* was work examining the temporal stages of thick film formation on iron in CO_2 under electrochemical control. The global corrosion status could be compared to *operando* observations on scaling.[11] The value of characterizing the colloidal properties of the solution, the amorphous gel formed on the surface followed by reorganization as a crystalline scale of siderite all while acquiring global electrochemical corrosion data was evident. Small and wide angle X-ray scattering experiments examined the iron CO_2 interface almost simultaneously for an unprecedented view of the events during scaling. It was found that small amounts of dissolved Cr critically effect scale formation. Lastly, a multi-faceted investigation into the effect of Mo on passivation in Fe–Cr–Ni and Fe–Cr–Ni–Mo alloys revealed the enormous value in applying several *in situ* and *ex situ* techniques simultaneously to unravel the long standing need to understand the beneficial role of major, trace and minor alloying elements on passivity and passivity breakdown.[12] Single crystal (111) surfaces were investigated. *Ex situ* X-ray photoelectron spectroscopy (XPS) and time of flight secondary ion mass spectrometry (ToF-SIMS) data were compared and contrasted.[12] Atomic scale morphological information was acquired in concert using *in situ* STM investigation of passivating terraces and ledges. To this end, the chemical, oxidation state, molecular and morphological behaviour of an electrochemically passivated Fe–Ni–Cr passivated surface was directly compared to the same alloy containing Mo. Several roles of Mo were identified and the fate of alloying elements at the metal/oxide, oxide layer, and at terrace sites was elucidated.[12]

With the age of environmental microscopes for *operando* interrogation, progress in this area of corrosion science is expected to produce unprecedented new findings, with some perhaps being transformational.

3.2 Solid–liquid interfaces and interactions of molecules at surfaces

In the corrosion field, information is sought about the underlying substrates, each interface, each film, scale, or layer, each synthetic coating or pre-treatment, as well as the interaction of water and the electrolyte both near the electrode surface at both double layer length scales and further away in the bulk of solution. Often corrosion research has been piecemeal where only one or two of these aspects are characterized *ex situ* upon air transfers. Challenges in corrosion include the need for real time *in situ* spatial characterization, depth profiling as well as the need for multi-channel information or "corrosion in 4-D".

New nanoscale understandings of the structure and fundamental chemical processes (*e.g.* ion adsorption) at solid–fluid interfaces were discussed. Particular attention was given to solid/aqueous solution interface, as they are of central importance to corrosion.[1] Ideally a layer-by-layer approach enables simultaneous characterization of each layer involved in a corroding interface.[1] Often, there is a metal conductor, semi-conductor oxide film, an amorphous highly porous outer film, and a colloidal solution within the overlying ionic conducting phase. It is clear that much is to be learned about the double layer, the structure of water near various oxide and metal interfaces as a function of potential, as well as the nature of non-traditional scales that may fall somewhere in between compact oxides and colloidal solutions near such interfaces.

The concept of obtaining *in situ* information at the oxide/liquid buried interface was exemplified in the study of radiolysis driven dissolution and the UO_2–water interface. Intense X-ray beams were utilized for *in situ* X-ray reflectivity and high angle diffraction (XRD). Information on layer thickness and thus rate of dissolution, roughening and electron density was obtained.[13] Moreover, the crystalline nature of the layers and dislocation defects could be examined. A layered UO_2/UO_x and a UO_x–water interface model was constructed and new nanoscale detail was revealed, possibly altering the understanding of current models addressing long term storage of spent nuclear fuel. The study of CO_2 corrosion also interrogated both the scale and the colloidal solution properties of the solution above the scale, concurrently, while collecting electrochemical information. Together these studies provide an unprecedented view of how a corroding interface behaves and responds to various environmental factors during corrosion.[11]

The interactions of special molecules at aqueous interfaces were also explored. In corrosion, inhibitor molecules and certain proteins can limit corrosion. In contrast, film disrupting ions can replace metal cations in oxides or otherwise damage oxides and reacting species such as thiosulfate can react on corroding surfaces leading to detrimental reduced species such as sulphur. Sulphur can in turn catalyse dissolution of certain metals (*e.g.* nickel).[14] One example of an interacting naturally occurring molecule is bovine serum albumin (BSA) and its interactions with pure Mg and Fe during corrosion is of great interest in *in vivo* studies such as those related to implant materials in humans. BSA has a strong interacting effect with Mg which leads to corrosion inhibition whereas selected accelerated corrosion is induced by interactions with Fe.[15] Moreover there was complex time dependence with this interaction. Favoured BSA adsorption and subsequent corrosion inhibition occurred in the case of Mg. This was traced to electrostatic interactions between the protein and the metal oxide which depended on the bond strength between metal cations and protein molecules.[15]

Chemical inhibitors are a key area of need for corrosion protection but also a fertile ground for rapid science advance. The first step concerns the interactions of molecules with surfaces while the second should consider the broader scope of the processes that conspire to regulate corrosion.[16–18] Computational studies with focus on surface interaction and adsorption phenomena have blossomed, perhaps led by the catalysis field. An ever increasing number of synthesized molecules and natural organic species may have potential as corrosion inhibitors. A major issue in the corrosion field is how to sort, categorize, and otherwise "screen" molecules for good replacement inhibitors which are at the same time more chemically friendly to the environment. *Ab initio* studies (mainly using Density Function Theory or DFT) have come of age and can begin to address the interaction of molecules with surfaces. In the Faraday Discussion it was pointed out that this task is much more difficult in the field of corrosion compared to reaction catalysis where relatively straight forward attributes of molecule–solid catalyst interactions may shed light on the qualities of the most favourable catalysts. The corrosion challenge is somewhat more complex partly due to the number of important steps involved in the overall rate determining process and the number of physical processes conspiring to regulate corrosion rate. Moreover, the complexity of many corrosion environments is substantial; with environments often containing ionic, molecular, polymeric, and colloidal species together.[18]

A common starting point in this issue is adsorption which is generally viewed as a key process by which corrosion inhibiting molecules improve corrosion resistance. The issue of bond strength of deprotonated and protonated inhibitor molecules in different orientations was directly addressed in *ab initio* studies of benzotriazole or $C_6H_5N_3$, hereafter referred to as BTAH. This is a well-known corrosion inhibitor of copper.[16,17] It was shown that the inhibitor molecule was able to bond to both oxide covered and oxide free surfaces. Moreover, BTAH may passivate under-coordinated copper sites that are "plausible sites for Cu oxidation". Organic metal complexes are also formed suggesting that "at risk" copper ad-atoms could be stabilized by inhibiting molecules.[17] The good qualities in an inhibiting molecule and the bad qualities were subsequently debated. No single metric accessed by *ab initio* methods seem to readily describe this; the consensus was that a combination of inhibitor qualities is needed and a good starting place is to understand adsorption. A soluble inhibitor molecule–metal complex was regarded as bad; strong adsorption could be good if electron donating events could be reduced by the presence of an absorbed species; for instance, metal oxidation and Cl^- adsorption.[17] Indeed, Cl^- and water adsorption energies were compared to the BTAH and BTA^- adsorption energies on copper crystals of different orientation and fairly strong competitive adsorption was suggested. Subtleties such as the how the BTAH and BTA^- are oriented with respect to the surface plane were shown to be very important with strong chemi-sorption *via* the azole moiety with coverage, protonation, and van der Waals forces all playing a role. Coverage changes the molecule arrangement ranging from flat lying towards end-to-surface configurations with inhibitor concentration which enabled close packing of BTAH. However, the attributes of a bad inhibitor in terms of a calculated metric were not entirely clear. Moreover, competitive adsorption needs to be considered in the complex electrolyte. Studies to date consider the bond strength of Cl^- *versus* the bond strengths of BTAH or BTA^- in isolation from each other and in a vacuum without lateral interactions. Studies at the next level should include inhibitor, competitive adsorbing ion and water molecules at the interface. Moreover, these bond strength arguments are weakened by limited insights regarding how the dissolution rate of first, second and third nearest neighbour adsorption sites might be affected by proximate adsorption. This has been considered elsewhere.[7] The next step in modelling might extend to copper oxidation with Monte Carlo simulation of half-cell Cu oxidation reactions (for an anodic inhibitor) for various copper sites with adsorbed molecules as well as for first, second and third nearest neighbours.[7] This will further enhance understanding of the role(s) of inhibiting ions and in order to move beyond the notion that strong adsorption is desired followed by the subsequent assumption that sites of adsorption are blocked while unblocked sites corrode as if no inhibitor were in the system at all; with each site behaving independently.

These studies are a significant advance over semi-empirical quantum chemistry approaches where the physicochemical properties of molecules in isolation are correlated with corrosion inhibition.[19] Relationships are sought between a number of molecular properties such as orbital energies, dipole moment, charge density, and other aspects.[19] This has been called the quantitative structure activity relationship or QSAR approach. For instance, in this approach the energies of the highest and lowest density occupied states of the inhibiting molecule in isolation are often used as a metric. A correlation between inferred coverage due to these attributes and corrosion inhibition is often attempted.

Finally, not only must the quantification of the various types and strengths of interactions of molecules with surfaces be considered. The modelling of dissolution rate with proximate inhibitor molecules is a good starting point. However, many other aspects such as electrolyte flow rate, colloidal properties and multilayering corrosion scales, wetting, biocide capability, and inhibiting films must be considered to provide a comprehensive view of how corrosion inhibitors operate.[18]

4 Corrosion science and engineering

Complex dissolution processes experienced during corrosion of commercial and model alloys in complex environments were discussed. Mirroring the technological needs of the field, some papers "took on this issue" in all of its complexity.[14,20,21] Specialized *in situ* mapping techniques, combination of complementary techniques and/or control of single corrosion sites were some of the different strategies used.[20–23] A recent advance includes various scanning probe techniques (SPTs) available for enhanced spatial resolution of corrosion in real time. Moreover, *operando* optical microscopy with long focal length microscopes remains of great utility.[20,21,23,24] New approaches such as the scanning electrochemical microscope (SECM) and scanning vibrating electrode technique (SVET) also were used in combination with more traditional DC electrochemistry methods. A continued challenge is the inability to acquire simultaneous data on multiple aspects of alloy corrosion at the same time and over the same length scale such as chemistry changes at surfaces while monitoring element-by-element dissolution and filming. Assessment of all types of anodic and cathodic reactions differentiated by ½ cell type with high spatial resolution continues to be a highly desirable capability needed to unravel complex alloy dissolution.

4.1 Localized corrosion and complex alloy dissolution

Spatially confined corrosion phenomena (*e.g.*, pitting), which can lead to rapid failure of otherwise corrosion resistant substrates, were addressed.[1] Studies aiming to understand at the nanoscale the combined influences of substrate microstructure/geometry, and fluid chemistry on complex alloy and local corrosion initiation/progression were presented.[1] One of the enduring themes was how complex alloy corrosion remains. Major, minor, and trace alloying elements impart a significant impact on corrosion and certain elements or phases often interact critically with specific solution chemistries.[14,20–25]

Anodic polarization of light alloys such as magnesium and aluminium can exhibit extraordinary rates of the hydrogen evolution reaction (HER) often observed in anodic areas.[26] This observation is counter to basic mixed potential theory which assumes that the HER rate decreases as applied potentials become more anodic and the hydrogen over potential is lowered. Hydrogen production reaches a minimum near open circuit conditions and is often intensified at more positive applied anodic currents with increasing metal dissolution rates. In the Mg corrosion literature this has been called the negative difference effect.[26] Anodically enhanced hydrogen evolution (EHE) has been observed in many alloy systems and modelled based on hydrolytic acidification and ohmic drop in pits and cracks.[27] However, the opening lecture posited that at high anodic reaction rates, hydrogen production may become "super-catalysed" as verified by hydrogen gas collection by mechanisms under current debate.[26] The notion is that the exchange current density for HER increases during anodic polarization to a much greater extent than the anodic overpotential; thereby offsetting a decrease in hydrogen overpotential. The concept could apply to a number of light metals which oxidize during open circuit corrosion and at anodic potentials far below reversible hydrogen. The various contributions to the effect (for instance the contributions from hydrolytic acidification and ohmic voltage drop,[27] impurities,[28,29] films, *versus* special sites which may affect the activation barrier of facile HE have not been clarified). Moreover, the study highlights some of the difficulty of gaining corrosion information from DC electrochemical methods which only measure net current and are not spatially resolved. Ramifications of EHE towards hydrogen embrittlement of Al were discussed.[26,30–34] Hydrogen embrittlement during freely corroding and anodic polarization in occluded sites is certainly observed during light alloy corrosion as well as steel and local hydrogen production has been shown to play a strong role.[27,32,35,36] High hydrogen levels local to anodically polarized crack tips are observed in numerous systems such as Al and Ti.[32,34,36,37] Additional study of hydrogen interactions with surfaces and precise effects of surface atom defects and coordination chemistry on reduction and oxidation processes are clearly warranted.[3,7,8,38]

Mg dissolution was discussed further, considering the effects of various trace levels of iron on localized corrosion morphology.[23] The theme of complex corrosion patterns was covered. For instance, the morphology of Mg corrosion changes with various Fe impurity levels in solutions of various ionic strength (conductivity).[23] Here anodically induced cathodic enrichment was observed and the formation of dark areas which created various corrosion patterns.[29,39] It was rationalized that these were regulated by Cl^- and Fe content.[23]

Complex interaction between thiosulfate and halides with three distinctly different alloys illustrates how radically different pitting morphologies may result from differences in alloy chemistry, concurrent electro-migration of differently charged ions and the presence of reacting species. The oxide breakdown process was short circuited or otherwise circumvented by scratching anodes. Studies were conducted in complex solutions containing several anions, some which attack the metal and support hydrolytic acidification, as well as others which react to form very detrimental species such as sulphur.[14] The localized (*e.g.*, pitting corrosion process) involved propagation of individual pits interacting dynamically with the chemistry of the solution and each other. Strong potential, solution chemistry and alloy tendencies were observed pointing to the complexity of the issue. For instance a lack of alloyed Mo contributed to pitting at low potentials in sulphate. The debate over the relative roles and contributions of alloyed Mo in connection with the passive film *versus* the propagation and stabilization stage of pitting was continued. The morphologies of both local and multiphase alloy attack were also discussed. Scanning probe techniques (SPTs) of various types have come of age in studying corrosion patterns on microstructural complex alloys. Methods discussed included SVET, SECM, and a novel shear force mounted scanning micro-capillary method.[20,21,23] There is a current focus within complex alloy corrosion research on light alloys such as micro-galvanically coupled Mg alloy ZEK100, zinc magnesium aluminium alloys, (ZMA), precipitation age hardened aluminium, and emerging steels.[20,21,23] Stainless steel and nickel based alloys continue to be of interest given their immense technological importance.[14,22,24] Future directions are likely to include high entropy alloys, 2-D materials, and others produced by high energy processes far from equilibrium.

Complicated patterns of corrosion develop as a function of alloy, solution chemistry, and electrolyte geometry (*i.e.*, droplets) and these patterns are controlled by a variety of factors such as active corrosion *versus* passive break-down,[24] electromigration of differently charge ions and anion ratios,[14] alloy microstructure including anodic and cathodic phases,[20,21] proximate cathodic reactions, and cathodic limited growth *versus* ohmically limited growth.[23] In another strategy, a single isolated pit was produced and controlled using a capillary drop method.[22] Critical concentrations, critical ratios of chemical species and critical depths were all envisioned to produce "windows of susceptibility" *versus* "kinetic immunity".[14,22,24]

In summary, the SPTs provided a strategy for *operando* studies and dealt with the incredible complexity of alloy dissolution. The dissolution morphologies observed are infrequently on the length scale of the microstructural features in the alloy illustrating the complexity of the issue. Progress to elucidate complex alloy corrosion continues fuelled often by (i) alloy "*du jour*" technological needs, and (ii) the adaptation of new or emerging techniques often developed in other fields such as biology, chemistry, surface science and applied electrochemistry.

4.2 Modelling of defined aspects of corrosion processes

Just as fundamental parameters and the details of specific unit processes must be understood, these processes must be assembled in meaningful ways to provide a more complete view of complex corrosion phenomena. Several papers were presented which modelled defined aspects of corrosion processes. Further progress will likely accelerate in this area.[18,40,41] Once models can be developed which can capture as many unit processes as necessary to faithfully describe a corrosion process then they can be exercised over a broad range of plausible input conditions to gain insight on the importance of variables to various outcomes, identify potential regulating factors during corrosion and even guide more focused experiments directed towards deeper insights. This differs from Edisonian experimental approaches where high throughput methods are used to rapidly map out response surfaces for corrosion processes under a broad range of conditions without much fundamental insight. High throughput methods are a worthy starting point for corrosion discovery when there is no model and little idea as to what is important. However, this approach can be improved upon as a corrosion process becomes better understood.

A mechanistic model of the oxide growth and dissolution of Cr-containing alloys was one example.[40] The model considered both electrochemical and chemical reactions at the metal/oxide and oxide solution interfaces. These included metal oxidation, oxide growth, and oxide dissolution. Charge as well as mass balances were maintained. Selected unit processes that operate together to regulate film formation were collected and dealt with together in one model.[40] The cation flux was quantified, while charge as well as mass balances were imposed. Noteworthy was that the metal cation flux at the metal/oxide interface must equal the sum of oxide growth and oxide dissolution rates at the oxide/solution interface. The effects of temperature, environmental chemistry and oxide thickness were all incorporated. Time dependent behaviour was predicted for pure iron, Co–Cr and Fe–Ni–Cr alloys. In another example, the synergistic influences of hydrogen peroxide and H_2 on simulated spent nuclear fuel corrosion were considered in a model framework where variations in corrosion potential, redox conditions, HCO_3^-/CO_3^{2-} and convective conditions all mediated U oxidation by H_2O_2 that could be reversed by H_2.[41] In this way, these various factors were all captured and were shown to operate in concert to govern spent fuel corrosion controlled by the chemical dissolution rate of the U(VI) layer. For instance, HCO_3^-/CO_3^{2-} could prevent formation of insulating U(VI) layers.

The need for multi-physics, multi-domain type approaches that collect a wide range of different processes and incorporate them all in a multi-length scale, multi-time scale model that captures the physical, chemical and electrochemical processes necessary to describe corrosion was clear.[18] Finally, risk assessment was mentioned as a capstone to the models.[18] The future of corrosion modelling will likely involve collection of the fundamental information discussed above, elucidation of the unit processes, plausible model frameworks, and then multi-scale models that can accurately meet the challenge of accurately forecasting corrosion.

4.3 Corrosion control and mitigation by coatings, pigments, and inhibitors

The largest practical area of corrosion control is by coatings, paints, and corrosion inhibitors. Nanoscale understanding of corrosion control methods was emphasized.[1] In particular, situations were addressed where altered surface chemistry, often of a multi-layered nature, played a key role in corrosion mitigation. Topics of discussion centred on those critical to obtaining mechanistic insight into corrosion control, including substrate bonding, structure/morphology, interfacial transport, and degradation mechanisms of coatings and treatments.[1] The future of corrosion control likely involves multi-functional solutions guided by the science behind individual steps in the coating degradation process.

Some topics of focus included replacement pigments for chromate, novel graphene and electro-reactive coatings, water uptake in intact organic coatings and film forming or molecular adsorbing inhibitors.[16–18,25,42–45] The benefits of atomistic scale observation were evident from the study of adsorbing inhibitors in the case of dealloying. Here, adsorbing inhibitors which formed a film and limited surface diffusion were effective towards limiting dealloying from solid solution alloys where one element was displaced far from its equilibrium oxidation potential. However, the inhibitor offered no long range protection since absence of the inhibitor on metal sites led to rapid dealloying.

Concerning, traditional engineered coatings, the behaviour of "intact coatings" which are a barrier to corrosion is of great importance. Here, degradation is driven by electrolyte uptake, transport and the development of microscopic defects and fast transport pathways in polymer coatings. Another important focus is corrosion control at "macro defects" in coatings.[25,44] In the former, long term organic coating failure was envisioned to occur through the development of ionic water channels; moisture sorption, and hydrogen bonding between the coating and water.[44] Nanoscale softening occurred as the coatings became plasticized. Increased wetting and a drop in the glass transition temperature both occurred as well. In the case of an artificial defect such as a physical scratch exposing bare metal, the challenge is to protect the bare area enabled by the functional response of the adjacent pigmented coating. Lithium bearing salts were incorporated in the coating as a leachable pigment in one example. Focus was on the ability of the pigment to supply Li forming compounds to create a deposit or conversion coating over the artificial scratch which exposed the engineering alloy substrate alloy (2024 or an alternative other precipitation age hardened high strength Al alloy). The morphology and chemical nature of the scale formed over the scratch were characterized. It is interesting to consider these two contrasting areas of study; one focused on the bare metal and the adjacent pigmented polymer and the ability to release, transport and form a protective layer over the bare scratch.[25,44] The other focused on the details of polymer degradation with respect to the "intact coating". Radically different aspects of time dependent polymer degradation, as well as storage/release processes are at play and many of the properties of interest differ from those typical of corrosion science. Clearly, both aspects must be studied in order to advance the understanding of coatings for corrosion control. There is the need for tools to observe and characterize both processes.[44] Corrosion science in this area could be furthered by capturing inhibitor release, transport and deposition in developing models, as well as to understand the attributes of so called smart coatings for successful protection of coating defects. Fortunately recent advances in continuum scale computational modelling can now be incorporated to address the efficacy of a pigment; for instance, regarding its ability to release and deliver the needed chemicals to protect a scratch.[46–52] Computational design of "scratch" protection schemes for coating defects is now possible including electrochemical scratch protection.[53] This type of approach could be used to study corrosion inhibition at coating defects using all kinds of strategies such as by partially soluble pigments or electroactive coatings where release might be triggered in response to environmental stimulation.[47,52,54–59]

Novel coatings are also a key part of future corrosion control strategies.[42,43] A conducting polymer with good adhesion was discussed. A closely related challenge is interface engineering to optimize adhesion and corrosion protection by producing favourable interface chemistries. The behaviour of graphene as a single molecular layer which is impermeable, transparent, flexible, and stable provides an interesting possibility. The impenetrable graphene layer is noteworthy but the edge exposed regions and subsequent lateral transport under the graphene over layer is an important challenge as it presents a weakness or potential flaw in the successful utilization of this mitigation strategy.[43] Graphene may not ultimately provide the answer to corrosion control, but is symbolic of the novel thinking that must be a part of an "all of the above" approach to corrosion control. It is also clear that no one aspect – neither the intact coating nor the defect can be considered in isolation. New technologies such as graphene and a mix of traditional methods and approaches must all be considered to improve corrosion. It was clear from these studies that, as in the case of passive films, the combination of detailed morphological, chemical and electrochemical information, preferably *in situ* will be of immense value towards future progress. In terms of modelling, a multi-scale approach must also be considered using first principle tools, continuum scale models combined with consideration of multiple physical phenomena. Since coatings like other materials contain a variable population of defects such modelling should include uncertainty analysis.[18]

5 Conclusions

Recent advances in characterization tools, computational capabilities and theory have created opportunities for advancement in understanding of solid–fluid interfaces at the nanoscale in corroding systems. The Faraday Discussion on Corrosion Chemistry highlighted some of the current and future needs, gaps and opportunities in the field of corrosion science. Themes could be organized into several hierarchical categories that provide a framework for inspiring further progress in corrosion science. These include foundational areas such as the need and opportunity to access fundamental physical and chemical data to further progress in thermodynamic and kinetic modelling, further improvements and new ways to understand unit processes that govern corrosion at the nanoscale such as a better understanding and control of scales, films and oxides, fluid–surface and molecular–surface interactions, as well as selected topics in corrosion science and engineering. Challenges were identified in each of these theme areas. Corrosion involves multiple processes, occurring at different lengths and time scales; these in combination and in certain hierarchical relationships with one another ultimately govern engineering scale behaviour. Multiple simultaneous tools to characterise corrosion at a sub-micrometre scale and on an *in situ* or *operando* basis are required. At the same time information is necessary on a larger length scale that factors in smaller length scale defects and features. More comprehensive and complete analysis of corrosion processes by modelling requires multi-physics based approaches which must consider these many processes – often together. The question arises, is the corrosion field making progress? The answer is yes, but progress is fragmented for a wide variety of reasons. Scientific advancement often occurs by piecing together isolated pockets of progress. Researchers are encouraged to find value in many of the studies discussed herein even if they are in an environment or on a different material that may not seem to be of immediate interest. Further advances in *operando* characterizations and computational modelling will enhance progress in the field, especially if coupling across length and time scales incorporating the various phenomena encountered in corrosion can be achieved and successful multi-scale approaches are found.

Acknowledgements

Support from the National Science Foundation DMR-1309999 under the direction of Dr Gary Shiflet is gratefully acknowledged. This work was also partially supported by an ONR MURI Program under the direction of Dr David Shifler titled "Understanding Atomic Scale Structure in Four Dimensions to Design and Control Corrosion Resistant Alloys on Grant number N00014-14-1-0675".

References

1 R. Lindsay, Corrosion Chemistry Faraday Discussion Web Site, accessed April, 2015, 2015.

2 M. Todorova and J. Neugebauer, *Faraday Discuss.*, 2015, DOI: 10.1039/C4FD00238E.

3 J. K. Nørskov, T. Bligaard, B. Hvolbæk, F. Abild-Pedersen, I. Chorkendorff and C. H. Christensen, *Chem. Soc. Rev.*, 2008, **37**, 2163–2171.

4 D. Kramer, Y. Wang and J. A. Wharton, *Faraday Discuss.*, 2015, DOI: 10.1039/C4FD00247D.

5 H. A. Hansen, J. Rossmeisl and J. K. Norskov, *Phys. Chem. Chem. Phys.*, 2008, **10**, 3722–3730.

6 S. R. Taylor, G. J. Shiflet, J. R. Scully, R. G. Buchheit, W. J. VanOoij, K. Sieradzki, R. E. Diaz, C. J. Brinker and A. L. Moran, *Corros. Rev.*, 2007, **25**, 491–522.

7 C. D. Taylor, M. Neurock and J. R. Scully, *J. Electrochem. Soc.*, 2011, **158**, F36–F44.

8 D. J. Horton, A. W. Zhu, J. R. Scully and M. Neurock, *MRS Commun.*, 2014, **4**, 113–119.

9 J. Lützenkirchen, F. Heberling, F. Supljika, T. Preocanin, N. Kallay, F. Johann, L. Weisser and P. Eng, *Faraday Discuss.*, 2015, DOI: 10.1039/C4FD00260A.

10 O. Karslioğlu, S. Nemšák, I. Zegkinoglou, A. Shavorskiy, M. Hartl, F. Salmassi, E. Gullikson, M.-L. Ng, C. Rameshan, B. Rude, D. Bianculli, A. Cordones-Hahn, S. Axnanda, E. Crumlin, P. N. Ross, C. M. Schneider, Z. Hussain, Z. Liu, C. S. Fadley and H. Bluhm, *Faraday Discuss.*, 2015, DOI: 10.1039/C5FD00003C.

11 B. Ingham, M. Ko, N. Laycock, N. M. Kirby and D. E. Williams, *Faraday Discuss.*, 2015, DOI: 10.1039/C4FD00218K.

12 V. Maurice, H. Peng, L. H. Klein, A. Seyeux, S. Zanna and P. Marcus, *Faraday Discuss.*, 2015, DOI: 10.1039/C4FD00231H.

13 R. Springell, S. Rennie, L. Costelle, J. Darnbrough, C. A. Stitt, E. Cocklin, C. Lucas, R. Burrows, H. E. Sims, D. Wermeille, J. Rawle, C. Nicklin, W. Nuttall, T. Scott and G. Lander, *Faraday Discuss.*, 2015, DOI: 10.1039/C4FD00254G.

14 W. Zhang, A. G. Carcea and R. C. Newman, *Faraday Discuss.*, 2015, DOI: 10.1039/C5FD00008D.

15 V. Wagener, A.-S. Faltz, M. S. Killian, P. Schmuki and S. Virtanen, *Faraday Discuss.*, 2015, DOI: 10.1039/C4FD00253A.

16 C. Gattinoni and A. Michaelides, *Faraday Discuss.*, 2015, DOI: 10.1039/C4FD00273C.

17 A. Kokalj, *Faraday Discuss.*, 2015, DOI: 10.1039/C4FD00257A.

18 C. Taylor, A. Chandra, J. Vera and N. Sridhar, *Faraday Discuss.*, 2015, DOI: 10.1039/C4FD00220B.

19 N. Khalil, *Electrochim. Acta*, 2003, **48**, 2635–2640.

20 P. Dauphin-Ducharme, W. J. Binns, M. E. Snowden, D. W. Shoesmith and J. Mauzeroll, *Faraday Discuss.*, 2015, DOI: 10.1039/C4FD00276H.

21 J. Sullivan, N. Cooze, C. Gallagher, T. Lewis, T. Prosek and D. Thierry, *Faraday Discuss.*, 2015, DOI: 10.1039/C4FD00251B.

22 S. Heurtault, R. Robin, F. Rouillard and V. Vivier, *Faraday Discuss.*, 2015, DOI: 10.1039/C4FD00252K.

23 G. Williams, N. Birbilis and N. McMurray, *Faraday Discuss.*, 2015, DOI: 10.1039/C4FD00268G.

24 T. Rayment, S. R. Street, N. Mi, A. J. M. C. Cook, H. B. Mohammed-Ali, L. Guo and A. Davenport, *Faraday Discuss.*, 2015, DOI: 10.1039/C4FD00246F.

25 P. Visser, Y. Lui, X. Zhou, T. Hashimoto, G. Thompson, S. Lyon, L. G. J. van der Ven, A. J. M. C. Mol and H. Terryn, *Faraday Discuss.*, 2015, DOI: 10.1039/C4FD00237G.

26 G. S. Frankel, S. Fajardo and B. M. Lynch, *Faraday Discuss.*, 2015, DOI: 10.1039/C5FD00066A.

27 B. A. Kehler and J. R. Scully, *Corrosion*, 2008, **64**, 465–477.

28 N. Birbilis, A. D. King, S. Thomas, G. S. Frankel and J. R. Scully, *Electrochim. Acta*, 2014, **132**, 277–283.

29 T. Cain, S. Madden, N. Birbilis and J. Scully, *J. Electrochem. Soc.*, 2015, **162**, C228–C237.

30 J. H. Ai, M. L. C. Lim and J. R. Scully, *Corrosion*, 2013, **69**, 1225–1239.

31 J. H. Ai and J. R. Scully, *Corrosion*, 2013, **69**, 752–767.

32 G. A. Young and J. R. Scully, *Metall. Mater. Trans. A*, 2002, **33**, 1167–1181.

33 Hydrogen solubility, diffusion and trapping in high purity aluminum and selected Al-base alloys, *Gaseous Hydrogen Embrittlement of Materials in Energy Technologies*, ed. J. R. Scully, G. A. Young and S. W. Smith, Woodhead Publishing, Philadelphia, PA 19102–3406, USA, 2012, vol. 1.

34 G. A. Young and J. R. Scully, *Acta Mater.*, 1998, **46**, 6337–6349.

35 D. G. Kolman, M. A. Gaudett and J. R. Scully, *J. Electrochem. Soc.*, 1998, **145**, 1829–1840.

36 D. G. Kolman and J. R. Scully, *Metall. Mater. Trans. A*, 1997, **28**, 2645–2656.

37 D. G. Kolman and J. R. Scully, *Corros. Sci.*, 2000, **42**, 1863–1879.

38 K. S. Williams, J. P. Labukas, V. Rodriguez-Santiago and J. W. Andzelm, *Corrosion*, 2015, **71**, 209–223.

39 L. G. Bland, A. D. King, N. Birbilis and J. R. Scully, *Corrosion*, 2015, **71**, 128–145.

40 M. Momeni and J. C. Wren, *Faraday Discuss.*, 2015, DOI: 10.1039/C4FD00244J.

41 M. Razdan and D. W. Shoesmith, *Faraday Discuss.*, 2015, DOI: 10.1039/C4FD00221K.

42 Y. Konno, E. Tsuji, Y. Aoki, T. Ohtsuka and H. Habazaki, *Faraday Discuss.*, 2015, DOI: 10.1039/C4FD00232F.

43 L. Kyhl, S. F. Nielsen, A. G. Cabo, A. Cassidy, J. A. Miwa and L. Hornekaer, *Faraday Discuss.*, 2015, DOI: 10.1039/C4FD00259H.

44 S. Morsch, S. Lyon, P. Greensmith, D. Smith and S. R. Gibbon, *Faraday Discuss.*, 2015, DOI: 10.1039/C4FD00229F.

45 B. R. Shrestha, A. Bashir, G. N. Ankah, M. Valtiner and F. U. Renner, *Faraday Discuss.*, 2015, DOI: 10.1039/C4FD00256C.

46 M. A. Jakab, F. Presuel-Moreno and J. R. Scully, *Corrosion*, 2005, **61**, 246–263.

47 M. A. Jakab and J. R. Scully, *Nat. Mater.*, 2005, **4**, 667–670.

48 M. A. Jakab and J. R. Scully, *Corrosion*, 2008, **64**, 198–209.

49 F. Presuel-Moreno, M. A. Jakab, N. Tailleart, M. Goldman and J. R. Scully, *Mater. Today*, 2008, **11**, 14–23.

50 F. J. Presuel-Moreno, M. E. Goldman, R. G. Kelly and J. R. Scully, *J. Electrochem. Soc.*, 2005, **152**, B302–B310.

51 F. J. Presuel-Moreno, H. Wang, M. A. Jakab, R. G. Kelly and J. R. Scully, *J. Electrochem. Soc.*, 2006, **153**, B486–B498.

52 J. R. Scully, F. Presuel-Moreno, M. Goldman, R. G. Kelly and N. Tailleart, *Corrosion*, 2008, **64**, 210–229.

53 A. King, J. Lee and J. Scully, *J. Electrochem. Soc.*, 2015, **162**, C12–C23.

54 M. Kendig and M. Hon, *Corrosion*, 2004, **60**, 1024–1030.

55 M. Kendig, M. Hon and L. Warren, *Prog. Org. Coat.*, 2003, **47**, 183–189.

56 M. Kendig and P. Kinlen, *J. Electrochem. Soc.*, 2007, **154**, C195–C201.

57 A. D. King, B. Kannan and J. R. Scully, *Corrosion*, 2014, **70**, 536–557.

58 A. D. King, B. Kannan and J. R. Scully, *Corrosion*, 2014, **70**, 512–535.

59 A. D. King and J. R. Scully, *Corrosion*, 2011, **67**, 055004-1, DOI: 10.5006/1.3590330.

60 *Research Opportunities in Corrosion Science and Engineering*, National Academy Press, Washington, DC, 2011.

Poster titles

Star-shaped crystallographic cracking of localized nanoporous gold defects, **Patricia Losada-Pérez, Buddha R. Shrestha, Asif Bashir, Genesis N. Ankah, Markus Valtiner and Frank U. Renner**, *Institute for Materials Research, Belgium*

Insight into oilfield corrosion by UHV-STM, **Matthew J. Acres, Hadeel Hussain, Christopher A. Muryn and Robert Lindsay**, *The University of Manchester, UK*

To provide a mechanistic understanding to in-pond corrosion of AGR fuel cladding, **Choen May Chan and Dirk Engelberg**, *The University of Manchester, UK*

Time-resolved IR spectroscopy during interfacial electrochemical reactions of organic corrosion inhibitors, **Ying-Hsuan Chen and Andreas Erbe**, *Max-Planck-Institut für Eisenforschung GmbH, Germany*

Ab initio study on the corrosion of pure Mg and Mg–Zn systems, **Su-Ting Cheng, Mira Todorova and Jörg Neugebauer**, *Max-Planck-Institute for Iron Research, Germany*

Investigating intergranular attack (IGA) in sensitised AGR cladding alloys, **Ronald Clark, Steve Walters and Geraint Williams**, *Swansea University, UK*

Controlled comparative study of imidazole and related corrosion inhibitors, **Colleen Richardson and Steven L. Bernasek**, *Princeton University, USA*

Atmospheric corrosion: The effect of NaCl precipitates, **Angus Cook, Trevor Rayment and Alison Davenport**, *University of Birmingham, UK*

Elucidating sour corrosion scales, **Karyn Cooper, Gaurav Joshi, Dirk Engelberg and Robert Lindsay**, *The University of Manchester, UK*

Corrosion resistance of ultra-light Mg–Li alloys. The influence of microstructural features, **Anna Dobkowska, Boguslawa Adamczyk-Cieslak, Jaroslaw Mizera, Jiri Kubasek and Dalibor Vojtech**, *Warsaw University of Technology, Poland*

The thickness of native oxides on aluminum alloys and single crystals, **Jonas Evertsson, Florian Bertram, Fan Zhang, Lisa Rullik, Lindsay Merte, Mikhail Shipilin, Markus Soldemo, Sarah Ahmadi, Nikolay Vinogradov, Francesco Carla, Jonas Weissenrieder, Mats Götelid, Anders Mikkelsen, Jan-Olov Nilsson, Jinshan Pan and Edvin Lundgren**, *Lund University, Sweden*

How can chemistry help National Grid to keep our lights on? A surface chemistry investigation on the anticorrosion additive Irgamet™39, **Marco Facciotti, Richard C. D. Brown, Paul L. Lewin, Gordon Wilson and Paul N. Jarman**, *University of Southampton, UK*

On the application of X-ray computed tomography to investigate the re-activation behaviour of corrosion pits in austenitic stainless steel, **Fahd Almuaili, S. A. McDonalds, P. J. Withers and D. L. Engelberg**, *The University of Manchester, UK*

In situ X-ray tomographic investigation of initiation of atmospheric corrosion pits in aluminium alloy AA2024, **Sarah Glanvill, Andrew du Plessis, Steven R. Street, Mark Basham, Trevor Rayment and Alison J. Davenport**, *University of Birmingham, UK*

The hematite (001)–water interface, insights from X-ray surface diffraction, **Frank Heberling, Johannes Lützenkirchen and Peter Eng**, *Karlsruhe Institute of Technology, Germany*

Lubricant additive effects on the interpretation of simple corrosion tests, **Gregory Hunt, Emma Ravenhill, Kris Sivagnanam and Michael Gahagan**, *Lubrizol, UK*

Corrosion of stainless steels in high temperature water, **Jonathan Morrison, Nathan Johnson, Thomas Caswell, Thomas Hutchinson, Clive Ponton, Brian Connolly and Andy Banks**, *University of Birmingham, UK*

Development of protective sweet corrosion scale, **Gaurav R. Joshi, Karyn Cooper, Xiangli Zhong, Dirk L. Engelberg and Robert Lindsay**, *The University of Manchester, UK*

Localised corrosion of surface treated type 316 stainless steel, **Benjamin Krawczyk and Dirk L. Engelberg**, *The University of Manchester, UK*

The corrosion behaviour of sputtered nanocrystalline stainless steel coating, **Ying Li, Tianshu Li, Bin Zhang and Li Liu**, *Chinese Academy of Science, China*

Phenylphosphonic acid as a corrosion inhibitor for hot dip galvanised steel, **Carol F. Glover and Geraint Williams**, *Swansea University, UK*

Impact of the microstructure of Cu–Zn alloys on the pore formation by chemical de-alloying, **Bao Lin, Huan Ma, Lingxue Kong, Peter D. Hodgson, Ralph Spolenak and Ludovic F. Dumée**, *Deakin University, Australia*

Aluminum oxide formation and stability, **Florian Bertram, F. Zhang, Jonas Evertsson, Lisa Rullik, Anders Mikkelsen, L. Ahl, J.-O. Nilsson, Francesco Carla, Jinshan Pan** and **Edvin Lundgren**, *Lund University, Sweden*

Minimising Pt nanoparticle corrosion in PEFCs *via* support engineering, **Stephen M. Lyth, Jianfeng Liu, Takeshi Daio and Kazunari Sasaki**, *Kyushu University, Japan*

Long-term corrosion of a magnesium alloy: are passive films really passive? **Tom P. Majchrowski, Hugh Godfrey and Andrew Diggle**, *National Nuclear Laboratory, UK*

Localised corrosion of a 7075 aluminium alloy exposed to a 3.5% KCl solution, **Christopher Mallinson, Mark Baker, James E. Castle, Ann Harvey and John Watts**, *University of Surrey, UK*

Comparison of the observed effects of pH, temperature and γ-radiolysis on oxide growth kinetics during corrosion of Fe–Ni–Cr alloys, **Ahmed Y. Musa, Mojtaba Momeni and J. Clara Wren**, *Western University, Canada*

Modeling of potentiostatic polarization and oxide film formation on stellite-6, **Mojtaba Momeni, Mehran Behazin and J. Clara Wren**, *Western University, Canada*

Improving industrial coatings by mapping electrochemical properties, **Sara Muhler, Julian A. Wharton and Jurgita Zekonyte**, *Winn & Coales (Denso) Ltd and University of Southampton, UK*

Mechanisms of CRUD deposition in PWR power plants, **Nathan Johnson, Jie-Jie Wu, Jonathan Morrison, Brian Connolly and Andy Banks**, *University of Birmingham, UK*

Impact of hot/wet conditions and aluminium alloys on titanium oxide layers, **Beatriz Rico Oller, Tobias Mertens, Max Kolb, Jürgen Wehr, Theo Hack and Mikhail Zheludkevich**, *Airbus Group Innovations, Germany*

Investigating the role of irradiation in corrosion simulated light water reactor conditions, **Elizabeth Parker-Quaife, Aliasander Biadak, Jonathon Duff, Fabio Scenini and Simon M. Pimblott**, *The University of Manchester, UK*

Specially treated super duplex stainless steel surfaces studied by X-PEEM and μ-XAS, **Lisa Rullik, Yuran Niu, Eleonora Bettini, Alexei A. Zakharov, Jinshan Pan and Edvin Lundgren**, *Lund University, Sweden*

Determination of local hydrogen concentrations and local effective hydrogen diffusivity in UHSS using the scanning Kelvin probe and scanning electrochemical microscope, **Rebecca Schaller and John R. Scully**, *University of Virginia, USA*

Electrochemical conditions that induce pitting corrosions on high-strength prestressed steel in high alkaline deaerated chloride electrolyte, **Siaw Foon Lee, Carmen Andrade, José Fullea and Javier Sanchez**, *Eduardo Torroja Institute for Construction Science, Madrid, Spain*

High-temperature electrochemical polarized resistance flow-type sensor to monitor RBMK NPP pipeline uniform corrosion rate, **Michail V. Sof'in, Vladimir G. Kritskj, Fiodor V. Nikolaeva, Natalia R. Nikolaeva and Natalia V. Mantserova**, *JSC "ATOMPROEKT" Enterprise of State Corporation "Rosatom", Russia*

On the origin of hydrogen gas evolution from copper in anoxic water, **Joakim Halldin Stenlid, Adam Johannes Johansson and Tore Brinck**, *KTH Royal Institute of Technology, Sweden*

Study of corrosion scale formation from quantum mechanics simulations to a continuum model, **Markus P. Tautschnig, Ariadna Blanca Romero and Nicholas M. Harrison**, *Imperial College London, UK*

Synthesis and characterization of corrosion inhibitor from waste PET and fatty acids, **Yesim Teke, Cansu Seckin and Esra Yonel-Gumruk**, *Turkish Petroleum Refineries Corporation, Turkey*

Oxide film growth and zinc uptake measurements in simulated pressurised water reactor environments, **Thomas Caswell, Jonathan Morrison, Brian Connolly and Sam Tulloch**, *University of Birmingham, UK*

Studying the inhibition of iron by a quaternary ammonium salt in an acidic environment, **Monika S. Walczak, Turia Belashehr and Robert Lindsay**, *The University of Manchester, UK*

Green corrosion inhibition: evaluation of aqueous *theobroma cacao* (cocoa) shell extracts as corrosion inhibitor for mild steel under acidic and basic conditions, **Ann M. Wilson, Tricia A. Bandoo and Anthony Guiseppi-Elie**, *The University of The West Indies, Trinidad and Tobago*

Use of biopolymer–polyelectrolyte membranes to control Mg corrosion, **Benjamin P. Wilson, Kirsi Yliniemi, Ferdinand Singer, Sarah Höhn, Eero Kontturi, Lasse Murtomäki and Sannakaisa Virtanen**, *Aalto University, Finland*

The kinetics and mechanism of filiform corrosion occurring on tin and iron–tin intermetallic coated steels, **Natalie Wint, Neil Hamilton McMurray, Arnoud C. A. de Vooys and Steven Geary**, *Swansea University, UK*

Characterising nickel corrosion and its inhibition, **Mary Wood, R. Welbourn, P. Gutfreund and Stuart Clarke**, *University of Cambridge, UK*

Effect of Zr additions on corrosion of Ti under simulated physiological conditions, **Yue Zhang, Owen Addison and Alison Davenport**, *University of Birmingham, UK*

Effect of duplex steel elemental composition and microstructure on the adsorption behaviour and efficiency of acid corrosion inhibitors, **Man Yi Ho, Evgeny Barmatov, Jill Geddes, Lynne Crawford and Trevor Hughes**, *Schlumberger Gould Research, UK*

Acid corrosion inhibitor film formation kinetics: experimental methods & effects of inhibitor chemistry, **Trevor Hughes, Evgeny Barmatov, Jill Geddes and Michaela Nagl**, *Schlumberger Gould Research, UK*

Effect of corrosion products on efficiency of film-forming corrosion inhibitors in hydrochloric acid, **Evgeny Barmatov and Trevor Hughes**, *Schlumberger Gould Research, UK*

Performance characterisation of duplex stainless steel for application in nuclear waste storage environment, **Cem Örnek and Dirk L. Engelberg**, *The University of Manchester, UK*

The Faraday Poster Prize sponsored by Cogent for the best poster was jointly awarded to Marco Facciotti of the University of Southampton, UK, for his poster on How can chemistry help National Grid to keep our lights on? A surface chemistry investigation on the anticorrosion additive Irgamet™39 and Su-Ting Cheng of the Max-Planck-Institute for Iron Research, Germany, for her poster on *Ab initio* study on the corrosion of pure Mg and Mg–Zn systems.

List of participants

Matthew Acres, *The University of Manchester, UK*
Fahd Almuaili, *The University of Manchester, UK*
Tricia Bandoo, *The University of the West Indies, Trinidad and Tobago*
Evgeny Barmatov, *Schlumberger Gould Research, UK*
Ariadna Blanca Romero, *Imperial College London, UK*
Hendrik Bluhm, *Lawrence Berkeley National Laboratory, USA*
Thomas Caswell, *University of Birmingham, UK*
Choen May Chan, *The University of Manchester, UK*
Ying-Hsuan Chen, *Max-Planck-Institut für Eisenforschung GmbH, Germany*
Su-Ting Cheng, *Max-Planck-Institute for Iron Research, Germany*
Ronald Clark, *Swansea University, UK*
Angus Cook, *University of Birmingham, UK*
Karyn Cooper, *The University of Manchester, UK*
Nathan Cooze, *Swansea University, UK*
Matthew Coward, *AWE plc, UK*
Alison Davenport, *University of Birmingham, UK*
Keith Dawes, *EC Lab Ltd, UK*
Mike Dawson, *Uniscan Instruments, UK*
Anna Dobkowska, *Warsaw University of Technology, Poland*
Dirk Engelberg, *The University of Manchester, UK*
Jonas Evertsson, *Lund University, Sweden*
Marco Facciotti, *University of Southampton, UK*
Gerald Frankel, *The Ohio State University, USA*
Callum Gallagher, *Swansea University, UK*
Chiara Gattinoni, *London Centre for Nanotechnology, UK*
Simon Gibbon, *AkzoNobel RD and I, UK*
Vivi Gkanavara, *Imperial College, UK*
Sarah Glanvill, *University of Birmingham, UK*
Carol Glover, *Swansea University, UK*
Hiroki Habazaki, *Hokkaido University, Japan*
James Hall, *Imperial College, UK*
Joakim Halldin Stenlid, *KTH Royal Institute of Technology, Sweden*
Frank Heberling, *Institute for Nuclear Waste Disposal, Karlsruhe Institute of Technology, Germany*
Patric Helbig, *Royal Navy, UK*
Stephane Heurtault, *CEA Saclay / DEN / DPC / SCCME / LECNA & CNRS, UMR 8235, LISE, France*
John Hewett, *University of Birmingham, UK*
Man Yi Ho, *Schlumberger Gould Research, UK*
Isobel Hogg, *Royal Society of Chemistry, UK*
Trevor Hughes, *Schlumberger Gould Research, UK*
Gregory Hunt, *Lubrizol, UK*
Hadeel Hussain, *The University of Manchester, UK*

Nathan Johnson, *University of Birmingham, UK*
Gaurav Joshi, *The University of Manchester, UK*
Anton Kokalj, *Jožef Stefan Institute, Slovenia*
Denis Kramer, *University of Southampton, UK*
Benjamin Krawczyk, *The University of Manchester, UK*
Line Kyhl, *Aarhus University, Denmark*
Siaw Foon Lee, *Eduardo Torroja Institute for Construction Science, Spain*
Tom Lewis, *Swansea University, UK*
Ying Li, *Institute of Metal Research, Chinese Academy of Science, China*
Yuandi Li, *Royal Society of Chemistry, UK*
Bao Lin, *Institute of Frontier Materials, Deakin University, Australia*
Rob Lindsay, *The University of Manchester, UK*
Patricia Losada-Pérez, *Institute for Materials Research, Belgium*
Edvin Lundgren, *Lund University, Sweden*
Helen Lunn, *Royal Society of Chemistry, UK*
Stuart Lyon, *The University of Manchester, UK*
Stephen Lyth, *Kyushu University, Japan*
Johannes Lützenkirchen, *Karlsruhe Institute of Technology, Germany*
Tom Majchrowski, *National Nuclear Laboratory, UK*
Christopher Mallinson, *University of Surrey, UK*
Philippe Marcus, *Chimie ParisTech-CNRS, France*
Vincent Maurice, *CNRS - Chimie ParisTech, France*
Janine Mauzeroll, *McGill University, Canada*
Angelos Michaelides, *UCL, UK*
Sofin Michail, *JSC "ATOMPROEKT" Enterprise of State Corporation "Rosatom", Russia*
Arjan Mol, *Delft University of Technology, Netherlands*
Mojtaba Momeni, *Western University, Canada*
Ella Moore, *Ashford Environmental Services Limited, UK*
Jonathan Morrison, *University of Birmingham, UK*
Suzanne Morsch, *The University of Manchester, UK*
Sara Muhler, *Winn & Coales (Denso) Ltd, UK*
David Munoz Ramo, *University of Cambridge, UK*
Fiona Nalden, *Royal Society of Chemistry, UK*
Roger Newman, *University of Toronto, Canada*
Sonja Nixon, *Airbus Group Innovations, Germany*
Cem Örnek, *The University of Manchester, UK*
Natasha Page, *B&V Water Treatment, UK*
Elizabeth Parker-Quaife, *The University of Manchester, UK*
Monica Patel, *Imperial College London, UK*
Bianca Provost, *University of Cambridge, UK*
Trevor Rayment, *Diamond Light Source Limited, UK*
Sandra Reisinger, *Hoffmann Elektrokohle, Austria*
Frank Renner, *IMO, Belgium*
Colleen Richardson, *Princeton University, USA*
Beatriz Rico Oller, *Airbus Group Innovations, Germany*
Lisa Rullik, *Lund University, Sweden*
Bethan Scadden, *University of Leeds, UK*
John Scully, *University of Virginia, USA*
Kyra Sedransk Campbell, *Imperial College, UK*
David Shoesmith, *Western University, Canada*
Thiago Simoes, *University of Leeds, UK*
Ross Springell, *University of Bristol, UK*

Viktoria Stelzhammer, *Royal Society of Chemistry, UK*
Steven Street, *University of Birmingham, UK*
James Sullivan, *Swansea University, UK*
Markus Tautschnig, *Imperial College London, UK*
Christopher Taylor, *DNV GL and Ohio State University, USA*
Yesim Teke, *Turkish Petroleum Refineries Corporation, Turkey*
Geoffrey Thornton, *UCL, UK*
Mira Todorova, *Max-Planck-Institut fuer Eisenforschung, Germany*
Sannakaisa Virtanen, *University of Erlangen, Germany*
Peter Visser, *Delft University of Technology, Netherlands*
Vincent Vivier, *Laboratoire Interfaces et Systèmes Electrochimiques - UMR8235, France*
Monika Walczak, *The University of Manchester, UK*
Becky Welbourn, *University of Cambridge, UK*
Julian Wharton, *University of Southampton, UK*
Gemma Wilkins, *Royal Society of Chemistry, UK*
Geraint Williams, *Swansea University, UK*
David Williams, *University of Auckland, New Zealand*
Ann Wilson, *The University of The West Indies, Trinidad and Tobago*
Benjamin Wilson, *Aalto University, Finland*
Natalie Wint, *Swansea University, UK*
Mary Wood, *University of Cambridge, UK*
Clara Wren, *Western University, Canada*
Jiejie Wu, *University of Birmingham, UK*
Jiwen Yin, *Chifeng University, China*
Kirsi Yliniemi, *Aalto University, Finland*
Louis Yu, *Imperial College, UK*
Yue Zhang, *University of Birmingham, UK*
Lijuan Zhao, *Nankai University, China*